Shoncair

W9-CII-200

PRINCIPLES OF TOXICOLOGY

PRINCIPLES OF TOXICOLOGY

Environmental and Industrial Applications

SECOND EDITION

Edited by

Phillip L. Williams, Ph.D.
Associate Professor
Department of Environmental Health Science
University of Georgia
Athens, Georgia

Robert C. James, Ph.D.
President, TERRA, Inc.
Tallahassee, Florida
Associate Scientist, Interdisciplinary Toxicology
Center for Environmental and Human Toxicology
University of Florida
Gainesville, Florida

Stephen M. Roberts, Ph.D.
Professor and Program Director
Center for Environmental and Human Toxicology
University of Florida
Gainesville, Florida

A Wiley-Interscience Publication

JOHN WILEY & SONS, INC.

New York • Chichester • Weinheim • Brisbane • Singapore • Toronto

This book is printed on acid-free paper. ♾

Copyright © 2000 by John Wiley & Sons, Inc. All rights reserved.

Published simultaneously in Canada.

No part of this publication may be reproduced, stored in a retrieval system or transmitted in any form or by any means, electronic, mechanical, photocopying, recording, scanning or otherwise, except as permitted under Sections 107 or 108 of the 1976 United States Copyright Act, without either the prior written permission of the Publisher, or authorization through payment of the appropriate per-copy fee to the Copyright Clearance Center, 222 Rosewood Drive, Danvers, MA 01923, (978) 750-8400, fax (978) 750-4744. Requests to the Publisher for permission should be addressed to the Permissions Department, John Wiley & Sons, Inc., 605 Third Avenue, New York, NY 10158-0012, (212) 850-6011, fax (212) 850-6008, E-Mail: PERMREQ@WILEY.COM.

For ordering and customer service, call 1-800-CALL-WILEY.

Library of Congress Cataloging in Publication Data:

Principles of toxicology: environmental and industrial applications / edited by Phillip L. Williams, Robert C. James, Stephen M. Roberts.—2nd ed.
p. cm.
Update and expansion on a previous text entitled: Industrial toxicology: safety and health applications in the workplace.
Includes bibliographical references and index.
ISBN 0-471-29321-0 (cloth: alk. paper)
1. Toxicology. 2. Industrial toxicology. 3. Environmental toxicology. I. Williams, Phillip L., 1952- II. James, Robert C., 1947- III. Roberts, Stephen M., 1950-

RA1211 .P746 2000
615.9'02—dc21 99-042196

Printed in the United States of America.

10 9 8 7 6 5 4 3 2 1

CONTRIBUTORS

LOUIS ADAMS, PH.D. Professor, Department of Medicine, University of Cincinnati, Cincinnati, Ohio

JUDY A. BEAN, PH.D., Director, Biostatistics Program, Children's Hospital, Cincinnati, Ohio

CHRISTOPHER J. BORGERT, PH.D., President and Principal Scientist, Appied Pharmacology and Toxicology, Inc.; Assistant Scientist, Department of Physiological Sciences, University of Florida College of Veterinary Medicine, Alachua, Florida

JANICE K. BRITT, PH.D., Senior Toxicologist, TERRA, Inc., Tallahassee, Florida

ROBERT A. BUDINSKY, JR., PH.D., Senior Toxicologist, ATRA, Inc., Tallahassee, Florida

CHAM E. DALLAS, PH.D., Associate Professor and Director, Interdisciplinary Toxicology Program, University of Georgia, Athens, Georgia

ROBERT P. DEMOTT, PH.D., Chemical Risk Group Manager, GeoSyntec Consultants, Inc., Tampa, Florida

STEVEN G. DONKIN, PH.D., Senior Scientist, Sciences International, Inc., Alexandria, Virginia

LORA E. FLEMING, M.D., PH.D., MPH, Associate Professor, Department of Epidemiology and Public Health, University of Miami, Miami, Florida

MICHAEL R. FRANKLIN, PH.D., Interim Chair and Professor, Department of Pharmacology and Toxicology, University of Utah, Salt Lake City, Utah

HOWARD FRUMKIN, M.D., DR.P.H., Chair and Associate Professor, Department of Environmental and Occupational Health, The Rollins School of Public Health, Emory University, Atlanta, Georgia

EDWARD I. GALAID, M.D., MPH, Clinical Assistant Professor, Department of Environmental and Occupational Health, The Rollins School of Public Health, Emory University, Atlanta

JAY GANDY, PH.D., Senior Toxicologist, Center for Toxicology and Environmental Health, Little Rock, Arkansas

FREDRIC GERR, M.D., Associate Professor, Department of Environmental and Occupational Health, The Rollins School of Public Health, Emory University, Atlanta, Georgia

PHILLIP T. GOAD, PH.D., President, Center for Toxicology and Environmental Health, Little Rock, Arkansas

CHRISTINE HALMES, PH.D., Toxicologist, TERRA, Inc., Denver, Colorado

DAVID E. JACOBS, PH.D., Director, Office of Lead Hazard Control, U.S. Department of Housing and Urban Development, Washington, D.C.

ROBERT C. JAMES, PH.D., President, TERRA, Inc., Tallahassee, Florida; Associate Scientist, Interdisciplinary Toxicology, Center for Environmental and Human Toxicology, University of Florida, Gainesville, Florida

WILLIAM R. KERN, PH.D., Professor, Department of Pharmacology and Therapeutics, University of Florida, Gainesville, Florida

PAUL J. MIDDENDORF, PH.D., Principal Research Scientist, Georgia Tech Research Institute, Atlanta, Georgia

GLENN C. MILLNER, PH.D., Vice President, Center for Toxicology and Environmental Health, Little Rock, Arkansas

ALAN C. NYE, PH.D., Vice President, Center for Toxicology and Environmental Health, Little Rock, Arkansas

ELLEN J. O'FLAHERTY, PH.D., Professor, Department of Environmental Health, University of Cincinnati, Cincinnati, Ohio

DANNY L. OHLSON, PH.D., Toxicologist, Hazardous Substances and Waste Management Research, Tallahassee, Florida

STEPHEN M. ROBERTS, PH.D., Professor and Program Director, Center for Environmental and Human Toxicology, University of Florida, Gainesville, Florida

WILLIAM R. SALMINEN, PH.D., Consulting Toxicologist, Toxicology Division, Exxon Biomedical Sciences, Inc., East Millstone, New Jersey

CHRISTOPER J. SARANKO, PH.D., Post Doctoral Fellow, Center for Environmental and Human Toxicology, University of Florida, Gainesville, Florida

CHRISTOPER M. TEAF, PH.D., President, Hazardous Substances and Waste Management Research, Tallahassee, Florida; Associate Director, Center for Biochemical and Toxicological Research and Hazardous Waste Management, Florida State University, Tallahassee, Florida

D. ALAN WARREN, PH.D., Toxicologist, TERRA, Inc., Tallahassee, Florida

PHILLIP L. WILLIAMS, PH.D., Associate Professor, Department of Environmental Health Science, University of Georgia, Athens, Georgia

GAROLD S. YOST, PH.D., Professor, Department of Pharmacology and Toxicology, University of Utah, Salt Lake City, Utah

CONTENTS

PREFACE

Purpose of This Book

Principles of Toxicology: Environmental and Industrial Applications presents compactly and efficiently the scientific basis to toxicology as it applies to the workplace and the environment. The book covers the diverse chemical hazards encountered in the modern work and natural environment and provides a practical understanding of these hazards for those concerned with protecting the health of humans and ecosystems.

Intended Audience

This book represents an update and expansion on a previous, very successful text entitled *Industrial Toxicology: Satety and Health Applications in the Workplace*. It retains the emphasis on applied aspects of toxicology, while extending its scope beyond the industrial setting to include environmental toxicology. The book was written for those health professionals who need toxicological information and assistance beyond that of an introductory text in general toxicology, yet more practical than that in advanced scientific works on toxicology. In particular, we have in mind industrial hygienists, occupational physicians, safety engineers, environmental health practitioners, occupational health nurses, safety directors, and environmental scientists.

Organization of the Book

This volume consists of three parts. Part I establishes the scientific basis to toxicology, which is then applied through the rest of the book. This part discusses concepts such as absorption, distribution, and elimination of toxic agents from the body. Chapters 4–10 discuss the effects of toxic agents on specific physiological organs or systems, including the blood, liver, kidneys, nerves, skin, lungs, and the immune system.

Part II addresses specific areas of concern in the occupational and environmental—both toxic agents and their manifestations. Chapters 11–13 examine areas of great research interest—reproductive toxicology, mutagenesis, and carcinogenesis. Chapters 14–17 examine toxic effects of metals, pesticides, organic solvents, and natural toxins and venoms.

Part III is devoted to specific applications of the toxicological principles from both the environmental and occupational settings. Chapters 18 and 19 cover risk assessment and provide specific case studies that allow the reader to visualize the application of risk assessment process. Chapters 20 and 21 discuss occupational medicine and epidemiologic issues. The final chapter is devoted to hazard control.

Features

The following features from *Principles of Toxicology: Environmental and Industrial Applications* will be especially useful to our readers:

- The book is compact and practical, and the information is structured for easy use by the health professional in both industry and government.

- The approach is scientific, but applied, rather than theoretical. In this it differs from more general works in toxicology, which fail to emphasize the information pertinent to the industrial environment.
- The book consistently stresses evaluation and control of toxic hazards.
- Numerous illustrations and figures clarify and summarize key points.
- Case histories and examples demonstrate the application of toxicological principles.
- Chapters include annotated bibliographies to provide the reader with additional useful information.
- A comprehensive glossary of toxicological terms is included.

Phillip L. Williams
Robert C. James
Stephen M. Roberts

ACKNOWLEDGMENTS

A text of this undertaking on the broad topic of toxicology would not be possible except for the contributions made by each of the authors in their field(s) of speciality. We especially appreciate the contributors patience during the many years it took to complete this revision. In addition, such an undertaking would not have been possible without the support provided by each of our employers—The University of Georgia, TERRA, Inc., and The University of Florida. We also owe a thank you to Valerie Rocchi for her administrative assistance throughout the effort and to Dr. Kelly McDonald for her editorial assistance.

Phillip L. Williams
Robert C. James
Stephen M. Roberts

PART I
Conceptual Aspects

1 General Principles of Toxicology

ROBERT C. JAMES, STEPHEN M. ROBERTS, and PHILLIP L. WILLIAMS

The intent of this chapter is to provide a concise description of the basic principles of toxicology and to illustrate how these principles are used to make reasonable judgments about the potential health hazards and the risks associated with chemical exposures. This chapter explains

- Some basic definitions and terminology
- What toxicologists study, the scientific disciplines they draw upon, and specialized areas of interest within toxicology
- Descriptive toxicology and the use of animal studies as the primary basis for hazard identification, the importance of dose, and the generation of dose–response relationships
- How dose–response data might be used to assess safety or risk
- Factors that might alter a chemical's toxicity or the dose–response relationship
- The basic methods for extrapolating dose–response data when developing exposure guidelines of public health interest

1.1 BASIC DEFINITIONS AND TERMINOLOGY

The literal meaning of the term *toxicology* is "the study of poisons." The root word toxic entered the English language around 1655 from the Late Latin word *toxicus* (which meant poisonous), itself derived from *toxikón*, an ancient Greek term for poisons into which arrows were dipped. The early history of toxicology focused on the understanding and uses of different poisons, and even today most people tend to think of poisons as a deadly potion that when ingested causes almost immediate harm or death. As toxicology has evolved into a modern science, however, it has expanded to encompass all forms of adverse health effects that substances might produce, not just acutely harmful or lethal effects. The following definitions reflect this expanded scope of the science of toxicology:

Toxic—having the characteristic of producing an undesirable or adverse health effect.

Toxicity—any toxic (adverse) effect that a chemical or physical agent might produce within a living organism.

Toxicology—the science that deals with the study of the adverse effects (toxicities) chemicals or physical agents may produce in living organisms under specific conditions of exposure. It is a science that attempts to qualitatively identify all the hazards (i.e., organ toxicities) associated with a substance, as well as to quantitatively determine the exposure conditions under which those hazards/toxicities are induced. Toxicology is the science that experimentally investigates the occurrence, nature, incidence, mechanism, and risk factors for the adverse effects of toxic substances.

Principles of Toxicology: Environmental and Industrial Applications, Second Edition, Edited by Phillip L. Williams, Robert C. James, and Stephen M. Roberts.
ISBN 0-471-29321-0 © 2000 John Wiley & Sons, Inc.

As these definitions indicate, the toxic responses that form the study of toxicology span a broad biologic and physiologic spectrum. Effects of interest may range from something relatively minor such as irritation or tearing, to a more serious response like acute and reversible liver or kidney damage, to an even more serious and permanent disability such as cirrhosis of the liver or liver cancer. Given this broad range of potentially adverse effects to consider, it is perhaps useful for those unfamiliar with toxicology to define some additional terms, listed in order of relevance to topics that might be discussed in Chapters 2–22 of this book.

Exposure—to cause an adverse effect, a toxicant must first come in contact with an organism. The means by which an organism comes in contact with the substance is the route of exposure (e.g., in the air, water, soil, food, medication) for that chemical.

Dose—the total amount of a toxicant administered to an organism at specific time intervals. The quantity can be further defined in terms of quantity per unit body weight or per body surface area.

Internal/absorbed dose—the actual quantity of a toxicant that is absorbed into the organism and distributed systemically throughout the body.

Delivered/effective/target organ dose—the amount of toxicant reaching the organ (known as the *target organ*) that is adversely affected by the toxicant.

Acute exposure—exposure over a brief period of time (generally less than 24 h). Often it is considered to be a single exposure (or dose) but may consist of repeated exposures within a short time period.

Subacute exposure—resembles acute exposure except that the exposure duration is greater, from several days to one month.

Subchronic exposure—exposures repeated or spread over an intermediate time range. For animal testing, this time range is generally considered to be 1–3 months.

Chronic exposure—exposures (either repeated or continuous) over a long (greater than 3 months) period of time. With animal testing this exposure often continues for the majority of the experimental animal's life, and within occupational settings it is generally considered to be for a number of years.

Acute toxicity—an adverse or undesirable effect that is manifested within a relatively short time interval ranging from almost immediately to within several days following exposure (or dosing). An example would be chemical asphyxiation from exposure to a high concentration of carbon monoxide (CO).

Chronic toxicity—a permanent or lasting adverse effect that is manifested after exposure to a toxicant. An example would be the development of silicosis following a long-term exposure to silica in workplaces such as foundries.

Local toxicity—an adverse or undesirable effect that is manifested at the toxicant's site of contact with the organism. Examples include an acid's ability to cause burning of the eyes, upper respiratory tract irritation, and skin burns.

Systemic toxicity—an adverse or undesirable effect that can be seen throughout the organism or in an organ with selective vulnerability distant from the point of entry of the toxicant (i.e., toxicant requires absorption and distribution within the organism to produce the toxic effect). Examples would be adverse effects on the kidney or central nervous system resulting from the chronic ingestion of mercury.

Reversible toxicity—an adverse or undesirable effect that can be reversed once exposure is stopped. Reversibility of toxicity depends on a number of factors, including the extent of exposure (time and amount of toxicant) and the ability of the affected tissue to repair or regenerate. An example includes hepatic toxicity from acute acetaminophen exposure and liver regeneration.

Delayed or latent toxicity—an adverse or undesirable effect appearing long after the initiation and/or cessation of exposure to the toxicant. An example is cervical cancer during adulthood resulting from in utero exposure to diethylstilbestrol (DES).

Allergic reaction—a reaction to a toxicant caused by an altered state of the normal immune response. The outcome of the exposure can be immediate (anaphylaxis) or delayed (cell-mediated).

Idiosyncratic reaction—a response to a toxicant occurring at exposure levels much lower than those generally required to cause the same effect in most individuals within the population. This response is genetically determined, and a good example would be sensitivity to nitrates due to deficiency in NADH (reduced-form nicotinamide adenine dinucleotide phosphate)–methemoglobin reductase.

Mechanism of toxicity—the necessary biologic interactions by which a toxicant exerts its toxic effect on an organism. An example is carbon monoxide (CO) asphyxiation due to the binding of CO to hemoglobin, thus preventing the transport of oxygen within the blood.

Toxicant—any substance that causes a harmful (or adverse) effect when in contact with a living organism at a sufficiently high concentration.

Toxin—any toxicant produced by an organism (floral or faunal, including bacteria); that is, naturally produced toxicants. An example would be the pyrethrins, which are natural pesticides produced by pyrethrum flowers (i.e., certain chrysanthemums) that serve as the model for the man made insecticide class pyrethroids.

Hazard—the qualitative nature of the adverse or undesirable effect (i.e., the type of adverse effect) resulting from exposure to a particular toxicant or physical agent. For example, asphyxiation is the hazard from acute exposures to carbon monoxide (CO).

Safety—the measure or mathematical probability that a specific exposure situation or dose will not produce a toxic effect.

Risk—the measure or probability that a specific exposure situation or dose will produce a toxic effect.

Risk assessment—the process by which the potential (or probability of) adverse health effects of exposure are characterized.

1.2 WHAT TOXICOLOGISTS STUDY

Toxicology has become a science that builds on and uses knowledge developed in other related medical sciences, such as physiology, biochemistry, pathology, pharmacology, medicine, and epidemiology, to name only a few. Given its broad and diverse nature, toxicology is also a science where a number of areas of specialization have evolved as a result of the different applications of toxicological information that exist within society today. It might be argued, however, that the professional activities of all toxicologists fall into three main areas of endeavor: descriptive toxicology, research/mechanistic toxicology, and applied toxicology.

Descriptive toxicologists are scientists whose work focuses on the toxicity testing of chemicals. This work is done primarily at commercial and governmental toxicity testing laboratories, and the studies performed at these facilities are designed to generate basic toxicity information that can be used to identify the various organ toxicities (hazards) that the test agent is capable of inducing under a wide range of exposure conditions. A thorough "descriptive toxicological" analysis would identify all possible acute and chronic toxicities, including the genotoxic, reproductive, teratogenic (developmental), and carcinogenic potential of the test agent. It would also identify important metabolites of the chemical that are generated as the body attempts to break down and eliminate the chemical, as well as analyze the manner in which the chemical is absorbed into the body, distributed throughout the body and accumulated by various tissues and organs, and then ultimately excreted from the body. Hopefully, appropriate dose–response test data are generated for those toxicities of greatest concern during the completion of the descriptive studies so that the relative safety of any given exposure or dose level that humans might typically encounter can be determined.

Basic research or *mechanistic* toxicologists are scientists who study the chemical or agent in depth for the purpose of gaining an understanding of how the chemical or agent initiates those biochemical or physiological changes within the cell or tissue that result in the toxicity (adverse effect). They identify the critical biological processes within the organism that must be affected by the chemical to produce the toxic properties that are ultimately observed. Or, to state it another way, the goal of mechanistic studies is to understand the specific biological reactions (i.e., the adverse chain of events) within the affected organism that ultimately result in the toxicity under investigation. These experiments may be performed at the molecular, biochemical, cellular, or tissue level of the affected organism, and thus incorporate and apply the knowledge of a number of many other related scientific disciplines within the biological and medical sciences (e.g., physiology, biochemistry, genetics, molecular biology). Mechanistic studies ultimately are the bridge of knowledge that connects functional observations made during descriptive toxicological studies to the extrapolations of dose–response information that is used as the basis of risk assessment and exposure guideline development (e.g., occupational health guidelines or governmental regulations) by applied toxicologists.

Applied toxicologists are scientists concerned with the use of chemicals in a "real world" or nonlaboratory setting. For example, one goal of applied toxicologists is to control the use of the chemical in a manner that limits the probable human exposure level to one in which the dose any individual might receive is a safe one. Toxicologists who work in this area of toxicology, whether they work for a state or federal agency, a company, or as consultants, use descriptive and mechanistic toxicity studies to develop some identifiable measure of the safe dose of the chemical. The process whereby this safe dose or level of exposure is derived is generally referred to as the area of *risk assessment*. Within applied toxicology a number of subspecialties occur. These are: forensic toxicology, clinical toxicology, environmental toxicology, and occupational toxicology. *Forensic toxicology* is that unique combination of analytical chemistry, pharmacology, and toxicology concerned with the medical and legal aspects of drugs and poisons; it is concerned with the determination of which chemicals are present and responsible in exposure situations of abuse, overdose, poisoning, and death that become of interest to the police, medical examiners, and coroners. *Clinical toxicology* specializes in ways to treat poisoned individuals and focuses on determining and understanding the toxic effects of medicines and simple over-the-counter (nonprescription) drugs. *Environmental toxicology* is the subdiscipline concerned with those chemical exposure situations found in our general living environment. These exposures may stem from the agricultural application of chemicals (e.g., pesticides, growth regulators, fertilizers), the release of chemicals during modern-day living (e.g., chemicals released by household products), regulated and unintentional industrial discharges into air or waterways (e.g., spills, stack emissions, NPDES discharges, etc.), and various nonpoint emission sources (e.g., the combustion byproducts of cars). This specialty largely focuses on those chemical exposures referred to as environmental contamination or pollution. Within this area there may be even further subspecialization (e.g., ecotoxicology, aquatic toxicology, mammalian toxicology, avian toxicology). *Occupational toxicology* is the subdiscipline concerned with the chemical exposures and diseases found in the workplace.

Regardless of the specialization within toxicology, or the types of toxicities of major interest to the toxicologist, essentially every toxicologist performs one or both of the two basic functions of toxicology, which are to (1) examine the nature of the adverse effects produced by a chemical or physical agent (*hazard identification* function) and (2) assess the probability of these toxicities occurring under specific conditions of exposure (*risk assessment* function). Ultimately, the goal and basic purpose of toxicology is to understand the toxic properties of a chemical so that these adverse effects can be prevented by the development of appropriate handling or exposure guidelines.

1.3 THE IMPORTANCE OF DOSE AND THE DOSE–RESPONSE RELATIONSHIP

It is probably safe to say that among lay individuals there exists considerable confusion between the terms poisonous and toxic. If asked, most lay individuals would probably define a toxic substance using the same definition that one would apply to highly poisonous chemicals, that is, chemicals capable of producing a serious injury or death quickly and at very low doses. However, this is not a particularly useful definition because all chemicals may induce some type of adverse effect at some dose, so all chemicals may be described as toxic. As we have defined toxicants (toxic chemicals) as agents capable of producing an adverse effect in a biological system, a reasonable question for one to ask becomes "Which group of chemicals do we consider to be toxic?" or "Which chemicals do we consider safe?" The short answer to both questions, of course, is all chemicals; for even relatively safe chemicals can become toxic if the dose is high enough, and even potent, highly toxic chemicals may be used safely if exposure is kept low enough. As toxicology evolved from the study of just those substances or practices that were poisonous, dangerous, or unsafe, and instead became a more general study of the adverse effects of all chemicals, the conditions under which chemicals express toxicity became as important as, if not more important than, the kind of adverse effect produced. The importance of understanding the dose at which a chemical becomes toxic (harmful) was recognized centuries ago by Paracelsus (1493–1541), who essentially stated this concept as "All substances are poisons; there is none which is not a poison. The right dose differentiates a poison and a remedy." In a sense this statement serves to emphasize the second function of toxicology, or risk assessment, as it indicates that concern for a substance's toxicity is a function of one's exposure to it. Thus, the evaluation of those circumstances and conditions under which an adverse effect can be produced is key to considering whether the exposure is safe or hazardous. All chemicals are toxic at some dose and may produce harm if the exposure is sufficient, but all chemicals produce their harm (toxicities) under prescribed conditions of dose or usage. Consequently, another way of viewing all chemicals is that provided by Emil Mrak, who said "There are no harmless substances, only harmless ways of using substances."

These two statements serve to remind us that describing a chemical exposure as being either harmless or hazardous is a function of the magnitude of the exposure (dose), not the types of toxicities that a chemical might be capable of producing at some dose. For example, vitamins, which we consciously take to improve our health and well-being, continue to rank as a major cause of accidental poisoning among children, and essentially all the types of toxicities that we associate with the term "hazardous chemicals" may be produced by many of the prescription medicines in use today. To help illustrate this point, and to begin to emphasize the fact that the dose makes the poison, the reader is invited to take the following pop quiz. First, cross-match the doses listed in column A of Table 1.1, doses that produce lethality in 50 percent of the animals (LD_{50}), to the correct chemical listed in column B. The chemicals listed in column B are a collection of food additives, medicines, drugs of abuse, poisons, pesticides, and hazardous substances for which the correct LD_{50} is listed somewhere in column A. To perform this cross-matching, first photocopy Table 1.1 and simply mark the ranking of the dose (i.e., the number corresponding next to the dose in column A) you believe correctly corresponds to the chemical it has been measured for in column B. [*Note*: The doses are listed in descending order, and the chemicals have been listed alphabetically. So, the three chemicals you believe to be the safest, should have the three largest doses (you should rank them as 1, 2, and 3), and the more unsafe or dangerous you perceive the chemical to be, the higher the numerical ranking you should give it. After testing yourself with the chemicals listed in Tables 1.1, review the correct answers in tables found at the end of this chapter.]

According to the ranking scheme that you selected for these chemicals, were the least potent chemicals common table salt, vitamin K (which is required for normal blood clotting times), the iron supplement dosage added to vitamins for individuals that might be slightly anemic, or a common pain relief medication you can buy at a local drugstore? What were the three most potently toxic chemicals (most dangerous at the lowest single dose) in your opinion? Were they natural or synthetic (human-made) chemicals? How toxic did you rate the nicotine that provides the stimulant properties of tobacco products? How did the potency ranking of prescription medicines like the sedative phenobarbital or

TABLE 1.1 Cross-Matching Exercise: Comparative Acutely Lethal Doses

The chemicals listed in this table are *not* correctly matched with their acute median lethal doses (LD_{50}'s). Rearrange the list so that they correctly match. The correct order can be found in the answer table at the end of the chapter.

	A	B	
N	LD_{50} (mg/kg)	Toxic Chemical	Correct Order
1	15,000	Alcohol (ethanol)	________
2	10,000	Arrow poison (curare)	________
3	4,000	Dioxin or 2,3,7,8-TCDD	________
4	1,500	(PCBs)—an electrical insulation fluid	________
5	1,375	Food poison (botulinum toxin)	________
6	900	Iron supplement (ferrous sulfate)	________
7	150	Morphine	________
8	142	Nicotine	________
9	2	Insecticide (malathion)	________
10	1	Rat poison (strychnine)	________
11	0.5	Sedative/sleep aid (phenobarbital)	________
12	0.001	Tylenol (acetaminophen)	________
13	0.00001	Table salt (sodium chloride)	________

the pain killer morphine compare to the acutely lethal potency of a poison such as strychnine or the pesticide malathion?

Now take the allowable workplace chronic exposure levels for the following chemicals—aspirin, gasoline, iodine, several different organic solvents, and vegetable oil mists—and again rank these substances going from the highest to lowest allowable workplace air concentration (listed in Table 1.2). Remember that the lower (numerically) the allowable air concentration, the more potently toxic the substance is per unit of exposure. Review the correct answers in the table found at the end of this chapter.

Defining Dose and Response

Because all chemicals are toxic at some dose, what judgments determine their use? To answer this, one must first understand the use of the dose–response relationship because this provides the basis for estimating the safe exposure level for a chemical. A dose–response relationship is said to exist when changes in dose produce consistent, nonrandom changes in effect, either in the magnitude of effect or in the percent of individuals responding at a particular level of effect. For example, the number of animals dying increases as the dose of strychnine is increased, or with therapeutic agents the number of patients recovering from an infection increases as the dosage is increased. In other instances, the severity of the response seen in each animal increases with an increase in dose once the threshold for toxicity has been exceeded.

The Basic Components of Tests Generating Dose–Response Data

The design of any toxicity test essentially incorporates the following five basic components:

1. The selection of a test organism
2. The selection of a response to measure (and the method for measuring that response)
3. An exposure period

TABLE 1.2 Cross-Matching Exercise: Occupational Exposure Limits—Aspirin and Vegetable Oil Versus Industrial Solvents

The chemicals listed in this table are *not* correctly matched with their allowable workplace exposure levels. Rearrange the list so that they correctly match. The correct order can be found in the answer table at the end of the chapter.

N	Allowable Workplace Exposure Level (mg/m^3)	Chemical (use)	Correct Order
1	0.1	Aspirin (pain reliever)	________
2	5	Gasoline (fuel)	________
3	10	Iodine (antiseptic)	________
4	55	Naphtha (rubber solvent)	________
5	170	Perchloroethylene (dry-cleaning fluid)	________
6	188	Tetrahydrofuran (organic solvent)	________
7	269	Trichloroethylene (solvent/degreaser)	________
8	590	1,1,1-Trichloroethane (solvent/degreaser)	________
9	890	1,1,2-Trichloroethane (solvent/degreaser)	________
10	1590	Toluene (organic solvent)	________
11	1910	Vegetable oil mists (cooking oil)	________

4. The test duration (observation period)
5. A series of doses to test

Possible test organisms range from isolated cellular material or selected strains of bacteria through higher-order plants and animals. The response or biological endpoint can range from subtle changes in organism physiology or behavior to death of the organism, and exposure periods may vary from a few hours to several years. Clearly, tests are sought (1) for which the response is not subjective and can be consistently determined, (2) that are conclusive even when the exposure period is relatively short, and (3) (for predicting effects in humans) for which the test species responds in a manner that mimics or relates to the likely human response. However, some tests are selected because they yield indirect measurements or special kinds of responses that are useful because they correlate well with another response of interest; for example, the determination of mutagenic potential is often used as one measure of a chemical's carcinogenic potential.

Fortunately or unfortunately, each of the five basic components of a toxicity test protocol may contribute to the uniqueness of the dose–response curve that is generated. In other words, as one changes the species, dose, toxicity of interest, dosage rate, or duration of exposure, the dose–response relationship may change significantly. So, the less comparable the animal test conditions are to the exposure situation you wish to extrapolate to, the greater the potential uncertainty that will exist in the extrapolation you are attempting to make. For example, as can be seen in Table 1.3, the organ toxicity observed in the mouse and the severity of that toxic response change with the air concentration of chloroform to which the animals are exposed. Both of these characteristics of the response—organ type and severity—also change as one changes the species being tested from the mouse to the rat.

In the mouse the liver is apparently the most sensitive organ to chloroform-induced systemic toxicity; therefore, selecting an air concentration of 3 ppm to prevent liver toxicity would also eliminate the possibility of kidney or respiratory toxicity. If the concentration of chloroform being tested is increased to 100 ppm, severe liver injury is observed, but still no injury occurs in the kidneys or respiratory tract of the mouse. If test data existed only for the renal and respiratory systems, an exposure level of 100 ppm might be selected as a no-effect level with the assumption that an exposure limit at this concentration would provide complete safety for the mouse. In this case the assumption would be incorrect, and this allowable exposure level would produce an adverse exposure condition for the mouse in the form of severe liver injury.

Note also that a safe exposure level for kidney toxicity in the mouse, 100 ppm, would not prevent kidney injury in a closely related species like the rat. This illustrates the problem in assuming that two

TABLE 1.3 Chloroform Toxicity: Inhalation Studies

Species	Toxicity of Interest	Duration of Exposure	Exposure/Dose (ppm)
Mouse	No effect—liver	6 h/day for 7 days	3
Mouse	Mild liver damage	6 h/day for 7 days	10
Mouse	Severe liver damage	6 h/day for 7 days	100
Mouse	No effect—kidneys	6 h/day for 7 days	100
Mouse	Mild kidney injury	6 h/day for 7 days	300
Mouse	No effect—respiratory	6 h/day for 7 days	300
Rat	No effect—respiratory	6 h/day for 7 days	3
Rat	Nasal injury	6 h/day for 7 days	10
Rat	No effect—kidneys	6 h/day for 7 days	10
Rat	Mild kidney injury	6 h/day for 7 days	30
Rat	No effect—liver	6 h/day for 7 days	100
Rat	Mild liver damage	6 h/day for 7 days	300

Source: Adapted from ATSDR (1996), *Toxicant Profile for Chloroform.*

similar rodent species like the mouse and rat have very similar dose–response curves and the same relative organ sensitivities to chloroform. For example, an investigator assuming both species have the same dose–response relationships might, after identifying liver toxicity as the most sensitive target organ in the mouse, use only clinical tests for liver toxicity as the biomarker for safe concentrations in the rat. Following this logic, the investigator might erroneously conclude that chloroform concentrations of 100 ppm were completely protective for this species (because no liver toxicity was apparent), although this level would be capable of producing nasal and kidney injury.

This simple illustration emphasizes two points. First, it emphasizes the fact that dose–response relationships are sensitive to, and dependent on, the conditions under which the toxicity test was performed. Second, given the variety of the test conditions that might be tested or considered and the variety of dose–response curves that might ultimately be generated with each new test system, the uncertainty inherent in any extrapolation of animal data for the purpose of setting safe exposure limits for humans is clearly dependent on the breadth of toxicity studies performed and the number of different species tested in those studies. This underscores the need for a toxicologist, when attempting to apply animal data for risk assessment purposes, to seek test data where the response is not subjective, has been consistently determined, and has been measured in a species that is known to, or can reasonably be expected to, respond qualitatively and quantitatively the way humans do.

Because the dose–response relationship may vary depending on the components of the test, it is, of course, best to rely on human data that have been generated for the same exposure conditions of interest. Unfortunately, such data are rarely available. The human data that are most typically available are generated from human populations in some occupational or clinical setting in which the exposure was believed at least initially, to be safe. The exceptions, of course, are those infrequent, unintended poisonings or environmental releases. This means that the toxicologist usually must attempt to extrapolate data from as many as four or five different categories of toxicity testing (dose–response) information for the safety evaluation of a particular chemical. These categories are: occupational epidemiology (mortality and morbidity) studies, clinical exposure studies, accidental acute poisonings, chronic environmental epidemiology studies, basic animal toxicology tests, and the less traditional alternative testing data (e.g., invertebrates, in vitro data). Each type or category of toxicology study has its own advantages and disadvantages when used to assess the potential human hazard or safety of a particular chemical. These have been summarized in Table 1.4, which lists some of the advantages and disadvantages of toxicity data by category:

Part *a*—occupational epidemiology (human) studies

TABLE 1.4 Some Advantages and Disadvantages of Toxicity Data by Category

Advantages	Disadvantages
a. Occupational Epidemiology (Human) Studies	
May have relevant exposure conditions for the intended use of the chemical	Exposures (especially past exposures) may have been poorly documented
As these exposure levels are usually far higher than those found in the general environment, these studies generally allow for a realistic extrapolation of a safe level for environmental exposures	Difficult to properly control; many potential confounding influences (lifestyle, concurrent diseases, genetic, etc.) are inherent in most work populations; these potential confounders are often difficult to identify
The chance to study the interactive effects of other chemicals that might be present; again at high doses relative to most environmental situations	Post facto—not necessarily designed to be protective of health
Avoid uncertainties inherent in extrapolating toxicities and dose–response relationships across species	The increase in disease incidence may have to be large or the measured response severe to be able to demonstrate the existence of the effect being monitored (e.g., cancer)
The full range of human susceptibility (sensitivity) may be measurable if sufficiently large and diverse populations can be examined	The full range of human sensitivity for the toxicity of interest may not be measurable because some potentially sensitive populations (young, elderly, infirm) are not represented
May help identify gender, race, or genetically controlled differences in responses	Effects must be confirmed by multiple studies as heterogeneous populations are examined, and confounders cannot always be excluded
The potential to study human effects is inherent in almost all industrial uses of chemicals; thus, a large number of different possible exposure/chemical regimens are available for study	Often costly and time-consuming; cost/benefit may be low if confounders or other factors limit the range of exposures, toxicities, confounders, or population variations that might occur with the chemical's toxicity
b. Clinical (Human) Exposure Studies	
The toxicities identified and the dose–response relationship measured are reported for the most relevant species to study (humans)	The most sensitive group (e.g., young, elderly, infirm) may often be inappropriate for study
Typically the components of these studies are better defined and controlled than occupational epidemiology studies	May be costly to perform
The chance to study the interactive effects of other chemicals	Usually limited to shorter exposure intervals than occupational epidemiologic studies
The dose–response relationship is measured in humans; exposure conditions may be altered during the exposure interval in response to the presence or lack of an effect making NOAELs or LOAELs easier to obtain	Only NOAELs are targeted for study; these studies are primarily limited to examining safe exposure levels or effects of minimal severity; more serious effects caused by the chemical cannot intentionally be examined by this type of study
Better than occupational studies for detecting relatively subtle effects; greater chance to control for the many confounding factors that might be found in occupational studies	Chronic effects are generally not identifiable by this type of study
Allows the investigator to test for and identify possible confounders or potential treatments	Requires study participant compliance
Allows one to test specific subpopulations of interest	May require confirmation by another study
May help identify gender, race or genetically controlled differences in responses	May raise ethical questions about intentionally exposing humans to toxicants
May be the best method for allowing initial human exposure to the chemical, particularly if medical monitoring is a prominent feature of the study	—

(continued)

TABLE 1.4 Continued

Advantages	Disadvantages
c. Environmentally Exposed Epidemiologic Studies	
The toxicities identified and the dose–response relationship measured are reported for the most relevant species to study (humans)	Exposures to the chemical are typically low relative to other types of human exposure to the chemical in question, or to chemicals causing related toxicities (e.g., exposure to other environmental carcinogens); thus, attributing the effects observed in a large population may be difficult if many confounding risk factors are present and uncontrolled for in the exposed population
Exposure conditions are relevant to understanding or preventing significant environmentally caused health effects from occurring	The exposure of interest may be so low that it is nontoxic and only acting as a surrogate indicator for another risk factor that is present but not identified by the study
The chance to study the effects of interactive chemicals may be possible	The number of chemicals with interactive effects may be numerous and their exposure heavy relative to the chemical of interest; this will confound interpretations of the data
The full range of human susceptibility may be present	The full range of human susceptibility may not be present, depending upon the study population
May allow one to test specific subpopulations of interest for differences in thresholds, response rates, and other important features of the dose–response relationship	The full complement of relevant environmental exposure associated with the population are not necessarily identified or considered
May help identify gender, race or genetically controlled differences in responses	Large populations may be so heterogeneous in their makeup that when compared to control responses, differences in confounders, gender, age, race. etc., may weaken the ability to discriminate real disease associations with chemical exposure from other causes of the disease
d. Acute Accidental Poisonings	
Exposure conditions are realistic for this particular safety extrapolation	If the exposure is accidental, or related to a suicide, accurate exposure information may be lacking and difficult to determine
These studies often provide a temporal description indicating how the disease will develop in an exposed individual	The knowledge gained from these studies may be of limited relevance to other human exposure situations
Inexpensive relative to other types of human studies	Confounding factors affecting the magnitude of the response may be difficult to identify as exposure conditions will not be recreated to identify modifying factors
Identifies the target organs affected by high, acute exposures; these organs may become candidate targets for chronic toxicity studies	Acute toxicities may not mimic those seen with chronic exposure; this may mislead efforts to characterize the effects seen under chronic exposure situations
Requires very few individuals to perform these studies	These studies are typically case reports or a small case series, and so measures of individual variations in response may be difficult to estimate
The clinical response requires no planning as the information gathering typically consists of responding to and treating the organ injuries present as they develop	These chance observations develop without warning, a feature that prevents the development of a systematic study by interested scientists who are knowledgeable about the chemical

(continued)

TABLE 1.4 Continued

Advantages	Disadvantages
e. Animal Toxicity Tests	
Easily manipulated and controlled	Test species response is of uncertain human relevance; thus, the predictive value is lower than that of human studies
Best ability to measure subtle responses	Species responses may vary significantly both qualitatively and quantitatively; thus, a number of different species should be tested
Widest range of potential toxicities to study	Exposures levels may not be relevant to (they may far exceed) the human exposure level
Chance to identify and elucidate mechanisms of toxicity that allow for more accurate risk extrapolations to be made using all five categories of toxicity test data	Selecting the best animal species to study, i.e., the species with the most accurate surrogate responses, is always unknown and is difficult to determine a priori (without a certain amount of human test data); thus, animal data poses somewhat of a catch-22 situation, i.e., you are testing animals to predict human responses to the chemical but must know the human response to that chemical to accurately select the proper animal test species
Cheaper to perform than full-scale epidemiology studies	May be a poor measure of the variability inherent to human exposures because animal studies are so well controlled for genetics, doses, observation periods, etc.
No risk of producing adverse human health effects during the study	The reproducibility of the animal response may create a false sense of precision when attempting human extrapolations

Source: Adapted from Beck et al. (1989).

f. Alternatives to Traditional Animal Testing

Type of Toxicity Test	Advantages	Disadvantages
Structure–activity relationships (SARs)	Does not require the use of any experimental animals Quick to perform	Many toxicants with very similar chemical properties have very different toxicities
In vitro testing	Reduces the number of experimental animals needed Allows for better control of the toxicant concentration at the target site Allows for the study of isolated functions such as nerve-muscle interaction and release of neurotransmitter Easier to control for host factors such as age dependency, nutritional status, and concurrent disease Possible to use human tissue	Cannot fully approximate the complexities that take place in whole organisms (i.e., absorption, distribution, biotransformation, and elimination)
Alternative animal testing (nonmammalian and nonavian species)	Less expensive and quicker (due to shorter lifespans) than using higher animals Since a whole organisms is used allows for absorption, distribution, biotransformation, and elimination of the toxicant	Since the animal is far removed from humans, the effect of a toxicant can be very different from that found with higher animals

Part *b*—clinical (human) exposure studies
Part *c*—environmentally exposed epidemiology studies
Part *d*—acute accidental poisonings
Part *e*—animal toxicity tests
Part *f*—alternative animal test systems

Frequency-Response and Cumulative-Response Graphs

Not only does response to a chemical vary among different species; response also varies within a group of test subjects of the same species. Experience has shown that typically this intraspecies variation follows a normal (Gaussian) distribution when a plot is made relating the frequency of response of the organisms and the magnitude of the response for a given dose (see Figure 1.1*a*). Well-established statistical techniques exist for this distribution and reveal that two-thirds of the test population will exhibit a response within one standard deviation of the mean response, while approximately 95 and 99 percent, respectively, lie within two and three standard deviations of the mean. Thus, after testing a relatively small number of animals at a specific dose, statistical techniques can be used to define the most probable response (the mean) of that animal species to that dose and the likely range of responses one would see if all animals were tested at that dose (about one or two standard deviations about the

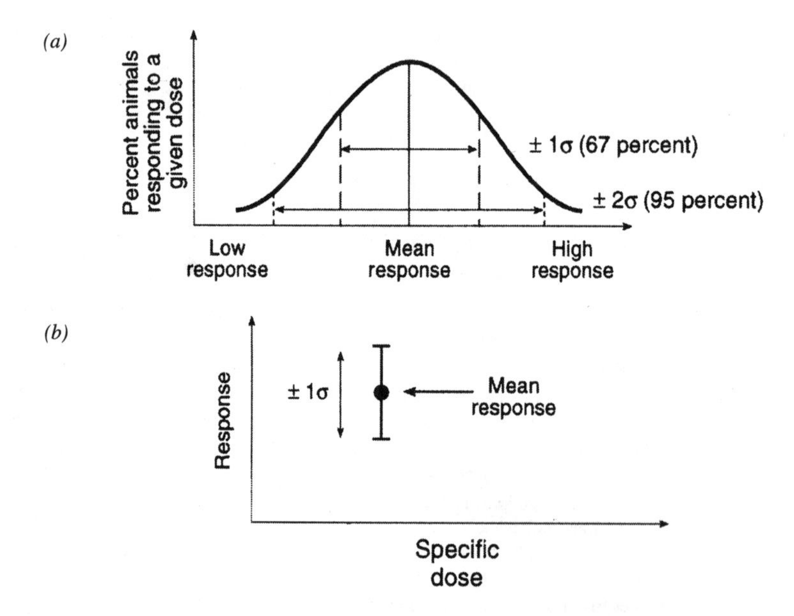

Figure 1.1 *(a)* When the response of test animals is plotted for a given dose, we see that some may show a minimal effect while others are more affected by the same dose. Plotting the percent of animals showing a particular magnitude of response gives a bell-shaped curve about the mean response. One standard deviation in either direction from the mean should encompass the range of responses for about two-thirds (67 percent) of the animals. Two standard deviations in both directions encompasses 95 percent of the animals. *(b)* The probable response for a test animal can therefore be easily predicted by testing *n* animals at a dose. By plotting the average of the *n* values as a point bracketed by one standard deviation, the probable response of an animal should fall within the area bracketed about the mean at least two-thirds of the time. *(c)* By plotting the cumulative dose-response (the probable responses for various doses), we generate a curve that is representative of the probable response for any given dose. *(d)* By plotting the cumulative dose–response curve, using the logarithm of the dose, we transform the hyperbolic shape of the curve to a sigmoid curve. This curve is nearly linear over a large portion of the curve, and it is easier to see or estimate values from this curve.

mean). Typically, a frequency–response curve for each dose of interest is not used to illustrate the dose–response relationship; instead, cumulative dose–response curves are generally used because they depict the summation of several frequency–response curves over a range of different dosages. Graphically, the separate results for each dose are depicted as a point (the average response) with bars extending above and below it to exhibit one standard deviation greater and less than this average response (see Figure 1.1*b*). A further refinement is then made by plotting the cumulative response in relation to the logarithm of the dose, to yield plots that are typically linear for most responses between 0 and 100 percent, and it is from this curve that several basic features of the dose–response relationship can be most readily identified (see Figures 1.1*c*,*d*).

In Figure 1.2, a cumulative dose–response curve is featured with a dotted line falling through the highest dose that produces no response in the test animals. Because this dose, and all doses lower than it, fail to produce a toxic response, each of these doses might be referred to as no-observable-effect levels (NOELs), which are useful to identify because they represent safe doses of the chemical. The highest of these NOELs is commonly referred to as the "threshold" dose, which may be simply defined as the dose below which no toxicity is observed (or occurs). For all doses that are larger than the threshold dose, the response increases with an increase in the dose until the dose is high enough to produce a 100 percent response rate (i.e., all subjects respond). All doses larger than the lowest dose producing a 100 percent response will also produce a 100 percent response and so the curve becomes flat as increasing dose no longer changes the response rate. For therapeutic effects, this region of the dose–response curve is typically the region physicians seek when they prescribe medicines. Because physicians are seeking a beneficial (therapeutic) effect, typically they would select a dose in this region

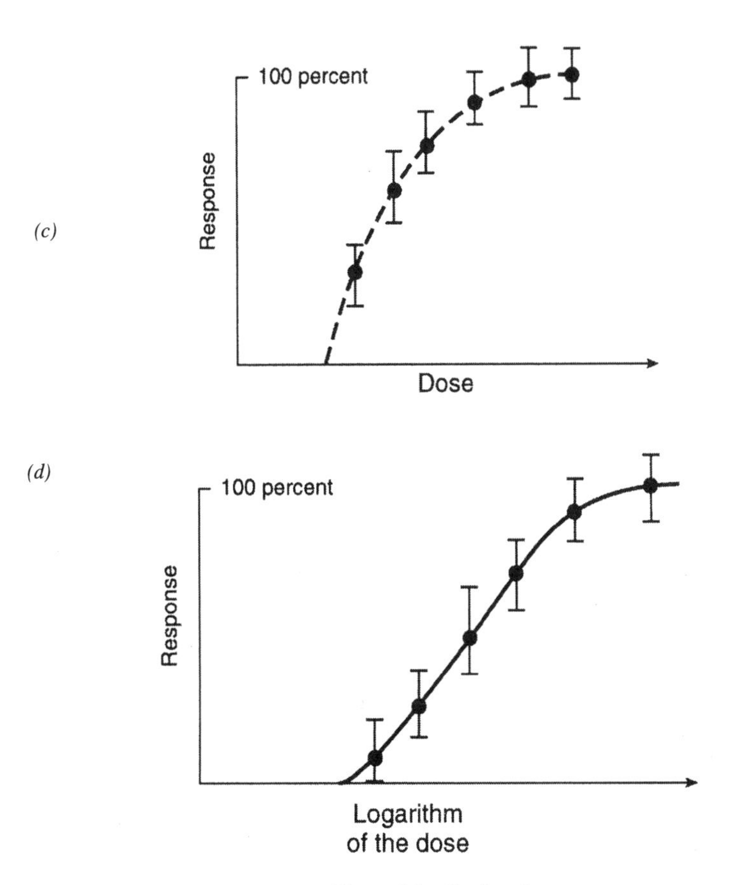

Figure 1.1 Continued

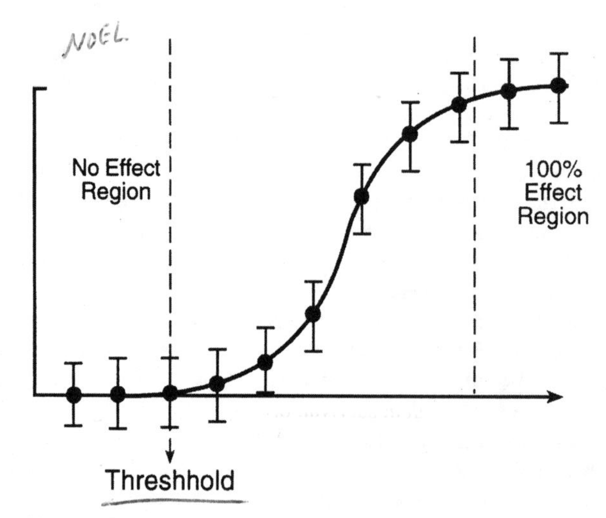

Figure 1.2 The no effect region is the range of doses that falls below the threshold dose. The threshold is the highest dose which elicits no effect (or the dose below which a response is not observed).

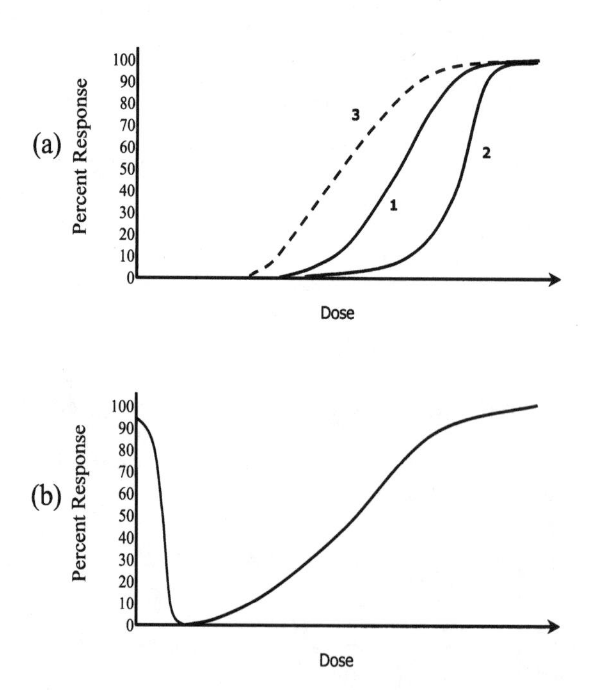

Figure 1.3 (a) The dose-response curve can have a variety of shapes, including line 1, which is linear; line 2, which is sublinear; and line 3, which is supralinear. (b) U-shaped curve representing a dose-response relationship for a chemical with beneficial, as well as adverse effects. At very low doses, a beneficial effect occurs, which is lost with increasing dose. Even higher doses produce a toxic effect. Other variations on the shape of the dose-response curve are possible, depending upon how toxic and beneficial effects are portrayed.

that is just large enough so that individual variations in response to the dose would still result in a 100 percent response, so as to ensure the efficacy of the drug. In contrast, a toxicologist is generally seeking those doses that produce no response because the effect induced by the chemical is an undesirable one. Thus, toxicologists seek the threshold dose and no effect region of the dose–response curve.

Before discussing other ways in which dose–response data can be used to assess safety, it will be useful to briefly discuss the various shapes a dose–response curve might take. Although the schematic shapes illustrated in the Figures 1.1 and 1.2 are the most common shapes, the dose–response curve could have either a *supralinear* or *sublinear* shape to it (see Figure 1.3). In Figure 1.3, the normal linear sigmoid curve is illustrated by line 1, line 2 is an example of a sublinear relationship, and line 3 depicts a supralinear relationship. In addition, some chemicals, while toxic at high doses, produce beneficial effects at low doses. Graphical presentation of this somewhat more complicated dose-response relationship results in a so-called U-shaped curve (Figure 1.3). The phenomenon of low dose stimulation (e.g., of growth, reproduction, survival, or longevity) and high dose inhibition is termed *hormesis*, and the most obvious examples of chemicals that exhibit this phenomenon are vitamins and essential nutrients. There are other agents that display hormesis for which the benefit of low doses is less intuitive. For example, a number of studies on animals and humans have suggested that low doses of ionizing radiation decrease cancer incidence and mortality while high doses lead to increased cancer risk. There is some evidence that hormesis may be applicable to a variety of types of chemical toxicants, but a careful assessment of the extent to which this represents a generalized phenomenon has been hampered by the limited availability of response data below the toxic range for most chemicals.

1.4 HOW DOSE–RESPONSE DATA CAN BE USED

Dosages are often described as *lethal doses* (LD), where the response being measured is mortality; *toxic doses* (TD), where the response is a serious adverse effect other than lethality; and *sentinel doses* (SD), where the response being measured is a non- or minimally-adverse effect. Sentinel effects (e.g., minor irritation, headaches, drowsiness) serve as a warning that greater exposure may result in more serious effects. Construction of the cumulative dose–response curve enables one to identify doses that affect a specific percent of the exposed population. For example, the LD_{50} is the dosage lethal to 50 percent of the test organisms (see Figure 1.4), or one may choose to identify a less hazardous dose, such as LD_{10} or LD_{01}.

Dose–response data allow the toxicologist to make several useful comparisons or calculations. As Figure 1.4 shows, comparisons of the LD_{50} doses of toxicants A, B, and C indicate the potency (toxicity relative to the dose used) of each chemical. Knowing this difference in potency may allow comparisons among chemicals to determine which is the least toxic per unit of dose (least potent), and therefore the safest of the chemicals for a given dose. This type of comparison may be particularly informative when there is familiarity with at least one of the substances being compared. In this way, the relative human risk or safety of a specific exposure may be approximated by comparing the relative potency of the unknown chemical to the familiar one, and in this manner one may approximate a safe exposure level for humans to the new chemical. For toxic effects, it is typically assumed that humans are as sensitive to the toxicity as the test species. Given this assumption, the test dose producing the response of interest [in units of milligrams per kilogram of body weight (mg/kg)], when multiplied by the average human weight (about 70 kg for a man and 60 kg for a woman), will give an approximation of the toxic human dose.

A relative ranking system developed years ago uses this approach to categorize the acute toxicity of a chemical, and is shown in Table 1.5. Using this ranking system, an industrial hygienist might get some idea of the acute danger posed by a workplace exposure. Similarly, if chronic toxicity is of greatest concern, that is, if the toxicity occurring at the lowest average daily dose is chronic in nature, combining a measure of this toxic dose (e.g., TD_{50}) and appropriate safety factors might generate an acceptable workplace air concentration for the chemical. Often the dose–response curve for a relatively minor acute toxicity such as odor, tearing, or irritation involves lower doses than more severe toxicities such

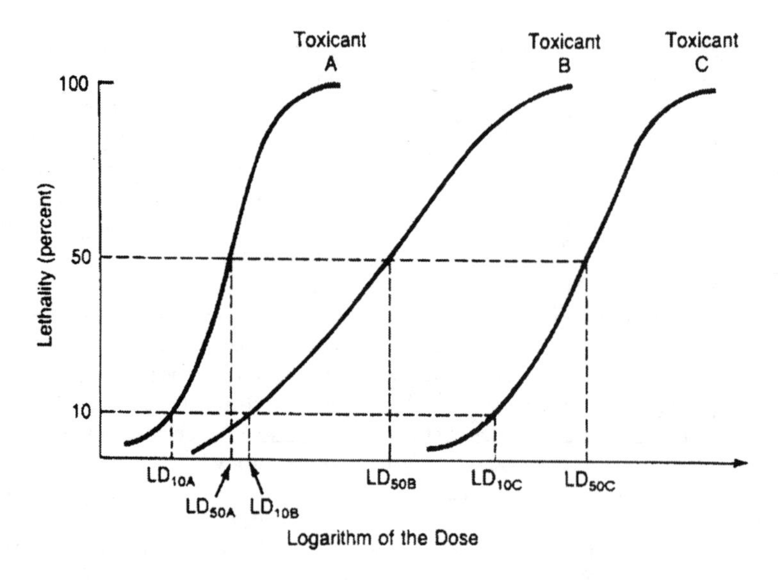

Figure 1.4 By plotting the cumulative dose–response curves (log dose), one can identify those doses of a toxicant or toxicants that affect a given percentage of the exposed population. Comparing the values of LD_{50A} to LD_{50B} or LD_{50C} ranks the toxicants according to relative potency for the response monitored.

as coma or liver injury, and much lower doses than fatal exposures. This situation is shown in Figure 1.5, and it can be easily seen that understanding the relationship of the three dose–response curves might allow the use of sentinel effects (represented in Figure 1.5 by the SD curve) to prevent overexposure and the occurrence of more serious toxicities.

The difference in dose between the toxicity curve and a sentinel effect represents the *margin of safety*. Typically, the margin of safety is calculated from data like that shown in Figure 1.5, by dividing TD_{50} by the SD_{50}. The higher the margin of safety, the safer the chemical is to use (i.e., greater room for error). However, one may also want to use a more protective definition of the margin of safety (for example, TD_{10}/SD_{50} or TD_{01}/SD_{100}) depending on the circumstances of the substance's use and the ease of identifying and monitoring either the sentinel response or the seriousness of the toxicity produced. Changing the definition to include a higher percentile of the sentinel dose–response curve (e.g., the SD_{100}) and correspondingly lower percentile of the toxic dose–response curve (e.g., the TD_{10} or the TD_{01}) forces the margin of safety to be protective for the vast majority of a population.

TABLE 1.5 A Relative Ranking System for Categorization of the Acute Toxicity of a Chemical

	Probable Oral Lethal Dose for Humans	
Toxicity Rating or Class	Dose (mg/kg)	For Average Adult
1. Practically nontoxic	> 15,000	> 1 quart
2. Slightly toxic	5000–15,000	1 pint to 1 quart
3. Moderately toxic	50–5000	1 ounce to 1 pint
4. Very toxic	50–500	1 teaspoonful to 1 ounce
5. Extremely toxic	5–50	7 drops to 1 teaspoonful
6. Supertoxic	< 5	< 7 drops

Source: Reproduced with permission of the *American Industrial Hygiene Association Journal*.

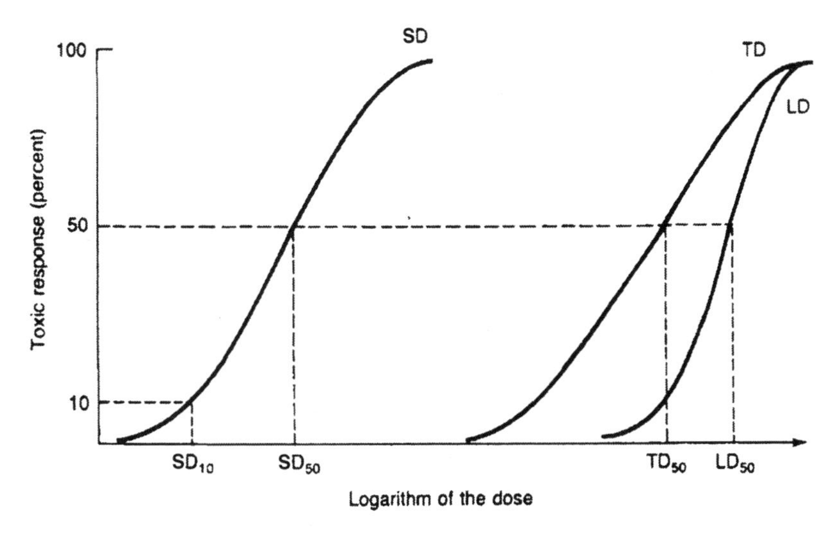

Figure 1.5 By plotting or comparing several dose–response curves for a toxicant, one can see the relationship which exists for several responses the chemical might produce. For example, the sentinel response (SD curve) might represent a relatively safe acute toxicity, such as odor or minor irritation to the eyes or nose. The toxic response (TD curve) might represent a serious toxicity, such as organ injury or coma. The lethal response (LD curve), of course, represents the doses producing death. Thus, finding symptoms of minor toxicity in a few people at sentinel response (SD_{10}) would be sufficient warning to prevent a serious or hazardous exposure from occurring.

$$\text{Margin of safety} = \frac{TD_{50}}{SD_{50}}$$

$$\text{Or redefine it as} = \frac{TD_{01}}{SD_{100}}$$

Finally, the use of dose–response curves allows for the estimation of the threshold dose or exposure (see Figure 1.2). The threshold is the lowest point on the dose–response curve, or that dose below which an effect by a given agent is not detectable. Thus, all doses, or exposures producing doses, less than the threshold dose should represent safe doses and exposures. As explained in more detail later in this chapter, the safety of extrapolating from the threshold dose is enhanced by dividing it by uncertainty factors, a procedure that is equivalent to selecting a lower dose from the no-effect region of the dose–response curve shown in Figure 1.2.

1.5 AVOIDING INCORRECT CONCLUSIONS FROM DOSE–RESPONSE DATA

While the dose–response relationship can be determined for each adverse health effect of a toxicant, one must be cognizant of certain limitations when using these data:

1. If only single values from the dose–response curves are available, it must be kept in mind that those values will not provide any information about the shape of the curve. So, while toxicant A in Figure 1.6 would appear to be more toxic than toxicant B chemical at higher doses, this is not true at lower doses. Toxicant B has a lower threshold and actually begins to cause adverse effects at lower doses than toxicant A. Once someone is exposed to a toxicant, the shape of the dose–response curve may be as important as the dose at which toxicity first begins (the threshold dose). Actually, in this regard toxicant A is of greater concern, not necessarily because of its lower LD_{50} and LD_{100} but rather

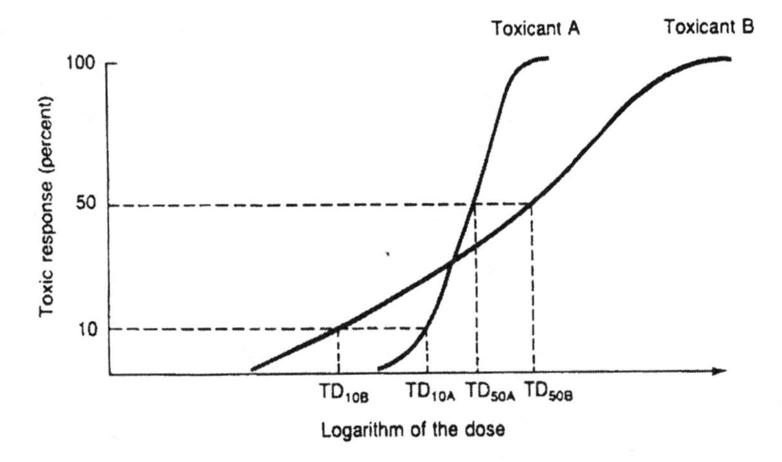

Figure 1.6 The shape of the dose–response curve is important. By finding the LD_{50} values for toxicants A and B from a table, one would erroneously assume that A is (always) more toxic than B. The figure demonstrates that this is not true at low doses.

because of its steeper dose–response curve. Once individuals become overexposed (exceed the threshold dose), the increase in response occurs with much smaller increases in dose, and more persons are affected with subsequent increases in dose. In other words, once the toxic level is reached, the margin of error for substance A decreases more rapidly than for substance B, because each incremental increase in exposure greatly increases the percent of individuals affected.

2. Acute toxicity, which is often generated in tests because of the savings in time and expense, may not accurately reflect chronic toxicity dose–response relationships. The type of adverse response generated by a substance may differ significantly as the exposure duration increases in time. Chronic toxicities are often not the same as acute adverse responses. For example, both toluene and benzene cause depression of the central nervous system, and for this acute effect toluene is the more potently toxic of the two compounds. However, benzene is of greater concern to those with chronic, long-term exposure, because it is carcinogenic while toluene is not.

3. There is usually little information for guidance in deciding what animal data will best mimic the human response. For example, a question that often arises initially in the study of a chemical is the following: Is the test species less sensitive or more sensitive than humans? As shown in Table 1.6, the dose of chloroform that is lethal to 50 percent of the test animals (i.e., the LD_{50}) varies depending on the species and strain of animal tested. Estimation of the fatal human dose based on the animal results shown in Table 1.6 would overstate the toxicity of chloroform when using the rabbit or CD-1 mouse data, and underestimate the toxicity of chloroform if projecting lethality using data from the two remaining mouse strains or the two rat strains tested.

Unfortunately, there are anatomical, physiological, and biochemical differences among animal species. These differences may confound the animal to human extrapolation by increasing the uncertainty and concern we have for the accuracy of the extrapolation being made. For example, some laboratory animals possess certain anatomical features that humans lack, such as the Zymbal gland and a forestomach. So, when a chemical produces organ toxicity or cancer within these structures, the relevance to humans is unknown. Similarly, male rats produce a protein known as α-2-microglobulin, which has been shown to interact with the metabolites of certain chemicals in a manner that results in repeated cellular injury within the kidney. This reaction is believed to be responsible for the kidney tumors seen in the male rat after chronic exposure to these chemicals. Because this unique protein from these animals does not occur to any appreciable extent in female rats or in mice, kidney tumors are not

TABLE 1.6 Oral LD_{50} Data for Chloroform

Species	LD_{50} (mg/kg/day)
Rabbit (Dutch Belted)	100[a]
Mouse (CD-1)	250
Human	602
Rat (Sprague–Dawley)	908
Mouse (Swiss)	1100
Mouse (ICR-Swiss)	1400[b]
Rat (Wistar)	2180

Source: Adapted from ATSDR (1996), *Toxicant Profile for Chloroform.*
[a]Based on 13 days of dosing.
[b]Female mice.

seen in female rats or male and female mice. From these important sex and species differences, regulatory agencies have concluded the male rat kidney tumors are of limited relevance to humans, a species which is also deficient in α-2-microglobulin. Finally, certain animal strains are uniquely sensitive to certain types of cancer. For example, a large proportion of B6C3F1 mice develop liver tumors before they die, and this sensitivity appears to be due in part to the fact that the H-*ras* oncogene in this mouse strain is hypomethylated, allowing this oncogene to be expressed more easily, especially during recurrent hepatocellular injury. Similarly, 100 percent of strain A mice typically develop lung tumors before these animals die, and so a chemical that promotes the early development of lung tumors in this strain of mice may not produce any lung tumors in other strains. To summarize, then, there are a number of important species differences that may cause changes in (1) basal metabolic rates; (2) anatomy and structure; (3) physiology and cellular biochemistry; (4) the distribution of chemicals to certain tissues and pharmacokinetics of the chemical in the animal; (5) the metabolism, bioactivation, and detoxification of the chemical; and (6) ultimately the cellular, tissue, or organ response to actions of the chemical at the biochemical, cellular, tissue, or organ level.

This problem of species-specific responses to chemicals creates somewhat of a paradox in toxicological research. We use animals as models to study the toxicities of many chemicals; yet, the proper selection of the animal to serve as the test system ideally requires prior knowledge of which animal species most closely resembles humans with respect to the chemical interaction of interest. Thus, the toxicologist is almost always faced with a dilemma. The goal of the toxicologist's study is the prediction of chemical effects on humans by using animal studies. However, selection of the right animal for that study requires a prior knowledge of the fate and effects of the chemical in humans (the goal), as well as its fate and effects in various animals. Thus, once data are generated in a test species, there are always inherent limitations to extrapolating the observed effects to humans. This is especially problematic when, as sometimes happens, one of the species tested is susceptible to a very undesirable effect, such as cancer or birth defects, yet several other species show no such effects. In that situation, determining or choosing which species represents the human response most accurately has, of course, a great impact on the estimated risk.

1.6 FACTORS INFLUENCING DOSE–RESPONSE CURVES

Organism-Related Factors

Characteristics of the test species or the human population may alter the dose–response curve or limit its usefulness. The following variables should be considered when extrapolating toxicity data:

Route of Exposure The exposure pathway by which a substance comes in contact with the body determines how much of it enters (rate and extent of absorption) and which organs are initially exposed to the largest concentration of the substance. For example, the water and lipid solubility characteristics of a chemical affect its absorption across the lungs (after inhalation), the skin (after dermal application), or the gastrointestinal (GI) tract (after oral ingestion), and the effect differs for each organ. The rate and site of absorption (organ) also may in turn determine the rate of metabolism and excretion of the chemical. So, changing the route of exposure may alter the dose required to produce toxicity. It may also alter the organ toxicity that is observed. For example, the organ with generally the greatest capacity for the metabolism and breakdown of chemicals is the liver. Therefore, a chemical may be more or less toxic per unit of dosage when the chemical is given orally or peritoneally, routes of administration that ensure the chemical absorbed into the bloodstream passes through the liver before it perfuses other organs within the animal. If the capacity of the liver to metabolize the chemical within the bloodstream is great, this leads to what is referred to as a *first-pass effect,* in which the liver metabolizes a large proportion of the chemical as it is absorbed and before it can be distributed to other tissues. If the metabolism of this chemical is strictly a detoxification process, then the toxic potency of the chemical (i.e., toxicity observed per unit of dose administered) may be reduced relative to its potency when administered by other routes (e.g., intravenously). On the other hand, if the metabolism of that dose generates toxic, reactive metabolites, then a greater toxic potency may be observed when the chemical is given orally relative to inhalation, dermal, or intramuscular administrations of the chemical. (See also discussion in Chapters 2 and 3.)

As an illustration that the route of exposure may or may not affect the toxic potency of the chemical, Table 1.7 lists LD_{50} data for various routes of exposure for three different chemicals. All of these chemicals were administered to the same test species so that differences relating to the route of exposure may be compared. As this table shows, in some instances the potency changes very little with a change in the route of administration (e.g., potency is similar for the pesticide DFP for all routes except dermal), in other instances—DDT, for example—the potency decreases 10-fold when changing the route of administration from intravenous to oral, and another 10-fold when moving from oral to dermal.

Sex Gender characteristics may affect the toxicity of some substances. Women have a larger percent of fat in their total body weight than men, and women also have different susceptibilities to reproduction system disorders and teratogenic effects. Some cancers and disease states are sex-linked. Large sex-linked differences are also present in animal data. One well-known pathway for sex-related differences occurs in rodents where the male animals of many rodent strains have a significantly greater capacity for the liver metabolism and breakdown of chemicals (they have more cytochrome P450; see Chapter 3). This greater capacity for oxidative metabolism can cause the male animals of certain rodent strains to be more or less susceptible to toxicity from a chemical depending on whether oxidative

TABLE 1.7 Effect of Route of Administration on Response (LD_{50})[a]

Route of Administration	Methadone[b]	Strychnine[b]	DDT[b]	DFP[c]
Oral	90	16.2	420	4
Subcutaneous	48	3	1500	1
Intramuscular	—	4	—	0.75
Intraperitoneal	33	1.4	100	1
Intravenous	10	1.1	40	0.3
Intraocular	—		—	1.15
Dermal	—		3000	117

Source: Adapted from *Handbook of Toxicology*, 1956, Vol. I.

[a]All doses are in units of mg/kg.

[b]Rat.

[c]Rabbit.

metabolism represents a bioactivation or detoxification pathway for the chemical at the dose it is administered. For example, in the rat, strychnine is less toxic to male rats when administered orally because their greater liver metabolism allows them to break down and clear more of this poison before it reaches the systemic circulation. This allows them to survive a dose that is lethal to their female counterparts. Alternatively, this greater capacity for oxidative metabolism renders male rodents more susceptible to the liver toxicity and carcinogenicity of a number of chemicals that are bioactivated to a toxic, reactive intermediate during oxidative metabolism.

Age Older people have differences in their musculature and metabolism, which change the disposition of chemicals within the body and therefore the levels required to induce toxicity. At the other end of the spectrum, children have higher respiration rates and different organ susceptibilities [generally they are less sensitive to central nervous system (CNS) stimulants and more sensitive to CNS depressants], differences in the metabolism and elimination of chemicals, and many other biological characteristics that distinguish them from adults in the consideration of risks or chemical hazards. For example, the acute LD_{50} dose of chloroform is 446 mg/kg in 14-day-old Sprague–Dawley rats, but this dose increases to 1188 mg/kg in the adult animal.

Effects of Chemical Interaction (Synergism, Potentiation, and Antagonism) Mixtures represent a challenge because the response of one chemical might be altered by the presence of another chemical in the mixture. A synergistic reaction between two chemicals occurs when both chemicals produce the toxicity of interest, and when combined, the presence of both chemicals causes a greater-than-additive effect in the anticipated response. Potentiation describes that situation when a chemical that does not produce a specific toxicity nevertheless increases the toxicity caused by another chemical when both are present. Antagonists are chemicals that diminish another chemical's measured effect. Table 1.8 provides simple mathematical illustrations of how the effect of one or two chemicals changes if their combination causes synergism, potentiation, additivity or antagonism, and gives a well-known example of a chemical combination that produces each type of interaction.

Modes of Chemical Interaction Chemical interactions can be increased or decreased in one of four ways

1. *Functional*—both chemicals affect the same physiologic function.
2. *Chemical*—a chemical interaction between the two compounds affects the toxicity of one of the chemicals.
3. *Dispositional*—the absorption, metabolism, distribution, or excretion of one of the chemicals is altered by the second chemical.
4. *Receptor-mediated*—when two chemicals bind to the same tissue receptor, the second chemical, which differs in activity, competes for the receptor and thereby alters the effect produced by the first chemical.

TABLE 1.8 Mathematical Representations of Chemical Interactions

Effect	Relative Toxicity (hypothetical)	Example
Additive	$2 + 3 = 5$	Organophosphate pesticides
Synergistic	$2 + 3 = 20$	Cigarette smoking + asbestos
Potentiation	$2 + 0 = 10$	Alcohol + carbon tetrachloride
Antagonism	$6 + 6 = 8$ or $5 + (-5) = 0$ or $10 + 0 = 2$	Toluene + benzene or caffeine + alcohol or BAL + mercury

TABLE 1.9 Chemical Interactions with Ethanol

Agent	Toxic Interaction	Mode: Mechanism
Aspirin	Increased gastritis	Functional—both agents irritate the GI tract
Barbiturates	Increased barbiturate toxicity	Functional/Dispositional—both agents are CNS depressants; altered pharmacokinetics and pharmacodynamics of the barbiturates
Benzene	Increased benzene-induced hematotoxicity	Dispositional—enhanced benzene bioactivation to toxic metabolites
Caffeine	Caffeine antagonizes the CNS depressant effects of ethanol	Functional—both agents affect the CNS, but one is a stimulant and one is a depressant
Carbon disulfide	Enhanced CS_2 toxicity	Dispositional—increased CS_2 bioactivation and retention in critical tissues
Chloral hydrate	Increased CNS sedative effects of chloral hydrate	Functional/dispositional—both agents are CNS depressants; ethanol also alters the metabolism of chloral hydrate, leading to greater trichloroethanol accumulation
Ethylene glycol	Decreased ethylene glycol toxicity	Dispositional—ethanol inhibits the metabolism of ethylene glycol to its toxic metabolites
Nitrosamines	Increase in formation of extrahepatic tumors induced by nitrosamines	Dispositional—ethanol alters the tissue distribution of nitrosamines by inhibiting hepatic metabolism

Source: Adapted from Calabrese (1991).

To help illustrate the ways in which chemical interactions are increased (additive, potentiation, synergism) or decreased (antagonism), Tables 1.9 and 1.10, adapted from a textbook on chemical interactions by Edward Calabrese, are provided. Table 1.9 summarizes a few of the chemical interactions identified for drinking alcohol (ethanol) and other chemical agents that might be found in home or occupational environments.

Like alcohol, smoking may also alter the effects of other chemicals, and the incidence of some minor drug-induced side effects have been reported to be lower in individuals who smoke. For example,

TABLE 1.10 Aquatic Toxicity Interactions between Ammonia and Other Chemicals

Chemicals	Toxic Endpoint	Ratio of Chemical EC_{50}s	Interaction
Ammonia + cyanide	96-h LC_{50}	1 : 1	Additive
Ammonia + sulfide	24-h LC_{50}	1 : 2.2	Antagonism
Ammonia + copper	48-h LC_{50}	1 : 1	Additive
	48-h LC_{25}	1 : 1	Synergism
	48-hr LC_{10}	1 : 1	Synergism
Ammonia + phenol	24-h LC_{50}	1 : 0.1	Antagonism
		1 : 0.7	Additive
Ammonia + phenol + zinc	48-h LC_{50}	1 : 1 : 0.5	Additive
		1 : 7 : 1	Synergism
		1 : 1 : 6	Antagonism

Source: Adapted from Calabrese (1991).

smoking seems to diminish the effectiveness of propoxyphene (Darvon) to relieve pain, and it lowers the CNS depressant effects of sedatives from the benzodiazepine and barbiturate families. Smoking also increases certain metabolic pathways in the liver and so enhances the metabolism of a number of drugs. Examples of drugs whose metabolism is increased by smoking include antipyrine, imipramine, nicotine, pentazocine, and theophylline.

Table 1.10 summarizes a few of the chemical interactions that have been reported in aquatic toxicity studies. Note that when the same chemicals are present but the ratio of components present in the mixture is changed, the type of interaction observed may change.

Genetic Makeup We are not all born physiologically equal, and this provides both advantages and disadvantages. For example, people deficient in glucose-6-phosphate dehydrogenase (G6PD deficiency) are more susceptible than others to the hemolysis of blood by aspirin or certain antibiotics, and people who are genetically slow acetylators are more susceptible to neuropathy and hepatotoxicity from isoniazid. Table 1.11 lists some of the genetic differences that have been identified in humans and some of the agents that may trigger an abnormal response in an affected individual.

Health Status In addition to the genetic status, the general well-being of an individual, specifically, their immunologic status, nutritional status, hormonal status, and the absence or presence of concurrent diseases, are features that may alter the dose–response relationship.

Chemical-Specific Factors

We have seen that a number of factors inherent in the organism may affect the predicted response; certain chemical and physical factors associated with the form of the chemical or the exposure conditions also may influence toxic potency (i.e., toxicity per unit of dose) of a chemical.

Chemical Composition The physical (particle size, liquid or solid, etc.) and chemical (volatility, solubility, etc.) properties of the toxic substance may affect its absorption or alter the probability of

TABLE 1.11 Pharmacogenetic Differences in Humans

Condition	Enzyme Affected	Some Chemicals Provoking Abnormal Responses
Acatalasia	Catalase—red blood cells	Hydrogen peroxide
Atypical cholinesterase	Plasma cholinesterase	Succinyl choline
Acetylation deficiency	Isoniazid acetylase	Isoniazid, sulfamethazine, procainamide, dapsone, hydralazine
Acetophenetidin-induced methemaglobinemia	Cytochrome P450	Acetophenetidin
Polymorphic hydroxylation of debrisoquine	Cytochrome P450	Encainide, metoprolol, debrisoquine, perphenazine
Polymorphic hydroxylation of mephenytoin	CYP 2C19	Mephenytoin
Glucose-6-phosphate dehydrogenase deficiency	Glucose-6-phosphate dehydrogenase	*Hemolytic anemia*: aspirin, acetanilide, aminosalicylic acid, antipyrine, aminopyrine, chloroquine, dapsone, dimercaprol, Gantrasin, methylene blue, naphthalene, nitrofurantoin, probenecid, pamaquin, primaquine, phenacetin, phenylhydrazine, potassium perchlorate, quinacrine, quinine, quinidine, sulfanilamide, sulfapyridine, sulfacetamide, trinitrotoluene

Source: Adapted from Vesell (1987).

exposure. For example, the lead pigments that were used in paints decades ago were not an inhalation hazard when applied because they were encapsulated in the paints. However, as the paint aged, peeled, and chipped, the lead became a hazard when the paint chips were ingested by small children. Similarly, the hazards of certain dusts can be reduced in the workplace with the use of water to keep finely granulated solids clumped together.

Exposure Conditions The conditions under which exposure occurs may affect the applied dose of the toxicant, and as a result, the amount of chemical that becomes absorbed. For example, chemicals bound to soils may be absorbed through the skin poorly compared to absorption when a neat solution is applied because the chemical may have affinity for, and be bound by, the organic materials in soil. Concentration, type of exposure (dermal, oral, inhalation, etc.), exposure medium (soil, water, air, food, surfaces, etc.), and duration (acute or chronic) are all factors associated with the exposure conditions that might alter the applied or absorbed dose.

1.7 DESCRIPTIVE TOXICOLOGY: TESTING ADVERSE EFFECTS OF CHEMICALS AND GENERATING DOSE–RESPONSE DATA

Since the dose–response relationship aids both basic tasks of toxicologists—namely, identifying the hazards associated with a toxicant and assessing the conditions of its usage—it is appropriate to summarize toxicity testing, or descriptive toxicology. While a number of tests may be used to assess toxic responses, each toxicity test rests on two assumptions:

1. *The Hazard Is Qualitatively the Same.* The effects produced by the toxicant in the laboratory test are assumed to be the same effects that the chemical will produce in humans. Therefore, the test species or organisms are useful surrogates for identifying the hazards (qualitative toxicities) in humans.
2. *The Hazard Is Quantitatively the Same.* The dose producing toxicity in animal tests is assumed to be the same as the dose required to produce toxicity in humans. Therefore, animal dose–response data provide a reliable surrogate for evaluating the risks associated with different doses or exposure levels in humans.

Which tests or testing scheme to follow depends on the use of the chemical and the likelihood of human exposure. In general, part or all of the following scheme might be required in a descriptive toxicology testing program.

Level 1: Testing for acute exposure

- a. Plot dose–response curves for lethality and possible organ injuries.
- b. Test eyes and skin for irritation.
- c. Make a first screen for mutagenic activity.

Level 2: Testing for subchronic exposure

- a. Plot dose–response curves (for 90-day exposure) in two species; the test should use the expected human route of exposure.
- b. Test organ toxicity; note mortality, body weight changes, hematology, and clinical chemistry; make microscopic examinations for tissue injury.
- c. Conduct a second screen for mutagenic activity.
- d. Test for reproductive problems and birth defects (teratology).
- e. Examine the pharmacokinetics of the test species: the absorption, distribution, metabolism, and elimination of chemicals from the body.
- f. Conduct behavioral tests.
- g. Test for synergism, potentiation, and antagonism.

Level 3: Test for chronic exposure

a. Conduct mammalian mutagenicity tests.
b. Conduct a 2-year carcinogenesis test in rodents.
c. Examine pharmacokinetics in humans.
d. Conduct human clinical trials.
e. Compile the epidemiologic data of acute and chronic exposure.

Establishing the safety and hazard of a chemical is a costly and time-consuming effort. For example, the rodent bioassay for carcinogenic potential requires 2–3 years to obtain results, at a cost of between \$3,000,000–\$7,000,000 and when completed the results, if positive, may in the end severely limit or prohibit the use of the chemical in question. Thus, this final test may entail additional costs if now a replacement chemical must be sought that does not have significant carcinogenic activity. Figure 1.7 outlines the approximate time required to test and develop the safety of chemicals assumed to have widespread human impact.

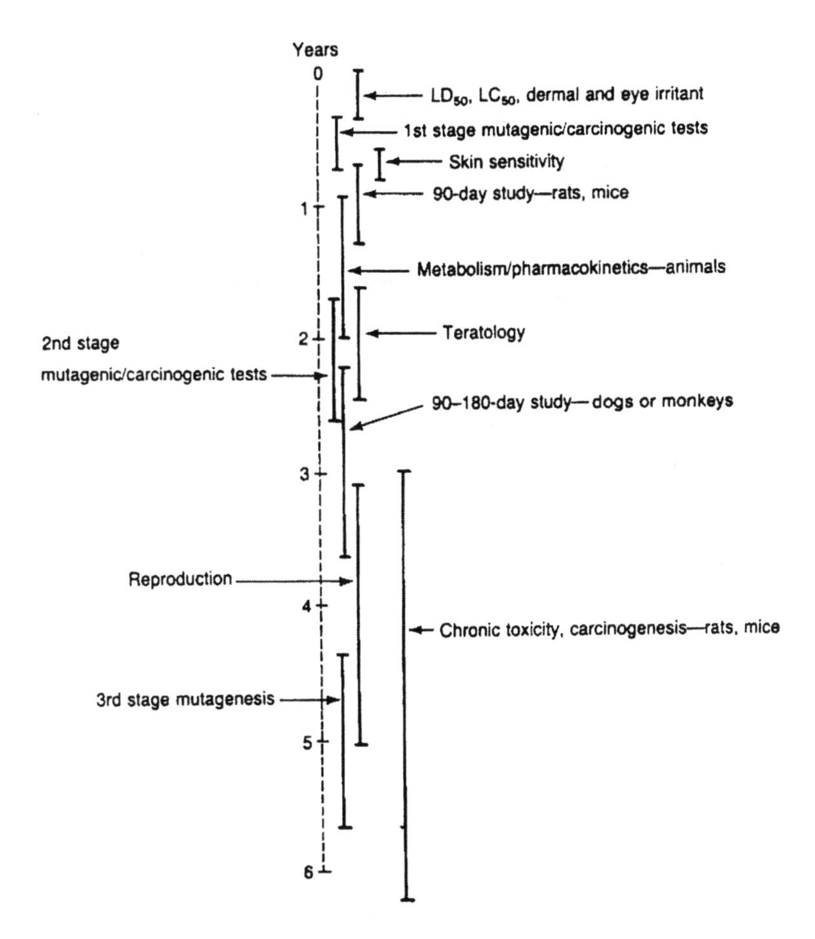

Figure 1.7 A timeline showing the approximate time that it might take to test a chemical having a broad exposure to the human population. The bars represent the approximate time required to complete the tests and suggest when testing might be initiated and completed.

1.8 EXTRAPOLATION OF ANIMAL TEST DATA TO HUMAN EXPOSURE

Several models can be used to extrapolate the human risks from chemical exposure on the basis of toxicity tests in animals. The model chosen is primarily determined by the health hazard of most concern. In the past, however, two basic methods for extrapolation were used. The first type consisted of extrapolating the human risk directly from either the threshold dose or some no-observable-effect-level (NOEL) dose. This method was applied to most toxicities or health hazards (except cancer), since thresholds were assumed to be present for all of these health hazards. The second type of model was generally used to assess the risk associated with carcinogens. Since the regulatory approach to carcinogens has been to assume that no identifiable threshold exists for this type of toxicity, any exposure was assumed to involve some quantifiable amount of risk. This concept dictated that the mathematical models used to extrapolate to exposures far below the dosages that induce observable responses in the test animal population involve some form of linear extrapolation at low doses. For noncancer-causing toxicants (those with threshold toxicity), the models for extrapolating risk are relatively simple and similar to the methods that have been suggested or used by the National Academy of Sciences (NAS) and various governmental agencies such as the Food and Drug Administration (FDA) or the Environmental Protection Agency (USEPA). These models derive a safe dosage by dividing the threshold (or NOEL/LOEL) by uncertainty factors. The purpose of adding these uncertainty factors is to ensure that the allowable human dose is one that falls within the no-effect region of the human dose–response curve. Basically, this type of calculation assumes that humans are as sensitive as the test species used; so, the amount of a chemical ingested by the test animal that gives no toxic response is considered the safe upper limit of exposure for humans (especially after inclusion of appropriate safety factors).

Calculating Safety for Threshold Toxicities: The Safe Human Dose Approach

The calculation of a safe human dose essentially makes an extrapolation on the basis of the size differential between humans and the test species. Usually this is a straightforward body-weight extrapolation, but a surface area scalar for dose could also be used. The calculation is similar to the following:

$$\text{SHD} = \frac{\text{NOAEL} = (\text{mg/kg per day}) \times 70\ \text{kg}}{\text{UF}} = N\ \text{mg/day}$$

where NOAEL = threshold dose or some other no-observable-adverse-effect-level selected from the no-effect region of the dose–response curve
SHD = safe human dose
UF = the total uncertainty factor, which depends on the nature and reliability of the animal data used for the extrapolation
N = number of milligrams consumed per day

(*Note*: In this example we are extrapolating for an average adult male, and so we have assumed a 70 kg body weight.)

Typically, the uncertainty factor used varies from 10 to 10,000 and is dependant on the confidence placed in the animal database as well as whether there are human data to substantiate the reliability of the animal no-effect levels that have been reported. Of course, the number calculated should use chronic exposure data if chronic exposures are expected. This type of model calculates one value, the expected safe human dosage, which regulatory agencies have referred to as either the *acceptable daily intake* (ADI) or the *reference dose* (RfD). Exposures which produce human doses that are at or below these safe human dosages (ADIs or RfDs) are considered safe.

Routes of Exposure and the SHD (Safe Human Dose)

Once the safe human dose has been estimated, it may be necessary to convert the dose into a concentration of the chemical in a specific environmental medium (air, water, food, soil, etc.) that corresponds to a safe exposure level for that particular route of exposure. That is, while some dose (in mg/day) may be the total safe daily intake for a chemical, the allowable exposure level of that chemical will differ depending on the route of exposure and the environmental medium in which it is found.

Exposure by Inhalation Inhalation is usually a major route for occupational exposures and safe levels are determined by the comparison of airborne concentrations to established standards. For converting the safe daily dose into a safe air concentration, the following formula may be used:

$$\text{Dosage} = \frac{(\alpha)(\text{BR})(C)(t)}{\text{BW}} = N \text{ mg/kg}$$

where

α = percent of the chemical absorbed by the lungs (if not known, considered to be 100 percent)
BR = breathing rate of the individual (which, for a normal worker, can be estimated as 2 h of heavy breathing at 1.47 m^3/h or as 6 h of moderate breathing at 0.98 m^3/h), depending on the size and physical activity of the individual
C = concentration of the chemical in the air (mg/m^3)
t = time of exposure in hours (usually considered to be 8 h)
BW = body weight in kilograms (usually considered to be 70 kg for men and 60 kg for women)

Thus, using the animal data, the preceding formula can be converted to calculate the safe air concentration if the SHD is known:

$$C = \frac{\text{SHD}}{(\alpha)(\text{BR})(t)} = N \text{ mg/m}^3$$

[*Note*: SHD = (threshold dosage divided by the uncertainty factor) × BW.]

This type of calculation can be used in two important ways:

- To predict a safe occupational airborne concentration for a chemical when there are no established airborne standards
- To compare an established occupational airborne standard (such as the TLV®—the threshold limit value established by the ACGIH—or an OSHA standard) to newly derived animal toxicity data

For many environmental exposures it may be assumed that α = 100 percent, and for adults that daily inhalation volume, equal to (BR)(t), is 20–30 m^3 for a 24-h period. To calculate an environmental air concentration for a chemical, the safe human daily dose (in units of mg/day) is divided by this total inhalation volume (in units of m^3/day). So, the acceptable air concentration (C) mg/m^3 = SHD ÷ 20 m^3/day (or 30 m^3/day). Should it be desirable to express the safe air concentration in parts of toxicant per million parts of air, the value of C [where the air concentration is in units of milligrams per cubic meter of air (mg/m^3)] may be converted to a ppm level by the following relationship:

$$\text{ppm} = \frac{C\ (\text{mg/m}^3) \times 24.5}{\text{MW}}$$

where MW is the molecular weight of the chemical (g/mol) and 24.5 is the amount (liters) of vapor per mole of contaminant at 25 °C and 760 mm Hg.

Example Calculations—Pentachlorophenol

Occupational Exposure Guidelines Pentachlorophenol (PCP), a general-purpose biocide, will be used as an example of how to derive various occupational and environmental exposure guideline extrapolations from an estimate of the safe human dosage. A literature review of the noncarcinogenic effects of PCP has shown that the toxicological effect of greatest concern is its teratogenic and fetotoxic effects in test animals. The PCP NOAEL for these effects has been reported to be as great as 5.8 mg/kg daily. Using the formulas shown above, an occupational exposure limit could be calculated as follows:

$$\mathrm{OEL} = \frac{\dfrac{5.8\ \mathrm{mg/kg\ daily\ (60\ kg)}}{100}}{1.0[(0.98\ \mathrm{m^3/h})\ 6\ \mathrm{h/day} + (1.47\ \mathrm{m^3/h})\ 2\ \mathrm{h/day}]}$$

$$\mathrm{OEL} = \frac{3.48\ \mathrm{mg/day}}{8.82\ \mathrm{m^3/day}} = 0.39\ \mathrm{mg/m^3}$$

where OEL = occupational exposure limit.

In this example, an uncertainty factor (UF) of 100 was chosen because there is extensive animal testing data for PCP; a BW of 60 kg was chosen since this type of OEL would be used to protect pregnant women; an α value of 100 percent was chosen because the amount of PCP that may be absorbed through the lungs is not known so this assumption is the most conservative; and the BR value is a standard estimate of the amount of air breathed daily by a worker performing moderately strenuous activity. This calculated level could then become a guideline for evaluating the occupational exposure of females to PCP. In using this approach, dermal exposure was not considered, but it is expected that the proper precautions (e.g., personal protective equipment and strict personal hygiene) could be used to limit these exposure pathways.

Another approach to the data would be to rearrange the formula to enable one to compare established OELs to animal toxicity data. Again, using PCP as an example, the ACGIH TLV® and OSHA PEL for PCP is an 8-h time-weighted average (TWA) exposure of 0.5 mg/m^3. This value can be compared to the animal daily NOAEL of 5.8 mg/kg by solving the following:

$$\text{Calculated daily dose} = \frac{[1.0(0.98\ \mathrm{m^3/h})\ 6\ \mathrm{h/day} + (1.47\ \mathrm{m^3/h})2\ \mathrm{h/day})]0.5\ \mathrm{mg/m^3}}{60\ \mathrm{kg}}$$

$$= 0.0735\ \mathrm{mg/kg}$$

The calculated daily dose of 0.0735 mg/kg can then be compared with the safe human dosage (SHD) of 0.058 mg/kg per day, which was generated by dividing the NOEL dosage of 5.8 mg/kg by a total uncertainty factor (UF) of 100. As one can see, if the present ACGIH TLV® and OSHA PEL are reached or exceeded, an occupationally exposed female may receive a dosage rate that exceeds the calculated or estimated SHD. Additionally, workers handling PCP would likely have some dermal exposure that will add to the daily dose calculation presented here causing the total female worker dosage to be even higher.

Environmental Exposure Guidelines

A similar approach can be used to set an acceptable ambient-air level (AAAL) or an environmental exposure guideline for other sources of exposure, such as water consumption and ingestion of foodstuffs.

Here again, it may be assumed that $\alpha = 100$ percent, and that (BR) × (t) is 20 m^3 for a 24-h period (the USEPA has recommended this value). Since environmental exposures include a more diverse population than the workplace (e.g., the old, the sick, the young), one might choose to use a UF larger than 100, one possibly as high as 1000. Thus, for a constant daily exposure the formula reduces to

$$\text{AAAL} = \frac{\text{SHD}}{20\ \text{m}^3/\text{day}} = N\ \text{mg}/\text{m}^3$$

Again using PCP as an example, the following calculation can be made:

$$\text{AAAL} = \frac{\frac{5.8\ \text{mg/kg per day}}{1000}\ 60\ \text{kg}}{20\ \text{m}^3/\text{day}}$$

where AAAL = 1.7×10^{-2} mg/m^3 per day, or 17 μg/m^3. This value could be used as an acceptable 24-h concentration of PCP in the ambient air.

Another approach to establishing an AAAL is to use the *estimated permissible concentration* (EPC). This approach uses an established OEL and applies two factors: one to take into account the potential increased exposure time for environmental exposures (i.e., 24 h per day for 7 days per week versus 8 h per day for 5 days per week); and an increased UF for the differences in populations between the workplace and the general community. The EPC can be calculated as follows:

$$\text{EPC} = \frac{\text{OEL}}{100(4.2)} = \frac{\text{OEL}}{420}$$

The value of 100 is used as an UF and the 4.2 value is used simply for the increased exposure time of 168 h per week (24 h per day for 7 days per week) versus a 40-h workweek (i.e., 168/40 =4.2). Using the PCP example, the following can be calculated:

$$\text{AAAL} = \frac{0.5\ \text{mg}/\text{m}^3}{420} = 1.19 \times 10^{-3}\ \text{mg}/\text{m}^3,\ \text{or}\ 1.2\ \mu\text{g/m}^3$$

Both of these approaches could be used for environmental exposures, but the first approach is preferable, assuming that the NOEL data for the most significant adverse effect (in this case, that occurring at the lowest dose) of a chemical are known.

For water consumption, one might adopt a 1000-fold UF and assume the average individual ingests 2 L of water per day. In this scenario, the safe water concentration for PCP becomes 174 μg/L or 174 ppb [(5.8 mg/kg per day × 60 kg) ÷ 1,000 = 348 μg/day of PCP is the SHD, which when divided by the water ingestion rate of 2 L/day of water becomes 174 μg/L.)]

If the route of environmental exposure to PCP were via the ingestion of food, then the level of PCP considered safe for a particular food item would be dependent on how much of the item is consumed each day. For this example let us assume that the fish ingestion rate is 20.1 g/day for the average fish consumer and 63 g/day for the high-end consumer (assuming this represents the 95th percentile). A safe fish concentration for PCP could be calculated as follows:

1. For the average ingestion rate:

$$\text{Fish concentration} = \frac{\frac{5.8\ \text{mg/kg daily}}{1000}\ 60\ \text{kg}}{20.1\ \text{g/day}}$$

Fish concentration (for average consumption rate) = 0.0173 mg/g = 17.3 μg/g or 17.3 ppm.

2. For the 95th percentile ingestion rate:

$$\text{Fish concentration} = \frac{\dfrac{5.8\ \text{mg/kg per day}}{1000}\ 60\ \text{kg}}{63\ \text{g/day}}$$

Fish concentration (95th) = 0.0055 mg/g = 5.5 μg/g or 5.5 ppm.

For the intake of fruits and vegetables, if we assume a daily mean consumption rate of 5.28 g/kg and a 95th percentile daily consumption rate of 22.44 g/kg, once again, using the PCP as an example, a safe vegetable and fruit concentration mean consumption rate could be calculated as follows:

Safe fruit–vegetable concentration = 0.001 mg/g = 1 μg/g = 1 ppm

To calculate the safe exposure levels for those individuals consuming fruits and vegetables at the 95th percentile consumption rate we simply divide by 22.44 g/kg per day rather than 5.28 g/kg per day and the safe fruit–vegetable concentration becomes 0.00025 mg/g = 0.25 μg/g = 0.25 ppm.

1.9 SUMMARY

Toxicology is a broad scientific field that utilizes basic knowledge of many other scientific disciplines.

- A toxicologist must understand these disciplines in order to discover and examine the variety of adverse effects produced by any toxicant.
- A toxicologist must utilize an understanding of each particular toxicant's adverse effects, and the dose–response curves for these toxicities, to develop either antidotal therapies or guidelines for risk prediction and prevention.
- A toxicologist uses dose–response relationships as a basic means of identifying the potency and toxicities that determine a chemical's relative hazards. Ultimately the dose–response curve for the toxicity of greatest concern is used to develop exposure guidelines for the human populations exposed to the chemical. These exposure levels may be dependent on the route of exposure and the perceived sensitivity of the population exposed.

Many types of toxicity tests and different factors can affect the outcome of a test or create uncertainty about its extrapolation to a heterogeneous human population.

- Often the inherent toxicity of a compound cannot be altered; in such cases the only way to lower the risk is to lower the exposure.
- Likewise, when unknown compounds are suspected of posing a hazard, or when our confidence in the estimate of their toxicity is poor, the only way to limit the risk and its liability is to limit exposure.

REFERENCES AND SUGGESTED READING

Ballantyne, B., T. C. Marrs, and P. Turner. "Fundamentals of toxicology," in *General and Applied Toxicology*, B. Ballantyne, T. Marrs, and P. Turner, eds., M. Stockton Press, New York, 1993, pp. 3–38.

Ballantine, B., "Exposure-dose-response relationships," in *Hazardous Materials Toxicology: Clinical Principles of Environmental Health*, J. B. Sullivan and G. R. Krieger, eds., Williams & Wilkins, Baltimore, 1992, pp. 24–30.

Ballantine, B., and J. B. Sullivan, "Basic principles of toxicology," in *Hazardous Materials Toxicology: Clinical Principles of Environmental Health*, J. B. Sullivan and G. R. Krieger, eds., Williams & Wilkins, Baltimore, 1992, pp. 9–23.

Beck, B. D., E. J. Calabrese, and P. D. Anderson, "The use of toxicology in the regulatory process," in *Principles and Methods of Toxicology,* 2nd ed., A. W. Hayes, ed., Raven Press, New York, 1989, pp. 1–28.

Calabrese, E. J., in *Multiple Chemical Interactions*, E. J. Calabrese, ed., Lewis Publishers, Chelsea, MI, 1991, pp. 467–544, 585–600.

Deschamps, J., and D. Morgan, "Information resources for toxicology," in *General and Applied Toxicology*, B. Ballantyne, T. Marrs, and P. Turner, eds., M. Stockton Press, New York, 1993, pp. 217–230.

Eaton, D. L., and C. D. Klassen, "Principles of toxicology," in *Casarett and Doull's Toxicology: The Basic Science of Poisons*, 5th ed., C. D. Klassen, ed., McGraw-Hill, New York, 1996, pp 13–34.

Gallo, M. A., "History and scope of toxicology," in *Casarett and Doull's Toxicology: The Basic Science of Poisons,* 5th ed., C. D. Klassen, ed., McGraw-Hill, New York, 1996, pp. 3–12.

Koeman, J. H., "Toxicology: History and scope of the field," in *Toxicology: Principles and Applications*, R. J. M. Niesink, J. deVries, and M. A. Hollinger, eds., CRC Press, New York, 1996, pp. 2–15.

Musch, A., "Exposure: Qualitative and quantitative aspects," in *Toxicology: Principles and Applications*, R. J. M. Niesink, J. deVries, and M. A. Hollinger, eds., CRC Press, New York, 1996, pp. 16–39.

Ottobani, M. A., "How chemicals cause harm," in *The Dose Makes the Poison*, M. A. Ottobani, ed., Van Nostrand-Rheinhold, New York, 1991, pp. 19–28.

Ottobani, M. A., "Toxiology—a brief history," in *The Dose Makes the Poison*, M. A. Ottobani, ed., Van Nostrand-Rheinhold, New York, 1991, pp. 29–38.

Ottobani, M. A., "Factors that influence toxicity: How much—how often," in *The Dose Makes the Poison*, M. A. Ottobani, ed., Van Nostrand-Rheinhold, New York, 1991, pp. 39–54.

Ottobani, M. A., "Other factors that influence toxicity," in *The Dose Makes the Poison,* M. A. Ottobani, ed., Van Nostrand-Rheinhold, New York, 1991, pp. 55–68.

Rhodes, C., M. Thomas, and J. Athis, "Principles of testing for acute effects," in *General and Applied Toxicology*, B. Ballantyne, T. Marrs, and P. Turner, eds., M. Stockton Press, New York, 1993, pp. 49–88.

Sullivan, J. B., and G. R. Krieger, "Introduction to hazardous material toxicology," in *Hazardous Materials Toxicology: Clinical Principles of Environmental Health*, J. B. Sullivan and G. R. Krieger, eds., Williams & Wilkins, Baltimore, 1992, pp. 2–8.

Vesell, E. S., "Pharmacogenetic differences between humans and laboratory animals: Implications for modeling," in *Human Risk Assessment: The Role of the Animal Selection and Extrapolation*, V. M. Roloff, ed., Taylor & Francis, 1987, pp. 229–238.

Williams, C. A., H. D. Jones, R. W. Freeman, M. J. Wernke, P. L. Williams, S. M. Roberts, and R. C. James, "The EPC approach to estimating safety from exposure to environmental chemicals," *Regul. Pharmacol. Toxicol.* **20**: 259–280 (1994).

Williams, P. L., "Pentachlorophenol: An assessment of the occupational hazard," *Am. Ind. Hyg. Assoc. J.* **43**: 799–810 (1982).

Answers to Table 1.1 Comparative Acutely Lethal Doses

Actual Ranking No.	LD_{50} (mg/kg)	Toxic Chemical
1	15,000	PCBs
2	10,000	Alcohol (ethanol)
3	4,000	Table salt—sodium chloride
4	1,500	Ferrous sulfate—an iron supplement
5	1,375	Malathion—pesticide
6	900	Morphine
7	150	Phenobarbital—a sedative
8	142	Tylenol (acetaminophen)
9	2	Strychnine—a rat poison
10	1	Nicotine
11	0.5	Curare—an arrow poison
12	0.001	2,3,7,8-TCDD (dioxin)
13	0.00001	Botulinum toxin (food poison)

Adapted from Loomis's Essentials of Toxicology, Fourth Edition, T.A. Loomis and A.W. Hayes, Academic Press, San Diego, CA, 1996.

Answers to Table 1.2 Occupational Exposure Limits: Aspirin and Vegetable Oil Versus Industrial Solvents

N	Allowable Workplace Exposure Level (mg/m^3)	Chemical (use)
1	0.1	Iodine
2	5	Aspirin
3	10	Vegetable oil mists (cooking oil)
4	55	1,1,2-Trichloroethane (solvent/degreaser)
5	188	Perchloroethylene (dry-cleaning fluid)
6	170	Toluene (organic solvent)
7	269	Trichloroethylene (solvent/degreaser)
8	590	Tetrahydrofuran (organic solvent)
9	890	Gasoline (fuel)
10	1590	Naphtha (rubber solvent)
11	1910	1,1,1-Trichloroethane (solvent/degreaser)

Source: American Conference of Government Industrial Hygienists (ACGIH), 1996.

2 Absorption, Distribution, and Elimination of Toxic Agents

ELLEN J. O'FLAHERTY

This chapter explains fundamental principles of toxicology, and discusses

- The broad principles that govern transfer of molecules across membranes
- The factors that influence absorption of foreign compounds from the GI tract and the lung, and across the skin
- Simple kinetic models that describe disposition (distribution and elimination)
- Mechanisms of elimination (biotransformation and excretion)

2.1 TOXICOLOGY AND THE SAFETY AND HEALTH PROFESSIONS

Occupational health specialists, including toxicologists, rely upon human and animal data to determine safe exposure levels. If effects observed in workers can be reproduced in a laboratory animal, it becomes possible to investigate the mechanisms that might reasonably be expected to produce such effects. On the other hand, shedding light on the mechanism by which a designated effect is produced in a test animal species may make it easier to find ways to prevent such effects from occurring in humans. Such an understanding may also help to identify subtle or delayed effects that have not been observed in workers, but to which health professionals should be alerted.

The uncertainties associated with converting test results in small animals to predictions relevant to humans are frequently stressed. Quantitative differences among species do exist. Occasionally, these differences can be so great that they obscure fundamental similarities. Physiological and biochemical attributes characteristic of a particular species can shift patterns of absorption, distribution, metabolism, excretion, or effect in significant ways. Nonetheless, the basic principles that control these processes are the same for all mammalian species. These principles will be surveyed in this chapter.

Toxic Agents and the Body

Figure 2.1 will be referred to several times during this chapter. It is a schematic overview of the behavior of a foreign compound as it enters the body, is distributed into tissues, exerts an effect, and is eliminated. A toxicant is absorbed into the body and then into the blood. From the blood, it is simultaneously eliminated and distributed to various tissues, including the target tissue. The target tissue is the tissue on which the toxicant exerts its effect.

Principles of Toxicology: Environmental and Industrial Applications, Second Edition, Edited by Phillip L. Williams, Robert C. James, and Stephen M. Roberts.
ISBN 0-471-29321-0 © 2000 John Wiley & Sons, Inc.

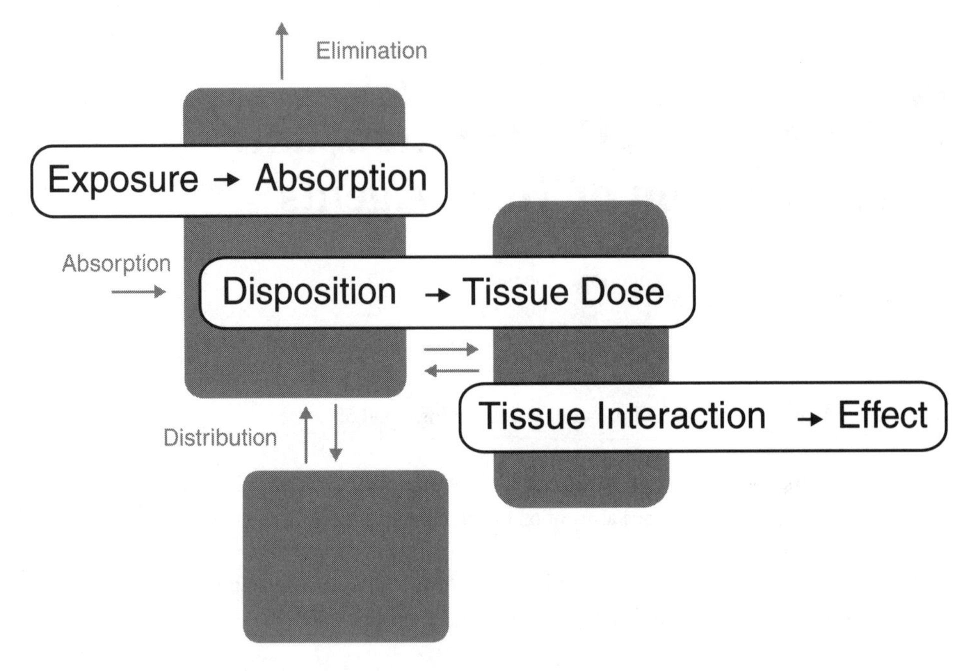

Figure 2.1 An overview of the absorption and disposition of a foreign compound. From the blood, the chemical is both eliminated and distributed to the target tissue, where it exerts its effect.

Generally, a toxicant must be considered absorbed in order to have an effect, but this is not always true. Some toxicants are locally toxic or irritating. For example, acid can cause serious damage to the skin even though it is not absorbed through the skin.

Although a distinction is made in Figure 2.1 between the target tissue and the central compartment that includes the blood, in some instances the blood itself represents the target tissue. Carbon monoxide, for example, combines with hemoglobin to form carboxyhemoglobin, whose presence in the blood reduces the availability of oxygen to the tissues. Hemolytic agents such as arsine are also active in the blood compartment, and blood is their target tissue. But most often the target tissue is a tissue other than the blood.

Significance of the Target Tissue

The target tissue or target organ is not necessarily the tissue in which the toxicant is most highly concentrated. For example, over 90 percent of the lead in the adult human body is in the skeleton, but lead exerts its effects on the kidney, the central and peripheral nervous systems, and the hematopoietic system. It is well known that chlorinated hydrocarbons tend to become concentrated in body fat stores, but they are not known to exert any effects in these tissues. Whether distribution and/or storage processes such as these are actually protective—that is, whether they act to lower the concentration of toxicant at its site of action—is not always clear. There is experimental support for the idea that certain highly localized and specialized sequestration mechanisms, such as incorporation of lead into intranuclear lead inclusion bodies or binding of cadmium to the tissue protein metallothionein, do indeed function as protective mechanisms. Whatever the case with regard to their function, however, the existence of sequestration mechanisms for many compounds means that the bulk movement of a toxicant through the body, or its kinetic behavior as reflected in plasma and tissue concentrations, must be interpreted with care. The concentration or amount of the biologically active form of the toxicant at sites in the target tissue controls the action—the dynamic behavior—of the toxicant.

2.2 TRANSFER ACROSS MEMBRANE BARRIERS

Every compound that reaches the systemic circulation and has not been intravenously injected has had to cross membrane barriers. Therefore, the first topic to be considered is the membrane itself and what enables a toxicant to cross it.

All membranes are similar in structure. They consist of a phospholipid bilayer, toward the interior of which are positioned the long hydrocarbon or fatty acid tails of the phospholipids, and toward the outside of which are the more polar and hydrophilic portions of the phospholipid molecules. The fatty acid tails align themselves in the interior of the membrane in a formation that is relatively fluid at body temperatures. The polar portions of the phospholipid molecules maintain a relatively rigid outer structure. Proteins embedded throughout the lipid bilayer have specific functions that will be considered later.

Molecules can traverse membranes by three principal mechanisms:

- Passive diffusion
- Facilitated diffusion
- Active transport

Passive Diffusion

Passive transfer does not involve the participation of any membrane proteins. Two factors determine the rate at which passive diffusion takes place across a membrane: (1) the difference between the concentrations of the chemical on the two sides of the membrane and (2) the ease with which a molecule of the chemical can move through the lipophilic interior of the membrane. Three major factors affect ease of passage: lipid solubility, or lipophilicity; molecular size; and degree of ionization.

The Partition Coefficient The lipid solubility of a compound is frequently expressed by its partition coefficient. The partition coefficient is defined as the concentration of the chemical in an organic phase divided by its concentration in water at equilibrium between the two phases. The organic phase is often chloroform, hexane or heptane, or octanol. The partition coefficient is determined by shaking the chemical with water and the organic solvent, and measuring the concentration of the chemical in each phase when equilibrium has been reached.

Although the partition coefficient does not have much meaning as an absolute value, it is very useful as an expression of the relative lipophilicities of a series of compounds. It is the rank order that is meaningful in most cases. For example, it has been shown that the partition coefficients of the nonionized forms of several series of representative drugs can be correlated with their rates of transfer across a number of biological membrane systems—from intestinal lumen into blood, from plasma into brain and into cerebrospinal fluid, and from lung into blood. In general, as lipophilicity increases, the partition coefficient increases, and so does ease of movement through the membrane (Table 2.1).

Molecular Size The second important feature of a molecule determining ease of movement across a membrane is molecular size. As the cylindrical radius of the molecule increases, with lipophilicity remaining approximately constant, rate of movement across the membrane decreases. This is because the transfer of larger molecules is slowed by frictional resistance and, depending on the structure of the molecule, may also be slowed by steric hindrance. Figure 2.2 illustrates the dependence of the permeability coefficient/partition coefficient ratio on molecular size in a series of lipophilic amides. The ratio would be constant if molecular size were not important. In this set of amides, both molecular size and steric hindrance (the branched-chain forms) are factors in slowing the diffusion of the larger molecules. Very small molecules, in contrast, may move across the membrane more rapidly than would be predicted on the basis of their partition coefficients alone. Small molecules are likely to be more

TABLE 2.1 Partition Coefficients and Rates of Transfer of Selected Drugs from Plasma into Cerebrospinal Fluid of Dogs

Drug	Heptane–Water Partition Coefficient of Nonionized Form of Drug	Half-Life of Transfer Process (min)
Thiopental	3.3	1.4
Aniline	1.1	1.7
Aminopyrine	0.21	2.8
Pentobarbital	0.05	4.1
Antipyrine	0.005	5.8
Barbital	0.002	27
N-Acetyl-4-aminoantipyrine	0.001	58
Sulfaguanidine	< 0.001	230

Source: Adapted from Brodie et al. (1960), Table 2.

water-soluble than their larger homologs. If this is the case, they may be able to move through membrane pores.

Pores are features of all membranes. Their size varies with the nature and function of the membrane. Cell membranes will not allow passage of water-soluble molecules larger than about 4×10^{-4} μm in diameter, while blood capillary walls allow passage of water-soluble molecules up to about 30×10^{-4} μm in diameter. Even within this size range of large water-soluble compounds, the rate of transcapillary movement is inversely proportional to molecular radius. Note that the cutoff of 30×10^{-4} μm excludes plasma proteins, so that they are retained within the plasma fluid volume.

Degree of Ionization The third important feature of the molecule determining ease of movement through membranes is its degree of ionization. Electrolytes are ionized at the pH values of body fluids. With the exception of very small ionized molecules that can pass through membrane pores, only the nonionized forms of most electrolytes are able to cross membranes. The ionized forms are generally too large to pass through the aqueous pores, and are insufficiently lipophilic to be transferred by passive diffusion. The rate of diffusion therefore will depend not only on the amount of an electrolyte present in the nonionized form but also on the ease with which the nonionized form of the molecule can cross the membrane, that is, on its molecular size and lipophilicity.

All ionizable acids and bases have a pK_a value related to the dissociation constant. The dissociation constant is always expressed for either acids or bases as an acid dissociation constant, K_a:

For acids:

$$K_a = \frac{(H^+)(A^-)}{(HA)}$$

For bases:

$$K_a = \frac{(H^+)(B)}{(HB^+)}$$

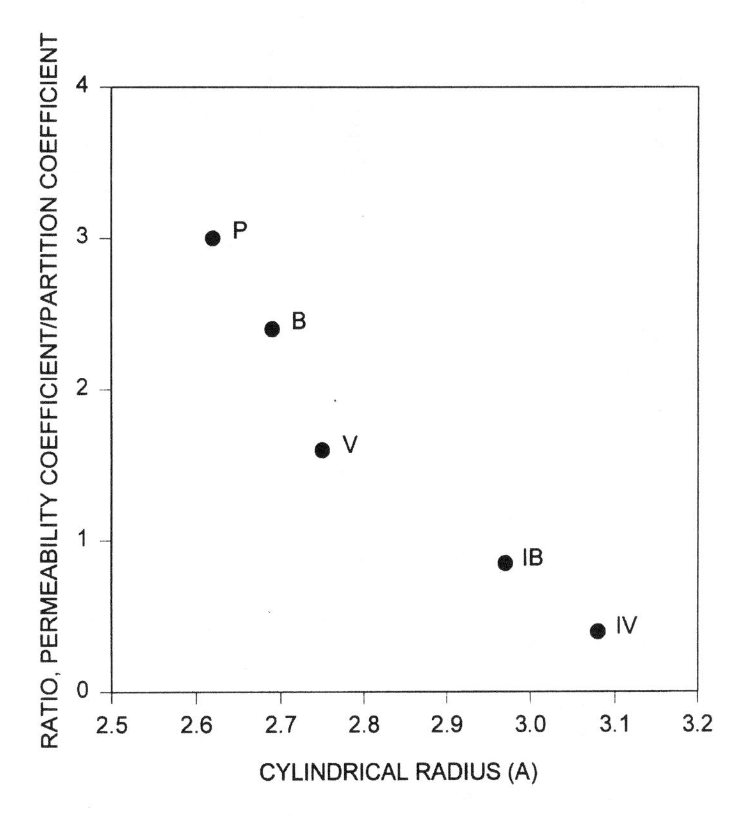

Figure 2.2 Dependence of the ratio of permeability coefficient to partition coefficient on the cylindrical radius of the diffusing molecule for diffusion of a series of lipophilic amides across the human red cell membrane. The ordinate scale is in relative, not absolute, values. (P—propionamide; B—butyramide; IB—isobutyramide; V—valeramide; IV—isovaleramide). The partition coefficients are P—0.01; B,IB—0.05; and V,IV—0.175. [Data from Sha'afi et al. (1971).]

The pK_a is the negative logarithm of the acid dissociation constant. If, for example, the acid dissociation constant K_a is 10^{-3}, then the pK_a is 3.

The degree of ionization in body fluids depends on the pH of the medium as well as on the pK_a of the acid or base. This relationship can be expressed by the Henderson–Hasselbalch equations:

For acids:

$$\mathrm{p}Ka - \mathrm{pH} = \log \frac{\text{(nonionized form)}}{\text{(ionized form)}} = \log \frac{(\mathrm{HA})}{(\mathrm{A}^-)}$$

For bases:

$$\mathrm{p}Ka - \mathrm{pH} = \log \frac{\text{(ionized form)}}{\text{(nonionized form)}} = \log \frac{(\mathrm{HB}^+)}{(\mathrm{B})}$$

When the pH is equal to the pK_a, half of the acid or base is present in the ionized form and half in the nonionized form. At pH values less than the pK_a, acids are less completely ionized. At pH greater than the pK_a, bases are less completely ionized. Another way of stating these relationships is that at a given pH, acids having a large pK_a (weak acids) are not as fully ionized as strong acids, while bases having a large pK_a (strong bases) are more fully ionized (associated with a hydrogen ion) than weak bases.

Schanker and his co-workers (Hogben et al., 1959) studied the intestinal absorption of a series of acids and bases in the rat, finding that the percentage of the drug absorbed depended on its pK_a and on whether it was an acid or a base. Weak acids and bases, largely nonionized at the intestinal surface pH of 5.3, were readily absorbed, while strong acids (pK_a < 3) and strong bases (pK_a > 7) were not. When the pH of the intestinal contents was increased from about 4 to about 8 by dissolving the test compounds in strongly buffered solutions rather than in water, the percentage absorption of the acids was decreased and that of the bases was increased.

Facilitated Diffusion

The second type of passage across a membrane is facilitated diffusion. Facilitated diffusion requires the participation of a carrier protein molecule. The carrier proteins are a subgroup of the proteins embedded in the membrane lipid bilayer. Because the number of carrier molecules is limited, facilitated diffusion has a maximum rate. It can also be inhibited selectively, competitively by agents that compete for binding sites on the carrier protein because they are structurally similar to the diffusing material, or noncompetitively by agents that affect the carrier in some other way. If diffusion equilibrium is reached, concentrations on both sides of the membrane are equal. In other words, with regard to the fraction of the substance that is absorbed, facilitated diffusion gives the same net results as passive diffusion, but facilitated diffusion is faster.

Several carrier protein systems have been studied in reasonable detail. Although many of the proteins incorporated into membranes appear at only one membrane face, all membrane carrier proteins that have been studied are known to span the membrane and to surface at both faces. It is probable that carrier proteins undergo a conformational change, associated with the binding of the diffusing molecule to the carrier, that facilitates their transport across the membrane. The requirement that the diffusing molecule must be able to bind to the carrier protein confers a certain degree of specificity on this mechanism. As might be expected, facilitated mechanisms have evolved to transport essential nutrients across membrane barriers. For example, the transport of essential nutrients such as sugars and amino acids into red blood cells and into the central nervous system takes place by facilitated diffusion.

Active Transport

Active transport is the third general process by which molecules can traverse cell membranes. In addition to its requirement for a carrier molecule, active transport requires controlled energy input. It maintains transport against a concentration gradient so that, when equilibrium is reached, the concentrations on the two sides of the membrane are not equal. Adenosine triphosphate (ATP) is the source of the energy required to maintain this concentration gradient.

Active transport processes are critical to conservation and regulation of the body's supply of essential nutrients. These are important functions of the kidney and the liver. Another function of these organs is the excretion of toxic chemicals and their metabolites. Accordingly, there are at least two active transport processes in the kidney for secretion into the urine (one for organic acids and one for organic bases) and at least four in the liver for elimination into the bile (one for acids, one for bases, one for neutral compounds, and one for metals). Other specialized active transport processes are found in the placenta and the intestine, and in the kidney for reabsorption of essential nutrients. Of course, other processes may operate simultaneously with active transport processes. Facilitated diffusion can occur along with active transport, and passive diffusion will take place in the presence of a concentration gradient whenever physical factors are sufficiently favorable.

Specialized Transport Processes

Certain specialized processes also transfer molecules across membrane barriers. Phagocytosis takes place in the alveoli of the lung and in the reticuloendothelial system of the liver and spleen. In phagocytosis, the cell membrane surrounds a particle to form a vesicle that detaches itself and moves into the cell interior. For macromolecules such as proteins, phagocytosis is particularly important. It is thought to be biologically significant in interiorizing certain kinds of specialized enzyme systems. Phagocytosis is also important because it is the mechanism used by the alveolar macrophage to scavenge particulates in the alveoli of the lung. A similar process, pinocytosis, is responsible for the cellular internalization of liquids.

2.3 ABSORPTION

For most practical purposes, we consider absorption to be absorption into the systemic circulation. Figure 2.3 is a schematic diagram drawn to show that the lungs and the lumen of the gastrointestinal (GI) tract, along with its contents, are exterior to the body. Toxicants may be absorbed from the GI tract, from the lung, or through the skin. In experimental studies, toxicants may also be injected. Injections are commonly given intravenously, intraperitoneally, subcutaneously, or intramuscularly.

Gastrointestinal Tract

The GI tract is a very important route for absorption of toxicants. Toxicants may be present in food or drinking water or, if they have been inhaled but are in the form of relatively large particles, they may

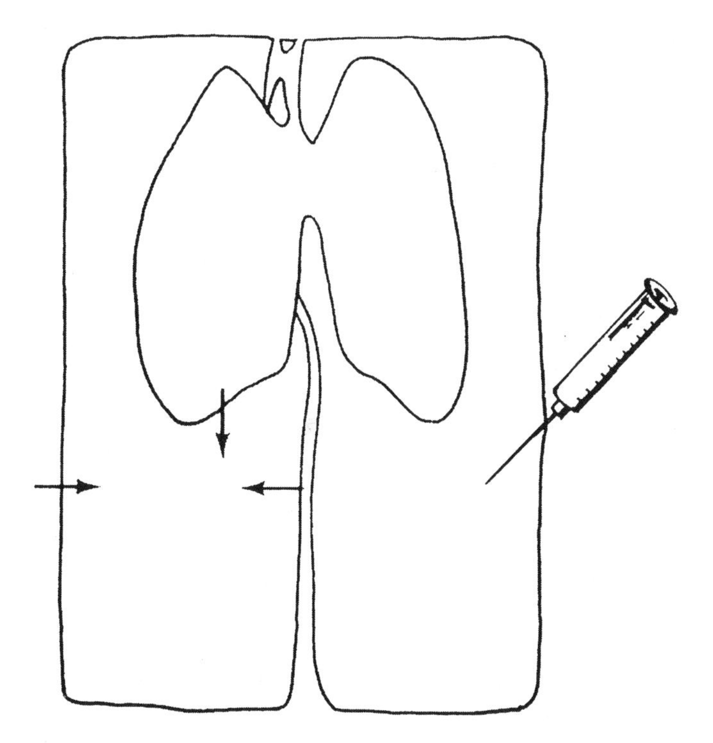

Figure 2.3 Schematic diagram of the entry of a chemical into the human body.

have been collected in the nasopharyngeal area and swallowed. Essentially only one cell thick, the epithelial wall of the GI tract is specialized not only for absorption but also for elimination.

Absorption from the GI tract is strongly site-dependent, since the pH varies from the very acidic range of about 1–3 in the stomach (depending on the amount and quality of the food and when it was eaten) to around 5–8 in the small intestine and colon (depending on location, food, and intestinal microflora). The intestinal contents can therefore be neutral or even slightly basic.

Absorption of Organic Acids and Bases Application of the Henderson-Hasselbalch equation to organic acids, which have pK_a values of 3–5, suggests that they should be relatively well absorbed from the acidic pH of the stomach. Salicylic acid is shown as an example in Figure 2.4. Its pK_a is about 2. The efficiency of its transfer across the gastric mucosa is dependent on the concentration gradient of the nonionized form across the mucosa as well as on the physical features of salicylic acid that control its rate of diffusion. As Figure 2.4 shows, in the stomach there are 100 nonionized molecules of salicylic acid for every salicylate ion. On the plasma side of the mucosal cell, however, there is relatively little salicylic acid; salicylate ion is overwhelmingly the dominant species. These calculations were carried out for steady-state conditions. In fact, once salicylic acid molecules have entered the plasma, they will be both ionized to a large extent and carried away from the absorption site by the plasma flow. These factors should combine to promote efficient absorption of organic acids from the stomach. However, organic acids are actually absorbed only moderately well in the stomach, perhaps because of its relatively small absorbing surface. Organic bases, in contrast, are largely ionized at the pH of the stomach contents, and so are much more efficiently absorbed from the intestine.

Determinants of GI Absorption A number of other factors are important in determining whether, and how rapidly, a compound will be absorbed from the GI tract. The physical factors, such as lipid solubility and molecular size, which determine the rate of diffusion of nonionized species, have already been discussed. Diffusion is also favored by the presence of villi and microvilli in the intestine. These greatly increase the surface area available for diffusion. Thus, even though absorption may not be particularly efficient per unit surface area, the very large total surface area helps to promote intestinal absorption.

Facilitated and active transport mechanisms present in the GI tract provide specialized transport for essential nutrients and electrolytes, including sugars, amino acids, sodium, and calcium. A toxicant that mimics the molecular size, configuration, and charge distribution of an essential nutrient sufficiently well may be transported by the carrier process already in place for absorption of that nutrient. Known examples of such mimicry are rare. 5-Fluorouracil has been shown to be absorbed by a pyrimidine transport mechanism. Interaction among metal ions with respect to their use of common transport mechanisms has also been documented.

Other factors affect absorption from the GI tract. Compounds that are chemically unstable at the acid pH of the stomach will not even reach the intestine to be absorbed there. Other compounds are

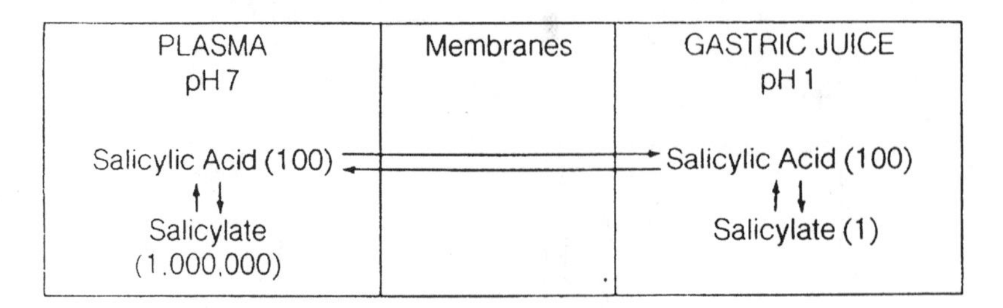

Figure 2.4 Partitioning of salicylic acid across the gastric mucosa. The numbers in parentheses are the numbers of molecules of the ionized or nonionized species present on either side of the membrane. (Reproduced with permission from O'Flaherty, 1981, Figure 2.9.)

susceptible to alteration by the actions of intestinal microflora, which are important for digestion of plant materials resistant to the action of mammalian enzymes. Enzyme systems of the intestinal wall and/or the liver may metabolize chemicals before they reach the systemic circulation, the intestinal and/or hepatic first-pass effect, which can result in significant reduction in bioavailability. For example, compared to its 100 percent availability on intravenous injection, the systemic bioavailability in rats of buprenorphine, an opiate analgesic, was found to be 49 percent when the drug was given intrahepatoportally and 10 percent when it was given intraduodenally. It can be calculated that after 80 percent of the intraduodenal dose had been inactivated in the intestine, half of the surviving 20 percent was inactivated in the liver.

Another determinant of gastrointestinal absorption is the rate at which foodstuffs pass through the GI tract. If the rate of passage is slowed, the length of time during which the compound is available for absorption is increased. Absorption also tends to increase during short periods of fasting but may fall off after a lengthy fast, probably consequent to a decrease in intestinal blood flow. Other important influences on absorption include the chemical and physical characteristics of the compound, its solubility under the conditions present in the GI tract, and its interactions with other compounds. Age and nutritional status of the individual may also affect absorption from the GI tract.

Skin

The second major pathway for absorption is the skin. The skin is a very effective barrier to absorption, primarily because of the outermost keratinized layer of thick-walled epidermal cells, the stratum corneum, which in general is not very permeable to toxicants, although its permeability varies from location to location. Compared with the total thickness of the epidermis and dermis together, the thickness of the stratum corneum is relatively slight, but this barrier is rate-limiting in the process of absorption through the skin. There may be slight absorption through sweat glands or hair follicles, but these structures represent a very small percentage of the total surface area and are not ordinarily important in the process of dermal absorption.

All toxicants that penetrate the skin appear to do so by passive diffusion. Lipophilic chemicals are much better absorbed through the skin than are hydrophilic chemicals, and the ease with which a compound penetrates the skin is correlated with its partition coefficient.

Dermal absorption can be increased in various ways. An increase in capillary blood flow, as in response to the demand of a warm environment for efficient heat loss, is associated with increased percutaneous absorption. Abrasion, which damages or removes the stratum corneum, greatly increases the permeability of the damaged area. The skin is normally partially hydrated; an increase in the degree of hydration increases permeability and promotes absorption. Certain solvents, such as dimethyl sulfoxide (DMSO), also increase skin permeability and facilitate absorption of toxicants.

Lipophilic drugs that would suffer extensive first-pass metabolism if given orally can be administered dermally. The glyceryl trinitrate patch used in treatment and prevention of angina is a good example.

Certain toxicants can produce systemic injury by percutaneous absorption. Hydrocarbon solvents, such as hexane, can produce a peripheral neurotoxicity, and carbon tetrachloride can produce liver injury. Organophosphate insecticides such as parathion and malathion have caused toxicity and deaths in industrial and field workers after absorption through the skin.

Lung

The third major site of toxicant absorption is the lung. In occupation-linked toxicology, the lung is a very important route of uptake. Gases and vapors such as carbon monoxide, sulfur dioxide, and volatile hydrocarbons are absorbed through the lung, and liquid or particulate aerosols, such as sulfuric acid aerosols or silica dust, are also deposited and/or absorbed in the lung. With solid and liquid particulates, the site of deposition is critical to the degree of absorption of a compound.

Solid and Liquid Particulates The lung can be thought of as consisting of three basic regions: the nasopharyngeal region, the tracheobronchiolar region, and the distal or alveolar region. Particles that are roughly 5 μm or greater in diameter are generally deposited in the nasopharyngeal region. If they are deposited very close to the surface, they can be sneezed out, blown out, or wiped away. If they are deposited slightly farther back, they may be picked up by the mucus-blanketed cilia lining the lung in this region (the mucociliary "escalator") and moved back up into the nasopharyngeal region, where they may be swallowed and absorbed in the GI tract in accordance with their solubility and absorption characteristics. Particles that fall into the size range of 2–5 μm generally reach the tracheobronchial region before they impact the lung surface. Most of these particles are also cleared by the mucociliary escalator back up to the nasopharyngeal region, where they are either eliminated directly or swallowed and absorbed or excreted in the GI tract. Particles smaller than 1 μm in diameter may reach the alveolar regions of the lung. Absorption in the lung, if it takes place at all, will most likely take place in the alveolar region, although there may be some absorption in the tracheobronchiolar region, particularly if the material is soluble in the mucus.

Size is probably the most important single characteristic determining the efficiency of particulate absorption in the lung. Size determines the region of the lung in which the aerosol is likely to be deposited. Even within the range of very small particles that reach the alveolar region and may be absorbed there, size is inversely proportional to the magnitude of particle deposition.

Figure 2.5 shows the dependence of lead deposition in the human lung on the size of the lead particles in an artificially generated lead sesquioxide aerosol that was inhaled by the subjects. Size is expressed as diffusion mean equivalent diameter (DMED), a measure of mean particle diameter. The amount deposited was calculated as the difference between the amount of lead that entered the lung and the amount that the subject exhaled. Thus, the lung regions in which the particles were actually deposited were not identified. However, the DMEDs for all three aerosols were less than 1 μm. For a standard subject, with a breathing cycle of about 4 s, the lung deposition of lead varied from about 24 percent for particles with a DMED of 0.09 μm to 68 percent for very small particles with a DMED of

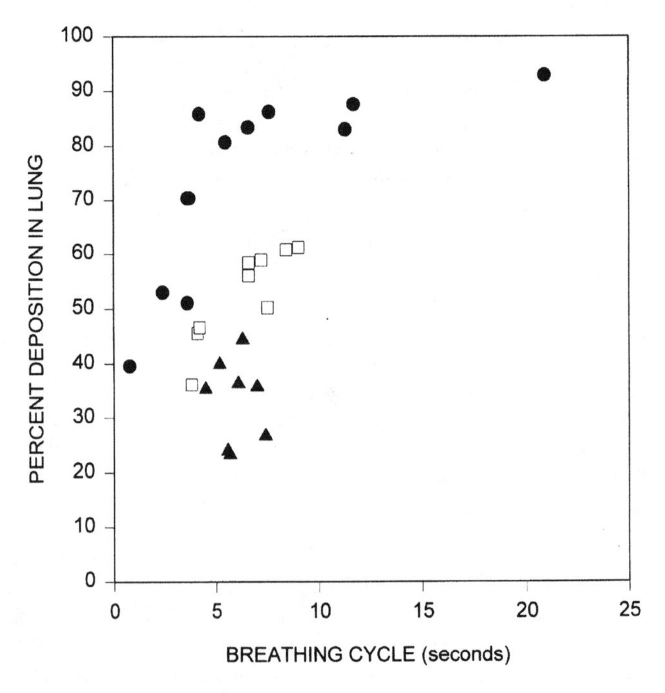

Figure 2.5 Deposition in the human lung of lead particles of various sizes: • DMED, 0.02 μm; □ DMED, 0.04 μm; ▲ DMED, 0.09 μm. Data from Chamberlain et al., 1978, Table 5.2.

0.02 μm. Note that deposition also depends on the breathing cycle. Slow, deep breathing was associated with greater percentage deposition of lead in each particle size range.

After being deposited in the alveolar region, particulates may be dissolved and absorbed into the bloodstream, reaching the systemic circulation directly. If they are not readily soluble, they may be phagocytized by alveolar macrophages and then either transferred to the lymphatic system, where they may remain for a considerable time period, or moved together with the macrophage to the mucociliary escalator for clearance by that route. They may also occasionally remain in the alveolus for an extended period of time. Absorption of particulates tends to be slower than absorption of gases and vapors, and appears to be controlled primarily by the solubility of the particulate. For example, the systemic bioavailability of chromium(VI) and nickel salts from the lung has been shown to parallel their solubility.

The absorption of water-soluble chemicals in the lung, and even in the nasal cavity, can be quite high. For example, aspirin was found to be 100 percent bioavailable from the rat nasal cavity but only 59 percent bioavailable when given orally. Nicotine was fully absorbed from intratracheal, bronchial, and distal sites in the dog lung, although absorption was not equally rapid from all three sites.

Gases and Vapors Absorption of gases and vapors in the lung depends on their solubility in the blood perfusing the lung. Very soluble compounds will be almost completely cleared from inhaled air and transferred to pulmonary blood in a single respiration. For such compounds, increasing the rate of pulmonary blood flow makes very little difference in absorption rate. The only way to increase absorption is to increase the rate of respiration; that is, to increase ventilation. Absorption of these compounds is said to be ventilation-limited. If they are also lipid-soluble, they will find their way rapidly to the lipid depots of the body.

Chloroform is a good example of such a compound. It is very highly lipid-soluble, and is readily cleared from inspired air. As the blood circulates through the body, the chloroform is transferred to fat, so that the blood is also effectively cleared and during its next pass through the lung is able to pick up more chloroform. The absorption of chloroform is ventilation-limited.

For poorly soluble gases, the capacity for absorption is rather limited. Little of the compound will be transferred to pulmonary blood in a single respiration. Often these are not compounds that are readily cleared from the blood. If this is the case, the blood will become saturated quickly, and the only way to increase absorption then is to increase the rate of pulmonary blood flow. Such compounds are said to be flow-limited in their absorption characteristics. Of course, there is a range of transition between these two extremes of pulmonary absorption behavior.

2.4 DISPOSITION: DISTRIBUTION AND ELIMINATION

Unlike absorption, disposition consists not just of one kind of process but, rather, of a number of different kinds of processes taking place simultaneously. Disposition includes both distribution and elimination, which occur in parallel in almost all cases (Figure 2.1). Elimination is also made up of two kinds of processes, excretion and biotransformation, which usually take place simultaneously.

Distribution and elimination are often considered independently of each other. While it is convenient to separate them for discussion, it is important to remember that they are taking place at the same time. If a substance is effectively excreted, it will not be distributed into peripheral tissues to any great extent. On the other hand, wide distribution of the compound may impede its excretion.

Kinetic Models

Before discussing some of the specific mechanisms for distribution, excretion, and biotransformation, it is useful to consider some simple kinetic disposition models. Rates of distribution are related to kinetic distribution constants. Rates of metabolism, or biotransformation, and of excretion are related to kinetic constants of elimination. It is possible to integrate all the essential information about distribution, biotransformation, and excretion of a chemical into a single kinetic model. Such models can be used to help formulate predictions about the kinetic behavior of a toxicant under different exposure conditions and, if used carefully and in conjunction with effect data, can sometimes also help to suggest a toxicant's site of action or mechanism of action. Kinetic models are finding increasing use in the interpretation of human exposure and response experience through an understanding of the dose–distribution/action–effect sequence gained from animal studies. In this connection, they are essential to the growing field of human health risk assessment.

Kinetic models can be assigned to two general groups: classical models and physiologically based models. The classical descriptive pharmacokinetic models were developed largely by and for the pharmaceutical industry. Their applications are often to situations in which at least some human data are available. Physiologically based models have been intensively developed and used in recent years by the toxicology community. These are based on actual anatomic and physiologic characteristics of the species and on physicochemical and metabolic characteristics of the chemical, and so lend themselves to the cross-species extrapolations with which the toxicologist must frequently deal.

Classical Kinetic Models The simplest classical kinetic model is the one-compartment open model (Figure 2.6), in which the compound is assumed to have been introduced instantaneously into the body, distributed instantaneously and homogeneously, and eliminated at a rate that is at all times directly proportional to the amount left in the body, that is, a first-order rate. The constant of proportionality between the rate of elimination and the amount in the body is the elimination rate constant (k_e), which has dimensions of reciprocal time, or time^{-1}. For this model, the logarithm of concentration in the blood is a linear function of time, as shown in Figure 2.7. The elimination rate constant is the absolute value of the slope of this line. The half-life is the time required for half the compound to be cleared from the plasma or, in this simple model, from the body. The nature of the half-life is such that it can be measured at any point on a concentration–time curve. Thus, it is a constant value, independent of dose, that characterizes the kinetic behavior of the chemical. The half-life is inversely proportional to the elimination rate constant: $t_{1/2} = \ln 2/k_e$.

Another useful concept derived from classical kinetics is the clearance. In first-order kinetics, clearance is defined as the rate constant times the volume of distribution. Therefore, clearance has dimensions of volume per unit time, or flow rate. It represents a volume of fluid cleared of the chemical per unit time by metabolism or excretion.

Few chemicals obey simple first-order one-compartment kinetics. Most compounds require at least a two-compartment model (Figure 2.8). In the two-compartment model, the chemical is assumed to enter the first or central compartment, which includes the blood; to be distributed instantaneously and homogeneously throughout this compartment; and then to be subject to the parallel processes of elimination and distribution to a second or peripheral compartment from which the chemical can return to the central compartment.

Concentration in the central compartment declines smoothly as a function of time. Concentration in the peripheral compartment rises, peaks, and subsequently declines (Figure 2.9). Assuming first-order kinetics, a half-life may also be calculated for a compound whose kinetic behavior fits a two-compartment model or, for that matter, a model with any number of compartments.

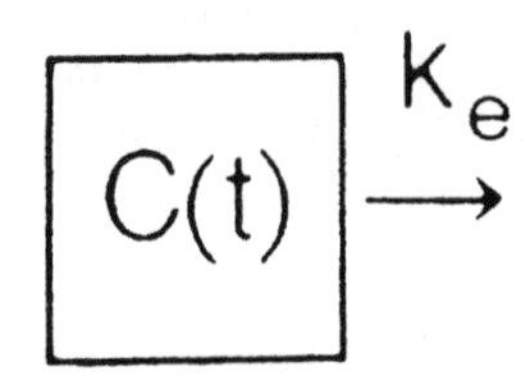

Figure 2.6 The linear one-compartment open model. $C(t)$ is the concentration, which is a function of time; and k_e is the elimination rate constant. (Reproduced with permission from O'Flaherty, 1981, Figure 2.12.)

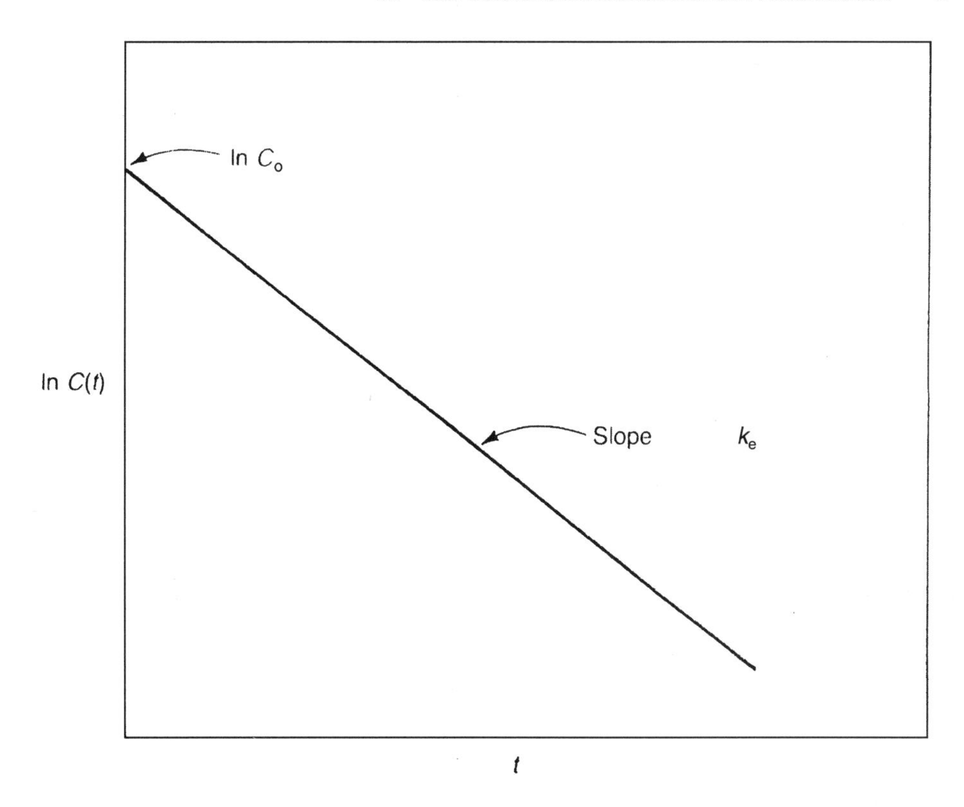

Figure 2.7 Plot of the logarithm of the concentration versus time for the linear one-compartment open model. C_0 is the concentration at time $t = 0$, assuming instantaneous distribution. (Reproduced with permission from O'Flaherty, 1981, Figure 2.15*a*.)

Calculated from the terminal slope of a plot of the natural logarithm of the concentration in the central compartment as a function of time, this half-life is designated the biological half-life. It is the parameter most frequently used to characterize the in vivo kinetic behavior of an exogenous compound.

Other features of chemical kinetic behavior or of mode of administration may be incorporated into the model as appropriate. For example, there may be more than one peripheral tissue compartment, as in Figure 2.1; or absorption, which is never truly instantaneous even for intravenous injection, may be first-order instead. An oral exposure, in which the rate of absorption is usually considered to be directly proportional to the amount remaining available in the GI tract, is an example of first-order uptake.

The important group of models that incorporate non-first-order kinetics should also be mentioned. Absorption and distribution are conventionally considered to be passive, first-order processes unless observation dictates otherwise. However, elimination often is not first-order. Frequently this is because excretion or metabolism is saturable, or capacity-limited, due to a limitation on the maximum number of active transport sites in organs of excretion or the maximum number of active sites on metabolizing enzymes. When all active elimination sites are occupied, the elimination process is said to be saturated. Kinetically it is a zero-order process, operating at a constant maximum rate independent of the amount or concentration of the chemical in the body. At very low concentrations at which relatively few elimination sites are occupied, capacity-limited kinetics reduces to pseudo-first-order kinetics. Capacity-limited kinetics is often referred to as *Michaelis–Menten kinetics*, after the authors of an early paper analyzing and interpreting this type of kinetic behavior. Classical kinetic models incorporating Michaelis–Menten elimination have been developed.

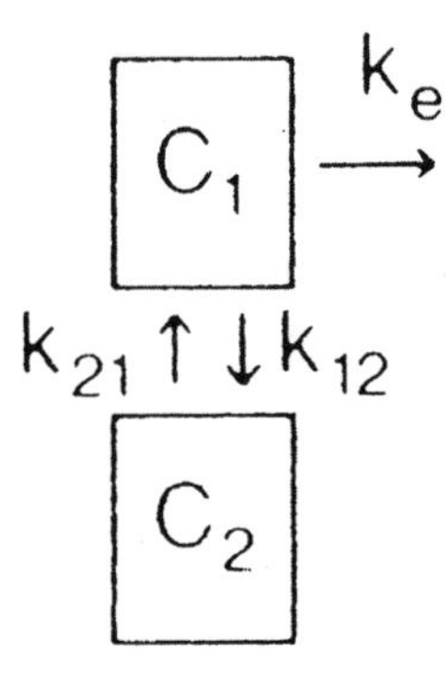

Figure 2.8 The linear two-compartment open model, where C_1 and C_2 are the concentrations in the central and peripheral compartments, respectively, and k_{12} and k_{21} are the rate constants for transfer between the two compartments. (Reproduced with permission from O'Flaherty, 1981, Figure 2.22.)

Most industrial or environmental exposures are not acute. Acute exposures do occur, but chronic exposures are much more frequent in both industrial and environmental settings. When exposure is approximately constant and continuous over a long period of time (e.g., if a contaminant is widely dispersed in ambient air), a steady state or "plateau" level will eventually be reached in all tissues. As long as elimination processes remain first-order (typical, e.g., of excretion by glomerular filtration in

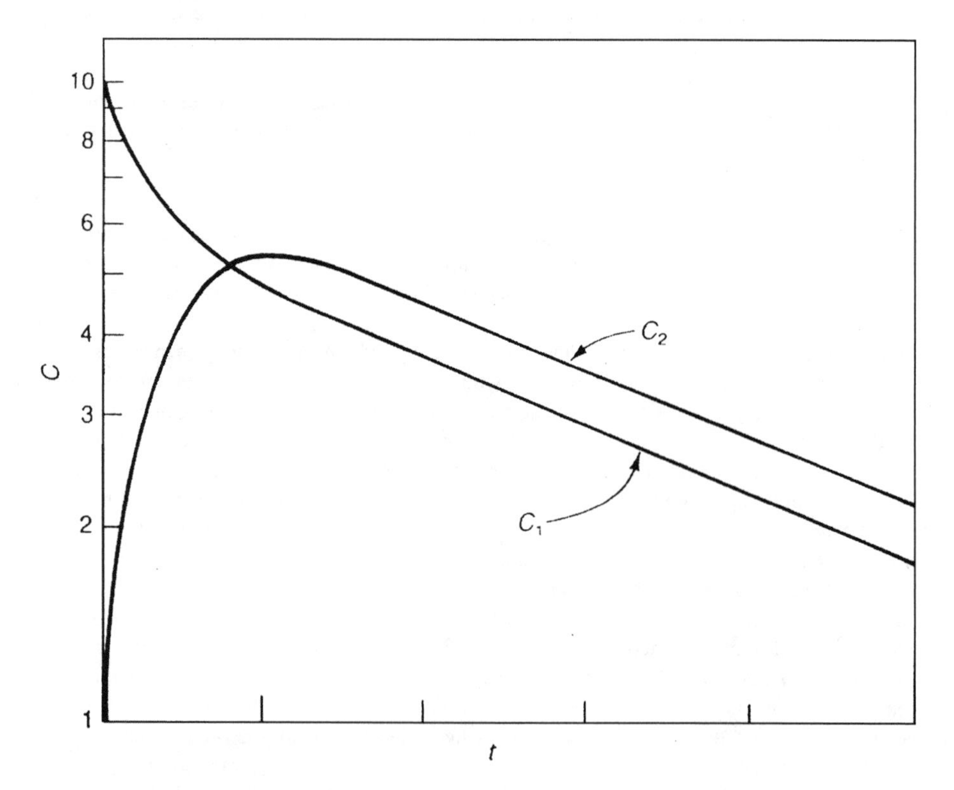

Figure 2.9 Plot of the logarithm of the concentration versus time for the linear two-compartment open model, showing ln C as a function of time for the central (C_1) and peripheral (C_2) compartments. (Reproduced with permission from O'Flaherty, 1981, Figure 3-24*b*.)

the kidney, or of loss of a volatile chemical in expired air), this steady state should be directly proportional to both the magnitude of exposure and the biological half-life.

If exposure were truly constant, the plateau level would be constant also. More commonly, exposure is intermittent, in which case blood concentrations at steady state will cycle in a way that reflects the absorption and elimination characteristics of the compound as well as the exposure pattern (Figure 2.10). However, on a larger timescale this cycling will take place about a constant mean that is predictable from the equivalent constant exposure rate and the biological half-life. This is one of the reasons why biological half-life is such an important attribute. Together with exposure rate, it determines mean steady-state blood level irrespective of whether exposure is continuous or intermittent. However, the individual exposed to large amounts of a substance at wide intervals will experience greater peak concentrations in blood and tissues following each new exposure than will an individual exposed to the same total amount as frequent small exposures. If the large peak concentrations are associated with toxicity or with saturation of elimination processes, then it becomes important to consider the pattern of administration as well as the equivalent mean exposure rate.

Physiologically Based Kinetic Models Physiologically based kinetic (PBK) models are simplified but anatomically and physiologically reasonable models of the body. Tissues are selected or grouped according to their perfusion (blood flow) characteristics and whether they are sites of absorption or elimination (by excretion or metabolism). The model design process is facilitated by reference to compilations of anatomic and physiologic data, including tissue and organ perfusion rates, that are now widely available.

Within this general structural framework, the kinetic behavior of the selected chemical is modeled. A key question is how the chemical is taken up into tissues. When flow-limited kinetics are assumed, the chemical is presumed to be in equilibrium between each tissue group and the venous blood leaving

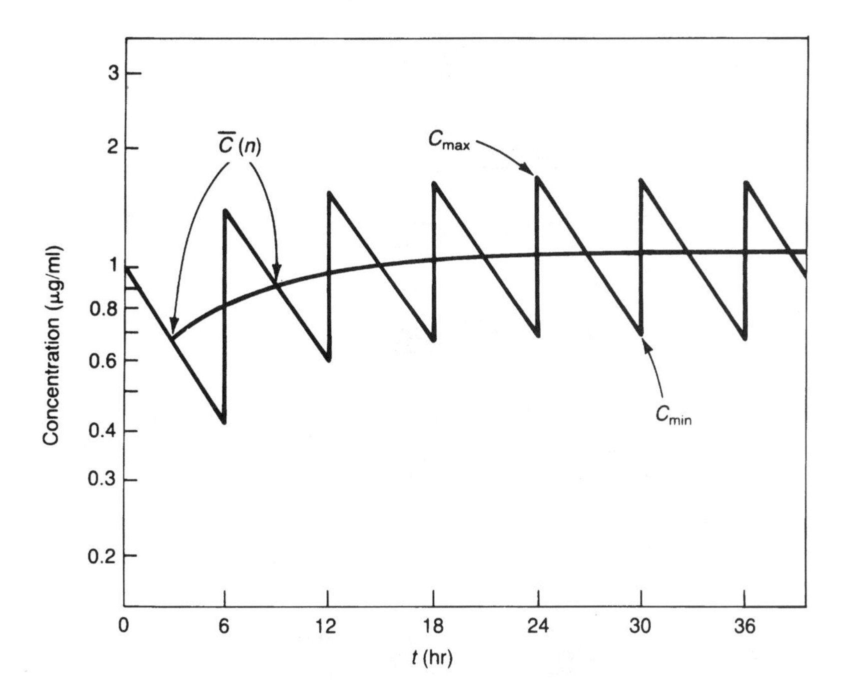

Figure 2.10 The relationship between average concentration $\overline{C}(n)$, calculated for repetitive administation, and the time course of concentration change during continuous administration of a hypothetical compound. C_{max} and C_{min} are the maximum and minimum concentrations in each time interval between doses, assuming instantaneous distribution of each successive dose. (Reproduced with permission from O'Flaherty, 1981. Figure 5-4.)

the tissue. This equilibrium will vary from tissue to tissue and may also vary from species to species. Simple partitioning phenomena, such as into body lipid stores, can be described by defining partition coefficients, whose values can be determined experimentally at steady state in vivo or in vial equilibration experiments in vitro. More complex partitioning, such as capacity-limited binding of a metal to specific binding sites in tissues, must be defined appropriately. Estimates of dissociation constants may be required.

Diffusion-limited kinetics can also be accommodated within the framework of PBK models. In diffusion-limited kinetics, the process of transfer across the membrane separating tissue from blood is the rate-limiting step in tissue uptake. The distinction between flow-limited and diffusion-limited tissue-uptake kinetics is roughly analogous to the distinction between ventilation-limited and flow-limited absorption in the lung.

The metabolism of the compound must also be known. Metabolic parameters are more likely than anatomic or physiologic parameters to be species-specific or even tissue-specific. The differences may be quantitative or qualitative. Capacity-limited metabolism, absorption, and/or excretion can be incorporated into PBK models as needed.

Figure 2.11 is a schematic diagram of a PBK model that might be designed for a volatile lipophilic chemical. Arrows designate the direction of blood flow, with arterial blood entering the organs and tissue groups and mixed venous blood returning to the lung to be reoxygenated. Organs of entry (lung, liver), excretion (kidney, intestine, lung), and metabolism (liver), and tissue of accumulation (fat) for this chemical class are explicitly included in the model. Other tissues are lumped into well-perfused and poorly perfused groups. Note that uptake into the liver is considered to take place both by way of the portal vein coming from the intestine and by way of the hepatic artery. An enterohepatic recycling

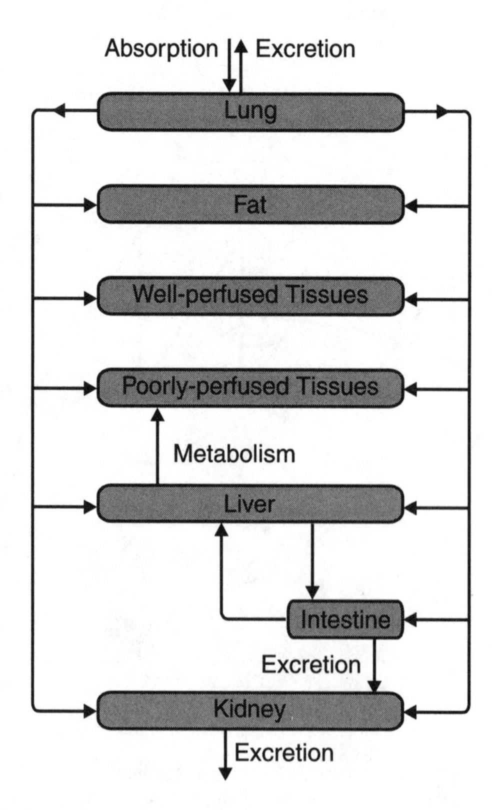

Figure 2.11 Schematic diagram of a physiologically-based model of the kinetic behavior of a volatile chemical compound.

between liver and intestine is also included in the model. These features of the model are choices made by the model developer, and reflect the known physicochemical behavior of the agent whose kinetics are being modeled. Models for other chemicals will be quite different. A model for a nonvolatile chemical would not include an explicit lung compartment, while models for bone-seeking elements like lead and uranium include bone as a distinct tissue.

In a sense, classical and PBK models work in opposite directions. In classical descriptive kinetics, model compartments having no necessary relationship to actual tissue volumes and clearances having no necessary relationship to tissue blood flow are inferred from a set of concentration data. In contrast, the PBK model is constructed from basic anatomic, physiologic, physicochemical, and metabolic building blocks. It is then used to simulate concentrations under a defined set of conditions, and its predictions are compared with observations. If the predictions are not accurate, some premise of the model is at fault. The need for model revision can afford insight into the processes that control the kinetic behavior of the chemical.

A PBK model for dichloromethane (DCM) forms the basis of a current human health risk assessment. DCM is metabolized by two pathways, a capacity-limited oxidative pathway and first-order conjugation with glutathione (for descriptions of these biotransformation processes, see Chapter 3). Either pathway was thought potentially capable of generating reactive intermediates involved in the tumorigenicity of DCM in mice. Andersen et al. (1987) demonstrated that tumorigenicity correlated well with the activity of the glutathione pathway, but not with the activity of the oxidative pathway. These investigators scaled a PBK model developed for DCM from mouse to human and from high dose to low dose in order to predict, based on studies carried out at high doses in mice, the risk associated with human environmental exposure to DCM. The mouse-to-human scaling of metabolism relied on experimentally-determined human metabolic parameter values.

Their physiologic foundation and the inclusion of species-specific physiologic and metabolic mechanisms, when these are known, confer on PBK models a flexibility that allows their use for route-to-route, dose-to-dose, and species-to-species extrapolations such as this one, for which classical models would be wholly inappropriate.

Biotransformation

Biotransformation is one of the two general elimination mechanisms. Biotransformation reactions in general can be divided into two classes: phase I and phase II reactions. Phase I reactions are catabolic or breakdown reactions (oxidation, reduction, and hydrolysis) that generate or free up a polar functional group. They produce metabolites that may be excreted directly or may become substrates for phase II reactions. Phase II reactions, which are often coordinated with phase I activity, are synthetic reactions in which an additional molecule is covalently bound to the parent or the metabolite, which usually results in a more water-soluble conjugate. Biotransformation reactions, and the factors that influence them, are discussed in detail in Chapter 3.

Excretion

Excretion takes place simultaneously with biotransformation and, of course, with distribution. The kidney is probably the single most important excretory organ in terms of the number of compounds excreted, but the liver and lung are of greater importance for certain classes of compounds. The lung is active in excretion of volatile compounds and gases. The liver, because it is a key biotransforming organ as well as an organ of excretion, is in a unique position with regard to the elimination of foreign chemicals.

Excretion in the Kidney About 20 percent of all dissolved compounds of less than protein size are filtered by the kidney in the glomerular filtration process. Glomerular filtration is a passive process; it does not require energy input. Filtered compounds may be either excreted or reabsorbed. Passive reabsorption in the kidney, as elsewhere, is a diffusion process. It is governed by the usual principles.

Thus, lipid-soluble compounds are subject to reabsorption after having been filtered by the kidney. The degree of reabsorption of electrolytes will be strongly influenced by the pH of the urine, which determines the amount of the chemical present in a nonionized form.

It is to be expected that some control could be exerted over the rate of excretion of weak acids and bases by adjusting urine pH. This type of treatment can be used very effectively in some cases. Alkalinization of the urine by administration of bicarbonate has been used to treat salicylic acid poisoning in humans. Alkalinization causes the weak acid to become more fully ionized; the ionized molecule is excreted in the urine rather than reabsorbed.

There are also active secretory and reabsorptive processes in the renal tubules of the kidney. These processes are specialized to handle endogenous compounds; active reabsorption helps to conserve the essential nutrients, glucose and amino acids. These pathways can also be used by exogenous compounds, provided the compounds have the structural and electronic configurations required by the carrier molecules.

The renal clearance represents a hypothetical plasma volume cleared of solute by the net action of all renal mechanisms during the specified period of time. A compound such as creatinine that is filtered but not secreted or reabsorbed is cleared in adult humans at a rate of about 125 mL/min. Compounds that are reabsorbed as well as filtered have clearances less than the creatinine clearance. Compounds that are actively secreted can have clearances as large as the renal plasma flow, about 600 mL/min.

The presence of disease in the kidney can affect the half-life of a compound eliminated via the kidney, just as the presence of disease in the liver can affect the half-life of a compound that is largely biotransformed.

Excretion in the Liver The liver is both the major metabolizing organ and a major excretory organ. Large fractions of many toxicants absorbed from the gastrointestinal tract are eliminated in the liver by metabolism or excretion before they can reach the systemic circulation, the hepatic first-pass effect. In addition, metabolites formed in the liver may be excreted into the bile before they themselves have had a chance to circulate. Although it does not excrete as many different compounds as the kidney does, the liver is in an advantageous position with regard to excretion, particularly of metabolites.

There are at least three active systems for transport of organic compounds from liver into bile: one for acids, one for bases, and one for neutral compounds. Certain metals are also excreted into bile against a concentration gradient. These transport processes are efficient and can extract protein-bound as well as free chemicals. The characteristics that determine whether a compound will be excreted in the bile or in the urine include its molecular weight, charge, and charge distribution. In general, highly polar and larger compounds are more frequently found in the bile. The threshold molecular weight for biliary excretion is species-dependent. In the rat, compounds with molecular weights greater than about 350 can be excreted in the bile. Those having molecular weights greater than about 450 are excreted predominantly in the bile, while compounds with molecular weights between 350 and 450 are frequently found in both urine and bile.

Once a compound has been excreted by the liver into the bile, and thereby into the intestinal tract, it can either be excreted in the feces or reabsorbed. Most frequently the excreted compound itself, being water-soluble, is not likely to be reabsorbed directly. However, glucuronidase enzymes of the intestinal microflora are capable of hydrolyzing glucuronides, releasing less polar compounds that may then be reabsorbed. The process is termed *enterohepatic circulation*. It can result in extended retention of compounds recycled in this manner. Techniques have been developed to interrupt the enterohepatic cycle by introducing an adsorbent that will bind the excreted chemical and carry it through the gastrointestinal tract.

Certain factors influence the efficiency of liver excretion. Liver disease can reduce the excretory as well as the metabolic capacity of the liver. On the other hand, a number of drugs increase the rate of hepatic excretion by increasing bile flow rate. For example, phenobarbital produces an increase in bile flow that is not related to its ability to induce metabolizing enzymes. Whether the increased rate of bile flow will increase the rate of elimination of a compound that is both metabolized and excreted by the liver depends on whether the rate-limiting step is the enzyme-catalyzed biotransformation or the transfer from liver to bile. If transfer from liver to bile is the rate-limiting step, enhancement of the rate of bile flow will enhance the rate of excretion.

Excretion in the Lung The third major organ of elimination is the lung, the key organ for the excretion of volatile chemical compounds. Pulmonary excretion, like pulmonary absorption, is by passive diffusion. For example, the rate of transfer of chloroform out of pulmonary blood is directly proportional to its concentration in the blood. Essentially, pulmonary excretion is the reverse of the uptake process, in that compounds with low solubility in the blood are perfusion-limited in their rate of excretion, whereas those with high solubility are ventilation-limited. Highly lipophilic chemicals that have accumulated in lipid depots may be present in expired air for a very long time after exposure.

Other Routes of Excretion Skin, hair, sweat, nails, and milk are other, usually minor routes of excretion. Hair can be a significant route of excretion for furred animals, and indeed the amount of a metal in hair, like the amount of a volatile compound in exhaled air, can be used as an index of exposure in both laboratory animals and humans. Hair is not quantitatively an important route of excretion in humans, however. Sweat and nails are only rarely of interest as routes of excretion, simply because loss by these routes is quantitatively so slight.

Milk may be a major route of excretion for some compounds. Milk has a relatively high fat content, 3–5 percent or even higher, and therefore compounds that are lipophilic may be excreted in milk to a significant extent. Some of the toxicants known to be present in milk are the highly lipid-soluble chlorinated hydrocarbons: for example, the polychlorinated biphenyls (PCBs) and DDT. Certain heavy metals may also be excreted in milk. Lead is thought to be secreted into milk by the calcium transport process.

2.5 SUMMARY

This chapter has conveyed some of the general biochemical and physiological principles that govern absorption, distribution, and elimination of toxic agents, in particular

- The importance of lipid solubility, molecular size, and degree of ionization to the rate at which a molecule moves through a membrane by passive transfer or diffusion.
- The characteristics of other transfer processes such as facilitated diffusion, active transport, phagocytosis, and pinocytosis.
- Absorption from the gastrointestinal tract with particular emphasis on the importance of pH as a determinant of absorption of ionizable organic acids and bases as well as on compound-specific and host-related factors such as lipid solubility and molecular size, the presence of villi and microvilli in the intestine, the possibility that the compound can be absorbed by facilitated or active transport mechanisms, and the action of gastrointestinal enzymes or intestinal microflora.
- Factors determining the rate of diffusion across the skin.
- Absorption of solid and liquid particulates and of gases and vapors in the lung.
- Simple classical and physiologically based kinetic models describing disposition (distribution, metabolism, and excretion).
- Excretion from kidney, liver (including enterohepatic circulation), and lung, and by less general routes such as skin, hair, sweat, nails, or milk.

REFERENCES AND SUGGESTED READING

Abernethy, D. R., and D. J. Greenblatt, "Drug disposition in obese humans: An update," *Clin. Pharmacokinet.* **11:** 199–212 (1986).

Andersen, M. E., H. J. Clewell, M. L. Gargas III, F. A. Smith, and R. H. Reitz, "Physiologically-based pharmacokinetics and the risk assessment process for methylene chloride," *Toxicol. Appl. Pharmacol.* **87:** 185–205 (1987).

Bragt, P. C., and E. A. van Dura, "Toxicokinetics of hexavalent chromium in the rat after intratracheal administration of chromates of different solubilities," *Ann. Occup. Hyg.* **27:** 315–322 (1983).

Brewster, D., M. J. Humphrey, and M. A. McLeavy, "The systemic bioavailability of buprenorphine by various routes of administration," *J. Pharm. Pharmacol.* **33:** 500–506 (1981).

Brodie, B. B., H. Kurz, and L. S. Shanker, "The importance of dissociation constant and lipid-solubility in influencing the passage of drugs into the cerebrospinal fluid," *J. Pharmacol. Exp. Therap.* **130:** 20–25 (1960).

Chamberlain, A. C., M. J. Heard, P. Little, D. Newton, A. C. Wells, and R. D. Wiffen. *Investigations into Lead from Motor Vehicles*, AERE. Publication N2R9198, Harwell, England, 1978.

Crouthamel, W. G., J. T. Doluisio, R. E. Johnson, and L. Diamond, "Effect of mesenteric blood flow on intestinal drug absorption," *J. Pharm. Sci.* **59:** 878–879 (1970).

English, J. C., R. D. R. Parker, R. P. Sharma, and S. G. Oberg, "Toxicokinetics of nickel in rats after intratracheal administration of a soluble and insoluble form," *Am. Ind. Hyg. Assoc. J.* **42:** 486–492 (1981).

Gariépy, L., D. Fenyves, and J.-P. Villeneuve, "Propranolol disposition in the rat: Variation in hepatic extraction with unbound drug fraction," *J. Pharm. Sci.* **81:** 255–258 (1992).

Gregus, Z., and C. D. Klaassen, "Disposition of metals in rats: A comparative study of fecal, urinary, and biliary excretion and tissue distribution of eighteen metals," *Toxicol. Appl. Pharmacol.* **85:** 24–38 (1986).

Guidotti, G., "The structure of membrane transport systems," *Trends Biochem. Sci.* **1:** 11–12 (1976).

Hamilton, D. L., and M. W. Smith, "Inhibition of intestinal calcium uptake by cadmium and the effect of a low calcium diet on cadmium retention," *Environ. Res.* **15:** 175–184 (1978).

Herrmann, D. R., K. M. Olsen, and F. C. Hiller, "Nicotine absorption after pulmonary instillation," *J. Pharm. Sci.* **81:** 1055–1058 (1992).

Hirom, P. C., P. Millburn, and R. L. Smith, "Bile and urine as complementary pathways for the excretion of foreign organic compounds," *Xenobiotica* **6:** 55–64 (1976).

Hogben, C. A. M., D. J. Tocco, B. B. Brodie, and L. S. Shanker, "On the mechanism of intestinal absorption of drugs," *J. Pharmacol. Exp. Therap.* **125:** 275–282 (1959).

Hussain, A. A., K. Iseki, M. Kagoshima, L. W. Dittert, "Absorption of acetylsalicylic acid from the rat nasal cavity," *J. Pharm. Sci.* **81:** 348–349 (1992).

King, F. G., R. L. Dedrick, J. M. Collins, H. B. Matthews, and L. S. Birnbaum, "Physiological model for the pharmacokinetics of 2,3,7,8-tetrachlorodibenzofuran in several species," *Toxicol. Appl. Pharmacol.* **67:** 390–400 (1983).

Lien, E. J., and G. L. Tong, "Physicochemical properties and percutaneous absorption of drugs," *J. Soc. Cosmet. Chem.* **24:** 371–384 (1973).

Nebert, D. W., A. Puga, and V. Vasiliou, "Role of the Ah receptor and the dioxin-inducible [Ah] gene battery in toxicity, cancer, and signal transduction," *Ann. NY Acad. Sci.* **685:** 624–640 (1993).

Nelson, D. R., T. Kamataki, D. J. Waxman, F. P. Guengerich, R. W. Estabrook, R. Feyereisen, F. J. Gonzalez, M. J. Coon, I. C. Gunsalus, O. Gotoh, K. Okuda, and D. W. Nebert, "The P450 superfamily: Update on new sequences, gene mapping, accession numbers, early trivial names of enzymes, and nomenclature," *DNA Cell Biol.* **12:** 1–51 (1993).

O'Flaherty, E. J., *Toxicants and Drugs: Kinetics and Dynamics,* Wiley, New York, 1981.

O'Flaherty, E. J., "Physiologically based models for bone-seeking elements. IV. Kinetics of lead disposition in humans," *Toxicol. Appl. Pharmacol.* **118:** 16–29 (1993).

Rollins, D. E., and C. D. Klaassen, "Biliary excretion of drugs in man," *Clin. Pharmacokinet.* **4:** 368–379 (1979).

Schanker, L. S., and J. J. Jeffrey, "Active transport of foreign pyrimidines across the intestinal epithelium," *Nature* **190:** 727–728 (1961).

Sha'afi, R. I., C. M. Gary-Bobo, and A. K. Solomon, "Permeability of red cell membranes to small hydrophilic and lipophilic solutes," *J. Gen. Physiol.* **58:** 238–258 (1971).

U.S. Environmental Protection Agency, *Update to the Health Risk Assessment Document and Addendum for Dichloromethane: Pharmacokinetics, Mechanism of Action and Epidemiology,* EPA 600/8-87/030A (1987).

Wagner, J. G., "Properties of the Michaelis-Menten equation and its integrated form which are useful in pharmacokinetics," *J. Pharmacokinet. Biopharmaceut.* **1:** 103–121 (1973).

Williams, R. T., "Interspecies scaling," in T. Teorell, R. L. Dedrick, and P. G. Condliffe, eds., *Pharmacology and Pharmacokinetics*, Plenum, New York, 1974, Table IV, p. 108.

3 Biotransformation: A Balance between Bioactivation and Detoxification

MICHAEL R. FRANKLIN and GAROLD S. YOST

This chapter identifies the fundamental principles of foreign compound (xenobiotic) modification by the body and discusses

- How xenobiotics enter, circulate, and leave the body
- The sites of metabolism of the xenobiotic within the body
- The chemistry and enzymology of xenobiotic metabolism
- The bioactivation as well as inactivation of xenobiotics during metabolism
- The variations in xenobiotic metabolism resulting from prior or concomitant exposure to xenobiotics and from physiological factors

The body is continuously exposed to chemicals, both naturally occurring and synthetic, which have little or no value in sustaining normal biochemistry and cell function. These chemical substances (xenobiotics) can be absorbed from the environment following inhalation, ingestion in food or water, or simple exposure to the skin (Figure 3.1). Biotransformation or metabolism of the chemicals allows the elimination of the absorbed chemicals to occur. Without this process, chemicals that were readily absorbed through lipid membranes because of a high octanol/water partition coefficients would fail to leave the body. They would be passively reabsorbed through the lipid membrane of the kidney tubule instead of remaining in, and passing out with, the urine (Figure 3.2). In addition, they would not be subject to active transport mechanisms capable of actively secreting many xenobiotic metabolites. Thus, an important objective of biotransformation is to promote the excretion of chemicals by the formation of water-soluble metabolites or products. Biotransformation can also alter the biological activity of chemicals, including endogenous chemicals released in the body, such as steroids and catecholamines, both by structural alteration and by enhancing their partition away from cellular compartments, membranes, and receptors. Thus biotransformation helps to both terminate the biological activity of chemicals and increase their ease of elimination.

Biotransformation is defined as the chemical alteration of substances by reactions in the living organism. For convenience, the conversion of xenobiotics is divided into two phases: metabolic transformations (phase I reactions) and conjugation with natural body constituents (phase II reactions) (Figure 3.3). The reactions of both of these phases are predominantly enzyme-catalyzed. A xenobiotic does not necessarily undergo metabolism by a sequential combination of phase I followed by phase II reactions for successful elimination. It may undergo phase I metabolism alone, phase II alone, and occasionally, phase I reactions subsequent to phase II conjugations are encountered.

An important objective of biotransformation is to promote the excretion of absorbed chemicals by the formation of water-soluble drug metabolites or products (*p* in Figure 3.1). Increased water solubility is derived primarily from the phase II reactions since most conjugates exist in the ionized state at physiological pH levels. This promotes excretion (*e* in Figure 3.1) by decreasing xenobiotic reabsorp-

Principles of Toxicology: Environmental and Industrial Applications, Second Edition, Edited by Phillip L. Williams, Robert C. James, and Stephen M. Roberts.
ISBN 0-471-29321-0 © 2000 John Wiley & Sons, Inc.

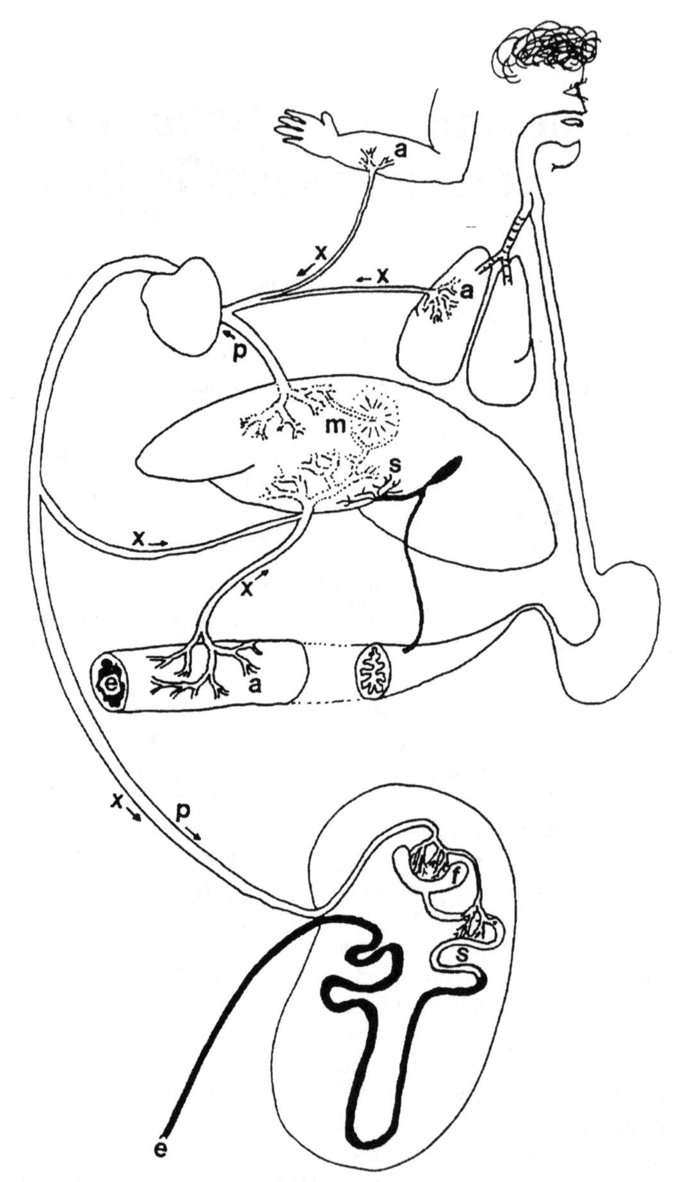

Abbreviations: a = major absorption sites; e = excretion sites; f = filtration sites; m = major metabolism site; p = metabolic product; s = secretion sites; x = xenobiotic.

Figure 3.1 Diagram of major sites of xenobiotic absorption, metabolism, and excretion.

tion from the renal tubule following glomerular filtration or active secretion (*f* and *s*, respectively, in Figures 3.1 and 3.2) and from the gastrointestinal tract following biliary secretion. Biotransformation also decreases the entry of xenobiotics into cells of all organs and makes them more suitable for secretion by active transport mechanisms into the bile and urine. Active secretion requires both energy and a carrier protein and is capable of forcing molecules up a chemical gradient. Of the carrier molecules, those that recognize and transport organic acids have particular importance for drug conjugates since they can carry glucuronides, sulfate esters, and amino acid conjugates. While

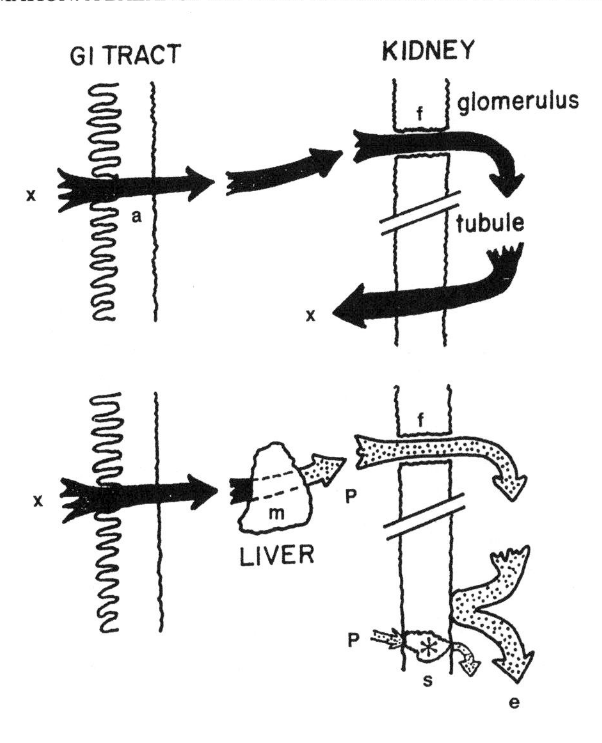

Figure 3.2 The role of metabolism in increasing urinary excretion.

excretion of xenobiotics into the urine largely terminates the exposure of the body to the chemical, excretion in the bile may not always result in efficient drug elimination because enterohepatic recirculation may occur. This can result in the prolonged effects and persistence seen with some drugs and chemicals. Enterohepatic recirculation most often involves the secretion of xenobiotic conjugates in the bile and their hydrolysis by enzymes from the host or microorganisms in the gastrointestinal tract. This deconjugation releases the free xenobiotic, which is often sufficiently lipid soluble (high octanol/water partition coefficient), to be reabsorbed. The reabsorbed xenobiotic returns in the portal circulation to the liver where it is reconjugated, resecreted, and so on. The same reabsorption can also occur if an unmetabolized lipid soluble xenobiotic is secreted in the bile.

As stated above, the conversion of xenobiotics is divided into the two phases of metabolic transformation and conjugation (Figure 3.3). The main chemical reactions involved in phase I or metabolic transformation, in approximate order of capacity or importance, are oxidation, hydrolysis, and reduction. Of the phase II or conjugation reactions, glucuronidations are generally the most prevalent in mammals, with the other conjugations having lesser overall capacity. All conjugation reactions, except with glutathione, involve the participation of energy-rich or activated cosubstrates. Conjugation with the cellular nucleophile, glutathione, is an especially important mechanism for the sequestering of electrophilic intermediates generated during phase I metabolism, and it can occur, albeit less efficiently, in the absence of enzyme.

As mentioned above, with reference to the generation of electophilic metabolites, biotransformation can have a variety of effects on the biological reactivity of the xenobiotic. The chemical can be inactivated or detoxified, can be changed into a more toxic substance (bioactivated), or can be changed into other chemical entities having effects that differ both quantitatively and qualitatively from the parent compound (Table 3.1).

Generally, phase II metabolites are inactive, but important exceptions exist. Phase I metabolites may or may not be inactive, and many are more reactive than the original xenobiotic. The greater reactivity can be viewed as an unfortunate necessary prerequisite to conjugation, which is the step contributing most to the facilitation of excretion (Figure 3.4).

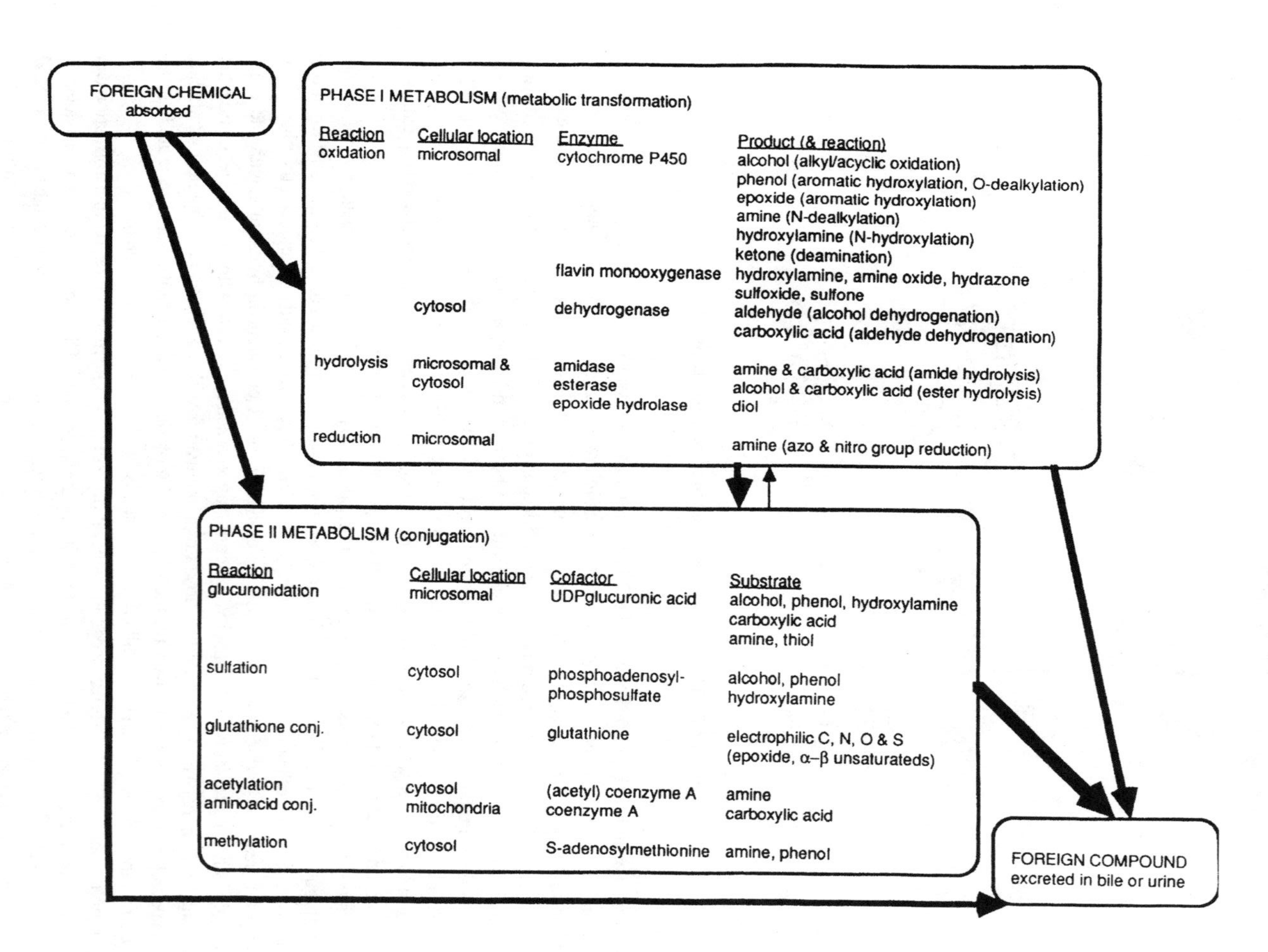

Figure 3.3 Xenobiotic metabolism summary; reaction characteristics and flowchart.

TABLE 3.1 Pharmacologic Effects with Xenobiotic Metabolism

Phase I	Phase II
Active to Inactive	
Amphetamine —P450→ phenylacetone	Acetaminophen —UGT/ST→
Cocaine —esterase→ benzoylecgonine	Aflatoxin 2,3-epoxide —GST→ 8 glutathionyl-9 hydroxyaflatoxin
Hexobarbital —P450→	Morphine —UGT→ morphine-3-glucuronide
Phenytoin —P450→	Testosterone —ST→
Active to Active	
Acetylsalicylic acid —esterase→ salicylic acid	
Codeine —P450→ morphine	Morphine —UGT→ morphine-6-glucuronide
Heroin —esterase→ morphine	Procainamide —AT → *N*-acetylprocainamide
Primidone-P450→ phenobarbital	Thiobarbital —P450→ barbital
Inactive to Active	
Chloral hydrate —reductase→ trichloroethanol	
Prontosil —reductase→ sulfanilamide	
Sulindac —reductase→ sulfide	
Inactive to Toxic	
Acetaminophen —P450→ *N*-acetyl-*p*-benzoquinine imine	*N*-Hydroxyacetylaminofluorene —ST→
Acetylhydrazine —P450→ acetylcarbonium ion	*N*-Hydroxymethylaminoazobenzene —ST→
Aflatoxin —P450→ aflatoxin-8,9 epoxide	Tetrachloroethylene —GST→
Malathion —P450→ malaoxon	Tolmetin —UGT→
Nitrofurantoin —reductase→ hydroxylamine	
Benzo(*a*)pyrene 7,8-diol —P450→ benzo(*a*)pyrene 7,8-diol 9,10-epoxide	
Dimethylnitrosamine —P450→ methyldiazohydroxide	

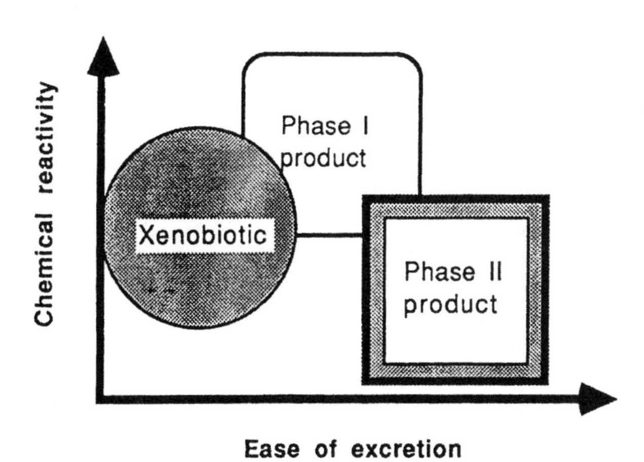

Figure 3.4 The balance of reactivity and excretability in xenobiotic metabolism.

3.1 SITES OF BIOTRANSFORMATION

Xenobiotic metabolism occurs in all organs and tissues in the body. Because many of the chemicals metabolized can have deleterious effects on the body, xenobiotic metabolism can be considered a defense mechanism that hastens the elimination of a toxic chemical and thus terminates the exposure. When viewed as a defense mechanism, it is not surprising that the exposure is best terminated at the point of exposure. These are the so-called portals of entry (shown as sites of absorption [*a*] for xenobiotics [*X*] in Figure 3.1), and constitute mainly the skin, lung, and intestinal mucosa. While drug metabolizing enzymes are present in all these tissues (Table 3.2), and at relatively high activity in some, particularly intestine and lung, the liver is by far the most important tissue for xenobiotic metabolism (site [*m*] in Figure 3.1).

Although it is not the first tissue of the body to be exposed to chemicals, the liver receives the entire chemical load absorbed from the gastrointestinal tract, which is the predominant portal of entry for most xenobiotics (Figure 3.1). The xenobiotic metabolizing enzymes are present in high concentrations and the organ itself has large bulk, approximately 5 percent of the total body weight. Xenobiotics absorbed from the lungs and skin can also quickly move to the liver for metabolism. Once in the liver, the highly vascular nature of the tissue and the intimate contact between blood and hepatocytes, which contain the xenobiotic metabolizing enzymes, allows for the rapid diffusion of chemicals in and metabolites out (Figure 3.5).

Although not a portal of entry, the kidney is an organ where xenobiotics are likely to be concentrated during the excretion process, and this may be the reason for the relatively high level of xenobiotic metabolizing enzymes in this tissue. Although the data presented in Table 3.2 are from laboratory animals, there is little evidence to contraindicate the existence of a similar distribution pattern in humans.

Within the liver, hepatocytes or parenchymal cells are the major site of drug biotransformation, and within these cells it is the endoplasmic reticulum, which occupies about 15 percent of the hepatocyte volume and contains 20 percent of the hepatocyte protein, which houses the bulk of the critical drug metabolizing enzyme activity. (The nonparenchymal cells, including endothelial and Kupffer cells, constitute 35 percent of liver cell number but only contribute 5–10 percent of liver mass. Their drug metabolizing enzyme activities are typically less than 20 percent of that in hepatocytes).

When liver is carefully homogenized, fragments of the endoplasmic reticulum are converted to microsomes (an artifact of cell disruption). The drug-metabolizing enzymes located in the endoplasmic reticulum are often referred to as *microsomal enzymes*, and it is often stated that chemicals are metabolized by liver microsomes. Enriched microsomal fractions are usually obtained by differential sedimentation, either as a suspension with cytoplasm (10,000*g* supernatant) or as a sediment free of cytosol (105,000*g* precipitate) (Table 3.3).

Many important xenobiotic metabolizing enzymes reside in the cytoplasm and microsomal fractions (Figures 3.3 and 3.6).

Oxidations and glucuronidations are the most common reactions occurring in microsomes. The terminal oxidase responsible for many of the oxidations, cytochrome P450, represents about 5 percent of the microsomal protein under normal conditions; more if induction has occurred (see text below). Other flavoproteins necessary for cytochrome P450 function and epoxide hydrolase, an enzyme important in the further metabolism of epoxides formed by cytochrome P450–dependent oxidation, are also conveniently located in the endoplasmic reticulum (Figure 3.6). Microsomal metabolism in tissues other than liver is seldom quantitatively important in overall drug elimination, but local generation of active metabolites may be important in drug-induced tissue damage, carcinogenesis, and other effects. Enzymes located in the cytoplasm of the hepatocyte catalyze a wide variety of both phase I and phase II reactions. Dehydrogenases and esterases are examples of phase I enzymes found predominantly in the cytosol. The sulfotransferase and glutathione transferase enzymes also depicted in Figure 3.6 serve as examples of phase II enzymes that are similarly located.

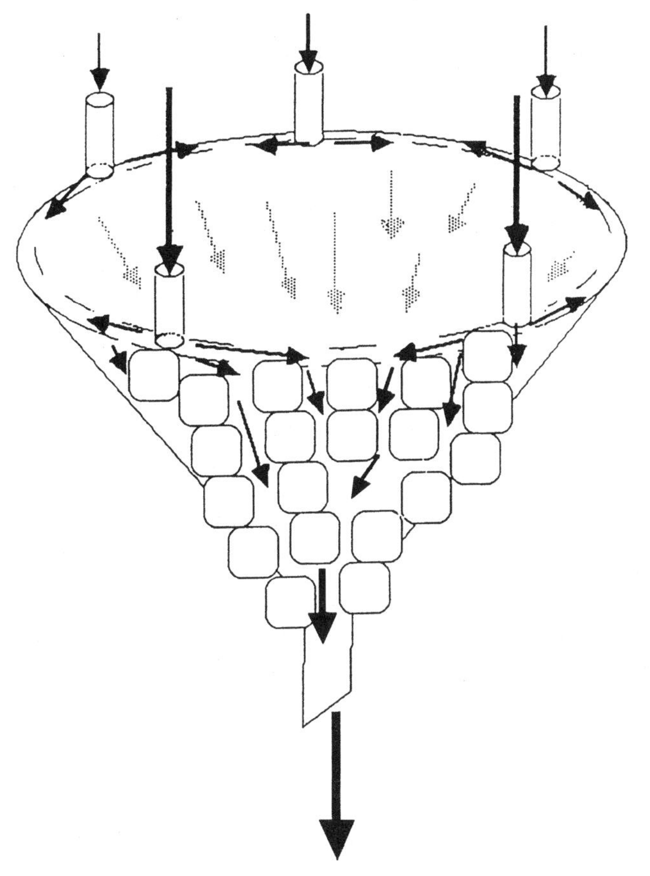

Figure 3.5 Diagrammatic rendition of hepatic lobule blood flow.

TABLE 3.2 Drug-Metabolizing Enzyme Activities[a] in Various Organs

	Lung	Intestine Mucosa	Liver	Kidney Cortex	Brain
Rabbit					
Cytochrome P450	0.4	0.34	1.45	0.33	0.02
UDP-glucuronosyltransferase (*p*-nitrophenol)[b]	0.4	—	6.6	2.9	
Glutathione *S*-transferase (DCNB)[b]	5.3	—	21.9	7.4	
Rat					
Cytochrome P450	0.09	0.05	0.84	0.12	0.01
Ethoxyresorufin demethylase (P4501A)	0.003	0.001	0.034	0.001	
Erythromycin demethylase (P4503A)	—	0.12	0.47	0.06	
UDP-glucuronosyltransferase (*p*-nitrophenol)[b]	0.8	—	4.4	3.3	
Glutathione *S*-transferase (DCNB)[b]	2.1	—	76.4	3.8	

[a]All activities are expressed on a per milligram of protein basis (DCNB = 1,2-dichloro 4-nitobenzene).

[b]Litterst CL, Mimnaugh EG, Reagan RL, Gram TE, *Drug Metab. Disp.* **3:** 259 (1975).

TABLE 3.3 Preparation of Subcellular Fractions for Xenobiotic Metabolism Studies

Step	Procedure	Result
1	Liver pieces homogenized in 4 volumes of 0.25 M sucrose in Potter Elvehjem glass–Teflon homogenizer	Tissue structure disrupted and hepatocytes sheared.
2	Homogenate centrifuged at 2000*g* for 10 min	Unbroken cells, connective tissue, and nucleii sedimented
3	2000*g* supernatant centrifuged at 10,000*g* for 15 min	Heavy mitochondria sedimented as pellet
4	10,000*g* supernatant centrifuged at 18,000*g* for 15 min	Light mitochondria sedimented as pellet
5	18,000*g* supernatant centrifuged at 105,000*g* for 60 min	Microsomes sedimented as pellet leaving nonturbid cytosol in 0.2 M sucrose supernatant

Without exception, the xenobiotic metabolizing enzymes occur in multiple forms (isozymes), often with differing substrate selectivities. The presence of specialized isozymes, which can more efficiently metabolize a specific range of chemicals, may enable those specific chemical challenges to be met more effectively. With differing substrate selectivities, often comes different sensitivity to inhibitors. The presence of multiple forms thus carries the advantage of not having all the metabolism of all compounds metabolized by that route or chemical reaction being subject to inhibitory influences at the same time. It has also been found that the synthesis of individual isozymes can be under different regulatory influences. The body can thus meet a chemical challenge with a finely tuned response to increase the production of only that enzyme best equipped to counter or neutralize the challenge.

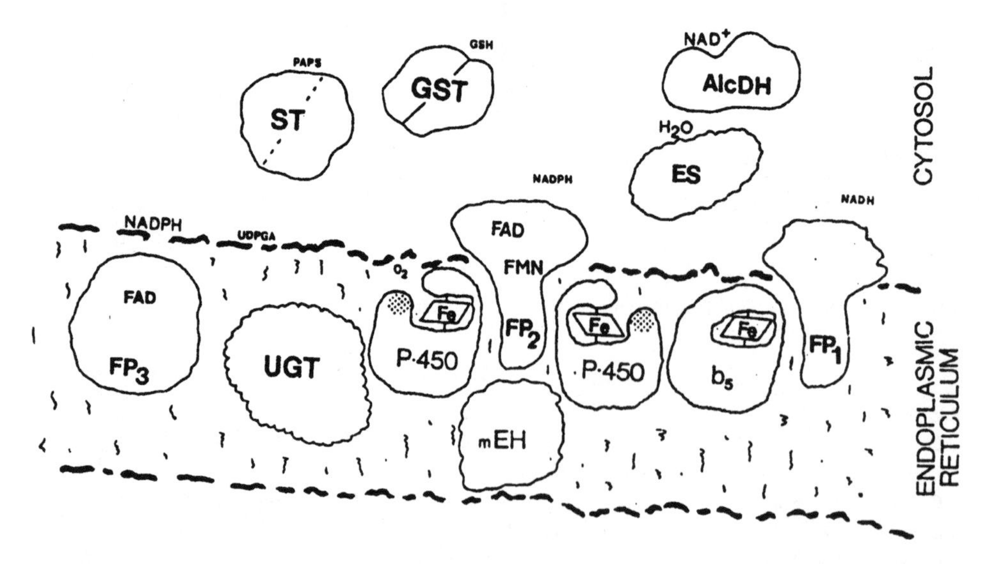

Abbreviations (clockwise) are ST = sulfotransferase; PAPS = adenosine 3-phosphate 5′-phosphosulfate; GST = glutathione *S*-transferase; GSH = glutathione; AlcDH = alcohol dehydrogenase; ES = esterase; FP_1 = NADH cytochrome b_5 reductase; b_5 = cytochrome b_5; P450 = cytochrome P450; mEH = microsomal epoxide hydrolase; FP_2 = NADPH-cytochrome P450/*c* reductase; UGT = UDP-glucuronosyltransferase; UDPGA = uridine 5-diphosphoglucuronic acid; FP_3 = flavin-dependent monooxygenase.

Figure 3.6 Diagram of the subcellular localization and organization of major xenobiotic metabolizing enzymes and necessary cofactors.

3.2 BIOTRANSFORMATION REACTIONS

There is multiple redundancy in metabolism. There may be more than one site of attack on a xenobiotic (e.g., amine and ester group of cocaine), there may be more than one metabolic reaction at a single site (e.g., sulfation and glucuronidation of the phenolic group of acetaminophen), and more than one enzyme/isozyme capable of catalyzing a single reaction at a single site. An example of the complexity of possible metabolism of a relatively simple hypothetical chemical is shown in Figure 3.7. From considerations in this chapter so far, it can be seen that the subcellular location of a metabolic reaction does not dictate the nature of the reaction. Both oxidations and hydolyses, albeit by different enzymes, occur in the cytoplasm and endoplasmic reticulum. Likewise, so do conjugations when considered collectively, but a specific form of conjugation may occur only in a single fraction (e.g., sulfation in the cytoplasm). The enzymes are therefore considered in the following paragraphs by the nature of the chemical reaction that they catalyze, and only for phase I oxidations is the subcellular location used as a convenient subdivision.

Phase I; Oxidations

Microsomal Microsomal oxidations are predominantly catalyzed by a group of enzymes called *mixed-function oxidases* or *monooxygenases*. The terminal oxidase is generally a hemoprotein called *cytochrome P450* but can be a flavoprotein.

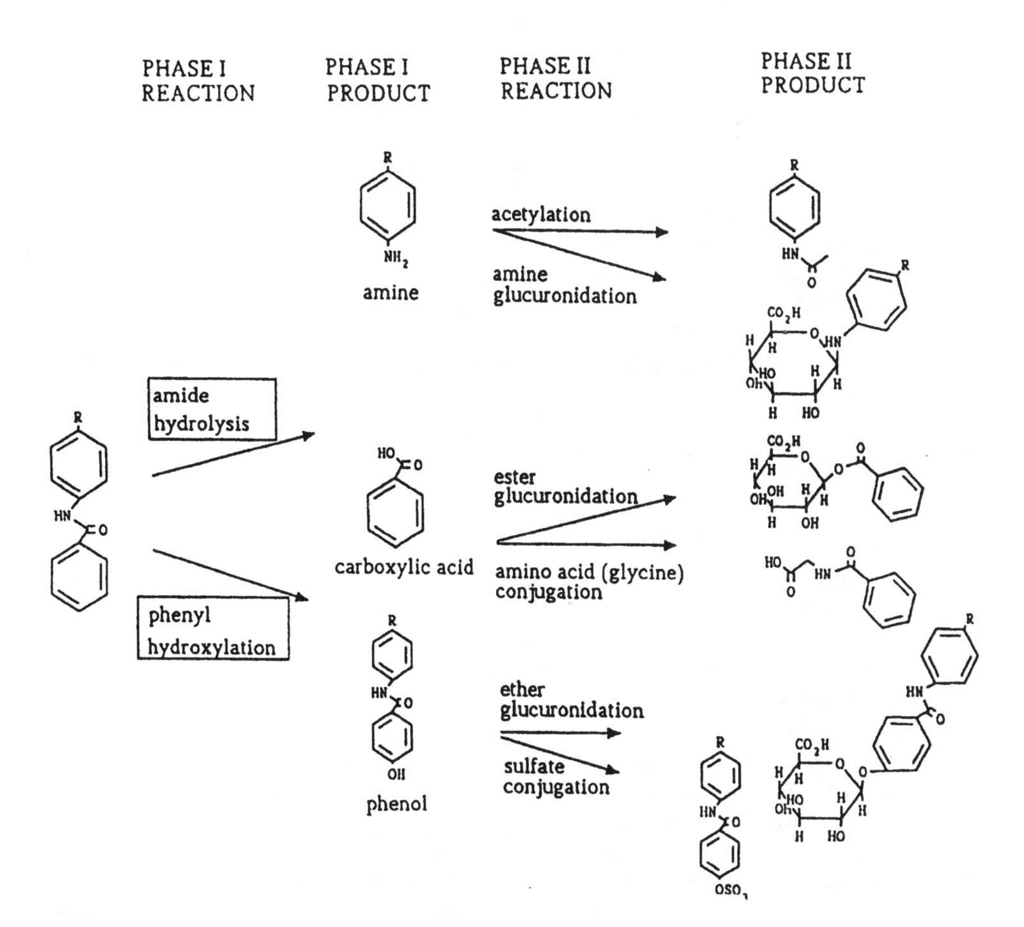

Figure 3.7 Possible metabolic conversions of a simple hypothetical xenobiotic.

Cytochrome P450 is a collective term for a group of related hemoproteins, all with a molecular weight (MW) around 50,000 daltons, which as will be seen later, differ in their substrate selectivity and in their ability to be induced and inhibited by drugs and chemicals (Table 3.4).

Cytochrome P450–catalyzed oxidations are categorized by the nature of the atom that is oxidized (see Figure 3.8). Subsequent to the oxidation, the oxygen atom from molecular oxygen may be retained within the major fragment of the chemical or it may be eliminated by molecular rearrangement (e.g., *O* and *N* dealkylations).

Whatever the atom oxidized, or the name given to the reaction, the cytochrome P450–mediated oxidation involves the same cyclic three-step series (Figure 3.9).

Step 1. The xenobiotic [*X*] first binds to the cytochrome at a substrate binding site on the protein and alters the conformation sufficiently to enable the efficient transfer of electrons to the heme from NADPH via a nearby (see Figure 3.6) flavoprotein, NADPH cytochrome P450 reductase. (The activity of this FAD- and FMN-containing flavoprotein is often determined experimentally using exogenously added mitochondrial cytochrome *c* rather than microsomal cytochrome P450 as the electron acceptor and so is often identified as NADPH cytochrome *c* reductase). The conformational change can sometimes be seen in vitro (in the absence of electron transfer) as an alteration of the heme from a low-spin to a high-spin state, which results in a blue shift in the absorbance maximum of the hemoprotein. The gain at 390 nm and loss at 420 nm, when seen by difference spectroscopy, is termed a *type I binding spectrum* (not to be confused with phase I metabolism).

Step 2. The reduction of the heme iron from its normal ferric state to the ferrous state allows a molecule of oxygen (O–O) to bind (the binding of CO rather than oxygen to ferrous cytochrome P450 in the in vitro situation provides a characteristic absorbance maximum around 450 nm, which gives this cytochrome its name).

Step 3. The ternary complex of xenobiotic, cytochrome, and oxygen receives another electron, either through the same flavoprotein as before or through an alternative path involving a different flavoprotein in which the electron is first passed through cytochrome b_5, another cytochrome present in the endoplasmic reticulum (see Figure 3.6). This alternate pathway for the second electron can also use NADH as the pyridine nucleotide electron donor. The addition of the second electron to the ternary complex results in a eventual splitting of the molecular oxygen, one atom of which oxidizes the chemical, the other atom picking up protons to form water, returning ferric cytochrome P450 to repeat the cycle.

Flavoprotein-catalyzed oxidations differ from cytochrome P450–catalyzed oxidations in mechanism and in substrate selectivity. For the flavoproteins (a 65,000-dalton protein containing only FAD), the enzyme forms an activated oxygen complex ("cocked gun") and the addition of a metabolizable chemical discharges this, in the process of becoming oxidized. The electrophilic oxygenated species attacks nucleophilic centers. A wide range of chemicals can thus be metabolized by this flavoprotein; the important feature for metabolism being a heteroatom (nitrogen, sulfur) presenting a lone pair of electrons (Table 3.5).

Some compounds are metabolized both by flavin-containing monooxygenases and cytochrome P450 but to different products. An example is dimethylaniline, which is metabolized to the *N*-oxide by the flavoprotein and is *N*-demethylated by cytochrome P450.

Nonmicrosomal Oxidations in other subcellular organelles can be catalyzed by flavoproteins (e.g., monoamine oxidase in mitochondria) or pyridine nucleotide linked dehydrogenases (e.g., alcohol and aldehyde dehydrogenases in cytoplasm).

Dehydrogenase-catalyzed oxidations do not involve molecular oxygen. The oxidation of the chemicals or drugs occurs through electron transfer to a pyridine nucleotide, usually NAD^+. Most of the dehydrogenases are cytoplasmic in location. The most noteworthy of this class of enzymes in humans is the dehydrogenase responsible for the metabolism of ethanol. In contrast to the major

TABLE 3.4 Important Cytochrome P450s[a]

Subfamily	Forms Present in: Rat	Rabbit	Human	Mouse	Tissue	Substrates/ Reactions	Inducers	Inhibitors
1A	1,2	1,2	1,2	1,2	liver, EH	ethoxyresorufin, phenacetin deE, caffeine 3 deM, benzopyrene, aflatoxin, cooked-food heterocyclic amines, NBI	PCB, PAH, TCDD, 3MC, isosafrole, omeprazole	7,8-benzoflavone, ellipticine, furafylline
2A	1,2			4,5			PAH (2A1)	
2B	1,2	4		9,10,13		pentoxyresorufin, benzphetamine	PB	
2C	6,7				liver, GI		PB	
			8,9,10			tolbutamide	PB, RIF	sulfaphenazole
	11,12,13						sex, maturation,	
			17,18,19			S-mephenytoin 4-OH, debrisoquine 4-OH, RIF	tranylcypromine quinidine	
2D			6	9,10,11, 12,13	liver, EH	sparteine, dextromethorphan, bufuralol 1′-OH		
2E	1	1,2	1	1	liver, EH	chlozoxazone, ethanol dimethylnitrosamine, acetaminophen, CCl_4, 4-nitrophenol-OH	ethanol, disulfiram ketones, pyridine, isoniazid	
2F			1	2	lung	3-methylindole naphthalene		

TABLE 3.4 Important Cytochrome P450s[a]

Subfamily	Forms Present in: Rat	Rabbit	Human	Mouse	Tissue	Substrates/ Reactions	Inducers	Inhibitors
3A	1,2	6	3,4,5,7	11,13		erythromycin *N* deM, TAO MI complex, cyclosporine, quinidine, testosterone, and cortisol 6β-OH, mephenytoin (rat), benzphetamine, nifedipine, and other dihydropyridines	glucocorticoids DEX, PCN, macrolides, TAO clotrimazole, phenobarbital	cimetidine, naringenin
4A	1,2,3,8	4,5,6,7	9,11	10,12	liver, kidney	Lauric acid and prostaglandin ω-OH	peroxisome proliferators, DEHP, clofibrate pregnancy (4A4)	
4B	1	1	1		lung	valproic acid		

[a]*Abbreviations*: deE = deethylation
deM = demethylation
DEHP = di(2-ethylhexyl)phthalate
DEX = dexamethasone
EH = extrahepatic
GI = gastrointestinal tract
Kid = kidney
MI = metabolic-intermediate
3MC = 3-methylcholanthrene
NBI = N-benzylimidazole
OH = hydroxylation
PAH = polycyclic aromatic hydrocarbons
PCB = polychlorinated biphenyls
PCN = pregeneolone 16α carbonitrile
PB = phenobarbital
RIF = rifampicin
TAO = troleandomycin
TCDD = 2,3,7,8-tetrachlorodibenzo-p-dioxin

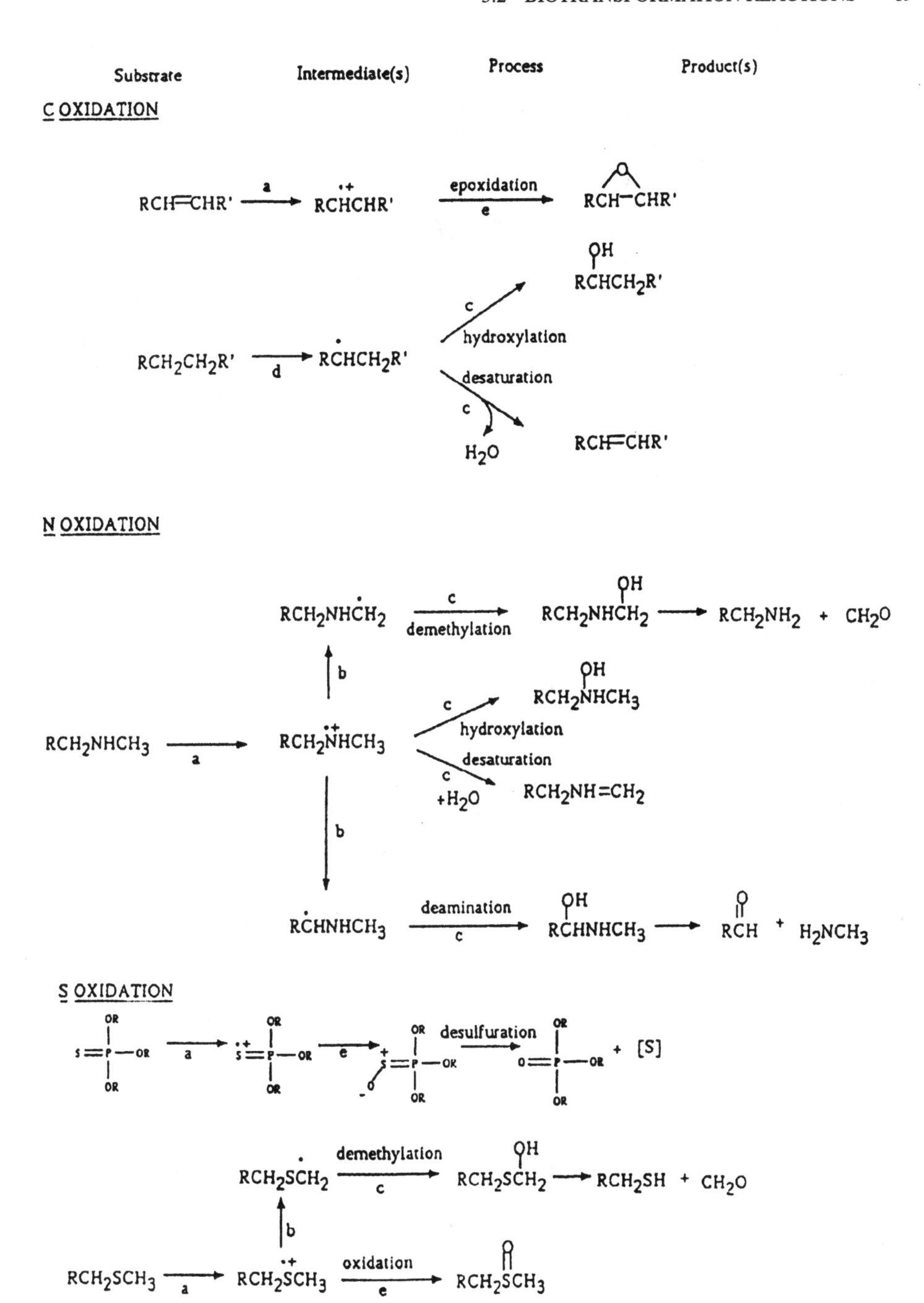

Figure 3.8 Cytochrome P450–catalyzed oxidations.

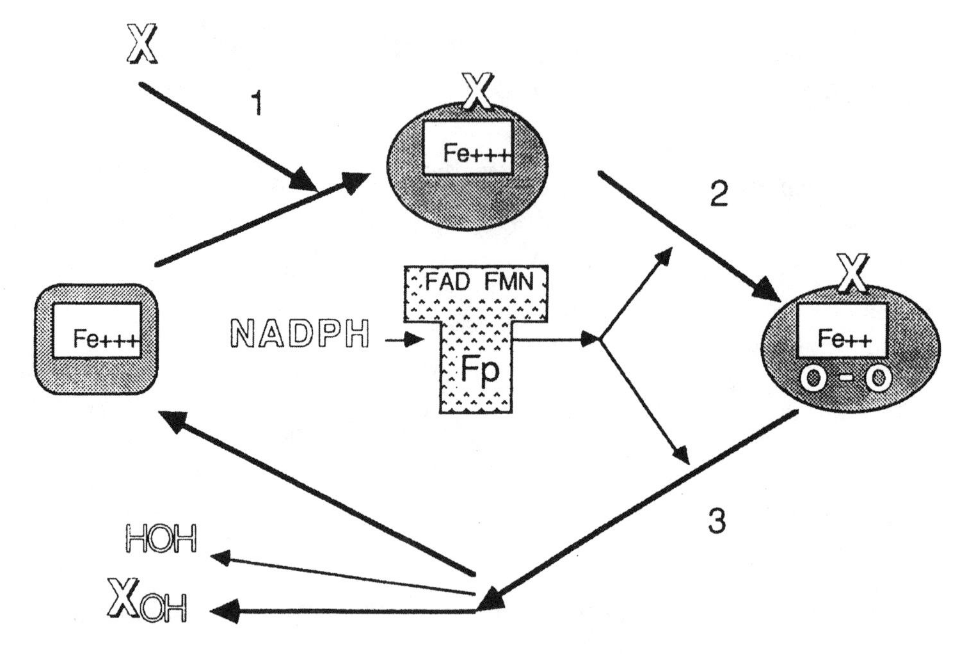

Figure 3.9 The cytochrome P450 oxidation cycle.

microsomal oxidizing enzyme, these enzymes are not subject to extensive induction (see discussion later).

Monoamine oxidases, which are usually mitochondrial in location, oxidize by electron transfer to a flavin group. Monoamine oxidases are responsible for the normal metabolism of neurotransmitters, and exposure to agents, which are also metabolized by this enzyme, (e.g., tyramine) can result in toxicities or pharmacological effects arising from accumulation of the unmetabolized neurotransmitter. A neurotoxin of much recent interest, 1-methyl-4-phenyl-1,2,3,6-tetrahydropyridine (MPTP), which leads to Parkinson's syndrome, is bioactivated by monoamine oxidase B (a form selectively inhibited by deprenyl and located in serotonergic neurons in the brain). Environmental compounds or drugs that are also tetrahydropyridines have been speculated to be causative agents in Parkinson's disease in the elderly.

Phase I; Hydrolyses

Hydrolysis reactions are catalyzed by esterases and amidases. While both can be microsomal, esterases are predominantly cytosolic in location. Hydrolysis of amides and esters produces two reactive centers,

TABLE 3.5 Compounds Metabolized by the Flavin-Containing Monooxygenases

Heteroatom	Class	Examples
Nitrogen	Tertiary amine	*N*-Dimethylaniline, imipramine, amitryptyline
	Secondary amine	*N*-Methylaniline, desipramine, nortryptyline
Sulfur	Thiocarbamides	Thiourea, propylthiouracil, methimazole
	Thioamides	Thioacetamide
	Thiols	Dithiothreitol, β-mercaptoethanol
	Sulfides	Dimethylsulfide

both of which are suitable for conjugation, if the metabolites are not first excreted as the phase I products (Figure 3.10).

Epoxide hydrolase activity is predominantly microsomal, but an enzyme is also present in the cytosol.

Most hydrolyses occur to a significant extent in tissues other than liver. Their quantitative importance is variable, depending on the chemical challenge. One significant extrahepatic location of esterases is in the blood (plasma and erythrocytes), and of great concern is the enzyme normally responsible for the hydrolysis of acetylcholine. Blockade of this enzyme is the mode of action of many insecticides and "nerve gases."

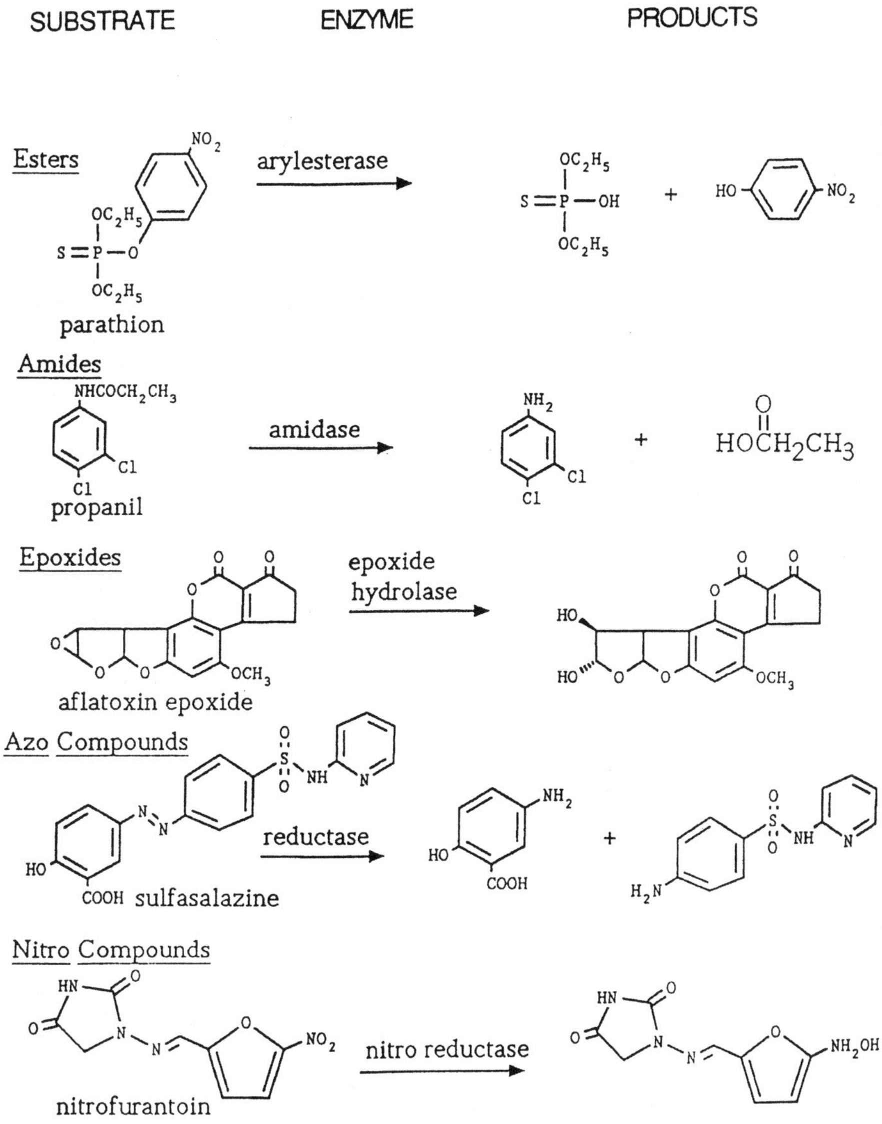

SUBSTRATE ENZYME PRODUCTS

Esters — parathion — arylesterase

Amides — propanil — amidase

Epoxides — aflatoxin epoxide — epoxide hydrolase

Azo Compounds — sulfasalazine — reductase

Nitro Compounds — nitrofurantoin — nitro reductase

Figure 3.10 Hydrolytic and reductive phase I reactions.

Phase I; Reductions

Reductive metabolism in the liver endoplasmic reticulum can occur through the mediation of both hemoprotein (cytochrome P450) and flavoproteins. Reductions of azo and nitro groups are the most commonly encountered (Figure 3.10), but reduction of disulfides, sulfoxides, epoxides, and *N*-oxides can also occur. In many instances, the products of reductive metabolism can be reoxidized under aerobic conditions.

Phase II; Glucuronidation

Glucuronidations are catalyzed by a group of closely related 55,000-dalton isozymes, termed *UDP-glucuronosyltransferases,* located within the endoplasmic reticulum. They catalyze the transfer of glucuronic acid from a uridinediphosphoglucuronic acid (UDPGA) cofactor to a carboxyl or hydroxyl (phenol), or less often an amine group on the xenobiotic (or phase I metabolite) (Figure 3.3). The UDPGA is generated from the abundant carbohydrate supply in the liver as glucose-1-phosphate, and following the reaction with UTP, the resultant UDP-glucose is oxidized. The formation of the glucuronide does not involve the acid group of glucuronic acid, so the conjugate retains acid and ionized character at physiological pH, providing dramatic enhancement of water solubility and excretability to the xenobiotic. Glucuronides are actively secreted into bile and in the proximal tubule of the kidney. Xenobiotics conjugated as glucuronides can be released as either a phase I metabolite or the original molecule by the action of glucuronidases of both mammalian and microbial origin.

UDP-glucuronosyltransferases occur in multiple forms. The most common classification utilized for the enzymes responsible for the metabolism of xenobiotics are those (GT1) that conjugate planar phenols (e.g., 1-naphthol, 4-nitrophenol) and are induced by polycyclic hydrocarbon-like molecules (see Table 3.6) and those (GT2) that conjugate nonplanar phenols (e.g. morphine, chloramphenicol) and are induced by phenobarbital and similar compounds. There are other forms which appear to be more selective for endogenous substrates, notably those for the 17 hydroxysteroids (testosterone), the 3 hydroxysteroids (androsterone) and bilirubin. More recent studies using the powerful techniques of molecular biology have provided a more rational classification system, but to aid the reader in understanding the bulk of existing literature, the old system has been used in this chapter. Like cytochrome P450s, UDP-glucuronosyltransferases are often substrate selective rather than substrate specific, being able to metabolize a wide range of compounds poorly (e.g., 4-nitrophenol is conjugated by almost all isozymes) while metabolizing substrates with particular characteristics very efficiently. Also like cytochrome P450s, more than one form may be induced by a xenobiotic inducing agent (both bilirubin and testosterone as well as morphine conjugations are induced by phenobarbital).

Phase II; Sulfation

Sulfate conjugation is an important alternative to glucuronidation for phenolic compounds and occasionally arylamines. Sulfate availability within the cell may be limited, so this conjugation pathway decreases in importance with higher xenobiotic or phenolic metabolite concentrations. The 3′-phosphoadenosine-5′-phosphosulfate (PAPS) cofactor from which the sulfate group is transferred is generated from ATP and inorganic sulfate. The sulfate can be derived from the sulfur containing amino acids, cysteine and methionine. The enzymes catalyzing the sulfate conjugations are a family of cytosolic 64,000-dalton enzymes, termed *sulfotransferases*, and are one of the exceptions to the major groups of drug metabolizing enzymes in that they appear to not be induced by xenobiotic compounds (see Table 3.6). The sulfates are completely ionized at physiological pH and easily eliminated. Much like glucuronides, enzymes exist (termed *sulfatases*) that can break the conjugate and return the xenobiotic, if it is phenolic, or the phase I metabolite of a xenobiotic, if it was oxidized or hydrolyzed to that functional group.

TABLE 3.6 Changes in Rat Hepatic Drug-Metabolizing Enzymes Following Xenobiotic Administration [a]

Agent (use)	Dose (mg/kg × days)	CYT P450	UGT (pNP)	UGT GT1 (N)	UGT GT2 (M)	GST (CDNB)	ST (pNP)
		(% control)					
SKF 525-A	80 × 4	400[b]	—	65	150	—	—
1-Benzylmidazole	75 × 3	320[c]	305	240	240	225	70
Troleandomycin (antibiotic)	500 × 4	300[b]	—	100	200	—	—
Clotrimazole (antifungal)	75 × 3	290[b]	120	120	225	265	85
Phenobarbital (anticonvulsant)	80 × 4	265	130	127	455	155	50
Dexamethasone (glucocorticoid)	100 × 3	245	70	80	155	125	60
3-Methylcholanthrene[d]	20 × 4	215	185	300	155	145	90
PCBs (Aroclor 1254) (transformer fluid)	25 × 6	205	265	270	280	140	110
2,3,6,7-Tetrachlorodibenzodioxin (TCDD)[d]	0.01 × 1	185	185	295	115	140	90
4,4″-Dipyridyl	75 × 3	190	245	130	225	130	90
Fluconazole (antifungal)	75 × 3	180[c]	180	175	170	130	115
Pregnenolone 16α carbonitrile[d]	75 × 4	180	85	100	150	140	120
Clofibrate (antihypertriglyceridemic)	400 × 3	180	65	80	85	95	—
trans-Stibene oxide[d]	400 × 4	175	170	195	260	215	100
5,6-Naphthoflavone [BNF]	80 × 3	165	495	250	130	185	85
Benzo(*a*)pyrene[d]	20 × 4	—	165	165	110	—	—
Miconazole (antifungal)	150 × 3	150	165	140	95	155	100
Phenytoin (anticonvulsant)	100 × 7	150	110	—	—	—	—
Carbamazepine (anticonvulsant)	100 × 7	145	125	—	—	—	—
Tioconazole (antifungal)	150 × 3	130[c]	300	215	330	170	70
Ketoconazole (antifungal)	150 × 4	130[c]	150	110	175	100	70
Isosafrole[d]	150 × 4	120	150	160	205	180	100
7,8-Benzoflavone[d]	100 × 4		125	120	110	—	—
Isoniazid (antitubercular)	100 × 4	105[e]	60	80	120	95	90
1,10-Phenanthroline	75 × 3	105	105	100	105	85	105
Cimetidine (antiulcer)	350 × 3	100	95	85	95	95	120
3,4 Benzoquinoline	75 × 3	90	230	240	320	140	80
Butylated hydroxyanisole (antioxidant)[d]	500 × 10	85	150	145	145	155	90
2,2′-Dipyridyl	75 × 3	85	310	205	235	145	80
Chloramphenicol (antibiotic)	300 × 3	80	95	90	125	105	70
Cyclosporine (immunosuppressant)	25 × 10	80	—	130	100	90	100
4,7-Phenanthroline	75 × 3	80	390	250	275	130	65

[a]*Abbreviations*: Enzymes: UGT = UDP-glucuronosyltransferase, GST = glutathione *S*-transferase, ST = sulfotransferase. Substrates: pNP = *p*-nitrophenol, N = 1-naphthol, M = morphine, CDNB = 1-chloro-2,4-dinitrobenzene.

[b]Full detection requires prior destruction of metabolic intermediate complex.

[c]Full detection requires time-dependent displacement of azole ligand by CO.

[d]From Watkins JB, Gregus Z, Thompson TN, Klaassen CD, *Toxicol. Appl. Pharmacol.* **64:** 439 (1982); Thompson TN, Watkins JB, Gregus Z, Klaassen CD, *Toxicol. Appl. Pharmacol.* **66:** 400 (1982).

[e]Induction of P4502E isozyme (see Table 3.3) obscured by decreases in other forms.

Phase II; Glutathione Conjugation

Conjugation with glutathione (γ-glutamylcysteinylglycine) is an important reaction for sequestering reactive (toxic) metabolites, which may be generated by cytochrome P450 oxidations. Glutathione

S-transferases are located predominantly in the cytosol, and hepatic concentrations of the necessary nucleophilic glutathione cosubstrate are high (> 5 mM). The major transferases consist of homo- or heterodimers of a limited number of forms of approximately 25,000-dalton subunits. The different subunit combinations confer different but overlapping substrate selectivity and isoelectric points and are expressed differently in different organs within an animal species. The subunits also respond differently to xenobiotic-inducing agents. In addition to cytosolic enzymes, a glutathione transferase unrelated to the cytosolic proteins is present in the endoplasmic reticulum.

Further metabolic products of glutathione conjugations include mercapturic acids (acetylated cysteine derivatives), which are the common excretory product. They are formed by sequential removal of glutamate and then glycine from the glutathione portion followed by acetylation of the amino group of the residual cysteine. Other metabolic products are methylated thiols and sulfones. Episulfonium ions and thioketenes can be formed from glutathione adducts and are reactive enough to form adducts with cell macromolecules and cause toxicity.

Phase II; Acetylation, Amino Acid Conjugation, and Methylation

The conjugations, involving acetylation of xenobiotics containing sulfonamide or amine groups, peptide conjugation of xenobiotics containing carboxylic acid groups, and methylation of xenobiotics containing amine or catechol groups (Figure 3.3), do not contribute much to enhanced excretability through an increase in water solubility, but serve to mask reactive centers. A problem with some early sulfonamides was that the acetylated metabolites were sufficiently less water-soluble that, they precipitated in the urine, resulting in renal damage. Both acetylations and amino acid conjugations utilize coenzyme A as a cofactor and require the formation of a thioester with the carboxylic acid group, either of acetate or of the xenobiotic. The thioester then reacts with an amine, either on the xenobiotic (acetylation) or amino acid (amino acid conjugation). In mammals, glycine and glutamate are the amino acids most commonly employed in xenobiotic conjugation, but taurine and aspartic acid conjugates are occasionally used, and in birds, ornithine is often used. Methylation reactions require the formation of *S*-adenosylmethionine (SAM) from ATP and the amino acid, methionine.

All the abovementioned conjugates can be deconjugated; deacetylases can remove acetyl groups, cytochrome P450 can remove methyl groups, and peptidases can split amino acid conjugates.

Most conjugations occur to varying degrees in tissues other than the liver. Quantitatively they are often minor, but can be very important for protection from reactive metabolites generated in extrahepatic tissues.

Factors Affecting Drug Metabolizing Capabilities With all that has been documented in this chapter so far, it is easy to overlook the fact that as in most biological systems, xenobiotic metabolism is a dynamic situation undergoing constant change. Numerous factors affect the ability to catalyze xenobiotic metabolism. Many are an inherent property of the animal species or strain. In addition, these genetic differences may be further altered by such physiological factors as gender or age. Xenobiotic metabolism in different animal species differs quantitatively and qualitatively from that in human. Extrapolation from animals to human and the selection of the most appropriate animal model is difficult unless the role of species and physiological factors in modulating metabolism is clearly delineated. The contribution of these various factors is also an important consideration within experimental research when there is a need to compare or reproduce findings generated in different laboratories.

Another factor of major concern is modification of xenobiotic metabolism by temporary stimuli, particularly chemical exposure. Typical human situations of chemical exposure can involve toxic accidental exposures but originate most often from ingestion of prescribed medications or ingestion of chemicals in the food, either as contaminants or as naturally occurring dietary constituents. The changes in xenobiotic metabolizing capability can be in either positive or negative directions, and each can occur by more than one mechanism. The response can be generalized over many enzymes catalyzing many different reactions or can be specific for a single isozyme and a single reaction.

Stimulation of Xenobiotic-Metabolizing Enzyme Activities by Xenobiotics. The activity of many microsomal and some cytoplasmic drug metabolizing enzymes can be increased by exposure to a wide range of drugs and other chemicals (Table 3.6). Generally, inducing agents possess two features in common: lipid solubility and a relatively long biological half-life (i.e., they gain access to the liver and remain there for a considerable period of time).

The stimulation of enzyme activity, called *induction*, is most often the result of the increased amount of enzyme present. If it is the result of an increased efficiency of existing enzyme it is termed *activation*, a phenomenon seen under some conditions with UDP-glucuronosyltransferases. Although not currently well documented for xenobiotic metabolizing enzymes, the activity of many enzymes can be altered by structural modification from processes such as phosphorylation by kinases and dephosphorylation by phosphatases.

Induction occurs by the inducing substance stimulating the synthesis of new enzyme. Because new protein (enzyme) synthesis requires time, the increase in activity is not an immediate event, and occurs over a period of many hours or days. Returning to a normal state following induction also takes a similar time course. The pattern of enzymes induced (both phase I and phase II) and the time course of induction varies with the agent. Induction is not open-ended, but rather, there appear to be limits to changes in each individual enzyme. Increases in liver microsomal enzyme activities determined in in vitro assays are often magnified for the metabolism of the xenobiotic in the whole animal because accompanying the increased enzyme activity per milligram of membrane protein are increased amounts of membrane per cell and increased overall size (most often an increased number of cells) of the liver.

The mechanism of induction is best understood for one group of compounds, the polycyclic aromatic hydrocarbon type of inducers, although this receptor-mediated induction (Figure 3.11) may not be the only mechanism by which these agents induce.

The cytosol contains a protein that has a high affinity for polycyclic aromatic hydrocarbon-like molecules. One of the chemicals most extensively utilized for these investigations has been 2,3,7,8,-tetrachlorodibenzo-*p*-dioxin (TCDD). When the agent binds to this "Ah receptor," it displaces a heat-shock protein (hsp90), which enables the receptor to enter the nucleus. Through an interaction

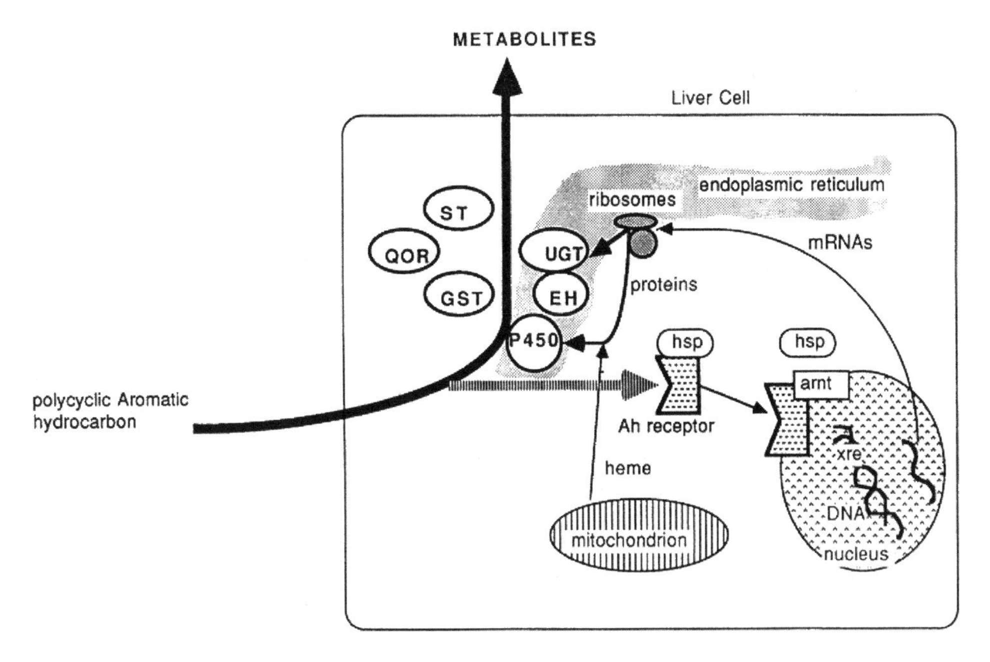

Figure 3.11 Induction of xenobiotic-metabolizing enzymes by polycyclic aromatic hydrocarbons and related compounds.

with a transporter protein (ARNT) in the nucleus, it initiates the transcription of mRNA to a limited number of proteins, including certain isozymes of cytochrome P450 (e.g., CYP1A isozymes) and UDP-glucuronosyltransferase (GT1), by binding to a regulatory region of these genes. The region of DNA to which it binds has been termed a xenobiotic response element (XRE). These mRNA molecules move out of the nucleus and are translated into new proteins on the ribosomes attached to the endoplasmic reticulum. The burst of mRNA production is usually seen within hours of exposure to the inducing agent. For increased amounts of active cytochrome P450, a coordinate induction of additional heme in the mitochondrion is also needed. Much of the information on this induction mechanism arose from work with the "nonresponsive" strains of mice (e.g., D2, CF-1; see Table 3.6) in which the Ah receptor appears defective with respect to its affinity for the polycyclic aromatic hydrocarbon. No such well-defined deficiency has yet been found in rat strains or humans.

The list of compounds that induce drug-metabolizing enzymes in a manner different from that of polycyclic hydrocarbons is much more extensive and includes chemicals of diverse chemical structure and biological effect. For some of these groups of chemicals (e.g., phenobarbital), no receptor has so far been identified. Different isozymes of the chemical/drug-metabolizing enzymes are induced (see Tables 3.4 and 3.6), and in contrast to the polycyclic hydrocarbons, many cause a marked proliferation of the endoplasmic reticulum and increase in liver size. Some of the induction seen with many of these agents has been attributed to a stabilization of existing enzyme in addition to the formation of new enzyme either via enhanced mRNA production (transcription) or changes in the translation rate of basal amounts of mRNA.

Nonmicrosomal enzymes, including sulfotransferases, are not induced as extensively as are microsomal enzymes. Exceptions are the cytosolic GSH transferases, which are induced by a wide range of agents (see Table 3.6). Extrahepatic microsomal enzymes are induced by a more restricted number of compounds compared to those that are able to induce liver enzymes, and polycyclic aromatic hydrocarbon-type induction predominates.

A similar degree of induction of both phase I and phase II enzymes does not always occur and can result in an imbalance in the ability of phase II reactions to conjugate all the reactive centers generated by the enhanced phase I activity (e.g., dexamethasone and pregnenolone 16α carbonitrile; Table 3.6). Sometimes, Phase II enzyme activities are increased with little (e.g., tioconazole, isosafrole; Table 3.6) or no (e.g., 2,2′-dipyridyl, 3,4-benzoquinoline; Table 3.6) effect on phase I enzymes. Changes in UDP-glucuronosyltransferases may be preferential for one or the other major isozyme (e.g., GT1 > GT2 for 5,6-naphthoflavone, 3-methylcholanthrene, and 2,3,6,7- tetrachlorodibenzodioxin; GT2 > GT1 for troleandomycin, phenobarbital, clotrimazole, and isosafrole). Changes in microsomal UDP-glucuronosyltransferase enzymes may (e.g., clotrimazole, isosafrole, and β-naphthoflavone) or may not (e.g., fluconazole) be accompanied by major induction of the cytosolic glutathione *S*-transferase activity.

The consequences of induction can be diverse. An inducing substance may increase the metabolism of one or more other xenobiotics and can even increase its own metabolism. Induction of microsomal enzymes can also enhance the metabolism of endogenous substrates such as steroids and bilirubin. Thus, induction may be important to consider in multiple drug therapy, chronic toxicity tests, crossover drug testing, and environmental toxicology. Some drug tolerance is explained by increased inactivation of the drug by induced enzymes. When major increases in phase I enzymes producing reactive intermediates are not matched by similar increases in the phase II enzymes responsible for their sequestration, increased toxicity may result.

Induction is qualitatively, if not quantitatively, similar in most common laboratory animal species, although the rat is perhaps the most responsive (see Table 3.7). Induction is known to occur in humans, often necessitating a change in the therapeutic dosage regimen of drugs. However, for some agents (e.g., peroxisome proliferators), the inductive response seen in experimental animals is absent in humans at therapeutic doses.

Although small differences are evident, the effects of inducers are also similar between strains of a species and between species. Thus, information derived from studies in one laboratory animal species can generally be assumed to occur in another. From the examples given in Table 3.7, the phenobarbi-

TABLE 3.7 Induction of Xenobiotic-Metabolizing Enzymes in Males of Various Animal Species [a]

		Phase I	Phase II			
			UGT			
		P450	GT1	GT2	GST	ST
Inducing Agent	Species and Strain		(% of naive animal)			
Ethanol	Rat: Fischer	125	140	115	175	90
	Rabbit: NZW	—	147	172	—	—
Phenobarbital	Rat: Sprague–Dawley	235	125	420	210	105
	Hamster: Syrian	160	120	220	120	80
	Mouse: CF-1	185	—	—	135	—
	Rabbit: NZW	—	110	155	—	—
B-Naphthoflavone	Rat: Sprague–Dawley	160	155	130	150	85
	Mouse: CF-1	85	—	—	75	—
	Mouse: D2	105	105	115	110	—
	Mouse B6	270	145	140	125	—
	Hamster: Syrian	145	—	—	65	—
	Rabbit: NZW	—	115	170	—	—
Dexamethasone	Rat: Sprague–Dawley	245	80	155	140	30
	Mouse: CF-1	240	—	—	100	—
	Mouse D2	280	100	60	80	160
	Hamster: Syrian	75	—	—	155	—

[a]*Abbreviations:* UGT = UDP-glucuronosyltransferase (two isozymes, GT1 and GT2); GST = glutathione *S*-transferase; ST = sulfotransferase; NZW = New Zealand White.

tal-induced increases in cytochrome P450, glutathione *S*-transferase, and preferential increase in GT2 UDP-glucuronosyltransferase activity over GT1 UDP-glucuronosyltransferase activity are similar in hamster and rat. Similarly, phenobarbital does not increase sulfotransferase activity in either species. β-Naphthoflavone, a polycyclic hydrocarbon-type inducer, has a similar effect in rat, mouse, and hamster, although the effect in the mouse depends on the strain employed. Two strains (CF-1 and D2) are considered nonresponsive with respect to induction by polycyclic hydrocarbon induction, and for these, in comparison with a B6 strain, there is no increase in cytochrome P450 nor induction of the GT1 UDP-glucuronosyltransferase. Dexamethasone produces large increases in cytochrome P450 with only minor increases in GT1 and GT2 UDP-glucuronosyltransferases and glutathione-*S*-transferases in either rat or mouse.

Inhibition of Xenobiotic-Metabolizing Enzymes

Since the body contains numerous but relatively nonspecific enzymes to metabolize xenobiotics, many chemicals compete for the same enzymes and mutually inhibit the metabolism of each. This may or may not be of great consequence, depending on whether the activity of xenobiotic-metabolizing enzymes is rate limiting. In considering inhibitors, and their beneficial or adverse effects, it is important to consider the perspective from which it is viewed. Piperonyl butoxide is used to inhibit insect cytochrome P450 so that the insect does not metabolize and rid itself of the pesticide, thus increasing (synergizing) the effectiveness of the pesticide. *N*-substituted imidazoles (e.g., clotrimazole) inhibit cytochrome P450-dependent ergosterol biosynthesis in fungi and prevent growth. These beneficial agents, if they inhibit human hepatic cytochrome P450 and slow the metabolism of other xenobiotics (usually labeled as drug–drug interactions), are considered as acting in an adverse manner. Inhibition of acetylcholine esterase by organophosphates is beneficial if it is being used as a pesticide, but not if it is directed against humans.

Most studies of inhibition of xenobiotic metabolism have centered on cytochrome P450. Early studies identified a compound, SKF 525A, as one of the first cytochrome P450 inhibitors, and although used extensively in laboratory investigations, it has no therapeutic use. Like most cytochrome P450 inhibitors, much of its effect can be attributed to it being an alternative substrate, namely, a competitive inhibitor. Some compounds exhibit noncompetitive characteristics. Many of these are heme ligands, which do not bind to the apoprotein "substrate site," and this class includes many nitrogenous heterocyclic compounds such as substituted pyridines, *N*-substituted imidazoles, and triazoles. Carbon monoxide, although an inhibitor of cytochrome P450 via heme binding, does not do so in vivo because it is sequestered in the blood before reaching the liver. In addition to the two classes described above, a third group of inhibitors that exhibit both of the abovementioned characteristics has been described. Their dual nature arises from their being substrates for metabolism initially, but the products of that metabolism either disrupt the protein structure (e.g., chloramphenicol, cyclophosphamide) or heme function. Heme function can be compromised by alkylation of the heme (e.g., dihydropyridines, unsaturated compounds such as olefins), which produces green pigments, by covalent linking of the heme to the protein (carbon tetrachloride), or by binding as ligands to the heme iron. This latter subgroup includes methylenedioxybenzene derivatives such as isosafrole and piperonyl butoxide and many amines including SKF 525A, troleandomycin, and related compounds. The complexes they form are classified as cytochrome P450 metabolic-intermediate complexes, and these can be detected by their characteristic ferrous state absorbance spectrum around the same wavelength as seen with the carbon monoxide, about 450 nm.

Because both competitive and suicide (mechanism-based) inhibitors require active-site recognition, inhibitors can be extremely selective for the enzyme or isozyme they inhibit. Some such selective cytochrome P450 isozyme inhibitors are given in Table 3.4. For some compounds, the exact nature of their inhibition of cytochrome P450 remains obscure; ethanol is one such example. Despite all the information available on drug interactions and toxic episodes resulting from inhibition, it is likely that the mechanism(s) of many of them have yet to be fully elucidated.

The biological consequences of inhibition of metabolism are two fold. In the acute phase, interactions can manifest themselves as either the potentiation of the biological effect of each, if metabolism results in inactivation, or protection from toxicity if toxicity arises from the bioactivation of the parent molecule. With chronic exposure, many agents generally considered as inhibitors (e.g., SKF 525A and clotrimazole) are also inducing agents (see Table 3.6). It appears that the compensation for long-term cytochrome P450 inhibition can be induction, perhaps as a response designed to circumvent the block. It should be noted, however, that more xenobiotic metabolizing enzymes than cytochrome P450 are induced by cytochrome P450 inhibitors. The induction seen with chronic exposure to inhibitors can thus result in drug interactions that are opposite to those listed as acute effects.

In addition to substrate-binding (active) site inhibition, drug metabolizing capability can be reduced by cosubstrate or cofactor depletion (e.g., glutathione, SO_4, NAD^+), by their diversion to other biochemical pathways, or by an inhibition of enzymes responsible for their formation. In laboratory investigations, glutathione conjugation can be inhibited by either buthionine sulfoximine, which inhibits the synthesis of glutathione; or diethylmaleate, which sequesters available glutathione. Galactosamine, prior to its hepatotoxic effect can deplete UDPGA by sequestering UTP. For multi-component reactions, the xenobiotic metabolism reaction can be inhibited at a distance (e.g., cytochrome P450 oxidations can be inhibited by the interruption of electron flow by heavy-metal ions, such as mercury, because the flavoprotein contains a more susceptible sulfhydryl group). Since xenobiotic metabolism is catalyzed by enzymes, many of the reactions can be inhibited nonselectively by protein denaturants such as heavy-metal ions and detergents, the degree of inhibition depending on the concentration. For enzymes that require a suitable membrane environment for activity, xenobiotics with lipid solvent properties can inhibit activity by destroying that necessary environment. Changes in lipid often lead to conformational changes that alter activity.

Not all inhibitors relate to enzymes located in membranes. There are inhibitors of nonmicrosomal xenobiotic-metabolizing enzyme activities that have toxicological importance and clinical usefulness. Disulfiram, an inhibitor of aldehyde dehydrogenase, is used as an adjunct to behavioral modification in the treatment of alcoholism since the unpleasant symptoms elicited by the accumulating acetaldehyde are sufficient to dissuade further ethanol ingestion. Monoamine oxidase inhibitors are available as drugs for the treatment of depression. If chemicals (e.g., tyramine) normally adequately metabolized by these enzymes are ingested simultaneously, they may accumulate to a sufficient concentration to cause severe toxicity (hypertensive crisis). Esterases where the active center contains a serine residue are readily inhibited by organophosphates and carbamates. Such inhibition results in the accumulation of other chemicals undergoing hydrolysis, particularly the endogenous substrate, acetylcholine, to the point of toxicity.

Other Factors Responsible for Variations in Xenobiotic Metabolizing Enzymes

Animal Species and Strain Much has been made of species differences in xenobiotic metabolism, both for the purposes of extrapolation to humans and for exploiting differences in the understanding of species selective toxicities. Rodents have higher cytochrome P450 concentrations than other mammalian species, birds, and fish. Among mammals, cats are particularly deficient in UDP-glucuronosyltransferase activities and fish are deficient in all conjugations. This latter point has been attributed to the lesser need of aquatic animals to render foreign compounds to their most water-soluble form, since the volume of water that xenobiotics can diffuse into via the gills compensates for the lower partition coefficient. In comparison to most laboratory animal species, the rat is well endowed with sulfotransferase activity, a little lower in cytochrome P450 concentration, and relatively deficient in glutathione transferase activity (Table 3.8).

TABLE 3.8 Species and Strain Variations in Xenobiotic-Metabolizing Enzymes[a]

	Phase I		Phase II			
			UGT			
Species and Strain	P450	pNA deM	GT1	GT2	GST	ST
			(vs. Male Sprague–Dawley Rat)			
Rabbit[b]	140	—	250	275	575	140
Hamster	160	300	155	235	470	25
Rat: Fischer	90	125	115	115	55	—
Rat: Gunn	125	—	30	120	—	—
Mouse: D2	85	245	60	170	225	320
Mouse: B6	—	85	325	90	325	235
Mouse: CF-1	120	490	—	—	265	—
Mouse: SW[b]	105	—	65	220	200	50
Guinea pig[b]	105	—	180	95	415	30
Cat[b]	60	—	5	50	75	50
Dog[b]	70	—	335	355	85	30
Quail[b]	45	—	220	25	75	35
Trout[b]	68	—	5	20	60	10

[a]*Abbreviations*: pNA deM—*p*-nitroanisole demethylase; UGT = UDP-glucuronosyltransferase (two isozymes: GT1 and GT2); GST = glutathione *S*-transferase; ST = sulfotransferase; SW = Swiss Webster.

[b]Gregus Z, Watkins JB, Thompson TN, Harvey MJ, Rozman K, Klaassen CD, *Toxicol. Appl. Pharmacol.* **67:** 430 (1983).

In some species, while the rate of metabolism of a compound may be similar, the products may be different. Amphetamines are oxidatively metabolized by cytochrome P450 in both rabbits and rats but deamination products predominate in rabbits and phenyl hydroxylation products predominate in rats.

Within an animal species there are also strain differences in xenobiotic-metabolizing capabilities. While a Fischer rat is very similar to a Sprague–Dawley rat, the Gunn strain of rat exhibits a deficiency in one of the two major classes of UDP-glucuronosyltransferase activity (GT1), yet its cytochrome P450 concentration is very similar to that in other strains. Two mouse strains (D2 and B6) that are very different in their response to induction by polycyclic hydrocarbons are very similar in their drug metabolizing enzymes activities before such exposure.

Clear examples of human genetic polymorphisms are known. The best documented is the variation in acetylator phenotype and its consequences for the use of the antituberculosis drug, isoniazid (see Figure 3.12).

Fast acetylators, occurring as a higher percent in Asian and Eskimo populations, tend to metabolize the drug and show liver toxicity from the reactive metabolite generated. Slow acetylators show the toxicity associated with the accumulation of unchanged drug, a peripheral neuropathy resulting from

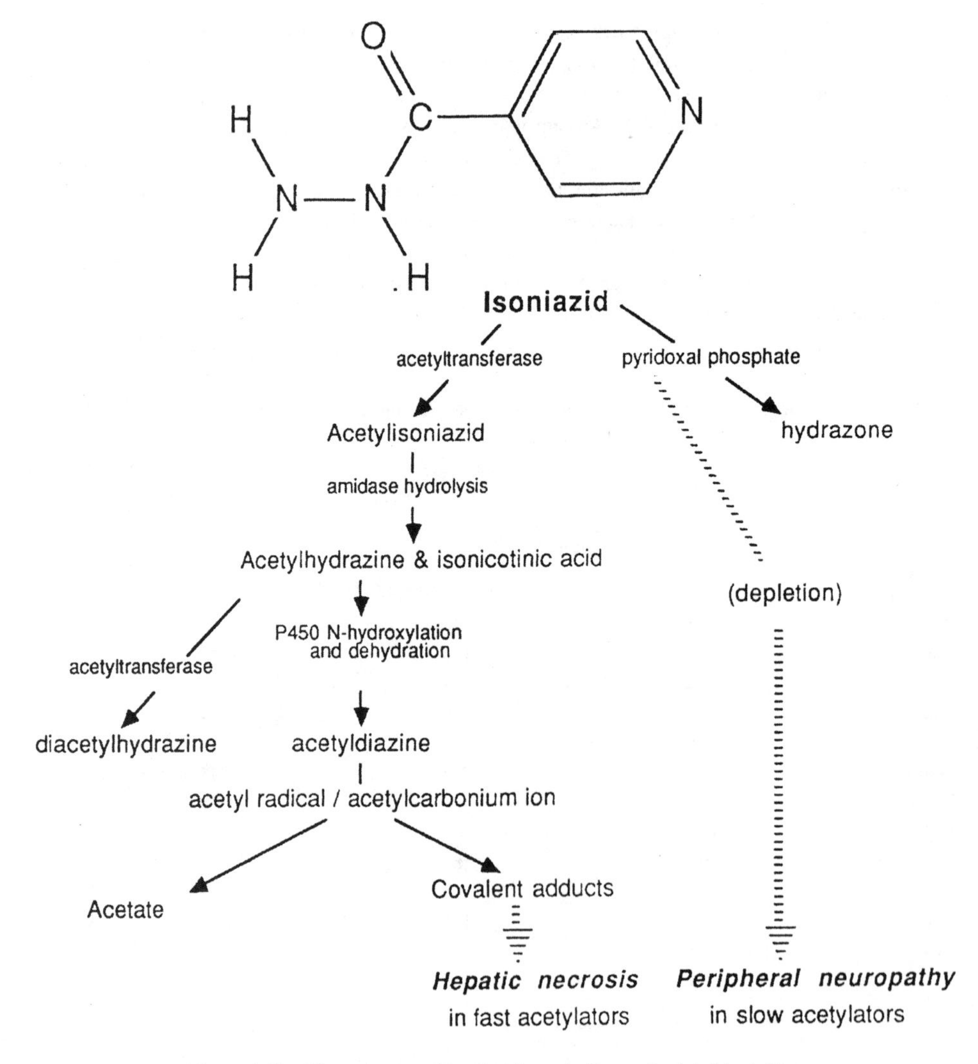

Figure 3.12 The pathways of isoniazid metabolism and related toxicities.

pyridoxal phosphate depletion by the unmetabolized isoniazid. Slow acetylators have an increased risk of developing bladder cancer when exposed to arylamine compounds but are less represented than the overall population among colorectal cancer patients.

Recently much attention has been directed at the number of isozymes, of human cytochrome P450 since variations in the amounts of the various isozymes that have some degree of substrate selectivity, could explain the variations in responses observed to standard doses of drugs. It could also partly explain the susceptibility of certain individuals to toxicity by chemicals that are bioactivated via oxidative metabolism. The cytochrome P450 2D6 polymorphism divides the Caucasian population into poor (5–10 percent) and extensive metabolizers of over 40 drugs and has been implicated in the development of some forms of cancer. Interestingly, extensive metabolizers are overrepresented in tobacco-smoke-associated lung cancer patients and underrepresented in leukaemia and melanoma patients. Polymorphism in the 2C19 form shows interracial differences, with an incidence of poor metabolizers of > 5 percent in Caucasian populations and 20 percent in Asian populations, although to date such differences have not been implicated in toxicities other than those arising from the slow metabolism of drugs.

Gender Although there is no evidence of major gender differences in hepatic xenobiotic metabolism in humans, a major difference has been well documented for rats, particularly with respect to cytochrome P450. (Limited studies in humans suggest that females have slightly greater oxidative metabolism rates than do males.) It is also well documented that there is gender-dependent expression of certain cytochrome P450 isozymes in the rat (see Table 3.4). Sex differences would appear to be independent of the strain of rat and also apparently occur in at least one other rodent species, the hamster (Table 3.9).

The mouse generally displays a higher cytochrome P450 concentration and activity in females. Phase II conjugations can also show sex differences, and these, like cytochrome P450, may be isozyme specific. Most of the gender-related differences in cytochrome P450 expression in rodents have been related to gender differences in growth hormone secretion.

Although small differences are evident, the effects of inducers are similar between sexes of a species. From the examples given in Table 3.9, the phenobarbital-induced increases in cytochrome P450, glutathione *S*-transferase, and preferential increase in GT2 UDP-glucuronosyltransferase activity over GT1 UDP-glucuronosyltransferase activity are similar in males and females of both hamster and rat. Similarly, phenobarbital does not increase sulfotransferase activity in either sex of either species.

TABLE 3.9 Gender Differences in Xenobiotic-Metabolizing Enzymes[a]

		Phase I		Phase II			
			pNA	UGT			
Status	Species and Strain/Gender	P450	deM	GT1	GT2	GST	ST
			(Female vs. Male)				
Naive	Rat: SD	0.85	0.75	0.70	1.00	1.05	0.30
	Rat: Fischer	0.70	0.85	0.35	0.90	1.15	—
	Hamster	0.90	0.80	0.80	0.80	1.00	0.85
			(Induced vs. Naive)				
Phenobarbital-induced	Rat: SD male	2.35	6.60	1.25	4.20	2.10	1.05
	Rat: SD female	1.65	3.75	1.50	3.30	1.55	0.90
	Hamster: male	1.60	3.26	1.20	2.20	1.20	0.80
	Hamster: female	1.40	3.35	1.30	1.95	1.15	0.60

[a]*Abbreviations:* pNA deM = *p*-nitroanisole demethylase; UGT = UDPglucuronosyltransferase (two isozymes: GT1 and GT2); GST = glutathione *S*-transferase; ST = sulfotransferase; SD = Sprague–Dawley.

Maturation The age of a rat can also cause changes in its complement of drug metabolizing enzymes. Old age decreases the cytochrome P450 concentration and activity, particularly in the male. Of the phase II enzymes, a decline in sulfotransferase activity is apparent. Glutathione *S*-transferase appears marginally increased in old rats and lower in immature rats as compared to the mature adult. Neonatal animals generally exhibit lower drug-metabolizing activities than adults (Table 3.10).

In humans, the activity of microsomal and perhaps nonmicrosomal xenobiotic-metabolizing enzymes is low in premature and neonatal infants. The effective glucuronidation of the bulky-type molecules, (e.g., morphine and chloramphenicol) appears to develop much later than does that for the planar phenol-type compounds. The activity of microsomal enzymes in neonates can be induced. Although there is evidence that the elderly have a decreased rate of hepatic microsomal metabolism of some drugs, the clinical importance of this is not clear because drug clearance remains unchanged as a consequence of changes in the volume of distribution of many drugs. The metabolism of many drugs and xenobiotic chemicals is fastest in adolescents.

Environment Human chemical drug metabolism can be influenced by the environment and diet. All diets contain naturally occurring nutrients and may also contain pesticide residues and food additives that are capable of altering the activity of chemical/drug-metabolizing enzymes. Among the more recent nutrient interactions is an inhibition of cytochrome P450 3A by grapefruit juice flavonoids, naringin and quercetin. Other flavonoids (catechin, myricetin, rutin, etc.) are able to induce phase II enzymes and protect against bioactivated intermediates (see discussion below).

Even the quality of the diet can have an effect. Protein deficiency or diets deficient in essential fatty acids or certain vitamins (e.g., A, C, E) can decrease xenobiotic metabolism. Supplementation of diets with these nutrients (e.g., high protein) can increase chemical metabolism above normal. Administration of drugs and exposure to toxic compounds in burned carbonaceous material such as cigarette smoke and charcoal-broiled foods are among the best known modifiers of xenobiotic metabolizing capabilities.

It is evident from the foregoing illustrations concerning the complex mixture of factors responsible for variations in xenobiotic metabolism that while a basic understanding of chemical metabolism can provide guidelines, prediction of actual situations is infinitely more difficult. While genetic differences between species may be obvious, subtle differences in physiology and diet all tend to confound extrapolations between experimental animals and, more critically, between experimental animal models and humans. These differences may even confound extrapolation between subgroups within the human population.

TABLE 3.10 Influence of Maturity on Rat Xenobiotic-Metabolizing Enzymes[a]

		Phase I		Phase II			
			pNA	UGT			
Age	Gender and Strain	P450	deM	GT1	GT2	GST	ST
				(vs. Young Adult)			
Immature	Male Sprague–Dawley[b]	1.00	1.05	0.50	0.90	—	—
	Female Sprague–Dawley[b]	1.40	1.00	0.55	1.05	—	—
Old	Male Fischer	0.65	0.70	0.95	1.00	1.10	0.60
	Female Fischer	0.90	0.80	1.00	1.10	1.05	0.70

[a]*Abbreviations:* pNA deM = *p*-nitroanisole demethylase; UGT = UDP-glucuronosyltransferase (two isozymes; GT1 and GT2); GST = glutathione *S*-transferase; ST = sulfotransferase.

[b]Chengelis CP, *Xenobiotica* **18:** 1225 (1988).

Biotransformation: A Balance between Bioactivation and Detoxification The balance between bioactivation and inactivation (detoxification) can often determine whether a chemical is "toxic" to cellular systems; toxic in this situation is usually defined as any cell damage leading to modified cell function, not necessarily cell and tissue death. Reactive intermediates may cause enzyme and protein modification and inactivation, membrane lipid peroxidation, and changes in DNA. Reactive intermediates are therefore implicated in carcinogenesis and tissue allergic responses in addition to tissue necrosis. Although enzymatic bioactivation of many environmental chemicals is a necessary step in the process of detoxification and elimination of the xenobiotic (Figure 3.4), the chemistry of the enzymatic product often precludes the elimination of the chemical without damage to critical cellular targets. When such reactive metabolites are produced in sufficient quantities, it is then that they cause cell and tissue damage. Thus, while the metabolism of certain toxic chemicals should represent a protective biological process, it frequently becomes a bioactivation process that is highly effective in the production of toxins within the organism.

The reactive intermediates produced by the enzymes are often electrophiles, free radicals, or chemicals that can rearrange nonenzymatically to such intermediates. The chemical nature of the enzymatically generated reactive intermediates can be useful in providing a classification of toxic compounds (Table 3.11).

The best documented example of enzymatic processes that produce reactive intermediates is the oxidation of chemicals by members of the cytochrome P450 superfamily; however, one should not overlook the fact that many xenobiotics are oxidized to nontoxic products such as phenols, *N*-oxides, alcohols, amines, aldehydes, and carboxylic acids. Thus, the same enzymes that detoxify one chemical can be responsible for the bioactivation of another.

The reactive electrophilic intermediates produced by cytochrome P450 range from epoxides to iminium ions and include the formation of free radicals. In addition to cytochrome P450, reactive intermediate generation by a variety of other enzymes also occurs (Figure 3.13).

The conjugation of xenobiotics with glutathione, glucuronic acid, sulfate, or acetate (phase II reactions) was originally thought to embody solely a detoxication process for drugs and environmental chemicals. In the vast majority of the examples that have been studied, these products of conjugation reactions are, in fact, detoxication metabolites. However, a significant number of studies with a variety of toxicants have shown that many of the conjugates are not innocuous. Glucuronidation has recently

TABLE 3.11 Classification of Toxicants by Reactive Intermediates

Acyl glucuronides of
Bilirubin, clofibric acid, diflunisal, indomethacin, tolmetin, valproic acid, zomipirac
Carbonium ions
2-Acetylaminofluorene, dimethylnitrosamine, nitrosonornicotine, procarbazine, pyrolizi dine alkaloids
Epoxides
Aflatoxins B_1 and B_2, benzo[*a*]pyrene, benzo[*e*]pyrene, benzy[*b*]fluoranthene, chrysene, 7,12-dimethylbenz[*a*]anthracene, 3-methylcholanthrene
Glutathione adducts of
Chlorotrifluoroethylene, 1,2-dibromo-3-chloropropane, dibromoethane, *N*-(3,5-dichlorophenyl)succinimide, hexachlorobutadiene, tetrachloroethylene, tetrafluoroethylene, trichlorethylene, tris(2,3-dibromopropyl)phosphate
Imines
Acetaminophen, amodiaquine, 3-methylindole, 2,6-dimethylaniline, ellipticine acetate, nico tine, phencyclidine
Nitrenium
2-Acetylaminofluorene, 4-aminobiphenyl, 2-aminonaphthalene, 2-aminophenanthrene, benzidi ne
Quinones
Adriamycin, *o*-benzoquinone, *p*-benzoquinone, bleomycin, menadione, mitomycin *c*, 1,2-naphthoquinone, streptonigrin

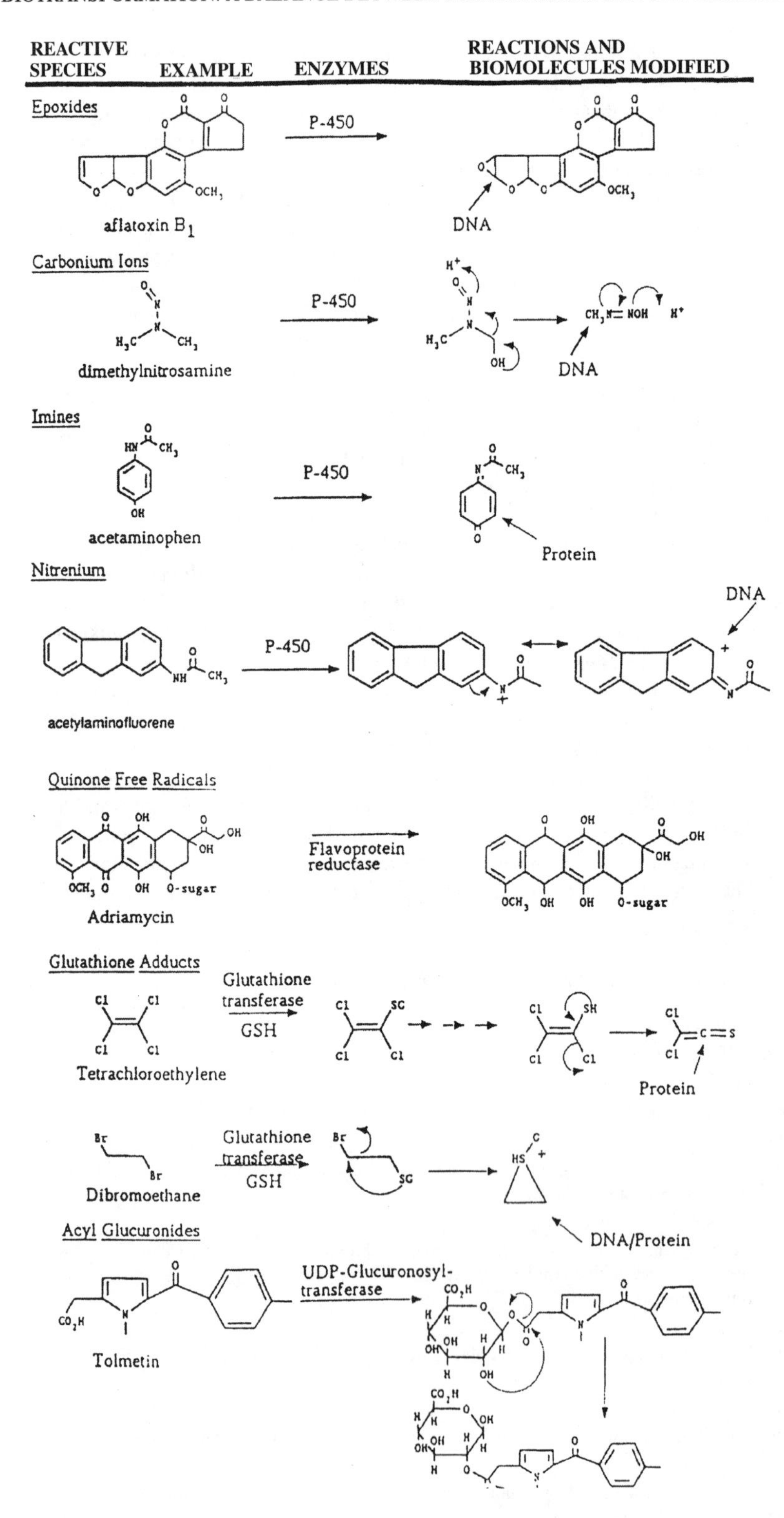

Figure 3.13 The enzymatic generation of reactive intermediates.

been shown to be responsible for the production of glucuronides, which, via protein alkylation, can result in the formation of immunogens. The immune response mounted to these aberrant molecules can be highly toxic to organisms. Examples of toxic conjugates (acyl-linked glucuronides and glutathione adducts) are shown in Figure 3.13 and Table 3.11. Glutathione conjugates can also serve as transport forms of reactive intermediates. Methyl isocyanate, a highly reactive electrophile, is such an example. The glutathione conjugate is transported to sites distant from the initial absorption site to cause toxicity to other organs.

The balance between detoxication and bioactivation of xenobiotics by metabolism enzymes can be dramatically changed by the induction or inhibition of the enzymes. Enzymes that are normally present at low levels, and therefore do not bioactivate toxicants to reactive intermediates, can become active participants in the toxicity of chemicals when the levels and activities of the enzymes are increased. Many examples of this situation exist. For example, induction of CYP2E1 by ethanol results in the greater bioactivation of hepatotoxins like CCl_4 and acetaminophen or carcinogens such as dimethylnitrosamine. Although the toxicants can produce damage normally, their potency is greatly increased after induction of CYP2E1; specifically, toxicity is elicited at much lower doses because more of the chemical is oxidized to a reactive intermediate.

Conversely, the toxicity of many chemicals can be ameliorated by induction of enzymes that are responsible for the detoxication of the compound. Bilirubin can cause significant central nervous system damage in neonates where the UDP-glucuronosyltransferase(s) that detoxify this naturally occurring heme breakdown product are present in low amounts. Inducing the levels of the necessary UDP-glucuronosyltransferase by drugs such as phenobarbital increases the glucuronidation of bilirubin and decrease its toxicity. In the same way that induction of bioactivation enzymes can increase toxicity and induction of detoxification enzymes can decrease toxicity, the inhibition of bioactivation enzymes or the inhibition of detoxification enzymes should decrease or increase toxicity, respectively. The carcinogenicity of complex mixtures of polycyclic aromatic hydrocarbons is sometimes found to be less than one would expect if the relative carcinogenicity of each component were summed. A probable reason for this decrease in toxicity lies in the inhibition, by components of the mixture, of the cytochrome P450 enzymes that bioactivate the carcinogens to their DNA-reactive intermediates.

The mechanisms by which xenobiotics cause toxicity can be highly diverse, and elucidating the precise biochemical and chemical mechanisms that induce toxicity can be a difficult process. There are many tools available that can be used to evaluate toxic mechanisms. They include the use of animal species, gender, or cellular differences that vary widely in their response to the toxin. For example, naphthalene is highly toxic to mice when administered by the intraperitoneal route or by inhalation but is much less toxic to rats. Investigators have used this species differences to provide vital information about a cytochrome P450 (CYP2F2) that is highly expressed only in murine lung and is responsible for the bioactivation of this toxicant. Limonene causes severe renal toxicity to male rats but not female rats. The primary cause for the toxicity was eventually linked to the expression of a globulin that is not expressed to a significant degree in female rats. An example of the use of specific cellular targets is with the nephrotoxic glutathione conjugates of halogenated hydrocarbons, such as hexachlorobutadiene, which are selective for the proximal tubule cells of the nephron. Analysis showed that these cells contain high amounts of the enzyme C-S lyase, and it is this enzyme which is responsible for the production of the electrophilic intermediates from these toxicants. Similarly, it is the high content of monoamine oxidase B in dopamine-containing neurons linked with the cellular selectivity of the toxicity of MPTP that has enabled the mechanism of bioactivation of this toxicant to be elucidated.

3.3 SUMMARY

By altering a portion of a chemical or by adding another molecule to it, drug-metabolizing enzymes can alter the toxicity of the chemical, its tissue-binding properties, and its distribution and duration within the body.

While the main benefit of biotransformation is to protect the body from attaining high chemical levels within the various tissues, the lack of specificity and predictability in the process sometimes leads to bioactivation or an increase in the toxicity of a chemical. Some such products are mutagenic and/or carcinogenic. Thus, metabolism of a toxic chemical has three possible outcomes, and it is the balance between these possible occurrences that determines the eventual outcome of the chemical exposure in question.

1. It may form a nontoxic metabolite.
2. It may generate a toxic metabolite that is subsequently detoxified.
3. It may generate a toxic metabolite that is not rendered harmless by detoxification before cellular and tissue injury have ensued.

The primary organ functioning to metabolize chemicals is the liver. It is the first organ to be exposed to chemicals absorbed from the gut, the major portal of entry of xenobiotics.

The characteristics of a person exposed to a toxic chemical that may alter the metabolism can include diet or nutritional status, age, gender, and/or hormonal status, and the genetic makeup of the individual. These characteristics may account for the interindividual variations observed in the human response to chemical exposure and the extent or type of toxicity observed in various model animals. A major determinant of metabolizing capabilities is prior exposure to chemicals that induce or inhibit the enzymes involved.

SUGGESTED READING

Gibson, G., and P. Skett, *Introduction to Drug Metabolism*, 2nd ed., Chapman & Hall, London, 1993.

Mulder, G. J., ed., *Conjugation Reactions in Drug Metabolism: An Integrated Approach*, Taylor & Francis, Bristol, PA, 1990.

Ortiz de Montellano, P. R., ed., *Cytochrome P450. Structure, Mechanism and Biochemistry*, 2nd ed., Plenum Press, New York, 1995.

Pacifici, G. M., and G. N. Fracchia, eds., *Advances in Drug Metabolism in Man*, European Commision, Luxembourg, 1995.

Timbrell, J. A., *Principles of Biochemical Toxicology*, 2nd ed., Taylor & Francis, Bristol, PA, 1991.

Witmer, C. M., R. R. Snyder, D. J. Jollow, G. F. Kalf, J. J. Kocsis, and I. G. Sipes, eds., *Biological Reactive Intermediates IV; Molecular and Cellular Effects and Their Impact on Human Health*, Plenum Press, New York, 1991.

4 Hematotoxicity: Chemically Induced Toxicity of the Blood

ROBERT A. BUDINSKY JR.

This chapter describes toxicities affecting blood. The following subjects are covered:

- The origin, formation, and differentiation of blood cells
- Clinical tests used to evaluate hematotoxicity
- Oxygen transport by erythrocytes (red blood cells or RBCs) and interference with oxygen transport by drugs and chemicals
- Chemicals that affect the formation of red blood cells, platelets, and white blood cells (bone marrow suppression)
- Leukemias and lymphomas (cancers of the white blood cells)
- Neurological and cardiovascular toxicities caused by interference with oxygen utilization (e.g., cyanide, hydrogen disulfide)
- Medical treatment of hematotoxicity

This chapter describes common occupational and environmental chemicals and drugs that affect blood formation and function. Only the recognized chemically induced blood toxicities are included in this chapter, although there are many examples of anecdotal reports linking a large number of chemicals and drugs with hematotoxicity.

4.1 HEMATOTOXICITY: BASIC CONCEPTS AND BACKGROUND

Hematotoxicity essentially involves two basic homeostatic functions: (1) RBC-mediated oxygen transport and (2) the production of red and white blood cells and platelets. Perhaps the earliest experiences with hematotoxicity involved the consumption of fava beans and the development of favism among people in the Mediterranean region. Favism is the development of acute hemolytic anemia in individuals with a deficiency in the red blood cell enzyme glucose-6-phosphate dehydrogenase following the ingestion of fava beans. It is also likely that methemoglobinemia, a blood condition characterized by cyanosis, was observed when individuals consumed well water containing large amounts of nitrates and nitrates.

Chemically induced hematotoxicity has been reported in the medical literature for over a century. For example, the 1919 publication by Dr. Alice Hamilton, entitled *Industrial Poisoning by Compounds of the Aromatic Series,* described a number of blood toxicities that were commonly encountered in occupational settings such as benzene-induced bone marrow suppression, aniline and nitrobenzene-induced methemoglobinemia, and hydrogen sulfide–induced effects. The major hematotoxicities encountered in the workplace involve benzene (bone marrow suppression and acute myelogenous leukemia), carbon monoxide (impairment of oxygen transport), aniline/nitrobenzene analogs

Principles of Toxicology: Environmental and Industrial Applications, Second Edition Edited by Phillip L. Williams, Robert C. James, and Stephen M. Roberts.
ISBN 0-471-29321-0 © 2000 John Wiley & Sons, Inc.

(hemolytic anemia, which reduces the oxygen transport capacity of blood), and hydrogen sulfide. Exposure to hydrogen sulfide can be a significant industrial hygiene concern in the refining of petroleum products and the biological degradation of silage (fermented corn, grain, etc., used to feed livestock) and sewage. As for the number of workers affected, benzene and hydrogen sulfide probably constitute the most significant risk factors for toxicity.

Hematotoxicity is also an important concern in the administration of pharmaceuticals. For example, dapsone (used to treat leprosy) and primaquine (used to treat malaria) can produce a fatal hemolytic anemia in certain genetically predisposed individuals (those with a deficiency in glucose-6-phosphate dehydrogenase). Unfortunately, individuals most likely to require primaquine or dapsone therapy live in tropical areas of Africa, Asia, and the Mediterranean and are most likely to inherit a deficiency in glucose-6-phosphate dehydrogenase. Of widespread concern are the risks of bone marrow injury and suppression caused by cancer chemotherapeutics, complications that can often limit the administration of cancer-curing drugs. Another longstanding problem involves carbon monoxide poisoning, which results from exposure to improperly ventilated combustion products. Outside the workplace, the most common occurrences of hematotoxicity involve carbon monoxide poisoning, due to faulty gas heating, and adverse hematologic effects due to prescription medications.

Fortunately, hematotoxicity is rarely encountered due to the resiliency of bone marrow, the redundancy of various hematologic controls and functions, and the implementation of more conservative occupational hygiene standards. However, when it occurs it is often life threatening. Likewise, examples of hematotoxicity resulting from exposure to environmental chemicals are relatively rare and generally involve foods or medications. Although hematotoxicity is not prevalent, it is useful for industrial hygienists, toxicologists, and occupational physicians to be aware of the chemicals that cause hematotoxicity, relevant signs and symptoms, and any antidotes and treatments that are available.

4.2 BASIC HEMATOPOIESIS: THE FORMATION OF BLOOD CELLS AND THEIR DIFFERENTIATION

All blood cells originate from undifferentiated mesenchymal cells, which are located in the bone marrow. The various stages of blood cell formation are depicted in Figure 4.1. From stem cells, clones of immature blood cells differentiate along one of two pathways: the myelogenous series or the lymphocytic series. The myeloid series gives rise to erythrocytes, macrophages, platelets, neutrophils, eosinophils, and basophils. The lymphoid series gives rise to T (thymus) and B (Bursa) lymphocytes.

Bone marrow production of blood cells is highly dependent on, and controlled by, a number of growth factors. Erythropoietin, a glycoprotein growth factor produced in the peritubular cells of the kidney, is essential for the differentiation and maturation of red blood cells. Under conditions of hypoxia (low oxygen), such as that occurring at higher altitudes or during anemia (a reduction in red blood cells or hemoglobin content) affliction, the release of erythropoietin by the kidney is enhanced. Conversely, the release of erythropoietin is inhibited by polycythemia (the increased number of circulating red blood cells) or hyperoxia. Other important glycoproteins that act alone or in conjunction with erythropoietin to control red blood cell formation include interleukins such as IL3, IL1, and IL2; granulocyte-macrophage colony stimulating factor (GM-CSF); insulin-like growth factor; and granulocyte colony stimulating factor.

White blood cell formation also depends upon stimulation and control by various growth factors. IL3 stimulates all of the myeloid series cells. GM-CSF stimulates the formation of granulocytes and macrophages. Additionally, specific G-CSF and M-CSF proteins stimulate the granulocyte series or the macrophage series, respectively. These growth factors, unlike erythropoietin, are produced by various cells including T lymphocytes, macrophages, fibroblasts, and endothelial cells. All the growth factors work in concert to regulate different stages of myeloid and lymphoid differentiation and replication.

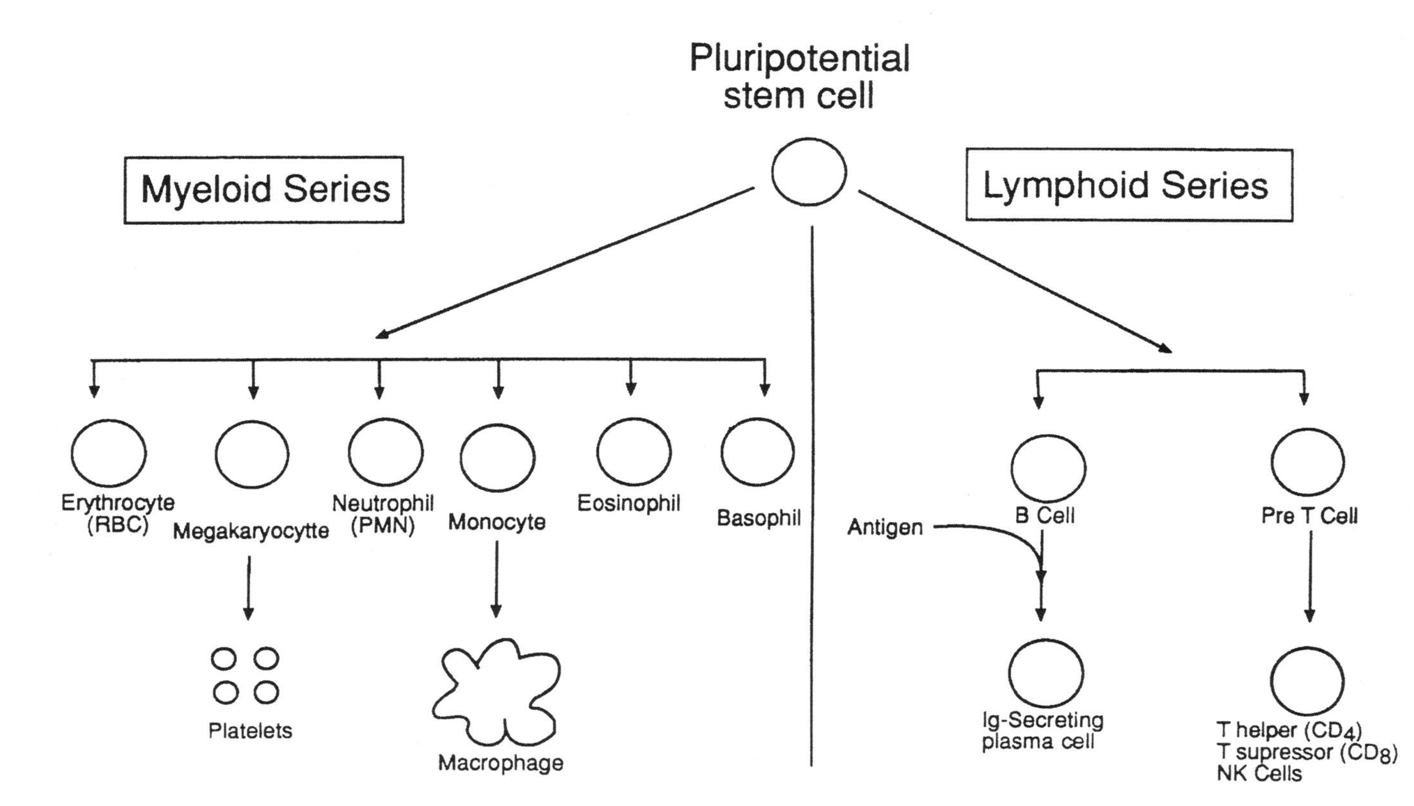

Figure 4.1 Formation of blood. Mature differentiated blood cells arise from a pluripotential stem cell via sequential complex interactions of regulatory molecules and cellular controls within the bone marrow environment. Disruption of the differentiation and maturation process gives rise to the various types of chemically induced blood deficits and/or leukemias/lymphomas.

When bone marrow is injured or suppressed, the number of specific types of blood cells (or all blood cells) may decline or even disappear. The decline in the number of cells from a specific blood cell lineage has its own diagnostic term and is based on the expected normal range that exists in healthy individuals. For example, a decrease in the normal number of circulating red blood cells leads to the clinical condition of anemia, a decrease in circulating platelets is known as *thrombocytopenia,* and a decrease in white blood cells is called *leukopenia.* Table 4.1 provides definitions for various clinical terms used to describe the abnormal number of circulating red blood cells, neutrophils, lymphocytes, and platelets. Some terms describe the same condition and may create confusion when used interchangeably. The suffix *penia* means an abnormal reduction, and the suffix *cytosis* refers to an abnormal excess.

Changes in the number of circulating white blood cells provide an important diagnostic parameter for many diseases. Granulocytopenia, when the granulocyte count (primarily the neutrophils) falls to less than 1000 cells/mm^3, may arise from chemical-induced bone marrow damage following administration of cancer chemotherapeutic drugs or the antibiotic chloramphenicol, antiinflammatory agents such as butazolidin, or exposure to benzene. When this occurs, an insufficient number of granulocytic cells are available to maintain the first line of defense against infectious agents, and recurrent infection is likely. Granulocytosis, or an increased number of circulating granulocytes (exceeding 10,000 cells/mm^3), often occurs in patients with leukemia or can be triggered by an underlying infection. In cancers of the myeloid series, such as myelogenous leukemia, neutrophil numbers may exceed 30,000 cells/mm^3. The leukemias generally consist of cells that lack the normal morphology and function of mature white blood cells (e.g., they resemble precursors of immature white blood cells found in the bone marrow).

The cancer biology of the various types of leukemias (myeloid and lymphoid) and lymphomas, as well as multiple myeloma is quite complex. The types of leukemias and lymphomas are described in Table 4.2 according to the International Classification of Disease codes (frequently abbreviated the ICD). These cancers can differ in morbidity as well as clinical presentation, symptoms, management, and long-term survival.

TABLE 4.1 Definitions of Hematological Clinical Terms (Normal Adult)

Clinical Term	Definition/Description
Anemia	A reduction in either the number or the volume of red blood cells, i.e., less than 3,500,000 RBC/mm^3 or 14 g of hemoglobin per 100 mL of blood
Aplastic anemia	A cessation of the normal regenerative production of red blood cells in the bone marrow
Agranulocytosis	A reduction in the number of polymorphonuclear leukocytes (PMNs) less than 500/mm^3
Granulocytopenia, neutropenia	A reduction in the normal number of granulocytic leukocytes in the blood; normal granulocytes number around 3000–4000/mm^3
Granulocytosis	An increase above normal of the normal number of circulating granulocytic leukocytes
Leukopenia	A reduction of the number of circulating leukocytes (white blood cells) below 5000 cells/mm^3
Leukocytosis, neutrophilia	An increase in the number of leukocytes, typically PMNs, above 10,000 cells/mm^3
Lymphopenia	A reduction in the number of circulating lymphocytes less than the normal 2500/mm^3
Lymphocytosis	An increase in the number of circulating lymphocytes from their normal number of around 2500/mm^3
Eosinophilia	Increase number of eosinophils above 200/mm^3
Thrombocytopenia	Abnormal number of circulating platelets less than 250,000–500,000/mm^3

TABLE 4.2 Leukemias and Lymphomas

ICD No.	General Description
ICD 200: lymphosarcoma and reticulosarcoma	Lymphocytic lymphomas involving resident lymphocytes in various tissues such as the liver, spleen, lung, skin, bone marrow, and gastrointestinal tract
ICD 201: Hodgkin's disease	Lymph node cancers exhibiting Reed–Sternberg cells
ICD 202: other malignant neoplasms of lymphoid and histiocytic tissue	Involves resident macrophage cell types located in the peripheral tissues such as lymph nodes, spleen, skin, liver, and connective tissue
ICD 203: multiple myeloma	Cancer of a specific plasma cell (antibody producing B lymphocyte)
ICD 204: lymphoid leukemia	Cancer of circulating B and T lymphocytes
ICD 205: myeloid leukemia	Cancer of the myeloid series, usually granulocytes (neutrophils or PMNs)
ICD 206: monocytic leukemia	Cancers of the monocytic series involving the bone marrow–derived monoblast
ICD 207: erythroleukemia	Cancer of RBC precursors
ICD 208: leukemia not otherwise specified	

Source: International Classification of Disease (ICD), 9th ed.

4.3 THE MYELOID SERIES: ERYTHROCYTES, PLATELETS, GRANULOCYTES (NEUTROPHILS), MACROPHAGES, EOSINOPHILS, AND BASOPHILS

Red Blood Cells (RBC): Erythrocytes

During erythrocyte maturation, the blast forming unit-erythrocyte or BFU-E matures into the colony forming unit-erythrocyte (CFU-E), which finally evolves into a mature erythrocyte after extrusion of the nucleus and acquisition of hemoglobin and a mature cytoskeletal support system. Accelerated erythrocyte formation due to anemia, may result in the release of RBCs prior to the loss of the nucleus. These immature erythrocytes known as reticulocytes can be found circulating in the blood in increased numbers (i.e., reticulocytosis). Physicians may suspect anemia or a bleeding disorder if the number of reticulocytes exceeds their normal blood level (normal circulating reticulocyte levels are 5 percent of the total number of RBCs) or if other clinical parameters are abnormal, specifically, the number of RBCs, the percent volume of packed RBCs or the hemoglobin content. Once the red blood cell is released into circulation, it has a normal lifespan of 120 days. Damaged or senescent RBCs are sequestered in the spleen and destroyed by splenic macrophages.

Mature RBCs are discoid-shaped and devoid of a nucleus and mitochondria. They are rich in the heme-containing protein, hemoglobin, which transports oxygen to peripheral tissues and carries carbon dioxide from the periphery to the lungs, where it is exhaled. Erythrocytes also provide a pH buffering function by converting carbon dioxide to carbonic acid via the enzymatic activity of carbonic anhydrase. Erythrocytes possess an outer membrane supported by a complex cytoskeletal system of various proteins that are essential for carrying out the normal functions of the RBC. Erythrocytes normally constitute around 40 percent by volume of whole blood and number around 3,500,000–5,000,000 per cubic millimeter. The normal hemoglobin content of 100 mL of blood is 14 g. When the 120-day lifespan of erythrocytes is shortened to the extent that stimulated erythrocyte replacement cannot keep pace, varying degrees of anemia may result.

There are numerous causes of anemia, including abnormal hemoglobin inherited in the form of sickle cell anemia or thalassemia; congenital nonspherocytic anemias involving defects in biochemical pathways in the RBC (e.g., glucose-6-phosphate dehydrogenase deficiency); congenital spherocytosis, in which inherited defects in cytoskeletal membrane proteins predispose RBCs to hemolysis and destruction by the spleen; proxysmal nocturnal hemoglobinuria involving nighttime episodes of hemolysis; autoimmune-induced RBC destruction; chemical-induced RBC damage or destruction; infectious diseases such as malaria; and hypersplenism. Many of the underlying pathological conditions that give rise to the various types of anemia can be quite complex and require careful clinical assessment. Anemias may present with different RBC morphology (shape or appearance). The types of anemia and information regarding specific anemias are listed in Table 4.3.

An essential characteristic of RBCs is their ability to undergo reversible shape changes or deformations as they travel through the narrow capillary beds. The reversible deformability involves a highly organized submembrane cytoskeletal scaffolding system. In this system, spectrin chains are attached to the membrane via ankyrin and further controlled by actin and numerous other cytoskeletal membrane proteins. Not surprisingly, if these cytoskeletal membranes are damaged by active oxygen (the high oxygen environment of the erythrocyte or redox cycling caused by chemicals, as discussed below) or by some other chemically induced perturbation or hereditary disease, the deformability of the erythrocyte is lost, and bizarre shape changes may ensue. Alterations in the normal cytoskeletal protein structure may also impart fragility to the RBC, inducing it to burst (hemolyze) under stressful conditions. In hereditary hemolytic anemias, familial mutations in various cytoskeletal membrane proteins induce premature destruction of the red blood cells. Chemical-induced hemolytic anemia is a similar condition in which exposure to certain chemicals results in damage to the cytoskeletal membrane proteins and the ultimate destruction of RBCs.

Platelets (or Thrombocytes)

Platelets are formed in a similar manner to erythrocytes, originating from the same myeloid stem cells. Platelets are actually cell fragments that pinch off the megakaryocyte. Mature platelets are the smallest cells found in circulation, and once a platelet is formed, its circulating life span is only 9 days.

Circulating platelets provide the first line of defense against blood loss by monitoring the integrity of the endothelial lining of arteries and veins. If a blood vessel ruptures, the connective tissue underneath the normally smooth, continuous, and nonreactive endothelial lining is exposed. The exposed collagen has a strong negative charge and is one of the most powerful inducers of platelet aggregation. Platelets bind to the collagen and to other platelets, and then the aggregated platelets begin

TABLE 4.3 Anemias

Types	Description
Microcytic hypochromic anemia; small RBCs	Results when red blood cells are formed having a low hemoglobin content (this occurs when red cells are produced rapidly in response to rapid blood loss); frequently an iron deficiency coincides with an inability to meet the demand of increased erythrocyte production, thereby causing a reduction in hemoglobin, hence hypochromic or reduced red pigmentation; iron-deficient diets may also produce hypochromic anemia; genetic diseases such as thalassemia result in abnormal hemoglobin production and microcytic anemia
Macrocytic; large RBCs	This anemia involves two different types: (1) megaloblastic and (2) nonmegaloblastic—it is an anemia resulting from a defect in DNA synthesis possibly secondary to folic acid or vitamin B_{12} deficiency; other etiologies may include liver disease and hypothyroidism that result in destruction of recticulocytes (reticulocytosis), or the myelodysplastic syndrome
Normocytic; normal-sized RBCs	Early stages of microcytic and macrocytic anemia may present as normocytic; causes of normocytic anemia may be primary bone marrow failure such as aplastic anemia, hemorrhage, hemolytic anemia (immune or drug-induced), and mild hypothyroidism

to contract. The contraction stimulates the release of ADP (adenine diphosphate), serotonin, and other locally active compounds that initiate the clotting cascade mechanism, which culminates with the formation of a fibrin clot. Thus, platelets serve two purposes: (1) they create a physical barrier via formation of a plug to seal a vascular break, and (2) they initiate the intrinsic and extrinsic clotting mechanism involving proteins (clotting factors) synthesized by the liver. Platelet activation is normally a life-saving process. However, in the presence of atherosclerotic plaques (fatty, fibrous, calcified lesions in the arterial endothelium), repeated rupture of the blood vessel may lead to thickening and closure of the arterial lumen (strokes and hypertension). This can result in sudden death if a platelet clot occludes the coronary artery (myocardial infarction, starving downstream heart tissue of oxygen).

Cigarette smoke–induced arteriosclerosis is a chemical toxicity that indirectly involves the platelets. In atherosclerosis accelerated by cigarette smoking, plaques, composed of a complex mixture of lipids (e.g., cholesterol), form underneath the normal smooth endothelial lining of the artery/arterioles. If the plaque ruptures, which it can do unpredictably, the connective tissue underneath the endothelial lining becomes exposed, and this event triggers platelet aggregation. If the platelet aggregation and fibrin deposition progresses to the point of occluding the blood vessel, an infarct occurs, and all tissue distal to the occlusion (platelet clot) dies from anoxia. The muscle that dies downstream of the clot (infarct or inclusion) is replaced by scar tissue, and hence the efficiency of the cardiac muscle is compromised. The scar may also impact normal electrical conduction pathways in the heart. Hence, platelets are involved in the final stages of cigarette-induced atherosclerosis, a condition that can lead to vascular disease, strokes, angina, and heart attacks.

Overall, few toxicologically effects result from direct stimulation of platelet aggregation. On the other hand, aspirin and inhibitors of prostaglandin synthetase, such as ibuprofen, inhibit platelet aggregation and can prolong bleeding times. In individuals with bleeding disorders, inhibition of platelet aggregation can lead to excessive blood loss. However, the impairment of clotting is actually beneficial to individuals who are at risk of strokes, angina (ischemia of the cardiac muscle, causing pain and potential arrhythmias), and heart attacks (myocardial infarction).

The most common platelets toxicity involves suppression of normal platelet number (thrombocytopenia). Alkylating agents used in the treatment of cancer are notorious for causing thrombocytopenia. For many of these cancer chemotherapeutics, their dosages are limited by potentially life-threatening thrombocytopenia, which can lead to hemorrhaging and death. The oncologist will frequently regulate the dose of alkylating agents based on the patient's platelet count. As one would anticipate, hemorrhaging from the mucous membranes of the mouth, nose, and kidneys often reveals the onset of a deficiency in platelet-controlled clotting. Sometimes it is necessary to administer platelets or whole blood to treat a cancer patient who develops serious thrombocytopenia.

Granulocytes or Neutrophils, Monocytes or Macrophages, Eosinophils, and Basophils

Differentiation pathways similar to RBCs exist for neutrophils, basophils, eosinophils, and monocytes (which mature into macrophages after leaving the bloodstream). Neutrophils provide the first line of defense against bacterial invasion. They are also commonly referred to as polymononuclear neutrophils (or PMNs) because they possess a multilobed nucleus. They generally amplify in number in response to infectious agents, which gives the physician a diagnostic endpoint to suspect an infection; specifically, PMNs exceeding 10,000 mm^3 in blood may indicate an infectious process.

Neutrophils spend less than a day in circulation before attaching to vascular epithelial cells and migrating to extravascular locations in response to foreign invasion. Monocytes circulate in the blood for 3–4 days before migrating to reticuloendothelial tissues, such as the liver, spleen, and bone marrow, where they set up residence as fixed macrophages. Macrophages are recruited into an area of infection or injury and remove cellular debris and phagocytize pathogens. Macrophages may also act as antigen presenting cells (APCs) by presenting a digested antigen to T lymphocytes in order to activate cellular immunity (the immune response mediated by T lymphocytes). Overall, granulocytes (including neutrophils, eosinophils, and basophils) and monocytes or macrophages can act as phagocytic cells by physically attaching to foreign particles via receptor recognition. Following attachment, the macrophage engulfs (phagocytizes) the particle or foreign cell, and enzymatic processes within these cells facilitate the digestion of the engulfed particle. Eosinophils provide protection against infectious organisms by releasing proteolytic enzymes and active oxygen and conducting phagocytotic activities. An increased number of eosinophils in the blood and tissue is normally observed in allergic (atopic) individuals who suffer from chronic hay fever or asthma. However, in certain toxicities, such as the L-tryptophan eosinophilia myalgia syndrome (LTEMS), eosinophil excess resulted from contaminants that were present an over-the-counter amino acid sleep aid. In this case, the increase in eosinophils constituted a harmful autoimmune response. Basophils, when stimulated, release histamine, proteolytic enzymes, and inflammatory mediators. Toxicities involving basophils are almost non-existent.

4.4 THE LYMPHOID SERIES: LYMPHOCYTES (B AND T CELLS)

The lymphoid series gives rise to cells involved in both humoral (B cells) and cellular (T cells) immunity. B cells function to produce antibodies, while T cells kill virus-infected cells and mediate the actions of other white blood cells. In the last 15–20 years (at the time of writing), considerable progress has been made toward understanding (1) the various types of T cells and how they differ in

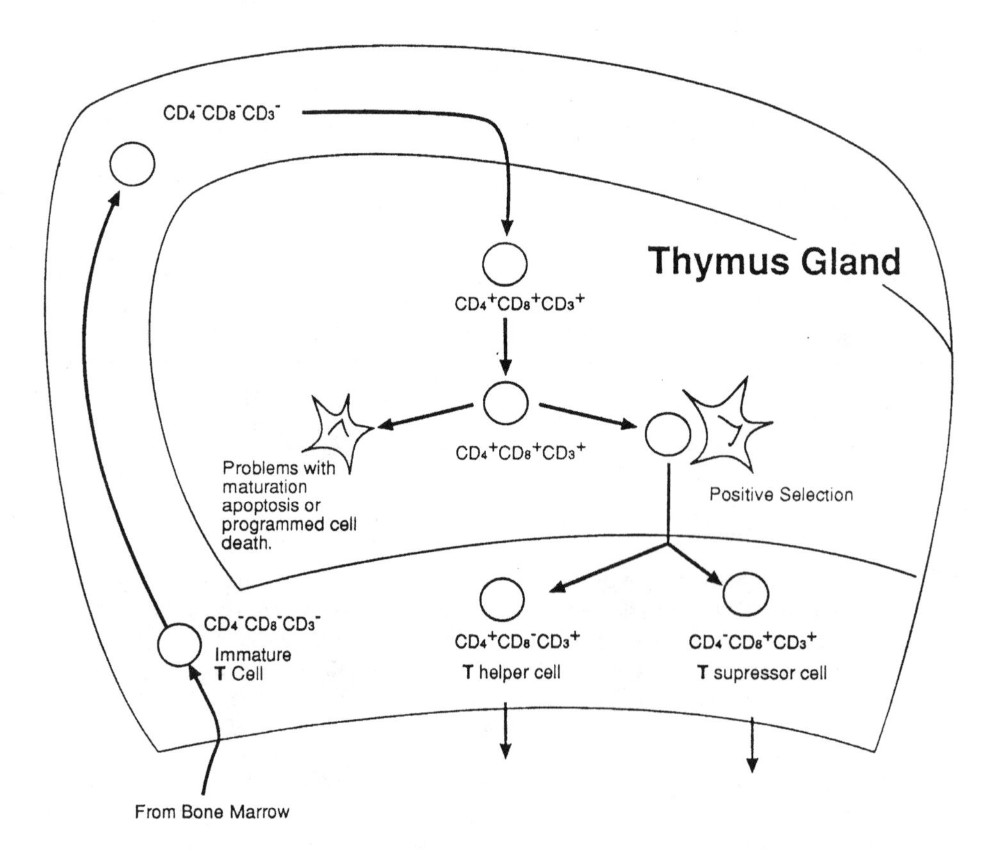

Figure 4.2 Thymic Maturation of T-Lymphocytes. Immature T-lymphocytes pass through the various layers and cavities of the thymus gland while acquiring specific functional capabilities. These capabilities result, in part, from the acquisition of receptors expressed on the cell surface of the T-lymphocytes. T-lymphocytes are identified by the phenotype expression of specific receptors such as T-suppressor cells which express the CD8 receptor while not expressing the CD4 receptor. Conversely, T-helper cells are defined by expression of the CD4 receptor protein while lacking CD8 expression.

function and response to stimuli, (2) the unique membrane-bound T cell receptors responsible for antigen recognition, and (3) many of the complex events that regulate T-cell maturation. Although the process of T-cell production begins in the bone marrow, the immature pre–T cell must migrate to the thymus gland, via the bloodstream, for further development and differentiation. The thymus-dependent differentiation of T cells into specific subpopulations is governed by the expression of unique cell surface proteins or receptors known as *cluster determinants.* Specific types of T cells are defined by their cluster determinant repertoire, namely, CD4 for T-helper cells, CD8 for T-suppressor cells, and CD3 as a marker for all T cells. The cluster determinant expression (phenotype of the mature T cell) ultimately determines the precise function of the mature T cell that leaves the thymus (T helper, suppressor, memory, and killer cells for example). Thymic maturation of T cells involving the acquisition and deletion of specific cluster determinants is depicted in Figure 4.2.

The post–bone marrow maturation of B cells in humans is not well understood. Like T cells, B lymphocytes may also be defined by their own distinct repertoire of cluster determinants (membrane proteins and protein receptors). Chemicals that affect T and B lymphocyte function are more appropriately discussed under the topic of immunotoxicity.

4.5 DIRECT TOXICOLOGICAL EFFECTS ON THE RBC: IMPAIRMENT OF OXYGEN TRANSPORT AND DESTRUCTION OF THE RED BLOOD CELL

Two types of toxicities essentially affect red blood cells: (1) competitive inhibition of oxygen binding to hemoglobin and (2) chemically induced anemia in which the number of circulating erythrocytes is reduced in response to red blood cell damage. Inhibition of oxygen transport is the more commonly observed toxicity directly affecting the RBC.

Carbon monoxide, cyanide, and hydrogen sulfide bind to hemoglobin and can potentially interfere with its ability to transport oxygen. Carbon monoxide directly inhibits oxygen binding to hemoglobin, which can result in a spectrum of adverse effects ranging from mild subjective complaints to life-threatening hypoxia. The mechanism underlying carbon monoxide toxicity is one of the simpler toxicological phenomena, in terms of its binding to the iron molecule in hemoglobin. However, some of the consequences of carbon monoxide poisoning, such as cardiovascular and neurological effects, are much more complex and occasionally are associated with somewhat controversial outcomes (i.e., delayed neurological injury, such as memory loss, purportedly expressed as a reduction in neuropsychological test performance). While cyanide and hydrogen sulfide can also bind to the heme iron in hemoglobin, their significant toxic effects relate to inhibition of mitochondrial energy production.

Chemically induced methemoglobin and methemoglobinemia associated with hemolytic anemia occur by two different mechanisms. The first mechanism involves oxidation of hemoglobin (methemoglobin formation). The second mechanism involves oxidation of hemoglobin coupled to modification of RBC membrane proteins causing the RBC to be recognized as foreign by the immune system. The ultimate outcome of either type of toxicity is hypoxia.

Oxygen Transport: Hemoglobin

An understanding of hemoglobin's protein structure is necessary to fully appreciate how carbon monoxide, cyanide, and hydrogen sulfide bind to the heme iron of hemoglobin and prevent oxygen from binding or being released. Hemoglobin (Hb) consists of four separate peptide chains (two alpha and two beta peptides). Each peptide chain is irregularly folded and surrounds a porphyrin molecule (protoporphyrin) located in a hydrophobic pocket. An iron molecule is located in the center of the protoporphyrin ring and forms a coordinate–ligand bond with oxygen. The oxidation state of the iron atom is an important factor in oxygen binding. Oxygen can only bind to iron when it is in its ferrous state (+2 oxidation state). Oxidation of the iron atom to its ferric state (+3 oxidation state) produces methemoglobin, a derivative of hemoglobin that does not form a coordinated ligand bond with oxygen.

The binding of one molecule of the iron molecule induces a conformational change in the tertiary structure of hemoglobin. The resulting shape change increases hemoglobin's affinity for subsequent oxygen binding. Thus, the binding of each oxygen molecule facilitates the binding of the next in a process known as *positive cooperativity.* Positive cooperativity produces a characteristic sigmoidal shaped oxygen binding curve (Figure 4.3), demonstrating that a disproportionately greater increase in oxygen binding to hemoglobin occurs as the oxygen concentration (P_{O_2}) of blood increases by only a small amount.

The release of oxygen from hemoglobin is caused by the tissue P_{O_2} gradient from the arteriole to the venous side. The release of the first oxygen molecule facilitates the release of the second oxygen molecule, and so on. The first oxygen is released in an area of relatively higher tissue oxygen content whereas the remaining oxygen is released in areas further down the capillary bed where the tissue oxygen content is lower. The transit of oxygenated blood from the arteriole to the venous side results in a loss of approximately 5 mL of oxygen from each 100 mL of blood.

Hypoxia

Hypoxia is defined as a decreased concentration of oxygen in inspired air, oxygen content in arterial blood, or oxygen content in tissue. *Anoxia,* on the other hand, is the complete absence of oxygen.

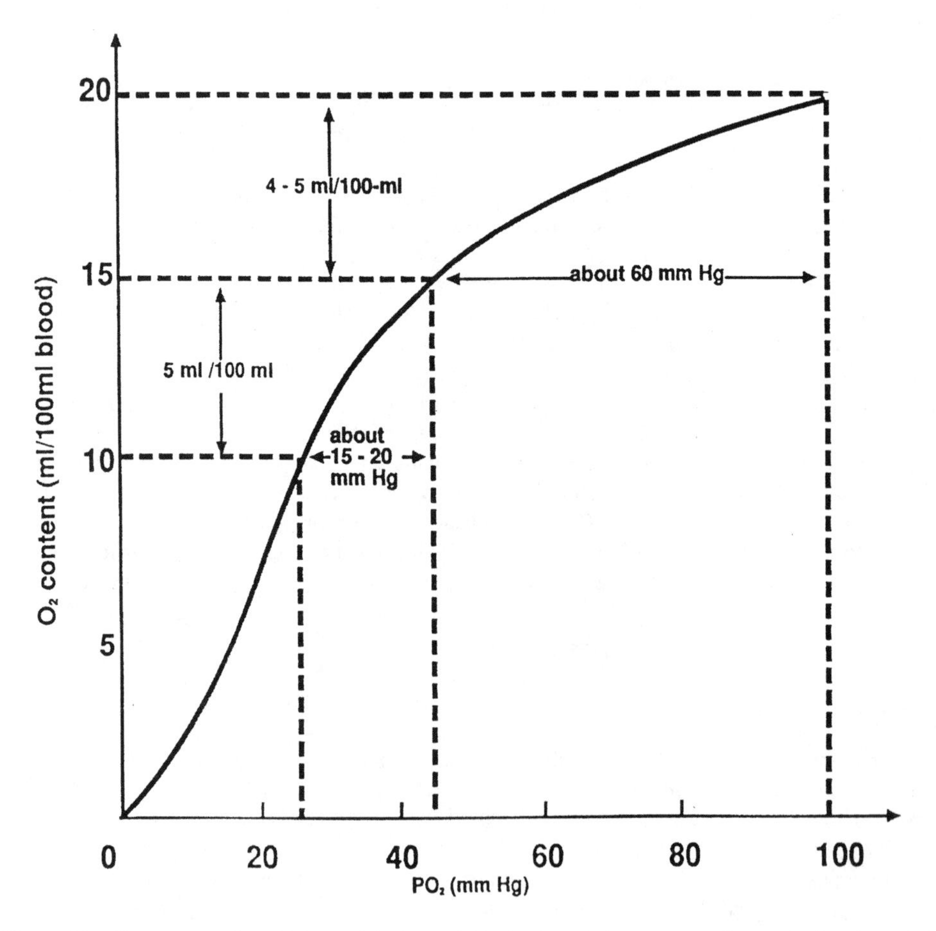

Figure 4.3 Characteristic sigmoid shape of the hemoglobin-O_2 dissociation curve. To liberate the first 4–5 ml of oxygen, the partial pressure of oxygen within the blood must drop about 60 mm Hg. The second 5 ml of oxygen per 100 ml of blood is liberated with a drop in pressure of only 15 to 20 mm Hg.

Hypoxia can result from a variety of conditions including anemia; a reduction in the iron carried by the RBC; ischemia (physical barrier to blood flow) caused by occlusion or vasoconstriction of an artery; or by an increased oxygen affinity (shift to the left of the oxygen-hemoglobin binding curve), which reduces the release of oxygen. In situations involving oxygen-deficient atmospheres, the blood oxygen concentration can drop to a level in which the central nervous system and cardiovascular system risk impairment.

Hypoxia typically occurs when workers enter confined spaces where the atmospheric oxygen (normally at 21 percent) is too low to sustain the oxygen saturation of hemoglobin above 80 percent. Under circumstances of reduced oxygen delivery to the lungs, serious cardiovascular and central nervous system impairment can develop. The symptoms range in severity from euphoria to loss of consciousness, seizures, and cardiac arrhythmias. Hemoglobin oxygen saturation less than 80 percent results in a sense of euphoria, impaired judgment, and memory loss. As the oxygen desaturation of hemoglobin worsens, the extent of central nervous system effects increase. If oxygen pressure drops to 30 mm Hg, a level corresponding to approximately 55–60 percent oxygen saturation, consciousness may be lost. Individuals with ischemic heart disease, such as atherosclerotic coronary vascular disease, may be more sensitive to hypoxic conditions than in healthy individuals. Individuals with atherosclerosis may be more prone to hypoxia-induced ischemia, which may lead to arrhythmias (irregular electrical conduction in the heart) or ischemia-like pains (i.e., chest pain encountered during angina or a myocardial infarction). Subjects with serious atherosclerosis of the cerebral vasculature are more likely to develop CNS impairment related to hypoxia than are healthy subjects. Hence, hypoxia resulting from either low oxygen concentrations or interference with oxygen transport must be assessed according the subject's cardiovascular status.

Physiological adaptations can affect oxygen's affinity for hemoglobin, especially when chronic low levels of hypoxia are present. 2,3-Diphosphoglycerate (or 2,3-bisphosphoglycerate) concentrations increase within RBCs under conditions of chronic hypoxia (e.g., high altitudes, various anemias). By complexing with deoxygenated hemoglobin, 2,3-diphosphoglycerate decreases hemoglobin's affinity for oxygen and facilitates oxygen release in peripheral tissues. This is illustrated by a shift to the right in the oxygen–hemoglobin binding curve. An increase in hydrogen ions (acidity of blood) also causes the hemoglobin–oxygen binding curve to shift to the right. Hydrogen ions are generated when carbon dioxide (formed during respiration or oxygen consumption) is converted to bicarbonate. When the hydrogen ions are then taken up by hemoglobin, oxygen is released. Consequently, ischemic tissue, where the oxygen tension is low and carbon dioxide is high, is benefited by the increased oxygen release that occurs in the presence of hydrogen ions. Conversely, if the oxygen–hemoglobin binding curve is shifted to the left, oxygen binds more avidly to hemoglobin. When this occurs, an even lower tissue oxygen concentration is required before oxygen can be released.

4.6 CHEMICALS THAT IMPAIR OXYGEN TRANSPORT

Carbon Monoxide

Carbon monoxide binds to hemoglobin, decreasing the available sites for oxygen while increasing the binding affinity of the oxygen that is already bound. The hemoglobin binding affinity of carbon monoxide is explained by the Haldane equation, named after the scientist who studied the effects of carbon monoxide in the late 1800s. The carbon monoxide binding affinity is denoted by M in the Haldane equation

$$\frac{[\mathrm{HbCO}]}{[\mathrm{HbO2}]} = \frac{M[P_{\mathrm{CO}}]}{[P_{\mathrm{O_2}}]}$$

where HbCO represents the percentage of carboxyhemoglobin (the carbon monoxide-hemoglobin complex), and HbO_2 represents the percentage of hemoglobin bound by oxygen. P_{CO} and P_{O_2} represent the carbon monoxide and oxygen tensions (percentages), respectively, in air. In humans, M is reported to be anywhere from 210 to 245, demonstrating that carbon monoxide binds to hemoglobin approximately 200 times more avidly than oxygen. To illustrate this further, consider the concentration of carbon monoxide that is required to decrease hemoglobin oxygenation by 50 percent. First, the concentrations of carboxyhemoglobin and oxyhemoglobin are equal so that the left side of the equation becomes one, that is, 50 percent of the blood exists as HbCO and 50 percent exists as HbO_2. The equation then simplifies to

$$[P\text{CO}] = \frac{[P\text{O}_2]}{M}$$

Since the normal oxygen concentration in air is 21 percent, solving the Haldane equation yields a carbon monoxide concentration in air of 0.1 percent or approximately 1000 ppm. When equilibrium is achieved, an individual inhaling 1000 ppm of CO will develop 50 percent carboxyhemoglobin and a serious hypoxic situation. Compounding this hypoxia is the increased binding affinity of oxygen caused by carbon monoxide inhibiting the release of oxygen to tissue. The ability of carbon monoxide to decrease oxygen's binding to hemoglobin and to increase oxygen's affinity for hemoglobin is called the *Haldane effect.*

Low level background carboxyhemoglobin concentrations of 1.0% or less normally exist in the blood as a result of porphyrin metabolism. Cigarette smoking increases carboxyhemoglobin concentrations to as much as 5–10 percent in heavy smokers—two packs per day, for example. If exposure to carbon monoxide from exogenous sources increases carboxyhemoglobin concentrations to around 20 percent, subjective complaints may be reported. As shown in Table 4.4, the adverse effects of carbon monoxide are concentration dependant.

Significant hypoxia caused by carboxyhemoglobin has been reported to produce brain injury resulting in a Parkinson's disease-like condition, cognitive impairment, and serious neurobehavioral changes. Some of these neurological sequelae may not be apparent for a number of days or even weeks following exposure. The more severe neurological effects generally occur in only a few individuals under circumstances of life-threatening hypoxia. Fortunately, most individuals with mild to moderate carbon monoxide poisoning experience complete recovery. Recovery is aided by the use of 100% oxygen or hyperbaric oxygen treatment along with supportive measures.

Assessment of carbon monoxide poisoning is typically performed in the emergency room. However, significant time may lapse between the exposure, emergency room arrival and the determination of carboxyhemoglobin. The time between loss of consciousness or serious clinical effects and drawing

TABLE 4.4 Carboxyhemoglobin and Effects

Carboxyhemoglobin (% Hemoglobin Saturation with Carbon Monoxide)	Effect
0.3–0.7	Background concentrations due to endogenous production of carbon monoxide
1–5	Increase in blood flow via compensating mechanisms such as increased heart rate or increased contractility (these concentrations are typically observed in cigarette smokers)
2–9	A reduction in exercise tolerance and an increase in the visual threshold for light awareness
16–20	Headache; abnormal visual responses
20–30	A throbbing headache accompanied by nausea, vomiting, and a decrease in fine-motor movement
30–40	Severe headaches, nausea, vomiting, and weakness
50+	Coma and convulsions
67–70	Lethal if not aggressively treated

a blood sample may lead to a significant decline in carboxyhemoglobin levels, especially if the patient is treated with oxygen. If breathing room air, the half-life of carboxyhemoglobin is approximately 4–5 hours; if 100 percent oxygen is administered, the half-life can be reduced by 4-fold. If hyperbaric oxygen treatment is implemented, the normal half-life can be shortened 10-fold. Hence, carboxyhemoglobin determinations at the time of medical intervention may not accurately gauge the extent of carboxyhemoglobin that occurred during exposure.

Carbon monoxide is generated by incomplete combustion; automobile fumes and cigarettes are among the most familiar sources. A common example of carbon monoxide poisoning occurs from heating with natural gas, especially natural gas of lesser quality, namely, individuals overcome by carbon monoxide heating homes with wet natural gas and without proper ventilation. Another potential occupational, as well as environmental, source of carbon monoxide results from methylene chloride exposure. Methylene chloride is metabolized to carbon monoxide by cytochrome P450 enzymes resulting in elevations in carboxyhemoglobin levels. Case reports have documented elevated carboxyhemoglobin levels in individuals stripping furniture with methylene chloride–based paint strippers. Physical activity, which increases the respiratory rate, will increase the amount of inhaled methylene chloride and the resulting carboxyhemoglobin levels. The current OSHA standard and ACGIH TWA for carbon monoxide is 50 and 25 ppm, respectively. By the time equilibrium is achieved, 50 ppm carbon monoxide will produce carboxyhemoglobin concentrations of approximately 5–6 percent after about 8 h of exposure. It should be noted that the binding equilibrium of carbon monoxide is not achieved instantaneously but requires time.

4.7 INORGANIC NITRATES/NITRITES AND CHLORATE SALTS

In blood, an equilibrium exists between ferrous and ferric hemoglobin. The oxygen-rich environment surrounding the RBC continually oxidizes hemoglobin to methemoglobin. Since methemoglobin does not bind and transport oxygen, the accumulation of methemoglobin is detrimental. Therefore, the accumulation of methemoglobin is prevented by the enzymatic reduction of ferric iron to ferrous iron via the enzyme methemoglobin reductase (also known as *diaphorase*). The normal concentration of methemoglobin is generally 0.5 percent or less, which produces no adverse health effects.

Methemoglobin formation results in a noticeable change in the color of blood from its normal red color to a brownish hue. In humans and animals, significant methemoglobinemia creates a bluish discoloration of the skin and mucous membranes. Mild to moderate concentrations of methemoglobin can be tolerated, and low levels of less than 10 percent may be asymptomatic, except for a slightly bluish color imparted to the mucous membranes. If blood methemoglobin concentrations achieve 15–20 percent of the total hemoglobin, clinical symptoms of hypoxia can develop, and above 20 percent, cardiovascular and neurological complications related to hypoxia may ensue. Methemoglobin concentrations exceeding 40 percent are often accompanied by headache, dizziness, nausea, and vomiting, and levels surpassing 60 percent may be lethal. Other than supportive care to maximize oxygen transport, such as oxygen administration, little can be done to treat methemoglobinemia. One available antidote is the intravenous administration of methylene blue, which provides reducing equivalents to methemoglobin reductase and thus facilitates the reduction of methemoglobin back to ferrous hemoglobin.

Inorganic nitrites such as sodium nitrite ($NaNO_2$) and chlorates (ClO_3^-) oxidize ferrous hemoglobin (Fe^{2+}) to ferric-hemoglobin (Fe^{3+} or methemoglobin). Nitrite and chlorate directly oxidize hemoglobin; nitrate, however, must first be reduced to nitrite by nitrifying bacteria in the gut. Exposures to nitrates, nitrites, and chlorates occur mostly in industrial settings or from contaminated drinking water. The typical concentrations of nitrate and nitrite found in foods and drinking water, however, do not present a risk in terms of methemoglobin production. If the rate of hemoglobin oxidation caused by nitrite/chlorate exceeds the capacity of methemoglobin reductases activity, a buildup in methemoglobin results. The oxidative conversion of hemoglobin to methemoglobin by nitrites and chlorates, combined with the reduction of methemoglobin back to ferrous-hemoglobin, is referred to as a redox cycle.

Nitrates, in addition to their conversion to methemoglobin-causing nitrite, can produce a complex array of vascular changes, such as venous pooling (reduced blood return to the right side of the heart). Episodes of as venous pooling aggravate the clinical complications of methemoglobinemia; cardiac output is reduced and tissue hypoxia is exacerbated. Thus, nitrate toxicity presents a complicated clinical picture that integrates the production of methemoglobin with a reduction in blood perfusion to tissues most in need of oxygen. The hematologic hazards regarding nitrite and chlorate, on the other hand, appear to be limited to the direct oxidation of hemoglobin to methemoglobin.

4.8 METHEMOGLOBIN LEADING TO HEMOLYTIC ANEMIA: AROMATIC AMINES AND AROMATIC NITRO COMPOUNDS

Aromatic amines and nitro compounds such as aniline and nitrobenzene cause methemoglobinemia by initiating a redox cycle in the RBC. The aromatic amines and nitro compounds are important building blocks in the dye, pharmaceutical, and agricultural chemical industries. Aromatic amines are also important structural components of numerous prescription medications. By in large, amine-induced methemoglobinemia and hemolytic anemia develop most often following treatment with antibiotics such as dapsone and primaquine, pharmaceuticals used to treat infectious diseases such as leprosy and malaria, respectively.

However, unlike those for nitrites and chlorates, the potential hazards of aromatic amines are not limited to methemoglobinemia. RBC changes occurring during or after methemoglobin formation may result in damage to the RBC membrane. The damaged RBCs are recognized by splenic macrophages, which remove and destroy them. Hemolytic anemia can result if the number of red blood cells destroyed exceeds the bone marrow's capacity to replenish them; for example, by amplification of RBC production in response to increased release of erythropoietin.

Reactive metabolite(s) of the parent aromatic amine compound, formed via cytochrome P450 metabolism, are also capable of causing methemoglobinemia and hemolytic anemia. Aromatic nitro compounds, like inorganic nitrate, must first be reduced to their respective aromatic amine by gut bacteria before being metabolized to an arylhydroxylamine. It is the *N*-hydroxyl metabolite that is directly responsible for initiating hemoglobin oxidation via a redox cycle. The redox cycle results in the formation of reactive oxygen species in the RBC (i.e., hydrogen peroxide). The reactive oxygen species oxidize proteins in the RBC cytoskeleton and damage the RBC membrane by crosslinking adjacent proteins. The crosslinked proteins can be visualized in the form of Heinz bodies, which consist of hemoglobin covalently linked to cytoskeletal proteins on the inner side of the red blood cell membrane. RBC membrane damage may alter the normal RBC discoid morphology, depicted in Figure 4.4 for dapsone *N*-hydroxylamine-induced RBC morphology alteration.

These spike-shaped RBCs produced by dapsone *N*-hydroxylamine are known as echinocytes. Other abnormally shaped RBCs that may result from exposure to various aromatic amines include anisocytes (asymetrically shaped RBCs); spherocytes (round RBCs); elliptocytes (ellipse or egg-shaped RBCs); sickle cell–shaped RBCs (known as *drepanocytes*); acanthocytes, which are round RBCs with irregular spiny projections; and stomatocytes, which are RBCs with a slit-like concavity. A senescent (aging) signal may appear on the membrane of the damaged red blood cell and serve as a recognition sign for the spleen. In effect, active oxygen species produced during redox cycles appear to cause premature aging and altered morphology of RBCs, leading to their early removal from circulation. Another name for redox cycle formation of reactive oxygen species and damage to the RBC is "oxidative stress."

Instances of aromatic amine-induced methemoglobinemia and hemolytic anemia are rather rare. This is due to their low volatility, which reduces inhalation exposure, and the fact that many of the amines are used in the form of salts, which reduces their potential for dermal absorption. The free amines, however, are dermally absorbed and can pose a potential hazard if directly contacted by the skin. Another serious concern with exposure to aromatic amines is their potential to induce hemorrhagic cystitis (bleeding from bladder damage) and bladder cancer.

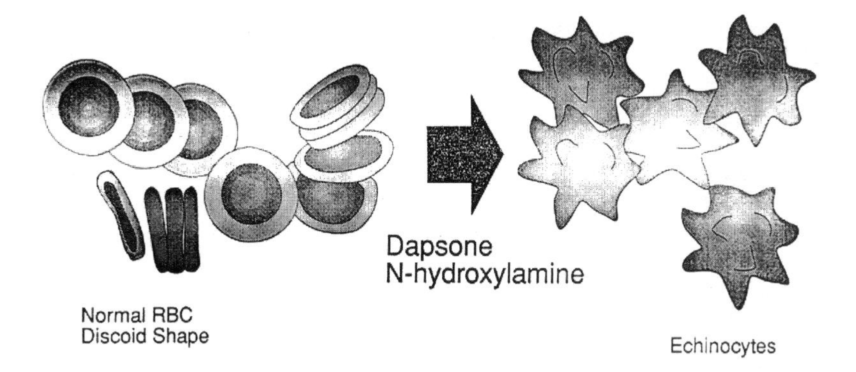

Figure 4.4 Dapsone N-hydroxylamine-induced Red Blood Cell Changes. Chemically induced damage to red blood cells is typically expressed as changes in red blood cell shape. The altered shape (morphology) results from damage to the cytoskeleton proteins or lipid membrane of the red blood cell.

Exposure to aromatic amines can be potentially life-threatening to individuals with a deficiency in the enzyme glucose-6-phosphate dehydrogenase (G6PDH). Individuals with deficiencies in G6PDH are limited in their ability to maintain sufficient levels of reduced glutathione (GSH) in their RBC. GSH acts as a scavenger of active oxygen species such as hydrogen peroxide that are formed during the redox cycle. In the event of oxidative stress caused by an activated redox cycle, these individuals cannot withstand the oxidations of GSH to GS-SG (glutathione disulfide) or GS-S-protein, and they will suffer oxidative damage to the RBC membrane proteins at lower blood concentrations of *N*-hydroxy metabolites than normal people. G6PDH deficiency exists primarily among individuals of Mediterranean, African, and Asian decent. It can be tested for prior to initiation of drug therapy that may cause hemolytic anemia.

Treatment modalities for chemically induced hemolytic anemia are limited. Methylene blue may be administered to maximize the ability of methemoglobin reductase, which reduces methemoglobin back to ferrous hemoglobin. Transfusions may be necessary to replace red blood cells prematurely sequestered and destroyed by the spleen. There is no information on the use of glutathione-related antidotes such as *N*-acetyl cysteine. Mild conditions of chemically induced hemolytic anemia are not fatal and can be treated supportively. The extent of hemolysis induced by aromatic amines is proportionate to the amount of methemoglobin produced. Therefore, low levels of methemoglobin, in the general range of 20–30 percent or less, do not typically lead to extensive removal of red blood cells and anemia.

4.9 AUTOIMMUNE HEMOLYTIC ANEMIA

Hemolysis mediated by the immune system occurs via a different mechanism than direct oxidative stress. In this instance, the drug or drug metabolites cause immunoglobulins (either IgG or IgM) to nonspecifically or specifically bind to the RBC. The IgG or IgM bound to the RBC attracts complement. Complement then binds to the surface of the RBC and initiates destruction of the RBC membrane. The damage to the RBC imparts fragility to the membrane, the RBC ruptures in the vasculature, and hemoglobin is released. The intravascular hemolysis can provoke disseminated intravascular coagulation (DIC), a serious consequence of autoimmune hemolytic anemia. Free circulating hemoglobin can also induce renal failure when it is excreted by the kidney. Hence, autoimmune hemolytic anemia, primarily caused by prescription drug use, can result in a battery of serious health effects. Fortunately, only a few drugs are known to provoke this adverse drug reaction, and most cases are considered idiosyncratic.

4.10 BONE MARROW SUPPRESSION AND LEUKEMIAS AND LYMPHOMAS

Bone Marrow Suppression

A variety of industrial chemicals and pharmaceuticals can cause partial or complete bone marrow suppression. Pancytopenia occurs when all cellular elements of the blood are reduced. Bone marrow suppression may be reversible or permanent depending on the chemical agent and the extent of exposure. Clinical signs of bone marrow suppression include bleeding, caused by a reduction in platelet counts; anemia, which leads to fatigue and altered cardiovascular/respiratory parameters; and a heightened susceptibility to various infectious processes. The cells with the shorter lifespans are the first to disappear, such as the platelets, which have a circulating lifespan of only 9 or 10 days. Therefore, if the bone marrow injury involves the myeloid series, thrombocytopenia (i.e., reduction in the number of blood platelets) bleeding is one of the first complications to be observed. Patients with this condition are at a high risk for life-threatening internal hemorrhaging. Examples of occupational chemicals and drugs reported to cause blood dyscrasias (e.g., thrombocytopenia, neutropenia, pancytopenia) are listed in Table 4.5.

Some of the examples listed in Table 4.5 are based solely on case reports and do not represent confirmed examples of chemically induced bone marrow suppression. For example, the evidence regarding the effects of pentachlorophenol-induced aplastic anemia is based on isolated case reports. However, larger clinical studies performed on wood-treatment workers and animal testing show no evidence of bone marrow suppression. Hence, concrete evidence that pentachlorophenol causes bone marrow suppression is lacking.

In contrast to the numerous single case reports weakly implicating specific chemicals with bone marrow suppression, there are a number of chemicals with undisputed bone marrow toxicity; benzene is the best-known example among industrial chemicals. A known marrow suppressant, benzene was experimentally used decades ago to inhibit the uncontrollable production of leukemia cells. Today, the cancer chemotherapeutics are the most frequently encountered causes of bone marrow suppression. The alkylating agents used in cancer chemotherapy are notorious for damaging the bone marrow and are often administered until the patient develops bone marrow suppression. In this event, the administration of further chemotherapy is discontinued, or more commonly, a reduction in the dose of the anticancer drug is attempted. Oncologists constantly monitor the patient's platelet and white blood cell count in order to evaluate the bone marrow suppressive effects of the cancer chemotherapy. Chloram-

TABLE 4.5 Chemicals Reported to Cause Bone Marrow Suppression

Benzene (an important industrial solvent and component of many refined petroleum products, e.g., gasoline)	Chloramphenicol (an important antibiotic used to treat resistant bacterial infections)	Phenylbutazone (antinflammatory used to treat arthritic conditions)
Procainamide (an antiarrhythmic used to control cardiac arrhythmias)	Allopurinol (a drug used to treat gout)	Tolbutamide (used to treat maturity onset or type II diabetes)
Methyldopa (an antihypertensive used to treat high blood pressure)	Sulindac (antiinflammatory agent)	Aminopyrine (analgesic and antipyretic)
Sulfasalazine (a drug used to treat inflammatory bowel disease)	Sodium valproate (used to treat certain epileptic conditions)	Alkylating and antimetabolite (cancer chemotherapy agents, e.g., nitrogen mustard, 5-fluorouracil, cytoxan)
Isoniazid (a mainstay antibiotic in treating tuberculosis)	Cephalothin (a cephalosporin antibiotic)	Gold (used as an antiflammatory agent in arthritic conditions)
Diphenylhydantoin (an important drug used in the treatment of epilepsy)	Pentachlorophenol (a chemical used to treat wood)	Carbamazepine (used to treat certain forms of epilepsy)

phenicol is an important antibiotic used to combat strains of bacteria that are resistant to first-line antibiotics; however, it bears a well-recognized risk of bone marrow suppression. The drug phenylbutazone, once commonly used as an antiinflammatory agent for treating arthritic conditions, is now conservatively prescribed for only a few weeks at a time in order to reduce the chance of developing bone marrow suppression.

The marrow suppressive effects of benzene were described long before benzene was established as a cause of acute myelogenous leukemia (AML). Benzene's suppressant effects range from mild and reversible to lethal, namely, life-threatening aplastic anemia or pancytopenia. Preleukemia or myelodysplasia, often viewed as a precursor to leukemia, is characterized by abnormal morphology of blood cells and may be associated with chronic bone marrow suppression. Evidence of benzene-induced bone marrow suppression in humans is based on many studies. One of the most highly publicized cases involved the Ohio Pliofilm workers of the 1940s and 1950s. The Pliofilm worker studies provided evidence that benzene exposures exceeding 50–75 ppm were associated with reductions in white blood cell counts. More recent evidence, using more sophisticated cell counting methods, suggest that lymphocytes may be the most sensitive target of benzene .

Metabolite(s) of benzene is (are) the actual cause(s) of marrow suppression. Benzene is metabolized by hepatic cytochrome P450 mixed function oxidases. Benzene is a substrate of cytochrome P450 IIE, which is one of the many isozymes among the family of cytochrome P450 mixed-function oxidases. Benzene oxide, the first intermediate in CYP 2EI-mediated metabolism, is converted into a number of metabolites including phenol, hydroquinone, and muconic acid/muconaldehyde (see Figure 4.5). Two benzene metabolites not shown in Figure 4.5 include catechol and trihydroxy benzene. In the bone marrow, myeloperoxidase further oxidizes phenolic metabolites of benzene to form free radicals capable of damaging the bone marrow.

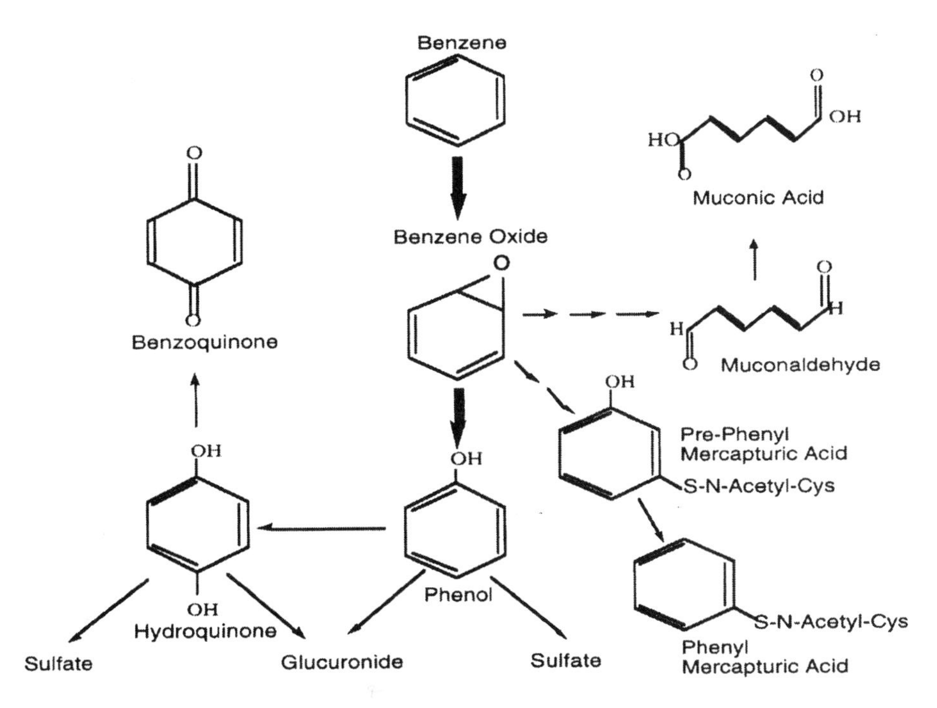

Figure 4.5 Benzene's Metabolism. Benzene is both bioactivated and detoxified via a number of different enzymatic-mediated steps. The bioactivated metabolites of benzene, such as hydroquinone and muconaldehyde, disrupt the various stages of blood formation in the bone marrow gives rise to any number of blood dyscrasias, myelodyplastic syndrome, and acute myelogenous leukemia.

Not all of benzene's metabolites cause bone marrow suppression. Phenol, hydroquinone, catechol, trihydroxy benzene, and muconaldehyde act in concert to cause bone marrow changes; by themselves these metabolites have less marrow toxicity. The precise mechanism by which these metabolites act alone or in concert to cause marrow suppression is uncertain, although these issues are among the topics of ongoing research.

4.11 CHEMICAL LEUKEMOGENESIS

Bone marrow injury may promote the development of myelodysplastic syndromes and acute myelogenous leukemia. Therefore, by damaging the bone marrow, benzene, chloramphenicol, and cancer chemotherapeutic agents increase an individual's risk of contracting bone marrow cancer. However, critical issues regarding exposure and dose, as well as the weight of evidence from epidemiologic and animal studies all influence the relative risk. The cancer biology of chemically induced leukemia is complex, and one or more of the following mechanisms may be involved in the progression toward myelodysplastic syndrome and possibly leukemia: bioactivation of the parent molecule to reactive intermediates, disruption of marrow physiology (e.g., interference with the mitotic spindle), inhibition of topoisomerases, formation of DNA adducts, chromosomal alterations, oncogene activation, and suppressor gene inactivation. As with any chemically induced cancer, benzene-induced AML follows a continuum or progression of events that includes repeated bone marrow injury and suppression, chromosomal changes, the development of dysplastic and metaplastic features, and the ultimate expression of AML.

Awareness of benzene's role in acute myelogenous leukemia came later. The mounting evidence of benzene-induced leukemias finally surfaced in the 1970s and 1980s with publication of NIOSH-conducted studies of Pliofilm workers from two plants in Ohio, the Turkish studies of shoemakers who used glues with high benzene content, and Italian rotogravure printers who used benzene-containing solvents, for example. The collective findings of these studies clearly implicated benzene in the development of AML. Recent Chinese studies suggest that other hematological tumors may occur at a higher incidence among benzene-exposed workers. However, the evidence for benzene-induced hematological cancers, other than AML, is still rather limited, and further investigations are needed. Industries with less benzene exposure (average benzene exposures of 1 part per million or less among refinery workers, rubber workers, and gasoline workers) and chemical workers exposed to benzene have not shown an increased incidence of AML. effects.

The Pliofilm studies have contributed information involving exposure estimates and dose–response relationships. For instance, Rinsky et al. (1988) first proposed a risk–exposure relationship:

$$\mathrm{OR} = e(0.0126 \times \mathrm{ppm{\cdot}year})$$

where OR stands for the odds ratio for leukemia relative to the unexposed workers in a worker who has acquired a specific cumulative ppm·year of benzene exposure. Based on this risk model a background exposure of 0.1 ppm·year generates a risk estimate no greater than background, that is, an odds ratio of 1.0. More recent studies of the Pliofilm workers have concluded that a threshold level of benzene as high as 50 ppm (or even higher) must be exceeded before a significant risk of developing AML exists.

In summary, epidemiologic evidence has established that high-level benzene exposure in the workplace is associated with an increased risk of acute myelogenous leukemia. Clear evidence that a causal relationship exists between benzene exposure and AML comes primarily from the studies on Pliofilm workers. When these studies are further evaluated for a dose–response relationship, the level of occupational exposure that bears a significant risk may be 50 ppm or greater. There is no sound evidence that benzene causes other types of cancer such as other types of leukemia, non-Hodgkin's lymphoma, or solid tumors such as lung cancer. Currently, the OSHA standard of 1.0 ppm should provide adequate protection against both benzene-induced bone marrow depression and a risk of AML.

4.12 TOXICITIES THAT INDIRECTLY INVOLVE THE RED BLOOD CELL

Two important chemicals interact with blood, and yet their toxicological effects directly involve the nervous and cardiovascular system. Both cyanide and hydrogen sulfide bind to the heme portion of hemoglobin. At toxic dosages, however, they first inhibit energy production by mitochondrial heme oxidase. Heme oxidase contains a porphyrin ring such as hemoglobin, which is essential for transporting electrons during oxidative phosphorylation. Cyanide and hydrogen sulfide are respiratory poisons that shut down energy production in cells carrying out aerobic metabolism. The selectivity of hydrogen sulfide and cyanide's apparent toxicity (on the nervous and cardiovascular system) is related to the high oxygen and energy demands of these two tissues. It has been suggested that carbon monoxide toxicity also affects the electron transport chain in the mitochondria.

4.13 CYANIDE (CN) POISONING

Cyanide inhibits cytochrome oxidase, thus halting electron transport, oxidative phosphorylation, and aerobic glucose metabolism. Inhibition of glucose metabolism results in the buildup of lactate (lactic acidemia) and the increase in the concentration of oxygenated hemoglobin in venous blood returning to the heart. Increased oxyhemoglobin in the venous circulation reflects the fact that oxygen is not being utilized in the peripheral tissues. The most serious consequences of oxidative phosphorylation inhibition are related to neurological and cardiovascular problems, including adverse neurological sequelae, respiratory arrest, arrhythmia, and cardiac failure. Cyanide exposure can occur via inhalation of hydrogen cyanide gas or through ingestion of sodium or potassium cyanide. Approximately 100 mg of sodium or potassium cyanide is lethal.

Sublethal doses of cyanide are quickly metabolized to thiocyanate via the enzyme rhodenase (a sulfurtransferase):

$$Na_2S_2O_3 + CN^- \rightarrow SCN^- + Na_2SO_3$$

The detoxification of cyanide to thiocyanate is facilitated by adding the substrate sodium thiosulfate, which reacts with cyanide through the action of rhodenase. Thiocyanate (SCN^-) is a relatively nontoxic substance eliminated in the urine.

4.14 HYDROGEN SULFIDE (H_2S) POISONING

Hydrogen sulfide also inhibits mitochondrial respiration by inhibiting cytochrome oxidase thus halting the production of adenosine triphosphate, or ATP. Central nervous system effects ranging from reversible CNS depression to loss of consciousness and death may occur. Cardiac effects may include alterations in the rhythm and contractility of the heart. Less serious consequences of hydrogen sulfide include irritation, inflammatory changes, and edema of the mucous membranes of the eyes, nose, throat, and respiratory tract. The ppb odor threshold for hydrogen sulfide (i.e., the rotten-egg odor) in normal individuals far precedes concentrations causing adverse health effects, and for a short period of time can serve as a warning signal.

Hydrogen sulfide exposure can occur around sewers and petroleum refinery wastestreams and in situations involving natural gas production or fermentation, such as with manure or silage (fodder for livestock stored in silos). Fortunately, most individuals are relatively sensitive to the odor of hydrogen sulfide and can detect it at ppb air concentrations, which provides an early warning. However, odor fatigue occurs with time and may result in a serious exposure if the individual remains in an area containing high or increasing concentrations of hydrogen sulfide. There are reports of individuals who are rapidly rendered unconscious and die from exposures to high levels of hydrogen sulfide, such as those exceeding 1000 ppm. For example, there are documented episodes of workers who collapse and died within minutes of entering silos storing silage. Table 4.6 lists increasing air concentrations of hydrogen sulfide and the effects that may result from exposure at each level.

The current OSHA acceptable ceiling concentration for hydrogen sulfide is 20 ppm, with maximum 10-min peak concentrations of 50 ppm allowed over an 8 h workshift (29 CFR, Part 1910). The American Conference of Governmental and Industrial Hygienists recommend a time-weighted average (TWA) exposure of 10 ppm.

Once absorbed from the lungs, hydrogen sulfide is rapidly metabolized in the blood and liver. A series of enzymatic and non-enzymatic pathways convert hydrogen sulfide (the sulfide anion) to thiosulfate and then sulfate, which is eliminated from the body. If a blood sample is drawn shortly after exposure, elevated blood concentrations of sulfide can be detected. However, in general, blood sulfide determinations are usually forgotten during an emergency since the immediate concern is to treat the patient. The delay between exposure and blood sampling is usually too long to determine the blood sulfide concentration that was responsible for the observed acute effects. Furthermore, blood sulfide determination is usually considered a specialty analysis that must be conducted by laboratories outside the hospital.

The term sulfhemoglobin has been used to describe hemoglobin with unique spectral characteristics distinguishable from simple methemoglobin. Sulfhemoglobin spectral changes were originally observed when hydrogen sulfide was bubbled through whole blood. The observation between high concentrations of hydrogen sulfide in the test tube and sulfhemoglobin formation has led to the misconception that hydrogen sulfide poisoning also produces measurable sulfhemoglobin (often used as a biomarker of hydrogen sulfide poisoning). However, this is an erroneous concept since sulfhemoglobin formation requires concentrations of hydrogen sulfide that far exceed those required to completely shut down oxidative phosphorylation. Thus, sulfhemoglobin determinations are not useful in verifying toxicity or lethality caused by hydrogen sulfide exposure.

TABLE 4.6 Dose-Response Relationship for Hydrogen Sulfide

Air Concentrations of Hydrogen Sulfide (parts per million)	Effect
0.022	No odor.
0.025–0.13	Noticeable to minimally detectable odor.
0.3	Distinct odor.
0.77	Generally perceptible.
3–6	Quite noticeable, offensive, moderately intense.
20	OSHA acceptable ceiling level.
20–30	Strong intense odor but not intolerable.
150	Olfactory nerve paralysis and mucous membrane irritation.
200	Less intense odor due to eventual sensory fatigue.
250	Prolonged exposure may cause pulmonary edema. Increasing mucous membrane irritation.
500	Dizziness over a few minutes to severe central nervous system impairment and unconsciousness if inhaled for more than a few minutes. Increasing mucous membrane irritation.
700–1000	Unconsciousness may develop rapidly followed by respiratory paralysis and death within minutes. Increasing mucous membrane irritation.
5,000	Imminent death.

4.15 ANTIDOTES FOR HYDROGEN SULFIDE AND CYANIDE POISONING

Unfortunately, there are no failproof antidotes for hydrogen sulfide poisoning, although methods that induce methemoglobinemia have been suggested. In instances of cyanide poisoning, and occasionally hydrogen sulfide exposure, the administration of nitrite in the form of amyl nitrite or intravenous sodium nitrite is recommended to purposely convert the patient's blood to a safe-level of methemoglobin. Methemoglobin has a very strong binding affinity for cyanide and hydrogen sulfide. The relative large amount of methemoglobin binds up and acts as a sink to remove cyanide or hydrogen sulfide from cellular spaces and the mitochondria. Once bound to methemoglobin, cyanide and hydrogen sulfide are no longer available to bind to (and thus inhibit) cytochrome oxidase, an mitochondrial enzyme essential to the aerobic metabolism of glucose. The chemicals are eventually released into the blood where they can be metabolized to thiocyanate (in the case of cyanide) and sulfite/sulfate (in the case of hydrogen sulfide). The ability of methemoglobin to trap cyanide and hydrogen sulfide is illustrated in Figure 4.6.

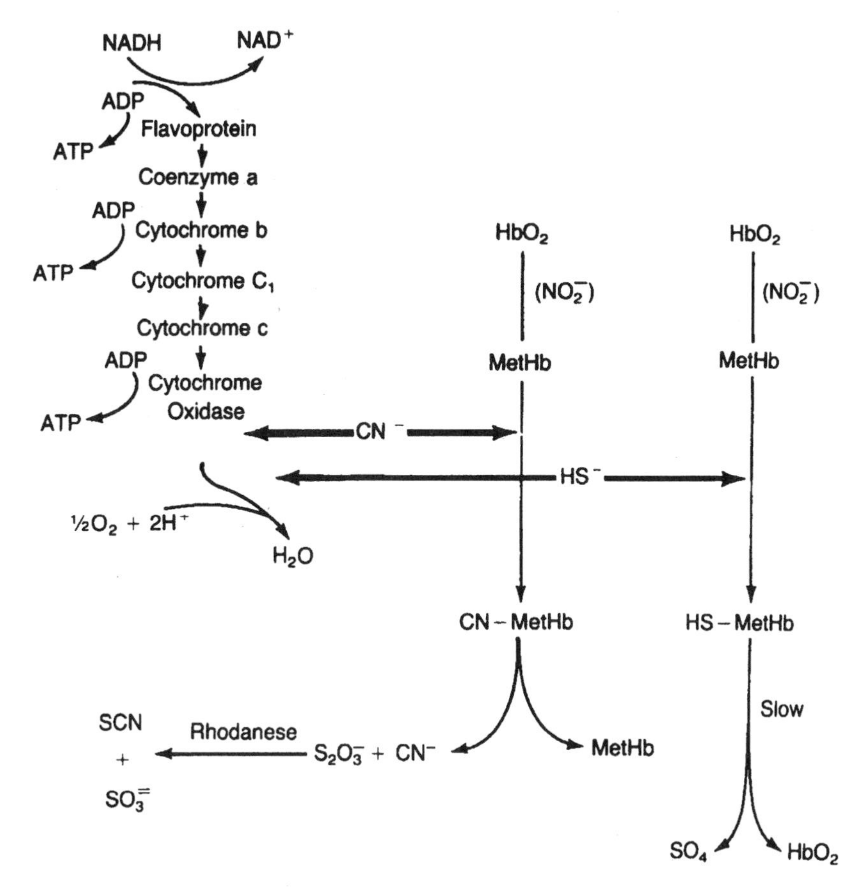

Figure 4.6 Schematic depiction of the electron transport chain through which the oxidation of NADH derived from sugar metabolism generates ATP. Both the cyanide (CN^-) and hydrogen sulfide (HS^-) anions bind to and inhibit cytochrome oxidase. However, both anions also bind the Fe^{+++} ion methemoglobin (MetHb) formed by the oxidation of hemoglobin with nitrate (NO_2^-). CN-MetHb denotes cyanmethemoglobin; HS-MetHb denotes sulfmethemoglobin.

4.16 MISCELLANEOUS TOXICITIES EXPRESSED IN THE BLOOD

Lead poisoning may affect normal red blood cell parameters. For one, lead interferes with heme synthesis in the liver, which can lead to anemias. This interference results in the accumulation of protoporphyrin, a heme precursor that is measurable in the form of zinc protoporphyrin in the blood. The term *basophilic stippling* is often associated with RBCs that are prematurely destroyed in response to lead-induced anemia. Basophilic stippling is characterized by various-sized purple granules that are microscopically observed within the RBC. The purple granules are comprised of pyrimidine compounds that accumulate because lead inhibits erythrocyte pyrimidine-5-nucleotidase, the enzyme responsible for the normal degradation of these pyrimidine nucleotides. The apparent blood lead threshold affecting porphyrin biochemistry is around 25–30 μg/dL and the threshold for affecting hemoglobin is around 50 μg/dL. Treatment of lead poisoning generally involves chelation therapy with drugs such as penicillamine, EDTA, Dimercarpol, or BAL (British anti-lewisite).

A number of chemicals affect the formation and action of clotting factors. Many of these chemicals inhibit clot formation and are extremely useful as anticoagulants in individuals with atherosclerotic cardiovascular and cerebrovascular disease. Thus, the anticoagulants aid in the prevention of heart attacks and strokes. For example, the drug warfarin effectively reduces circulating clotting factors within a few hours to days following treatment. Warfarin's mechanism of action involves the antagonism of vitamin K, which is involved in the carboxylation of clotting factor proteins. Anticoagulants such as warfarin are also used as pesticides. Several additional rodenticides include difenacoum, chlorphacinone, and brodifacoum. Unless ingested, these anticoagulants are relatively safe since they are nonvolatile and cannot be absorbed through the skin. Poisoning by the anticoagulants usually occurs in infants and suicide cases. In the clinical setting, physicians monitor the patient's clotting times to control for the desired therapeutic effect and to avoid excessive anticoagulation, which could result in a fatal hemorrhage. Vitamin K is the recommended antidote for treating individuals poisoned by anticoagulants.

4.17 SUMMARY

Hematotoxicity involves a wide range of effects ultimately affecting oxygen delivery, maintenance of a viable immune or clotting system, and cancer. It is fortunate that hematotoxicity is a relatively uncommon occurrence. Overall, the real concern regarding bone marrow injury is related to benzene exposure in occupational settings. Benzene exposure in the workplace has dramatically declined since the days of the Pliofilm workers and before, and currently, there is little evidence to suggest that existing occupational settings pose a risk of AML. The threshold for benzene-induced AML has been reported to range of 0.1–50 ppm, although the actual concentration which poses a serious threat is still heavily debated. On the other hand, benzene exposure resulting from ingestion of ppb concentrations in ambient air or drinking water does not pose a risk of AML or any other hematopoietic tumors. In general, hematotoxicity is an occupational concern since the exposures and doses of chemicals required to cause a toxic response cannot be achieved from the low levels found in the environment (i.e., ppb air concentrations). The exceptions, of course, are carbon monoxide poisonings, which frequently occur in home settings, or toxicities from medications, such as chemotherapeutic agents used to treat cancer.

REFERENCES AND SUGGESTED READING

Ellenhorn's Medical Toxicology Diagnosis and Treatment of Human Poisoning. Matthew J. E llenhorn editor, 2nd edition. Williams & Wilkins, Baltimore, (1997).

Fishbeck, W. A., J. C. Townsend, and M. G. Swank, "Effects of chronic occupational exposure to measured concentrations of benzene." *J. Occup Med.* **20**(8): 539–542 (1978).

Hamilton, A., "Industrial poisoning by compounds of the aromatic series." *J. Ind. Hygi.* 200–212 (1919).

Hancock, G., A. E. Moffitt, Jr., and E. B. Hay, "Hematological findings among workers exposed to benzene at a coke oven by-product recovery facility," *Arch. Environ. Health* **39**(6): 414–418 (1984).

Kipen, H. M., R. P. Cody, K. S. Crump, B. C. Allen, and B. D. Goldstein, "Hematological effects of benzene: A thirty-five year longitudinal study of rubber workers," *Toxicol. Ind. Health* **4**: 411–430 (1988).

Peterson, J. E., and R. D. Stewart, "Absorption and elimination of carbon monoxide by inactive young men." *Arch. Environ. Health* **21**: 165–171 (1970).

Rinsky, R. A., A. B. Smith, R. Hornung, T. G. Filloon, R. J. Young, A. H. Okun, and P. J. Landrigan, "Benzene and Leukemia. An epidemiologic risk assessment," *N. Engl. J. Med.* **316**: 1044–1050 (1987).

Stewart, R. D., "The effects of low concentrations of carbon monoxide in man," *Scand. J. Respir. Dis. Suppl.* **91**: 56–62 (1974).

Yin, S.-N., Q. Li, Y. Liu, F. Tian, C. Du, and C. Jin. "Occupational exposure to benzene in China," *Br. J. Ind. Med.* **44**: 192–195 (1987).

5 Hepatotoxicity: Toxic Effects on the Liver

STEPHEN M. ROBERTS, ROBERT C. JAMES, AND MICHAEL R. FRANKLIN

This chapter will familiarize the reader with

- The basis of liver injury
- Normal liver functions
- The role the liver plays in certain chemical-induced toxicities
- Types of liver injury
- Evaluation of liver injury
- Specific chemicals that are hepatotoxic

5.1 THE PHYSIOLOGIC AND MORPHOLOGIC BASES OF LIVER INJURY

Physiologic Considerations

The liver is the largest organ in the body, accounting for about 5 percent of total body mass. It is often the target organ of chemical-induced tissue injury, a fact recognized for over 100 years. While the chemicals toxic to the liver and the mechanisms of their toxicity are numerous and varied, several basic factors underlie the liver's susceptibility to chemical attack.

First, the liver maintains a unique position within the circulatory system. As Figure 5.1 shows, the liver effectively "filters" the blood coming from the gastrointestinal tract and abdominal space before this blood is pumped through the lungs and into the general circulation. This unique position in the circulatory system aids the liver in its normal functions, which include (1) carbohydrate storage and metabolism; (2) metabolism of hormones, endogenous wastes, and foreign chemicals; (3) synthesis of blood proteins; (4) urea formation; (5) metabolism of fats; and (6) bile formation. When drugs or chemicals are absorbed from the gastrointestinal tract, virtually all of the absorbed dose must pass through the liver before being distributed through the bloodstream to the rest of the body. Once a chemical reaches the general circulation, regardless of the route of absorption, it is still subject to extraction and metabolism by the liver. The liver receives nearly 30 percent of cardiac output and, at any given time, 10–15 percent of total blood volume is present in the liver. Consequently, it is difficult for any drug or chemical to escape contact with the liver, an important factor in the role of the liver in removing foreign chemicals.

The liver's prominence causes it to have increased vulnerability to toxic attack. The liver can particularly affect, or be affected by, chemicals ingested orally or administered intraperitoneally (i.e., into the abdominal cavity) because it is the first organ perfused by blood containing the chemical. As discussed in Chapter 2, rapid and extensive removal of the chemical by the liver can drastically reduce the amount of drug reaching the general circulation—termed the *first-pass effect*. Being the first organ

Principles of Toxicology: Environmental and Industrial Applications, Second Edition, Edited by Phillip L. Williams, Robert C. James, and Stephen M. Roberts.
ISBN 0-471-29321-0 © 2000 John Wiley & Sons, Inc.

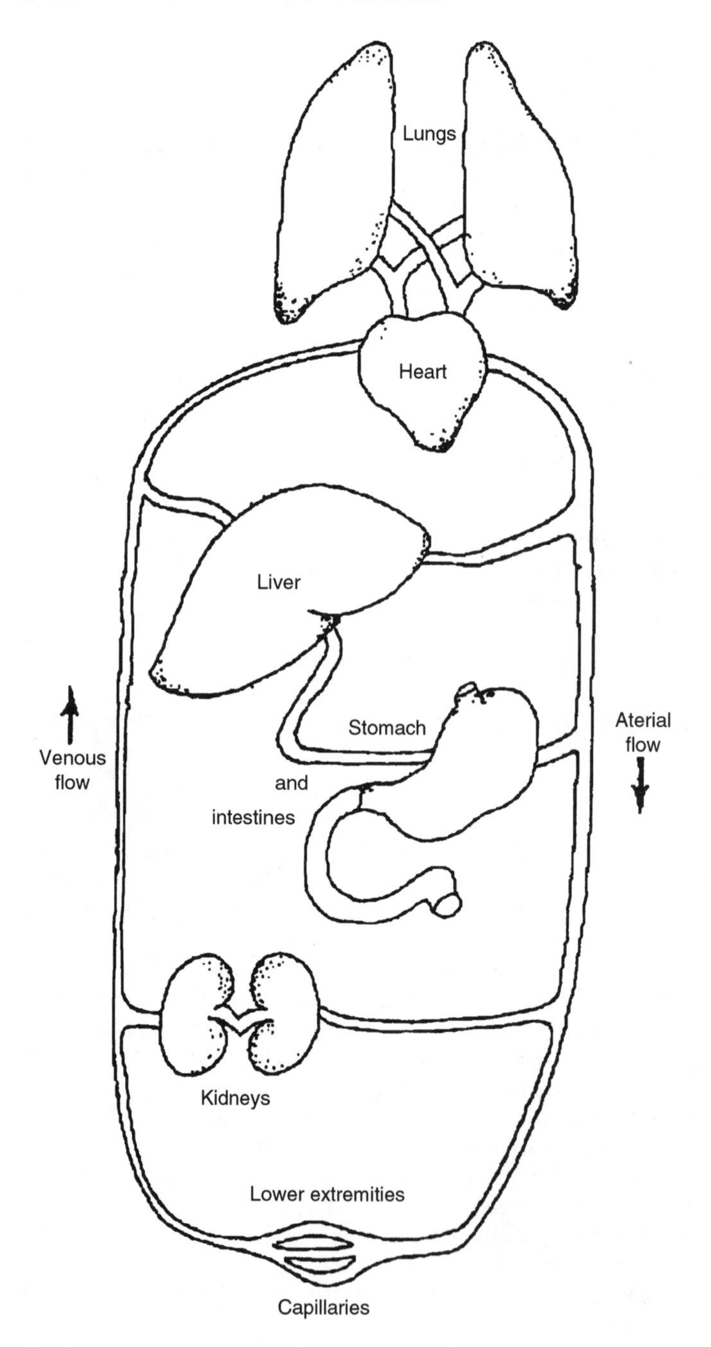

Figure 5.1 The liver maintains a unique position within the circulatory system.

encountered by a drug or chemical after absorption from the gastrointestinal tract or peritoneal space also means that the liver often sees potential toxicants at their highest concentrations. The same drug or chemical at the same dose absorbed from the lungs or through the skin, for example, may be less toxic to the liver because the concentrations in blood reaching the liver are lower, from both dilution and distribution to other organs and tissues.

A second reason for the susceptibility of the liver to chemical attack is that it is the primary organ for the biotransformation of chemicals within the body. As discussed in Chapter 3, the desired net outcome of the biotransformation process is generally to alter the chemical in such a way that it is (1) no longer biologically active within the body and (2) more polar and water-soluble and, consequently, more easily excreted from the body. Thus, in most instances, the liver acts as a *detoxification* organ. It lowers the biological activity and blood concentrations of a chemical that might otherwise accumulate to toxic levels within the body. For example, it has been estimated that the time required to excrete one-half of a single dose of benzene would be about 100 years if the liver did not metabolize it. The primary disadvantage of the liver's role as the main organ metabolizing chemicals, however, is that toxic reactive chemicals or short-lived intermediates can be formed during the biotransformation process. Of course, the liver, as the site of formation of these bioactivated forms of the chemical, usually receives the brunt of their effects.

Morphologic Considerations

The liver can be described as a large mass of cells packed around vascular trees of arteries and veins (see Figure 5.2). Blood supply to the liver comes from the hepatic artery and the portal vein, the former normally supplying about 20 percent of blood reaching the liver and the latter about 80 percent. Terminal branches of the hepatic artery and portal vein are found together with the bile duct (Figure 5.2). In cross section, these three vessels are called the *portal triad.* Blood is collected in the terminal hepatic venules, which drain into the hepatic vein. The functional microanatomy can be viewed in different ways. In one view, the basic unit of the liver is termed the *lobule.* Blood enters the lobule

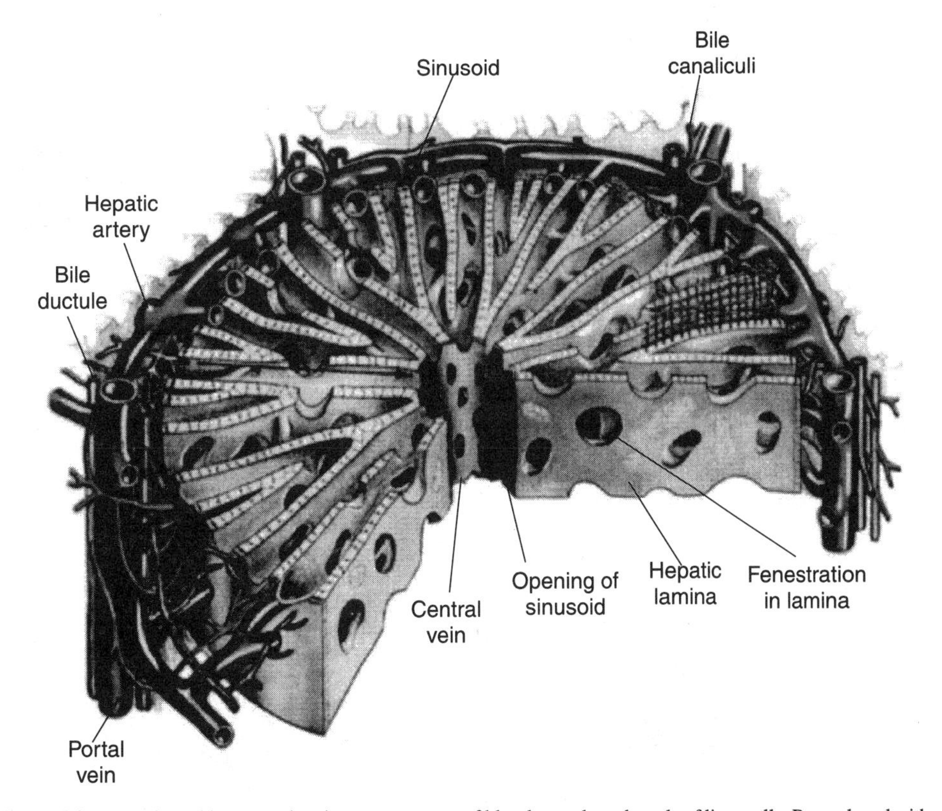

Figure 5.2 Hepatic architecture, showing arrangement of blood vessels and cords of liver cells. Reproduced with permission from Textbook of Human Anatomy, Second Edition, C.V. Mosby Co., St. Louis, MO, 1976.

from the hepatic artery and portal veins, traverses the lobule through hepatic sinusoids, and exits through a hepatic venule. In the typical lobule view, cells near the portal vein are termed *periportal*, while those near the hepatic venule are termed *perivenular*. The hepatic venule is visualized as occupying the center of the lobule, and cells surrounding the venule are sometimes termed *centrilobular*, while those farther away, near the portal triad, are called *peripheral lobular*. Rappaport proposed a different view of hepatic anatomy in which the basic anatomical unit is called the *simple liver acinus*. In this view (Figure 5.3, left), cells within the acinus are divided into zones. The area adjacent to small vessels radiating from the portal triad is zone 1. Cells in zone 1 are first to receive blood through the sinusoids. Blood then travels past cells in zones 2 and 3 before reaching the hepatic venule. As can be seen in Figure 5.3, zone 3 is roughly analogous to the centrilobular region of the classic lobule, since it is closest to the central vein. Zone 3 cells from adjacent acini form a star-shaped pattern around this vessel. Zone 1 cells surround the terminal afferent branches of the portal vein and hepatic artery, and are often stated as occupying the *periportal* region, while cells between zones 1 and 3 (i.e., in zone 2) are said to occupy the *midzonal* region. A modification of the typical lobule and acinar models has been provided by Lamers and colleagues (1989) (Figure 5.3, right). Based on histopathologic and immunohistochemical studies, they propose that zone 3 should be viewed as a circular, rather than star-shaped, region surrounding the central vein. Zone 1 cells surround the portal tracts, and zone 1 cells from adjacent acini merge to form a reticular pattern. As with the Rappaport (1979) model, cells in zone 3 may be described as centrilobular (matching closely the classic lobular terminology), cells in zone 1 as periportal, and the cells in zone 2 in between are called midzonal.

Each of these viewpoints has in common a recognition that the cells closest to the arterial blood supply receive the highest concentrations of oxygen and nutrients. As blood traverses the lobule, concentrations of oxygen and nutrients diminish. Differences in oxygen tension and nutrient levels are reflected in differing morphology and enzymatic content between cells in zones 1 and 3. Consistent with their greater access to oxygen, hepatocytes in zone 1 are better adapted to aerobic metabolism. They have greater respiratory activity, greater amino acid utilization, and higher levels of fatty acid oxidation. Glucose formation from gluconeogenesis and from breakdown of glycogen predominate in zone 1 cells, and most secretion of bile acids occurs here. On the other hand, most forms of the biotransformation enzyme cytochrome P450 are found in highest concentrations in zone 3 cells. As the site of biotransformation for most drugs and chemicals, zone 3 cells have greatest responsibility for their detoxification. This also means that zone 3 cells are often the primary targets for chemicals that are bioactivated by these enzymes to toxic metabolites in the liver.

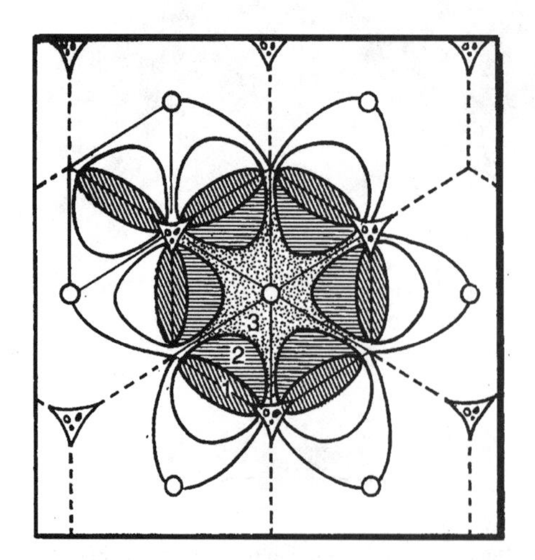

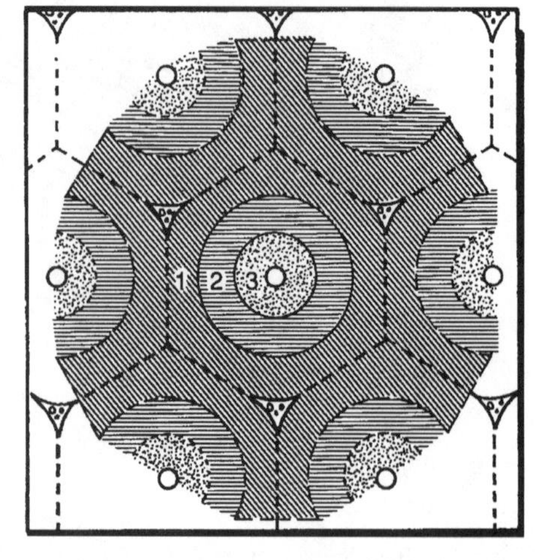

Figure 5.3 Alternative views of the liver acinus. Reproduced with permission from Lamers et al., 1989.

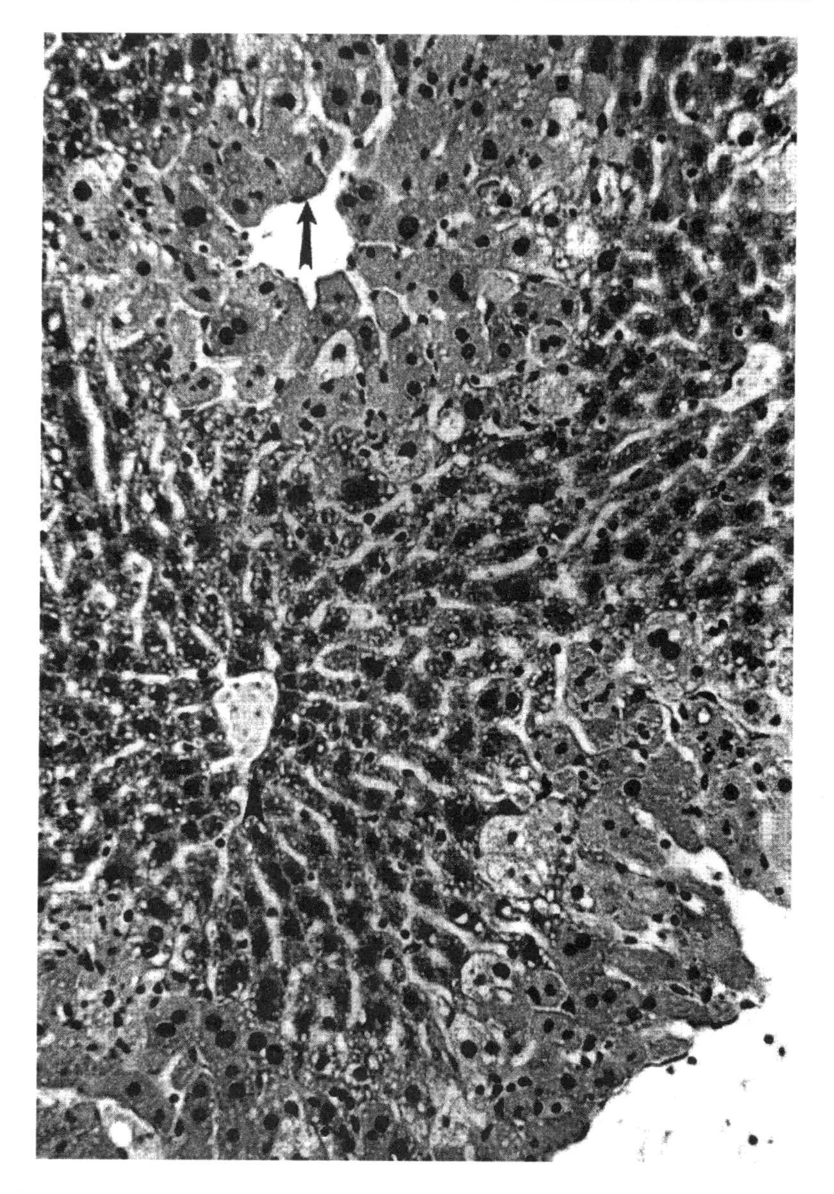

Figure 5.4 Liver section from mouse given an hepatotoxic dose of acetaminophen. With acetaminophen, liver cell swelling and death characteristically occurs in regions around the central vein (Zone 3, arrow); cells near the portal triad (Zone 1, arrow head) are spared.

There are several types of liver cells. *Hepatocytes*, or *parenchymal cells*, constitute approximately 75 percent of the total cells in the human liver. They are relatively large cells and make up the bulk of the hepatic lobule. By virtue of their numbers and their extensive xenobiotic metabolizing activity, these cells are the principal targets for hepatotoxic chemicals. The sinusoids are lined with endothelial cells. These cells are small but numerous, making up most of the remaining cells in the liver. The hepatic microvasculature also contains resident macrophages, called *Kupffer cells*. Although comparatively few in number, these cells play an important role in phagocytizing microorganisms and foreign particulates in the blood. While these cells are a part of the liver, they are also part of the immune system. They are capable of releasing reactive oxygen species and cytokines, and play an important role in inflammatory responses in the liver. The liver also contains *Ito cells* (also termed *fat-storing cells, parasinusoidal cells,* or *stellate cells*) which lie between parenchymal and endothelial cells. These cells appear to be important in producing collagen and in vitamin A storage and metabolism.

5.2 TYPES OF LIVER INJURY

All chemicals do not produce the same type of liver injury. Rather, the type of lesion or effect observed is dependent on the chemical involved, the dose, and the duration of exposure. Some types of injury are the result of acute toxicity to the liver, while others appear only after chronic exposure or treatment. Basic types of liver injury include the anomalies described in the following paragraphs.

Hepatocellular Degeneration and Death

Many hepatotoxicants are capable of injuring liver cells directly, leading to cellular degeneration and death. A variety of organelles and structures within the liver cell can be affected by chemicals. Principal targets include the following:

1. *Mitochondria.* These organelles are important for energy metabolism and synthesis of ATP. They also accumulate and release calcium, and play an important role in calcium homeostasis within the cell. When mitochondria become damaged, they often lose the ability to regulate solute and water balance, and undergo swelling that can be observed microscopically. Mitochondrial membranes can become distorted or rupture, and the density of the mitochondrial matrix is altered. Examples of chemicals that show damage to hepatic mitochondria include carbon tetrachloride, cocaine, dichloroethylene, ethionine, hydrazine, and phosphorus.

2. *Plasma Membrane.* The plasma membrane surrounds the hepatocyte and is critically important in maintaining the ion balance between the cytoplasm and the external environment. This ion balance can be disrupted by damage to plasma membrane ion pumps, or by loss of membrane integrity causing ions to leak in or out of the cell following their concentration gradients. Loss of ionic control can cause a net movement of water into the cell, resulting in cell swelling. Blisters or "blebs" in the plasma membrane may also occur in response to chemical toxicants. Examples of chemicals that show damage to plasma membrane include acetaminophen, ethanol, mercurials, and phalloidin.

3. *Endoplasmic Reticulum.* The endoplasmic reticulum is responsible for synthesis of proteins and phospholipids in the hepatocyte. It is the principal site of biotransformation of foreign chemicals and, along with the mitochondria, sequesters and releases calcium ions to promote calcium homeostasis. As discussed in Chapter 3, hepatic biotransformation enzyme activity is substantially increased in response to treatment or exposure to a variety of chemicals. Many of these enzymes, including cytochrome P450, are located in the endoplasmic reticulum, which undergoes proliferation as part of the enzyme induction process. Because the endoplasmic reticulum is the site within the cell of most oxidative metabolism of foreign (xenobiotic) chemicals, it is also the site where reactive metabolites from these chemicals are formed. This makes it a logical target for toxicity for chemicals that produce injury through this mechanism. Morphologically, damage to the endoplasmic reticulum often appears in the form of dilation. Examples of chemicals that show damage to endoplasmic reticulum include acetaminophen, bromobenzene, carbon tetrachloride, and cocaine.

4. *Nucleus.* There are several ways in which the nuclei can be damaged by chemical toxicants. Some chemicals or their metabolites can bind to DNA, producing mutations (see Chapter 12). These mutations can alter critical functions of the cell leading to cell death, or can contribute to malignant transformation of the cell to produce cancer. Some chemicals appear to cause activation of endonucleases, enzymes located in the nucleus that digest chromatin material. This leads to uncontrolled digestion of the cell's DNA—obviously not conducive to normal cell functioning. Some chemicals

cause disarrangement of chromatin material within the nucleus. Morphologically, damage to the nucleus appears as alterations in the nuclear envelope, in chromatin structure, and in arrangement of nucleoli. Examples of chemicals that produce nuclear alterations include aflatoxin B, beryllium, ethionine, galactosamine, and nitrosamines.

5. *Lysosomes.* These subcellular structures contain digestive enzymes (e.g., proteases) and are important in degrading damaged or aging cellular constituents. In hepatocytes injured by chemical toxicants, their numbers and size are often increased. Typically, this is not because they are a direct target for chemical attack, but rather reflects the response of the cell to the need to remove increased levels of damaged cellular materials caused by the chemical.

Not all hepatocellular toxicity leads to cell death. Cells may display a variety of morphologic abnormalities in response to chemical insult and still recover. These include cell swelling, dilated endoplasmic reticulum, condensed mitochondria and chromatin material in the nucleus, and blebs on the plasma membrane. More severe morphological changes are indicative that the cell will not recover, and will proceed to cell death, that is, undergo *necrosis*. Examples of morphological signs of necrosis are massive swelling of the cell, marked clumping of nuclear chromatin, extreme swelling of mitochondria, breaks in the plasma membrane, and the formation of cell fragments.

Necrosis from hepatotoxic chemicals can occur within distinct zones in the liver, be distributed diffusely, or occur massively. Many chemicals produce a zonal necrosis; that is, necrosis is confined to a specific zone of the hepatic acinus. Table 5.1 provides examples of drugs and chemicals that produce hepatic necrosis and the characteristic zone in which the lesion occurs. Figure 5.4 shows an example of zone 3 hepatic necrosis from acetaminophen. Confinement of the lesion to a specific zone is thought to be a consequence of the mechanism of toxicity of these agents and the balance of activating and inactivating enzymes or cofactors. Interestingly, there are a few chemicals for which the zone of necrosis can be altered by treatment with other chemicals. These include cocaine, which normally produces hepatic necrosis in zone 2 or 3 in mice, but in phenobarbital-pretreated animals causes necrosis in zone 1. Limited observations of liver sections from humans experiencing cocaine hepatotoxicity are consistent with this shift produced by barbiturates. The reason for the change in site of necrosis with these chemicals is unknown.

Necrotic cells produced by some chemicals are distributed diffusely throughout the liver, rather than being localized in acinar zones. Galactosamine and the drug methylphenidate are examples of chemicals that produce a diffuse necrosis. Diffuse necrosis is also seen in viral hepatitis and some forms of idiosyncratic liver injury. The extent of necrosis can vary considerably. When most of the cells of the liver are involved, this is termed *massive necrosis*. As the name implies, this involves destruction of most or all of the hepatic acinus. Not all the acini in the liver are necessarily affected to the same extent, but at least some acini will have necrosis that extends across the lobule from the portal triad to the hepatic vein, called *bridging necrosis*. Massive necrosis is not so much a characteristic of specific hepatotoxic chemicals as of their dose.

Because of the remarkable ability of the liver to regenerate itself, it is able to withstand moderate zonal or diffuse necrosis. Over a period of several days, necrotic cells are removed and replaced with new cells, restoring normal hepatic architecture and function. If the number of damaged cells is too great, however, the liver's capacity to restore itself becomes overwhelmed, leading to hepatic failure and death.

Another form of cell death is *apoptosis*, or programmed cell death. Apoptosis is a normal physiological process used by the body to remove cells when they are no longer needed or have become functionally abnormal. In apoptosis, the cell "commits suicide" through activation of its endonucleases, destroying its DNA. Apoptotic cells are morphologically distinct from cells undergoing necrosis as described above. Unlike cells undergoing necrosis, which swell and release their cellular contents, apoptotic cells generally retain plasma membrane integrity and shrink, resulting in condensed cytoplasm and dense chromatin in the nucleus. There are normally few apoptotic cells in liver, but the number may be increased in response to some hepatotoxic chemicals, notably thioacetamine and ethanol. Also, some chemicals produce hypertrophy, or growth of the liver beyond its normal size.

TABLE 5.1 Drugs and Chemicals that Produce Zonal Hepatic Necrosis

	Site of Necrosis		
Chemical	Zone1	Zone 2	Zone 3
Acetaminophen			X
Aflatoxin	X		X
Allyl alcohol	X		
Alloxan	X		
α-Amanitin			X
Arsenic, inorganic			X
Beryllium		X	
Botulinum toxin			X
Bromobenzene			X
Bromotrichoromethane			X
Carbon tetrachloride			X
Chlorobenzenes			X
Chloroform			X
Chloroprene			X
Cocaine[a]		X	X
Dichlorpropane			X
Dioxane			X
DDT			X
Dimethylnitrosamine			X
Dinitrobenzene			X
Dinitrotoluene			X
Divinyl ether			X
Ethylene dibromide			X
Ethylene dichloride			X
Ferrous sulfate	X		
Fluoroacetate			X
Iodobenzene			X
Iodoform			X
Manganese compounds	X		
Methylchloroform			X
Naphthalene			X
Ngaione		X	
Paraquat		X	X
Phalloidin			X
Pyridine			X
Pyrrolidizine alkaloids			X
Rubratoxin			X
Tannic acid			X
Thioacetamide			X
Urethane			X
Xylidine			X

Source: Adapted from Cullen and Reubner, 1991.

[a]Necrosis is shifted to zone 1 in phenobarbital-pretreated animals.

Examples include lead nitrate and phenobarbital. When exposure or treatment with these agents has ended, the liver will return to its normal size. During this phase, the number of apoptotic cells is increased, reflecting an effort by the liver to reduce its size, in part by eliminating some of its cells.

Drugs and chemicals can produce hepatocellular degeneration and death by many possible mechanisms. For some hepatotoxicants, the mechanism of toxicity is reasonably well established. For example, galactosamine is thought to cause cell death by depleting uridine triphosphate, which is essential for synthesis of membrane glycoproteins. For most hepatotoxicants, however, key biochemical effects responsible for hepatocellular necrosis remain uncertain. The search for a broadly applicable mechanism of hepatotoxicity has yielded several candidates:

Lipid Peroxidation Many hepatotoxicants generate free radicals in the liver. In some cases, such as carbon tetrachloride, the free radicals are breakdown products of the chemical generated by its cytochrome P450-mediated metabolism in the liver. In other cases, the chemical causes a disruption in oxidative metabolism within the cell, leading to the generation of reactive oxygen species. An important potential consequence of free-radical formation is the occurrence of lipid peroxidation in membranes within the cell. Lipid peroxidation occurs when free radicals attack the unsaturated bonds of fatty acids, particularly those in phospholipids. The free radical reacts with the fatty acid carbon chain, abstracting a hydrogen. This causes a fatty acid carbon to become a radical, with rearrangement of double bonds in the fatty acid carbon chain. This carbon radical in the fatty acid reacts with oxygen in a series of steps to produce a lipid hydroperoxide and a lipid radical that can then react with another fatty acid carbon. The peroxidation of the lipid becomes a chain reaction, resulting in fragmentation and destruction of the lipid. Because of the importance of phospholipids in membrane structure, the principal consequence of lipid peroxidation for the cell is loss of membrane function. The reactive products generated by lipid peroxidation can interact with other components of the cell as well, and this also could contribute to toxicity.

The list of chemicals that produce lipid peroxidation as part of their hepatotoxic effects is extensive, and includes halogenated hydrocarbons (e.g., carbon tetrachloride, chloroform, bromobenzene, tetrachloroethene), alcohols (e.g., ethanol, isopropanol), hydroperoxides (e.g., *tert*-butylhydroperoxide), herbicides (e.g., paraquat), and a variety of other compounds (e.g., acrylonitrile, cadmium, cocaine, iodoacetamide, chloroacetamide, sodium vanadate). Consequently, it is an attractive common mechanism of hepatotoxicity. There is some question, however, as to whether it is the most important mechanism of toxicity for these chemicals. For some of these hepatotoxic compounds, experiments have been conducted in which lipid peroxidation was blocked by concomitant-treatment with an antioxidant. In many cases, hepatotoxicity still occurred. This argues that for at least some agents, lipid peroxidation may contribute to their hepatotoxicity, but is not sufficient to explain all of their toxic effects on the liver.

Irreversible Binding to Macromolecules Most of the conventional hepatotoxicants must be metabolized in order to produce liver toxicity, producing one or more chemically reactive metabolites. These reactive metabolites bind irreversibly to cellular macromolecules—primarily proteins, but in some cases also lipids and DNA. This binding precedes most manifestations of toxicity, and the extent of binding often correlates well with toxicity. In fact, histopathology studies with some of these chemicals have found that only cells with detectable reactive metabolite binding undergo necrosis. Examples of hepatotoxic chemicals that produce reactive metabolites include acetaminophen, bromobenzene, carbon tetrachloride, chloroform, cocaine, and trichloroethylene.

It is certainly plausible that irreversible binding of a toxicant to a critical protein or other macromolecule in the cell could lead to loss of its function, and the fact that binding precedes most, if not all, toxic responses in the cell make it a logical initiating event. However, demonstrating precisely how irreversible binding causes cell death has been extremely challenging. Several studies have been conducted attempting to identify the macromolecular targets for binding and to determine whether this binding results in an effect that could lead to cell death. Acetaminophen, in particular, has been studied in this regard. While several proteins bound by the acetaminophen reactive metabolite, *N*-acetyl-*p*-

benzoquinone imine, have been identified, none as yet has been clearly shown to be instrumental in acetaminophen-induced hepatic necrosis. Without identification of the critical target(s) for irreversible binding for hepatotoxicants, this remains an attractive but unproven mechanism.

Loss of Calcium Homeostasis Intracellular calcium is important in regulating a variety of critical intracellular processes, and the concentration of calcium within the cell is normally tightly regulated. The plasma membrane actively extrudes calcium ion from the cell to maintain cytosolic concentrations at a low level compared with the external environment (the ratio of intracellular to extracellular concentration is about 1:10,000). Both the mitochondria and endoplasmic reticulum are capable of sequestering and releasing calcium ion as needed to modulate calcium concentrations for normal cell functioning. Loss of control of intracellular calcium can lead to a sustained rise in intracellular calcium levels, which, in turn, disrupts mitochondrial metabolism and ATP synthesis, damages microfilaments used to support cell structure, and activates degradative enzymes within the cell. These events could easily account for cell death from hepatotoxic chemicals.

Early studies of toxic effects of chemicals on liver cells in culture suggested that an influx of calcium from outside the cell (e.g., from plasma membrane failure) was responsible for their toxic effects. Later experiments showed that this was probably not the case, but nonetheless supported disregulation of intracellular calcium as a key event in toxicity. Intracellular calcium levels were observed to rise substantially in response to a number of hepatotoxicants, apparently due to chemical effects on mitochondria and/or the endoplasmic reticulum leading to loss of control of intracellular calcium stores. Impaired extrusion of calcium out of the cell by the plasma membrane might also be important, at least for some chemicals. In general, increases in intracellular calcium preceded losses of viability, suggesting a cause–effect relationship. It is sometimes difficult, however, to discern to what extent elevated calcium levels are the cause of, or merely the result of, cytotoxicity.

Immune Reactions This mechanism of hepatotoxicity is not common, but nonetheless important. Characteristically, an initial exposure is required that does not produce significant hepatotoxicity—a sensitizing event. Subsequent exposure to the drug or chemical can lead to profound liver toxicity that may be accompanied by hepatic inflammation. Consistent with a hypersensitivity reaction, there is little evidence of a dose–response relationship, and even small doses can trigger a reaction. This response is usually rare and difficult to predict; hence it is often considered an idiosyncratic reaction. Typically, this kind of hepatotoxicity for a drug or chemical is very difficult to demonstrate in laboratory animals, and unfortunately becomes known only after widespread use or exposure in humans.

Perhaps the most familiar example of a drug or chemical producing this type of hepatotoxicity is the general anesthetic halothane. Studies suggest that halothane is metabolized to a reactive metabolite that binds with proteins. These proteins become expressed on the cell surface where they are recognized by the immune system as being foreign. The immune system then mounts a cell-mediated response, resulting in destruction of the hepatocytes. This response, called *halothane hepatitis*, seldom occurs (only about 1 in 10,000 anesthetic administrations in adults) but has a 50 percent mortality rate. A similar phenomenon has been observed with other drugs, including diclofenac.

Fatty Liver

Many chemicals produce an accumulation of lipids in the liver, called *fatty liver* or *steatosis*. Examples of chemicals that produce fatty liver are provided in Table 5.2. Just as hepatocellular necrosis preferentially occurs in specific acinar zones in response to certain chemicals, so does fatty liver. For example, zone 1 is the primary site of lipid accumulation from white phosphorus, while zone 3 is where most of the lipid accumulation is observed with tetracycline and ethanol. The lipid accumulates in vacuoles within the cytoplasm, and these vacuoles are usually present as either one large, clear vacuole (called *macrovesicular steatosis*) or numerous small vacuoles (*microvesicular steatosis*). The type of steatosis (macro- or microvesicular) is characteristic of specific hepatotoxicants and, in some cases, of certain diseases or conditions. For example, microvesicular steatosis has been associated with

TABLE 5.2 Drugs and Chemicals that Produce Fatty Liver

Antimony	Ethyl chloride
Barium salts	Hydrazine
Borates	Methyl bromide
Carbon disulfide	Orotic acid
Chromates	Puromycin
Dichloroethylene	Safrole
Dimethylhydrazine	Tetracycline
Ethanol	Thallium compounds
Ethionine	Uranium compounds
Ethyl bromide	White phosphorus

tetracycline, valproic acid, salicylates, aflatoxin, dimethylformamide, and some of the antiviral nucleoside analogs used to treat HIV. It is also associated with Reye's syndrome and fatty liver of pregnancy. Macrovesicular steatosis has been associated with antimony, barium salts, carbon disulfide, dichloroethylene, ethanol, hydrazine, methyl and ethyl bromide, thallium, and uranium compounds.

There are several potential chemical effects that can give rise to accumulation of lipids in the cell. These include:

1. *Inhibition of Lipoprotein Synthesis.* A number of chemicals are capable of inhibiting synthesis of the protein moiety needed for synthesis of lipoproteins in the liver. These include carbon tetrachloride, ethionine, and puromycin.
2. *Decreased Conjugation of Triglycerides with Lipoproteins.* Another critical step in lipoprotein synthesis is conjugation of the protein moiety with triglyceride. Carbon tetrachloride, for example, can interfere with this step.
3. *Interference with Very-Low-Density Lipoprotein (VLDL) Transfer.* Inhibition of transfer of VLDL out of the cell results in its accumulation. Tetracycline is an example of an agent that interferes with this transfer.
4. *Impaired Oxidation of Lipids by Mitochondria.* Oxidation of nonesterified fatty acids is an important aspect of their hepatocellular metabolism, and decreased oxidation can contribute to their accumulation within the cell. Carbon tetrachloride, ethionine, and white phosphorus have been shown to inhibit this oxidation.
5. *Increased Synthesis of Fatty Acids.* The liver is capable of synthesizing fatty acids from acetyl-CoA (coenzyme A), and increased fatty acid synthesis can increase the lipid burden of the cells. Ethanol is an example of a chemical that produces this effect.

Other possible mechanisms might contribute to fatty liver, such as increased uptake of lipids from the blood by the liver, but the role of these processes in drug- or chemical-induced steatosis is less clear. The mechanisms listed above are not mutually exclusive. Indeed, it is likely that many of the chemicals that produce steatosis do so by producing more than one of these effects.

Fatty liver may occur by itself, or in conjunction with hepatocellular necrosis. Many chemicals produce a lesion that consists of both effects. Examples include: aflatoxins, amanitin, arsenic compounds, bromobenzene, carbon tetrachloride, chloroform, dimethylnitrosamine, dinitrotoluene, DDT, dichloropropane, naphthalene, pyrrolizidine alkaloids, and tetrachloroethane. Drug- or chemical-induced steatosis is reversible when exposure to the agent is stopped.

Phospholipidosis is a special form of steatosis. It results from accumulation of phospholipids in the hepatocyte, and can be caused by some drugs as well as by inborn errors in phospholipid metabolism. Liver sections from patients with phospholipidosis reveal enlarged hepatocytes with "foamy" cytoplasm. Often this condition progresses to cirrhosis. Examples of drugs associated with phospholipidosis include amiodarone, chlorphentermine, and 4,4′-diethylaminoethoxyhexoestrol.

Cholestasis

The term *cholestasis* refers to decreased or arrested bile flow. Many drugs and chemicals are able to produce cholestatic injury, and examples are listed in Table 5.3. There are several potential causes of impaired bile flow, many of which can become the basis for drug- or chemical-induced cholestasis. Some of these are related to loss of integrity of the canalicular system that collects bile and carries it to the gall bladder, while others are related to the formation and secretion of bile. For example, α-naphthylisothiocyanate disrupts the tight junctions between hepatocytes that help form the canaliculi, the smallest vessels of the bile collection system. This causes a leakage of bile contents out of the canaliculi into the sinusoids. Other toxicants, such as methylene dianiline and paraquat, impede bile flow by damaging the bile ducts. The primary driving force for bile formation is the secretion of bile acids into the canalicular lumen. This requires uptake of bile acids from the blood into hepatocytes, and then transport into the canaliculus. Anabolic steroids are an example of a class of compounds that produce cholestatic injury by inhibiting these transport processes.

Some cholestatic injury can be expected whenever there is severe hepatic injury of any type. This is because normal bile flow requires functioning hepatocytes as well as a reasonably intact cellular architecture in the liver. Whenever this is disrupted, some impairment of bile flow can be expected as a secondary consequence. Many agents produce primarily hepatic necrosis with perhaps limited cholestasis (see Table 5.1), others produce primarily cholestasis with some necrosis (chlorpromazine and erythromycin are examples), and still others are capable of producing cholestasis with little or no damage to the hepatocytes. The contraceptive and anabolic steroids are examples of this last category of agents.

Vascular Injury

Cells lining the vasculature within the liver are also potential targets for hepatotoxicants. Injury of vascular cells leads to occlusion (impaired blood flow), which in turn leads to hypoxia. Cells in zone 3 are most vulnerable, since the oxygenation of blood reaching these cells is low even under normal conditions. Typically, hypoxia results in necrosis, and continuing injury over time leads to fibrosis. Severe cases can result in fatal congestive cirrhosis. There are several examples of chemicals known

TABLE 5.3 Drugs and Chemicals that Produce Cholestasis

Amitryptyline	Ethanol
Ampicillin	Haloperidol
Arsenicals, organic	Imipramine
Barbiturates	Methylene dianiline
Carbamazepine	Methyltestosterone
Chlorpromazine	α-Naphthylisothiocyanate
Cimetidine	Norandrostenolone
Cyproheptadine	Paraquat
4,4-Diaminodiphenylmethane	Phalloidin
4,4-Diaminodiphenylamine	Phenytoin
1,1-Dichloroethylene	Prochlorperazine
Dinitrophenol	Tolbutamide
Erythromycin estolate	Troleandomycin
Estrogens	

to produce hepatic *venoocclusive disease*, including many of natural origin such as pyrrolizidine alkaloids in herbal teas. Oral contraceptives and some anticancer drugs have also been associated with this effect.

Peliosis hepatis is another vascular lesion characterized by the presence of large, blood-filled cavities. It is unclear why these cavities form, but there is reason to suspect that it may be due to a weakening of sinusoidal supporting membranes. Use of anabolic steroids has been associated with this effect. Although patients with peliosis hepatis are usually without symptoms, the cavities occasionally rupture causing bleeding into the abdominal cavity.

Cirrhosis

Chronic liver injury often results in the accumulation of collagen fibers within the liver, leading to fibrosis. Fibrotic tissue accumulates with repeated hepatic insult, making it difficult for the liver to replace damaged cells and still maintain normal hepatic architecture. Fibrous tissue begins to form walls separating cells. Distortions in hepatic microcirculation cause cells to become hypoxic and die, leading to more fibrotic scar tissue. Ultimately, the organization of the liver is reduced to nodules of regenerating hepatocytes surrounded by walls of fibrous tissue. This condition is called *cirrhosis*. Hepatic cirrhosis is irreversible and carries with it substantial medical risks. Blood flow through the liver becomes obstructed, leading to portal hypertension. To relieve this pressure, blood is diverted past the liver through various shunts not well suited for this purpose. It is common for vessels associated with these shunts to rupture, leading to internal hemorrhage. Even without hemorrhagic episodes, the liver may continue to decline until hepatic failure occurs.

The ability of chronic ethanol ingestion to produce cirrhosis is widely appreciated. Occupational exposures to carbon tetrachloride, trinitrotoluene, tetrachloroethane, and dimethylnitrosamine have also been implicated as causing cirrhosis, as well as the medical use of arsenicals and methotrexate. Some drugs (e.g., methyldopa, nitrofurantoin, isoniazid, diclofenac) produce an idiosyncratic reaction resembling viral hepatitis. This condition, termed *chronic active hepatitis*, can also lead to cirrhosis if the drug is not withdrawn.

Tumors

Many chemicals are capable of producing tumors in the liver, particularly in laboratory rodents. In fact, in cancer rodent bioassays for carcinogenicity, the liver is the most common site of tumorigenicity. Hepatic tumors may be benign or malignant. Conceptually, the distinction between them is that benign tumors are well circumscribed and do not metastasize (i.e., do not invade other tissues). Malignant tumors, on the other hand, are poorly circumscribed and are highly invasive (see Chapter 13 for additional discussion on benign and malignant tumors). Benign tumors, despite their name, are capable of producing morbidity and mortality. However, they are easier to manage and have a much better prognosis than malignant tumors.

Tumors are also classified by the tissue of origin, that is, whether they arise from epithelial or mesenchymal tissue, and by the specific cell type from which they originate. The nomenclature for naming tumors is complex, and the reader is referred elsewhere for a complete discussion of the topic. Basically, malignant tumors arising from epithelial tissue are termed *carcinomas*, while malignant tumors of mesenchymal origin are *sarcomas*. Thus, malignant tumors derived from hepatocytes, which are of epithelial origin, are termed *hepatocellular carcinomas*. Malignant tumors from bile duct cells, also of epithelial origin, are termed *cholangiocarcinomas* (the prefix *cholangio-* refers to the bile ducts). Cells of the vascular lining are of mesenchymal origin. Consequently, a malignant tumor in the liver arising from these cells may be called *hemangiosarcoma*. Benign tumors are also named on the basis of tissue of origin and their appearance. For example, benign tumors of epithelial origin with gland, or glandlike structures are called *adenomas*, and in the liver these can occur among hepatocytes or bile duct cells. Benign tumors of fibrotic cell origin are termed *fibromas*, and those in the bile ducts are called *cholangiofibromas*.

To make things more complicated, cells go through a series of morphological changes as they progress to become a benign or malignant tumor. Thus, groups of cells that represent proliferation of liver tissue, but are not (or not yet) tumors, may be described as nodular hyperplasia, focal hepatocellular hyperplasia, or foci of hepatocellular alteration, depending on their morphological characteristics. The foci of hepatocellular alteration represent the earliest stages that can be detected microscopically. These foci are small groups of cells that are abnormal, but have no distinct boundary separating them from adjacent cells. Their growth rate is such that they are producing little or no compression of surrounding cells. The abnormalities are subtle at this stage, and special stains and markers are sometimes used to help visualize them. Nodular hyperplasia is more readily observed; the group of cells is more circumscribed and compression of adjacent cells is apparent. These cells are thought to represent an intermediate step in tumor development. The significance of these lesions is not that they are associated with any clinical signs or symptoms of disease, but rather that they may represent an area from which a tumor may develop. Consequently, their appearance is important in the assessment of the ability of a drug or chemical to cause cancer. For most chemicals, only a very small percentage—or perhaps none—of the neoplastic areas will go on to produce a malignant tumor. Consequently, the issue of how to use data regarding the appearance of these lesions in the assessment of carcinogencity of a chemical is one of considerable discussion and debate among toxicologists.

Liver tumors from chemical exposure can arise through numerous mechanisms. Some hepatocarcinogens form DNA adducts leading to mutations. Nitrosoureas and nitrosamines are examples of hepatocarcinogens thought to produce tumors through this mechanism (see also Chapters 12 and 13 for further discussion of genotoxicity and carcinogenicity). Many chemicals that produce liver tumors are not genotoxic, however, and appear to work through epigenetic mechanisms. Nongenotoxic hepatocarcinogens are many and diverse, and include tetrachlorodibenzo-*p*-dioxin, sex steroids, synthetic antioxidants, some hepatic enzyme inducing agents (e.g., phenobarbital), and peroxisome proliferators (e.g., clofibrate). A discussion of the mechanisms underlying epigenetic carcinogenesis (e.g., inhibition of cell-to-cell communication, recurrent cellular injury, receptor interactions) is beyond the scope of this chapter, and the reader is referred to Chapter 12 for more information on this subject.

Despite the many chemicals found to produce benign and malignant liver tumors in mice and rats, relatively few have been clearly associated with liver tumors in humans. Adenomas have been associated with the use of contraceptive steroids, and clinical and epidemiologic studies implicate anabolic steroids, arsenic, and thorium dioxide as causing hepatocellular carcinoma in humans. Hemangiosarcoma is a rare tumor that has been strongly linked to occupational exposure to vinyl chloride, and has also been associated with arsenic and thorium dioxide exposure.

5.3 EVALUATION OF LIVER INJURY

Symptoms of Liver Toxicity

As discussed above, liver injury may be either acute or chronic, and may involve liver cell death, hepatic vascular injury, disruption of bile formation and/or flow, or the development of benign or malignant tumors. Obviously, the signs and symptoms that accompany this array of types of liver injury can vary significantly. There are some generalizations that can be made, however. Common symptoms of liver injury include anorexia (loss of appetite), nausea, vomiting, fatigue, and abdominal tenderness. Physical examination may reveal hepatomegaly (swelling of the liver) and ascites (the accumulation of fluid in the abdominal space). Patients whose liver toxicity involves impaired biliary function may develop *jaundice*, which results from the accumulation of bilirubin in the blood and tissues. Jaundice will appear as a yellowish tint to the skin, mucous membranes, and eyes. *Pruritis*, or an itching sensation in the skin, will often accompany the jaundice.

If the injury is particularly severe, it may lead to *fulminant hepatic failure*. When the liver fails, death can occur in as little as 10 days. There are several complications associated with fulminant hepatic failure. Because the liver is no longer able to produce clotting factor proteins, albumin, or glucose, hemorrhage and hypoglycemia are common. Also, failure of the liver leads to renal failure and deterioration of the central nervous system (*hepatic encephalopathy*). Inability to sustain blood pressure and accumulation of fluid in the lungs may also result. Prognosis is poor for patients with fulminant hepatic failure, with a mortality rate of about 90 percent.

Morphologic Evaluation

For laboratory animal studies of hepatotoxicity, histopathologic examination of liver tissue by light or electron microscopy can be extremely valuable. Histopathologic evaluation can provide information on the nature of the lesion and the regions of the liver affected. This, in turn, can provide insight as to the mechanism of toxicity. For example, the presence of fatty liver would suggest that the chemical may interfere with triglyceride metabolism and/or lipoprotein secretion by the liver. Hepatocellular necrosis confined to the centrilobular region might suggest bioactivation of the chemical by cytochrome P450, since most of the activity of this enzyme normally exists in centrilobular cells. Altered morphology of mitochondria as an early event in toxicity might suggest that mitochondrial toxicity is an important initiating event in the sequence of events leading up to cell death. Histopathologic observations alone cannot establish the mechanism of toxicity, and additional experimentation would be required to explore these hypotheses. Nevertheless, morphologic observation provides important clues, and is an integral part of any comprehensive study of potential hepatotoxicity of a chemical.

In humans, morphologic evaluation of liver biopsies is sometimes used in the diagnosis and management of chronic liver toxicity, particularly liver cancer. Also, noninvasive techniques such as computerized tomography (CT) or magnetic resonance imaging (MRI) scans are used to detect liver cancer, obstructive biliary injury, cirrhosis, and venoocclusive injury to the liver.

Blood Tests

A great deal of insight into the nature and extent of hepatic injury can often be gained through tests on blood samples. There are two fundamental types of blood tests that can be performed. One type is an assessment is based on measuring the functional capabilities of the liver. This can involve an evaluation of the liver's ability to carry out one or more of its basic physiological functions (e.g., glucose metabolism, synthesis of certain proteins, excretion of bilirubin) or its capacity to extract and metabolize foreign compounds from the blood. The second type of assessment involves a determination of whether there are abnormally high levels in the blood of intracellular hepatic proteins. The presence of elevated levels of these proteins in blood is presumptive evidence of liver cell destruction. Examples of these two types of tests are described below:

1. *Serum Albumin.* Albumin is synthesized in the liver and secreted into blood. Liver damage can impair the ability of the liver to synthesize albumin, and serum albumin levels may consequently decrease. The turnover time for albumin is slow, and as a result it takes a long time for impaired albumin synthesis to become evident as changes in serum albumin. For this reason, serum albumin measurements are not helpful in assessing acute hepatotoxicity. They may assist in the diagnosis of chronic liver injury, but certain other diseases can alter serum albumin levels, and the test is therefore not very specific.

2. *Prothrombin Time.* The liver is responsible for synthesis of most of the clotting factors, and a decrease in their synthesis due to liver injury results in prolonged clotting time. In terms of clinical tests, this appears as an increase in prothrombin time. Several drugs and certain diseases also increase prothrombin time. As with serum albumin measurement, this is a relatively insensitive and nonspecific tool for detecting or diagnosing chemical-induced liver injury.

3. *Serum Bilirubin.* The liver conjugates bilirubin, a normal breakdown product of the heme from red blood cells, and secretes the glucuronide conjugate into the bile. Impairment of normal conjugation

and excretion of bilirubin results in its accumulation in the blood, leading to jaundice. Serum bilirubin concentrations may be elevated from acute hepatocellular injury, cholestatic injury, or biliary obstruction. This test is always included among the battery of tests to assess liver function clinically, although it is not a particularly sensitive test for acute injury.

4. *Dye Clearance Tests.* These tests involve administration of a dye that is cleared by the liver and measurement of its rate of disappearance from the blood. Delayed clearance is interpreted as evidence of liver injury. One such dye is sulfobromophthalein (Bromsulphalein; or BSP). Clearance of BSP from the blood is dependent on its active transport into liver cells, conjugation with glutathione, and then active transport into the bile. Conceivably, disruption of any of these processes could result in delayed clearance, although the biliary excretion step is regarded as most critical. The test consists of administering a dose of the dye intravenously and measuring its concentration in blood spectrophotometrically over time. Another dye used for this purpose is indocyanine green (ICG). Unlike BSP, ICG is excreted into the bile without conjugation. Following an intravenous dose, the disappearance of ICG from blood can be measured with repeated blood samples or noninvasively by ear densitometry. The dye tests, although well established, are seldom used clinically.

5. *Drug Clearance Tests.* This test relies on the principle that liver injury will result in impaired biotransformation. The biotransformation capacity of the liver is assessed by following the rate of elimination of a test drug whose clearance from blood is dependent on hepatic metabolism (i.e., a drug for which other elimination processes, such as renal excretion, are insignificant). A test drug such as antipyrine, aminopyrine, or caffeine is administered, and its rate of disappearance from blood is followed over time through serial blood sampling. This rate is compared with a value considered "normal" to determine whether impaired biotransformation exists. This can also be used to test for hepatic enzyme induction, in which the rate of elimination from blood would be increased, rather than decreased as in liver injury. This test is primarily used for research purposes.

6. *Measurement of Hepatic Enzymes in Serum.* Cells undergoing acute degeneration and injury will often release intracellular proteins and other macromolecules into blood. The detection of these substances in blood above normal, baseline levels signals cytotoxicity. This is true for any cell type, and in order for the presence of intracellular proteins in blood to be diagnostic for any particular type of cell injury (e.g., liver toxicity versus renal toxicity versus cardiotoxicity), the proteins must be associated rather specifically with a target organ or tissue. Fortunately, several proteins are found primarily in hepatocytes, and their presence in blood in elevated levels is the basis for some of the most commonly used tests for hepatotoxicity. Table 5.4 shows many of the most common proteins measured in these tests. The reader will note that all of these proteins are enzymes. This is not a coincidence. While any intracellular protein specific to the liver would be useful theoretically, enzymes are proteins that can be measured specifically (by measuring the rate of their particular enzyme activity) using

TABLE 5.4 Serum Enzyme Indicators of Hepatotoxicity

Enzyme	Acronym	Comments
Alanine aminotransferase	ALT	Found mainly in the liver; increase reflects primarily hepatocellular damage
Aspartate aminotransferase	AST	Less specific to the liver than ALT; increase reflects primarily hepatocellular damage
Alkaline phosphatase	ALP	Increases reflect primarily cholestatic injury
γ-Glutamyl transferase; γ-glutamyltranspeptidase	GGTP	Increases reflect primarily cholestatic injury, although elevated in hepatocellular damage as well
5′-Nucleotidase	5′ND	Increases reflect primarily cholestatic injury
Sorbitol dehydrogenase	SDH	High specificity for liver; increase reflects primarily hepatocellular damage
Ornithine carbamoyltransferase	OCT	High specificity for liver; increase reflects primarily hepatocellular damage

assays that are rapid and inexpensive. In fact, the concentrations of each of these proteins are typically measured as an enzyme activity rate, rather than a true concentration per se.

Aminotransferase activities [alanine aminotransferase (ALT) and aspartate aminotransferase (AST)], alkaline phosphatase activity, and gamma glutamyltransferase transpeptidase (GGTP) are included in nearly all standard clinical test suites to assess potential hepatotoxicity. The value of performing a battery of these tests is that each test responds slightly differently in the various forms of liver injury, and evaluating the pattern of responses can offer insight into the type of injury that has occurred. For example, severe hepatic injury from acetaminophen can result in dramatic increases in serum ALT and ALT activities (up to 500 times normal values), but only modest increases in alkaline phosphatase activity. Pronounced increases in alkaline phosphatase is characteristic of cholestatic injury, where increases in ALT and AST may be limited or nonexistent. In alcoholic liver disease, AST activity is usually greater than ALT activity, but for most other forms of hepatocellular injury ALT activities are higher. Serum GGTP is an extremely sensitive indicator of hepatobiliary effects, and may be elevated simply by drinking alcoholic beverages. It is not a particularly specific indicator (it is increased by both hepatocellular and cholestatic injury) and is best utilized in combination with other tests. Serum levels of enzymes such as lactate dehydrogenase have been used to evaluate liver toxicity, but this enzyme has such low specificity for the liver that interpretation of these results is impossible without other confirming tests. Other enzymes such as sorbitol dehydrogenase (SDH) and ornithine carbamoyltransferase (OCT) are quite specific to the liver.

5.4 SUMMARY

Both the anatomic location and its role as a primary site for biotransformation make the liver uniquely susceptible to drug- and chemical-induced injury. Many chemicals encountered in the workplace and environment are capable of producing toxic effects in the liver:

- There are many types of liver injury, including hepatocellular degeneration and death (necrosis), fatty liver, cholestasis (decreased or arrested bile flow), vascular injury, cirrhosis, and tumor development.
- Hepatic injury from drugs and chemicals can arise from a variety of mechanisms. While the mechanism of toxicity for some chemicals is reasonably well established, many aspects of toxic mechanisms for most chemicals remain unclear.
- Hepatotoxic chemicals can attack a variety of subcellular targets. Principal organelles and structures affected include the plasma membrane, mitochondria, the endoplasmic reticulum, the nucleus, and lysosomes.
- Liver injury can be evaluated morphologically (microscopic examination of liver tissue) or through blood tests. Blood tests are designed to either measure the functional capacity of the liver or the appearance of intracellular hepatic contents in the blood.

REFERENCES AND SUGGESTED READING

Cullen, J. M., and B. H. Ruebner, "A histopathologic classification of chemical-induced inju ry of the liver," in *Hepatotoxicity,* R. G. Meeks, S. D. Harrison, and R. J. Bull, eds., CRC Press, Boca Raton, FL, 1991, pp. 67–92.

Delaney, K., "Hepatic principles," in *Goldfrank's Toxicologic Emergencies,* L. R. Goldfrank, N. E. Flomenbaum, N. A. Lewin, R. S. Weisman, M. A. Howland, and R. S. Hoffman, eds., Appleton & Lange, Stamfor d, CT, 1998, pp. 213–228.

Kedderis, G. L. "Biochemical Basis of Hepatocellular Injury." *Toxicologic Pathology,* **24** (1): 77–83 (1996).

Lamers, W. H., A. Hilberts, E. Furt, J. Smith, G. N. Jonges, C. J. F. von Noorden, J. W. G. Janzen, R. Charles, and A. F. M. Moorman, "Hepatic enzymic zonation: A reevaluation of the concept of the liver acinus," *Hepatology* **10**: 72–76 (1989).

Marzella, L., and B. F. Trump, "Pathology of the liver: Functional and structural alterations of hepatocyte organelles induced by cell injury" in *Hepatotoxicity,* R. G. Meeks, S. D. Harrison, and R. J. Bull, eds., CRC Press, Boca Raton, FL, 1991, pp. 93–138.

MacSween, R. N. M., and R. J. Scothorne, "Developmental anatomy and normal structure," in *Pathology of the Liver,* R. N. M. MacSween, P. P. Anthony, P. J. Scheuer, A. D. Burt, and B. C. Portmann, eds., Churchill Livingstone, Edinburgh, 1994, pp. 1–49.

Miyai, K., "Structural organization of the liver," in *Hepatotoxicity,* R. G. Meeks, S. D. Harrison, and R. J. Bull, eds., CRC Press, Boca Raton, FL, 1991, pp. 1–65.

Moslen, M. T., "Toxic responses of the liver," *Casarett and Doull's Toxicology. The Basic Science of Poisons,* 5th ed., C. D. Klaasen, M. O. Amdur, and J. Doull, eds., McGraw-Hill, New York, 1996, pp. 403–416.

Popper, H., "Hepatocellular degeneration and death," in *The Liver: Biology and Pathobiology,* I. M. Arias, W. B. Jakoby, H. Popper, D. Schachter, and D. A. Shafritz, eds., Raven Press, New York, 1988, pp. 1087–1103.

Rappaport, A. M., "Physioanatomical basis of toxic liver injury," in *Toxic Injury of the Liver, Part A,* E. Farber and M. M. Fisher, eds., Marcel Dekker, New York, 1979, pp. 1–57.

Zimmerman, H. J., and K. G. Ishak, "Hepatic injury due to drugs and toxins," in *Pathology of the Liver,* R. N. M. MacSween, P. P. Anthony, P. J. Scheuer, A. D. Burt, and B. C. Portmann, eds., Churchill Livingstone, Edinburgh, 1994, pp. 563–633.

6 Nephrotoxicity: Toxic Responses of the Kidney

PAUL J. MIDDENDORF and PHILLIP L. WILLIAMS

This chapter will give the environmental and occupational health professional information about

- The importance of kidney functions
- How toxic agents disrupt kidney functions
- Measurements performed to determine kidney dysfunctions
- Occupational and environmental agents that cause kidney toxicity

6.1 BASIC KIDNEY STRUCTURES AND FUNCTIONS

The principal excretory organs in all vertebrates are the two kidneys. The primary function of the kidney in humans is removing wastes from the blood and excreting the wastes in the form of urine. However, the kidney plays a key role in regulating total body homeostasis. These homeostatic functions include the regulation of extracellular volume, the regulation of calcium metabolism, the control of electrolyte balance, and the control of acid–base balance.

The adult kidneys of reptiles, birds, and mammals (including humans) are nonsegmental and drain wastes only from the blood (principally breakdown products of protein metabolism). The kidneys are paired organs that lie behind the peritoneum on each side of the spinal column in the posterior aspect of the abdomen. The adult human kidney is approximately 11 cm long, 6 cm broad, and 2.5 cm thick. In human adults individual kidneys weigh 125–170 g for males and 115–155 g for females. The renal artery and vein pass through the hilus, which is a slit in the medial or concave surface of each kidney (Figure 6.1*b*). From each kidney a common collecting duct, the ureter, carries the urine posteriorly to the bladder where it can be voided from the body.

Each human kidney consists of an outer cortex and an inner medulla (see Figures 6.1*b* and 6.2). The cortex constitutes the major portion of the kidney and receives about 85 percent of the total renal blood flow. Consequently, if a toxicant is delivered to the kidney in the blood, the cortex will be exposed to a very high proportion.

Blood Flow to the Kidneys

The kidneys represent approximately 0.5 percent of the total body weight, or approximately 300 g in a 70-kg human. Yet the kidneys receive just under 25 percent of the total cardiac output, which is about 1.2–1.3 L blood/min, or 400 mL/100 g tissue/min. The rate of blood flow through the kidneys is much greater than through other very well perfused tissues, including brain, heart, and liver. If the normal blood hematocrit (i.e., that proportion of blood that is red blood cells) is 0.45, then the normal renal plasma flow is approximately 660 to 715 mL/min. Yet only 125 mL/min of the total plasma flow is

Principles of Toxicology: Environmental and Industrial Applications, Second Edition, Edited by Phillip L. Williams, Robert C. James, and Stephen M. Roberts.
ISBN 0-471-29321-0 © 2000 John Wiley & Sons, Inc.

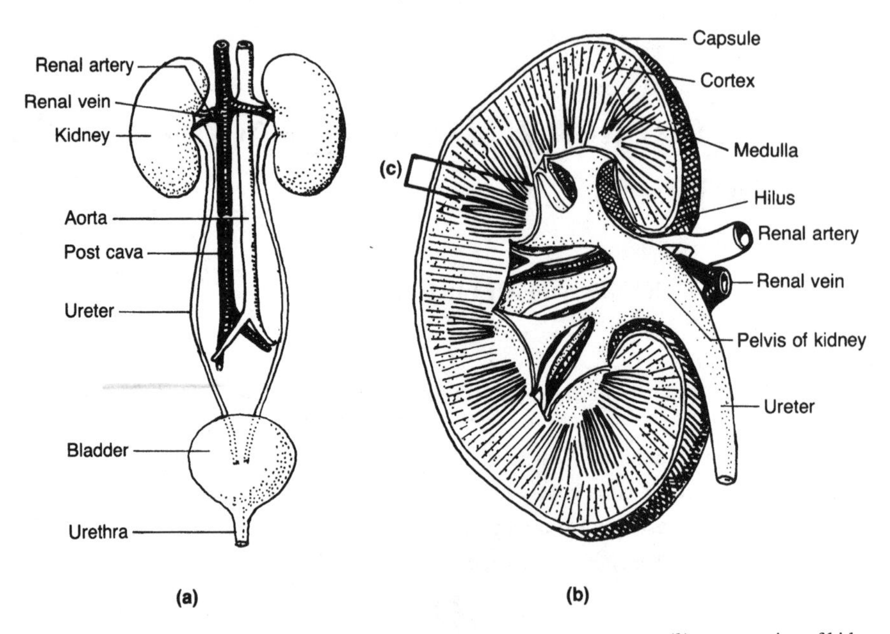

Figure 6.1 The human renal excretory system: (*a*) the complete excretory system; (*b*) cross section of kidney; (*c*) representative section for the enlargement in Figure 6.2.

actually filtered by the kidney. Of this, the kidney reabsorbs approximately 99 percent, resulting in a urine formation rate of only about 1.2 mL/min. Thus, the kidneys, which are perfused at approximately 1 L/min, form urine at approximately 1 mL/min or 0.1 percent of the perfusion. Because of the high volume of blood flow to the kidneys, a chemical in the blood is delivered to this organ in relatively large quantities.

The kidney requires large amounts of metabolic energy to remove wastes from the blood by tubular secretion and to return filtered nutrients back to the blood. Roughly 10 percent of the normal resting oxygen consumption is needed for the maintenance of proper kidney function. Therefore, the kidney is sensitive to agents, such as barbiturates, that induce *ischemia,* a lack of oxygen caused by a decrease in blood flow. Acute intoxication by barbiturates induces severe hypotension (i.e., low blood pressure) and shock. The severe decrease in blood pressure results in a decrease in filtration of the plasma, resulting in a decrease (oliguria) or cessation (anuria) of urine formation. At an early stage this is called *pre–renal failure,* and a reversal in the blood deficit to the kidney will restore normal renal function. However, a critical point is reached when renal sufficiency cannot be restored because of the cell death caused by ischemic anoxia, and the resultant renal failure is irreversible. In this situation, the accumulation in the blood of wastes normally excreted (uremia) results in death. It should be remembered, then, that any agent or physical trauma that causes severe hypotension and shock may produce acute renal failure and eventually death by a similar mechanism.

Nephrons: The Functional Units of the Kidney

The cortex of each kidney in humans contains approximately one million excretory units called nephrons. Agents toxic to the kidney generally injure these nephrons, and such agents are therefore referred to as nephrotoxicants. Degeneration, necrosis, or injury to the nephron elements is referred to as a *nephrosis* or *nephropathy.*

An individual nephron may be divided into three anatomic portions: (1) the vascular or blood-circulating portion, (2) the glomerulus, and (3) the tubular element (Figures 6.2 and 6.3). The glomerulus, which is about 200 μm in diameter, is formed by the invagination of a tuft of capillaries

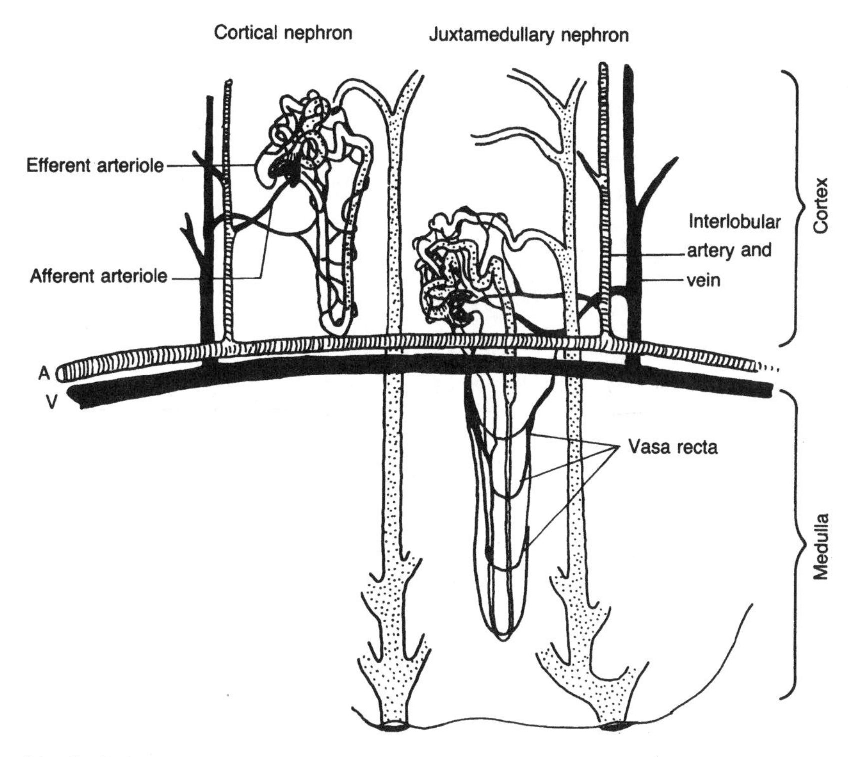

Figure 6.2 Cortical and juxtamedullary nephrons. Enlargement of representative kidney section in Figure 6.1*c*. (Based on B. Brenner and F. Rector, *The Kidney*, Saunders, Philadelphia, 1976.)

into the dilated, blind end of the nephron (Bowman's capsule). The capillaries are supplied by an afferent arteriole and drained by an efferent arteriole. These vascular elements deliver waste and other materials to the tubular element for excretion, return reabsorbed and synthesized materials from the tubular element to the blood circulation, and deliver oxygen and nourishment to the nephron.

The Glomerulus and Glomerular Filtration The glomerulus behaves as if it were a filter with pores 100 Å in diameter, or about 100 times more permeable than the capillaries in skeletal muscle. Substances as great as 70,000 daltons can appear in the glomerular filtrate, but most proteins in the plasma are still too large to pass through the glomerulus. Therefore, a substance that is, for example, 75 percent bound to plasma proteins has an effective filterable concentration of 25 percent its total plasma concentration. Small amounts of protein, principally the albumins, which are important chemical-binding proteins, may appear in the glomerular filtrate, but these are then normally reabsorbed. The glomerular filter can be made more permeable in certain disease states and by actions of certain nephrotoxicants. Both circumstances may result in the appearance of protein in the urine (proteinuria). If damage to the glomerular element is severe, the result is a loss of a large amount of the plasma proteins. If this occurs at a rate greater than the rate at which the liver can synthesize the plasma proteins, the result will be hypoproteinemia (lower than normal levels of proteins in the blood) and a concomitant edema due to the reduction in osmotic pressure. This clinical picture is sometimes referred to as the *nephrotic syndrome*. However, transient but significant proteinuria occurs normally after prolonged standing or strenuous exercise, so a single measurement of high protein levels in the urine may not indicate kidney damage.

Nephron Tubules and Tubular Reabsorption The tubular element of the nephron selectively reabsorbs 98–99 percent of the salts and water of the initial glomerular filtrate. The tubular element of the

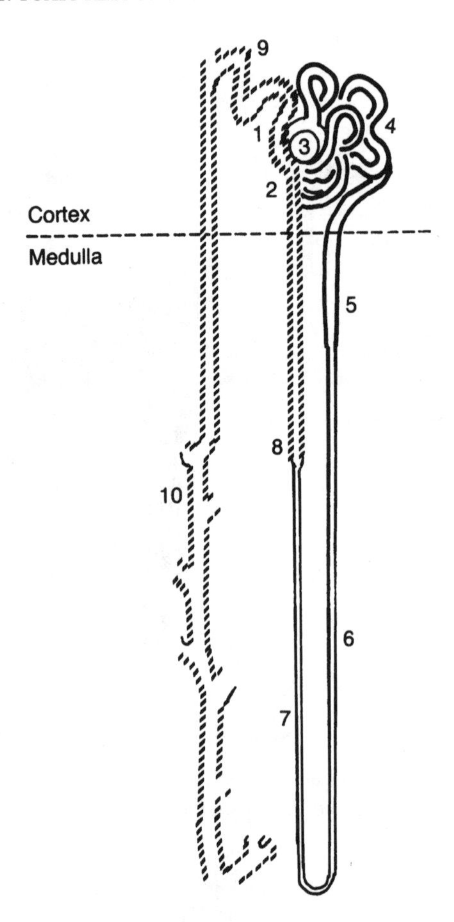

Figure 6.3 Juxtamedullary nephron: (1) afferent arteriole; (2) efferent arteriole; (3) glomerulus; (4) proximal convoluted tubule; (5) proximal straight tubule (pars recta); (6) descending limb of the loop of Henle; (7) thin ascending limb of the loop of Henle; (8) thick ascending limb of the loop of Henle; (9) distal convoluted tubule; (10) collecting duct. (Based on J. Doull, et al., eds., *Casarett and Doull's Toxicology: The Basic Science of Poisons,* 2nd ed., Macmillan, New York, 1980.)

nephron consists of the proximal tubule, the loop of Henle, the distal tubule, and the collecting duct (see Figure 6.3). The proximal tubule consists of a proximal convoluted section (pars convoluta) and a distal straight section (pars recta). Substances that are actively reabsorbed in the proximal tubule include glucose, sodium, potassium, phosphate, amino acids, sulfate, and uric acid. Essentially all amino acids and glucose are reabsorbed in the proximal tubule, and virtually none normally appear in the urine. Agents toxic to the proximal tubule cause amino acids and glucose to appear in the urine (aminoaciduria and glycosuria). Even though 250 g of glucose normally passes through the kidney daily, no more than 100 mg is usually excreted in 24 h. However, glucose does appear in excess quantities in the urine if high blood glucose levels produce a glucose load in the filtrate and this exceeds the resorptive capacity of the proximal tubule of the nephrons. This occurs in diabetes mellitus, in which excess glucose appears in urine because excessive amounts of glucose in the blood plasma filtrate have overwhelmed the glucose transport system in the nephron. Water is also reabsorbed in the proximal tubule because of an osmotic gradient between the filtrate in the tubule and the blood plasma. Thus, isotonicity is maintained in the proximal tubule even though there is a selective reabsorption of solutes. Approximately 75 percent of the glomerular filtrate fluid is reabsorbed in the proximal tubule.

If tubular reabsorption of substances is compromised, then less water is reabsorbed. The result is diuresis (increased urine flow) and polyuria (excess urine production). Toxic agents can cause polyuria by affecting active solute reabsorption.

Tubular Secretion Active transport of certain organic compounds into the tubular fluid also occurs in the proximal tubule. There are two separate active secretory systems in the proximal tubule: one for anionic (negatively charged) organic chemical species, and a similar but separate system for cationic (positively charged) organic chemical species. The organic anion secretory system is the better studied. Organic cations such as tetramethyl ammonium are actively secreted, but this system is not as well studied as the organic anion secretory system. The two secretory systems also have unique competitors and inhibitors. Penicillin and probenecid are actively secreted by the organic anion secretory system. As a consequence, they inhibit the excretion of PAH (*p*-amminohippuric acid) and each other. In fact, probenicid has been used to prolong the half-life of penicillin in the blood since probenicid inhibits secretion of penicillin into the proximal tubules and its subsequent excretion in the urine. These organic anions do not inhibit secretion of organic cations or compete with them for secretion. The reverse is also true. The result is that substances reabsorbed from the tubule will have a clearance significantly less than the glomerular filtration rate (approximately 125 mL/min), while those secreted into the tubules will have a clearance greater than the glomerular filtration rate in the adult human.

The Loop of Henle After the glomerular filtrate has passed the proximal tubule in the nephron, it moves into the loop of Henle. A nephron with a glomerulus in the outer portion of the renal cortex has a short loop of Henle, whereas a nephron with a glomerulus close to the border between the cortex and medulla (juxtamedullary nephrons) has a long loop of Henle extending into the medulla and papilla (Figures 6.2 and 6.3). Approximately 15 percent of the nephrons in humans are juxtamedullary. As the tubule descends into the medulla there is an increase in osmolality of the interstitial fluid. In the descending limb the tubular fluid becomes hypertonic (high in salt) as water leaves the tubule to maintain isoosmolality with the hypertonic interstitial fluid. However, in the thick segment of the ascending portion of the loop of Henle the tubule becomes impermeable to water, and sodium is actively transported out of the tubule with a decrease in the osmolality of the filtrate and an increase in the osmolality of the interstitial fluid. The sodium transport in the ascending limb is necessary for maintenance of the interstitial fluid concentration gradient. An additional 5 percent of the glomerular filtrate fluid is reabsorbed in the loop of Henle, making a total of 80 percent of the total water reabsorbed at this point.

Urine Formation Once the tubular fluid enters the distal convoluted tubule and collecting duct, it is hypotonic (low salt concentration) in comparison to blood plasma because of the active transport of sodium out of the tubule at the loop of Henle. In the presence of vasopressin, the antidiuretic hormone, the collecting duct becomes permeable to water, and the water moves from the tubular fluid in order to maintain isoosmolality. However, in the absence of vasopressin, the collecting duct is impermeable to water, which results in excretion of a large volume of hypotonic urine. Normally, another 19 percent of the original glomerular filtrate fluid is reabsorbed in the last portion of the nephron, so that a total of 99 percent of the fluid filtered at the glomerulus is reabsorbed—only 1 percent of the fluid entering the nephron is excreted in the urine. Thus, the normal flow of urine is only about 1 mL/min, while in the absence of vasopressin it can be increased to 16 mL/min. The kidney's ability to concentrate urine is determined by the measurement of urine osmolality. Urine osmolality can vary between 50 and 1400 mOsm/L. Certain nephrotoxicants compromise the kidney's ability to concentrate the urine. These changes occur early after the exposure to the nephrotoxicant and frequently foreshadow graver consequences.

The excretion of urea, a metabolic breakdown product of protein, is a special case. Urea passively diffuses out of the glomerular filtrate of the tubules as fluid volume decreases. At low urine flow, more urea has the opportunity to leave the tubule. Under these conditions only 10–20 percent of the urea is excreted. At conditions where the urine flow is high, the urea has less time to diffuse through membranes with the water; this results in a 50–70 percent excretion of urea. A second factor in urea excretion is that it accumulates in the medullary interstitial fluid along a concentration gradient. Since the walls of the collecting ducts are permeable to urea fluid where they pass through the medulla, the urea content of the urine is higher than it would be if they passed only through regions with low urea concentration.

Passive reabsorption occurs for all nonionic compounds, while ionic chemicals are not passively reabsorbed. For organic acids, a basic urine is desirable to maximize excretion since more of the acid will be ionized at higher pH (Haldane equation, Chapter 4). For organic bases, an acidic urine is desirable for maximal excretion, because more of the basic compound will be ionized.

Bladder

The urine that flows from the collecting ducts is deposited in the bladder. Little of the literature is devoted to the bladder and its functioning. However, some compounds are toxic to the bladder. Bladder cancer is thought to be caused by occupational exposure to bicyclic aromatic amines. The bladder epithelium contains high levels of an enzyme, prostaglandin H synthase (PHS), which can activate certain aromatic amines, such as benidine, 4-aminobiphenyl, and 2-aminonaphthalene, to compounds that can react with DNA. The normal metabolism of these compounds involves acetylation, and there are several genetic polymorphisms of the enzymes (*N*-acetyltransferases) responsible for acetylating them. Individuals with slow acetylating enzymes are more likely to develop bladder cancer after exposure.

Important Kidney Functions Seldom Considered as Toxic Endpoints

Renal Erythropoietic Factor The kidney synthesizes hormones essential for certain metabolic functions. For example, hypoxia stimulates the kidneys to secrete renal erythropoietic factor, which acts on a blood globulin (proerythropoietin) released from the liver to form erythropoietin, a circulating glycoprotein with a molecular weight of 60,000 daltons. The erythropoietin acts on erythropoietin-sensitive stem cells in the bone marrow, stimulating them to increase hemoglobin synthesis, produce more red blood cells, and release them into the circulating blood. The increased oxygen-carrying capacity of the blood reduces the effects of hypoxia. Thus, in chronic renal failure, anemia usually develops, in large part caused by decreased synthesis of erythropoietic factor because of damage to the kidney tissues responsible for its synthesis. In addition to hypoxia, androgens and cobalt salts also increase production of renal erythropoietic factor by the kidneys. In fact, administration of cobalt salts produces an overabundance of red cells in the blood (i.e., polycythemia) by this mechanism. Polycythemia has been observed in heavy drinkers of cobalt-contaminated beer.

Regulation of Blood Pressure The kidney is involved in regulating blood pressure in several ways. The kidney produces renin, a proteolytic enzyme, which cleaves a plasma protein globulin to form angiotensin I. Angiotensin I is converted to angiotensin II, a potent vasoconstrictor. The angiotensin II stimulates release of aldosterone from the adrenal cortex, and aldosterone increases reabsorption of sodium in the kidney, leading to an increase in blood plasma osmolality and an increase in extracellular volume. A decrease in the mean renal arterial pressure is the stimulus controlling kidney renin production and the compensatory increase in arterial pressure by the abovementioned mechanisms. In addition, renal disease and narrowing of the renal arteries are known to cause sustained hypertension in humans. It appears that the kidney produces vasodepressor substances that are thought to be important in the regulation of blood pressure. Thus, changes in the kidney that disturb the renin–angiotensin–aldosterone system and/or secretion of the vasodepressor substances are suspected of playing a key role in the etiology of certain forms of hypertension.

Metabolism of Vitamin D The kidney also plays a key role in the metabolism of vitamin D, thus performing a vital function in the hormonal regulation of calcium in the body. Vitamin D_3 (cholecalciferol) is relatively inactive. The liver hydroxylates vitamin D_3 to 25-hydroxycalciferol, and then, the kidney hydroxylates the 25-hydroxycalciferol to 1,25-dihydroxycalciferol, the most potent active form of vitamin D. The kidney is also the key to the metabolism of parathyroid hormone, another hormone important to calcium regulation. If the kidney is damaged, thereby disrupting its role in vitamin D and parathyroid hormone metabolism, the development of a renal osteodystrophy can occur, which is characterized by skeletal disease and hyperplasia of the parathyroid gland.

6.2 FUNCTIONAL MEASUREMENTS TO EVALUATE KIDNEY INJURY

From the preceding paragraphs it should be clear that the kidney plays an essential role in maintaining a number of vital body functions. Therefore, if a disruption of normal kidney function is caused by the action of a toxic agent, a number of serious sequelae can occur besides a disruption in blood waste elimination. However, for clinical purposes, alterations in the excretion of wastes are the principal endpoints for determining the action of nephrotoxicants. Nevertheless, it must be remembered that changes in the other functions may also be present, even if they are not conveniently or routinely measured as toxic endpoints.

Determining the excretion rate of certain drugs from the kidney is a useful clinical procedure for diagnosing the functional status of the kidney. This rate of elimination in the urine is the net result of three renal processes:

- Glomerular filtration
- Tubular reabsorption
- Tubular secretion

The rates of glomerular filtration and tubular secretion are dependent on the concentration of the drug in the plasma, and the rate of reabsorption by the tubules is dependent on the concentration of drug in the urine.

The Glomerular Filtration Rate

The glomerular filtration rate (GFR) can be measured in intact animals and humans by measuring both the excretion and plasma levels of those chemicals that are freely filtered through the glomeruli and neither secreted nor reabsorbed by the kidney tubules. The substance used should ideally be one that is freely filtered, not metabolized, not stored in the kidney, and not protein bound. Inulin, a polymer of fructose with a molecular weight of 5200 daltons, meets these criteria. For measuring the glomerular filtration rate the inulin is allowed to equilibrate within the body, and then accurately timed urine specimens and plasma samples are collected.

The following general formula is used to determine the clearance in this procedure:

$$\frac{U_a \times V}{P_a} = \text{Cl}$$

where
U_a = concentration of substance a per milliliter urine
V = urine volume excreted per unit time
P_a = concentration of substance a per milliliter of plasma
Cl = clearance of substance per unit of time

For clearance of inulin (in), the following values can be used to demonstrate a sample calculation:

$$U_{\text{in}} = 31 \text{ mg/mL}$$

$$V = 1.2 \text{ mL/min}$$

$$P_{in} = 0.30 \text{ mg/mL}$$

Thus,

$$\frac{(31 \text{ mg/ml}) \times (1.2 \text{ ml/min})}{0.30 \text{ mg/ml}} = 124 \text{ ml/min.}$$

The normal human glomerular filtration rate in adult humans is about 125 mL/min and inulin clearance is routinely used as a measure of glomerular function. The GFR is not only a measure of the functional capacity of the glomeruli but also indicates the kidney's ability to concentrate urine by removal of water. By comparing the amount (milliliters) of urine voided in one minute to the amount (milliliters) of plasma cleared, information can be gained about the amount of water reabsorbed during passage through the tubules.

Diseases or nephrotoxicants that affect the glomerulus or those that produce renal vascular disease have a profound effect on the glomerular filtration rate. Indeed, any significant renal disease or nephrotoxic compromise can reduce the glomerular filtration rate. It should also be realized that any agent inducing severe hypotension or shock will likewise reduce the glomerular filtration rate.

Measurement of certain natural endogenous substances in the blood can be used to assess glomerular function as well. The measurement of blood urea nitrogen (BUN) and plasma creatinine are two endogenous compounds routinely measured for the clinical assessment of glomerular function. As glomerular filtration decreases, BUN and plasma creatinine become more elevated. Normal BUN ranges from 5 to 25 mg/100 mL, while serum creatinine ranges from 0.5 to 0.95 mg/mL of serum.

Nephrotoxicants may also disrupt the selective permeability of the glomerular apparatus. Normally, the result is an increase in porosity in the glomerulus; protein enters the glomerular filtrate and subsequently the urine. Therefore, if a compound causes excretion of large amounts of protein into the urine it must be suspected as a nephrotoxicant, and measurement of protein in urine, particularly those of high molecular weight, is used to determine which chemicals produce toxic changes to the glomerulus. The normal excretion of protein in humans is no more than 150 mg in 24 h.

Renal Plasma Flow

Some organic acids, such as *p*-aminohippuric acid (PAH), can be used in clearance studies to obtain information about the total amount of plasma flowing through the kidneys. PAH is transported so effectively that it is almost completely removed from the plasma in a single passage through the kidney (i.e., 80–90 percent). Any chemically induced reduction in the PAH clearance may be caused by either a disruption of the active secretory process or by an alteration of the renal blood flow.

In a clinical setting, measurements can be made of the concentration of PAH per milliliter of plasma (P_{PAH}), of the concentration of PAH per milliliter of urine (U_{PAH}), and of the volume of urine excreted per minute (V). Using the formula that was previously discussed, the clearance of PAH in mL/min can be calculated. This calculation represents the rate of plasma flow through the kidneys (average renal plasma flow in the normal, healthy adult male is about 650 mL/min).

Excretion Ratio Another useful calculation for evaluating kidney injury is the excretion ratio:

$$\text{Excretion ratio} = \frac{\text{Renal plasma clearance of drugs (ml/min)}}{\text{Normal GFR (ml/min)}}$$

If the ratio is less than 1.0, it indicates that a drug has been partially filtered, perhaps also secreted, and then partially reabsorbed. A value greater than 1.0 indicates that secretion, in addition to filtration, is involved in the excretion. A substance that is completely reabsorbed, such as glucose, would have an excretion ratio of 0, and a substance such as PAH that is completely cleared can have a ratio of about 5.

Additional Clinical Test Alterations in renal function can be determined by a variety of other tests. A battery of such tests includes urinary pH, measurement of urine volume, and a determination of the excretion of sodium and potassium. An excess of protein or the appearance of sugar in the urine indicates abnormalities in renal function as would changes in urine sediments. These are all general tests, but they can provide information about the changes in total kidney function.

6.3 ADVERSE EFFECTS OF CHEMICALS ON THE KIDNEY

Frequently, exposure to large amounts of a chemical can cause kidney effects that are not observed at lesser exposures. Effects of kidney damage are frequently assessed in nonspecific terms such as changes in kidney weight (both increases and decreases) or increases in protein content of the urine (proteinuria) or changes in volume of urine (polyuria, oliguria, or anuria).

Acute renal failure (ARF) is one of the more common responses of the kidney to toxicants. ARF is characterized by a rapid decline in glomerular filtration rate and an increase in the concentration of nitrogenous compounds in the blood. Numerous mechanisms have been identified that lead to ARF. Compounds that cause renal vasoconstriction reduce the amount of blood that reaches the glomerulus and cause hypoperfusion, a reduction in the amount of blood filtered. When toxicants cause glomerular injury, they can reduce the amount of filtrate that enters the tubules, called *hypofiltration.*

When the tubular cells are injured by toxicants, the permeability of the tubule is increased and the filtrate is allowed to backleak into the interstitium and into the circulation, producing an apparent reduction of the GFR. Some toxicants may reduce the adhesion of tubular cells to each other, causing them to obstruct the pathway for filtrate to be reabsorbed and thus increasing the pressure within the tubule leading to a resistance of movement of filtrate into the tubule.

The kidney is capable of overcoming substantial loss of function. If a single kidney is lost, the remaining kidney can increase its GFR by 40–60 percent. Individual nephrons can increase the reabsorption of water and solutes so that the osmotic balance is maintained and there is no apparent difference in tests of kidney function. Although the compensatory mechanisms protect the whole organism in the short term, the compensatory responses may lead to chronic renal failure in the long term. The increase in glomerular pressure leads to sclerosis of the glomerulus and the degeneration of the capillary loops, among other changes in the nephron whose roles in compensatory nephron damage are not as well documented. The loss of additional nephrons and the capacity to remove wastes by this mechanism leads to additional compensation by other nephrons, which are subsequently damaged by similar mechanisms, eventually leading to chronic renal failure.

Other means of protecting the kidney from damage include the induction of metallothionein and heat-shock proteins. Heat-shock proteins play a housekeeping role to maintain normal protein structure and/or degrade damaged proteins. Metallothionein is a low-molecular-weight protein that binds heavy metals and prevents them from inducing toxic responses. The production of metallothionein is induced by the presence of heavy metals, and, when low doses of the heavy metal are given, the metallothionein is produced and can provide protection against larger doses given at a later time. If no exposure has occurred previously, no protection is provided because metallothionein is not present to bind the heavy metal.

In addition to the organ-level response of the kidney, many toxicants affect specific regions of the nephron. They may damage the glomerulus, the proximal tubule, or the further tubule elements such as the loop of Henle, distal tubule, or collecting duct. The most common site of injury for toxicants is the proximal tubule.

Nephrotoxic Agents

Many compounds are known to adversely affect kidney tissues at some exposure level, but the kidney is the tissue affected at the least lowest observed adverse effect levels for only a few compounds. The chemicals for which the American Conference of Governmental Industrial Hygienists (ACGIH) has established Threshold Limit Values (trademark) (TLVs) that are intended to protect against affects on the kidney are given in Table 6.1. For these compounds, however, the renal system may not be the only system the TLV is intended to protect.

Two classes of environmentally or occupationally relevant chemicals that damage the kidney are the heavy metals and halogenated hydrocarbons. The adverse effects of representative chemicals from each group are discussed below. Some occupations that have exposure to nephrotoxicants are given in Table 6.2.

Cadmium The kidney is the organ most sensitive to the toxic effects of cadmium. Numerous factors have been used as indicators of kidney damage by cadmium. One of the early indicators is the presence of 2-microglobulin, a low-molecular-weight protein that is usually reabsorbed by the proximal tubules. Proximal tubule damage of the nephrons caused by cadmium is also evidenced by glycosuria, aminoaciduria, and the diminished ability of the kidney to secrete PAH. As damage increases, there is an increase in urinary excretion of low- and high-molecular-weight proteins, which predicts an acceleration of the decline in glomerular filtration rate. Workers in factories where nickel/cadmium batteries are manufactured and who are exposed to excessive amounts of cadmium oxide exhibit

TABLE 6-1. Chemicals with ACGIH TLVs™ Specifically Set to Prevent Renal Effects

Arsine	Methyl tert-butyl ether
Cadmium	Methyl Chloride
Chloroform	Methyl Chloroform
1-chloro-1-nitropropane	Methylcyclohexanol
o-Chlorostyrene	2-methyl cyclopentadienyl manganese tricarbanol
Hexavalent Chromium Compounds (water soluble)	4,4′-methylene bis(2-chloroaniline)
Chromyl chloride	Methyl ethyl ketone peroxide
†Cyclohexane	Methyl isoamyl ketone
p-dichlorobenzene	Methyl isobutyl ketone
1,1-dichloroethane	Nickel, elemental
Diethanolamine	Nitrogen trifluoride
1,4-dioxane	Oxygen difluoride
Diphenylamine	Paraquat
Dipropylketone	Phenothiazine
Epichlorohydrin	Phosphorous (yellow)
Ethyl bromide	Picloram
Ethylene chlorohydrin	Pindone
Ethylene dibromide	Propargyl alcohol
Ethylene oxide	Propylene dichloride
Ethyl silicate	Pyridine
Hexachlorobutadiene	Stoddard Solvents
Hexachloroethane	4,4′-thiobis(6-tert-butyl-m-cresol)
Hexafluoroacetone	o-Tolidine
Indene	o-,m-, and p-toluidene
Iodoform	1,2,3-trichloropropane
Lead	Uranium
Lead Arsenate	Vinylidene chloride
Mercury, aryl, inorganic, elemental	Xylidene (mixed isomers)
Mesityl oxide	

†1998 Notice of Intended Changes includes kidney effects which were not listed previously

TABLE 6.2 Industrial Operation with Exposure to Nephrotoxicants[a]

Industrial operation	Nephrotoxicant
Amalgam manufacturers	Mercury
Chemists	Chloroform
Chloralkali	Mercury
Dry cleaning	Perchloroethylene
Manufacturing batteries	Mercury, Lead, Cadmium
Manufacturing cellulose acetate	Dioxane
Metal degreasing	Perchloroethylene
Paint manufacturers	Lead, Cadmium
Plumbers	Lead

[a]List in alphabetical order.

consistent proteinuria, and cadmium-induced kidney damage may appear years after workers are removed from exposure.

In Japan excessive cadmium intake was also linked to a peculiar form of renal osteodystrophy known as "ouch-ouch disease" or "itai-itai byo." It has been proposed that this disease is caused by excessive loss of cadmium and phosphorus in the urine, combined with dietary calcium deficiency.

The kidney naturally accumulates cadmium. Normally cadmium accumulates in the kidney over the lifetime of the individual until the age of 50. About 50 percent of the total burden of cadmium in the body is borne by the liver and kidney, with the kidney having 10 times the concentration of the liver. Cadmium induces synthesis in the liver of metallothionein, a protein with a high binding affinity for cadmium. While metallothionein acts to protect certain organs, such as the testes, from cadmium toxicity, it may play a role in cadmium toxicity in the kidney. After the available metallothionein in proximal tubule cells is overcome by high cadmium concentrations, the free cadmium exerts toxic effects on the cells in the proximal tubule.

Chronic cadmium exposure has also been implicated as a factor in hypertension. However, while the development of hypertension may involve the kidney, the role of cadmium in the etiology of hypertension in humans is far from conclusive.

Mercury Inorganic mercury (Hg^{2+}) is a classical nephrotoxicant. It is used as a model compound for producing kidney failure in animals, and massive doses of mercuric ion can damage the proximal tubule and cause acute renal failure. A brief polyuria is followed by oliguria or even anuria. The anuria (kidney failure) leads, of course, to a life-threatening accumulation of bodily wastes and may last many days. If recovery occurs, a polyuria follows, which is probably caused by a decreased sodium absorption in the proximal tubule. Such disturbances in tubular function may last several months.

Acute exposure to high concentrations of mercury is rare; usually mercury exposure occurs at lower dose rates. The part of the nephron most sensitive to mercuric ion toxicity is the pars recta or straight portion of the proximal tubule (Figure 6.3). Early damage is characterized by the presence of enzymes in the urine that are normally found in the brush border portion of the cells lining the tubule. Further damage results in the presence of intracellular enzymes from these cells in the urine. Longer-term exposure and damage can lead to the presence of glucose, amino acids, and proteins in the urine. Also associated with long-term exposure to mercury is a reduction in the GFR caused by vasoconstriction, tubular damage, and damage to the glomerulus.

Chloralkali workers exposed to mercury have increased glomerular dysfunction and elevated excretion of high-molecular-weight proteins. 2-Microglobulin has been found at elevated levels in the blood plasma of these workers, but levels in urine were not increased.

Lead Lead is a known nephrotoxicant in humans. Lead causes damage principally to the proximal tubule of the nephron. Reabsorption of glucose, phosphate, and amino acids is depressed in the proximal tubule. This leads to glycosuria, aminoaciduria, and a hyperphosphaturia with hypophosphatemia. These changes are reversible on treatment with a chelating agent such as ethylene–diamine tetraacetic acid (EDTA), but only when the lead exposure has been relatively short. Long-term, prolonged exposure to lead may cause irreversible dysfunction and morphologic changes. This is manifested by intense interstitial fibrosis accompanied by tubular atrophy and dilation. The glomeruli are involved in later stages of the disease. Eventually, long-term lead exposure syndrome results in renal failure and death. There has been linkage of the chronic renal damage to saturnine (lead-induced) gout, in which uric acid is increased in the kidney.

Other Toxic Metals

Table 6.3 lists those metals known to be toxic to the kidneys. As with most nephrotoxicities, the proximal tubule appears to be the most sensitive to toxic effects, with more extensive nephron involvement at higher dosages. In all animal and human exposures to uranium that are acutely injurious, the kidney is the main target of concern. Necrosis in the pars recta of proximal tubules and ascending limb of the loop of Henle and collecting tubules occurs, with accompanying loss of function. In the urine, there are increased casts, protein, glucose, catalase and other enzymes, amino acids, and 2-microglobulin; assays of urinary amino acid and 2-microglobulin appear to be the most sensitive in this regard. With recovery, areas of renal necrosis are at least partially replaced; however, the new cells may not be functionally equivalent to the original cells.

Halogenated Hydrocarbons

Carbon tetrachloride (CCl_4) and chloroform ($CHCl_3$) are nephrotoxicants. Again, the proximal tubule appears to be the portion of the nephron most sensitive to damage by these agents. However, lesions are seen in other parts of the nephrons as well. It should be noted that carbon tetrachloride causes severe blood hepatic necrosis in humans, but the ultimate cause of death is kidney failure. The mechanism of the injury to the kidney is not known. However, it has been reported that backdiffusion of glomerular filtrate was important in the early stages of oliguria, and decreased renal blood flow contributed in the later stages of oliguria following carbon tetrachloride inhalation in humans. It appears that chloroform and carbon tetrachloride are activated to a toxic chemical species in the kidney by a mixed-function oxidase system similar to that found in the liver. The toxic metabolite covalently binds to tissue macromolecules in the kidney, and this leads to nephrotoxicity.

The exposure levels leading to renal damage in humans have not been well defined. An increased incidence of proteinuria was reported in workers exposed to vapor concentrations of around 200 ppm, while the urine protein content was not changed after inhalation exposure to 50 ppm for 70 min or 10 ppm for 3 h.

Bromobenzene, tetrachloroethylene, and 1,1,2-trichloroethylene also produce toxic effects to the kidney similar to those of chloroform and carbon tetrachloride.

TABLE 6.3 Metal Nephrotoxic Agents

Metals of Principal Concern	Other Metals Having Nephrotoxicity
Cadmium	Arsenic
Lead	Bismuth
Mercury	Chromium
	Platinum
	Thallium
	Uranium

Methoxyflurane (1,1-difluoro-2,2-dichloromethyl ether) is a halogenated surgical anesthetic that causes renal failure in humans and animals. Its causes a polyuria and an increase in serum osmolality, serum sodium, and blood urea nitrogen. Methoxyflurane is metabolized to inorganic fluoride anion and oxalate. The fluoride anion has been shown to be responsible for acting on the collecting tubules, which results in vasopressin resistance and causes polyuria.

Bromomethane Humans exposed to high levels of bromomethane vapor commonly suffer from renal congestion, anuria or oliguria, and proteinuria; however, renal effects after exposure are frequently minimal or absent. Animal studies report similar signs of renal injury such as swelling, edema, nephrosis, and tubular necrosis.

Hexachloroethane Hexachlorethane has been found to cause tubular atrophy, degeneration, hypertrophy, and/or dilation in rats.

Other Nephrotoxic Compounds

Methyl Isobutyl Ketone In rats exposed to as low as 100 ppm of MIBK, microscopic examination showed toxic nephrosis of the proximal tubules. The exposed rats also showed hyaline droplet degeneration of the proximal renal tubules and occasionally tubular necrosis. The tubular damage was considered transient and reversible.

Dioxane Dioxane exposure can be significant from both an oral and inhalation exposure route. It has a selective action on the convoluted tubules of the kidneys, and causes renal obstruction. In several cases of fatal industrial exposure, injuries to the kidney were identified as causing the deaths. Signs of severe hemorrhagic nephritis and central hepatic necrosis occurred after about 2 months of what were considered to be heavy exposures to dioxane vapor. No cases of jaundice were observed, and death occurred within 1 week after onset of illness from acute renal failure.

Phenol In subchronic toxicity tests with guinea pigs phenol causes renal proximal tubule swelling and edema and glomerular degeneration.

Agents Causing Obstructive Uropathies

A number of agents cause nephrotoxicity through physical deposition in the tubular sections of the nephron. Certain chemical agents can be concentrated in the tubular fluid to levels well above their solubility limit in water. The result is that crystals are deposited in the kidney tubules, causing physical damage. Methotrexate and sulfonamide drugs can cause nephrotoxicity by this mechanism.

Acute renal failure of this type is also associated with the ingestion of ethylene glycol. Ethylene glycol is metabolized to oxalic acid by the body; the acid, in turn, is deposited in the lumen of the tubule of the nephron as well as within the cell of the tubule as insoluble calcium oxalate salt. However, ethylene glycol additionally appears to cause a nephrotoxicity to the proximal tubule, which is independent of oxalate deposition. The deposition of large quantities of oxalate crystals in the tubular elements of the nephrons probably contributes to the nephrotoxicity observed. Oxalate found in the leaves of rhubarb is of a sufficient quantity that it can cause deposition of oxalate crystals in the tubular elements of the nephron, and can lead to nephrotoxicity. Part of the nephrotoxicity caused by methoxyflurane is likewise believed to be caused by deposition of calcium oxalate crystals in the tubular elements of the nephrons.

Agent Producing Pigment-Induced Nephropathies

A number of chemicals can cause the release of certain pigments such as methemoglobin, hemoglobin, and myoglobin into the blood. When this occurs, an associated acute renal failure may develop. Arsine

TABLE 6.4 Therapeutic Agents Known to Cause Nephrotoxicity

Acetaminophen (analgesic)
Aminoglycoside antibiotics
Amphotericin B (antibiotic)
Cephalosporadine (antibiotic)
Colistimethate (antibiotic)
Gentamycin
Kanamycin
Neomycin
Polymyxin B (antibiotic)
Streptomycin
Tetracyclines (particularly outdated formulations) (antibiotics)

gas causes massive hemolysis of red blood cells, which results in hemoglobinuria and associated renal failure (see Chapter 4 for a general listing of hemolytic agents, methemoglobin formers, etc.).

Heroin overdosage can result in a prolonged pressure on dependent muscles and a lysis of the muscle cell, leading to a release of myoglobin into the blood. Heroin may also cause some direct lysis of the muscle cells. The result can be myoglobinuria and ultimately acute renal failure. Aniline dyes are another group of chemicals that have been shown to release methemoglobin, with an associated renal failure.

Therapeutic Agents

Table 6.4 lists a number of therapeutic agents known to cause nephrotoxicity. Acetaminophen is oxidized by the microsomal P450 oxygenase system in the renal cortex to a toxic metabolite. The microsomal P450 oxygenase system of the kidney is similar to that of the liver (see Chapter 5).

Cephalosporadine reaches high toxic concentrations in the nephron because the organic ion transport system of the proximal tubule secretes it into the tubule. The nephrotoxicity of cephalosporadine can be diminished by compounds that compete with the organic anion secretion system in the proximal tubule, such as probenicid. The resulting decrease in tubular concentration of cephalosporidine in tubular fluid results in elimination of toxicity.

Other therapeutic agents can, in certain individuals, elicit a nephrotoxicity by an allergic type of reaction. However, such nephrotoxicities are usually only rarely encountered.

6.4 SUMMARY

The kidney performs a number of functions essential for the maintenance of life:

- Elimination of waste products (particularly nitrogen-containing wastes from the metabolism of proteins) from the blood
- Regulation of acid-base balance, extracellular volume, and electrolyte balance

Toxic agents that disrupt these key functions can be life-threatening.

The kidney is a highly metabolic organ sensitive to deprivation of oxygen, and any agent that significantly impedes renal flow will cause two adverse sequelae; acute renal failure that can result in death:

- First, less blood plasma will reach the kidney, resulting in a decrease in removal of blood wastes with a resulting increase of wastes in the blood (i.e., uremia).

- Second, if blood flow is compromised long enough, tissue ischemia will result in irreversible organ damage.

Nephrotoxicants, agents toxic to the nephron, the principal excretory unit of the kidney, also disrupt key life-preserving functions.

- The glomerulus normally filters out the high-molecular-weight proteins from the blood. However, toxic agents will increase its permeability, allowing these proteins to appear in urine.
- Agents that damage the tubular element of the nephron will compromise its ability to reabsorb solutes such as glucose and amino acids, which are necessary for normal maintenance of the body, or disrupt sodium transport out of the nephron tubule, which could result in diuresis or excess urine formation or an unbalancing of the body's ionic (salt) homeostasis.
- If damage to the nephron is excessive, renal failure can decrease or completely stop urine flow, and cause death by poisoning from the body's own products.

Many agents directly toxic to the nephron are commercially or industrially important.

- Mercury, lead, and cadmium are industrially the most important nephrotoxic metals.
- Halogenated hydrocarbons, particularly carbon tetrachloride and chloroform, are nephrotoxic.
- Certain therapeutic agents, such as phenacetin, aspirin, and the aminoglycoside antibiotics are directly nephrotoxic.

Chemicals can cause nephrotoxicity indirectly:

- Some agents deposit crystals in the tubular element of the nephron, resulting in physical damage.
- Hemolytic agents such as arsine gas are capable of pigment neuropathy by releasing hemoglobin into the blood.

REFERENCES AND SUGGESTED READING

American Conference of Governmental Industrial Hygienists, *TLVs and Other Occupational Exposure Values—1998,* ACGIH, Cincinnati, OH, 1997.

Agency for Toxic Substances and Disease Registry, *Toxicological Profiles on CD-ROM,* CRC Press, Boca Raton, FL, 1997.

Berndt, W. O., "Renal Methods of Toxicology," in *Principles and Methods of Toxicology,* A. W. Hayes, eds., Raven Press, New York, 1982, pp. 447–474.

Brenner, B., and F. Rector, *The Kidney,* Saunders, Philadelphia, 1976. Ganong, F., *Review of Medical Physiology,* Lange Medical Publications, Los Altos, CA, 1973, pp. 510–532.

Goldstein, R., and R. Schnellman, "Toxic responses of the kidney," in *Casarett and Doull's Toxicology: The Basic Science of Poisons,* 5th ed., C. D. Klaassen, ed., McGraw-Hill, New York, 1996, pp. 417–442.

National Institute for Occupational Safety and Health, *NIOSH Criteria Documents on CD-ROM,* CDC-NIOSH, Cincinnati, 1996.

Pitts, R., *The Physiology of the Kidney and Body Fluids,* 2nd ed., Year Book Medical Publications, Chicago, 1968.

Porter, G. A., and W. A. Bennett, "Toxic nephropathies," in B. M. Brenner, and F. C. Rector, e ds., *The Kidney,* 2nd ed., Vol. II, Saunders, Philadelphia, 1981.

Ullrich, K., and D. Marsh, "Kidney, water, and electrolyte metabolism," *Ann. Rev. Physiol.,* **25**: 91 (1963).

Weiner, I., "Mechanisms of drug absorption and excretion: The renal excretion of drugs and related compounds," *Ann. Rev. Pharmacol.* **7**: 39 (1967).

7 Neurotoxicity: Toxic Responses of the Nervous System

STEVEN G. DONKIN and PHILLIP L. WILLIAMS

Neurotoxic chemicals are significant contributors to the human health effects that result from environmental and workplace chemical exposures. The National Institute for Occupational Safety and Health (NIOSH) reports that exposure to neurotoxic chemicals is one of the 10 leading causes of work-related disease and injury and that over 25 percent of the chemicals for which the American Conference of Governmental Industrial Hygienists (ACGIH) has established Threshold Limit Values (TLV) (trademarks) have demonstrated nervous system effects. Other sources have estimated that of the 400 or so commonly used chemicals (primarily solvents and various pesticides), 42 percent are neurotoxic. Thus neurotoxicity is an important consequence of human exposure to industrial chemicals.

Studying toxic effects in the nervous system presents many challenges not usually encountered when working with other systems. Foremost is the sheer complexity of the human nervous system. Indeed, it is this complexity that in large part distinguishes us from other organisms and accounts for the exceedingly diverse spectrum of human behavior.

As a result of its complexity, the nervous system displays a variety of responses to toxicant exposure. These may include changes in heart rate, breathing rate, sensory perception, coordination, mood, and many other physiological, behavioral, cognitive, and emotional effects. Quantitating these effects is sometimes difficult enough, but even when it is possible, the significance for human health may not be clear. For example, are feelings of euphoria or drowsiness toxic effects? Also, while a temporary decrease in reaction time may not in itself be life-threatening, in an industrial setting where the worker is surrounded by other hazards, a loss in the ability to react may result in disaster. Considerations such as these must become a part of the overall neurotoxicity assessment.

Not all industrial chemicals are neurotoxicants, but for those that are, neurotoxic effects are often extremely sensitive indicators of low-level exposure. This, of course, depends on developing appropriate methods for monitoring such subtle effects. Changes in behavior are commonly used as sensitive and easily measured neurotoxic end points, although they may present some difficulty in terms of their objective quantification and baseline variability among individuals. Some of the standard tests for neurotoxicity are described later in this chapter.

Besides being complex, the nervous system is ubiquitous, its network extending throughout the body. We can conveniently divide this network into the central nervous system, comprising mainly the brain and spinal cord, and the peripheral nervous system, comprising all other components, including sensory and motor nerves. This distinction is important for the purposes of neurotoxicity because, as described later, some neurotoxicants appear to target only the central or peripheral nervous systems, but not both. The brief overview of the nervous system in the next section will suggest several different ways in which neurotoxic chemicals may impair nervous system function:

- Outright neuronal destruction may result in permanent damage since neurons do not usually regenerate.

Principles of Toxicology: Environmental and Industrial Applications, Second Edition, Edited by Phillip L. Williams, Robert C. James, and Stephen M. Roberts.
ISBN 0-471-29321-0 © 2000 John Wiley & Sons, Inc.

- Chemicals may disrupt the electrical impulse along the axon, either by harming the myelin sheath or membrane integrity, or by impairing the synthesis or functioning of proteins essential to axonal transport.
- Chemicals may also inhibit the neurotransmitters by blocking their synthesis, release, or binding to receptors.
- General protein synthesis impairment may have an effect not only on neurotransmitter production, but also the production of important enzymes which break down neurotransmitters when they are no longer needed.

We will next consider the mechanisms of electrical and chemical signal transmission through the nervous system in more detail. The reader should keep in mind that proper nervous system function depends on all steps in signal transmission working properly, and a disruption in any step may result in what would be described as a neurotoxic effect.

7.1 MECHANISMS OF NEURONAL TRANSMISSION

In one sense, the nervous system is little more than an enormous network of interconnected nerve cells, or neurons, supported by various other auxilliary cell types. However, this description is deceptive in its simplicity. Neurons come in many shapes, sizes, and functions, but may be generically described as having dendrites, a cell body, and an axon. The dendrites receive chemical signals from an adjacent neuron. These signals then trigger electrical impulses along the axon and in turn stimulate the release of more chemical signals at the terminal boutons. In this way, a stimulus may travel the entire length of the human body. The electrical impulse is often maintained along the length of the neuron with the aid of the myelin sheath, which acts as an insulator surrounding the axon. Successive neurons meet at a gap called the *synapse,* and it is across this gap that the chemical signals, or neurotransmitters, diffuse from one neuron to the dendrites of the next. Alternatively, neurons may terminate at muscles or glands, releasing neurotransmitters to specialized receptors at these sites.

It has been found that these basic features of the nervous system are similar throughout a wide taxonomic range. Most multicellular organisms possess some form of nervous system which includes neurons, neurotransmitters, and electrical signal conduction. This similarity provides us with substantial confidence in using neurotoxicity test results in animals to predict neurotoxic effects in humans.

The Action Potential

Electrical signals are initiated and propagated along the axon by what is called an *action potential.* The source of this potential is a charge difference across the nerve membrane, created by the movement of sodium (Na^+), potassium (K^+), and chloride (Cl^-) ions. This charge difference is determined by the selective permeability of the membrane, as well as concentration and potential gradients, and active transport. When the membrane is at rest, the concentration of K^+ ions is greater inside the cell than outside, while the concentrations of Na^+ and Cl^- are greater outside the cell. The concentrations of K^+ and Cl^- ions counterbalance each other, and this balance is maintained against their concentration gradients by the resulting potential gradient. Thus, in this equilibrium state, the tendency of either ion to diffuse across the membrane and down its concentration gradient is controlled by the imbalance caused by the potential gradient. Meanwhile, the membrane is relatively impermeable to Na^+, and therefore a net positive charge exists on the outside of the cell relative to the inside (Figure 7.1*a*).

When the cell is stimulated and an action potential is created, the membrane becomes locally permeable to sodium, and an influx of positive charge occurs. The result is a depolarization of the membrane that is propagated down the axon as current flows ahead of the action potential, depolarizing the membrane further (Figure 7.1*b*). Behind the action potential, the membrane permeability again shifts to favor K^+ movement and to decrease Na^+ movement. The resulting repolarization from the K^+

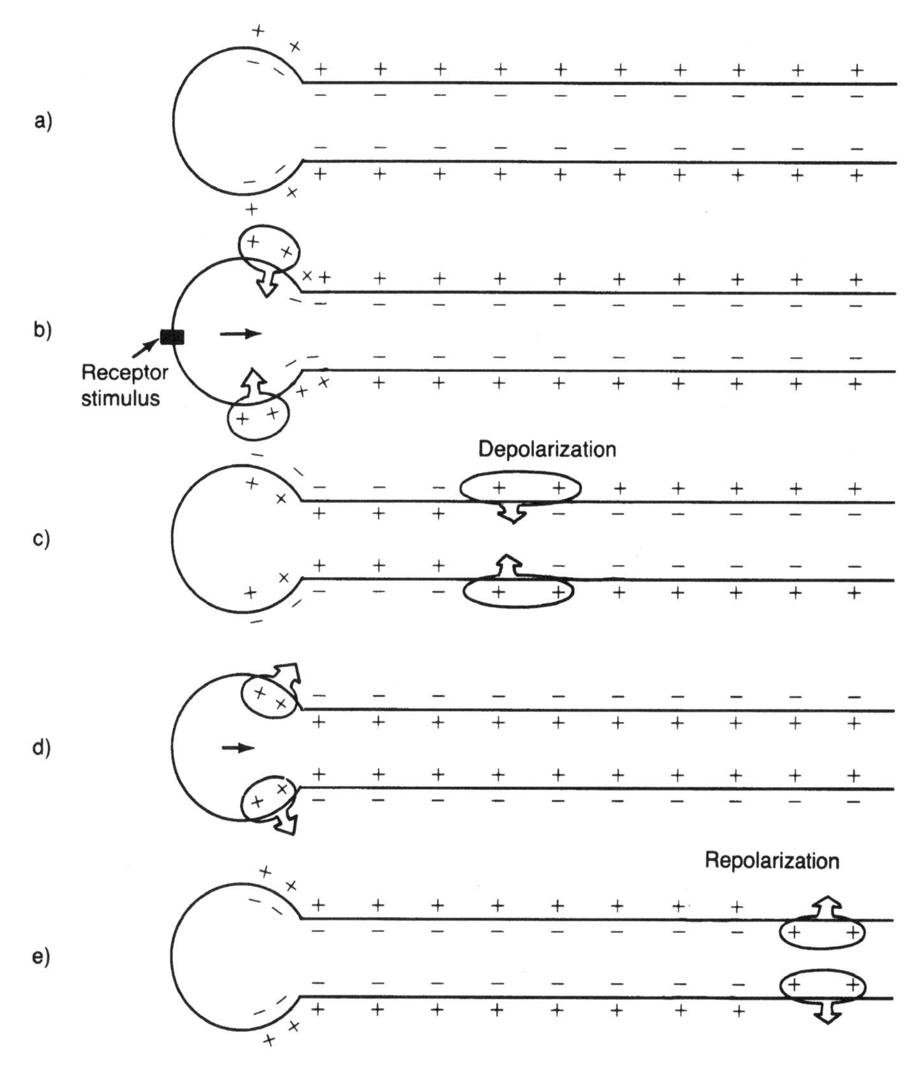

Figure 7.1 Propagation of an action potential along a nerve fiber. (a) The resting electrochemical potential of a nerve. (b) Stimulation alters the sodium permeability of the nerve. (c) As sodium ions rush in, the adjacent gradient begins to depolarize, which increases sodium permeability and allows sodium to enter this part of the nerve as well. This action propagates down the nerve as a small, local current. (d) Repolarization begins in the same place the impulse started. The high positive charge inside the cell increases the potassium permeability. Potassium ions flow out of the cell and reestablish the resting potential. (e) Repolarization travels down the nerve until it is complete.

ions moving out of the cell brings the membrane state back to its original resting potential (Figure 7.1*c*).

Neurotransmitter Activity

Communication between neurons and other cells occurs by both electrical and chemical signals. Electrical signals, the fastest means of communication, are transmitted between tightly packed neurons through membrane pores called *gap junctions*. The slightly slower chemical signals consist of neurotransmitters released at the synapse, which then bind to receptors on the postsynaptic cell, triggering a response in that cell. The neurotransmitters include several types of molecules whose release from the neurotransmitter vesicles is stimulated by the action potential in the presynaptic cell. This process is illustrated in Figure 7.2.

The most common neurotransmitter at synapses in the human body is acetylcholine (ACh), a component primarily of the peripheral nervous system. This chemical usually functions in an excitatory fashion, meaning that it initiates an action potential in the postsynaptic neuron, although it may also serve to inhibit signal propagation at some synapses. Whatever its role, ACh is hydrolyzed and its activity terminated by the enzyme acetylcholinesterase (AChE) after it has triggered a response in the postsynaptic cell, and in this way the signal is sent only once. Future signals are transmitted by newly synthesized ACh molecules released by the presynaptic neuron. Failure to hydrolyze older molecules,

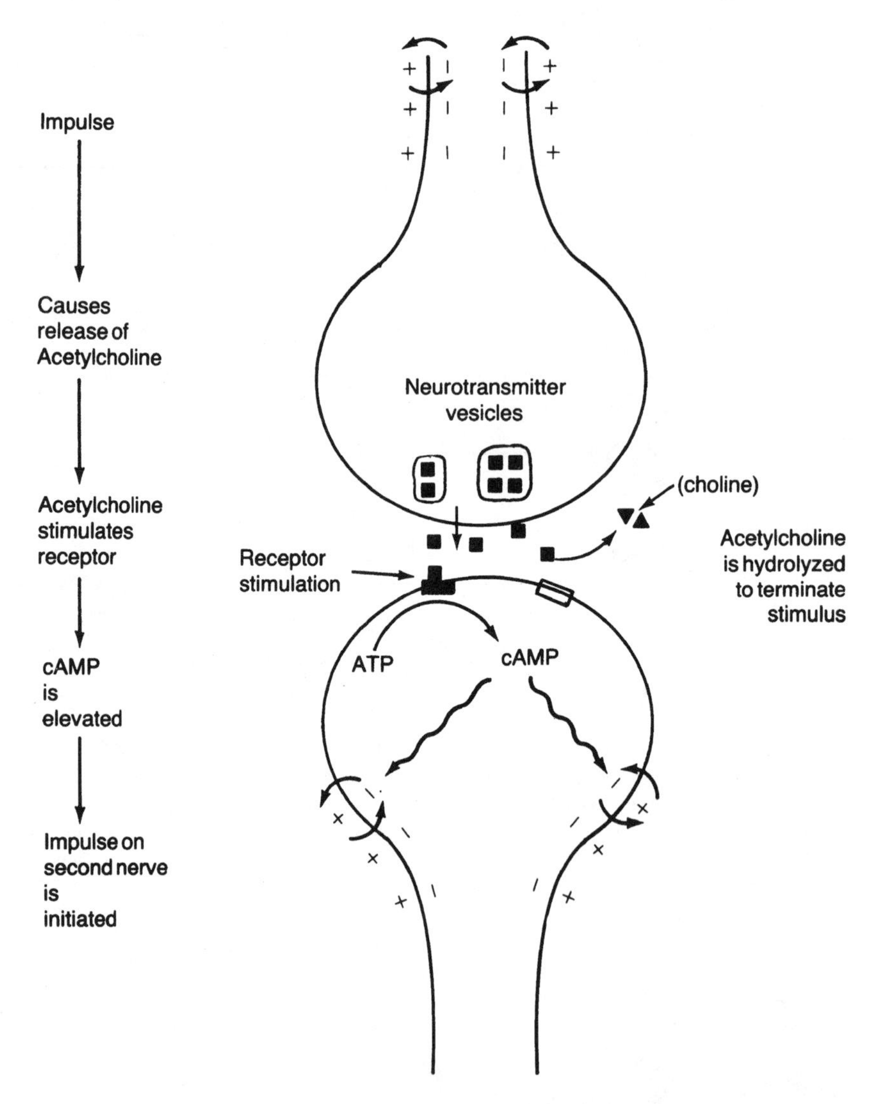

Figure 7.2 Schematic model of a nerve synapse and the use of neurotransmitters to pass the stimulus from nerve to nerve. Neurotransmitters are released and stimulate receptors. Receptor stimulation increases cAMP levels, which affect sodium/potassium ATPase activity and hence the electrochemical gradient, membrane permeabilities, etc. Stimulus is ended either by (1) the breakdown of acetylcholinesterase, or (2) the reuptake of epinephrine in adrenergic nerves.

an effect seen with some neurotoxicants described later, will result in the inappropriate and continuous stimulation of the postsynaptic cell by the accumulated ACh molecules.

Other neurotransmitters are known and include γ-aminobutyric acid (GABA—a component of the central nervous system), amines (epinephrine, norepinephrine, dopamine, serotonin), amino acids (glycine, glutamate), and peptides (enkephalins, endorphins). Their actions on the postsynaptic cell may be either excitatory or inhibitory, and may be directed toward another neuron, a muscle fiber, or a glandular cell.

7.2 AGENTS THAT ACT ON THE NEURON

Membrane Disruption

Effective transmission of neuronal signal depends on intact membranes, both along the length of the axon and at the terminus where neurotransmitter is released. In addition, the axons of many neurons are typically wrapped with myelin, which aids in the propagation of the action potential by minimizing any loss of potential across the membrane to the outside of the cell. Chemicals that disrupt the integrity of this membrane system can seriously impair nervous system function.

Many commonly used industrial solvents, such as methanol, trichloroethylene, and tetrachloroethylene, are excellent cleaning agents and degreasers because of their lipophilicity. However, this same property also makes them destructive to the lipids in cell membranes. Coupled with the potential for substantial inhalation exposure resulting from their volatility, these membrane disrupters may pose a serious threat to the nervous system.

Some metals are disruptive of the myelin sheath that surrounds neurons of the central nervous system and some of the peripheral nervous system. In an industrial setting, some of these metal compounds, such as lead, thallium, and triethyltin, may be readily inhaled in the course of smelting or soldering operations. They may then directly attack the myelin sheath, or disrupt the functioning of the accessory cells that myelinate neurons, namely, the Schwann cells and oligodendrocytes. The results of such damage may vary from the vision loss commonly seen with thallium poisoning to the impaired cognition associated with lead exposure.

Many insecticides impair membrane function by interfering with the ion channels responsible for maintaining the proper balance of sodium and potassium ions across the membrane. An example is the organochlorine insecticide DDT (dichlorodiphenyltrichloroethane), which blocks ion channels and inhibits active transport, thus impeding the repolarization of the membrane after propagation of the action potential. The resulting symptoms, which include tremors, seizures, and increased sensitivity to stimuli, are seen in both the poisoned target insect and the accidentally exposed human. The use of DDT has been banned in the United States (though more because of its environmental persistence than its neurotoxicity), but many organochlorine insecticides, as well as some similarly acting pyrethroid esters, remain in use worldwide and are thus a potential hazard for workers involved in both their manufacture and application.

Peripheral Sensory and Motor Nerves

Effects on peripheral sensorimotor nerves are manifested as abnormal sensation and impaired motor control in the distal regions of the body, primarily hands and feet. Peripheral nerves often have the capacity to regenerate if damage is limited to the axonal region, so that quick action to limit exposure to the toxicant may result in complete reversal of the toxic effects. This is a fortunate contrast to CNS neuropathy, which usually results in permanent and irreversible damage.

Some peripheral neurotoxicants, such as the solvents methyl *n*-butyl ketone and *n*-hexane, are thought to exert their effects mainly through the formation of toxic metabolites formed in the liver. The observed pathology is distal axonal swelling in both motor and sensory neurons, and is thought to be caused mainly by the metabolite 2, 5-hexanedione. A similar effect is seen as a result of exposure to carbon disulfide, another common industrial solvent, although the role of its metabolites is less clear.

Other chemicals may be transformed outside of the body into products that remain closely associated with the parent compound, leading to confusion about what is actually causing the observed effect. An interesting example is the effect on the cranial nerves that has long been seen with trichloroethylene inhalation. Trichloroethylene is known to target the central nervous system, while the cranial nerves are part of the peripheral nervous system controlling sensory and motor functions in the face and head. Recent studies suggest that the cranial neuropathy previously attributed to trichloroethylene inhalation may be caused by dichloroacetylene, an abiologically formed breakdown product of trichloroethylene that may occur in some industrial settings. Dichloroacetylene has clearly been shown to target the cranial nerves, whereas this has not yet been demonstrated with pure trichloroethylene.

Permanent Brain Lesions

Damage to the neurons of the brain may produce varying results, depending on the affected area. Sensory, cognitive, or motor skills may be impaired, and the degree of effect may range from slightly debilitating to severe or even fatal. As with other CNS effects, those resulting from lesions in the brain are likely to be irreversible.

Due to the high metabolic rate of neuronal tissue, the brain is by necessity highly perfused with blood vessels, and this presents a problem when a neurotoxic chemical finds its way into the bloodstream. Humans have evolved an important protection against potential brain toxicants known as the blood–brain barrier. This consists of several anatomical adaptations, such as tightly joined cells with few transport vesicles, which serve to decrease the permeability of membranes to many blood-borne chemicals. While it is quite effective at minimizing brain exposure to most large or hydrophilic molecules, the blood–brain barrier may still be traversed by some highly lipophilic molecules, as discussed below.

The classic example is the neurotoxic metal mercury. When it exists in its ionized form or as an inorganic salt, mercury is water-soluble and, although it may circulate in the bloodstream of an exposed individual, it is not likely to cross the blood–brain barrier and cause damage to the brain. The neurotoxic symptoms of mercury poisoning, which may include tremors, mood disorders, psychosis, and possibly death, are manifested when the lipophilic elemental mercury or an organic mercury species is formed. The transformation of inorganic mercury to organic mercury is commonly performed by bacteria in the environment, but may also occur as a result of bacterial activity within the human gut. Since elemental mercury and organic mercury can cross membranes more easily than the ionized form, the blood–brain barrier presents a less formidable obstacle. Once inside the brain, elemental or organic form of mercury may be transformed into the ionized mercury, and thus remain there for a long time, producing severe brain lesions.

Anoxia

In addition to those mentioned above, the nervous system depends on other important physiological functions that, if impaired by toxic agents, may result in symptoms of neurotoxicity. Of particular importance is the relationship between the nervous system and the respiratory system.

The high metabolic rate of neurons requires that they be well supplied with oxygen and a rapid waste transport system. Compounds like carbon monoxide, which compete with oxygen for hemoglobin binding sites, may severely reduce the oxygen supply to neurons and eventually cause their deaths by anoxia. The most serious case would be destruction of neurons in the brain, leading to functional damage or death of the individual. Other compounds may produce the same result but through a different mechanism, as in the case of cyanide or hydrogen sulfide, which irreversibly bind to cytochrome oxidase, an essential component of the respiratory electron transfer chain.

When the affected neurons are those of the brain, the results are usually serious, as described in the previous section. However, life-threatening effects may also result from damage to the autonomic nerves, such as those controlling breathing and heart rate. This is often the effect of cyanide or hydrogen sulfide, which leads to death.

7.3 AGENTS THAT ACT ON THE SYNAPSE

Anticholinesterase Agents

When the enzyme acetylcholinesterase (AChE) is prevented from hydrolyzing acetylcholine (ACh), overstimulation of the postsynaptic cell results. This is an important mode of action for a variety of insecticides, two groups of which are the organophosphorus and the carbamate esters. However, as is the case with the organochlorines, what makes these anticholinesterase compounds effective insecticides also makes them potentially hazardous to humans who come into contact with them.

Organophosphorus esters, or organophosphates, include malathion and parathion. These compounds bind to acetylcholinesterase (AChE), rendering it inactive in what is generally considered an irreversible reaction. Carbamate esters, such as sevin and aldicarb, also inactivate AChE by binding to it, although this reaction is considered reversible. Thus, carbamate poisoning is usually less severe than organophosphate poisoning.

Since ACh is such a ubiquitous neurotransmitter in the body, the symptoms of anticholinesterase toxicity may take on a variety of forms. These may include decreased cognitive and motor skills and loss of autonomic nervous function, resulting in vomiting and diarrhea, seizures, tremors, and fatigue.

Neurotransmitter Inhibitors and Receptor Antagonists

Although many neurotoxic insecticides target AChE and thus the function of the neurotransmitter ACh, other insecticides exert their effects on other neurotransmitters. An example is the chlorinated cyclodienes, such as chlordane and endosulfan, which block the binding of the inhibitory neurotransmitter γ-aminobutyric acid (GABA) to its postsynaptic receptor. The indicators of toxicity are again the generalized symptoms of seizures, nausea, dizziness, and mood swings.

Other chemicals that are implicated in neurotransmitter inhibition include carbon disulfide and DDT, which inhibit norepinephrine function, and manganese, which inhibits serotonin, norepinephrine, and dopamine function. Another commonly encountered chemical, nicotine, which is found in tobacco products and some insecticides, binds to a subset of ACh receptors that bear its name. These "nicotinic receptors" are found throughout the central and autonomic nervous systems and at neuromuscular junctions. Their increased stimulation can lead to the well-known and often contrary symptoms of nicotine poisoning, such as excitability, nausea, and increased heart rate, followed by muscle relaxation, decreased heart rate, and sometimes coma or death.

7.4 INTERACTIONS OF INDUSTRIAL CHEMICAL WITH OTHER SUBSTANCES

The effects of nicotine on the nervous system have already been discussed. While truly severe effects are expected only at high acute exposures, the lower but chronic exposure of a smoker to nicotine may combine with workplace exposure to other chemicals to produce an additive or even synergistic effect on the nervous system. The same is true for other neurotoxic chemicals to which a worker may be exposed outside the workplace, but that nonetheless serve to exacerbate the symptoms of neurotoxicity resulting from workplace chemical exposure.

Perhaps the most common neurotoxicant of this sort is ethanol, which impairs axonal signal transmission by disrupting the sodium and potassium channels. This results in general CNS depression and uncoordination. It may also have serious consequences when combined with other chemicals with depressive effects. Prescribed sedatives, such as barbiturates, may produce similar results.

Caffeine is another commonly encountered substance with mild stimulatory effects resulting from the inhibition of cyclic AMP metabolism, a second messenger in the response cascade of postsynaptic nerve cells. However, the effects of caffeine on the nervous system are relatively mild, and are not thought to be particularly hazardous. More potent, and often illegal, stimulatory drugs, such as amphetamines and cocaine, may produce additive effects when taken in conjunction with workplace exposure to other neurotoxicants. The resulting symptoms may be life-threatening alterations in breathing and heart rates, or violent mood swings.

7.5 GENERAL POPULATION EXPOSURE TO ENVIRONMENTAL NEUROTOXICANTS

Besides being workplace hazards, many neurotoxicants find their way into the environment through either deliberate or inadvertent release. Thus the general population may become exposed to these chemicals through the air, food, or drinking water. The infamous Minamata incident of the 1950s, in which residents of a coastal Japanese town suffered severe and occasionally fatal neurotoxicity from methyl mercury poisoning is only one example. In this case, exposure was due to ingestion of fish contaminated by discharge from a local acetaldehyde plant.

Because humans are exposed to a variety of environmental chemicals, pinpointing the culprit causing a particular toxic effect can be difficult. Also, environmental exposure levels are usually, except in extreme cases, not so high as to cause blatant symptomology in all individuals. More often, the situation is such that only some members of an exposed population (presumably those that are most highly exposed or sensitive) show symptoms of varying degrees. In the case of neurotoxic effects, these often appear as changes in behavior that may serve as early warning signs of further toxicity possibly occurring if exposure is continued.

Neurotoxic effects may be found among individuals exposed to dissolved metals such as lead and arsenic in their drinking water. Likewise, organic solvents, such as trichloroethylene and carbon tetrachloride, are now seen at various concentrations in most groundwater and surface water supplies. Often, these levels are not considered high enough to pose a serious risk, although we still do not know enough about the effects of low-level, chronic exposure to these chemicals on the developing nervous systems of fetuses and infants. Several epidemiological studies of populations living near water supplies contaminated with industrial solvents have reported elevated symptoms of neurotoxicity (e.g., decreased reaction times, reduced cognitive skills, mood disorders), although usually not at levels high enough as to be incontrovertible.

Exposure to neurotoxicants through inhalation is seldom considered to be a serious threat outside an industrial setting. This is because volatile chemicals are easily dispersed in the atmosphere, so that only people living near industrial sources of air pollution may be subject to breathing elevated levels of neurotoxic chemicals. Usually, the primary concern is exposure within a confined space, such as the potential inhalation of volatile organic compounds from showering with contaminated water. Substantially increased levels of tetrachloroethylene have been measured in houses or automobiles in which newly dry-cleaned clothes are being stored or transported. However, the correlation between exposure to such levels and possible neurotoxic effects is as yet unknown.

7.6 EVALUATION OF INJURY TO THE NERVOUS SYSTEM

Clinical Signs

Taken as a whole, the evaluation of an individual for possible neurotoxicant exposure can generally be described as a series of steps, such as

- A study of the patient's history, including diseases, chronic health problems, drug use, and exposure to other industrial or environmental chemicals.
- An evaluation of the patient's mental status, as determined by various intelligence, memory, or mood tests.
- An evaluation of the patient's sensory, motor, reflex, and cranial nerve function. These are assessed by simple, noninvasive tests, some of which are described below.
- An inspection of the patient's work environment, which includes monitoring for the neurotoxic chemicals that may be implicated on the basis of results from the patient's evaluative tests.

Many clinical symptoms can often be indicators of CNS disturbance. These may include dizziness, vertigo, headache, mood swings, fatigue, memory loss, and various other cognitive disorders. Effects on the peripheral nervous system, specifically the peripheral sensory and motor neurons, are manifested by changes in breathing rate, heart rate, tendon reflex, perspiration, and gastrointestinal function.

Standardized tests for cognition include IQ tests and performance with discriminatory tasks. An example is the Wechsler Adult Intelligence Scale (WAIS), in which the subject is presented with a list of words of increasing difficulty and is asked to provide a definition. Since the results obtained often depend strongly on the way in which the test is administered, and may be particularly vulnerable to hidden biases, evaluations of cognitive skills are not without controversy. Similar tests, which rely on the subject's answers to certain questions, may be used to measure mood and memory effects. A difficulty with using such tests to evaluate possible neurotoxicant exposure in the workplace is that, in order to measure a change in cognitive skills, the individual's preexposure level of skill must be available for comparison. This is rarely the case.

Less subjective than cognitive tests are the standard physiological measurements, such as heart and breathing rate for autonomic nervous system effects, and sensory effects such as impaired hearing or vision. Decreased reaction time in response to a stimulus may indicate peripheral nervous system effects as well, and electroencephalograph (EEG), or "brain wave," measurements present a noninvasive method for monitoring the central nervous system. With some large, easily accessible neurons, as exist in the legs or arms, changes in the conduction velocity along the axon may be measured directly.

Behavioral Tests

There is a vast array of available behavioral tests, which may be performed on workers to indicate potential neurotoxicity. These include measurements of reaction time to a stimulus, changes in dexterity as measured while performing various tasks, perception, motor steadiness, and general coordination. Sometimes cognitive and mood factors will be involved in determining the outcome of these tests as well. Several batteries of standard neurobehavioral tests have been developed, such as the World Health Organization Neurobehavioral Core Test Battery and the Finnish Test Battery, and are routinely used in industry around the world.

An example of the type of neurobehavioral tests that may be administered is the Luria test for acoustic-motor function. The subject listens to a sequence of high and low tones, then repeats the pattern by knocking on a table with a fist for high tones and a flat hand for low tones. This test measures both acoustic perception and motor skills. Visual perception and hand dexterity may be measured with the Santa Ana dexterity test, in which the subject must rotate, within a given time, a pattern created with moveable colored pegs.

Because behavior is determined by many factors, neurobehavioral tests are useful only as a first-step screening procedure for neurotoxicity. Taking preventative steps on behalf of workers requires knowledge of the neurotoxicants present, as well as their mechanisms of action which may result in the observed behavioral effects.

7.7 SUMMARY

Evaluating neurotoxicity can be problematic for several reasons:

- The proper functioning of the nervous system depends upon many complicated steps occurring in a precisely controlled fashion, and thus an agent that is disruptive of any of these steps may be potentially neurotoxic.
- The operation of the nervous system is intimately involved with that of other systems, such as the respiratory, gastrointestinal, and endocrine systems, so that impairment of respiration or hormone function may occur in conjunction with, or as a result of, neurotoxic symptoms.

TABLE 7.1 Some Common Neurotoxic Chemicals

Chemical	Symptom(s)	Site(s) of Action
Metals		
Arsenic	Seizures, tremors	Peripheral motor neurons
Barium	Muscle spasms	Ion channels
Lead	Insomnia, tremors	Myelin, synapse, axon
Manganese	Insomnia, confusion	Synapse
Mercury (organic)	Ataxia, tremors, confusion	Peripheral motor neurons, axon
Thallium	Seizures, psychosis	Myelin, axon
Tin (organic)	Headache, psychosis	Myelin
Organic solvents		
Acetone	CNS depression	Neuronal membrane
Benzene	Giddiness, ataxia	Neuronal membrane
Carbon tetrachloride	CNS depression, giddiness	Neuronal membrane
Carbon disulfide	Dizziness, psychosis	Peripheral sensory and motor neurons, synapse
n-Hexane	Numbness, giddiness	Peripheral sensory and motor neurons, axon
Methanol	Blindness, mild inebriation	Neuronal membrane, axon
Tetrachloroethylene	Dizziness, ataxia	Neuronal membrane
Toluene	Dizziness, euphoria	Neuronal membrane
Trichloroethylene	Giddiness, tremors	Neuronal membrane, axon
Organic pesticides		
Chlordane	Blurred vision, ataxia	Synapse
Cyanide	Confusion, labored breathing	Cytochrome oxidase
2,4-D	Muscle spasms, convulsions	Neuronal membrane
DDT	Dizziness, convulsions	Ion channels, synapse
Endosulfan	CNS depression, convulsions	Synapse
Lindane	Headache, convulsions	Synapse
Malathion	Blurred vision, headache	Synapse, axon
Parathion	Ataxia, convulsions	Synapse, axon
Rotenone	Tremors, convulsions	Electron transport chain
Other		
Carbon monoxide	Headache, dizziness	Hemoglobin
Ethanol	CNS depression, giddiness	Ion channels, neuronal membrane
Hydrogen sulfide	Convulsions, coma	Cytochrome oxidase
Nicotine	Excitability, nausea	Synapse
Petroleum distillates	CNS depression, dizziness	Neuronal membrane

- Although much of the damage which may occur in the nervous system is irreversible, some protective adaptations and redundancy of function exist, which sometimes makes predicting the effects of neurotoxicant exposure less than straightforward.

The chemicals cited in the sections above serve only as examples and are by no means a comprehensive list of neurotoxic hazards in the workplace. In addition, their symptoms of exposure are often varied and may be attributed to several mechanisms of action, which makes categorizing them by effect somewhat difficult. Recommended exposure limits for chemicals commonly encountered in an occupational setting are published by various agencies in the United States, such as the Occupational Safety and Health Administration (OSHA), the National Institute for Occupational Safety and Health (NIOSH), and the American Conference of Governmental Industrial Hygienists (ACGIH). In addition, several other countries have organizations with similar purposes. The publications of these groups should be consulted for specific recommended exposure levels of neurotoxic chemicals. Table 7.1 presents a few of the more common industrial neurotoxicants, along with some general symptoms of exposure and their primary sites of action.

It should be clear to the reader by now that, as the result of multiple and overlapping effects of many neurotoxic chemicals, any listing of effects must necessarily be a simplified representation. Also, the science of neurotoxicity is continually evolving, so that revisions of such lists are to be expected as new information is obtained. Nonetheless, considerable progress has been made in recent years toward developing reliable methods of neurotoxicity evaluation and minimizing exposure to potential neurotoxicants.

REFERENCES AND SUGGESTED READING

Annau, Z., ed., *Neurobehavioral Toxicology,* Johns Hopkins Univ. Press, Baltimore, 1986.

Anthony, D. C., T. J. Montine, and D. G. Graham. "Toxic responses of the nervous system," in *Casarett and Doull's Toxicology: The Basic Science of Poisons,* 5th ed., C. D. Klaassen, ed., McGraw-Hill, New York, 1966, pp. 463–486.

Anthony, D. C., and D. G. Graham. "Toxic responses of the nervous system," in *Casarett and Doull's Toxicology: The Basic Science of Poisons,* 4th ed., M. O. Amdur, J. Doull, and C. D. Klaassen, eds., McGraw-Hill, New York, 1991.

Araki, S., ed., *Neurobehavioral Methods and Effects in Occupational and Environmental Health,* Academic Press, London, 1995.

Baker, E. L. Jr., "Neurologic and behavioral disorders," in *Occupational Health: Recognizing and Preventing Work-Related Disease,* 2nd ed., B. S. Levy and D. H. Wegman, eds., Little, Brown, Boston, 1988.

Chang, L. W., and W. Slikker, Jr., eds., *Neurotoxicology: Approaches and Methods,* Academic Press, London, 1995.

Feldman, R. G., *Occupational and Environmental Neurotoxicology*, Lippincott, Williams and Wilkins Publishers, Philadelphia, 1998.

Johnson, B. L., ed., *Advances in Neurobehavioral Toxicology: Applications in Environmental and Occupational Health,* Lewis Publishers, Chelsea, MI, 1990.

Kilburn, K. H., *Chemical Brain Injury*, Van Nostrand-Reinhold, New York, 1998.

Norton, S., "Toxic responses of the central nervous system," in *Casarett and Doull's Toxicology: The Basic Science of Poisons,* 3rd ed., C. D. Klassen, M. O. Amdur, and J. Doull, eds., Macmillan, New York, 1986.

Office of Technology Assessment, Congressional Board of the 101st Congress, *Neurotoxicity: Identifying and Controlling Poisons of the Nervous System,* Van Nostrand-, New York, 1990.

Tilson, H. A., and G. J. Harry, eds., *Neurotoxicology* (*Target Organ Toxicology Series*), Taylor and Francis, London, 1999.

Tilson, H. A., and C. L. Mitchell, *Neurotoxicology,* Raven Press, New York, 1992.

Weiss, B., and J. L. O'Donoghue, eds., *Neurobehavioral Toxicity: Analysis and Interpretation,* Raven Press, New York, 1994.

8 Dermal and Ocular Toxicology: Toxic Effects of the Skin and Eyes

WILLIAM F. SALMINEN and STEPHEN M. ROBERTS

The skin is the body's first line of defense against external toxicant exposure. Normal skin is an excellent barrier to many substances, but because of its enormous surface area (1.5–2.0 m^2), it can act as a portal of entry for many diverse chemicals that come into contact with it, causing local and/or systemic effects. Understanding the composition of the skin and factors that influence the migration of chemicals across it are prerequisites to understanding the various manifestations of toxicant injury of the skin. Ocular toxicity will also be touched on in this chapter since many aspects pertaining to skin toxicity are relevant to ocular toxicity; the main difference is that the eye seldom serves as a significant portal of entry because of its small surface area. In this chapter you will learn about the

- Composition of the skin
- Ability of the skin to defend against toxicants
- Types of skin maladies
- Commonly used tests to determine skin disorders
- Composition of the eye and exposures pertaining to ocular toxicity

8.1 SKIN HISTOLOGY

The skin is composed of two layers: the outer epidermis and the underlying dermis. The two layers are firmly associated and together form a barrier that ranges in thickness from 0.5–4 mm or more in different parts of the body. The epidermis and dermis are separated by a basement membrane, which has an undulating appearance. The uneven interface gives rise to dermal ridges and provides the basis for the fingerprints used in personal identification since the patterns of ridges are unique for each individual. Hair follicles, sebaceous glands, and eccrine (i.e., sweat) glands span the epidermis and are embedded in the dermis. A third subcutaneous layer lays below the dermis and is composed mainly of adipocytes. Even though this layer is not technically part of the skin it plays an integral role by acting as a heat insulator and shock absorber. (See Figure 8.1.)

The epidermis is composed of several layers of cells—some living and some dead. The majority of the epidermis is composed of keratinocytes, which undergo keratinization, a process during which they migrate upward from the lowest layers of the epidermis and accumulate large amounts of keratin (80 percent once fully mature and nonviable). By the time they reach the outer layer of the epidermis, the stratum corneum, the cells are no longer viable. They have become flattened and have lost their aqueous environment, which is replaced by lipids. The superficial cells of the stratum corneum are continuously lost and must be replaced by new cells migrating from the lower layers of the epidermis. The lowest layers of the epidermis immediately adjacent to the dermis (stratum germinativum and stratum spinosum) are responsible for the continual supply of new keratinocytes and initiation of the keratinization process. The migration and differentiation of keratinocytes from the lower viable layers

Principles of Toxicology: Environmental and Industrial Applications, Second Edition, Edited by Phillip L. Williams, Robert C. James, and Stephen M. Roberts.
ISBN 0-471-29321-0 © 2000 John Wiley & Sons, Inc.

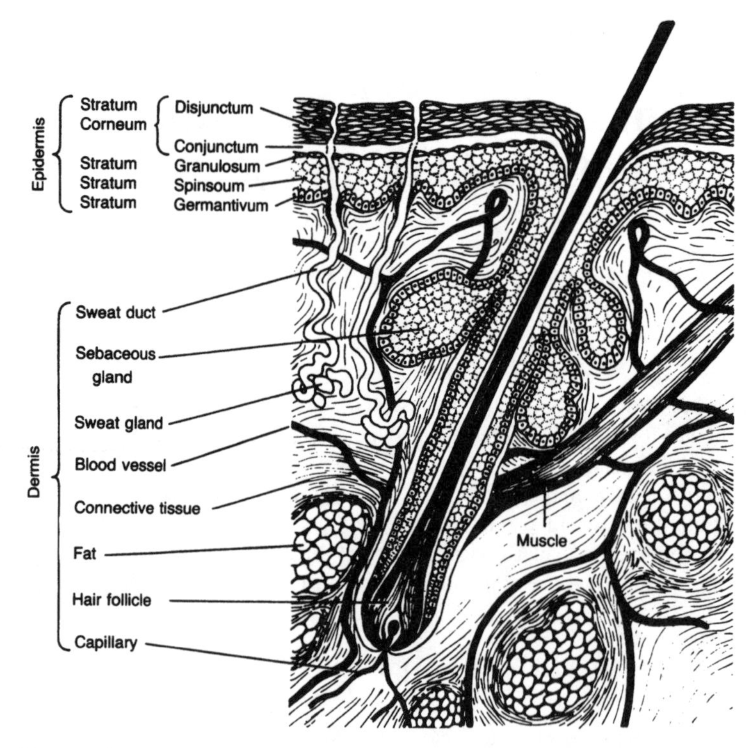

Figure 8.1 Diagram of a cross-section of skin. (Based on Doull, J., *et al.*, eds., *Casarett and Doull's Toxicology: The Basic Science of Poisons,* 2nd Ed. New York: Macmillan Publishing Company, 1980). Reprinted with permission.

to the upper stratum corneum take approximately 2 weeks, with another 2 weeks elapsing before the keratinocytes are shed from the surface. The lowest two layers of the epidermis also contain melanocytes, which produce the pigment melanin. Melanocytes extrude melanin, which is taken up by the surrounding epidermal cells, giving them their characteristic brown color. Langerhans' cells are also found in these layers and play a role in the skin's immune response to foreign agents.

The dermis has a largely supportive function and represents about 90 percent of the skin in thickness. The predominant cells of the dermis are fibroblasts, macrophages, and adipocytes. Fibroblasts secrete collagen and elastin, thereby providing the skin with elastic properties. The dermis is well supplied with lymph and blood capillaries. The capillaries terminate in the dermis without extending into the epidermis. A toxicant must penetrate the epidermis and dermis in order to enter the systemic circulation; however, once the stratum corneum is breached, the remaining layers pose little resistance to toxicant penetration. Hair follicles are embedded within the dermis and have a capillary at the bulb of the follicle. In some instances, hair can enhance toxicant absorption across skin by providing a shunt to the blood supply at the base of the follicle. Eccrine glands are embedded deep within the dermis, and coiled sweat ducts wind upward through the epidermis and out through the stratum corneum.

8.2 FUNCTIONS

The skin is an effective barrier to many substances, but it is a perfect barrier to very few. This is an important concept, since even though relatively small amounts of chemicals cross the skin, it can be sufficient to cause local and/or systemic toxicity. The passage of chemicals through the skin appears to be by passive diffusion with no evidence of active transport of compounds. The outer stratum corneum is the primary layer governing the rate of diffusion, which is very slow for most chemicals. This layer also prevents water loss by diffusion and evaporation from the body except, of course, at the sweat glands, which helps regulate body temperature. The viable layers of the epidermis and the dermis are poor barriers to toxicants, since hydrophilic agents readily diffuse into the intercellular water and hydrophobic agents can embed in cell membranes, eventually reaching the blood supply in the dermis.

Several factors influence the rate of diffusion of chemicals across the stratum corneum. In general, hydrophobic agents of low molecular weight can permeate the skin better than can those that are hydrophilic and of high molecular weight. This is due to the low water and high lipid content of the stratum corneum, which allows hydrophobic agents to penetrate more readily. However, if the skin becomes hydrated on prolonged exposure to water, its effectiveness as a barrier to hydrophilic substances is reduced. Often the skin of lab animals is covered with plastic wrap to enhance the hydration of the skin and increase the rate of uptake of agents applied to the surface of the skin. For compounds with the same hydrophobicity, the smaller compound will diffuse across the skin fastest since its rate of diffusion is quickest. A good example of the diffusion of a class of toxicants across the skin that can cause systemic toxicity is the organophosphate pesticides (e.g., parathion) used in agriculture. These compounds are hydrophobic, are very potent, and can lead to systemic effects such as peripheral neuropathy (i.e., nerve damage) and lethality after exposure to only the skin.

The property of diffusion of agents across the skin and the reservoir capacity of the skin can be useful in delivering drugs to the systemic circulation over a prolonged period (typically 1–7 days). Transdermal drug delivery using specially designed skin patches is used to deliver nicotine, estradiol, and nitroglycerin. This approach provides a steady dose, avoids large peak plasma concentrations from loading doses, and prevents first-pass metabolism by the liver for agents that are sensitive to metabolism such as nitroglycerin.

The rate of diffusion through the epidermis varies among anatomical sites and is not solely a function of skin thickness. In fact, the skin on the sole of the foot has a higher rate of diffusion than the skin of the forehead or abdomen, even though it is much thicker. Therefore, skin thickness is not a useful indicator of how much chemical will reach the systemic circulation in a given amount of time. If the skin is wounded, the barrier to chemicals is compromised and a shorter or direct route to the systemic circulation is available since the skin can no longer repel the chemicals. In addition, diseases (e.g., psoriasis) can compromise the ability of skin to repel chemicals.

The skin also provides protection against microorganisms and ultraviolet (UV) radiation. Hydrated skin has a greater risk of becoming infected by microorganisms than does dry skin, which is why soldiers in Vietnam often suffered from foot infections. The stratum corneum and epidermis, but primarily melanin pigmentation, provide protection against UV radiation by absorbing the energy before it reaches more sensitive cells and causes adverse effects such as DNA damage. (See Table 8.1.)

Another important aspect of the skin's barrier function is its ability to metabolize chemicals that cross the stratum corneum and enter the viable layers of the skin. Even though the metabolic activity of the skin on a body weight basis is not nearly as great as that of the liver, it does play a crucial role in determining the ultimate effects of some chemicals. The epidermis and pilosebaceous units of the skin contain the highest levels of metabolic activity, which includes phase I (e.g., cytochrome

TABLE 8.1 Defense Roles of the Skin

Prevent water loss
Act as a barrier for physical trauma
Retard chemical penetration
Prevent ultraviolet light penetration and damage
Inhibit microorganism growth and penetration
Regulate body temperature and electrolyte homeostasis

P450-mediated) and phase II enzymes (e.g., epoxide hydrolase, UDP glucuronosyl transferase, glutathione transferase). Some chemicals that cross the skin are simply degraded and eliminated as innocuous metabolites. For others such as benzo(*a*)pyrene or crude coal tar (the latter is often used in dermatological therapy), metabolism of the parent compound can produce a metabolite that is a skin sensitizer or carcinogen. In addition to metabolizing foreign agents, the skin also has anabolic and catabolic metabolic activity important to its maintenance.

8.3 CONTACT DERMATITIS

Irritants

Irritant contact dermatitis is one of the most common occupational diseases. The highest incidence of chronic irritant dermatitis of the hands occurs in food handlers, janitorial workers, construction workers, mechanics, metal workers, horticulturists, and those exposed to wet working environments, such as hairdressers, nurses, and domestic workers. Contact irritant dermatitis is confined to the area of irritant exposure, and since it is not immunity-related, it can occur in anyone given a sufficient exposure to a chemical. Previous exposure to the chemical is not required to elicit a response as is needed for allergic contact dermatitis, since contact irritant dermatitis is not a hypersensitivity reaction (discussed below). A range of responses can occur after exposure to an irritant, including, but not limited to, hives (wheals), reddening of the skin (erythema), blistering, eczemas or rashes that weep and ooze, hyperkeratosis (thickening of the skin), pustules, and dryness and roughness of the skin. Unlike corrosive chemicals (e.g., strong acids and bases), the ultimate skin damage from irritant contact dermatitis is not due to the primary actions of the chemicals but to the secondary inflammatory response elicited by the chemical. It is important to note that even though the ultimate inflammatory response elicited by different chemicals may appear the same, they often occur through different mechanisms.

A wide array of factors influence the ability of an irritant to elicit an inflammatory response. As discussed in Section 8.2, factors affecting skin permeability and chemical composition of the irritant determine the rate of percutaneous penetration and how much chemical reaches the viable layers of the skin. A variety of other factors determine whether irritant dermatitis occurs and to what magnitude. Higher concentrations and greater amounts of a given agent contacting the skin surface are more likely to elicit a response than lower concentrations and smaller quantities. The surface area of skin exposed to an irritant can also be important. For some irritants, a certain area of skin exposure is required to trigger a response, and below that threshold dermatitis does not occur. The genetic makeup and age of the individual plays a critical role in the sensitivity to a particular agent since the same chemical can cause no response in one individual and a dramatic response in another. The genetic factors influencing sensitivity are unknown, however. In general, children appear to be more, and the elderly less, susceptible to irritants. Concomitant disease may increase or decrease sensitivity to an irritant, and certain medications such as corticosteroids can suppress the irritant response to some agents. Extremes in temperature, humidity, sweating, and occlusion can lower the threshold of irritation for a given compound.

The range of agents that can cause irritant dermatitis is extensive and diverse, and all cannot be touched on in this section. Table 8.2 lists some of the most commonly encountered classes of irritants. All of these agents have the potential of causing irritation on primary exposure; however, in the workplace, exposure to a potential irritant often occurs repeatedly and to relatively low quantities. Since the response is dependent on the amount of irritant to which the individual is exposed, repeated exposure may be required before clinical signs of dermatitis appear. Management of contact irritant dermatitis is based on reducing or avoiding the amount of exposure to the irritant. Wearing gloves to provide protection against wetness or chemicals and minimizing wet working conditions and hand washing can be very helpful. Complete healing of lesions may take several weeks, and the likelihood of a flare-up is often increased for months.

TABLE 8.2 Potential Inducers of Irritant Contact Dermatitis

Agent	Examples
Water	—
Cleansers	Soaps and detergents
Bases	Epoxy resin hardeners, lime, cement, and ammonium
Acids	Hydrochloric acid and citric acid
Organic solvents	Many petroleum-based products
Oxidants	Peroxides, benzoyl peroxide, and cyclohexanone
Reducing agents	Thioglycolates
Plants	Orange peel, asparagus, and cucumbers

Source: Adapted from Rietschel (1985).

Extremely corrosive and reactive chemicals can cause immediate coagulative necrosis at the site of contact resulting in substantial tissue damage. These chemicals, called *primary irritants,* differ from those that cause irritant contact dermatitis in that they cause nonselective damage at the site of contact, which is not a result of the secondary inflammatory response. Primary irritants cause damage resulting from their reactivity, such as acids precipitating proteins and solvents dissolving cell membranes, both resulting in cell damage, death, or disruption of the keratin ultrastructure. The resulting damage is in direct proportion to the concentration of chemical in contact with the tissue. It is important to realize that primary irritants are not always in a liquid form. Many primary irritants are solid chemicals that become hydrated on contact with the skin, and gaseous agents are often converted to acids on contact with water available on the skin and mucous membranes. Ammonia, hydrogen chloride, hydrogen peroxide, phenol, chlorine, sodium hydroxide, and a variety of antiseptic or germicidal agents (e.g., cresol, iodine, boric acid, hexachlorophene, thimerosal) are some of the many commonly encountered primary irritants that can cause skin burns.

Allergic Contact Dermatitis

Allergic contact dermatitis is a delayed type IV hypersensitivity reaction that is mediated by a triggered immune response. Typical of a true immune reaction, minute quantities of the allergenic agent can trigger a response. This differentiates it from irritant dermatitis, which is proportional to the dose applied. Allergic contact dermatitis can be very similar to irritant contact dermatitis clinically, but allergic contact dermatitis tends to be more severe and is not always restricted to the part of the body exposed to the chemical.

On first exposure to the allergenic chemical, little or no response occurs. After this first exposure, the individual becomes sensitized to the chemical, and subsequent exposures elicit the typical delayed type IV hypersensitivity reaction. The allergenic agents (haptens) are typically low-molecular-weight chemicals that are electrophilic or hydrophilic. These agents are seldom allergenic alone and must be linked with a carrier protein to form a complete allergen. Some chemicals must be metabolically activated in order to form an allergen, which can occur within the skin as a result of the skin's phase I and phase II metabolic activities.

Sensitization occurs when the hapten/carrier protein (antigen) is engulfed by an antigen-presenting cell (e.g., macrophages and Langerhans cells) and the processed antigen is presented to a helper T cell ($CD4^+$). The T cell produces cytokines that activate and cause the proliferation of additional T cells that specifically recognize the antigen. The secretion of cytokines also causes inflammation of the contact area and activation of monocytes into macrophages. The active macrophages are the ultimate effector cells of the reaction. They act to eliminate the foreign antigen and, through secretion of additional chemical mediators, enhance the inflammation of the contact site. Keratinocytes also play a role in the hypersensitivity reaction. They are capable of producing many different cytokines and can act as antigen-presenting cells under certain circumstances. After the sensitization process occurs, subsequent exposure to the allergenic chemical triggers the same cascade of events as described above. However, the prior sensitization reaction resulting in a population of T cells specific for the antigen allows the cascade of events to proceed much faster.

Table 8.3 lists some of the most common agents that trigger contact dermatitis. The actual number of potential allergenic agents is almost limitless. Individual sensitivity to a particular allergen varies greatly and is dependent on many factors, as discussed for irritant contact dermatitis. The genetic makeup of the person probably plays the greatest role in determining whether a response occurs. This is similar to the variability noticed among individuals for their sensitivity to IgE-mediated allergies, such as hay fever, in which some people respond while others do not.

Patch testing is used to try to determine to which agent a person with suspected allergic contact dermatitis may be sensitive. Unfortunately, the test is usually limited to agents that are the most frequent causes of allergic contact dermatitis. As such, identifying sensitivity to an agent unique to a given occupation may be impossible. Patch testing should be performed by physicians trained and experienced in the technique, its pitfalls, and the subtleties of interpretation. If a compound is identified as allergenic, the sensitive individual can attempt to avoid exposure to that agent. The distribution of the allergic response on the body can also provide clues as to what the allergenic compound is. For example, linear stripes may indicate plant-induced dermatitis while a rash on the lower abdomen may indicate an allergy to a nickel-containing pants button. A variety of treatments are used to help alleviate contact dermatitis. The best treatment, however, is avoidance of the allergen or irritant. Baths and wet compresses, antibiotics, antihistamines, and corticosteroids are used in various combinations to treat contact dermatitis.

A unique situation arises when a contact allergen is ingested or enters the systemic circulation. The most serious effects include generalized skin eruption, headache, malaise, and arthralgia. Flaring of a previous contact dermatitis, vesicular hand eruptions, and eczema in flexor areas of the body may be less dramatic disturbances. Systemic exposure can trigger a delayed type IV hypersensitivity reaction with subsequent deposition of immunoglobulins and complement in the skin, which are potent inducers of the secondary inflammatory response. Therefore, systemic exposure to a contact allergen may induce a widespread delayed type IV hypersensitivity reaction that is not localized to one area of the body.

Ulcers

Some chemicals can cause ulceration of the skin. This involves sloughing of the epidermis and damage to the exposed dermis. Ulcers are commonly triggered by acids, burns, and trauma and can occur on

TABLE 8.3 Commonly Encountered Contact Allergens

Source	Allergen(s)	Examples
Plants and trees	Rhus	Poison oak and ivy
Metals	Nickel and chromium	Earrings, coins, and watches
Glues and bonding agents	Bisphenol A, formaldehyde, acrylic monomers	Glues, building materials, and paints
Hygiene products and topical medications	Bacitracin, neomycin, benzalkonium chloride, lanolin, benzocaine, and propylene glycol	Creams, shampoos, and topical medications
Antiseptics	Chloramine, glutaraldehyde, thimerosal, and mercurials	Betadine
Leather	Formaldehyde and glutaraldehyde	
Rubber products	Hydroquinone, diphenylguanidine, and *p*-phenylenediamine	Rubber gloves and boots

mucous membranes and the skin. Two commonly encountered compounds that induce ulcers are cement and chrome.

Urticaria

Like allergic contact dermatitis, urticaria can be triggered by immunity-related mechanisms, and minute quantities of allergen can therefore trigger the reaction. Urticaria results in the typical hives, which are pruritic red wheals that erupt on the skin. Asthma is also a common occurrence after exposure to an inducer of urticaria. The symptoms often last less than 24 h. In severe cases, however, anaphylaxis and/or death may occur. The reaction is an immediate type I hypersensitivity reaction that is mediated by activated mast cells. The mast cells may be activated directly by the chemical (nonimmune), or activation may occur when the chemical acts as an allergen (immunity-mediated) and binds to the IgE immunoglobulins located on mast cells. When sufficient quantities of IgE become bound by the allergen or the mast cell is directly activated, the mast cell releases vasoactive peptides and histamine that cause the ultimate hive through activation of additional cellular components of the reaction.

Most compounds that induce urticaria must enter the systemic circulation. Often urticaria is triggered by compounds to which the responder has a specific allergy, but induction by completely idiopathic mechanisms is also seen. Some potential nonimmune inducers of urticaria (i.e., direct activators of mast cells) are curare, aspirin, azo dyes, and toxins from plants and animals. A smaller number of agents may cause contact urticaria on exposure only to the epidermis. Cobalt chloride, benzoic acid, butylhydroxyanisol (BHA), and methanol have been reported to cause this form of urticaria. One of the most common inducers of contact urticaria seen in the medical community is caused by latex rubber products such as gloves. Natural latex rubber contains a protein that is capable of inducing an immediate type I hypersensitivity reaction. Simple contact with latex rubber products, such as gloves, can trigger the hypersensitivity reaction and cause hiving, asthma, anaphylaxis, and sometimes death.

Toxic Epidermal Necrolysis

Toxic epidermal necrolysis (TEN) is one of the most immediate life-threatening skin diseases caused by chemicals or drugs. Mortality is usually 25–30 percent, but can be as high as 75 percent in the elderly. Luckily, the incidence of TEN is fairly low, with approximately one person per million per year becoming affected. The disease is characterized by a sudden onset of large, red, tender areas involving a large percentage of the total body surface area. As the disease progresses, necrosis of the epidermis with widespread detachment occurs at the affected areas. Once the epidermis is lost, only the dermis remains, severely compromising the ability of the skin to regulate temperature, fluid, and electrolyte homeostasis. Since the epidermis is lost, the remaining dermis posses little resistance to chemicals entering the systemic circulation and to infection from microorganisms.

The ultimate mechanism of drug or chemical induction of the disease has remained elusive. Recent evidence has implicated immunologic and metabolic mechanisms, but they are far from conclusive. TEN has been associated with graft–host disease, and, even though it is a controversial area, TEN is believed to be part of the same spectrum of disease as the Stevens–Johnson syndrome (erythema multiforme major), which is another serious reaction to drugs and infection.

Acneiform Dermatoses

Acne is a very disfiguring ailment, but fairly innocuous in terms of producing long-lasting damage to the skin. In the workplace, the most common causes of acne are petroleum, coal tar, and cutting oil products. They are termed comedogenic since they induce the characteristic comedo, which is either open (blackhead) or closed (whitehead). The black color of open comedones is due to pigmentary changes resulting in accumulation of melanin. The comedogenic agents produce biochemical and physiological alterations in the hair follicle and cell structure that cause accumulation of compacted keratinocytes in the hair follicles and sebaceous glands. The keratinocytes clog the hair follicles and sebaceous glands and become bathed in sebum (released from the sebaceous glands).

Halogenated chemicals—especially polyhalogenated naphthalenes, biphenyls, dibenzofurans, and contaminants of herbicides such as polychlorophenol and dichloroaniline—cause a very disfiguring and recalcitrant form of acne called *chloracne*. Chloracne is typically characterized by the presence of many comedones and straw-colored cysts behind the ears, around the eyes, and on the shoulders, back, and genitalia. Other symptoms that may or may not occur include conjunctivitis and eye discharge due to hypersecretion of the Meibomian glands around the eyelids, hyperpigmentation, and increased hair in atypical locations. Since chloracne is a very persistent disease, the best method of treatment is to prevent exposure to the halogenated chemicals. This could involve putting up splash guards and other devices to prevent the chemicals from coming into contact with the skin along with changing chemical soaked clothing frequently.

Pigment Disturbances

Some chemicals can cause either an increase or decrease in pigmentation. These compounds often cause hyperpigmentation (darkening of the skin) by enhancing the production of melanin or by causing deposition of endogenous or exogenous pigment in the upper epidermis. Hypopigmentation (loss of pigment from the skin) can be caused by decreased melanin production and/or loss, melanocyte damage, or vascular abnormalities. Some common hyperpigment inducers are coal tar compounds, metals (e.g., mercury, lead, arsenic), petroleum oils, and a variety of drugs. Phenols and catechols are potent depigmentors that act by killing melanocytes.

Photosensitivity

Photosensitivity is an abnormal sensitivity to ultraviolet (UV) and visible light and can be caused by endogenous and exogenous factors. Wavelengths outside the UV and visible light ranges are seldom involved, since the earth's atmosphere significantly filters those wavelengths or they are not sufficiently energetic to cause skin damage. In order for any form of electromagnetic radiation to produce an effect, it must first be absorbed. Chromophores, epidermal thickness, and water content all affect the ability of light to penetrate the skin, and those parameters vary from region to region on the body. Melanin is the most significant chromophore, since it can absorb a wide range of radiation from UVB (290–320 nm) through the visible spectrum.

Exposure to intense sunlight causes erythema (redness or sunburn) due to vasodilation of the exposed areas. Inflammatory mediators may be released at these areas and have been implicated in the systemic symptoms of sunburn such as fever, chills, and malaise. UVB is the most important radiation band in causing erythema. Sunlight has up to 100-fold greater UVA (320–400 nm), but UVA is 1000 times less potent than UVB in causing erythema. UVB exposure causes darkening of the skin through enhanced melanin production or through oxidation of melanin. Oxidation of melanin occurs immediately, but offers no additional protection against sun damage. Enhanced melanin production is noticeable within 3 days of exposure. UV exposure also enhances thickening of the skin, primarily in the stratum corneum, which further retards subsequent UV absorption. Chronic exposure to UV light can induce a number of skin changes such as freckling, wrinkling, and precancerous and malignant skin lesions. UV light is not the only type of radiation that can induce skin changes. Depending on the dose delivered, ionizing radiation can cause acute changes such as redness, blistering, swelling, ulceration, and pain. Following a latent period or chronic exposure, epidermal thickening, freckling, nonhealing ulcerations, and malignancies may occur.

Phototoxicity results from systemic or topical exposure to exogenous chemicals. The symptoms are very similar to severe sunburn and include reddening and blistering of the skin. Chronic exposure can result in hyperpigmentation and thickening of the affected areas. Unlike sunburn, phototoxicity often results from exposure to the UVA band, but the UVB band is sometimes involved. Phototoxic chemicals are protoxicants (i.e., they are not toxic in their native form) and must be activated by UV light to a toxic form. Phototoxic chemicals readily absorb UV light and become excited to a higher-energy state. Once the excited chemical returns to the ground state, it releases its energy, which can lead to production of reactive oxygen species and other reactive products that damage cellular components and macromolecules, ultimately causing cell death. The resulting damage is similar to that caused by irritant chemicals (discussed in Section 8.3) that cause cell death. Phototoxicant-induced cell death triggers an inflammatory response that produces the clinical signs of phototoxicity. Dyes (eosin, acridine orange), polycyclic aromatic hydrocarbons (anthracene, fluoranthene), tetracyclines, sulfonamides, and furocoumarins (trimethoxypsoralen, 8-methoxypsoralen) are commonly encountered phototoxic drugs and chemicals.

Photoallergy is very similar to contact allergic dermatitis and is a delayed type IV hypersensitivity reaction. The difference between an allergenic chemical and a photoallergenic chemical is that the photoallergenic chemical must be activated by exposure to light—most often UVA. Once activated, the photoallergen complexes with cellular protein to form a complete allergen that triggers the delayed type IV hypersensitivity reaction. Since it is a hypersensitivity reaction, previous exposure to the phototoxic chemical is required for a response. Subsequent topical or systemic exposure to the photoallergen may induce the hypersensitivity reaction, which has clinical manifestations similar to allergic contact dermatitis (see the subsection on allergic contact dermatitis). Testing for photoallergy is similar to the patch testing used for regular allergens, but the potential allergens are tested in duplicate. One set of the patches is removed during the test and irradiated with UV light. By comparing duplicate samples, the physician can determine whether the compound is allergenic and is also a photoallergen.

Skin Cancer

Skin cancer is the most common neoplasm in humans with half a million new cases occurring per year in the United States. Even though exposure to UV light is the primary cause of skin cancer, chemicals can also induce malignancies. UV light and carcinogenic agents induce alterations in epidermal cell DNA. These alterations can lead to permanent mutations in critical genes that cause uncontrolled proliferation of the affected cells, ultimately leading to a cancerous lesion. UVB rays are the most potent inducers of DNA damage and work by inducing pyrimidine dimers. In addition to inducing DNA damage, UV light also has an immunosuppressive effect that may reduce the surveillance and elimination of cancerous cells by the immune system. Since UVB light is the most potent inducer of DNA damage, utilization of a sunscreen that blocks UVB radiation is critical in preventing skin cancer along with the other skin effects associated with UV light exposure. Ionizing radiation is also a potent inducer of skin cancer. Fortunately, ionizing radiation is no longer used for treatment of skin ailments such as acne and psoriasis, as was done in the recent past.

The best characterized chemical inducers of skin cancer are the polycyclic aromatic hydrocarbons (PAH). In the 1700s, scrotal cancer was found to be prevalent among chimney sweeps in England. The compounds that induced the cancer were later determined to be PAHs present in high concentrations in coal tar, creosote, pitch, and soot. PAHs must be bioactivated within the skin, often to a reactive epoxide, by cytochrome P450 metabolism (discussed in Section 8.2) in order to cause DNA damage. The epoxides are electrophilic and can form DNA adducts that may produce gene mutations. Other carcinogenic agents may cause DNA damage through different mechanisms, but the ultimate lesion is a gene mutation that leads to a cancerous lesion.

Eye Toxicity

The eye is a very complex organ composed of many different types of cells. Disease, drugs, and chemicals can injure various parts of the eye with many different manifestations of injury. The most common cause of injury in an occupational setting is exposure of the cornea and conjunctiva to agents that are splashed onto the eye. Many other effects can occur to other parts of the eye such as the retina and optic nerve (see Figure 8.2), but they are usually limited to effects caused by drugs and various diseases. This section therefore focuses on external exposure of the eye to chemicals.

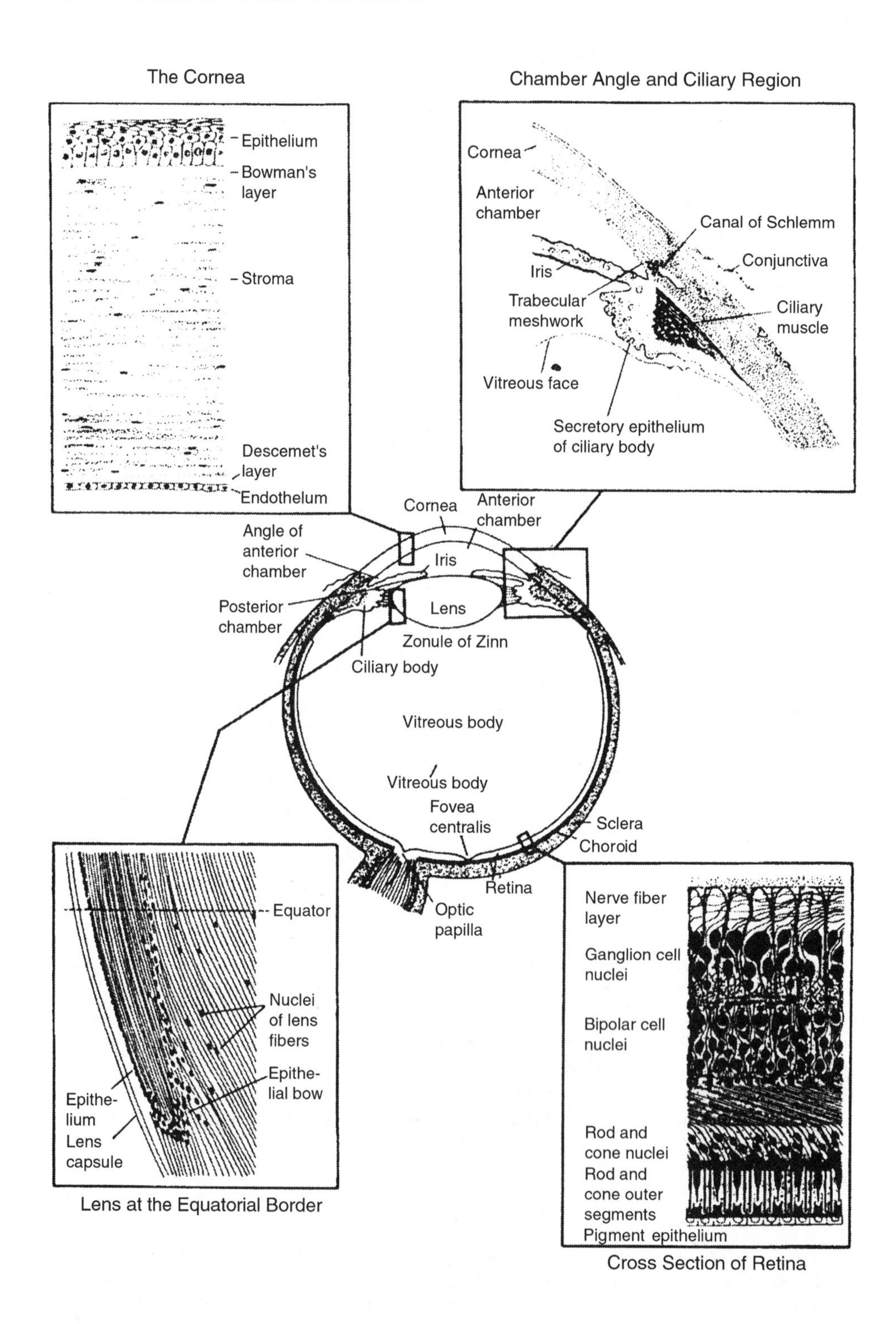

Figure 8.2 Diagrammatic cross section of the eye, with enlargement of details in cornea, chamber angle, lens, and retina. *Casarett & Doull's Toxicology: The Basic Science of Poisons*, 5th Ed. McGraw-Hill, 1996. Reprinted with permission.

The structure of the eye is shown in Figure 8.2. The cornea and conjunctiva are the first line of defense against chemicals that contact the eye. Acids and alkalis are the more commonly encountered agents that cause eye damage. Acids cause protein damage, which leads to eye injury that can range in severity from burns that heal completely to those that perforate the globe. Alkalis such as ammonia can also cause serious eye burns. Alkali burns differ from acid burns in that they may lead to additional damage as time elapses, even if the burn was relatively mild at the time of injury. The best treatment for both types of substances is irrigation with large volumes of water, which reduces the acid or alkali concentration. Some compounds such as unslaked lime, which is found in many commercial wall plasters, may stick to the eye and form clumps that are not readily diluted or washed away with water irrigation. In these cases, irrigation followed by debridement of any remaining particles is required to remove as much contamination as possible.

Two other agents that are frequent causes of eye damage are organic solvents and detergents. Organic solvents cause damage by dissolving fats in the eye. The damage is seldom extensive or long-lasting; however, if the solvent is hot, thermal burn may complicate the picture. Detergents act by disrupting proteins in the eye and lowering the surface tension of aqueous solutions. Detergents contain a nonpolar section and a polar section on the same molecule, allowing them to emulsify compounds with widely different hydrophobicities. They are commonly found in wetting agents, antifoaming agents, emulsifying agents, and solubilizers.

Other parts of the eye can be affected by chemicals, either directly or as a result of the ensuing immune response that follows chemical burns. Many corrosive chemicals can cause lid damage and scarring of the puncta or canaliculi. Normal tear flow enters the lacrimal canaliculi in the lid margin via the puncta. The tears flow through the common canaliculus, lacrimal sac, and nasolacrimal duct into the nasopharynx. Damage of the puncta or canaliculi obstructs tear flow and can cause the tears to run down the cheeks. If a corrosive chemical penetrates the cornea and reaches the anterior chamber, it may cause damage to the iris. Damage to the iris increases vascular permeability with ensuing liberation of protein into the normally low-protein aqueous humor. These proteins can clog the outflow of fluid from the interior of the eye and lead to pressure buildup and glaucoma. Leukocytes may also infiltrate the aqueous humor from the inflamed iris vasculature and contribute to the blockage of the outflow system.

Methanol is a unique eye toxicant since it affects the nerves of the eye and retinal and photoreceptor cells. Ingested methanol is metabolized to formaldehyde, then formate, then CO_2 and water, with formate considered the toxic metabolite. Methanol intoxication can lead to appreciable, and sometimes permanent, loss of vision. Since methanol is first metabolized by alcohol dehydrogenase, ethanol can be used to prevent the formation of formate. The ethanol successfully competes for the alcohol dehydrogenase enzyme preventing the metabolism of methanol. Ethanol must be administered for a sufficient length of time so that all the methanol is eliminated from the body.

8.4 SUMMARY

Toxicity of the skin and eye can occur after exposure to many different substances that cause injury through a variety of mechanisms. This chapter covered the major problems caused by chemicals encountered at home and at work, but a variety of other skin and eye diseases, including the ones mentioned in this chapter, can occur in reaction to systemically administered drugs. Whether a chemical can produce an effect after it comes into contact with the skin or eye depends on many factors, including genetic makeup, status of health, and efficiency of the skin's barrier function. The following are some important points about eye and skin toxicity.

- The most common skin disease is irritant and allergic contact dermatitis, with allergic contact dermatitis usually being more severe.

- A person must first be sensitized to a chemical before allergic contact dermatitis can occur. Since allergic contact dermatitis is an immune reaction, minute quantities of allergen can trigger the reaction, which makes management of future flare-ups difficult.
- Urticaria may or may not occur through immunity related mechanisms. The ultimate trigger of the hives associated with urticaria is due to the release of histamine and vasoactive agents from mast cells that are activated after chemical exposure.
- Phototoxicity and photoallergy are similar to irritant and allergic contact dermatitis, respectively. The difference is that the phototoxicant or photoallergen must be activated by exposure to UV light.
- Skin cancer is the most prevalent form of cancer. Its main cause is exposure to UV light, but many chemicals can induce cancerous lesions, too, such as polycyclic aromatic hydrocarbons and arsenic.
- The main cause of eye toxicity in the workplace is due to chemicals that are splashed onto the eye and cause corneal burns. Secondary events triggered by the burn can lead to further complications such as glaucoma.

REFERENCES AND SUGGESTED READING

Bradley, T., R. E. Brown, J. O. Kucan, E. C. Smoot, and J. Hussmann. "Toxic epidermal necrolys is: A review and report of the successful use of biobrane for early wound coverage," *Ann. Plastic Surg.* **35**: 124–132 (1995).

Goldstein, S. M., and B. U. Wintroub, *Adverse Cutaneous Reactions to Medication,* Williams & Wilkins, Baltimore, 1996.

Grandjean, P., *Skin Penetration: Hazardous Chemicals at Work,* Taylor & Francis, New York, 1990.

Haschek, W. M., and C. G. Rousseaux, *Handbook of Toxicologic Pathology,* Academic Press, San Diego, 1991.

Hogan, D. J., "Review of contact dermatitis for non-dermatologists," *J. Florida Med. Assoc.* **77**: 663–666 (1990).

Marzulli, F. N., and H. I. Maibach, *Dermatotoxicology,* Hemisphere Publishing, Washington, DC, 1987.

Potts, A. M., "Toxic responses of the eye," in *Casarett and Doull's Toxicology: The Basic Science of Poisons,* 5th ed., C. D. Klaassen, ed., McGraw-Hill, New York, 1996, pp. 583–615.

Rice, R. H., and D. E. Cohen, "Toxic responses of the skin," in *Casarett and Doull's Toxicology: The Basic Science of Poisons,* 5th ed., C. D. Klaassen ed., McGraw-Hill, New York, 1996, pp. 529–546.

Rietschel, R. L., "Dermatotoxicity: Toxic effects in the skin," in *Industrial Toxicology: Safety and Health Applications in the Workplace,* P. L. Williams and J. L. Burson, eds., Van Nostrand-Reinhold, New York, 1985, pp. 138–161.

Taylor, J. S., and P. Praditsuwan, "Latex allergy: Review of 44 cases including outcome an d frequent association with allergic hand eczema," *Arch. Dermatol.* **132**: 265–271 (1996).

Wang, R. G. M., J. B. Knaak, and H. I. Maiback, *Health Risk Assessment: Dermal and Inhalation Exposure and Absorption of Toxicants,* CRC Press, Ann Arbor, MI, 1993.

Zug, K. A., and M. McKay, "Eczematous dermatitis: A practical review," *Am. Family Phys* **54**: 1243–1250 (1996).

9 Pulmonotoxicity: Toxic Effects in the Lung

CHAM E. DALLAS

In this chapter the anatomy and physiology of the lung will be related to the most prevalent mechanisms of lung toxicity resulting from exposure to commonly encountered industrial toxins. Specifically, the chapter will discuss

- Lung anatomy and physiology
- Defense mechanisms of the lung
- Different classes of chemicals that are known to damage the lung
- Four basic mechanisms by which industrial chemicals exert toxic effects on the lung
- Clinical evaluation of occupational lung injuries

When considering toxicology and the lung, it is important to note that the lung is both a target organ for many toxins and a major port of entry into the body, providing toxins the opportunity to exert toxic effects in other organs.

9.1 LUNG ANATOMY AND PHYSIOLOGY

The lung and the rest of the respiratory system provide all the cells in the body with the ability to exchange oxygen and carbon dioxide. The same system can also provide many industrial toxins with entry to (and in some cases exit from) the body. Essentially, the respiratory system is an air pump, just as the heart is a blood pump for the circulatory system. Changes in the anatomy and physiology of the lung due to toxin exposure can often result in very severe health consequences for the exposed individual. An understanding of the structure and function of the respiratory tract is essential to understanding why so many individuals in occupational exposures suffer these toxicologic outcomes.

Upper Airway

The entry-level area into the respiratory system is known as the nasopharyngeal region. The upper airway is generally considered to extend from the nose down to approximately the area of the vocal cords. Air that is inhaled into the nose enters the nasal openings and goes initially upward, then takes an abrupt turn and goes downward into the throat. Of course, humans can also choose to breathe through the mouth, in which the nasopharyngeal portion of the "respiratory tree" is bypassed. In most instances, mouth breathing entails a calculated effort on the part of the individual and has been observed when the nasal pathway is blocked or obstructed and when the individual needs to dramatically increase the volume of breathing.

Principles of Toxicology: Environmental and Industrial Applications, Second Edition, Edited by Phillip L. Williams, Robert C. James, and Stephen M. Roberts.
ISBN 0-471-29321-0 © 2000 John Wiley & Sons, Inc.

Inhaled air is highly "conditioned" before it leaves the upper airway system. Relatively cold air, for instance, will be warmed to body temperature (37 °C) before it reaches the lung. In like manner, air that is at an elevated temperature will be cooled to body temperature within the nasopharyngeal system. The lung and the portion of the respiratory system below the upper airway is a very moist physiological system and is quite sensitive to humidity. The inhaled air is therefore highly humidified during its passage from the nares to the lung, and the air that enters the nares is cleared of the larger particles. The nose hairs function to some extent in this process, and the turbulent nature of the air passages in the nares also contributes to the deposition of the larger particles, preventing them from being inhaled into the lower passages of the respiratory system.

The lining of the nasal wall is known as the *mucosa* and is highly inundated with blood capillaries. Therefore, air that is inhaled through the nose comes immediately into contact with mucosal surfaces, which only thinly separate the air from these blood vessels. Deposition of toxic chemicals in the upper airway system can therefore result in both toxicity to the mucosal tissue and absorption of the agent into the systemic circulation by way of these capillaries.

Sinus Cavities

There are four pairs of hollow cavities within the skull that are lined with a mucosal lining that is similar to the lining of the nasopharyngeal region. In order to view these sinuses from different angles, Figure 9.1 shows a frontal view of the skull, while Figure 9.2 represents a sagittal view. Since the sinuses are connected to the nasopharyngeal airways through a number of small openings, inhaled air also enters the sinuses. Acute sinusitis can occur when inhaled airborne toxins irritate the surfaces of sinus mucosa. As in other parts of the respiratory system, irritation of the mucosal lining leads to an inflammatory response in these tissues. As a result of the inflammation, there is an accumulation of

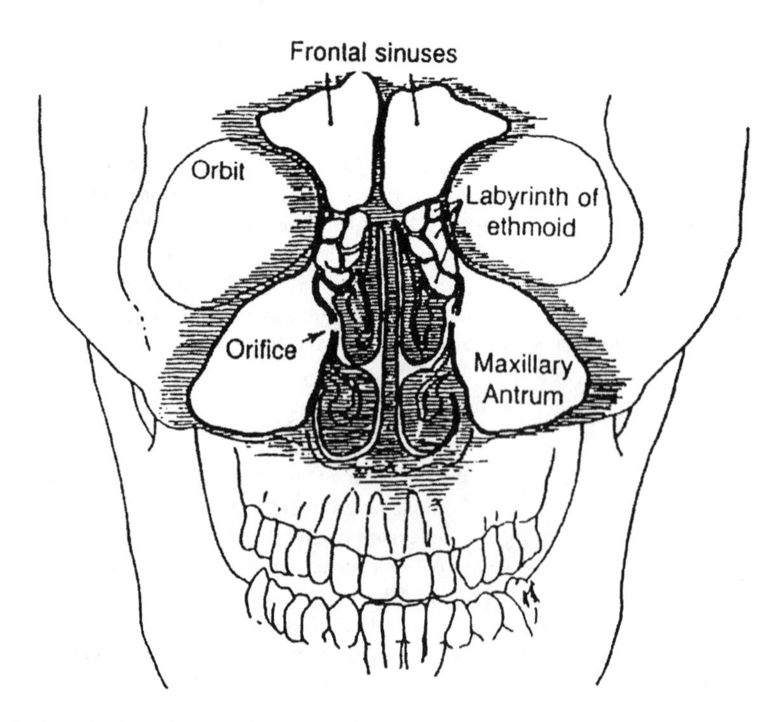

Figure 9.1 Frontal view of the skull, showing frontal, maxillary, and ethmoid sinuses. (Reproduced with permission from W. O. Fenn and H. Rahn, *Handbook of Physiology*. American Physiology Society, Washington, DC, 1964.)

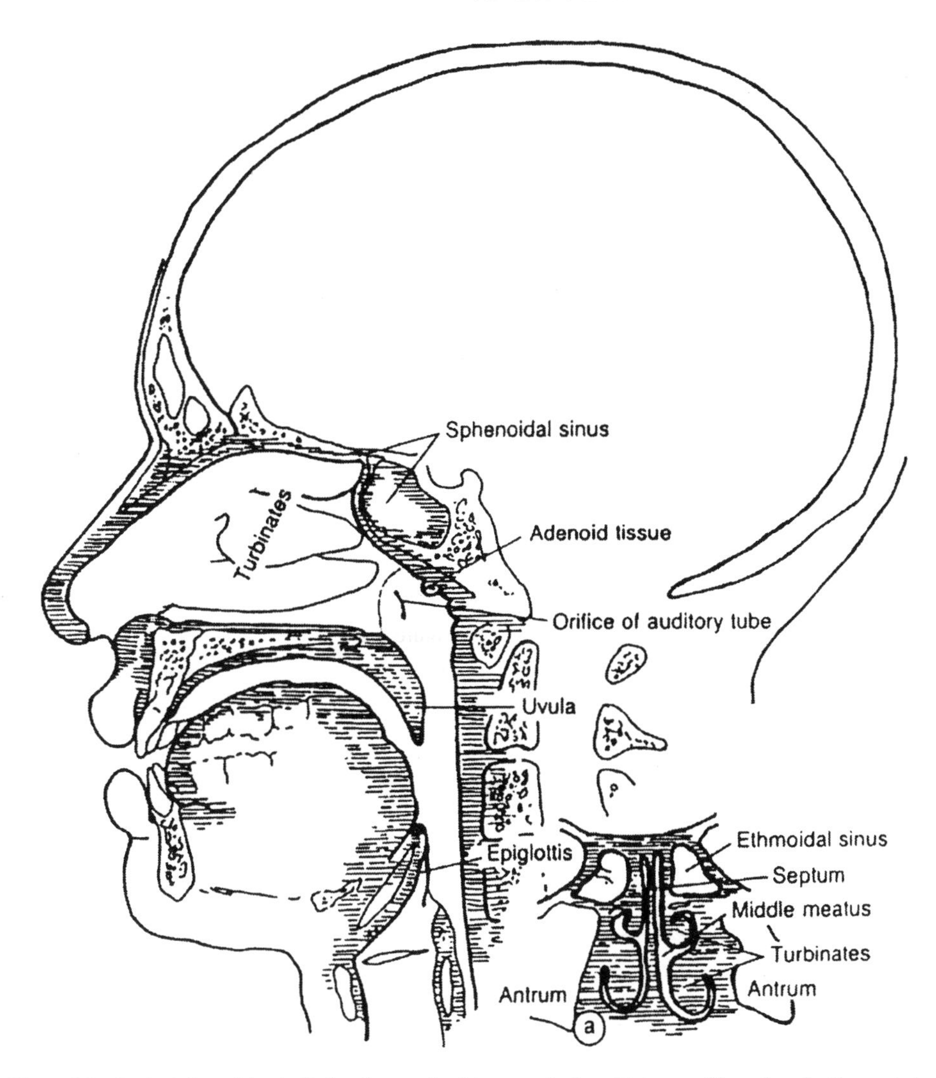

Figure 9.2 Sagittal view of the skull, showing nasal turbinates and sphenoid sinuses. [Reproduced with permission from Fenn and Rahn (1964) (see Figure 9.1 source note).]

mucous, and the poor drainage characteristics of the sinuses lead to the growth of bacteria. Some individuals who suffer some sinusitis have severe headaches while others may experience only a continuous "postnasal drip." Many factors can contribute to sinusitis, in addition to or in conjunction with inhaled toxins, such as allergic hypersensitivity, individual characteristics of the sinuses in each person, and climatic conditions.

Tracheobronchiolar Region

The trachea is a tube surrounded with cartilaginous rings that connects the nasopharyngeal region with the bronchioles. This region is essentially a conducting airway system to the lungs. The bronchi are a sequence of bifurcating branches of tubes. Each tube divides into two or three smaller tubes, and each successive branch then divides into smaller tubes, and so on. (see Figure 9.3). The bronchi themselves do not allow for the absorption of oxygen or carbon dioxide across their surfaces; they are merely

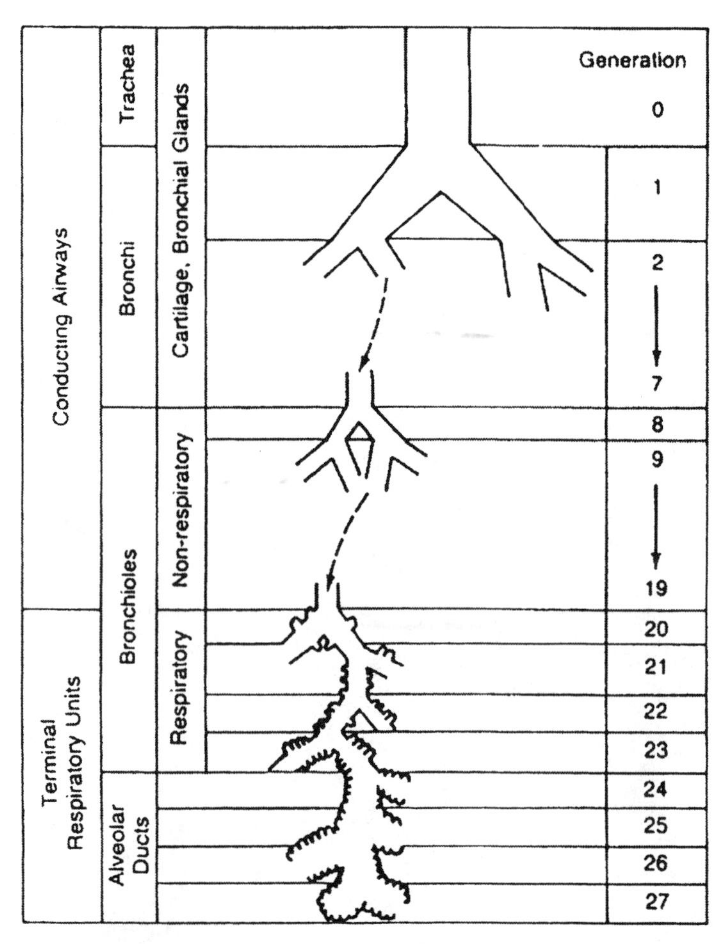

Figure 9.3 Schematic representation of the subdivisions of the conducting airways and terminal respiratory units. (Reproduced with permission from E. R. Weibel, *Morphometry of the Human Lung*, Springer-Verlag, New York, 1963.)

conducting airway tubes. In the bronchi near the lung itself, very small air sacs, or alveoli, begin to appear (at about the nineteenth or twentieth division) and increase in frequency with proximity to the lung. The bronchi in this region are known as respiratory bronchioles. It is in these alveoli that gas exchange between the inhaled air and the blood circulatory system occurs.

Pulmonary System and Gas Exchange

The number of alveoli in the lungs number in the hundreds of millions, although the size of each individual alveolus is quite small. The total surface area of the human lung, which results from the summation of these alveoli, approximates that of about one-third of the square footage of an average American home.

In each alveolus, a thin wall separates the blood in the capillary vessels from the inhaled air in the alveolus. In Figure 9.4, the terminal bronchiole and the many surrounding alveoli can be seen in relationship to the pulmonary blood supply. The wall between the blood vessel and the alveolus is a combination of the capillary endothelium, a basement membrane adjacent to the capillary, the space between the capillary and the alveolus (known as the interstitial space), a basement membrane adjacent

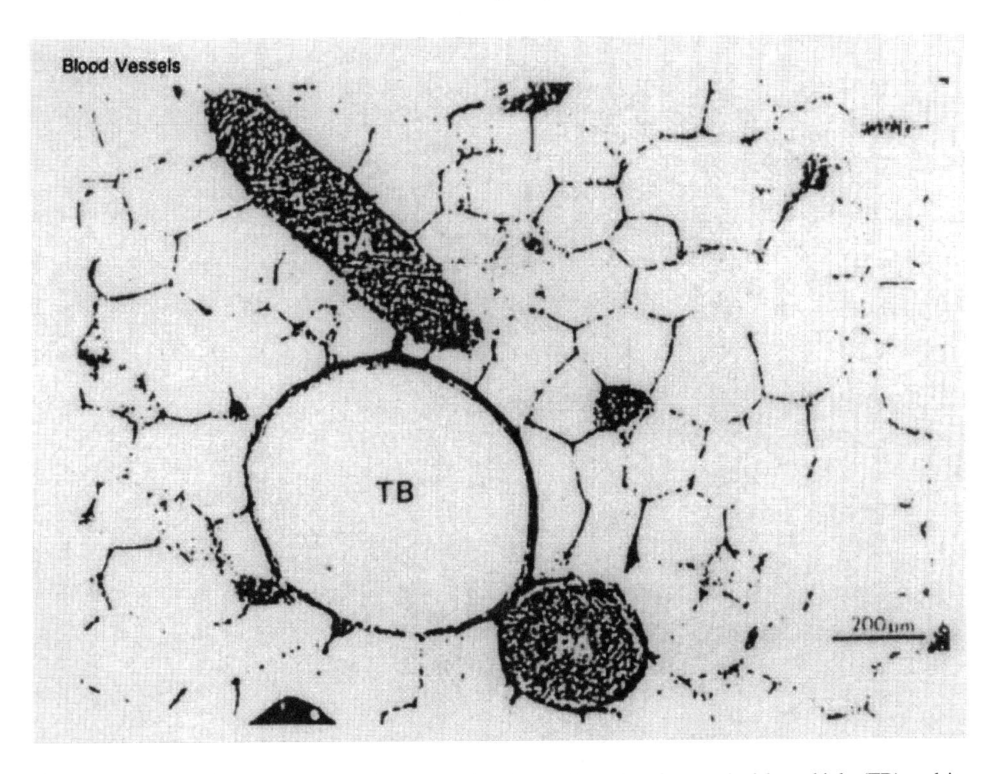

Figure 9.4 Photomicrograph of lung tissue, showing the relationship of a terminal bronchiole (TB) and its accompanying blood vessel, the pulmonary artery (PA), to the alveoli. (Reproduced with permission from J. F. Murray, *The Normal Lung. The Basis for Diagnosis and Treatment of Pulmonary Disease,* Saunders, Philadelphia, 1976.)

to the alveolus, and the alveolar epithelium. In many instances, the red blood cells are just barely able to fit through the small capillaries, so the blood cell wall is often in very close proximity to this membrane complex with the alveolus.

Figure 9.5 illustrates how the remarkable design discussed above facilitates gas exchange. Carbon dioxide and oxygen readily cross this membrane complex in a process of simple diffusion. Many inhaled airborne industrial chemicals will also readily cross this membrane and will enter the bloodstream. These potential toxins thus enter the blood circulatory system in a manner analogous to someone receiving an intravenous infusion of a drug. A unique view of the alveoli is provided in Figure 9.6. The small holes, called *pores of Kohn,* provide for some ventilation between adjacent alveoli.

Toxicologic insult to the lung as well as various disease states can result in a functional derangement of this membrane system. Exposure to some chemicals may result in an increase in fluid in the interstitial space. If sufficient fluid accumulates, a condition known as pulmonary edema, gas exchange can be hindered sufficiently to result in severe difficulty in breathing and even in death. Damage to the membrane itself can result in scarring, which may increase the thickness of the membrane or decrease the elasticity of the lung tissue, or both. As with pulmonary edema, an increase in the thickness of the membrane can deleteriously affect pulmonary gas exchange. Alterations in elasticity make the work of breathing harder, which can decrease the volume of respiration as the individual tires from the increased effort required. Of course, whenever gas exchange or the volume of respiration is sufficiently decreased, the amount of oxygen pressure in the circulatory system will also decline. If this decline proceeds to a sufficient extent, affected individuals can become seriously compromised in their health status.

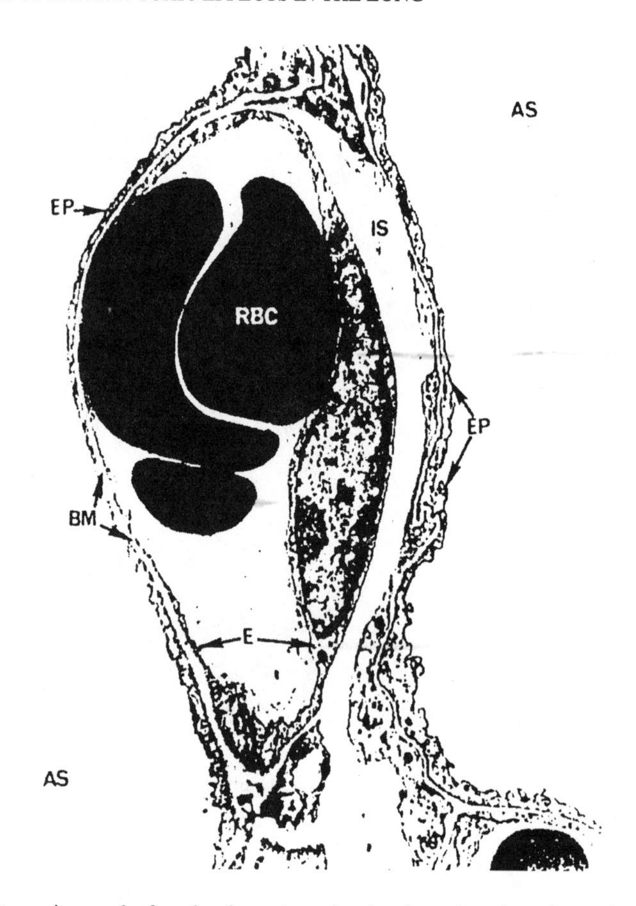

Figure 9.5 Electron micrograph of an alveolar septum, showing the various tissue layers through which oxygen and carbon dioxide must move during the process of diffusion. The surface of the alveolar spaces (AS) is lined by continuous epithelium (EP). The capillary containing red blood cells (RBC) is lined by endothelium (E). Both layers rest on basement membranes (BM) that appear fused over the "thin" portion of the membrane and that are separated by an interstitial space (IS) over the "thick" portion of the membrane. [Reproduced with permission from Murray (1976) (see Figure 9.4 source note).]

Physiologic Differences between Inhalation and Ingestion

Following inhalation, the chemical goes directly into the bloodstream without being first processed by the gastrointestinal system. This can result in an extremely rapid uptake of an industrial chemical from the air. For some chemicals, this also results in an extremely rapid onset of toxicity following inhalation of the agent.

Inhalation of a chemical might also result in a higher degree of toxicity than if the compound were ingested. This is because a chemical absorbed from the gut will go first to the liver, which is the primary metabolizing organ of the body. The liver thus has the opportunity to eliminate the compound before it exerts its effect in some other target organ. This is called the *first-pass effect.* When the chemical is inhaled, it bypasses the liver and the toxin has the opportunity to reach a specific target organ and exert some degree of toxicity before the liver has the opportunity to eliminate it.

Particulates

Many chemical and radionuclide agents are deposited in the respiratory tract in the form of solid particles or droplets, also referred to as *aerosols,* meaning a population of particles that remain

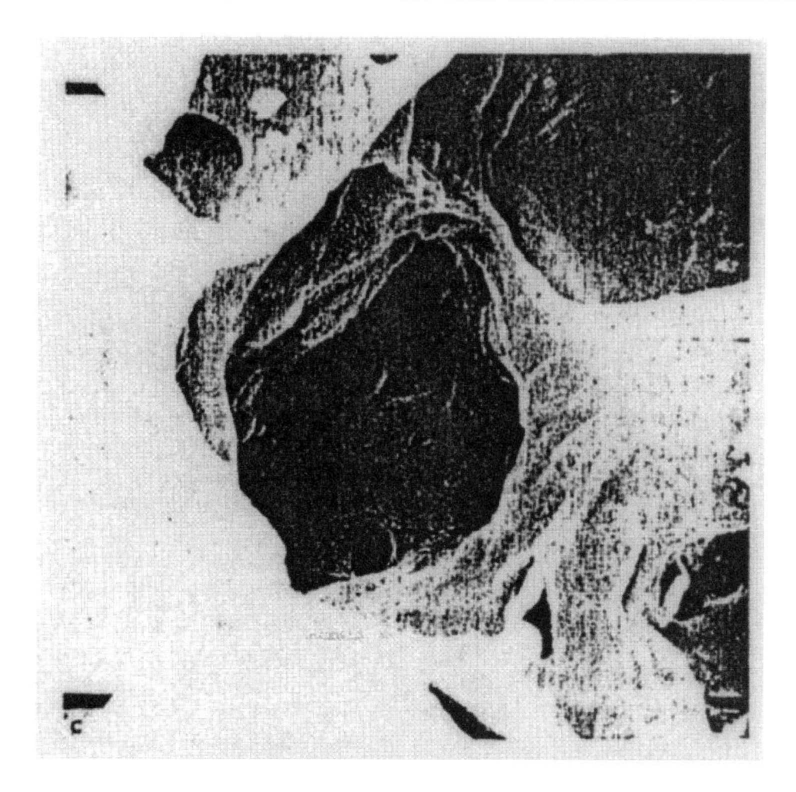

Figure 9.6 Scanning electron micrograph showing interior of an alveolus and its pores of Kohn. (Reproduced with permission from D. V. Bates *et al.*, *Respiratory Function in Disease. An Introduction to the Integrated Study of the Lung*. Saunders, Philadelphia, 1971.)

suspended in air over time. Some terms that are also used are dusts, fumes, smokes, mists, and smog. Dusts, which result from industrial processes such as sandblasting and grinding, are considered to be identical to the compounds from which they originated. In contrast, fumes usually result from a chemical change in compounds during processes such as welding, in which combustion or sublimation occurs. Smokes result when organic materials are burned; mists are aerosols composed of water condensing on other particles; and smog is a conglomerate mixture of particles and gases that is prevalent in certain environments such as areas with mountains, plenty of sunlight, and periodic temperature inversions. The toxicity of inhaled particulates has been known for a long time, especially in relation to occupational exposure. The early (1493–1541) famous toxicologist Paracelsus described the relationship between mining occupations and pulmonary toxicity in the sixteenth century.

Particle Size

In the case of particulates, size is the primary critical determinant of how much of and where the agents will be deposited. The range in particle size for various aerosols is generally as follows: dusts, up to 100 μm; fumes, from 10 Å to 0.1 μm; smokes, less than 0.5 μm. The pattern of airflow in the respiratory system and anatomic features of the exposed individual are also important.

Most inhaled particles are not spherical, but highly irregular in shape. In order to categorize the highly heterogeneous nature of inhaled particles, the aerodynamic diameter is calculated for the population of particulates of interest. This value is based on the settling velocity of the population of particles and roughly approximates what the particles' diameter would be if it were compared to a spherical particle in the time it takes the particles to settle in the air. This calculation is also referred to as the *mass median aerodynamic diameter.* If the number of particles is of primary interest (and not necessarily particle shape), the count median diameter is determined. Of course, the size of particles may change during the course of traversing the respiratory tract. Since the respiratory tract is highly humidified, particles that absorb water could be expected to undergo chemical reactions and increase in size as they descend.

Lung Deposition Mechanisms

Particles tend to deposit in the lung according to size, air velocity, and regional characteristics of the respiratory system. In the nares, nose hairs tend to block out the very large particles that enter the nose. Once inside the nares, the abrupt turn in the nasopharyngeal system of humans (from going up to going down) results in the impact of many of the larger particles on the walls of this region of the respiratory system.

This mechanism, referred to as *impaction,* results from the aerodynamic tendency of particles to travel in a linear direction, even when the respiratory system is turning and branching. An analogy would be a bifurcating freeway system, in which the safety department will often place barrels at the point of bifurcation since cars are most likely to strike this location. In a similar manner, particles are more likely to strike the points of bifurcation in the respiratory system.

A related mechanism of deposition is known as *interception*. This process occurs when a particle comes close enough to contact a respiratory surface and, subsequently, deposits there. Interception does not have to occur at the bifurcations or turns and is mostly a factor in the deposition of fibers, which are much longer than other forms of particles. It is not uncommon for a fiber to be only a few μms in diameter and several hundred μms in length, so the probability of contact with the respiratory surfaces is enhanced.

In the tracheobronchiolar region, the declining airflow allows gravitational influences to result in the deposition of particles in the 1–5 μm range. This process, referred to as *sedimentation*, increases in frequency as the particles in this size range descend lower into the bronchiolar tree. Sedimentation can also occur in the alveolar region, but the simple process of *diffusion* will result in the deposition of particles in the 1-μm range.

Clearance Mechanisms

The respiratory system has an extraordinary design for the clearance of particles and other toxins. Generally, the clearance mechanism is related to the site of deposition. This respiratory clearance should not be confused with total body clearance or systemic clearance in the pharmacokinetic sense. Respiratory clearance removes particles and other toxins from the respiratory tree; ultimate removal from the body is achieved through the gastrointestinal system, the lymphatics, and the pulmonary blood.

In the nasopharyngeal and tracheobronchial regions, there is a *mucociliary escalator* mechanism. In the respiratory wall, there are pseudostratified columnar epithelial cells together with specialized goblet cells, which produce a layer of mucous along the wall of epithelial cells. Hundreds of cilia, which resemble small hairs, protrude from the epithelial cells (Figure 9.7). The mucous itself is in two layers: the lower layer, known as *sol,* contains the cilia and is thin and watery so that cilia movement is not impeded; the upper layer, the *gel,* is thick and viscous. The cilia beat in unison and move the gel layer along like a continuous sheet (Figure 9.8). Inhaled particles and other toxins become trapped on the gel layer. In the tracheobronchial region, the cilia beat upward, and the entrapped particles in the gel are propelled up toward the mouth. Typically, an individual will solubilize the material in saliva, which is then eliminated via the gastrointestinal tract. Occasionally the material may be coughed out of the body. In the nasopharyngeal region, the cilia beat downward toward the mouth and rely on the same mechanisms of removal. Typically, mucociliary clearance will occur within hours of the deposition of most particles, and in healthy individuals, the process is usually completed within 48 h.

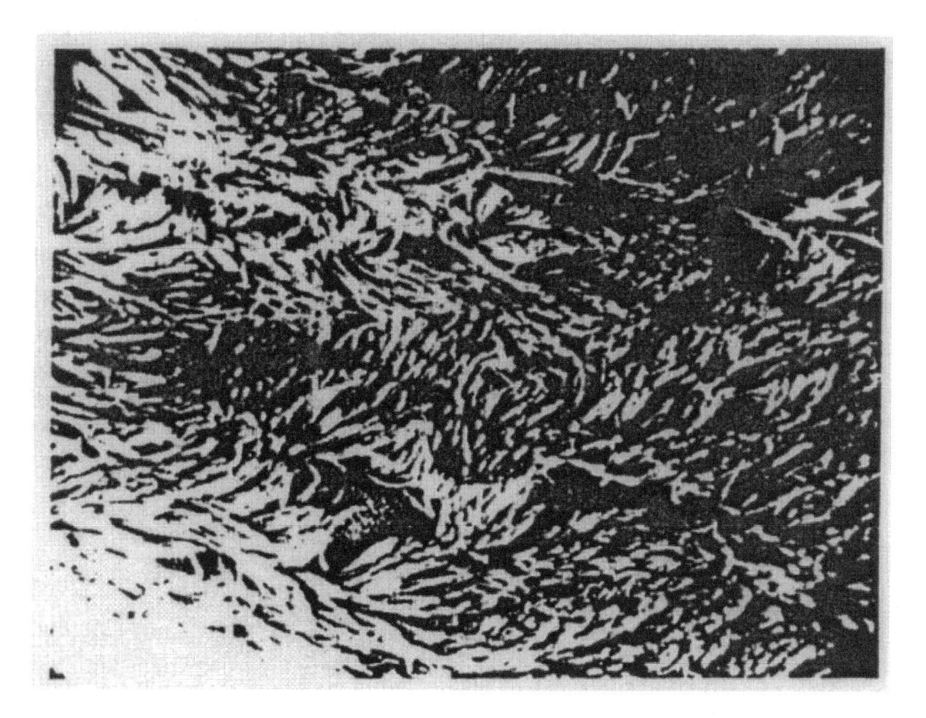

Figure 9.7 Scanning electron micrograph of the luminal surface of a bronchiole, showing the cilia. The mucous layer has been removed. [Reproduced with permission from Ebert and Terracio, "The Bronchiolar Epithelium in Cigarette Smokers," *Am. Rev. Resp. Disease* **111**, 6 (1975).]

In the alveolar region, *macrophages* provide a mobile and effective defense against particles, bacteria, and other offensive agents that reach the lower respiratory tree. Chemotactic factors are released when these inhaled agents deposit in the lung, and these factors alert the phagocytic cells to the location of the agents. The macrophages then engulf them and attempt to ingest them with proteolytic enzymes. An example of a macrophage moving from one alveolus to another through a connecting pore is shown in Figure 9.9. A very wide variety of potentially toxic agents, including viruses, bacteria, chemicals, and particles of many sizes, can be successfully broken down by macrophages. However, in certain situations, such as in unhealthy individuals (e.g., long-term smokers), the macrophages might be inefficient or in lower numbers, and this defense might be abrogated to a significant extent. Additionally, some particles are not particularly digestible by the macrophages. In such cases, as with tuberculosis infections and with some fibers, the macrophage may

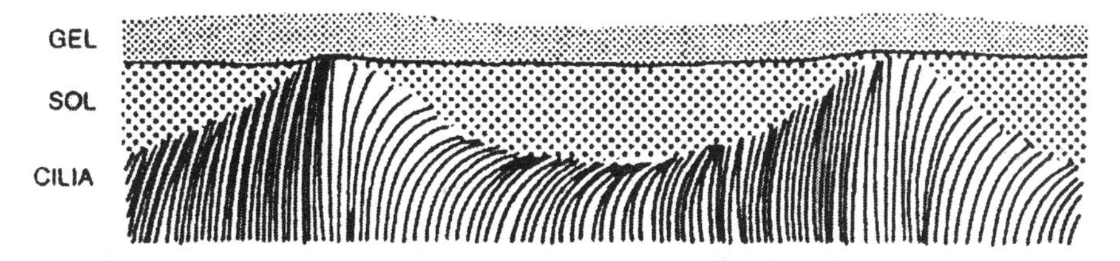

Figure 9.8 Schematic representation of the mucociliary blanket, showing the wavelike motion of the cilia within the sol layer. [Reproduced with permission from A. C. Hilding, "Experimental studies on some little-understood aspects of the physiology of the respiratory tract," *Trans. Am. Acad. Ophthalmol. And Otol.* (July–Aug. 1961).

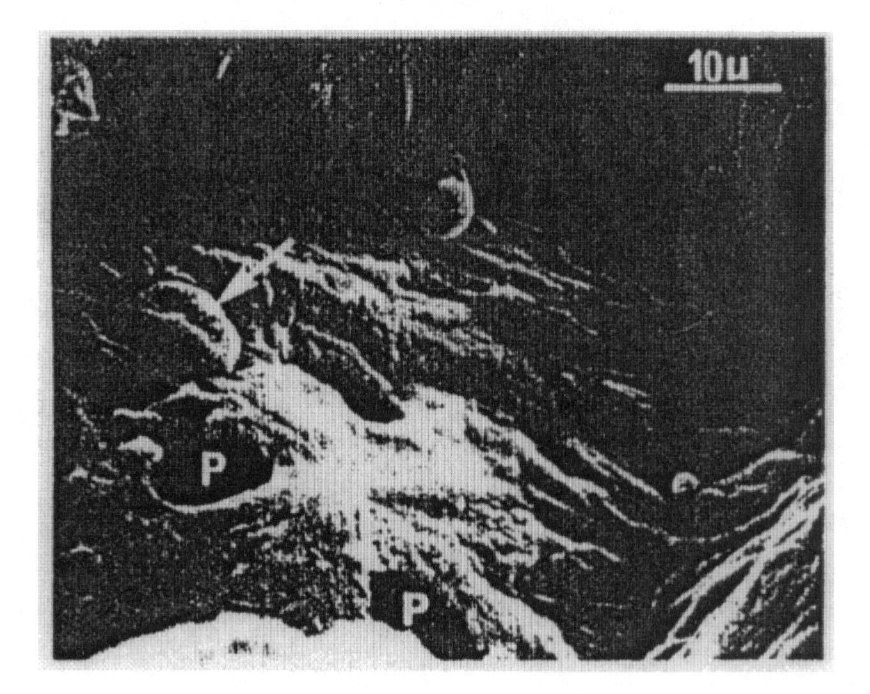

Figure 9.9 Scanning electron micrograph of the interior of an alveolus showing pores of Kohn (P) and a macrophage (arrow). [Reproduced with permission from Murray (1976) (see Figure 9.4 source note).]

rupture and spill the proteolytic enzymes into the lung tissues and damage them. If successful phagocytosis has occurred, the phagocytized material is then removed by either the mucociliary escalator or by lymphatic drainage. The action by the macrophages is initially very rapid, with inhaled particles engulfed by some macrophages within minutes of inhalation.

Gases and Vapors

Many injuries to the lung and to distant organs have been known to occur following inhalation exposure to gases and vapors, especially in the workplace. Most industrial chemicals can exist in the gas or vapor state under certain situations, and various industrial processes can create even the fairly extreme physicochemical conditions necessary to vaporize potentially toxic agents. Everyday in the workplace, millions of workers are exposed to countless potentially toxic chemicals in the form of gases and vapors.

The potential for highly toxic outcomes from inhalation exposures to gases and vapors is related to the fact that once they are inhaled into the lung, they can pass directly into the bloodstream. In a pharmacokinetic sense, inhaled gases and vapors are injected into the bloodstream as a patient would receive a drug through an intravenous (or intraarterial) infusion. Once a gaseous chemical enters the alveolar spaces of the lung, it can cross the relatively permeable *alveocapillary membrane* complex and enter the pulmonary blood. This complex consists primarily of the capillary and alveolar membranes, separated by an interstitial space (sometimes with fluid in it). The lining of the alveolar membrane also has a lining of surfactant (dipalmitoyl lecithin), which serves to equalize the inflation pressures of the heterogeneously sized alveolar sacs.

The passage of the inhaled gases and vapors across the alveocapillary membrane complex, or the diffusion efficiency, is influenced by several factors. The solubility of the inhaled compound is important, as highly water-soluble compounds are often more likely to deposit in the upper respiratory system, before reaching the alveolar regions of the lung. The condition of the alveocapillary membrane is also important. Poor health conditions in a patient might lead to the engorgement of the interstitial space with fluid, which would impair the diffusion of toxic chemicals across the alveocapillary membrane. While this protects the affected individual from the toxic effects of the inhaled chemical, it also prevents the free exchange of oxygen and carbon dioxide, which can have obvious life-threatening outcomes.

The degree of uptake of inhaled gases and vapors can be quite significant in workers in many occupations. Following the initiation of inhalation, rapid uptake of perchloroethylene, a commonly used dry cleaning solvent for which there are thousands of potential exposures, can be observed in many different tissues (Figure 9.10). In this case, the uptake of perchloroethylene in circulating blood and seven tissues was remarkably rapid, and for many industrial chemicals, it is often within minutes of exposure. It is often interesting to note that the levels of the inhaled solvent remained fairly constant throughout the inhalation exposure period. This can have important ramifications in occupational exposures, as workers who enter an environment with a potentially toxic gas can experience systemic toxic effects almost immediately, and these effects can persist for long periods of time (while the inhalation exposure period continues). For instance, many industrial solvents cause neurobehavioral depression following inhalation exposure, and workers have been known to be injured as a result of falls or mishaps with industrial machinery almost immediately after breathing the chemicals.

Obviously, the length of exposure affects the amount of chemical inhaled. However, for many gases and vapors a steady-state equilibrium can be established after a certain period of inhalation exposure. In this way, the level of chemical in the blood does not continue to increase, despite the continued inhalation exposure to the compound (Figure 9.10). This has important ramifications in industrial exposures because it helps explain why workers sometimes do not experience toxic effects to certain chemicals despite long-term exposure.

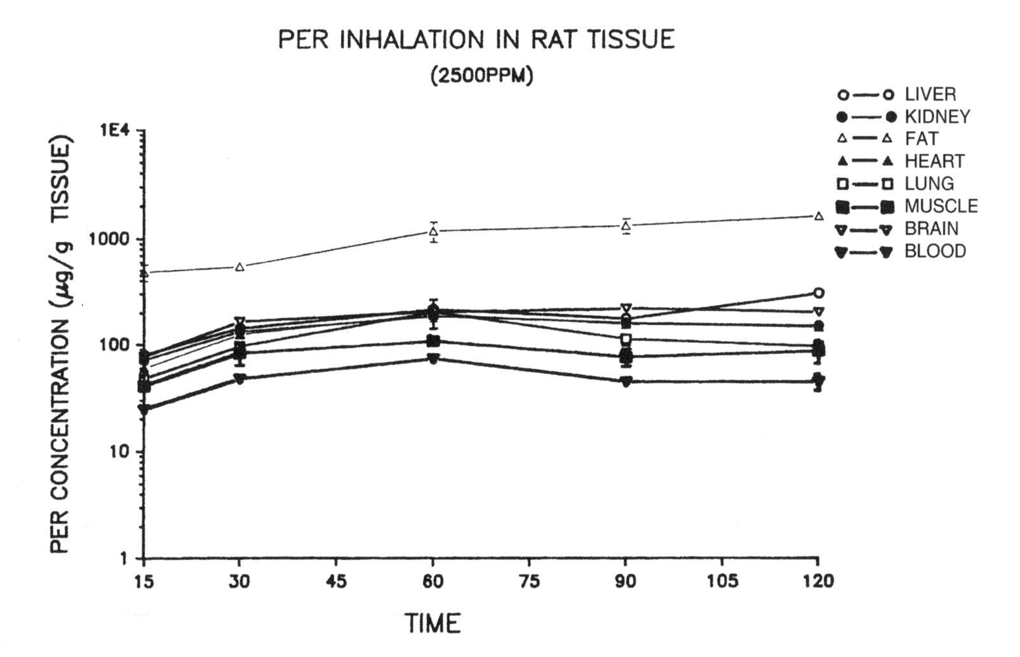

Figure 9.10 The uptake and disposition of perchloroethylene (PER) in the blood and seven tissues of laboratory rats is shown. The animals inhaled 2500 ppm of perchloroethylene for 120 min in dynamic inhalation exposure chambers, and blood and tissues were analyzed for the solvent by electron capture-gas chromatography. (Supported by US Air Force Grants AFOSR 870248 and 910356.)

Air-Pollutant Gases

Many of the air pollutants are inhaled as gases, such as carbon monoxide, sulfur dioxide, and the various oxides of nitrogen. By far, the number one killer as far as toxic gases are concerned is carbon monoxide. The incomplete burning of various fuels results in the emission of carbon monoxide, and every year there are many deaths and injuries from individuals who breathe this gas in an enclosed space. While some of these are suicides, there are also many industrial exposures to carbon monoxide and other combustion pollutants. A number of air pollutant gases are produced by a complex interaction of sunlight, humidity, temperature, hydrocarbons, and the oxides of nitrogen. These interactions generate smog, as well as other gases such as ozone and the aldehydes.

Tobacco Smoke

Toxicity resulting from the intentional and unintentional inhalation of tobacco smoke is an important consideration given its enormous magnitude of incidence, its interaction with the toxicity of other inhaled industrial pollutants, and its representation of the toxicity of both particulates and gases. The number of people who die and are significantly injured each year in the United States due to inhalation exposures to industrial chemicals cannot be stated with certainty; however, it is definitely much smaller than the number of people who die and are experiencing diminished health status as a result of tobacco smoke inhalation. The smoking of tobacco products causes pulmonary emphysema, chronic bronchitis, and lung cancer in many thousands of Americans each year.

Interference with Pulmonary Defense Tobacco smoke inhalation results in the derangement of the pulmonary defense mechanisms necessary to protect against the inhalation of industrial toxins. It has been shown that, following chronic cigarette smoking, the cilia in the mucociliary escalator become increasingly paralyzed. The decrease in ciliary activity slows or prevents the removal of deposited toxins from the nasopharyngeal and tracheobronchial regions, as the gel layer becomes more sedentary. Many of the more than 2000 components of tobacco smoke are known to be respiratory irritants, and these irritating properties lead to an increased production of mucous in the respiratory system. Therefore, there is a decreased movement (and removal) of mucous simultaneously with an increase in mucous production. Eventually, some of the airways can become impeded and even blocked, severely limiting the respiratory volume of the affected individual. Sometimes the overworked mucous glands will increase in size sufficiently to block the airways themselves, further impeding airflow and increasing resistance.

It has been shown that the cellular defense mechanisms of the lung, particularly the alveolar macrophages and the alveolar polymorphonuclear leukocytes, are significantly impacted by tobacco smoke inhalation. In many cases, these cells may be killed, causing the release of proteolytic enzymes, which come in contact with the respiratory membrane surfaces. Pulmonary emphysema can result, if this process is extensive, from the severe rupturing of the septa walls. Even short of cell death, these cells become less efficient in the removal of particulates and other toxins. Therefore, the inhalation of toxic agents in industrial environments has the potential to exert greater toxicity in smokers than in equally exposed nonsmokers. This has been shown repeatedly for many exposures to toxic chemicals in occupational studies, such as with asbestos. For this reason, occupational epidemiologists and physicians will often look for correlations between toxicity in an industrial worker population and tobacco use.

Lung Cancer and Tobacco Smoke Bronchogenic carcinoma data from the 1980s estimated that approximately 90 percent of the more than 100,000 lung cancer cases each year in the United States are due to tobacco smoke inhalation. A very distressing aspect of this unpleasant data is that the incidence of lung cancer, previously occurring more often in men, is growing rapidly in the female population. The increasing incidence of tobacco smoke inhalation by women has been followed in an appropriate timeframe by an explosion in lung cancer cases in women. Whereas breast cancer was previously the number one cause of cancer deaths in women, now this dubious honor is being replaced by lung cancer, as is the case in men. Women are also entering the industrial environment in increasing numbers, pursuing occupations previously held predominantly by men. This now incites the question of whether there will be a correlation between this increased smoking incidence among women and the incidence of cancer from industrial chemicals.

9.2 MECHANISMS OF INDUSTRIALLY RELATED PULMONARY DISEASES

Irritation of Respiratory Airways

One of the most common toxicity manifestations from inhaled agents in industrial exposures is the irritation of the airways, resulting in breathing difficulties and even death for the exposed individual. Often, this response results from bronchoconstriction, as the airways react to diminish the extent of the unwanted exposure. This can be a protective mechanism, if the affected person can quickly remove himself/herself or be removed from the offending agent. Of course, diminished inhalation over any extended period of time has obvious deleterious effects for the worker.

The chemical warfare agents, chlorine and phosgene, exert immediate toxicity by airway irritation. If the level of exposure is sufficient, the exposed individual can die within minutes of the initiation of exposure. Often a high dose exposure is accompanied by *dyspnea* (difficulty in breathing, either real or perceived), cough, lacrimation (tears), nasopharyngeal irritation, dizziness, and headache. The dose response for chlorine exposures is summarized in Table 9.1.

An interesting aspect of most industrial inhalation exposures involving the irritation of the airways is that the symptoms appear very serious at first, but seldom result in permanent respiratory damage. The coughing and choking are very alarming to both the affected individual and onlookers (including medical personnel), and at least should result in the injury being taken seriously (which is often a problem in industrial toxicity episodes). Chest X rays and pulmonary function tests should be conducted on these individuals, in case there are permanent or late onset toxicity manifestations such as pulmonary edema. Although most of these individuals will recover completely, many people have died from irritation of the airways following industrial chemical inhalation, and every incident must be treated as a serious episode. It is highly recommended that workers have a baseline pulmonary function test on file with which to compare after an irritant exposure.

Fibrosis and Pneumoconiosis

A variety of lung diseases resulting from the inhalation of dusts has been encountered in occupational environments. The disease mechanism, known as *fibrosis,* results when the lung gradually loses elasticity as a result of the pulmonary response to long-term dust inhalation. The disease condition is referred to as *pneumoconiosis,* derived from the Latin and Greek root words *pneumo,* which means breath or spirit, and *coniosis,* which means dust.

TABLE 9.1 Chlorine Dose–Response Relationships

<4 ppm	Can be tolerated up to 30 min
15 ppm	Severe respiratory symptoms begin
30 ppm	Coughing, choking, chest pain
>40 ppm	Pulmonary edema
>1000 ppm	Immediate death

Silicosis

Following long-term inhalation of silica-containing dusts, many workers have developed irreversible lung damage known as silicosis. One-half to two-thirds of the rocks in the crust of the planet contain silica, so it is to be expected that many industrial processes result in the production of silica-containing dusts. While some of the inhaled silica dioxide crystals will deposit in the nares and on the mucociliary escalator, a certain number will reach the alveolar regions of the respiratory system. Unfortunately, the alveolar macrophages that ingest the silica particles will be damaged by the silicic acid produced following phagocytosis. Damaged and killed macrophages will release phagocytic enzymes into the alveolar sacs, which will result in their progressive destruction over time. This eventually results in a "stiffening" of the lung tissues, which makes breathing more difficult for the affected patient. Over a long period of time, the body will try to wall off the area, resulting in the development of a silicotic nodule. Patients with advanced silicosis often have greater susceptibility to respiratory infections such as tuberculosis. In any one patient, one might find each of these stages located in the same lung. Even after an individual has been removed from the further inhalation of silica dust, this progressive deterioration will continue. Another negative aspect of the disease is that it is very difficult to treat, and currently, clinicians can do little more than alleviate symptomatic suffering.

Asbestosis

The highly effective flame retardant asbestosis has been used for centuries, and in the past few decades, it has been used in industry for a variety of purposes. Many thousands of workers have received very high doses of asbestos in the shipbuilding industry. Usually, insulation workers were exposed to asbestos dust in very enclosed spaces, which tended to increase the concentration of the inhaled fibers. Countless individuals have been exposed to asbestosis fibers while working with the brake linings of cars. Chrysotile, or "white" asbestos, accounts for about 90 percent of the asbestos in industrial applications; the amphiboles account for most of the other potential exposures, in which crocidolite, or "blue" asbestos, is the most important (and was the first form found to be carcinogenic).

The insidious nature of asbestosis is that major symptoms seldom appear until 5–10 years (or longer) after the inhalation of the asbestos fibers. As with silicosis, the inability of macrophages to digest the fibers leads to a progressive fibrosis of the lung tissue. However, with asbestosis there is also pleural thickening and calcification, which can be picked up by X-ray examination in the relatively early stages of the disease. Pleural calcification may exist in patients when there are no other symptoms present. Pulmonary function tests are often useful, in that decreases in compliance and total lung capacity are observed. A pathologic finding in asbestosis is the appearance of "asbestos bodies," which are structures formed by the protein encapsulation of asbestos fibers that resemble a "barbell" in weight lifting (the protein is thicker on the ends). Asbestosis eventually leads to the development of malignant neoplasms in the respiratory tract. One form of cancer, mesothelioma, is so rare in situations outside of asbestos exposure that many physicians consider it a "marker" disease for asbestosis. A higher incidence (up to an 80-fold increase) of bronchogenic carcinoma is distinctly correlated with tobacco smoke inhalation and asbestos exposure. These asbestos related cancer deaths generally occur from 25–40 years after the asbestos inhalation.

Excess Lung Collagen

Most types of pulmonary fibrosis involve distinct changes in the proportion of the types of lung collagen that is produced in the affected lung. Such information is used by pathologists today in determining the degree of pulmonary fibrosis that has occurred. In most normal lungs, the two most common collagen types, type I and type III, are observed at a ratio of approximately 2:1. When pulmonary fibrosis occurs, there is generally an increase in type I collagen in relation to type III collagen. Mechanistically, the presence of the fibers causes macrophages to release lymphokines and various growth factors, which leads to an increase in the production of certain collagen types. Since type III is considered to be more compliant than type I, this might be the cause of the "stiffening" of the lung tissue, but this is not known for certain.

Emphysema

Whenever inhaled toxins result in the progressive destruction of the alveolar walls of the lung tissue, there is an enlargement of the lung air spaces accompanied by a decrease in the surface area of the lung available for gas exchange. This is commonly referred to as *emphysema,* and it is a relatively common pulmonary disease condition in the United States. Although emphysema is due primarily to tobacco smoke inhalation, a number of inhaled industrial toxins may also be responsible for the development of emphysematic conditions. For instance, the inhalation of coal dust by miners over extended periods has been shown to result in both pulmonary fibrosis and emphysema.

Recent research has indicated that a genetically related deficiency in α-1-antiprotease, of a biochemical inhibitor of elastase, is clinically related to the relatively early onset of emphysema. It is believed that the breakdown of the alveolar walls is modulated by elastases, which are released by neutrophils and perhaps alveolar macrophages, and if the α-1-antiprotease enzyme is genetically absent or decreased, this results in a higher incidence of emphysema. In this scenario, if an inhaled toxin causes increased migration of the normally protective cells (neutrophils and macrophages) to the site of the inhaled toxin deposition, then these cells may end up damaging the lung tissue in addition to eliminating the toxins.

Pulmonary Edema

Many inhaled agents produce sufficient cellular toxicity to cause an increase in the membrane permeability of the alveocapillary membrane complex of the lung and other airway linings. This results in an increase in fluid, either in the interstitial space of the alveocapillary membrane complex or on the surface of the airways or alveolar sacs. This increase in fluid is called *edema,* and its presence impedes the exchange of oxygen and carbon dioxide between the alveolar air and the pulmonary blood. If the decrease in gas exchange proceeds sufficiently, the affected individual can die, literally in their own fluids.

Among the many agents that result in pulmonary edema are the air pollutant gases, such as nitrogen dioxide and ozone. These agents typically exert their lung toxicity at relatively low levels of exposure in air-pollution episodes, but in industrial exposures, workers may be exposed to considerably higher concentrations. Chlorine and phosgene, two of the more potent inducers of pulmonary edema, were shown to induce thousands of deaths when used as chemical warfare gases in World War I. Recently, it was reported that the Iraqi military has used one or both of these agents against the Kurdish minority in that country. Since chlorine is now the primary chemical used to keep water supplies clean, its industrial use has soared. Municipalities use chlorine for their drinking water treatment; therefore, its geographic distribution is widespread. Large-scale releases of chlorine have occurred during transport to these disparate localities, and there have been a number of fatalities from pulmonary edema following chlorine inhalation. Phosgene is also used frequently in industry; however, strict industrial hygiene controls, due to the extreme toxicity of the chemical, has resulted in a low frequency of worker injury. Other agents known to cause pulmonary edema include nickel oxide, paraquat, cadmium oxide, and some industrial solvents.

The delayed onset of pulmonary edema in most cases of chemical inhalation results in a significant hazard for exposed workers. Usually, the edema fluid is not readily detected by the exposed individual or by clinical examination for at least several hours after the termination of exposure.

In a typical occupational exposure, the worker may experience short-term symptoms involving irritation of the airway, which influences them to seek immediate medical assistance. Since the short-term symptoms usually have no immediate cytotoxic sequelae, the medical examination will result in no revelation of significant morbidity, and the patient will be released. Then, 4–24 h later, the pulmonary edema rapidly develops, usually while the patient is asleep. Often, when patients awake with difficulty breathing, they are already in an advanced stage of pulmonary decline, and the condition is difficult to treat. It is critical that individuals who have been exposed (or potentially exposed) to agents known to cause pulmonary edema, be kept overnight (or at least 24 h following the exposure) at a medical facility where they can be closely monitored. A series of chest X rays during the "critical period," when pulmonary edema could be initiated, should be taken and examined for the appearance of fluid in the lung.

Respiratory Allergic Responses

Among the potential allergic reactions of the respiratory system in industrial exposures, there are many well-characterized conditions, as well as somewhat mysterious and hard-to-define personnel histories. Many of the characterized diseases have historically involved certain occupations and are often named after the occupations in which they were first observed. The allergic reactions involve antibody formation against certain inhaled toxins or to dusts and organic particles. Subsequent exposure to the same agent then often results in a more severe reaction, which is understandably a real problem in the workplace where individuals often work in the same environment and receive repeated exposures. In the less characterized occurrences, it often appears that exposure to one agent might result in a nonspecific reaction to a multitude of other compounds inhaled at some later time.

Occupation-Related Inhaled Allergic Disorders

A very old disease, known as "farmers' lung," involves the allergic reaction to the Actinomycetes spores found in hay. Hay that is collected in the field is often damp, and the high temperatures that can arise inside damp hay over time may give rise to large numbers of the thermophilic Actinomycetes spores. When the farmers inhale these spores, IgG antibodies are produced (against the spores), and subsequent exposures result in potentially severe allergic reactions. An interesting aspect of the disease is that the time interval between the initial exposure and the expressed toxicity can be highly variable. Various aches and pains, fever, chills, cough, weight loss, and malaise accompany the condition, which is often confused with pneumonia. Over the long term, fibrosis can also materialize. "Malt worker's lung," contracted from the dust of bird droppings, presents with similar allergic alveolitis and has been reported in individuals in the whiskey industry. "Cheese washer's lung" has been reported in the widespread cheese industry. Ironically, this condition is due to *Penicillium* spores. In the lumber industry, "maple bark stripper's disease" results from the inhalation of fungus particles, particularly *Cryptostroma. Bagassosis* results from the inhalation of the bagasse dust left behind after the moisture has been removed from sugar cane stalks. Once the disease is in progress, the worker must be removed from any further contact with the bagasse dust, or the symptoms are likely to return and will usually get progressively worse.

In the textile industry, the inhalation of cotton dust and other organic fibers has long been associated with reactive airway disturbances known as *byssinosis*. Individuals with this condition complain of chest tightness, wheezing, and other respiratory difficulties. It should be noted that these symptoms might appear after a short, or even an extended, absence from the industrial setting. A particular pattern seems to be that the first day back at work after a break, such as a weekend, is the most likely time for an episode. Unlike the previously cited occupational diseases, byssinosis does not appear to be necessarily related to the presence of bacteria, fungus, or some other living organism; the cotton or textile dust is the only requirement. Bronchoconstriction results from the release of histamine and 5-hydroxytryptamine following inhalation of the cotton dust. If the affected workers are removed from the environment containing the offending dusts relatively early in the process (i.e., months or very few years), then the patients appear to recover without permanent lung decrements. Long-term development of the disease, however, has been shown to result in permanent injury. In addition, the symptoms associated with byssinosis are usually more severe in smokers than in nonsmokers.

Industrially Related or Occupational Asthma

Many individuals develop asthma following workplace exposure, and some asthmatics suffer additional provocation following the inhalation of certain industrial toxins. The inhalation of wood dusts, for instance, has been implicated in both situations. Some grocery workers have developed an asthmatic condition following the wrapping of meats with plastic film. Apparently, heating the plastic to seal it releases toluene diisocyanate, which is then inhaled. Subsequent exposure to even very low levels of the plastic, or its component, may result in a severe reoccurrence of symptoms.

It has been shown that the bronchiolar muscles of asthmatics will undergo constriction at a lower concentration of inhaled industrial chemicals than will those of nonasthmatics. Not surprisingly, these individuals often find themselves reacting in situations in which their co-workers do not respond. A further complication for these workers is that exercise tends to exacerbate the asthma symptoms. Physical exertion, obviously required in many industrial situations, along with the simultaneously chemical exposure can lead to severe complications for the affected worker.

Lung Cancer

Until the twentieth century, lung cancer was relatively rare. The rapid promotion of lung cancer to the number one cancer killer is directly related to the inhalation of tobacco smoke (probably 80–90 percent of all lung cancers) and industrial/atmospheric chemicals. The relationship between tobacco smoke inhalation and lung cancer was discussed previously. Many industrial chemicals have also been linked to lung cancer in workers and laboratory animal studies.

The dusts and fumes of many metals have been demonstrated to be carcinogenic in lung tissue. Epidemiologic studies conducted on worker populations in smelting operations have long shown definitive relationships between metal inhalation and lung cancer. Industrial metal carcinogens include nickel, arsenic, cadmium, chromium, and beryllium. Workers in mining operations, including metal recovery from ores, are at risk for developing lung cancers because of exposure to certain metals such as chromium and uranium. The inhalation of benzo(*a*)pyrene and other polycyclic aromatic hydrocarbons, from coke oven emissions, has also been linked to the development of lung cancer.

Radioactive materials have long been recognized as inducers of lung cancer. Uranium miners have an elevated incidence of lung cancers, as did the victims of the atomic bomb explosions at Hiroshima and Nagasaki. Recently, the potential for inhalation of radon gas has become a concern, due to the large population with the possibility for long-term exposure. Smoking has been shown to exacerbate the incidence of lung cancer when in conjunction with exposure to radioactive materials.

An important feature regarding the development of lung cancer in humans is the generally long latent period. Normally it takes 20–40 years following the inhalation of most toxins before lung tumors appear. For this reason, it is often difficult to establish the definitive etiology of the lung cancer. Cancer of the upper respiratory tract does occur and is associated with some professions, such as chromate and nickel industry workers. By far, though, the majority of respiratory system cancers occur in the bronchioles and the lung tissues.

9.3 SUMMARY

The lungs provide a unique pathway for industrial toxins and tobacco smoke to enter the body, since the interface between the alveolar air and the pulmonary blood can facilitate the diffusion of both life-giving air and life-threatening toxins. The beautiful design of the respiratory system provides a number of highly efficient methods of protection from commonly encountered potential toxins, including

- Humidification and temperature control
- The mucociliary escalator

- Alveolar macrophages

Many industrial toxins are encountered as particulates, which undergo characteristic deposition in certain regions of the respiratory system according to various physicochemical processes. The speed and mechanism by which particulates are cleared from the various respiratory regions vary significantly. Industrial chemicals that are inhaled as gases and vapors are often taken up very rapidly, and the effects in workers can be substantial, both in the lung and at distant sites.

Inhaled industrial toxins exert toxicity by several distinct physiological mechanisms, which have historically led to many deleterious disease states in workers. Specific mechanisms of respiratory-related toxicity include

- Irritation of respiratory airways
- Fibrosis and pneumoconiosis
- Pulmonary edema
- Respiratory allergic responses
- Lung cancer

Some inhaled agents exert toxic effects by more than one mechanism, and many workers may suffer from more than one lung-related disease condition. Potential interactions between different inhaled toxins, especially tobacco smoke and various industrial chemicals, pose an additional threat. There is a tremendous potential for inhalation exposure to toxic chemicals in the workplace; therefore, workers must be monitored thoroughly by vigorous programs in industrial hygiene, environmental monitoring, occupational physicals, and toxicology.

REFERENCES AND SUGGESTED READING

Church, D. F., and W. A. Pryor, "The oxidative stress placed on the lung by cigarette smoke," in *The Lung,* Vol II, R. G. Crystal, J. B. West, P. J. Barres, et al., eds., Raven Press, New York, 1991, pp. 1975–1979.

Dosman, J. A., and D. J. Cotton, eds., *Occupational Pulmonary Disease. Focus on Grain Dust and Health,* Academic Press, New York, 1980.

Duffell, G. M., "Pulmonotoxicity: Toxic effects in the lung," in *Industrial Toxicology,* 1st ed., P. L. Williams, and J. L. Burson, eds., Van Nostrand-Reinhold, New York, 1985.

Ebert, R. V., and M. J. Terracio, "The bronchiolar epithelium in cigarette smokers," *Am. Rev. Resp. Disease* **111**: 6 (1975).

Fenn, W. O., and H. Rahn, *Handbook of Physiology,* American Physiology Society, Washington, D.C., 1964.

Frazier, C. A., ed., *Occupational Asthma,* Van Nostrand-Reinhold, New York, 1980.

Guyton, A. V., *Textbook of Medical Physiology,* 8th ed. Saunders, Philadelphia, 1991.

Hahn, F. F., "Carcinogenic responses of the lung to inhaled materials," in *Concepts in Inhalation Toxicology,* R. O. McClellan, R. F. Henderson, eds., Hemisphere, New York, 1989, pp. 313–346.

Hatch, T., and P. Gross, *Pulmonary Deposition and Retention of Inhaled Aerosols,* Academic Press, New York, 1964.

Lippmann, M., "Biophysical factors affecting fiber toxicity," in *Fiber Toxicology,* D. B. Wahrheit, ed., Academic Press, San Diego, 1993, pp. 259–303.

Mauderly, J. L., "Effects of Inhaled Toxicants on Pulmonary Function," in *Concepts in Inhalation Toxicology,* R. O. McClellan, and R. F. Henderson, eds., Hemisphere, New York, 1989, pp. 347–402.

McClellan, R. O., and R. F. Henderson, eds., *Concepts in Inhalation Toxicology,* Hemisphere, New York, 1989.

Menzel, D. B., and M. O. Amdur, "Toxic responses of the respiratory system," in *Doull's Toxicology: The Basic Science of Poisons,* 3rd ed., Macmillan, New York, 1986.

Morgan, W. K. C., and A. Seaton, eds., *Occupational Lung Diseases.* Saunders, Philadelphia, 1975.

Morrow, P. E., "Dust overloading in the lungs: Update and appraisal," *Toxicol. Appl. Pharmacol.* **113**: 1–12 (1992).

Muir, D., ed., *Clinical Aspects of Inhaled Particles,* Davis, Philadelphia, 1972.

Parent, R. A., *Treatise on Pulmonary Toxicology,* Vol. I, *Comparative Biology of the Normal Lung.* CRC Press, Boca Raton, FL, 1991.

Parkes, W. R., *Occupational Lung Disorders,* 2nd ed., Butterworths, Woburn, MA, 1982.

Samet, J. M., "Epidemiology of lung cancer," in *Lung Biology in Health and Disease,* C. Lenfant, ed., Marcel Dekker, New York, 1994.

Shami, S. G., and M. J. Evans, "Kinetics of pulmonary cells," in *Comparative Biology of the Normal Lung,* Vol. 1. *Treatise on Pulmonary Toxicology,* R. A. Parent, ed., CRC Press, Boca Raton, FL, 1991, pp. 145–155.

Steele, R, "The pathology of silicosis," in *Medicine in the Mining Industries,* J. M. Rogan, ed., Davis, Philadelphia, 1972.

Tager, I. B., S. T. Weiss, A. Muñoz, B. Rosener, and F. E. Speizer, "Longitudinal study of the effects of maternal smoking on pulmonary function in children," *NEJM,* **309**: 699–703 (1983).

USEPA, *Respiratory Health Effects of Passive Smoking: Lung Cancer and Other Disorders,* USEPA/600/6-90/006, 1992.

Witschi, H. R., and J. A. Last, "Toxic responses of the respiratory system," in *Casarett and Doull's Toxicology: The Basic Science of Poisons,* 5th ed., C. D. Klaassen, ed., McGraw-Hill, New York, 1996.

10 Immunotoxicity: Toxic Effects on the Immune System

STEPHEN M. ROBERTS and LOUIS ADAMS

This chapter discusses

- Basic elements and functioning of the immune system
- Types of immune reactions and disorders
- Clinical tests to detect immunotoxicity
- Tests to detect immunotoxicity in animal models
- Specific chemicals that adversely affect the immune system
- Multiple chemical sensitivity

10.1 OVERVIEW OF IMMUNOTOXICITY

Exposure to a variety of chemicals and biological agents has been implicated in the onset of symptoms of immune origin, including acute and chronic respiratory distress, dermal reactions, and manifestations of autoimmune disease. The types of substances associated with immune system effects is extraordinarily diverse, and include chemicals found in occupational and environmental settings, infectious materials, certain foods and dietary supplements, and therapeutic agents. As discussed in this chapter, dysregulation of the immune system by toxicants can lead directly to adverse health effects, as well as rendering the body more susceptible to infectious disease and cancer.

The immune system is highly complex, with many facets poorly understood. Because of this, assessment of potential immunotoxic effects of drugs, chemicals, and other agents is not a simple task. Often, measurement of a variety of components of the immune system and/or their functionality is required to gain an appreciation of the likelihood of immune dysfunction from drug or chemical exposure. Increasingly, there is realization that the immune system may be among the most sensitive target organs for toxicity for many chemicals and, as a result, merits special attention.

10.2 BIOLOGY OF THE IMMUNE RESPONSE

The immune system has evolved primarily to defend the body against the invasion of microorganisms, although normal immune function is important in regulating and sustaining the internal environment as well, such as recognition and removal of malignant cells. There are two types of immunity: natural immunity (also termed *innate immunity*) and acquired immunity (also termed *specific immunity*). *Natural immunity* is nonspecific in that it is directed to a wide variety of foreign substances, and is rarely enhanced by prior exposure to these substances. Natural immunity arises from several mechanisms, including complement, natural-killer (NK) cells, mucosal barriers, and the unique activity of

Principles of Toxicology: Environmental and Industrial Applications, Second Edition, Edited by Phillip L. Williams, Robert C. James, and Stephen M. Roberts.
ISBN 0-471-29321-0 © 2000 John Wiley & Sons, Inc.

polymorphonuclear and mononuclear phagocytic cells. Parts of this nonspecific immune system may contribute to the pathogenesis of an inflammatory response, and certain aspects of this system may be important in the etiology of autoimmunity.

Acquired immunity, in contrast, is highly specific and increases in magnitude with successive exposure to foreign substances. Substances that trigger these specific immune responses are termed *immunogens,* and may be either foreign or endogenous. In many cases, immunogens are proteins, although a variety of macromolecules can be immunogenic under appropriate circumstances, including polysaccharides, nucleic acids, and ribonucleic acids. There are two types of acquired immune responses: humoral immunity and cell-mediated immunity. *Humoral immunity* involves the production of proteins capable of binding to foreign substances. These belong to a special class of proteins called *immunoglobulins*, and the proteins themselves are called *antibodies*. The substances to which the antibodies bind are called *antigens*. Antibody binding can neutralize toxins, cause agglutination of bacteria and other microorganisms, and lead to precipitation of soluble foreign proteins. Each of these is important in defense of the host. In *cell-mediated immunity,* specialized cells rather than antibodies are responsible for the destruction of foreign cells.

A critical function of the immune system is to effectively distinguish between macromolecules that belong, or do not belong, in the body. The specific immune response is believed to be highly individualistic, a process which defines "self" while also defending the organism against "nonself." This is evident by the response to certain environmental toxicants, to allergens or antigens, and the specific rejection of allografts. Recognition of "self" is known to be guided, in part, by genetic variations in proteins of the class I and II *major histocompatibility complex* (MHC). Initially, the ability of the immune system to differentiate "self" from "nonself" is an educational process. During maturation, the system must ignore an infinite variety of self-molecules and yet be primed and ready to respond to an array of exogenous antigens. Immunomodulatory control mechanisms lead to immune tolerance of self and carefully orchestrate the immune response to targets and removal of foreign macromolecules and cells. These control mechanisms arise from interactions among the several different cell types with roles in proper immune function.

Lymphocytes are considered to be the major cells involved in a specific immune response in humans. They are derived from pluripotent stem cells and undergo an orderly differentiation and maturation process to become T cells or B cells (see Figure 4.1 in the chapter on hematotoxicity), with critical functional roles in the host defense. T-cell development occurs primarily in the thymus, where cell surface protein markers are acquired during the selection and differentiation process. These protein markers are called CD antigens (for *cluster of differentiation*), and at least 78 different CD antigens have been identified in humans. The presence of certain CD antigens, detectable by immunofluorescence, has been used to positively identify immunocytes. In general, mature T cells are characterized by the presence of $CD3^+$ and $CD4^+$ or $CD8^+$ surface markers and are devoid of surface or cytoplasmic immunoglobulin. There are various subtypes of T cells, such as T-helper (T_H) cells, T-suppressor (T_S) cells, and cytotoxic cells (T_C). T_H lymphocytes carry the $CD4^+$ marker, while T_S and T_C lymphocytes have the $CD8^+$ marker. Together, these T-lymphocyte populations play a vital role in initiating and regulating the immune response.

Human B cells develop from stem cells in the fetal liver and, after birth, B-cell development occurs principally in the bone marrow. B-cell development and maturation are characterized by class-specific immunoglobulin (Ig) expression on the cell surface. Monoclonal reagents can identify the Ig expressed on the surface of B cells. Immunophenotypic characterization of cells via these markers has proved to be invaluable in certain clinical situations and a useful research tool. B cells play an important role in recognition of antigens and are responsible for antibody production.

Another important cell in the specific immune response is the *antigen-presenting cell* (APC). These cells make first contact with the antigen and may also process the antigen; that is, modify it in such a way as to enable its recognition by T cells. This category of cells is defined more by function than cell type. In general, the most important APCs are tissue macrophages and peripheral blood monocytes, although cells of other types (e.g., Langerhans cells in the skin, dendritic cells in lymphoid tissue) may also perform this function.

In the specific immune response, antigen may be taken up by APCs and presented to T or B cells. In order to present the antigen to T cells, the antigen must be processed, or partially digested by the APC and then presented on its cell surface bound to an MHC class II molecule. Presentation of antigen to B cells does not require this processing, and in fact B cells are capable of recognizing antigens directly, without APC presentation. Antigens, either presented by APCs or encountered independently, interact with immunoglobulins on the cell surface of B-cell clones. Different B-cell clones vary in the immunoglobulins expressed on their cell surface, and these immunoglobulins can be quite specific in terms of the antigens with which they will interact. Thus, a particular antigen may interact with only one or a few B cell clones, a critical aspect in creating a specific immune response. When the antigen binds to an immunoglobulin receptor on the B cell surface, the antigen–receptor complex migrates to one pole of the cell and is internalized within the cell. The B cell becomes activated, and the antigen is processed leading to display of antigenic peptides on the cell surface in conjunction with an MHC class II protein.

T-cell activation is postulated to require at least two signals. The first signal is thought to be an interaction between the $CD4^+$ T-cell receptor of T-helper (T_H) lymphocytes and antigenic peptides and MHC class II proteins presented by APCs or B cells. The second signal may be under the influence of other receptor–ligand pairs on the T cell and cognate interactions through adhesion molecules of APCs, MHC complex, and the various cytokines produced by T-cell subsets and accessory cells, such as macrophages. When activated, T_H cells proliferate, creating more cells for interaction with APC and B cells.

An effective immune response requires the activation of specific subsets of T_H cells (T_H1 and T_H2 cells) which secrete different cytokines. *Cytokines* are low-molecular-weight proteins that mediate communication between cell populations. A list and functional classification of cytokines is shown in Table 10.1. The T_H1 cells are involved in the activation of macrophages by INF-γ, secrete tumor necrosis factor (TNF), and mediate delayed-type hypersensitivity responses. The most critical function of T_H2 cells is to regulate B cells, but they also secrete cytokines (specifically, interleukins, designated IL) that may regulate mast cells (IL3, IL4, and IL10), eosinophils (IL5) and IgE (IL4) responses in allergic diseases. Of the several factors known to participate in immunomodulation, IL4 and IL10 are particularly noted to upregulate the humoral response while suppressing the cell-mediated response (see below for more discussion of humoral versus cell-mediated immunity). IL13, which is produced by activation of T cells (Table 10.1) and shares many of the properties of IL4, also suppresses cell-mediated immune responses and the production of proinflammatory cytokines (IL1, IL6, IL8, IL10, IL12, and TNF).

When an activated T_H cell binds to the antigenic peptide-MHC complex of a B cell, the B cell is stimulated to replicate and differentiate into an antibody secreting *plasma cell*. This B cell clonal expansion leads to increased production of antibody specific to that B cell, and this antibody, in turn, has reactivity directed rather specifically to the antigen initiating the response. Through this mechanism, the immune system is able to produce the necessary quantities of antibodies targeting specific molecules (antigens) regarded as foreign. The synthesis of the antibody is tightly regulated, however, and the proliferation of plasma cells and antibody synthesis are controlled by cytokines and interactions with T cells. T-amplifier cells (T_A) and T-suppressor cells (T_S), as their names imply, function to enhance or suppress the immune response, respectively. Control of the immune response is achieved by balancing the stimulatory and inhibitory effects of T cells and various cytokines.

After an encounter with an antigen, the immune system appears to retain "memory" of that antigen and is able to mount a more rapid and greater antibody response on subsequent contact, even if the period between exposures to the antigen span several years. The basis for this memory is still not well understood. Initial (*primary*) immune responses to T-dependent antigens require a proliferative response by naive T and B cells. As these cells mature, they differentiate and become *effector cells*. The elimination of effector T cells and the factors controlling the survival of memory cells is still controversial. Because immune responses to viruses or immunization encountered in childhood generally result in lifelong immunity, it has been presumed that memory

TABLE 10.1 Cytokines and their Functions

Cytokine	Produced by	Function(s)
IL1 (IL1-α and IL1-β)	Several cell types, including neutrophils and macrophages	Variety of effects, including neutrophil and macrophage activation, T- and B-cell chemotaxis, and increased IL2 and IL6 production
IL2	T cells	Stimulates replication of T cells, NK cells, and B cells
IL3	T cells	Involved in regulation of progenitor cells for several different cell types, including granulocytes, macrophages, T cells, and B cells
IL4	Activated T cells	Activates T and B cells; suppresses synthesis of IL1 and TNF
IL5	T cells and activated B cells	Increases secretion of immune globulins by B cells
IL6	Several cell types, including T and B cells	Important in inflammatory reactions and in differentiation of B cells into Ig-secreting cells
IL7	Bone marrow stromal cells	Important in regulating lymphocyte growth and differentiation
IL8	Activated monocytes and macrophages	Activates neutrophils; important for chemotaxis of neutrophils and lymphocytes
IL9	T_H cells	Stimulates growth of T_H cells
IL10	B cells	Stimulates growth of T cells in the presence of IL2 and IL4
TNF-α	Variety of cells, primarily activated macrophages	Important in inflammatory responses; effects similar to IL1
TGF-β	Variety of cells	Inhibits T-cell proliferation and suppresses inflammatory responses
TNF-β	Activated CD4+ cells (T_H)	Important in mediating cytotoxic immune responses, cell lysis
Interferons	Leukocytes (INF-α), fibroblasts (INF-β), and lymphocytes and NK cells (INF-γ)	(INF): Neoplastic growth inhibitor; activates macrophages; protects against viral infections by interfering with viral protein synthesis

is afforded by long-lived cells that become activated only following repeat exposure to the antigen or immunization. While it has been assumed that "memory cells" last indefinitely following a single antigen contact, recent evidence suggests the life-span of memory cells may be related to repeat contact with antigen.

In order to be recognized by the immune system, antigens must be of appreciable size. Some of the smallest antigens, for example, are natural substances with molecular weights in the low thousands. There are circumstances where much smaller molecules can elicit an immune response, but this requires the participation of a large molecule to serve as a carrier. For example, some metals, drugs, and organic environmental and occupational chemicals too small to be recognized by the immune system can become antigenic when bound to a macromolecule such as a protein. Once the immune response has been initiated, antibodies will recognize and bind the small molecule even when it is not bound to the carrier molecule. In situations such as this, the small molecule is called a *hapten*.

The antibodies themselves are glycoproteins, the basic unit of which consists of two pairs of peptide chains (see Figure 10.1) connected by disulfide bonds. The longer peptide chain is termed the *heavy* (or H) chain and the shorter is the *light* (or L) chain. There are five main types of antibodies, or immunoglobulins (Ig): IgG, IgM, IgA, IgE, and IgD. They differ both in structure and function. IgG is present in the greatest concentration in serum, has a molecular weight of around 150,000 (there are four subtypes of somewhat different sizes), and is important in secondary immune responses. IgM is a primary response antibody, meaning that it is increased

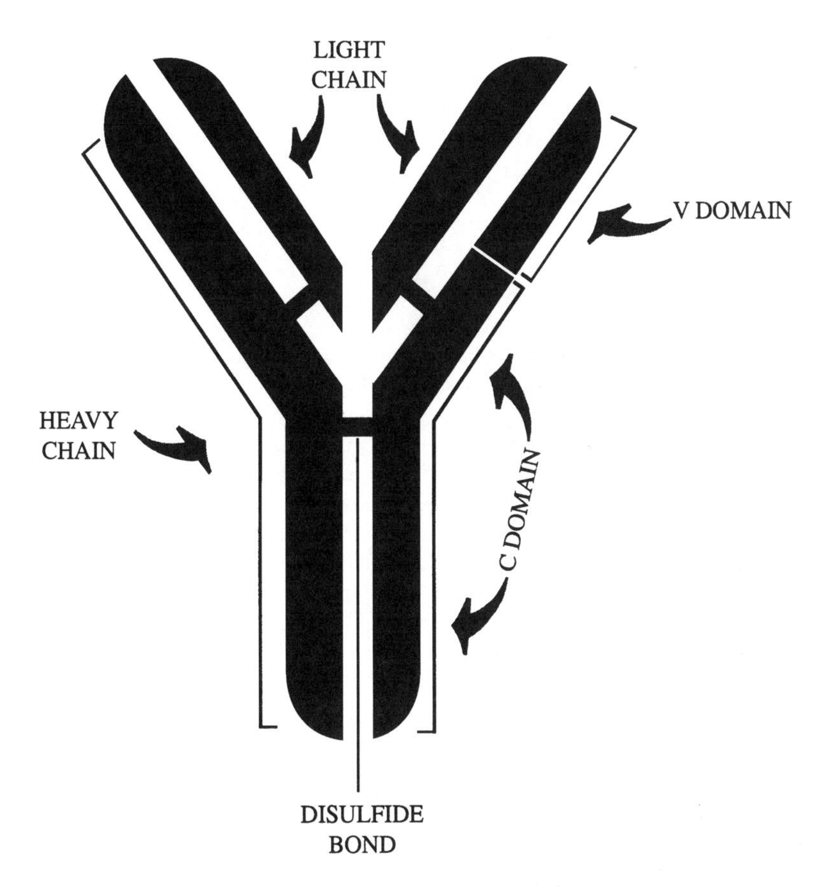

Figure 10.1 Light and Heavy Chain Structure of IgG. IgG illustrates the basic structure of antibody proteins, which consists of two long, heavy chains and two shorter, light chains held together by disulfide bonds. Composition of C domains is relatively constant, while V domain varies, creating the binding specificity characteristic of antibodies.

very early in an immune response. IgM is much larger than the other Igs, consisting of five sets of heavy/light-chain pairs bound together at a single point with another peptide (the J chain). Its molecular weight is about 970,000. IgA may exist as a monomer (one basic unit of two pairs of H and L chains) or as a dimer—two basic units bound together with a J chain. The monomeric IgA has a molecular weight of about 160,000 and is the predominant form of IgA found in serum. IgA is the primary Ig found in secretions (e.g., tears and saliva), mostly in the dimeric form with a molecular weight of 385,000. IgD has a molecular weight of about 184,000, and is present in very low concentrations in serum. Its function is unclear, but it may play a role in B-cell differentiation. IgE is slightly larger than IgG (molecular weight of 188,000), and is normally present in low concentrations in serum. It can attach itself to leukocytes and mast cells, and is the primary antibody involved in hypersensitivity reactions.

In cell-mediated immunity, cells carrying the antigen on their surface are attacked directly by cytotoxic T cells (T_C) or other cell types such as natural-killer (NK) cells. In the case of T_C cells, recognition of cells to be destroyed is through interaction between processed antigen in conjunction with MHC class I molecules on the target cell surface and an antigen receptor on the T_C. In order to be active, the T_C must also receive stimulation from $CD4^+$ cells, principally in the form of IL2. Mechanisms of target cell recognition by NK cells are not well understood.

10.3 TYPES OF IMMUNE REACTIONS AND DISORDERS

Interactions of toxicants with the immune system may result in undesirable effects of three principal types—those manifested as (1) a hypersensitivity reaction, (2) immunosuppression, or (3) autoimmunity. Each is discussed below.

Allergic Reactions

Allergic reactions are divided into four classes:

Type I. Type I immune response is limited to IgE-mediated hypersensitivity (allergic) reaction. This reaction involves an initial exposure in which immune symptoms are generally absent (sensitization), followed by reexposure that can elicit a strong allergic reaction. In type I immune responses, antigen interacts with IgE antibodies passively bound to mast cells. On binding of antigen to the IgE, the mast cells release histamine and serotonin, which are responsible for many of the immediate symptoms of an allergic reaction such as upper respiratory tract congestion and hives. In a severe reaction, termed *anaphylaxis*, histamine and serotonin release can cause vasodilation leading to vasomotor collapse, and bronchiolar constriction making breathing difficult. This type of reaction has occurred following the administration of a number of different drugs and diagnostic agents, hormones, and a variety of sulfiting agents (e.g., sodium bisulfite, sodium metabisulfite, etc.).

Type II. Type II reaction is believed to be the result of the binding of a drug or chemical to a cell surface, followed by a specific antibody-mediated cytotoxicity that is directed at the agent (drug or chemical) or at the cell membrane that has been altered by the compound. Under some circumstances, immune complexes may become adsorbed to a cell surface (erythrocytes, thrombocytes or granulocytes) resulting in a complement-mediated cytotoxic response, leading to induction of immune hemolytic anemia, thrombocytopenia or granulocytopenia.

Type III. Soluble immune complexes consisting of a drug or chemical hapten (plus carrier molecule) and its specific antibody plus complement components are primarily responsible for immune complex disease. A particular form of immune complex disease arising from injection of an antigen is called *serum sickness syndrome*. Clinically, a type III reaction may be characterized by the onset of fever and the occurrence of a rash that may include purpura and/or urticaria. The immunopathology includes the activation of complement and the deposition of immune complexes in areas such as blood vessel walls, joints, and renal glomeruli. Some of the signs and symptoms associated with drug-related lupus may be included under type III reactions.

Type IV. These reactions involve cell-mediated and/or delayed-type hypersensitivity responses. The expression of type IV reactions requires prior exposure to the agent and T-cell sensitization. A special subpopulation of T cells (T_D) appear to be responsible for this reaction. The T_D cells react with antigens in tissues and release lymphokines, attracting macrophages to the site and leading to an inflammatory response. The reaction is termed delayed because the inflammatory reaction may not peak for 24–48 h, as opposed to responses occurring within a few minutes to a few hours with other reaction types. These reactions are usually seen as contact dermatitis occurring after the use of certain drugs or exposure to some chemicals.

Immunosuppression

Impairment of one or more components of the immune system from drug or chemical exposure can lead to loss of immune function, or *immunosuppression*. Clinically, this is manifested primarily as increased susceptibility to infectious disease, although diminished immune function could conceivably increase vulnerability to cancer by impairing immune surveillance and removal of malignant cells. In certain situations immunosuppression is intentionally induced via drug therapy to prevent rejection of transplants. Agents employed for this purpose are diverse, and several potential mechanisms are involved, including inhibition of cytokine production (e.g., corticosteroids, cyclosporin) and lymphocyte proliferation (e.g., azothioprine). Most of the evidence that environmental and occupational chemicals suppress immune responses is derived from animal studies, and while the same principles likely apply to humans as well, there are few clear examples in the clinical literature of immunosuppression from chemical exposure other than that from intentional treatment with immunosuppressive drugs.

The opposite reaction, immunological enhancement, is also possible, and several natural and synthetic agents have been shown to increase immune responsiveness under experimental conditions. Examples of agents that increase immune reactivity include the bacillus Calmette-Guerin (BCG), alum (aluminum potassium sulfate or aluminum hydroxide), bacterial lipopolysaccharides and peptidoglycans, a variety of synthetic polymers, and the antiparasitic drug Levamisole (phenylimidazolethioazole). Difficulty in producing a controlled stimulation of the immune system and the enormous potential for undesirable side effects limit the therapeutic use of these agents. To date, there are no examples of environmental or occupational chemicals shown to produce immune stimulation in humans, other than in the context of allergic reactions.

Autoimmunity

Autoimmunity is defined as the induction and expression of antibodies to self-tissue, including nuclear macromolecules. Studies of drug-related autoimmunity in humans have provided some of the best examples of this type of reaction. Although there are many types of autoimmune disease, the most common autoimmune syndrome produced by drugs is one resembling systemic lupus erythematosus (SLE). Clinical signs and symptoms of so-called *drug lupus* are not identical to idiopathic SLE, however. Both are characterized by arthralgia and the appearance of antinuclear antibodies in the blood, but the pattern of antinuclear antibodies is somewhat different, and renal and CNS complications dominate idiopathic SLE while these are typically absent in drug-lupus. Symptoms of drug lupus generally subside after the drug is withdrawn. Demonstration of autoimmune responses from environmental exposure to chemicals (other than drugs) has been difficult, in part because of problems identifying etiologic agents in retrospective studies of patients developing autoimmune disease. One concern is that some chemicals may exacerbate underlying autoimmune disease (e.g., SLE), rendering symptomatic a patient with subclinical disease or increasing the duration or severity of symptoms in those with active disease. Unfortunately, differentiating the effects of chemical exposure from progression of the underlying disease is difficult or impossible in practice. Understanding of autoimmune consequences of chemical exposure is further hampered by the general lack of satisfactory animal models—the results obtained in laboratory animals seldom correspond exactly to observations in humans.

10.4 CLINICAL TESTS FOR DETECTING IMMUNOTOXICITY

In the clinical setting, the use and proper interpretation of immunologic laboratory tests can be important in establishing a differential diagnosis in a patient who has been exposed to an immunotoxic agent. Immune system testing for diagnostic purposes can be challenging, however, because of the complexity of the immune system and difficulty in establishing normal values for many of the tests. When immune dysfunction from chemical exposure is suspected, it is important to be sure that the patient is free from infectious disease and not taking medications that can influence immune function—obvious confounders to interpretation of any immune tests. Also, it is important to recognize that many immune parameters, such as lymphocyte subpopulation counts, can vary normally by age and gender, making the use of appropriate controls essential for proper interpretation of results. Finally, temporal variations in most tests are common. In order to demonstrate that an abnormality exists, it is usually advisable to repeat the test one or more times to insure that a consistent result is obtained.

Some of the laboratory tests available provide information relevant to assessing humoral immunity, others are useful in evaluating cellular immunity, and some can provide insight regarding both. Described below are examples of assays commonly used in the evaluation of individuals exposed to chemicals in the environment or workplace.

Immunoglobulin Concentrations The concentrations in serum of each immunoglobulin can be determined with the exception of IgD, which exists primarily on cell surfaces. Single-radial diffusion is commonly employed for most immunoglobulins, although enzyme linked immunosorbent assay (ELISA) or radioimmunoassay (RIA) is often needed to measure the low concentrations of IgE typically present. Diminished immunoglobulin concentrations, either in total or of specific classes, may suggest immunodeficiency, but is not sufficient to establish a diagnosis. Conversely, immunoglobulins within normal limits do not necessarily indicate immunocompetence. There may be defects in subtypes of immunoglobulins not quantified by the assay, and patients with normal or high values may nonetheless exhibit increased susceptibility to disease. Immunoglobulin values may be profoundly influenced by viral or bacterial infections and the presence of some drugs.

T- and B-Cell Concentrations Immunotyping of T- and B-cell subsets by ethidium bromide and cytofluorometry techniques is used by many laboratories for screening studies of chemical-related injury. Concentrations of B cells, either in absolute terms or as a percentage of peripheral blood lymphocytes, can be expressed, and the distribution of B cells expressing different immunoglobulin types (IgM, IgG, IgA) can be measured. Some studies have sought to evaluate a potential immunosuppressive effect through measurement of the ratio of T_H to T_S lymphocytes in peripheral blood, using the $CD4^+$ marker to indicate T_H cells and the $CD8^+$ marker for T_S cells. As discussed above, these markers are not specific for T_H and T_S cells, however, and interpretation of a decreased $CD4^+$ to $CD8^+$ ratio as a loss of T help relative to T suppression is an oversimplification. A significant reduction in $CD4^+$ cells is associated with several immunodeficient states (e.g., in patients with AIDS, undergoing radiotherapy, or chemotherapy), implying that diminished $CD4^+$ is indicative of impaired antibody production. This assumption is not infallible, however, because there are also circumstances in which $CD4^+$ cells may be reduced without loss of antibody production. Significant changes in absolute or relative concentrations of lymphocyte subsets may be suggestive of immunotoxic effects from chemical exposure, but are not, by themselves, reliable indicators of compromised function.

Cutaneous Anergy *Anergy* is a generalized clinical condition of non-responsiveness to ubiquitous skin test antigens that is frequently observed in patients who are immunosuppressed. Cutaneous anergy may suggest functional impairment or abnormalities of the cellular immune system. The most cost-effective method for evaluation of cutaneous anergy is the use of a battery of attenuated, premeasured and well-standardized ubiquitous antigens that are available from commercial sources. The assessment of a person who is thought to be immunologically suppressed due to exposure to an environmental chemical can be attained within 48 h through the use of these antigens. The intradermal skin test antigens frequently used to measure cellular delayed hypersensitivity are: tetanus toxoid, diphtheria toxoid, *Streptococcus* (group C), old tuberculin (PPD), *Candida albicans*, *Trichophyton mentagrophytes*, and *Proteus mirabilis*. Measurement of specific IgG antibodies to diphtheria and tetanus toxoids in serum at 2 weeks following booster immunization is also useful in assessing the ability to form antibodies to protein antigens.

In Vitro *Tests* Functional capabilities of lymphocytes can be evaluated by taking a blood sample and performing a variety of tests *in vitro*. In general, these tests involve isolating lymphocytes from a blood sample, placing them in culture, and exposing them to a stimulatory agent. The ability of the cells to proliferate in response to the stimulus and, in the case of B cells, to synthesize immunoglobulins, can be measured. For example, treatment of peripheral blood lymphocytes with pokeweed mitogen (PWM) normally produces cellular proliferation and increased immunoglobulin synthesis. This response requires both T_H and B cells, and provides an indication of the capability of these two cells to interact properly and of B cells to produce immunoglobulins. Lipopolysaccharide (LPS) is a mitogen effective selectively on B cells, while phytohemaglutinin (PHA) and concanavalin A (con A) are selective T-cell mitogens. Other stimulants to lymphocyte activation can be used, such as tetanus toxoid, diptheria toxoid, *Candida*, and PPD, if the subject has been previously exposed to these. The rapid cell division characteristic of a normal response to these mitogens is typically assessed by measuring incorporation of ^{3}H-thymidine into DNA of the cells. Other endpoints of stimulation, such as increased expression of IL2 receptors on T cells, can also be evaluated. The results of these tests are particularly prone to variability, and the tests should be repeated on several occasions in order to demonstrate an abnormal response.

In the mixed-lymphocyte reaction (MLR) test, lymphocytes from the test subject and another individual are mixed. Normally, contact with the allogenic lymphocytes will cause the test subject's lymphocytes to become activated and proliferate. To conduct this assay, the target lymphocytes are rendered incapable of replication, often by irradiation or by treatment with mitomycin C. Test subject lymphocytes are then added, and the rate of their replication is evaluated by measuring incorporation of ^{3}H-thymidine. The cytotoxic lymphocyte (CTL) assay takes the lymphocyte interactions one step further to evaluate the ability of cytotoxic T cells (T_C) to destroy target cells. After incubation of the test subject and target lymphocytes, the subject T_C are isolated, washed, and reincubated with target lymphocytes preloaded with ^{51}Cr. As the target cells are destroyed, ^{51}Cr is released into the medium and can be measured, providing an index of cytotoxic capabilities of the T_C lymphocytes.

Fluorescent Antinuclear Antibody Assay (FANA) The indirect immunofluorescence antinuclear antibody assay (FANA) may be the initial screening test used to show autoimmunity. However, several FANA patterns are recognized in various connective-tissue diseases and some low-titer staining patterns have also been reported in sera from persons exposed to environmental agents. The following staining patterns may be observed:

1. The *diffuse (homogenous) staining pattern,* which is usually associated with antibody directed to DNA-histone or histone subfractions. This staining pattern is frequently found in sera from patients receiving chronic treatment with procainamide, hydralazine, isoniazid, anticonvulsant drugs, and some environmental chemical agents.
2. A *peripheral (rim) pattern,* which is attributed to antibody reacting with native DNA and soluble DNA-histone complexes. This staining pattern is frequently seen in sera from patients with systemic lupus erythematosus (>95 percent).
3. *Speckled FANA staining,* which is usually attributed to antibodies reacting with saline-soluble antigens. These antibodies are directed to nonhistone antigens and include Sm, ribonucleoprotein, SS-A/Ro, SS-B/La, PM-1, and SCl-70. While these staining patterns frequently occur in patients with mixed connective tissue diseases, including Sjögren's syndrome, polymyositis and progressive systemic sclerosis, they have also been found in sera from persons exposed to immunotoxic agents.
4. The *nucleolar staining pattern,* which has been restricted to antibodies reactive with nucleolar RNA. This pattern is associated with a particular form of systemic sclerosis (progressive systemic sclerosis).

10.5 TESTS FOR DETECTING IMMUNOTOXICITY IN ANIMAL MODELS

For most chemicals, an assessment of their potential to produce immunotoxicity in humans is based on testing in animals. Many of the tests used in animal studies are the same as, or at least analogous to, those available for clinical assessment described above. However, studies in animals offer the opportunity to evaluate directly toxic endpoints difficult or impossible to assess clinically, such as the development of immunopathology or loss of resistance to infectious disease.

Currently, a tiered approach is recommended for standardized testing for immunotoxicity in animals. Tier I consists of a battery of tests intended to evaluate both humoral and cell-mediated immune system integrity. An assessment of immune system pathology is also included in tier I (see Table 10.2). If the results of tier I tests are negative, the chemical is considered not to possess significant immunotoxic potential at the dosages tested. If effects are observed in tier I tests, additional tests are conducted in tier II to better characterize the immunotoxic properties of the chemical. Tier II does not consist of a rigid battery of tests, but rather the opportunity to select more specific tests to follow up on observations made in tier I. Examples of tests that might be used in tier II are included in Table 10.2.

Many of the endpoints examined in tier I are basic. Total and differential white cell counts are obtained from blood, body and specific organ weights are recorded, and tissues of particular relevance for immune function (viz., spleen, thymus, and lymph nodes) are examined histologically for evidence of injury. Humoral immunity is assessed with a plaque-forming cell (PFC) assay. In this assay, the test animal is injected with sheep red blood cells (SRBCs) as the source of antigen. Four days later the spleen is removed, and cells isolated from the spleen are cultured with intact SRBCs. B cells producing IgM directed to SRBC antigens result in lysis of the red cells, producing clear areas in the culture called plaques. The number of plaques (per spleen or per million spleen cells) provides an indication of the ability of splenic cells to synthesize and secrete antigen-specific antibodies. This, in turn, offers information regarding the ability of the immune system to mount a primary (IgM-mediated) response. Cell-mediated immunity is evaluated by measuring the responsiveness of peripheral blood T and B lymphocytes to mitogens (such as concanavalin A), and through the MLR assay. Nonspecific immunity is evaluated in tier 1 by measuring NK cell function. These tests are essentially identical to the *in vitro* methods described above for clinical assessment of potential immunotoxicity in humans.

More detailed follow-up tests are available for tier II. For example, if disturbance in the numbers of immunocytes is suggested by tier I tests, the abundance of individual T- and B-cell types in the spleen or blood can be measured using reagents that detect specific cell surface antigens. In the assessment of humoral immunity, an abnormal primary response (IgM-mediated) to SRBCs detected in the PFC assay in tier I might lead to an evaluation of the secondary response (IgG-mediated) to SRBCs. Evidence of altered cell-mediated immunity could lead to expanded tests of T-lymphocyte cytotoxicity in tier II, commonly using tumor cells as targets. Tier II could also include an assessment of delayed-type hypersensitivity response. Evaluation of non-specific immunity may be extended in tier II to include enumeration of macrophages and tests of their function. For functional tests, macrophages are typically taken from the peritoneal or alveolar space of test animals, cultured, and examined for phagocytic activity, secretion of cytokines, and/or production of reactive oxygen or

TABLE 10.2 Tier I and Tier II Tests for Immunotoxicity

Tier I	Hematology, including CDC and differential counts
	Body and organ weights, including spleen, thymus, kidney, and liver
	Histology of lymphoid organs, including spleen, thymus, and lymph nodes
	Humoral immunity, assessed through IgM plaque-forming cell (PFC) response
	Cell-mediated immunity, assessed through T- and B-lymphocyte responses to mitogens, and the mixed-lymphocyte response (MLR)
	Nonspecific immunity, assessed through measurement of natural-killer (NK) cell activity
Tier II	Quantitation of individual T- and B-cell populations in blood and spleen
	Humoral immunity, assessed through IgG plaque-forming cell (PFC) response
	Cell-mediated immunity, assessed through cytotoxic T-cell (CTC) activity, as well as the delayed hypersensitivity (type IV) response
	Host resistance, assessed through challenge with pathogens or tumors

TABLE 10.3 Examples of Agents Used for Immune Challenge in Host Resistance Tests

Type of Agent	Name	Typical Exposure Route
Virus	Cytomegalovirus	Intraperitoneal or intratracheal administration
	Herpes simplex virus type 2	Intraperitoneal, intravenous, or intravaginal administration
	Influenza virus	Intranasal administration
Bacteria	*Corynebacterium parvum*	Injected intravenously
	Listeria monocytogenes	Injected intravenously
	Pseudomonas aeruginosa	Injected intravenously
	Streptococcus pneumoniae	Injected intravenously
Parasites	*Plasmodium* species	Intravenous or intraperitoneal injection of infected blood
	Trichinella spiralis	Intragastric administration
Tumor cells	B16-F10 melanoma	Cells are injected intravenously
	PYB6 fibrosarcoma	Cells are injected subcutaneously

nitrogen species. The ability of macrophages in culture to phagocytize foreign materials is typically examined using light microscopy, with either biological (e.g., SRBCs or bacteria) or nonbiological materials (e.g., fluorescent beads) as targets. On activation, macrophages normally release specific cytokines (e.g., TNF-α and IL2), as well as reactive oxygen and nitrogen. Cytokine production by activated macrophages in culture can be measured by ELISA (enzyme-linked immunosorbent assay) using antibodies directed to specific cytokines, or by ELISPOT, which is capable of identifying the numbers of cells producing specific cytokines. Several techniques are available for quantitating reactive oxygen and nitrogen species.

When immunosuppression (or, less commonly, immunostimulation) is suspected, one of the most direct means to test overall immune competence is through a *host resistance model* (also sometimes called a *host susceptibility model*). With this model, the ability of the animal to withstand an immune challenge is assessed with and without exposure to the drug or chemical. Immune challenge can take the form of an infectious microorganism or a syngeneic tumor. A variety of types of infectious microorganisms are used for these tests, including viruses, bacteria, yeast, fungi, and parasites. Syngeneic tumor lines are derived from the same strain and species as the test animal, requiring their recognition as tumor cells and not simply a source of foreign protein. Examples of microorganisms and tumor cell lines used for host resistance models are provided in Table 10.3. Many of these agents are human pathogens, and this type of test arguably provides the best direct evidence of the ability of a drug or chemical to produce clinically relevant immune suppression or stimulation.

10.6 SPECIFIC CHEMICALS THAT ADVERSELY AFFECT THE IMMUNE SYSTEM

The number of drugs and chemicals associated with immunotoxicity in humans is extensive. As discussed in Section 10.3, immunotoxicity typically occurs as a hypersensitivity reaction, immunosuppression, or autoimmunity. Several agents commonly encountered in occupational settings are capable of producing contact, cell-mediated hypersensitivity, with common symptoms of rash, itching, scaling, and the appearance of redness and vesicles on the skin. Examples of these agents are shown in Table 10.4. The respiratory tract is also a common site of allergic symptoms from drug or chemical exposure. Inhalation of respiratory allergens can cause an immediate-type reaction (an *early-phase reaction*, occurring and waning rapidly) or a delayed-type reaction (sometimes called a *late-phase reaction*), which may appear 6–8 h later and require 12 to 24 hours to resolve. Both reactions are

TABLE 10.4 Examples of Agents that Produce Dermal Contact Sensitivity

Drugs	Metals
Benzocaine	Beryllium
Thimerosal	Cadmium
Neomycin	Chromates
Resins	Gold
Acrylic resins	Mercury
Epoxy resins	Nickel
Formaldehyde resins	Silver
Phenolic resins	Zirconium
Other industrial chemicals	
Ethylenediamine	
Paraphenylenediamine and other dyes	
Antioxidants	
Chlorinated hydrocarbons	
Dinitrochlorobenzene	
Mercaptans	

IgE-mediated. Table 10.5 lists examples of common agents associated with respiratory allergy. *Occupational asthma* represents a special kind of inhalation disorder that is distinct from typical respiratory allergy. In general, a longer sensitization period is required, and symptoms may resemble an early-phase reaction, a late-phase reaction, or both. IgE may be responsible for some, but not all, of the manifestations of occupational asthma. In fact, the role of the immune system in occupational asthma may be different for asthma initiated or provoked by high-molecular-weight compounds, low-molecular-weight compounds, and irritatants.

The potential for immunosuppression from occupational and environmental exposure to chemicals has been suggested by numerous in vitro studies and experiments in laboratory animals. Direct evidence for clinical immunosuppression following workplace or environmental exposures is extremely limited. However, there are many well-documented examples of the development or exacerbation of autoimmunity from chemical exposure. Most of these examples (shown in Table 10.6) are drugs, and for agents such as procainamide, up to 80 percent of patients treated chronically will develop increased levels of autoimmune antibodies. Many of these drugs produce signs and symptoms resembling systemic lupus erythematosus, while others produce autoimmune disease of the kidney, liver, thyroid, and other organs; scleroderma; or autoimmune hemolytic anemia. Evidence suggests that several environmental contaminants may also have the ability to either produce or worsen autoimmune disease, although the association with autoimmune disease is often less well substantiated.

TABLE 10.5 Examples of Agents that Product Respiratory Allergy

Molds	Dusts and Small Particulates
Aspergillus	Coffee
Cladosporum	Enzymes
Hormodendrum	Flour
Penicillium	Mites
Rhizopus	Sawdust
Pollens (various)	Pet dander
	Cockroach proteins

TABLE 10.6 Examples of Agents Associated with Autoimmune Disease

Drugs
Acebutalol
Allopurinol
Alprenolol
Amiodarone
Ampicillin
Bleomycin
Captopril
Carbamazepine
Cephalosporin
Chlorpromazine
Chlorthalidone
Dapsone
Diphenylhydantoin
Ethosuximide
Fenoprofen
Iodine
Isoniazid
Lithium
Lovastatin
Mefenamic acid
Methyldopa
Minocycline
Nitrofurantoin
Penicillamine
Phenylbutazone
Propylthiouracil
Quinidine
Sulfonamides
Amino acids
L-Tryptophan
L-Canavanine
Environmental/industrial chemicals
Aromatic amines
Cadmium
Chlordane
Chorpyrifos
Chromium
Formaldehyde
Gold
Hydrazine
Mercury
Paraquat
Pentachlorophenol
Perchlorethylene
Silicon (silica)
Thallium
Trichloroethylene
Vinyl chloride

Some classes of chemicals or agents have, in particular, been associated with immunotoxic effects in humans. These are discussed briefly below.

Metals

Metals have been associated with various types of hypersensitivity reactions. Beryllium, nickel, chromium, cadmium, silver, and zirconium have all been found to produce contact dermatitis. Nearly 10 percent of women and 2 percent of men have sensitivity to nickel, and may develop rashes upon contact with nickel in jewelry, coins, and clothing fasteners. Sensitive individuals may also respond to chromium in tanned leather products. Metals are also associated with pulmonary hypersensitivity reactions and occupational asthma. One of the most serious of these diseases is *berylliosis*, a delayed hypersensitivity (type IV) reaction thought to result from beryllium acting as a hapten. Acutely, hypersensitivity to beryllium is manifested as pneumonitis and pulmonary edema. Chronically, workers exposed to beryllium develop a severe, debilitating granulomatous lung disease.

Studies in experimental animals have shown that metals such as lead, mercury, nickel, and cadmium are associated with activation of $CD4^+$ T cells or cause suppression of antibody responses and cell-mediated immunity, resulting in increased susceptibility to infection. There is some clinical and epidemiologic evidence that lead may decrease resistance to infectious disease, and the use of arsenic for medicinal purposes suggests that it, too, may have immunosuppressive effects. Arsenic was used in the early twentieth century to treat some inflammatory diseases, and currently appears to have some efficacy in treating leukemia. Also, patients treated with arsenicals were reported to have a relatively high incidence of the viral disease herpes zoster, suggesting some impairment of the immune system.

A number of studies have reported increased or unusual autoantibodies in association with exposure to some metals in the workplace, suggesting potential autoimmune toxicity. For example, there is evidence of immune complex glomerulonephritis in nephrotoxicity from cadmium and mercury. Iodine and lithium have been linked to autoimmune thyroid disease, and chromium and gold have been associated with systemic lupus erythematosus-like disease.

Polychlorinated Dibenzo(p)dioxins

Studies in rodents have shown that perinatal exposure to 2,3,7,8-tetrachlorodibenzodioxin (TCDD) appears to affect the developing thymus, leading to a persistent suppression of cellular immunity. The depression of T-cell function from perinatal exposure appears to be greater and more persistent than when exposure occurs in adults. The potential for TCDD immunotoxicity in humans is less clear. Individuals exposed to very high TCDD doses during an industrial explosion in Seveso, Italy in 1976 have not shown demonstrable loss of immune function. Studies of individuals exposed to TCDD chronically in Times Beach, Missouri have revealed a few differences from a control population in some parameters, but overall the observations do not suggest significantly altered immunocompetence. These studies have focused on humans exposed as adults to TCDD, and it is possible that perinatal exposure to TCDD may have more profound effects, as has been observed in laboratory animals. Increased antinuclear antibodies and immune complexes have been reported in blood of dioxin-exposed workers, but increases in clinical manifestations of autoimmunity have not been observed.

Dusts and Particulates

A number of occupations involve inhalation exposure to high-molecular-weight organic molecules or particles containing these molecules. Examples include flour and wood dust; enzymes (e.g., from *B. subtilis* and *A. niger* in the detergent industry); dusts from agricultural wastes; fungi and bacteria in moldy hay, feeds, and wood products; and dander, feces, pupae, and other residue from insect and rodent pests. These high-molecular-weight substances are capable of producing an IgE-mediated, type I allergic reaction. This reaction can manifest itself as eye and upper respiratory tract congestion, occupational asthma, and hypersensitivity pneumonitis. Acute inhalation of dusts from bacterial or animal origin have also been shown to produce a short-term flulike illness called *organic dust toxic syndrome*. This is not a type I allergic reaction because no prior sensitization is required, nor are antigen-specific antibodies present during the illness. Inhalation of silica dusts both activates and damages alveolar macrophages. Activation of these macrophages can lead to pulmonary inflammation. Reported effects on lymphocyte responsiveness are somewhat conflicting, but suggest that immune function may be impaired.

Pesticides

Dermal and pulmonary symptoms among workers handling pesticides are not uncommon, but most of these cases appear to be due to irritant rather than hypersensitivity reactions. Studies of workers exposed to pesticides have sometimes found changes in various specific immune parameters, but there is currently little evidence that host resistance is compromised in these individuals. Isolated reports suggest an association of pesticides (i.e., paraquat) with the development of renal autoimmune disease. Also, recent studies in animals suggest that some chlorinated pesticides may accelerate the development of autoimmunity, although no studies are yet available to assess whether this occurs in humans as well.

Solvents

Benzene is capable of producing bone marrow hypoplasia and pancytopenia. Along with other formed elements of the blood, peripheral blood lymphocyte counts are diminished, leading to impaired immune function. Immunotoxic effects of benzene may extend beyond individuals experiencing bone marrow toxicity from benzene, as humans exposed chronically to benzene have been observed to have diminished serum immunoglobulins and immune complement.

Immune abnormalities, such as alterations in serum immunoglobulin concentrations, immunocyte counts, or immunocyte ratios have been observed in workers exposed to solvents, either individually or as mixtures. The significance of these findings is unclear, however, as no deficits in host resistance or other clinical immune effects have been demonstrated. Exposure to vinyl chloride has been linked to the development of scleroderma, and there is epidemiologic evidence of an association between chronic exposure to trichloroethylene in groundwater and lupus syndromes.

Miscellaneous Agents

In 1981, thousands of individuals in Spain were poisoned with cooking oil adulterated with rapeseed oil containing aniline. The symptoms that developed were called *toxic oil syndrome*, and included pneumonitis, rash, gastrointestinal distress, and marked eosinophilia. These patients developed autoantibodies and a connective tissue disorder characterized by myalgia, neuropathy, myopathy, and cutaneous manifestations. Hundreds of poisoned patients died, attributed primarily to impairment of respiratory musculature.

Acid anhydrides are used to produce a number of commercial products, including paints and epoxy coatings. On inhalation exposure, acid anhydrides can become haptens, binding to carrier proteins in the respiratory tract to elicit an immune response. After sensitization, subsequent exposure leads to asthma-like symptoms or to a reaction resembling hypersensitivity pneumonitis. Chronic exposure may lead to severe restrictive lung disease.

10.7 MULTIPLE-CHEMICAL SENSITIVITY

Multiple-chemical sensitivity is a term applied to a subjective illness in individuals attributed to contact with a broad array of chemicals in the environment. Other terms for this condition include *environmental illness*, *total allergy syndrome*, *chemical-induced immune dysregulation*, *chemical hypersen-*

sitivity syndrome, and, more recently, *idiopathic environmental intolerances*. There is no defined symptomology for this condition; in fact, physical diagnostic and laboratory findings are typically normal. Complaints are almost always subjective, including fatigue, headache, nausea, irritability, and loss of concentration and memory. It appears almost exclusively in adults, primarily in women. Offending substances are commonly identified by odor, although symptoms can also be ascribed to substances in food, to drugs, and to electromagnetic fields. While sensitivity is thought to arise from a single initial exposure, perhaps to a single agent, it is visualized as progressing to eventually involve an expanded array of substances; hence the term *multiple-chemical sensitivity.* Diagnosis is made principally on the basis of history—the patient indicates intolerance to a variety of substances in the environment. The symptoms are triggered by exposure to these substances at levels generally well tolerated by the vast majority of the population. Improvement in symptoms is attributed to avoidance of these substances.

Tests are sometimes performed on these patients, including a provocation–neutralization test and a panel of immunologic tests. In the provocation–neutralization test, reaction to various substances is tested by administering small doses sublingually, intracutaneously, or subcutaneously. The test agents are not necessarily those thought to be causing the patient's illness, and the battery of agents tested can vary from practitioner to practitioner. After administration of the test substance, the patient records any symptoms that occur over the next 10 minutes. There is no standard for what constitutes a symptom in this testing. If no symptoms are recorded, the dose is increased until a positive response is obtained. Increased or diminished doses are then given until the symptom(s) abate. This becomes the "neutralization" dose that may be recommended to the patient for subsequent treatment of the condition. Immunologic tests usually consist of quantitation of serum immunoglobulins, complement components, lymphocyte counts, autoantibodies, and immune complexes.

The status of multiple-chemical sensitivity as a legitimate disease entity has been controversial. Mainstream medical organizations do not recognize it as a defined disease for several reasons: (1) there are no objective physical signs or symptoms, or clinical laboratory observations that characterize the disease; (2) there are no clearly defined diagnostic criteria—the provocation-neutralization test described above has no physiologic basis, and the diagnostic value of the immunologic tests typically performed has not been validated; (3) although several mechanisms have been proposed to explain symptoms in these patients, there are little data to support any of these, and many explanations are contrary to current understanding of immunology and toxicology; (4) objective evidence of any chemical agent as a specific cause of multiple chemical sensitivity is virtually nonexistent; and (5) there are no treatments of proven efficacy.

Many theories have been proposed to explain multiple-chemical sensitivity, most involving the immune system. Some have proposed that environmental chemicals may act as allergens or haptens, and that an IgG (rather than IgE) response to these agents leads to an immune complex disease. However, the symptoms of multiple-chemical sensitivity do not resemble serum sickness, and IgG antibodies to the postulated array of triggering substances have not been demonstrated in these patients. An autoimmune mechanism has also been proposed, but patients with multiple-chemical sensitivity typically do not have demonstrably elevated autoantibody titers, nor do they have the usual clinical manifestations of any of the autoimmune diseases. A more general concept of immune dysregulation has also been advanced, commonly including the notion that T suppression has been impaired by environmental chemical exposure. Compelling evidence that T suppression (or any other specific immune abnormality) is a consistent feature of multiple chemical sensitivity is lacking, however, as is an explanation as to how several different chemical substances, at low exposure levels, could produce this effect. A high prevalence of psychiatric disorders has been observed in patients claiming to have multiple chemical sensitivity. This suggests that somatization (i.e., symptoms of psychogenic origin) may be involved, although proponents of multiple chemical sensitivity argue that this is a manifestation, rather than a cause, of the disease. Credibility for multiple-chemical sensitivity within the medical and scientific community has also been impaired somewhat by the significant percentage of individuals with this condition using it as a basis for workers' compensation claims or other litigation.

The mainstay of treatment of multiple-chemical sensitivity involves avoidance of what are regarded as the inciting chemicals. In some cases, this can be taken to extremes, involving near isolation in specially controlled environments. Vitamins and mineral supplements are often recommended, as well as intravenous gammaglobulin, ostensibly to fortify the immune system. "Neutralization" doses of extracts identified positively in provocation–neutralization tests are sometimes recommended to relieve or prevent symptoms. Reports of efficacy of these treatments are either anecdotal or from poorly controlled studies. Objective evidence that any of these treatments leads to improvement in the patient's condition is generally considered to be absent.

10.8 SUMMARY

A fully functioning immune system is vital for defense against pathogenic microorganisms and to prevent the emergence of cancerous cells. It is a complex system, requiring the cooperation of many types of cells. The immune system is capable of both specific and nonspecific responses to insults. Specific responses are elicited by macromolecules recognized by the body as being foreign, termed antigens. The presence of an antigen can trigger a humoral response (i.e., the production of antibodies that bind rather specifically to that molecule) or a cell-mediated response in which cells carrying the antigen on their surface are attacked by specialized immune cells (e.g., natural-killer cells or cytotoxic lymphocytes).

Drugs and chemicals can produce adverse health effects by influencing the immune system in one of three ways:

1. *Causing a Hypersensitivity Reaction.* There are four basic types of hypersensitivity reactions (types I–IV), each with a different mechanism. Depending on the type of reaction, symptoms may be immediate or delayed, mild or severe, and involve different organs and tissues. Allergic reactions can cause considerable discomfort in the workplace, and some types (e.g., a severe type I reaction, or anaphylaxis) can be life-threatening.

2. *Suppressing the Immune System.* Normal function of the immune system requires participation by many components, and disruption of any of these could conceivably result in impaired capability. If impairment is sufficient, the individual is at increased risk of infection and cancer. This has been clearly demonstrated by patients on immunosuppressive therapy (e.g., transplant patients) and in animal studies involving a variety of chemicals. Although there are few clear examples of immunosuppression from occupational or environmental exposure in humans, there is no reason to expect that this effect cannot occur under these circumstances as well.

3. *Causing or Exacerbating Autoimmune Disease.* By producing a dysregulation of the immune system, drugs and chemicals are capable of causing the immune system to attack normal body constituents. This has been clearly demonstrated for several drugs, and a number of reports suggest that it may also occur from occupational and environmental exposures.

The potential for a chemical to produce immunotoxicity can be assessed through a variety of in vivo and in vitro tests. Most of these tests focus on effects on a very specific aspect of the immune system. The immune system possesses considerable functional redundancy and extra capacity, and alterations (or "abnormalities") in one or a few parameters may not necessarily result in diminished overall functional of the immune system. Consequently, the results of these tests must be interpreted carefully.

REFERENCES AND SUGGESTED READING

Burleson, G. R., J. H. Dean, and A. E. Munson, *Methods in Immunotoxicology,* Wiley-Liss, New York (1995).

Burrell, R., D. K. Flaherty, and L. J. Sauers, *Toxicology of the Immune System, A Human Approach,* Van Nostrand-Reinhold, New York (1992).

Farine, J-C., Animal models in autoimmune disease in immunotoxicity assessment. *Toxicology* **119**: 29–35 (1997).

Kammuller, M. E., N. Bloksma, and W. Seinen, eds., *Autoimmunity and Toxicology: Immune Disregulation Induced by Drugs and Chemicals,* Elsevier Science Publishing, New York (1989).

National Research Council, *Biologic Markers in Immunotoxicology,* National Academy Press, Washington, DC (1992).

Sell, S., *Basic Immunology: Immune Mechanisms in Health and Disease,* Elsevier Science Publishing, New York (1987).

Smialowicz, R. J., and M. P. Holsapple, *Experimental Immunotoxicology,* CRC Press, Boca Raton, FL (1996).

Vial T., B. Nicolas, and J. Descotes, Clinical immunotoxicology of pesticides, *Journal of Toxicology and Environmental Health* **48**: 215–229 (1996).

Vos, J. G., and H. Van Lovern, Experimental studies on immunosuppression: How do they predict for man? *Toxicology* **129**: 13–26 (1998).

PART II
Specific Areas of Concern

11 Reproductive Toxicology

ROBERT P. DEMOTT and CHRISTOPHER J. BORGERT

The possibility of disruptions in normal reproductive function or proper development is one of the potential health effects of chemical exposure that causes the most concern. Increases in the number of women in the workplace and the subsequent pregnancies that occur along with occupational exposure to various chemicals make the overall population especially aware of the potential for this type of toxicity. Also, our environment is frequently perceived to harbor more potential reproductive hazards due to growing awareness of the distribution and persistence of some human-made (or human-related) chemicals. How do these perceptions of increased reproductive risk correspond to currently available scientific information and, how can future scientific investigation be focused toward important issues? These are questions that can be addressed by considering how reproductive processes interact with chemical (and non-chemical) workplace or environmental exposure. Our understanding of reproductive toxicity has expanded rapidly over the last several decades, providing much more information on which to base critical evaluations of the potential for reproductive risk.

This chapter will review the established toxic responses of the human male and female reproductive systems and of human fetal development. The focus will be on potential human health effects from occupational and/or environmental exposure and examples will be drawn from this area. However, there is much additional mechanistic information from experimental systems and many more chemicals for which there are experimental indications of potential reproductive toxicity. In evaluating the importance of such chemicals or mechanisms with regard to human exposure, two basic tenets of toxicology must be considered: 1) what are the characteristics of a likely exposure, and 2) what could be a relevant dose. Understanding how reproductive processes respond to chemical challenges is the key to addressing these issues.

Topics to be covered include

- Male reproduction and the susceptibility of rapidly dividing germ cells
- Female reproduction and the regulation of endocrine status as a potential target for toxic responses
- Fetal development—the major opportunities for toxic responses during development and the established causes of developmental defects
- Current research concerns—hot, timely topics

Many of the toxicological principles pertaining to reproduction can be clearly illustrated with examples relating to the male. Also, many of the most fully characterized examples of toxic responses on human reproduction from occupational exposures are male-related. This state of knowledge may well reflect a historical bias toward interest in male-related effects due to the former predominance of men in the industrial setting. However, male reproductive physiology and function does provide certain susceptibilities to toxic agents that are useful for illustrating how toxicological responses relate to reproductive biology in general. Understanding the underlying mechanisms will help make sense out of the more complex, and less understood issues in female and developmental toxicology. As interest

Principles of Toxicology: Environmental and Industrial Applications, Second Edition, Edited by Phillip L. Williams, Robert C. James, and Stephen M. Roberts.
ISBN 0-471-29321-0 © 2000 John Wiley & Sons, Inc.

continues to shift toward non-industrial occupational exposures and general environmental exposures, many new discoveries about the mechanisms of toxic injury to the female reproductive system and developing offspring can be expected.

Most of the best described examples of reproductive toxicity rely on experimental results obtained with laboratory animals for their explanations. While inferring the actual human potential for an adverse occurrence from animal testing results must always be done critically and cautiously, there are certain factors relating to reproductive endpoints that add unique uncertainties. The relevance of the experimental dose level to potential human exposures is always an important factor for interpreting animal studies. For reproductive endpoints, not only is the dose level an issue, the duration and interval used for dosing is also critical since sequential, delicately timed progressions of physiological events are a hallmark of reproductive processes. Experimental testing for any given chemical must encompass the timeframe likely to be relevant for the mechanisms of toxicity involved. In evaluating developmental toxicity, concomitant maternal toxicity can be a problematic complication. In trying to demonstrate effects, dose levels are often pushed high enough to result in general wasting, nutritional problems, or other stresses on the animal. Thus, it can be difficult to distinguish between direct toxicity to the developing offspring and females that are simply too compromised to maintain a normal pregnancy. With such complications firmly in mind, animal testing remains a critical and valuable tool for characterizing reproductive toxicity. There is no way around the need for test systems that can be readily manipulated to tease out potential mechanisms of toxicity. The challenge is interpreting the implications of particular animal testing results and determining how they relate to potential concerns for humans.

11.1 MALE REPRODUCTIVE TOXICOLOGY

In the most basic sense, the functions of the male reproductive system are to produce and deliver the male germ cells, spermatozoa, in such condition that union with a female germ cell and subsequent development can occur. Toxic responses must interfere with either germ cell production or delivery. The reproductive organs are obvious targets for such responses, but damage in the nervous and endocrine systems can also be important due to their role in controlling reproductive function. This section will describe some of the toxic chemicals that affect germ cell production and their mechanisms of action. Additionally, information about some of the toxic responses that can cause problems with sperm delivery will be presented.

Susceptibility of Spermatogenesis

The process of germ cell production in the male, spermatogenesis, provides clear examples of how cells may have enhanced susceptibility to certain classes of chemicals at particular times. In spermatogenesis, germ cells are produced from a pool of progenitor cells, stem cells, through a series of mitotic and meiotic divisions that eventually produce a large number of spermatozoa from each original stem cell and provide replacement stem cells. Spermatogenesis occurs in specialized, thick walled tubules within the testis called seminiferous tubules. The germ cells are initially located at the outer edge of the tubule (Figure 11.1) and move progressively toward the center of the tubule. From here the spermatozoa move along the tubules and into the duct system that will carry them out of the body.

In a human, it takes about 64 days to produce a mature spermatozoon through this process, which continues throughout adult life. The rate of spermatogenesis increases dramatically following puberty until a hundred million or more sperm are produced each day. Spermatogenesis can be equated to a mass production process where constant high rates of production and high quantity of output are the focus. In a biological system, this requires an extremely active, rapidly dividing cellular environment within the testis.

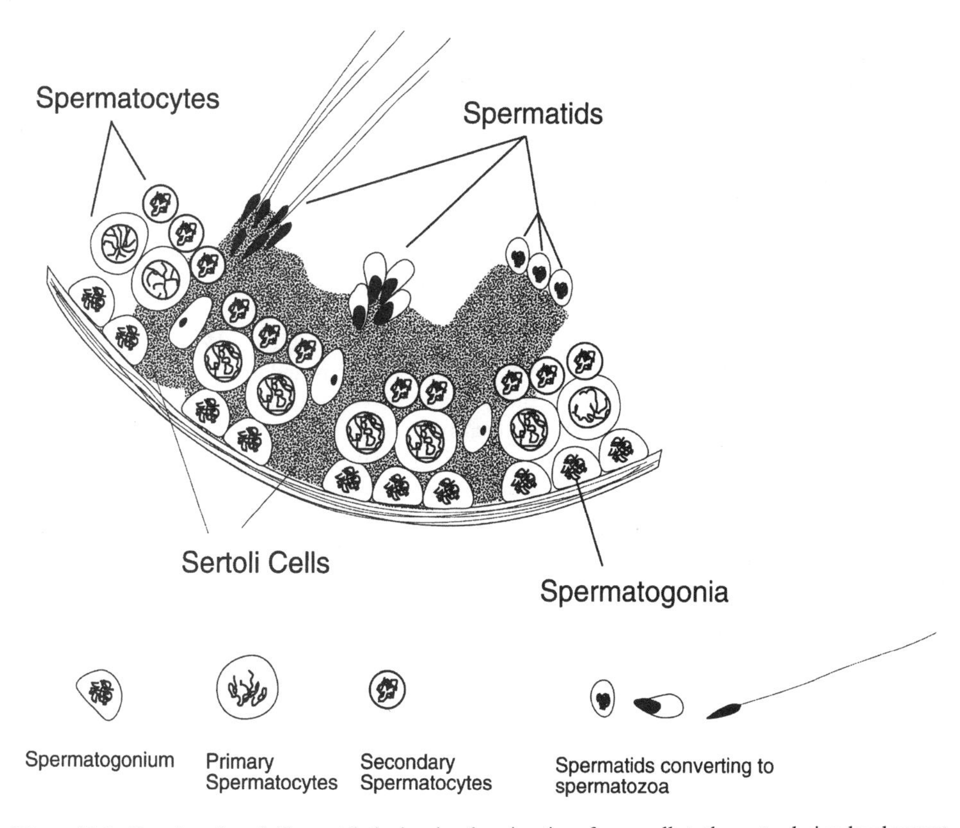

Figure 11.1 Drawing of seminiferous tubule showing the migration of germ cells to the center during development and stages of spermatogenesis.

The high rates of cellular division and metabolic activity associated with spermatogenesis are the basis for susceptibility to certain types of damage. During the duplication of genetic material and cell division, DNA is particularly vulnerable to damage. In addition, many specialized cellular proteins and enzymes are needed and a high level of cellular respiration is required. Therefore, chemicals that can cause DNA damage or interfere with cellular protein function or respiration are of particular concern in rapidly dividing tissues. Examples include reactive electrophilic chemicals such as alkylating agents and ionizing radiation.

Many chemicals or their metabolites that are considered to be relatively toxic have the ability to undergo chemical reactions with DNA or important cellular proteins. Depending on the particular chemical, DNA damage may result from direct interaction with the strands or with other cellular macromolecules involved in stabilizing the DNA. Reactions with DNA can affect base pairing and strand linkage. Protein damage can include modifying enzymes and carrier molecules such that they cannot participate in biochemical reactions.

Anti-neoplastic drugs used in chemotherapy, such as methotrexate, adriamycin, cyclophosphamide, vincristine, and vinblastine, are good examples of reactive compounds that can cause failures of germ cell production. Some examples of reactive chemicals with common occupational or environmental exposures and particular concerns with regard to rapidly dividing spermatogenic tissues include:

- Acrylamide & ethylene oxide—extensively industrial use
- Polynuclear aromatic hydrocarbons (PAHs)—combustion products
- Ethylene dibromide & dibromochloropropane—fumigants/pesticides

Understanding general mechanisms of action can help categorize chemicals as to the effects that are possible. It is important to remember that while a mechanism for cellular damage to spermatogenic cells exists for reactive chemicals in general, not every alkylating agent or reactive metabolite will actually act as a specific reproductive toxicant. Other factors control the susceptibility of spermatogenic tissues to particular chemicals. Among the critical factors are the dose level that is received, the extent of distribution to the target tissue that occurs, and how/where metabolism occurs for the particular chemical. Individual and species differences in such factors explain the differences in susceptibility and demonstrate the need to proceed carefully when predicting whether a chemical will be a human reproductive toxicant.

The energy associated with ionizing radiation, including x-rays, can also result in chemical modifications to DNA that affect its potential to be copied correctly and to direct cellular functions. While germ cells at all stages of spermatogenesis can be affected, it appears that some of the early stages are most susceptible to DNA strand breaks from irradiation. Such breaks can result in chromosomal malformations in germ cells. In addition, the death of damaged somatic cells (non-germ cells) can result in the collection of cellular debris in the duct system that carries the germ cells.

In addition to DNA damage that can lead to cell death, there is also the possibility that modifications of DNA can be repaired incorrectly, producing a mutation. There are cellular mechanisms available to remove and replace damaged segments of DNA, but there is a certain error rate, albeit low, associated with this type of repair. When such repair errors occur in a germ cell, there is the possibility that the resulting mutation could be passed to offspring and become heritable. While this is theoretically possible and can be demonstrated in some experimental systems, the generation of an inherited human mutation following chemical-induced DNA damage has not been documented.

Direct and Indirect Modes of Toxicity

Another important toxicological concept well illustrated in the male reproductive system is the distinction between direct acting toxicants and indirect acting toxicants. Direct acting toxicants may be reactive chemicals or chemicals with sufficient structural similarity to molecules used in cellular communication that they can interfere with signaling pathways. Indirect acting toxicants may eventually cause damage in the same ways, but must first be modified through metabolic reactions or bioactivated. Ironically, reactive metabolites result frequently from the chemical reactions also used to break down and eliminate foreign chemicals. The metabolism may take place in the cells or tissues that are eventually damaged, or may occur in other organs, such as the liver. In the latter case, the toxic metabolites must be transported to the target tissue.

Some of the best known reproductive toxicants have a direct mode of action on male reproductive tissues. Lead and cadmium are two examples of metals in this category. Lead can damage genetic material, disrupting cell division and resulting in cell death. While many different cell types are susceptible to damage from lead, the importance of continual division during spermatogenesis makes this process particularly vulnerable.

Cadmium has an interesting direct mode of action on the vasculature surrounding the testis and epididymis, the adjacent tubular organ in which sperm are stored and mature. Sperm must be kept at temperatures slightly below core body temperature. Extensive and specialized vascularization is provided to remove heat from the testis. This vasculature is extremely vulnerable to direct damage by cadmium. When such damage occurs, the remaining vessels are unable to carry away as much blood, and thus as much heat. The combination of reduced perfusion with oxygenated blood and higher temperatures can subsequently destroy the spermatogenic cells.

Many of the alkylating chemotherapeutic drugs have direct modes of action resulting in male reproductive toxicity, including busulfan and cyclophosphamide. Ethylene oxide, used extensively as a chemical intermediate in industry and for gas sterilization of medical devices and even foods, also has direct actions on cellular biomolecules and experimental studies suggest later stage germ cells are particularly susceptible to its damage. DNA is stabilized by extensive interactions with structural elements called protamines during later stages of spermatogenesis and protamines are particularly vulnerable to reactions with ethylene oxide.

These medically related examples indicate the delicate balance between therapeutic or beneficial uses and potential reproductive toxicity. The powerful alkylating properties of chemotherapy drugs allow them to work against rapidly dividing cancer cells and the risk of ancillary effects on other cells, even frank reproductive toxicity, may represent an acceptable tradeoff for the cancer patient. Also, the safety and benefits of dry, gas sterilization with ethylene oxide are clearly significant. While we can document the mechanisms of action producing male reproductive toxicity experimentally, there is no evidence that the ethylene oxide exposures associated with sterilization has produced reproductive toxicity in men. Again, the toxicology indicates what could happen if the right conditions existed, not what happens under the typical situations in which the chemicals are encountered.

Many of the compounds of interest in occupational or environmental toxicology require metabolic activation to produce reproductive effects. The testis has the enzymatic capabilities for oxidative metabolism, a pathway that frequently produces reactive intermediates. While this activity is low compared to the liver, it is sufficiently high to produce toxic amounts of metabolites for some compounds.

Metabolism of common industrial chemicals including the solvents *n*-hexane and the glycol ethers appears to contribute to their reproductive toxicity. Some of the phthalates, a chemical class used extensively as plasticizers and distributed widely in the environment, are also capable of affecting male reproductive tissues after metabolism. All of these examples are discussed further with regard to the specific cells they affect and have at least purportedly affected humans. There are also many other examples of indirect acting male reproductive toxicants where there is at least experimental or mechanistic information on toxic potential including the intermediate acrylamide and vinyl chloride, another common industrial intermediate also found in the environment, often as a breakdown product of dry cleaning solvent.

Cell-type Specific Toxicity

Another principle illustrated by examining male reproductive toxicology is the specificity of action on certain cell types due to the characteristics of the cells or their metabolic potential. For some reproductive toxicants, there is varying sensitivity among the somatic and germ cell types. While this may be explained in some cases by the high levels of cell division and activity among the germ cells, in some cases there appear to be more specific factors involved. Besides the germ cells, there are two major types of somatic cells in the testis required for spermatogenesis, Sertoli cells and Leydig cells. Both of these cell types may also be specific targets for some toxicants. In addition, the microvasculature of the testis can be a specific target and the functional consequences of impaired circulation in the testis have been described above.

Developing Sperm Cells As germ cells proceed through spermatogenesis, several different terms are used to distinguish the varying degrees of maturity. At the earliest stages are the spermatogonia, followed by the spermatocytes, the spermatids, and finally spermatozoa (Figure 11.1). Besides describing the developmental stage of the germ cells, these distinctions also correspond to some degree of toxicological specificity as certain stages are targeted by certain compounds. This specificity generally relates to which stages are the most sensitive to a particular agent. In most cases, as the dose increases or exposure conditions change, more than one stage can be affected.

One of the occupational episodes that stirred interest in effects on male reproductive function was reported sterility among workers handling the pesticide dibromochloropropane (DBCP). Subsequent investigations suggested a toxic effect that would certainly explain sterility. The most significant cell type damaged by DBCP is probably the spermatogonia. Since these progenitor germ cells are at the base of spermatogenic cellular expansion, their destruction precludes future cycles of spermatogenesis. Thus, the expected observation would be a depletion of all the later stages and an inability to recover spermatogenic potential after the toxicant was removed. This fits the observed effects on occupationally exposed humans.

Spermatocytes, particularly at the pachytene stages, are susceptible to damage from ethylene glycol monoethyl ether (EGME), one of the glycol ethers with considerable potential for human exposure. A metabolite of EGME, 2-methoxyacetic acid (MAA), may cause indirect damage to the spermatocyte by decreasing lactate production in the Sertoli cells. Lactate is a key metabolic substrate for developing spermatocytes.

Spermatids may be a particular target for ethylene dibromide (EDB) toxicity. This is another compound, used as a fumigant, for which there is a least some information that occupational exposures may have adverse effects on male reproduction. Effects on the later, spermatid, stages of spermatogenesis would be consistent with the abnormalities and deficits observed in some occupationally exposed workers. Experimentally, however, EDB turns out to be an example where the stage specificity breaks down as the dose increases.

Spermatozoa can be affected by various toxic mechanisms. Epichlorohydrin (widely used intermediate in plastics/rubbers) is an example of a compound that appears to affect sperm motility by interfering with metabolism. Motility is required for fertilization and the needed energy is produced through specialized metabolic pathways that operate within the sperm. Chlorpromazine, a drug used to treat psychosis, appears to cause metabolic effects on sperm secondary to permeabilizing their membranes. Agents such as mercury and lead may have a combined effect on both sperm cell membrane dynamics and on the epididymis. Sperm complete their maturation in the epididymis, and the secretory and absorptive functions of the epididymal epithelium are required for the maintenance of viable sperm.

Another important mechanism of toxicity may also be pertinent to mature sperm. Many biochemical reactions result in the formation of highly reactive radical groups. Cellular phospholipids, important to the structure of cell membranes, are particularly sensitive targets for reactions with these radicals. The resulting damage, called lipid peroxidation, can impair the integrity of cell membranes. Free radicals can be generated during the oxidative metabolism of many different compounds. Some of those that may be male reproductive toxicants are adriamycin, ethylene dibromide, and the herbicides paraquat and diquat.

Free radical-induced lipid peroxidation may be important to sperm for two reasons: 1) the detoxification pathways that typically keep free radicals in check are modified in reproductive tissues and appear to be especially limited in the sperm cells, and 2) sperm contain highly specialized membranes that can be easily compromised. Investigations of lipid peroxidation in sperm have only begun recently, and currently, the only clear occurrence of such damage in human sperm is found in frozen semen samples. Freezing seems to destroy one of the major anti-oxidant defense enzymes making the sperm especially susceptible. As further investigations of chemical-induced lipid peroxidation are carried out, some of the membrane disruption associated with spermatotoxicity may be better explained.

Sertoli Cells Sertoli cells are in direct contact with the germ cells and provide support for them, both structurally and functionally (Figure 11.1). By virtue of specialized junctions between Sertoli cells, which isolate the germ cells from any other somatic cells outside of the seminiferous tubule, the Sertoli cells create a barrier that provides a degree of insulation and protection from chemicals distributed through the circulatory system. Thus, when Sertoli cells are targeted by toxicants, not only is their support of germ cell production impaired, the blood/testis barrier may be disrupted, exposing the germ cells to more potential damage.

Due to their close relationship with the germ cells, it is not surprising that toxicants which specifically affect Sertoli cells have a subsequent effect on germ cell production. Some of the characteristics of Sertoli cell damage are that all stages of developing germ cells are impacted and the damage is frequently irreversible. This is due to the limited replacement of Sertoli cells; they divide relatively little in mature males. Also, part of the function of Sertoli cells is to initiate the sequence of germ cell development. Without Sertoli cells, remaining progenitor germ cells are unable to begin the spermatogenic cycle.

The solvent *n*-hexane is a demonstrated reproductive toxicant. Its testicular toxicity is related to interference with microtubule formation in Sertoli cells by the metabolite 2,5-hexanedione. The susceptibility of Sertoli cells to a microtubule poison is understandable since Sertoli cells form a scaffolding supporting multiple layers of germ cells and this function relies on the assembly of microtubules. This process involves extensive remodeling of the Sertoli cell architecture as the germ cells are moved through the seminiferous tubule, and such remodeling is heavily dependent on microtubule formation.

Some of the phthalate plasticizers also appear to affect Sertoli cells. The toxicity appears to occur in the Sertoli cells, involving a breakdown of the attachments between Sertoli cells and germ cells. Thus, all of the spermatogenic cells in the developmental sequence at the time of intoxication may be compromised.

Bioactivation of tri-o-cresyl phosphate (TOCP) is required prior to its effects on Sertoli cells. Interestingly, the metabolism occurs in the Leydig cells but does not appear to interfere with their function. The reactive metabolite subsequently reaches the Sertoli cells, which are damaged, and spermatogenesis is subsequently impaired.

Another Sertoli cell toxicant of interest in occupational toxicology is dinitrobenzene (DNB). This compound, and structurally related analogues, appears to disrupt Sertoli cell function, possibly through involvement in a metabolic reaction cycle that depletes the cells of important reducing equivalents, impairing their functional support for the spermatogenic cells. Subsequently, all of the stages of developing germ cells may be compromised.

The Sertoli cell tight junctions can also be affected by toxic agents, disrupting the blood/testis barrier. Platinum-based anti-neoplastic drugs, such as cisplatin, appear to operate in this manner. The germ cells divide improperly subsequent to this toxicity; however, it is not clear whether this is due directly to exposure through the disrupted barrier or if the Sertoli cells are incapable of directing germ cell development properly.

Leydig Cells Leydig cells are located outside, but surrounding the seminiferous tubules. Their major function is producing testosterone, a key to regulating spermatogenesis as well as male reproductive development and behavior. There are several toxicants that can be demonstrated to affect Leydig cells experimentally. It is not clear, however, whether any demonstrable human reproductive toxicity is primarily due to actions on Leydig cells. In part, this is because androgen regulation is so complex that it is difficult to determine which observations are primary toxic responses and which result secondarily from hormonal dysregulation. Also, several of the toxicants with specific actions on Leydig cells can also cause responses in other cell types, depending on the dose received.

A well-described experimental Leydig cell toxicant which does not appear to directly affect other testicular cell types is ethane-1,2-dimethanesulfonate (EDS). This compound affects androgen production by the Leydig cells and appears to interfere with specific early steps in the synthesis of steroid hormones. While the mechanism leading to cell death is not clear, EDS does kill the Leydig cells. The eventual result of EDS toxicity is, somewhat predictably, impaired spermatogenesis. In addition, Sertoli cells, which are dependent to some degree on Leydig cell secretions, may also be damaged.

Among the toxicants with significant human exposure that operate primarily on the testis, diethylhexyl phthalate and its metabolites are the only example with specific Leydig cell effects. In addition to their Sertoli cell effects, the phthalates also appear capable of interfering with steroid synthesis and Leydig cell function. It is not yet clear what the relative contribution of the Leydig cell damage is to the resulting spermatotoxicity.

Hormonal Regulation and the Hypothalamic-Pituitary-Gonadal Axis

Disruptions of male reproductive function can also occur secondary to toxic responses in the endocrine system. Androgen production in the testes is regulated primarily by luteinizing hormone (LH), a gonadotropin released by the pituitary gland. Gonadotropin secretion is in turn regulated by gonadotropin releasing hormone (GnRH), secreted by the hypothalamic portion of the brain. This hierarchical arrangement, where the hypothalamus regulates the pituitary which in turn regulates the gonads, is known as the hypothalamic-pituitary-gonadal axis. The hypothalamic-pituitary-gonadal axis is illustrated for both males and females in Figure 11.2. From a toxicological perspective, this arrangement creates even more sites where toxic responses may have an impact on reproduction. With this in mind, it is not surprising that some compounds generally considered to affect the central nervous system, can impact Leydig cell function and male reproduction.

The toxic effects of ethanol are wide ranging and complex. Experimental evidence for direct testicular toxicity is not clear; however, it is clear that alcoholics suffer decreased testosterone levels and subsequently, decreased gonadal function. In alcoholics, the ability to stimulate testosterone production appears to be impaired. Experimentally, it can be demonstrated that ethanol affects LH release. It is clear that alcohol interferes with male reproduction by affecting endocrine regulation, and ultimately, in part, Leydig cell production of testosterone, but the relative contributions of direct testicular action and toxic responses elsewhere in the regulatory axis are not known.

A variety of other drugs and industrial compounds also affect male reproduction through endocrine-related mechanisms. These include the anti-hypertensive drug propanolol, the opiates, and tetrahydrocannabinol (THC). The use of such drugs is pertinent to occupational toxicology, since their effects can confound observations on reproductive impairment related to direct occupational exposure. Carbon disulfide, the pesticide chlordecone, and the phthalates are examples of industrial chemicals that can disrupt the endocrine axis.

Besides the potential impact on spermatogenesis, there is another aspect of toxicity to the hypothalamic-pituitary-gonadal axis that affects male reproductive function. Both libido, or behavioral drive, and physical aspects, such as penile erection and ejaculation, are controlled by the central nervous system. Libido is controlled primarily by androgens and any of the drugs or industrial compounds that can dysregulate the endocrine axis and affect androgen production can affect libido. Alcohol and THC

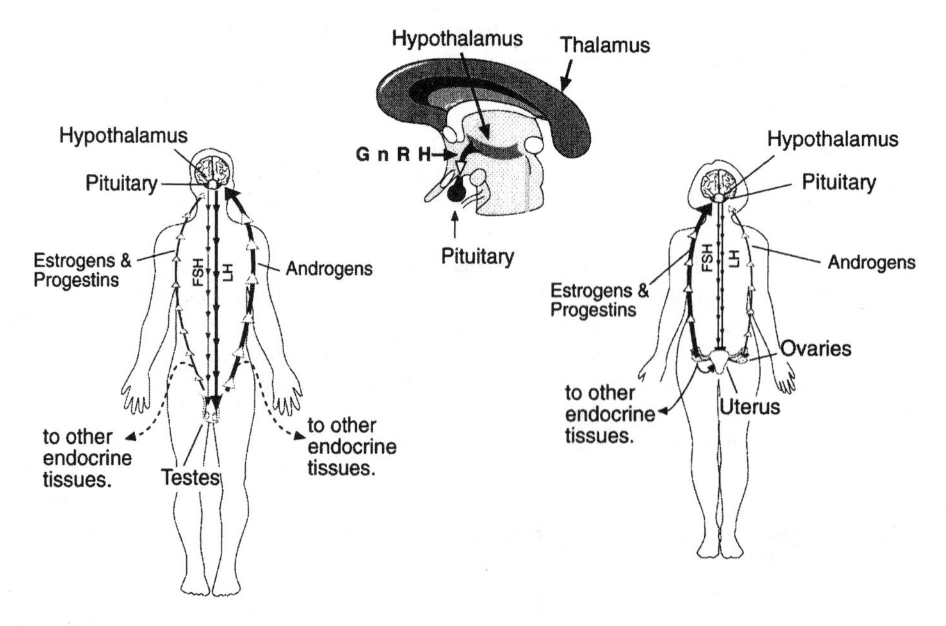

Figure 11.2 The hypothalamic-pituitary-gonadal axis for human males and females illustrating the sequence of control from the brain to the secretion of gonadotropins by the pituitary, to the production of steroid hormones in the gonads. The major hormonal products and their contribution to regulatory feedback loops are indicated. GnRH—Gonadotropin Releasing Hormone.

are common examples of compounds that decrease libido. Carbon disulfide and chlordecone-exposed workers also reported decreased libido. It is important to remember, however, that neurological and psychological factors play a tremendous role in libido and the relative role of chemical-induced mechanisms for decreasing libido is likely minor. Where there are demonstrable effects that relate to endocrine balance, it is not surprising to find reported effects on libido. However, where the claimed effect is simply a report of decreased libido, there is little likelihood of establishing a clear toxicological basis.

Other factors that can preclude sperm delivery are impotence and inability to ejaculate. These aspects of reproductive function are centrally controlled, but the autonomic nervous system also plays a key role in coordinating the physical events. The most common examples of chemical-induced impotence are some of the adrenergic antihypertensives, methyl DOPA, clonidine and guanethidine, and the opiates and ethanol. Psychoactive drugs, such as chlorpromazine and diazepam, can also produce impotence. As with effects on libido, psychological factors are major contributors to impotence. Clinical findings with humans suggest that the preponderance of reported problems with impotence are psychological and not related to toxic responses.

Endocrine Feedback and Potential Dysregulation

Some of the key signaling molecules essential for endocrine feedback loops are the steroid hormones and other protein-based hormones. Since both the absolute levels and the ratio between androgens and estrogens serve as signals to the hypothalamic-pituitary-gonadal axis, modifications of this balance can subsequently disrupt endocrine function and result in reproductive effects.

A highly publicized case of occupational exposure leading to reproductive impairment involves the pesticide chlordecone. Occupationally exposed men appeared to have reductions in fertility. Subsequent analysis provided a possible mechanism for such effects, as chlordecone turned out to have estrogenic activity. The endocrine dysregulation produced by elevated estrogenic feedback could explain the neurotoxic and reproductive effects that have been attributed to chlordecone.

The role of exogenous compounds with estrogenic, anti-estrogenic, and anti-androgenic activity is a current source of substantial controversy. The extremely potent toxicant dioxin has the potential to interfere with endocrine balance, and the contribution of this activity to its toxicity is hotly debated. A variety of pesticides, including DDT, the carbamates, and mirex, are reported to possess endocrine activity as are some of the polychlorinated biphenyls (PCBs). What is not yet clear is whether environmental levels of exposure to such compounds can actually produce endocrine-mediated toxic effects. Most synthetic steroid-like molecules can only displace the actual endogenous compounds in molecular interactions to a very limited degree. Also, typical exposure levels of the exogenous compounds are extremely low, and metabolic deactivation before they reach the target tissue further limits the potential activity. On the other hand, the endocrine system functions with very low effective concentrations of signal molecules at the target cells. The significance of endocrine-mediated toxicity in the male reproductive system is not yet clear, however, it is a topic of intense interest and potentially wide ranging ramifications for the future (see further discussion in Section 11.4).

Male Reproduction Summary

There are many industrial and pharmaceutical compounds that are potential male reproductive toxicants (Table 11.1). Such chemicals may interfere with the development of germ cells directly, or indirectly, by disrupting the cellular and endocrine factors involved in supporting and regulating spermatogenesis. The regulatory roles of the neuroendocrine system are also important to the delivery of sperm. The potential for various compounds to work through a particular toxic mechanism, however, does not necessarily indicate that there is a relevant human reproductive risk associated with that mechanism, or even with the compound at all. Mechanistic possibilities and experimental results must be carefully evaluated in terms of those effects that are actually observed in exposed humans. There

TABLE 11.1 Suspected Human Male Reproductive Toxicants

Industrial/Environmental	Pharmaceutical Agents and Drugs
Cadmium	Adriamycin
Carbon disulfide	Busulfan
Chlordecone	Chlorambucil
Dibromochloropropane (DBCP)	Chlorpromazine
DDT	Clonidine
Diethylhexyl phthalate	Cyclophosphamide
Dinitrobenzene	Diazepam
Epichlorohydrin	Ethanol
Ethylene dibromide	Guanethidine
Ethylene oxide	Methotrexate
n-Hexane	Methyl DOPA
2-Hexanedione	Opiates
Ionizing radiation	Propanolol
Lead	Tetrahydrocannabinol
Mercury	Vincristine
2-Methoxyethanol	Vinblastine
Tri-*o*-cresylphosphate	

have been relatively few instances of occupational exposure leading to demonstrable decreases in male fertility.

Cell type susceptibility to toxic injury can be roughly generalized with germ cells as the most sensitive, followed by Sertoli cells and then Leydig cells. The hierarchical regulation of spermatogenesis underlies this since development of germ cells is often affected by toxicants acting on the Sertoli and Leydig cells. In turn, Sertoli cells are often affected by both Sertoli cell and Leydig cell-specific toxicants. In addition, there need not be any specific site within the testis targeted by a reproductive toxicant. Reproductive function is susceptible to agents that interfere with the central nervous system and autonomic nervous function because of the importance of neuroendocrine regulation.

11.2 FEMALE REPRODUCTIVE TOXICOLOGY

For the sake of this chapter, female reproductive toxicology will only include toxic responses of mature females not directly affecting the fertilized egg or subsequent development. All post-fertilization toxicity relating to the developing offspring will be considered developmental toxicology (see Section 11.3). With this limitation, female reproductive function can be described by the same characteristics outlined for the male—the key features being the production of female germ cells, eggs, and transport of the germ cells, in this case the sperm and eggs, to the site of fertilization.

With reference to human occupational and environmental exposures, substantially less is known about female-specific toxicology compared to male or developmental toxicology. One of the reasons is that reproductive impairment of human females is difficult to both establish and analyze. Unless there is an observed prolonged inability to maintain a pregnancy there is often little reason to investigate whether toxic responses may have affected female reproductive function. While this may also be true of occupationally exposed men as well, fast and fairly sensitive means to test the reproductive capacity of men are available. Semen samples are easily obtained and evaluated, and, while such analysis cannot definitively determine fertility, abnormalities are obviously a sign of a potential problem. On the other hand, germ cell production by women is very difficult to monitor and potential indicators, such as failure to menstruate or irregular menstruation, occur frequently enough and for such a variety of reasons that associating these occurrences with a particular exposure is difficult. In addition, pregnancy failure occurs frequently as well, and can be due to problems with either the female or the developing embryo.

There are also some fundamental differences in female germ cell production that relate to the potential mechanisms of toxic injury. In contrast to the constant cellular division required to generate millions of new sperm each day, the ovary contains all of the eggs that will ever be ovulated, and then some, by the time of birth. All mitotic divisions and the initial stages of meiosis have been completed by the middle of fetal development and result in the generation of around 10 million primary oocytes arrested in the meiotic progression. Subsequently, a rapid process of degeneration (atresia) occurs, and there are around one million primary oocytes left at birth. Atresia continues somewhat more slowly throughout life, and the mature woman has around a half a million oocytes with the potential to develop into mature eggs.

In the human, a handful of the primary oocytes begins the process of folliculogenesis during each menstrual cycle, but typically only one forms a completely mature follicle and is ovulated. Estimating the number of menstrual cycles in a typical reproductive lifetime, only around 500 germ cells ever complete development. There is no further division of the germ cells prior to leaving the ovary and while there are somatic cells that divide and develop to support the maturing egg, this occurs over the course of about two weeks. Clearly, female germ cell production does not rely on mass production and rapidly dividing cell populations. Correspondingly, the overall process is less susceptible to cytotoxic compounds that operate against dividing cells.

Oocyte Toxicity

This being noted, however, some of the powerful antineoplastic drugs are still capable of disrupting oocyte development, including cyclophosphamide, chlorambucil, busulfan, and vinblastine. Experimental results indicate that busulfan may be capable of destroying the arrested, primordial oocytes, preventing further maturation and ovulation. Other alkylating agents appear to work only on the follicles developing at the time of exposure. For the reproductive process as a whole, this is advantageous, since other primordial oocytes are still available for future cycles of folliculogenesis.

While there is experimental evidence that ovarian-derived metabolites of some polycyclic aromatic hydrocarbons (PAHs–combustion byproducts) are also capable of destroying the primordial oocytes, there is limited information on ovotoxicity relating to occupational or environmental exposures. There are no good examples of human reproductive toxicants that appear to affect egg production under such circumstances. This is surely due in part to difficulties in analyzing ovarian toxicity that have slowed down the identification of toxicants with the ability to target the ovary. Experimental findings suggest that lead, mercury, and cadmium are capable of damaging oocytes, but in light of their generalized toxicity, this is hardly surprising and probably does not occur in the absence of their major toxic effects on other systems.

Somatic Ovarian and Reproductive Tract Toxicity

There are a few experimental examples of toxic compounds that can have a direct effect on the somatic ovarian cells or the reproductive tract. For the ovary, such toxicity relates to successful germ cell production since the ovarian cells differentiate into layers known as the granulosa and theca, both integral to the development of mature follicles. 4-Vinylcyclohexene, an industrial compound used in the production of epoxy resins, appears to be capable of producing generalized ovarian atrophy. Similar atrophy can be produced with the antibiotic nitrofurantoin, which also has some specific toxicity toward the cells lining the follicles. The granulosa layer in particular can be targeted by metabolites of some of the phthalates. It is not clear how such toxicity relates to other effects of these compounds.

The female reproductive tract is responsible for transport of the eggs and sperm to the site of fertilization and subsequent transport of any fertilized embryo to the site of implantation in the uterus. Fertilization occurs in the uterine tube, also known as the Fallopian tube or oviduct (Figure 11.3). The

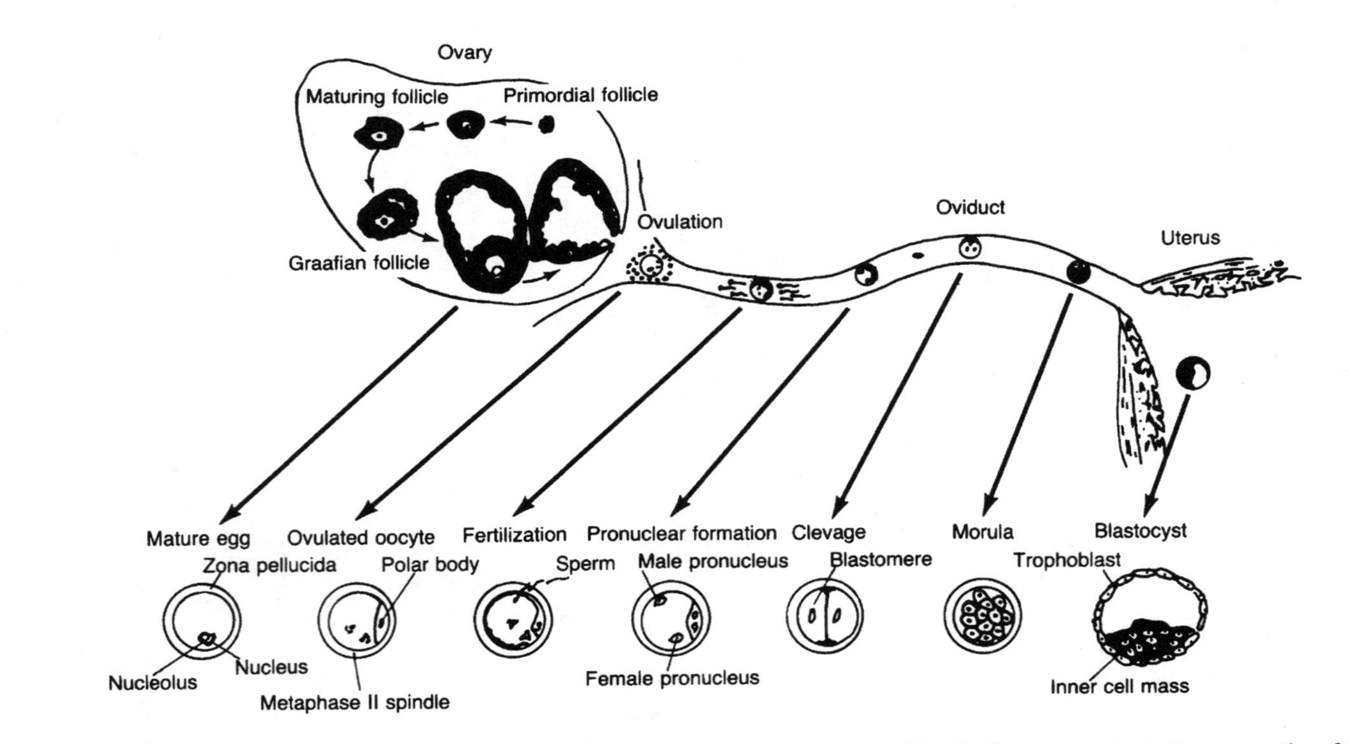

Figure 11.3 Cycle of follicular development and the progression through the human female reproductive tract. Note fertilization occurring in the upper portion of the oviduct and implantation occurring after the earliest cell divisions at the blastocyst stage. (Reproduced with permission from Dean J., American Journal of Industrial Medicine, 4 (1983) 3. pp. 31–39. Figure 1.)

early stages of embryonic development occur in the uterine tube, and then the embryo moves down to the uterus to implant. All of this transport requires a patent lumen in the oviduct, and the movement occurs due to a combination of ciliary beating and muscular action. In other words, the tract is more than a transport tube and appropriate biological function is required to support the early embryo in particular.

Atrophy of the oviduct or uterus can clearly prevent the transport of the germ cells and embryo. Cadmium can produce such atrophy, and presumably other metals which cause overall tissue degeneration could as well. For cadmium, the response is a general metabolic inhibition of the cells in the reproductive tract, leading to cell death and declining organ weight. Not surprisingly, uterine, as well as ovarian, cyclicity is impaired. Lead is another example of a metal that may directly affect the cells lining the uterus, subsequently interfering with proper uterine cyclicity.

The features responsible for moving the germ cells and early embryo appear to be a potential target for some components of cigarette smoke. Experimental exposures have resulted in both increases in muscular related oviductal and uterine motility and immobilization of the cilia lining the oviduct. The pertinence of these observations to human exposures and the observed effects of smoking are not clear. However, the potential outcomes, improper migration of the germ cells precluding fertilization, or the early embryo preventing proper implantation, are consistent with the overall decreased fertility and increases in irregular cyclicity reported for smokers.

Hormonal Regulation of Reproductive Function and Associated Toxicity

Endocrine regulation of female reproduction is even extremely complex. The hypothalamic-pituitary-gonadal axis is present in females, and GnRH release and the timing of changes in the relative levels of the two major gonadotropins, LH and follicle stimulating hormone (FSH) are linked to the ovarian follicular cycle (Figure 11.2). This is accomplished by endocrine feedback loops involving the steroid hormones estrogen and progesterone, as well as some protein hormones.

In a simplified form, the female endocrine cycle can be considered to start with increasing levels of FSH production by the pituitary during the early stages of folliculogenesis (Figure 11.4). As the follicles develop, the granulosa cells surrounding the oocyte are a major source of estrogens. As the estrogen levels increase, FSH production is shut down and production of the other gonadotropin, LH increases. When the follicle is fully mature, there is a surge of LH release, directing ovulation and the subsequent formation of a progesterone secreting tissue, the corpus luteum, at the site where the follicle had been. Progesterone levels rise and support the establishment of a pregnancy. Progesterone also causes the levels of both gonadotropins to drop. If there is no pregnancy the corpus luteum degenerates and progesterone levels decline, releasing the inhibition of gonadotropin secretion. FSH can again rise, starting the cycle over. For the human, this is the point where menses occurs, lasting through the early stages of the next follicular cycle.

Disruptions at the gonad, pituitary, or hypothalamus during the preovulatory stages can cause a failure of folliculogeneis, and there will be no ovulation for the affected cycle. Later disruptions can cause a failure of the corpus luteum maintenance, preventing the establishment of pregnancy if the egg had been fertilized, or causing a shortened cycle. Alternatively, interference with luteal degeneration can cause a cycle extension and this may be manifest as delayed menstruation. It is clear that there are plenty of opportunities for endocrine disrupting toxicants to interfere with both the follicular cycle and the ability to maintain a pregnancy.

There are examples of toxicants that can interfere with female hormonal regulation. Lead toxicity, for instance, is associated with decreased progesterone production. This may in part explain its historical use as an abortofacient, since progesterone is the key hormone for establishing and maintaining pregnancy. The actual mechanism of lead-induced progesterone inhibition is not clear. However, the established effects of lead on the neuroendocrine system could reasonably be expected to interfere with the hypothalamic or pituitary secretion patterns required for the luteal phase of the cycle.

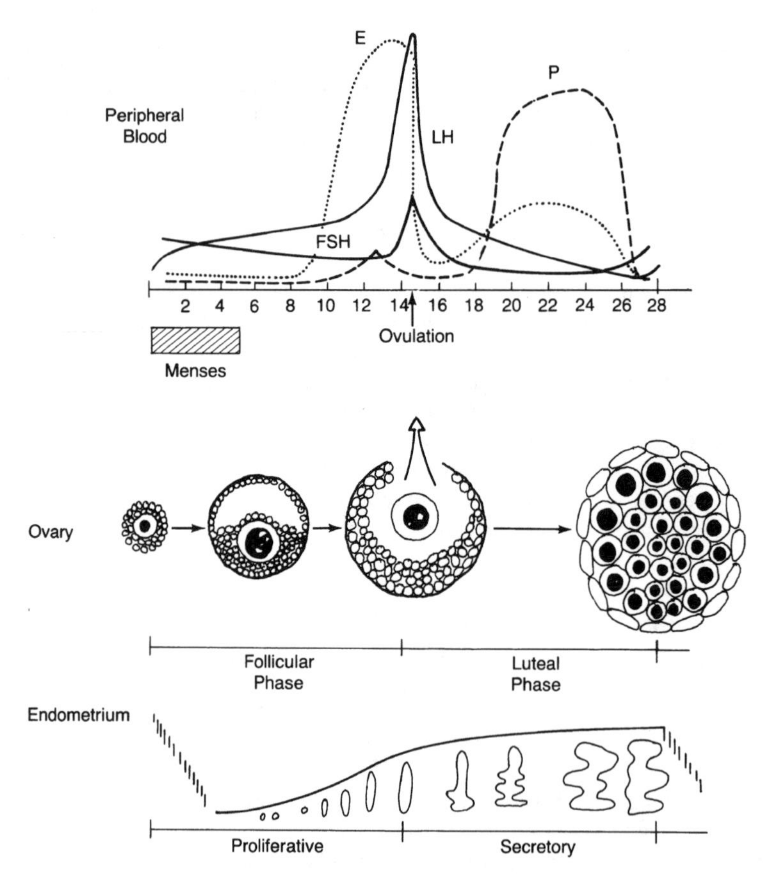

Figure 11.4 Sequence of hormonal peaks and hormone balance with concomitant follicle development during the human female reproductive cycle. Timeline indicates number of days beginning with the first day of menses; E—Estrogen; FSH—Follicle Stimulating Hormone; LH—Luteinizing Hormone; P—Progesterone. (Reproduced with permission from Mattison (Ed.), "Reproductive Toxicology," American Journal of Industrial Medicine, 4 (1983): 1–2. P. 20.)

Cigarette smoking, probably the nicotine, and alcohol are both capable of interfering with GnRH release. Alcoholics have been observed to be unable to produce the LH surge needed for ovulation, and tobacco use is associated with decreased estrogen levels. These observations are consistent with irregularities of the cycle that have been associated with alcohol and tobacco use. The decreased libido and lack of cyclicity associated with narcotics abuse may relate to the hypothalamic depression that these compounds cause. Clearly, drugs of abuse and smoking have wide ranging physiological effects, and their toxicity is not specifically tied to female reproductive function. However, the endocrine actions of such compounds clearly present a mechanism by which associated reproductive impairments could be explained.

Clomiphene citrate is an example of a compound that is used therapeutically to interfere with the female endocrine balance. This anti-estrogen is used to hyperstimulate ovulation in infertile women. It appears to work by increasing gonadotropin release, allowing more follicular development to occur. The anti-estrogenic properties prevent the estrogen-mediated shut down of FSH production. There are, however, other ramifications of the anti-estrogenic activity including decreased luteal function and decreased ability of the uterus to establish a pregnancy. The mixture of desired and undesired outcomes is a good example of the complex outcomes associated with endocrine interference.

Occupational and environmental exposures that could dysregulate the female endocrine pattern are a major topic of current investigation. Dioxins, other polycyclic chlorinated compounds and organochlorine and organophosphate pesticides are all potential compounds of concern. As described in the Male Reproductive Toxicology section, the concern is that many of these compounds have some type of estrogenic or androgenic activity because they are able to replace the endogenous compounds in cellular interactions. The theoretical potential for such compounds to affect female reproductive function is clear. However, demonstrating such an effect is extraordinarily difficult, and there are still no convincing examples of endocrinologically active compounds causing reproductive impairment in women through typical occupational or environmental exposures.

The endpoints that can generally be observed for women are menstrual interval, fertility as measured by time to pregnancy, and ability to carry a pregnancy to term. This last potential effect will be discussed in the Developmental Toxicology section. There are such extreme interpersonal differences in menstrual interval and regularity and so many established causes for missed or delayed menstruation that associating any variation with a particular chemical exposure is difficult. Many cultural and occupational factors are clearly relevant for affecting time to pregnancy, and difficulty achieving a pregnancy when desired may affect as much as 25 percent of couples in the United States at times. It is clear that this is not always due to female reproductive problems, but this "naturally" occurring background obscures potential toxicologically mediated effects.

The high degree of interest in endocrine disruption as a potential mechanism for female reproductive toxicity is driving extensive investigations of this hypothesis. In the future it should become clearer whether environmental estrogens and other endocrinologically active compounds can actually reach levels at which they can produce a significant endocrine disruption and subsequent reproductive impairment. Currently, we are left with a potential mechanism for reproductive effects and candidate compounds that could act through this mechanism, but no clear demonstration of any of the candidates posing an actual risk through such a mechanism for humans following occupational or environmental exposures.

The potential for exposure to chemicals that could alter endocrine processes and the need to use pharmacological agents known to cause reproductive toxicity opens up controversial occupational and societal issues about restricting women's chemical exposure. What types of data or experimental results should be sufficient to indicate the need to control occupational exposures? When considering whether women should be excluded from certain jobs during certain segments of their reproductive lives, suddenly, the need to get beyond the uncertainties of extrapolating doses and mechanisms of toxicity from animal testing becomes crystal clear. The associated issues are as widely disparate as the economic impacts of possibly needing to move employees in and out of certain jobs or requiring specialized exposure control equipment to the potential for claims of discrimination, should women be excluded from opportunities on the basis of concerns they do not believe are relevant for them.

Alternatively, when deciding a certain therapy is needed, what constitutes an adequate representation to the patient of the risks to herself or a developing fetus? Clearly, we cannot always discard effective treatments. The recent return of thalidomide, discussed below as the cause of one of the most notorious cases of human developmental toxicity, is a shining example. Thalidomide turns out to be a particularly effective treatment for patients suffering complications of leprosy or some complications of AIDS. It may further be an effective sedative for cancer patients and those suffering autoimmune diseases. These uses expand the patient population where reproductive effects are a possible concern. Effective patient education and carefully planned distribution policies may be relatively straightforward for thalidomide, where the toxic timing and dosage is established and the outcomes are readily documented, but what is the appropriate balance between protection and restrictiveness for other drugs?

Female Reproduction Summary

Compared to the male, there are relatively few female-specific reproductive toxicants that are not related to developmental toxicity (Table 11.2). (Developmental toxicants will be covered in the following section.) In part, this is due to the difficulties in analyzing oocyte production and determining

TABLE 11.2 Suspected Human Female Reproductive Toxicants

Industrial/Environmental	Pharmaceutical Agents and Drugs
Arsenic	Androgens, estrogen, and progestins
Cadmium	Busulfan
Diethylhexyl phthalate	Chlorambucil
Dioxins	Cyclophosphamide
Ionizing radiation	Ethanol
Lead	Opiates
Mercury	Vinblastine
PCBs—coplanar forms	

when there is a reproductive impairment. Also, the relatively small proliferative cell population in the ovary and the intermittent nature of the proliferative stage makes the ovary less susceptible to disruptions of cell division. The compartmentalization of the active germ cell and its supporting follicle is also pertinent to ovarian toxicology since this means that only a few germ cells are vulnerable during a given cycle. This decreases the likelihood that a toxic injury will cause permanent interference with oogenesis. While there are occasional toxicants, such as busulfan, that can wipe out the arrested population of primordial oocytes and prevent future follicular development, in general, the arrested cells are fairly resistant to damage.

Probably the most significant type of toxicants for female reproductive function are those that interfere with the dynamic endocrine balance required for folliculogenesis and ovulation. A wide array of toxicants have at least the potential to interfere with the female hormonal pattern, including heavy metals, drugs of abuse, and some chlorinated biphenyls. Chemicals that can structurally mimic the steroid hormones, or have a competing functional activity, can be found among a variety of therapeutic drugs, pesticides, and environmental contaminants. Though a mechanism for interfering with female reproductive function is suggested, the impact of endocrinologically active chemicals, especially through environmental exposures on human reproduction, is not yet clear.

11.3 DEVELOPMENTAL TOXICOLOGY

Recognition that environmental factors could cause congenital defects grew following the 1941 report by Gregg that there was an association between exposure to Rubella virus (German measles) during pregnancy and the occurrence of blindness and deafness in the offspring. Further analysis following a Rubella epidemic and the thalidomide incident solidified people's awareness of the potential for prenatal exposures to cause developmental defects. Part of the reason that recognition was so late in coming, even in the scientific community, is that the placenta was thought to serve as a barrier preventing any potentially harmful agents from reaching the fetus. Since the 1960s it has become clear that the placenta is actually quite porous to chemicals of the molecular size that encompasses all but the largest drugs and industrial compounds. Developmental toxicity testing has now become commonplace, and many agents that can affect development, chemical and biological, as well as physical phenomena, have been identified.

As we have seen in male and female reproductive toxicology, experimental demonstrations of toxic potential far outnumber demonstrated cases of human developmental toxicants. This may be due in part to species differences and the high doses used in experimental protocols; however, with developmental defects, it is also not always clear whether the experimentally demonstrable differences in structure or behavior can be extrapolated to humans and considered abnormal. Overall, the class of chemicals with demonstrated human effects are less important in the population than both biologically infective agents, such as Rubella virus, syphilis and cytomegalovirus, and maternal metabolic disorders, such as diabetes and phenylketonuria. In turn, genetically based developmental disorders still surpass all of the environmental phenomena that can affect development in terms of proportional importance within the population.

This section will cover the toxicology relevant to all stages of development from fertilization onward. This will include the action of toxicants on the mother that affect the ability to establish and maintain pregnancy, as well as direct actions on the fetus. The developmental stages can be divided into 1) the preimplantation stage, where toxicity generally affects the entire organism and there is typically an all or none response (i.e., there is repair or the developing organism is aborted), and 2) the later embryonic and fetal stages, where specific structural defects can occur. The most sensitive period for teratogenesis, the production of congenital defects, is during organogenesis in the embryonic period.

Mammalian development can be thought of as an expansive flow diagram (Figure 11.5). Following fertilization there is a very particular sequence of events that is followed, directed by the expression of certain genes at certain times. From a toxicological point of view, disruptions during this relatively linear phase generally derail the entire developmental sequence. Certain developmental steps serve as branch points and once particular branches are followed, the occurrence of events on that branch will not necessarily affect other branches. As development progresses, many smaller branches are reached, which may each relate to the development of certain tissues, cell types, or regions of a structure. Toxicologically, when specific sequences are affected, the response may be restricted to the features that develop out of that particular sequence. This illustrates how very specific defects can occur in response to toxicants or other environmental factors and exhibit clear time dependence.

Spontaneous Abortion and Embryonic Loss

Recent improvements in the ability to measure human chorionic gonadotropin, a very early indicator of the presence of an embryo, have allowed reasonable estimates of early pregnancy loss. Overall, more than 50 percent of fertilized eggs/embryos are lost through spontaneous abortion. Around 30 percent are lost after implantation but before the first menstrual period is missed. An additional 20–25 percent are lost after they have been clinically recognized as a pregnancy. There are also probably substantial pre-implantation losses, but these are much harder to accurately estimate. Presented another way, it appears that the chance of achieving a full-term pregnancy for any one menstrual cycle in which fertilization is likely to have occurred (based on non-contracepted intercourse with ovulation) is around 25 percent.

The preponderance of embryonic loss occurs during what was described as the linear phase of development. Based on chromosomal analysis of spontaneously aborted embryos, approximately two-thirds of this loss can be explained by gross genetic abnormalities. Around 10 percent can be attributed to a known environmental cause, and a cause cannot be determined for the remainder. While some of the genetically associated loss could be related to environmentally mediated DNA damage, it is clear that most is due to major chromosomal aberrations associated with germ cell production and fusion. Once again, the potentially chemically induced responses are hidden within an extremely high background rate of embryonic loss.

In general, the response to toxicological insult during the early embryonic stage is considered to be an all or none event, where damage up to minor cell death is completely repairable and major cellular disruption or death results in abortion of the pregnancy. This is based on the flexibility of the cells in the early embryo, which allows them to functionally replace a few lost cells. As development progresses, most cell lines become committed to a particular fate and such compensation is less likely.

The limited responses available during early embryogenesis mean that typically the endpoint of concern for toxic exposures is spontaneous abortion. While this criterion has held over the years for most toxicants, there are recent experimental results which suggest that very early exposure to ethylene oxide and some other mutagens may cause responses that are manifest much later in development. This implies that the exposure may result in a sublethal injury that is not repaired, nor are all of the

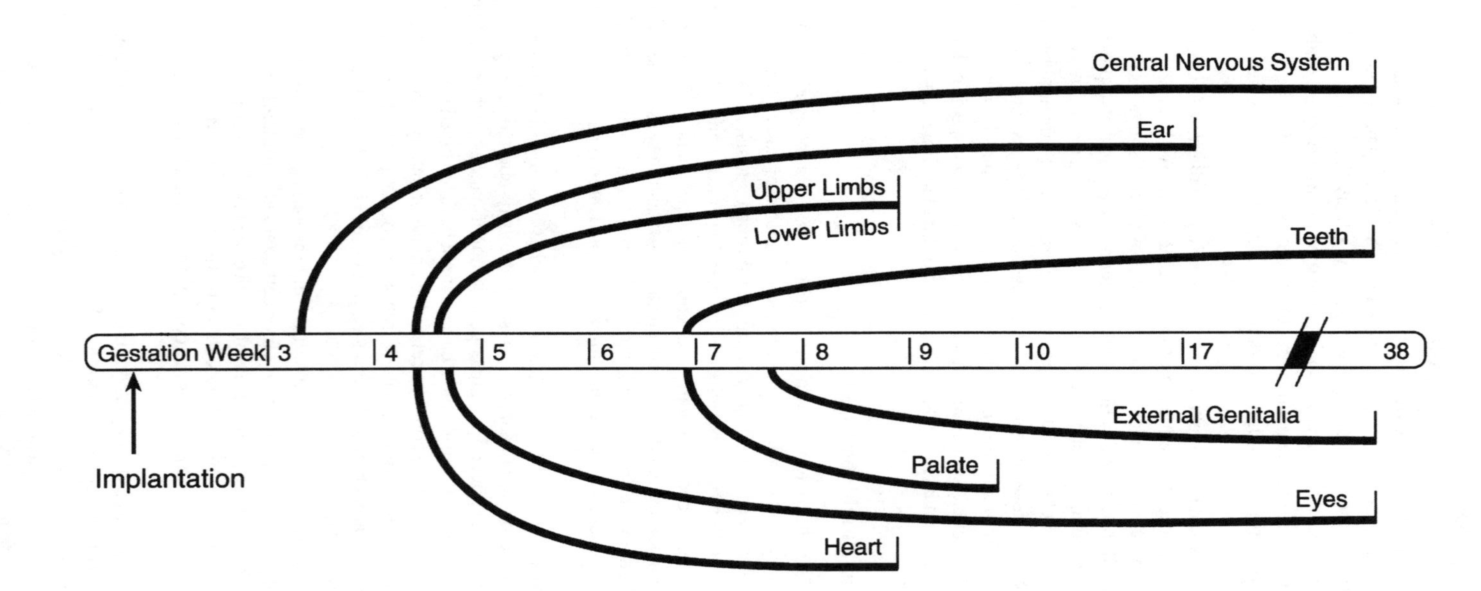

Figure 11.5 Developmental tree indicating the time during human gestation at which the development of various major organ systems becomes a separate progression from the rest of the organs and tissues. The branching corresponds roughly to the periods of tissue-specific teratogenic sensitivity.

affected cells replaced. It is not clear how many toxicants could create developmental defects in this manner, as they would likely have to be capable of producing specific, minor changes in DNA, nor is it clear that the experimental conditions are relevant to humans.

Spontaneous abortion remains the most useful indicator of early embryonic toxicity, and there are both experimental and occupational examples where chemical exposures appear to affect spontaneous abortion rates. One of the most investigated, and most controversial, occupational examples relates to anesthetic gases. Studies both large and small have reported both positive and negative results when looking for elevated rates of spontaneous abortion among health care professionals using gas anesthetics such as nitrous oxide and halothane. The potential effect does not appear to relate to paternal exposure since there is not an observable elevation in the spontaneous abortion rate among the wives of occupationally exposed men. Methodological flaws have called some of the positive results into question. At this point, the most defensible conclusion is that based on the evidence there is an elevated incidence of spontaneous abortion among women in such occupations; however, the association between anesthetic exposures and the spontaneous abortion rates cannot be reliably demonstrated. This suggests that other, unidentified factors present in the study populations could play an important role.

Carbon disulfide, dimethylformamide, and some of the phthalates are other examples in which investigations of spontaneous abortion rates have detected differences that may be attributable to occupational exposures of women. The data for carbon disulfide are the most convincing in terms of documenting an association, but even this conclusion is weakened because the proportionate increases are small and not clearly out of the expected background range.

Chloroprene, an industrial chemical used in polymer manufacture, is an example in which male exposure may have subsequent effects on spontaneous abortion. In this case, the wives of occupationally exposed men showed elevated spontaneous abortion rates. This could be classified as a male reproductive effect, if it results from an effect on the sperm that necessarily occurs prior to fertilization, and suggests that some types of sperm damage may still be compatible with the ability to fertilize an egg. While such a mechanism of toxicity could explain the observations, this is not recognized as a common mode of toxic action, and it is thus difficult to exclude some other explanation not directly linked to the chloroprene exposure of the men.

There are many compounds that can be shown experimentally to cause spontaneous abortions. Some of the classes might be expected based on their common toxic effects, such as the antineoplastic drugs and heavy metals. Their cytotoxicity is well known, so embryonic interference is hardly surprising. In addition, solvents such as benzene and toluene, many chlorinated pesticides and herbicides, PAH's, and aldehydes such as formaldehyde can all experimentally cause early pregnancy failure. In short, most cytotoxic chemicals have the ability to interfere with early development under experimental conditions. The relevance to human exposure conditions and potential dose levels, especially in the occupational setting, is not clear.

A common occupational chemical exposure that illustrates the difficulties in establishing embryotoxicity occurring in humans is the use of ethylene oxide. Large quantities of this chemical are used in manufacturing, especially for the production of ethylene glycol antifreezes. Female workers clearly have potential industrial exposures. In terms of the number of exposed workers, even more significant is the use of ethylene oxide as a sterilant of medical devices. Ethylene oxide exposure during unloading of sterilizers and in the area where the sterilized packages are aerated can be significant enough to produce toxic responses in other organs systems. This was especially true prior to interest in the potential long-term effects that grew during the 1980s. Also, ethylene oxide clearly causes embryotoxicity and death and structural abnormalities during the fetal stages in animal tests. These factors suggest that ethylene oxide could be a developmental concern for occupational exposure levels.

Epidemiological studies of ethylene oxide exposed workers are equivocal. Though occasional findings suggesting elevated spontaneous abortion rates among potentially exposed workers have been reported, this result has been inconsistently observed. Furthermore, the largest studies, best designed to account for exposure levels and potential biases, have been routinely negative. So, the database stacks up as follows: 1) the toxic potential from animal tests is clear, 2) the potential for human exposure is clear, 3) suggestive associations have been reported in relatively few worker studies that have been criticized for inability to clearly establish exposure levels and for sensitivity to possible biases, and 4) no association is supported by the biggest and best controlled studies.

Toxic Responses of the Embryo and Fetus

For many people, the possibility of congenital defects is the most alarming aspect of reproductive toxicology. Congenital defects are any morphological, biochemical, or functional abnormalities that result from an occurrence prior to birth. The defect may not be detected until later, as with some learning deficits, but the biological basis for the defect occurs during uterine development. Most congenital defects are not due to chemical exposures, but it is clear that some defects have been caused by drugs and environmental exposure.

The rate of major congenital structural malformations runs at about two to three percent of live births. When other more subtle congenital defects are added in, the rate reaches about seven percent of live births. Approximately 15 percent of these defects are linked to an inheritable disorder at a known gene locus, such as Tay Sachs disease or hemophilia. Another 10 percent are linked to major chromosomal malformations, such as monosomies or trisomies. Overall, genetic factors can account for up to 35 percent of congenital defects. Identifiable external factors (physical, biological, and chemical) account for around 10 percent of congenital defects. Estimates of additional, uncharacterized chemical and drug-induced defects account for between one and five percent. There remains a large proportion of defects without a well understood cause.

Teratogens are agents, chemical or otherwise, capable of creating congenital defects. They are generally considered to create specific defects during the period of organogenesis, which begins around five weeks after fertilization for humans, and continues for various organs through most of the second trimester of pregnancy. For many teratogens with specific structural targets, the fetus shows a period of sensitivity corresponding to the development of the target structure. This is significant to human concerns since it means that to cause a defect, exposure must generally occur in a particular window of time during pregnancy.

Disruptions of Tissue Organization The prototypical example of a teratogen with a narrow window for toxic potential is thalidomide. This drug was widely used in the late 1950s and into the 1960s to treat morning sickness and as a sedative. A sudden rise in children with limb deformities was associated with mothers who had taken thalidomide. The critical window of sensitivity was identified based on the severity of the deformities and period during which the mother had used the drug. A relatively narrow window in weeks 6–7 of gestation was identified in which exposure to thalidomide produced deformities in nearly all infants. Exposures after this time were associated with minor and less prevalent defects. In addition to the striking limb deformities, thalidomide exposed infants also exhibited congenital heart and renal defects along with ear deformities.

Despite extensive animal experimentation in the aftermath of the thalidomide incident, the mechanism of action has still not clearly been determined. The probable reactive metabolites have been identified, and mechanisms of actions, such as interference with vitamin or amino acid metabolism in the developing limb bud and direct disruption of DNA in this region, have been suggested. It seems unfortunately ironic that while so many chemicals have a clear potential mechanism of action, yet no clearly observable human effects, perhaps the best example of an actual human teratogen has been recalcitrant to the studies that might identify the mechanisms that actually operate to disrupt human development.

With thalidomide removed from use by pregnant women, the most significant teratogenic drug is isotretinoin, or Accutane, a highly effective agent against cystic acne. This drug is especially important in relation to teratogenicity, precisely because it is so effective and there is not a suitable replacement. Thus, despite its ability to effectively produce major fetal deformities, its use continues. Despite aggressive warnings by physicians and many exclusionary policies that attempt to prevent patients at risk for becoming pregnant from taking the drug, developmental deformities related to isotretinoin exposure continue to be reported.

The defects associated with isotretinoin are wide ranging and include craniofacial deformities, including cleft palate, and cardiac and central nervous system abnormalities. This can be understood on the basis of the role of the retinoids in normal development. Isotretinoin is a synthetic retinoid, or chemical relative of Vitamin A. Gradients of certain retinoids in tissues appear to play a major role in the organization and orientation of tissue growth. The direction of cellular growth needed to extend a limb, for instance, is guided by retinoid signals. Disrupting this road map with exogenous retinoids could clearly be a basis for inappropriate development. Exogenous exposure to most retinoids can produce developmental defects, at least experimentally. Isotretinoin is a good example of a teratogen that works by interfering with the chemical signaling used to guide development.

Fetal Hydantoin Syndrome Another class of teratogenic drugs is still used by pregnant women because the developmental risks are less than the risks of removing the drug. Diphenylhydantoin—phenytoin, valproic acid, and other anticonvulsants are used in epileptics to prevent seizures. They are also teratogenic. In these cases, however, the therapeutic regimens are not associated with substantial numbers of congenital defects, and the potential for injury to both the fetus and mother should a seizure occur is more of a concern.

A specific set of characteristic developmental features associated with anticonvulsant treatment has been classified as Fetal Hydantoin Syndrome. There are craniofacial features, limb alterations, as well as growth and learning deficits. While the structural effects may be mild, the growth and learning retardation are commonly permanent. The syndrome is not particularly common, however, and many studies conclude that the risk is low, especially when the potential for epileptic seizures is considered. In experimental protocols, phenytoin produces more severe, specific craniofacial defects, including cleft palate when given in the window of time associated with palatogenesis. Later exposures produce the limb effects.

Valproic acid is clearly teratogenic in experimental protocols as well. The primary effects appear to be on the central nervous system, however, skeletal and craniofacial defects can also be produced. There are reports of neural tube defects and spina bifida in humans exposed to valproic acid through maternal treatment, and the occurrence appears to be higher than expected in some studies. However, most of these studies have selected cases to examine, and it is not clear what the actual incidence rate of valproic acid-related human defects is. Again, the concurrent epilepsy confounds the situation.

DES: A Teratogen Associated with Cancer Endpoints Another type of teratogen is exemplified by another human tragedy. Diethylstilbestrol (DES) is a synthetic steroid hormone that was used to help prevent miscarriage in women with difficulty maintaining a pregnancy. In this case, rather than the half decade it took for thalidomide's effects to become clear, around a quarter of a century passed between the mid-1940s and 1970 before the teratogenicity of DES became clear. An extremely rare form of reproductive tract cancers, clear cell adenocarcinomas, was detected in a series of women whose mothers had taken DES during the first trimester of pregnancy. Though the cancer risk was first noted in some early teenage girls, it peaked around age 19–22. This explains the delay in discovering DES's effects and illustrates an example where the congenital defect was not immediately obvious. DES is also an interesting example of a teratogen that produces cancer as its congenital defect. It is currently the only established human carcinogen that acts transplacentally.

In addition to the cancer risk associated with DES, a variety of other reproductive disorders were noted as the exposed children grew up. Among the female children, these included an increased risk of ectopic pregnancy, spontaneous abortion, menstrual irregularities and infertility. For the male children, abnormalities of the genitals, decreased sperm production, cryptorchidism, and a possible increase in testicular cancer were observed. All of these reproductive tract defects help point out the likely actions of DES in the developing children. The development of the internal and external genitalia is coordinated by steroid hormone production, primarily by the fetal gonads. The hyperestrogen environment produced by DES is consistent with improper formation of the male internal and external genitalia and may have disrupted the development of the steroid feedback loops in the hypothalamic-pituitary-gonadal axis of the females.

Additional Human Teratogens Other drugs that are established human teratogens include lithium, tetracyclines, aminopterin, and the coumarin anticoagulants. Antineoplastic agents such as busulfan and cyclophosphamide are teratogenic in addition to their multiple other reproductive toxicities. Androgenic hormones, used to help maintain pregnancies, are also teratogenic and interfere with reproductive development in the fetus, somewhat like the estrogenic DES. With all of these therapeutic agents, the dose resulting in teratogenicity, the issues of leaving the underlying illness untreated, and the actual likelihood of teratogenic effects must be considered in characterizing their toxicological potential.

Drugs of Abuse and Maternal Nutrition Some of the drugs of abuse are teratogenic, but their effects typically relate more to generalized metabolic disruptions than to interference with specific features of development. Fetal alcohol syndrome (FAS) is the best example of this class of teratogens. The primary features of FAS include growth retardation, psychomotor dysfunction, and craniofacial anomalies. Growth retardation is the most sensitive and prevalent effect following alcohol consumption during pregnancy. This can be demonstrated at fairly low doses, around 1 drink per day, but the effects at low dose exposure are controversial. At higher doses, 5 or more drinks at a time and at least 2 each day, however, the risk of having an infant classified as small-for-gestational-age increases three-fold. Among children born to chronic alcoholics, one-third have been classified as FAS by some studies. Subsequent development is characterized by reduced height and an inability to catch up during postnatal development. Among children classified as FAS, the rate of mental retardation is 85 percent. Impulsiveness, attention disorders, and language deficits are commonly observed.

The mechanism by which FAS is produced is not yet clear. Especially for the chronic alcoholics, it seems most likely to relate primarily to the overall nutritional state, metabolic and endocrine imbalances of the mother. Nutrient deficits and interference with placental metabolism and transport are clearly mechanisms that can affect fetal growth and neural function. Infants born to heroin addicts also exhibit growth retardation that is probably related to overall maternal nutritional and metabolic status.

Cigarette smoking by the mother has also been associated with general developmental deficits. Growth retardation of the fetuses of smokers is clear. The array of chemicals in tobacco smoke, however, makes defining the key pathway difficult. It has been suggested that nicotine, carbon monoxide, and cyanide interfere with the transport of amino acids across the placenta. In addition, cadmium is capable of producing placental necrosis and could affect placental exchange. Also, the PAH's induce metabolic enzymes in the placenta that may create toxic reactive metabolites. Which of these mechanisms actually operate in humans, and which are the major causes of growth retardation is not clear.

Maternal deficiencies of specific nutrient factors have been reported to be associated with teratogenicity. Zinc, folic acid, and retinoic acid deficiencies have all been reported to have negative effects on development. It is clear that these nutrients are all required by the fetus, just as they are for any human, but the degree of deficiency that must be reached to cause developmental defects is not clear. Severe deficiencies of any of the vitamins, minerals, or amino acids could reasonably be expected to interfere with development. Declines in zinc, folic acid, and retinoic acid may not be tolerated very well because developmental growth processes are heavily dependent on them. Zinc and folic acid are utilized extensively in metabolism within the rapidly growing tissue. The role of retinoic acid as a major signal molecule during fetal growth has been described above.

Methylmercury Poisoning During the 1950s, Minamata disease was described and related to mercury contamination of the Minamata Bay in Japan by industrial facilities. Fish from these coastal waters, a major food source for women in the area, were accumulating elevated methylmercury levels. Mercury levels realized by the women were not high enough to produce obvious signs of mercury toxicity. However, methylmercury accumulation was sufficient to cause CNS abnormalities in developing fetuses, with cerebral palsy being the most common problem. It was later discovered that mercury levels increased in the placenta and fetal membranes of pregnant women that were exposed to metallic mercury during work. However, these exposures were apparently not sufficient to cause developmental toxicity since there was no increase in spontaneous abortion rates and no defects found in the offspring.

Developmental Toxicity Summary

Developmental toxicity can be separated into two categories. Early in pregnancy, the predominant effect of chemical and other stresses is spontaneous abortion. Later, when the specific differentiation of the various organs and structures is taking place, the response to some toxicants is congenital defects of structure or function. Determining whether spontaneous abortions have been caused by a particular chemical exposure is extremely difficult, primarily because there is such a high background rate and so many non-toxicological causes. Detecting a difference in the spontaneous abortion rate within a population is difficult for similar reasons.

Most congenital defects are due to inherited or developmental genetic factors rather than teratogenic chemicals. Though in many cases there is a desire to establish whether a defect is the result of external factors, clearly identifying relevant factors and isolating the definitive cause is frequently impossible. From a population perspective, finding a relevant, common factor to ascribe to a certain set of defects is difficult and establishing a causal role for such a factor is even more challenging. It is relevant to consider that the best examples of chemical-induced teratogenesis relate to therapeutic doses of chemicals given with the knowledge and documentation of a health professional.

Experimental results have clearly established mechanisms by which both early and later development can be affected by exogenous chemicals. This information is useful in prioritizing investigations of potential human health effects and judging whether reported effects could reasonably be expected

TABLE 11.3 Suspected Human Developmental Toxicants and Teratogens

Industrial/Environmental	Pharmaceutical Agents and Drugs
Anesthetic gases (e.g., Halothane)	Aminopterin
Benzene	Busulfan
Cadmium	Coumarin anticoagulants
Carbon disulfide	Cyclophosphamide
Chloroprene	Diethylstilbestrol (DES)
Chlorobiphenyls	Dimethylformamide
Diethylhexyl phthalate	Ethanol
Dioxins	Isotretinoin and other retinoids
Ethylene oxide	Lithium
Lead	Phenytoin
Methylmercury	Opiates
Toluene	Tetracyclines
	Thalidomide
	Valproic acid
Miscellaneous Nonchemical Agents	
Cytomegalovirus	Rubella
Diabetes	Syphilis
Ionizing radiation	Toxoplasmosis
Phenylketonuria	

to occur from certain candidate chemicals. However, development is a very finely regulated process that depends on small pools of cells to serve as the starting point for various structures, and the potential of many experimental protocols to interfere with this process is not surprising. The challenge is to decide which of these experimental sources of developmental effects should be of concern to humans. Table 11.3 provides a summarization of the suspected developmental toxicants.

11.4 CURRENT RESEARCH CONCERNS

This section will describe some areas of reproductive toxicology that are currently drawing intensive interest from researchers. Some of these research areas are likely to follow a progression into applied toxicology and become issues affecting the future regulatory framework in the United States and, consequently, affect industrial and environmental decisions and concerns. The goal is to point out both exciting areas for investigation and suggest areas of toxicology that may be important in the future for both government and industry.

Endocrine Disruption

Since passage of the 1996 Food Quality Protection Act, a law that required EPA to screen pesticides for the ability to produce estrogenic effects in humans, endocrine disruption has moved to the forefront in terms of toxicological research and regulatory controversy. The potential for some chemicals to manifest adverse effects through interactions with the endocrine system has been discussed for male, female, and developmental toxicology in earlier sections of this chapter. Although scientific and regulatory attention is recent, ecologists, agricultural scientists and farmers have known about the practical ramifications of hormonally active agents for many years. A large number of chemicals produced by one organism can affect the hormonal status of other organisms. On the ecological scale, this type of chemical signaling between species is critical to the functioning of certain communities. Farmers have long known that grazing sheep or cattle on rich, new growth clover reduces pregnancies. This affect is due to compounds produced by the clover that can mimic estrogen, called phytoestrogens. Many different classes of chemicals produced by plants or fungi can affect reproduction. One question drawing much toxicological and regulatory attention is whether exposure to synthetic chemicals at levels relevant in the environment or workplace can also have hormonal effects. A second question is whether synthetic chemicals with weak hormonal potency could adversely affect endocrine functioning given that the human diet already contains large amounts of naturally-derived hormonally active agents. A third critical question is whether it is practical to regulate chemicals based on presumed mechanisms of action—i.e., on the basis of a potential endocrine mechanism—rather than on production of adverse effects such as reproductive or developmental impairment.

There is little doubt that certain wildlife exposures to high concentrations of synthetic chemicals have produced reproductive and developmental effects. However, scientists disagree as to which chemicals or environmental factors may be responsible, and whether the effects are caused by hormonal mechanisms or by other types of toxicity. There is even greater controversy regarding whether adverse effects are also occurring in humans and other wildlife at lower exposure levels (the so-called endocrine disruptor hypothesis). Proponents of the hypothesis claim support from three main tenets: the effects observed in wildlife inhabiting highly contaminated environments; the chemical similarities among endogenous hormones, naturally occurring hormonally active agents, and certain synthetic chemicals; and, the fact that the endocrine system is responsive to minute levels of hormones. Those who find the hypothesis unsupportable point out different tenets: that synthetic chemicals are much less potent than natural hormones; that the human diet already contains many naturally-occurring hormonally active agents that may actually enhance health; that there is a propensity for weakly hormonal chemical signals to cancel each other by antagonistic actions; and, that the reproductive health of humans in industrialized societies has tended to improve rather than decline over recent decades.

The basis for presuming that exposure to hormonally active agents can lead to significant risks is mechanistically sound and clearly operates in certain high-dose situations, such as the birth control pills and other therapeutic uses of synthetic hormones such as DES. However, there are well-understood biological reasons to expect that the characteristics of the endocrine response differ under dramatically different levels of stimulation (i.e., dose). Since doses of DES prescribed during the first three months of pregnancy are equivalent to more than 150 years worth of a woman's natural estrogen production, this example is probably not relevant to potential risks from low levels of weakly estrogenic environmental contaminants. It is important to recognize that the potency of "environmental estrogens" typically range from hundreds to millions of times less than estradiol itself. Although there are factors that might tend to reduce these potency differences, such as binding to serum proteins, these factors are insufficient to answer the serious questions raised as to whether synthetic chemicals could affect endocrine signaling in humans at realistic exposure levels.

Other putative developmental effects of hormonally active agents in humans derive from environmental exposures, some of which occurred in accidental, high-dose poisoning incidents. Prenatal and postnatal exposure to PCBs and polychlorinated dibenzofurans (PCDFs) in high-dose accidental poisonings from contaminated rice oil in Yusho, Japan and Yucheng, Taiwan have resulted in various developmental defects. The syndrome of effects includes low birth weight, dark pigmentation of the skin and mucous membranes, gingival hyperplasia, exophthalmic edematous eyes, dentition at birth, abnormal calcification of the skull, rocker bottom heel, and low birth weight. Most of the affected infants were found to be shorter and had less total lean mass and soft-tissue mass. Follow-up studies on poisoned individuals suggest neurobehavioral effects and cognitive deficits. Gross developmental defects have not been observed in populations exposed at lower levels, but proponents of the endocrine disruptor hypothesis point out that typical body burdens of PCB's and dioxins are relatively close to the levels measured in epidemiological study groups where adverse effects on IQ and neuromuscular development have been reported. Nonetheless, these body burdens of PCBs have not been shown to cause any specific adverse effects, and the overall epidemiological evidence is equivocal and does not support a causal association between typical body burdens of PCBs and adverse developmental outcomes.

Another highly publicized putative consequence of endocrine disruption in humans is reduced sperm counts in men living in industrialized nations. Increasing background levels of a variety of persistent, estrogenic environmental chemicals have been identified as a potential cause. This theory has some mechanistic plausibility because sperm production is controlled by androgen levels, and some effects of androgens can be antagonized by estrogens. This theory also has some high-dose precedent from DES, a potent estrogenic compound that may have reduced sperm counts among males exposed to therapeutic levels *in utero*. However, several of the studies that report declining sperm counts have been criticized for methodological flaws, failing to account for alternative factors, and biases in data collection. The statistical tests used and the proper interpretation of the tests have also been called into question. Based on these difficulties and criticisms, many scientists question whether sperm counts have actually declined in men from industrialized nations.

It is important to recognize that suggested reduction in sperm numbers is not the type of readily apparent pathological condition observed in DBCP manufacturers (discussed above under Male Reproductive Toxicology). The sperm count decline suggested by some authors (up to around 50 percent) would not be expected to correspond to a general fertility reduction because of the large excess of sperm that are produced by most men.

Numerous methodological difficulties arise in evaluating sperm counts from different laboratories over a long time frame, and hence, the degree of change that is purportedly related to environmental exposures may be too subtle to be easily measured. Collection and preparation methods have varied over the years, and differing criteria have been used to categorize typical, or "normal" sperm counts. Different studies have handled samples from possibly infertile patients differently, some studies including and others excluding them according to differing criteria. These differences in methodology have confounded attempts to combine the results into a larger database for integrated analysis. For these reasons, many scientists are convinced that reported sperm count declines are an artifact of methodological and analytical flaws of the studies.

Many synthetic chemicals have been suggested as potential human endocrine disruptors based upon widespread human exposure and their hormone-like activity in certain laboratory assays. Various lists of putative endocrine disruptors have been published or otherwise publicized in the media or on the internet. It is important to recognize that the quality of data supporting inclusion of chemicals on these lists varies considerably, and there is no generally accepted scientific source providing an authoritative listing at this time. Most lists include chemicals from diverse chemical classes, many of which have produced a positive result in at least one of a variety of bioassays and receptor-binding methods devised to determine the potential interaction of a chemical with the endocrine system. Despite positive results in laboratory assays, few chemicals—e.g., those drugs and chemicals already discussed in this section—have been shown to produce adverse developmental outcomes in exposed humans. Some prominent examples of chemicals listed as endocrine disruptors include organochlorine pesticides (e.g., toxaphane, methoxychlor, chlordecone, DDT and metabolites), alkylphenol ethoxylates (detergents or dispersing agents in household cleaners), PAHs (combustion products) dioxins (TCDD), co-planar PCBs, phthalate and phenolic plasticizers (e.g., benzyl butyl phthalate, di-*n*-butyl phthalate, bisphenol A). However, more definitive laboratory studies and risk assessments developed for a number of such chemicals (e.g., alkylphenol ethoxylates, phthalate and phenolic plasticizers) indicate little or no potential for adverse effects in humans at environmentally relevant exposure levels.

Two particular issues have arisen in the controversy over endocrine disruption that deserve special mention. In 1996, just months before Congress passed the 1996 Food Quality Protection Act, Arnold and coworkers published a paper in *Science* that brought national attention to the subject of endocrine disruption. The report claimed that a combination of synthetic chlorinated pesticides were one-thousand times more potent than any of the chemicals individually in stimulating an estrogenic response. This so-called demonstration of estrogenic synergism was later shown to be in error, and the publication was retracted more than a year later. Despite its failure to demonstrate synergy, this study raised a debate within the scientific, regulatory, and regulated communities over the frequency with which synergistic interactions are likely to occur and their relevance to human and environmental health. Though the interest in synergy has subsided considerably since the retraction of the Arnold publication, a considerable amount of effort is still underway to determine whether such chemical interactions are important considerations for risk assessment.

The second issue of debate involves the dose-response function for endocrine active agents. First, is there a threshold for endocrine-mediated adverse effects and second, do toxic effects of high doses of hormonally active agents mask more subtle adverse effects that can only be detected at low doses using specialized assay systems? These issues arise from two publications suggesting that very low doses of plasticizing agents could produce subtle effects on the developing male reproductive tract not seen at higher doses, possibly because subtle effects are masked by more overt toxicity at higher doses. Neither study has been replicated, despite attempts that employed more comprehensive study designs. Nonetheless, the issue has lead to an outcry from consumer and environmental activist groups to cease the use of certain plastics in baby bottles and childrens' toys. Former Surgeon General of the United States Dr. C. Everett Koop has responded, calling this reaction irresponsible.

In summary, a number of critical questions have been raised with respect to the identification of hormonally active agents in general, and laboratory studies that purport to demonstrate potential hormonal activity in particular.

- Are positive results in short-term *in vivo* and *in vitro* laboratory assays predictive of adverse health effects in humans?
- Can measurements of hormonal potency in laboratory assays be extrapolated to human populations at environmentally relevant exposure levels?

- Given that the human diet contains high amounts of naturally-derived hormonally active agents, is it feasible that synthetic chemicals with weak hormonal potency could adversely affect human endocrine functioning?
- Do the dose-response curves of hormonally active agents lack a threshold for adverse effects?
- Do toxic effects of high doses of hormonally active agents mask more subtle adverse effects that can only be detected at low doses using specialized assay systems?
- Are hormonally active agents more prone to exhibiting interactive effects (synergism or antagonism) than chemicals that operate through other mechanisms?
- Is it practical to regulate chemicals based on presumed mechanisms of action—i.e., on the basis of a potential endocrine mechanism—rather than on production of adverse effects such as reproductive or developmental impairment?

The way that the scientific and regulatory communities answer these questions could have a profound impact on the risk assessment of hormonally active agents in the workplace and in the environment.

Lead Poisoning and the Lowering of the Threshold

Currently, a hot area of research is the sensitivity of the developing nervous system to low-dose lead exposure. Lead toxicity is apparent in a variety of organ systems. As mentioned above, lead effects on both male and female reproduction have been investigated and the use of lead salts for inducing abortion reaches back to antiquity. The neurological system is recognized as one of the key targets for toxic responses to lead. Some reports have recently suggested that the levels of environmental lead exposures received by large populations, especially in urban areas, could be sufficient to produce adverse cognitive effects. This has lead to substantial investigation of both lead toxicity mechanisms in animals and the occurrence of cognitive deficits in children. Though reports of low-dose lead effects have struck parental and societal chords, the body of research on intelligence and cognitive outcomes does not support a consistent association with today's common levels of environmental lead exposure.

Rather than the traditional applied dose, lead exposure is typically considered on the basis of a measured blood level. There is little dispute about the potential for lead toxicity in children when chronic blood levels reach the 30–50 μg/dl range or higher. A standard regulatory criterion of concern is 10 μg/dl. However, there are suggestions that cognitive effects may accrue even at this threshold, or perhaps even up to 10-fold lower. Unfortunately, the endpoints of intelligence and verbal ability that have been suggested as the most sensitive indicators are exceedingly difficult to measure in a repeatable, reliable, and objective manner. A further complication is the considerable plasticity in learning processes and the ability of children to "make up ground" as they develop.

Scientific arguments rage over the verbal abilities of two-year olds and the meaning of IQ differences of less than one or two points on the typical scale. Research has suggested that verbal development is a brain function particularly vulnerable to lead. However, despite claims of statistical significance in some studies, the uncertainty associated with evaluating these endpoints, which is not captured statistically, clearly makes definitive conclusions impossible. The testing methods for assessing cognitive development and verbal ability in infants and toddlers are not generally regarded as sensitive enough to reliably distinguish between inter-individual variability and exposure-associated effects at the required levels.

However, information from animal studies has begun to shed light on mechanisms by which lead could affect brain development. There does appear to be a heightened sensitivity of fetal and neonatal brain cells to lead effects compared to adults. This may relate to the much more active process of forming connections among neural cells and expansion of vascular, blood carrying elements during fetal and neonatal stages. It is not clear what degree of change in this process must occur to represent an adverse reaction to lead, however, since there is considerable variation and plasticity in the process anyway.

Two important areas for additional investigation are: 1) developing better tools for investigating human cognitive function and abilities and 2) characterizing the relationship between the dose-response characteristics of experimental animals and that of humans for lead. These questions are important in terms of both public health and economics. In general, the scientific and regulatory communities have regarded the clear and dramatic drop in children's blood lead levels since the 1970s as a real public health improvement realized through the control of lead from gasoline and paints. If neurocognitive development turns out to be as sensitive as some suggest to the effects of lead, much tougher questions about whether and how to address exposures, down to the range associated with naturally occurring lead, will be up for consideration. Without obvious and readily replaceable major exposure sources, like gasoline or paint, the costs associated with additional incremental reductions in lead exposure for the population as a whole may be dramatic.

11.5 SUMMARY

This chapter has outlined the toxic responses of the male and female reproductive systems and the developing fetus. Some of the mechanisms of toxicity, generally described using experimental toxicants, have been presented to illustrate the types of responses and effects that should be considered. In most cases, however, the experimental toxicants have limited direct application to human health effects. Especially for occupational exposures, the gap between toxic potential and demonstrated effects is large. Examples of actual human reproductive and developmental toxicants have been pointed out so that those chemicals, which are currently known to represent a risk to humans, can be identified.

Some of the key points in the chapter included:

- The differential sensitivity of various tissues and cell types in the male and female reproductive organs to certain types of toxicants.
- The functional and toxicological implications of the different patterns of cellular division and germ cell maturation used by males and females.
- The multiple interactions between the reproductive and endocrine systems and the balance of endocrine regulation that may be vulnerable during certain toxic responses.
- The relationship of the sequential course of developmental processes to toxic responses.
- The major difference in toxic responses between the embryonic and fetal periods of development.

REFERENCES AND SUGGESTED READING

Alvarez, J. G., and B. T. Storey, "Evidence for increased lipid peroxidative damage and loss of superoxide dismutase as a mode of sublethal damage to human sperm during cryopreservation." *Jo. of Androl.* **13:** 232–241 (1992).

Arnold, S. F., D. M. Klotz, B. M. Collins, P. M. Vonier, L. J. Jr., Guillette, and J. A. McLachl an, Synergistic activation of estrogen receptor with combinations of environmental chemicals [see comments] [retra cted by McLachlan JA. In: Science 1997 Jul 25; **277**(5325):462–463], *Science*, 1996; **272**: 1489–1492.

Ashby, J., J. Odum, H. Tinwell, and P. A. Lefevre, "Assessing the risks of adverse endocrine- mediated effects: where to from here?" Regulatory Toxicology and Pharmacology **26**: 80–93 (1997).

Ashby, J., H. Tinwell, P. A. Lefevre, J. Odum, D. Paton, S. W. Millward, S. Tittensor, and A. N. Brooks, "Normal sexual development of rats exposed to butyl benzyl phthalate from conception to weaning." Regulatory Toxicology and Pharmacology **26**(1 Pt 1):102–118 (1997).

Auger, J., J. M. Kunstmann, F. Czyglik, and P. Jouannet, "Decline in semen quality among ferti le men in Paris during the past 50 years." *New England Journal of Medicine* **332**: 281–285 (1995).

Bromwich, P., J. Cohen, I. Stewart, and A. Walker, "Decline in sperm counts: and artefact or c hanged reference range of 'normal'?" *British Medical Journal* **309**: 19–22 (1994).

Cagen, S. Z., J. M. Jr., Waechter, S. S. Dimond, W. J. Breslin, J. H. Butala, F. W. Jekat, R. L. Joiner, R. N. Shiotsuka, G. E. Veenstra, and L. R. Harris, "Normal reproductive organ development in CF-1 mice following prenatal exposure to bisphenol A." *Toxicological Sciences* **50**(1): 36–44 (1999).

Carney, E. W., A. M. Hoberman, D. R. Farmer, R. W. Jr., Kapp, A. I. Nikiforov, M. Bernstein, M. E. Hurtt, W. J. Breslin, S. Z. Cagen, and G. P. Daston. "Estrogen modulation: tiered testing for human hazard evaluation." American Industrial Health Council, Reproductive and Developmental Effects Subcommittee. Reproductive Toxicology **11**(6): 879–892 (1997).

Colborn, T., F. S. Vom Saal, and A. M. Soto, "Developmental effects of endocrine-disrupting chemicals in wildlife and humans." Environmental Health Perspectives **101**: 378–384 (1993).

Crisp, T. M., E. M. Clegg, R. L. Cooper, W. P. Wood, D. G. Anderson, K. P. Baetcke, J. L. Hoffman, M. S. Morrow, D. J. Rodier, J. E. Schaeffer, L. W. Touart, M. G. Zeeman, and Y. M. Patel, "Environmental endocrine disruption: An effects assessment and analysis." Environmental Health Perspectives **106** (Supplement 1): 11–56 (1998).

Endocrine Disruptor Screening and Testing Advisory Committee (EDSTAC), "Endocrine Disruptor Screening and Testing Advisory Commitee (EDSTAC) Final Report." Washington, D.C. USEPA, editor, (1998).

Faber, K. A., and C. L., Jr., Hughes, "Clinical Aspects of Reproductive Toxicology" in Witorsch, R. J., ed., Reproductive Toxicology. 2nd edition. New York: Raven Press, Ltd, (1995).

Gorospe, W. C., and M. Reinhard, "Toxic Effects on the Ovary of the Nonpregnant Female." in Witorsch, R. J., ed., Reproductive Toxicology. 2nd edition. New York: Raven Press, Ltd, (1995).

Koop, C. E., "The Latest Phoney Chemical Scare." *The Wall Street Journal*, June 22, 1999.

Manson, J. M., and L. D. Wise, "Teratogens." in Amdur, M. O., Doull, J., and Klaassen C. D., eds., Casarett and Doull's Toxicology: The Basic Science of Poisons. 4th edition. New York: Pergamon Press (1991).

Matt, D. W., and J. F. Borzelleca, "Toxic Effects on the Female Reproductive System During Pregnancy, Parturition, and Lactation." in Witorsch, R. J., ed., Reproductive Toxicology. 2nd edition. New York: Raven Press, Ltd, (1995).

Mattison, D. R., D. R. Plowchalk, M. J. Meadows, A. Z. Al-Juburi, J. Gandy, and A. Malek, "Reproductive Toxicity: Male and Female Reproductive Systems as Targets for Chemical Injury." Medical Clinics of North America **74**: 391–411 (1990).

McLachlan, J. A., Retraction: Synergistic activation of estrogen receptor with combinations of environmental chemicals, *Science* **277**: 462–463 (1997).

NagDas, S. K. "Effect of chlorpromazine on bovine sperm respiration." Archives of Andrology **28**: 195–200 (1992).

Nair, R. S., F. W. Jekat, D. H. Waalkens-Berendsen, R. Eiben, R. A. Barter, and M. A. Martens, "Lack of Developmental/Reproductive Effects with Low Concentrations of Butyl Benzyl Phthalate in Drinking Water in Rats." The Toxicologist, **48**(1-S): 218 (1999).

National Research Council, Committee on Hormonally Active Agents in the Environment, Board on Environmental Studies and Toxicology, Commission on Life Sciences, 1999. "Hormonally Active Agents in the Environment." National Academy Press, Washington.

Nimrod, A. C. and W. H. Benson, "Environmental estrogenic effects of alkylphenol ethoxylates." Critical Reviews in Toxicology **26**: 335–364 (1996).

Olsen, G. W., K. M. Bodner, J. M. Ramlow, C. E. Ross, and L. I. Lipshultz, "Have sperm counts been reduced 50 percent in 50 years? A statistical model revisited." Fertility and Sterility **63**: 887–893 (1995).

Peltola, V., E. Mantyla, I. Huhtaniemi, and M. Ahotupa, "Lipid peroxidation and antioxidant enzyme activities in the rat testis after cigarette smoke inhalation or administration of polychlorinated biphenyls or polychlorinated naphthalenes." *Jo. of Androl.* **15**: 353–361 (1994).

Safe, S. H., "Do environmental estrogens play a role in development of breast cancer in women and male reproductive problems?" Human and Ecological Risk Assessment **1**: 17–23 (1995).

Schardein, J. L. Chemically Induced Birth Defects. New York: Marcel Dekker, Inc (1985).

Schilling, K., C. Gembardt, and J. Hellwig, "Reproduction toxicity of di-2-ethylhexyl phthalate (DEHP)" The Toxicologist, **48**; (1-S): 692 (1985, 1999).

Sharpe, R. M., J. S. Fisher, M. M. Millar, S. Jobling, and J. P. Sumpter, "Gestational and lactational exposure of rats to xenoestrogens results in reduced testicular size and sperm production." Environmental Health Perspectives **103**(12): 1136–1143 (1995).

Shepard, T. H., Catalog of Teratogenic Agents. 6th edition. Baltimore: Johns Hopkins University Press (1989).

Sundaram, K., and R. J. Witorsch, "Toxic Effects on the Testes." in Witorsch, R. J., ed., Reproductive Toxicology. 2nd edition. New York: Raven Press, Ltd, (1995).

Thomas, J. A. "Toxic Responses of the Reproductive System." In Amdur, M. O., Doull, J., and Klaassen, C. D., eds., Casarett and Doull's Toxicology: The Basic Science of Poisons. 4th edition. New York: Pergamon Press (1991).

vom Saal, F. S., B. G. Timms, M. M. Montano, P. Palanza, K. A. Thayer, S. C. Nagel, M. D. Dhar, V. K. Ganjam, S. Parmigiani, and W. V. Welshons, "Prostate enlargement in mice due to fetal exposure to low doses of estradiol or diethylstilbestrol and opposite effects at high doses." *Proceedings of the National Academy of Sciences* **94**(5): 2056–2061 (1997).

12 Mutagenesis and Genetic Toxicology

CHRISTOPHER M. TEAF and PAUL J. MIDDENDORF

Genetic toxicology combines the study of physically or chemically induced changes in the hereditary material (deoxyribonucleic acid or DNA) with the prediction and the prevention of potential adverse effects. Modification of the human genetic material by chemical agents or physical agents (e.g., radiation) represents one of the most serious potential consequences of exposure to toxicants in the environment or the workplace. Nevertheless, despite increasing research interest in this area, the number of agents or processes that are known to cause such changes is quite limited. This chapter presents information regarding the following areas:

- Types and characteristics of genetic alteration
- Common research methods for the assessment of genetic change
- Practical significance of test results from animal and human studies in the identification of potential mutagens
- Theoretical relationships between mutagenesis and carcinogenesis

12.1 INDUCTION AND POTENTIAL CONSEQUENCES OF GENETIC CHANGE

Historical Perspective

The term *mutation* is defined as a transmissible change in the genetic material of an organism. This actual heritable change in the genetic constitution of a cell or an individual is referred to as a genotypic change because the genetic material has been altered. While all mutational changes result in alteration of the genetic material in the parent cells, not all are immediately expressed in cell progeny as functional, or phenotypic, changes. Thus, it is possible to have genetic change that is not associated with a transmissible change. These distinctions are discussed in greater detail in subsequent sections.

Potential environmental and occupational mutagens may be classified as physical, biological, or chemical agents. Ames and many subsequent researchers have identified representative chemical mutagens in at least 10 classes of compounds, including the following: cyclic aromatics, ethers, halogenated aliphatics, nitrosamines, selected pesticides, phthalate esters, selected phenols, selected polychlorinated biphenyls, and selected polycyclic aromatics (PAHs). Despite nearly 50 years of research concerning chemical-induced genetic change, ionizing radiation still represents the best described example of a dose-dependent mutagen and was first demonstrated in the 1920s. Chemical mutagenesis was first demonstrated in the 1940s, and many of the characteristics of radiation-induced mutation are believed to be common to chemically induced mutation. This is particularly true for molecules known as free radicals, which are formed in radiation events and some chemical toxic events. Radicals contain unpaired electrons, are strongly electrophilic, and extraordinarily reactive, features that are well correlated with both mutagenic and carcinogenic potency. Such reactive molecules

Principles of Toxicology: Environmental and Industrial Applications, Second Edition, Edited by Phillip L. Williams, Robert C. James, and Stephen M. Roberts.
ISBN 0-471-29321-0 © 2000 John Wiley & Sons, Inc.

probably are responsible for at least some of the alterations of nucleic acid sequences that are observed in genotoxic processes.

Over 3500 functional disorders or disease states have been linked to heritable changes in humans, and the ambient incidence of genetic disease may be as great as 10 percent in newborns. In the case of some cancers, a change in the genotype of a cell results in a change in phenotype that is grossly defined by rapid cellular division and a reversion of the cell to a less specialized type (dedifferentiation). The subsequent generations eventually may form a growing tumor mass within the affected tissue. This simplified sequence has been termed the *somatic cell mutation theory of cancer*. While not all chemically-induced cancers can be explained by this hypothesis, general applicability of the somatic cell mutation theory is supported by the following points:

- Most demonstrated chemical mutagens are demonstrably carcinogenic in animal studies
- Carcinogen-DNA complexes (adducts) often can be isolated from carcinogen-treated cells
- Heritable defects in DNA-repair capability, such as in the sunlight-induced disease xeroderma pigmentosum, predispose affected individuals to cancer
- Tumor cells can be "initiated" by carcinogens but may remain in a dormant state for many cell generations, an observation consistent with permanent DNA structural changes
- Cancer cells generally display chromosomal abnormalities
- Cancers display altered gene expression (i.e., a phenotypic change)

The issue of correlation between genotoxicity or mutagenicity assays and cancer is discussed in greater detail in subsequent sections of this chapter.

Although genetic changes in somatic cells are of concern because consequences such as cancer may be debilitating or lethal, mutational changes in germ cells (sperm or ovum) may have even more serious consequences because of the potential for effects on subsequent human generations. If a lethal and dominant mutation occurs in a germinal cell, the result is a nonviable offspring, and the change is not transmissible. On the other hand, a dominant but viable mutation can be transmitted to the next generation, and it need only be present in single form (heterozygous) to be expressed in the phenotype of the individual. If the phenotypic change confers evolutionary disadvantage to the individual (e.g., renders it less fit), it is unlikely to become established in the gene pool. In contrast, individuals that are heterozygous for recessive genes represent unaffected carriers that are essentially impossible to detect in a population. Thus, recessive mutations are of the greatest potential concern. These mutations may cause effects ranging from minor to lethal whenever two heterozygous carriers produce an offspring that is homozygous for the recessive trait (i.e., the genes are present in both copies). Figure 12.1 describes the potential consequences of mutagenic events.

Occupational Mutagens, Spontaneous Mutations, and Naturally Occurring Mutagens

In considering the potential adverse effects of chemicals, it is important to recognize that both physical and chemical mutagens occur naturally in the environment. Radiation is an ubiquitous feature of our lives, sunlight representing the most obvious example.

Incomplete combustion produces mutagens such as benzo[*a*]pyrene, and some mutagens occur naturally in the diet, or may be formed during normal cooking or food processing (e.g., nitrosamines). In addition, drinking water and swimming-pool water have been shown to contain potential mutagens that are formed during chlorination procedures. Thus, the genetic events that influence the human evolutionary process appropriately may be viewed as a combination of normal background incidence of spontaneous mutations that may be occurring during cellular division, coupled with the exposure to naturally occurring chemical or physical mutagens.

Mutagenic chemicals in the workplace, or those that are introduced into the environment via industrial operations, represent a potential contribution to the genetic burden, though the practical significance of this contribution is not known with precision. It is estimated that over 70,000 synthetic

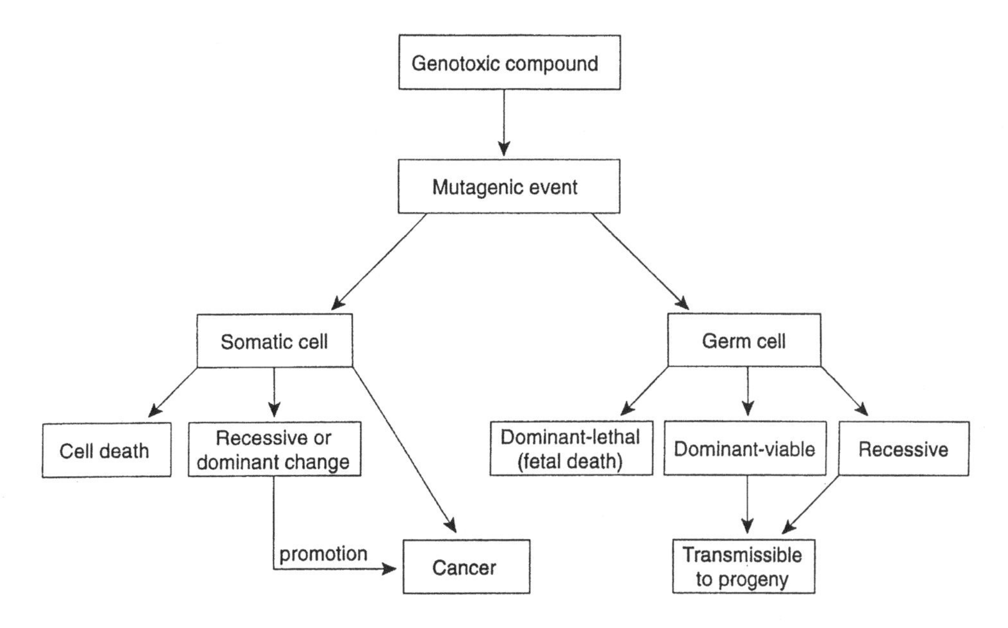

Figure 12.1 Possible consequences of mutagenic event in somatic and germinal cells.

organic compounds currently are in use, a number which increases annually. Only a very small fraction of these have been confirmed as human carcinogens (see Chapter 13), and no compound has been shown unequivocally to be mutagenic in humans. However, animal and bacterial tests have demonstrated a mutagenic potential for some occupational and environmental compounds at high exposure levels, and it is reasonable to consider human exposure to these compounds, particularly in occupational situations where contact may be frequent and/or intense. This is not to suggest that very small exposures common to environmental circumstances are likely to be associated with adverse effects.

12.2 GENETIC FUNDAMENTALS AND EVALUATION OF GENETIC CHANGE

Transcription and Translation

DNA (deoxyribonucleic acid) is the structural and biochemical unit on which heredity and genetics are based for all species. It is the only cellular macromolecule that is self-replicating, alterable, and transmissible. Subunits of the DNA molecule are grouped into genes that contain the information, which is necessary to produce a cellular product. An example of such a cellular product is a polypeptide or protein, which may have a structural, enzymatic, or regulatory function in the organism. Figure 12.2 illustrates how the sequence of messages on the DNA molecule is transcribed into the RNA (ribonucleic acid) molecule and ultimately is translated into the polypeptide or protein. The sequence of base pairs in the DNA molecule specifies the appropriate complementary ("mirror image") sequence that governs the formation of the messenger RNA (mRNA). Transfer RNAs (tRNA), each of which is specific for a single amino acid, are matched to the appropriate segment of the mRNA. When the amino acids are released from the tRNAs and are linked in a continuous string, the polypeptide (or protein) chain is formed.

Recognition of the mRNA regions by the tRNA-amino acid complex is accomplished by a system of triplet, or three-base, codons (in the mRNA) and complementary anticodons (in the tRNA). The critical features of this coding system are that it is simultaneously *unambiguous* and *degenerate*. In

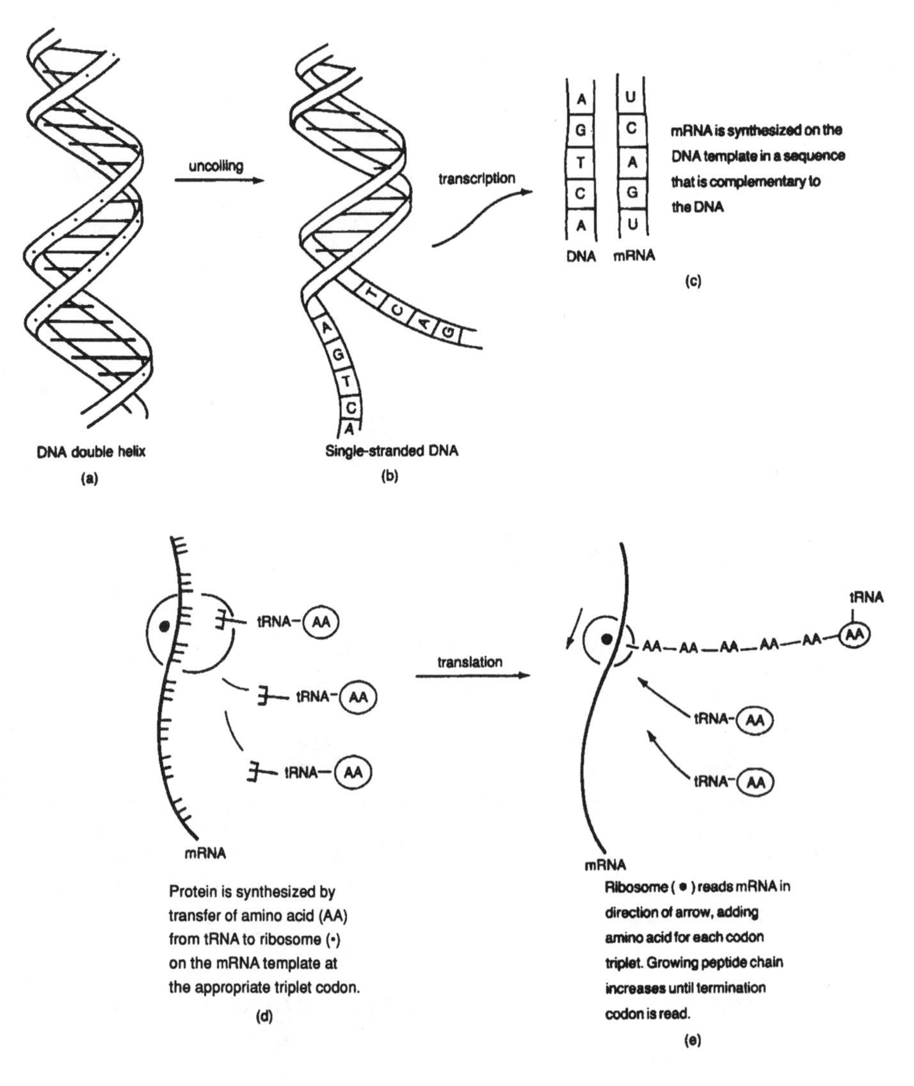

Figure 12.2 Schematic representation of transcription and translation.

other words, no triplet codon may call for more than a single specific tRNA-amino acid complex (unambiguous), but several triplets may call for the same tRNA-amino acid (degenerate). This results from the fact that four nucleotides, which form DNA (DNA is composed of adenine, cytosine, guanine, and thymine), and the nucleotides forming RNA (RNA is made up of A, C, G, and uracil) may be combined in triplet form in 64 different ways ($4 \times 4 \times 4$ or 4^3). The 20 amino acids and three terminal codes account for less than half of the available codons, leaving well over 30 codons of the possible 64. The biological significance of this degeneracy is that such a characteristic minimizes the influence of minor mutations (e.g., single basepair deletions or additions) because codons differing only in minor aspects may still code for the same amino acids. The significance of having an unambiguous code is clear; the formation of proteins must be perfectly reproducible and exact. Table 12.1 depicts the amino acids that are coded for by the various triplet codons of DNA, as well as the initiation and termination signal triplets.

The process of mutagenesis results from an alteration in the DNA sequence. If the alteration is not too radical, the rearrangement may be transmitted faithfully through the mRNA to protein synthesis,

TABLE 12-1. Correspondence of the Genetic Code with the Appropriate Amino Acids (Note Unambiguity and Degeneracy)

	Second position in triplet				
First position in triplet	U	C	A	G	Third position in triplet
A	Isoleucine	Threonine	Asparagine	Serine	U
	Isoleucine	Threonine	Asparagine	Serine	C
	Isoleucine	Threonine	Lysine	Arginine	A
	*Methionine	Threonine	Lysine	Arginine	G
C	Leucine	Proline	Histidine	Arginine	U
	Leucine	Proline	Histidine	Arginine	C
	Leucine	Proline	Glutamate	Arginine	A
	Leucine	Proline	Glutamate	Arginine	G
G	Valine	Alanine	Aspartate	Glycine	U
	Valine	Alanine	Aspartate	Glycine	C
	Valine	Alanine	Glutamate	Glycine	A
	Valine	Alanine	Glutamate	Glycine	G
U	Phenylalanine	Serine	Tyrosine	Cysteine	U
	Phenylalanine	Serine	Tyrosine	Cysteine	C
	Leucine	Serine	STOP	STOP	A
	Leucine	Serine	STOP	Tryptophan	G

*The sequence AUG, in addition to coding for methionine, is part of the initiator sequencehat starts the translation process by which mRNA is formed from the DNA template.

which results in a gene product that is partially or completely unable to perform its normal function. Such changes may be correlated with carcinogenesis, fetal death, fetal malformation, or biochemical dysfunction, depending on the cell type that has been affected. However, cause and effect relationships for such correlations typically are lacking.

Initiation and termination of DNA transcription are controlled by a separate set of regulatory genes. Most regulatory genes respond to chemical cues, so that only those genes that are needed at a given time are expressed or available. The remaining genes are in an inactive state. The processes of gene activation and inactivation are believed to be critical to cellular differentiation, and interruption of these processes may result in the expression of abnormal conditions such as tumors. This represents an example of a case in which a non-genetic event may result in tumorigenesis. Oncogenes represent an example of a situation where activation of a genetic phenomenon may act to initiate carcinogenicity. In contrast, loss of "tumor suppressor" genes may, by omission, result in initiation of the carcinogenic process.

Chromosome Structure and Function

The DNA of mammalian species, including humans, is packaged in combination with specialized proteins (predominantly histones) into units termed *chromosomes*, which are found in the nucleus of the cell. The proteins are thought to "cover" certain segments of the DNA and may act as inhibitors of expression for some regions. Each normal human cell (except germ cells) contains 46 chromosomes (23 pairs). Chromosomes may be present singly (haploid), as in germ cells (sperm or ovum), or in pairs (diploid), as in somatic cells or in fertilized ova. In haploid cells, all functional genes present in the cell can be expressed. In diploid cells, one allele may be dominant over the other, and in this case, only the dominant gene of each functional pair is expressed. The unexpressed allele is termed *recessive*, and recessive genes are expressed only when both copies of the recessive type are present. Some cell types in mammals are found in forms other than diploid. Functionally normal liver cells, for example, are occasionally found to be tetraploid (two chromosome pairs instead of one pair).

Some features and terminology that are important to cytogenetics, or the study of chromosomes, include:

- Karyotype—the array of chromosomes, typically taken at the point in the cell cycle known as *metaphase*, which is unique to a species and forms the basis for cellular taxonomy; may be used to detect physical or chemical damage
- *Centromere*—the primary constriction, which represents the site of attachment of the spindle fiber during cell division; useful in identifying specific chromosomes, as its location is relatively consistent
- *Nucleolar organizing region*—the secondary constriction, which represents the site of synthesis of RNA, subsequently used in ribosomes for protein synthesis
- *Satellite*—the segment terminal to the nucleolar organizing region; useful in specific chromosome identification
- *Heterochromatin*—tight-coiling region; relatively inactive
- *Euchromatin*—loose-coiling region; primary transcription site

Mitosis, Meiosis, and Fertilization

The process by which a somatic cell divides into two diploid daughter cells is called *mitosis*. The first stage of mitosis is called *prophase*, during which the spindle is formed and the chromatin material (DNA and protein) of the nucleus becomes shortened into well-defined chromosomes. During *metaphase*, the centriole pairs are pulled tightly by the attached microtubules to the very center of the cell, lining up in the equatorial plane of the mitotic spindle. With still further growth of the spindle, the chromatids in each pair of chromosomes are broken apart, a stage called *anaphase*. All 46 pairs (in humans) of chromatids are thus separated, forming pairs of daughter chromosomes that are pulled toward one mitotic spindle or the other. In *telophase*, the mitotic spindle grows longer, completing the separation of daughter chromosomes. A new nuclear membrane is formed, and shortly thereafter the cell constricts at the midpoint between the two nuclei, forming two new cells.

Meiosis is the term for the process by which immature germ cells produce gametes (sperm or ova) that are haploid. During meiosis, DNA is replicated, producing 46 chromosomes with sister chromatids. The 46 chromosomes arrange into 23 pairs at the center of the nucleus, and in the first division the pairs separate. In a second division, the sister chromatids separate, with one chromosome of each pair being incorporated into four gametes. At the time of fertilization, or zygote formation, the fusion of gametes once again forms a cell with a full complement of 46 chromosomes.

Genetic Alteration

Tests for genotoxicity in higher organisms may be placed into one of three basic categories: gene mutation tests, chromosomal aberration tests, and DNA damage tests. These tests are conducted individually or in combination to identify various types of mutagenic events (Figure 12.3) or other genotoxic effects. For the purpose of this discussion, the principles of each test category will be reviewed and specific tests will be discussed by broad phylogenetic classifications. Over 200 individual test methods have been developed to assess the extent and magnitude of genetic alteration; however, less than 20 have been validated or are in common use. Numerous mutagenic agents have the demonstrated capacity to cause genetic change in one or more of these test systems, but no well-documented cases of human mutation are available. This latter conclusion may change as a result of improvements in the ability to detect human genetic change. Nevertheless, as discussed in this section, use of a reasonable battery of tests is capable of identifying almost all of the known human carcinogens, consistent with the hypothesis that somatic cell mutations are, at least in part, responsible for a large proportion of human cancers.

A transmissible change in the linear sequence of DNA can result from any one of three basic events:

- Infidelity in DNA replication
- Point mutation
- Chromosomal aberration

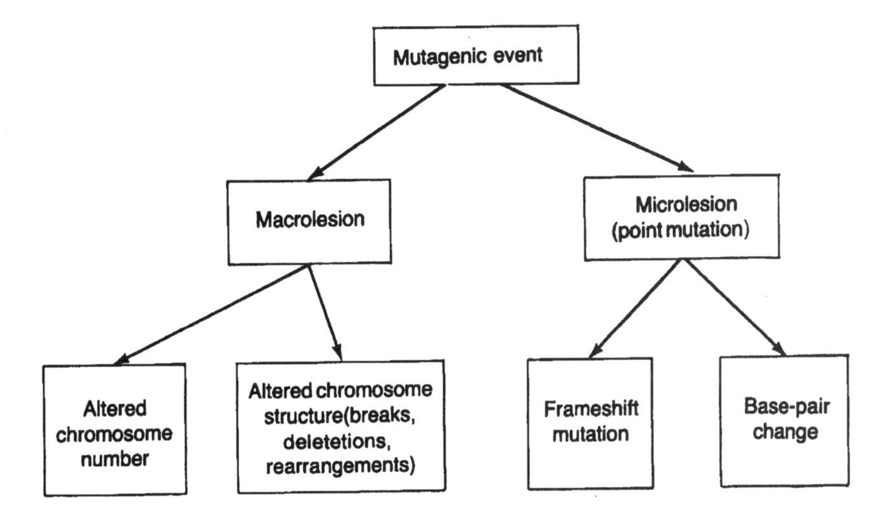

Figure 12.3 Types of mutagenic changes (Adapted from Brusick, 1980).

One possibility, infidelity or inexact copying of a DNA strand during normal cellular replication, may result from inaccurate initiation of replication, failure of the transcription enzymes to accurately "read" the DNA, or interruption of the transcription process by agents that interpose (intercalate) themselves within the DNA molecule or between the DNA and an enzyme.

Point mutations, as the second possibility, may be subdivided into basepair changes and frameshift mutations (Figure 12.4). The former result from transition or transversion of DNA base pairs so that

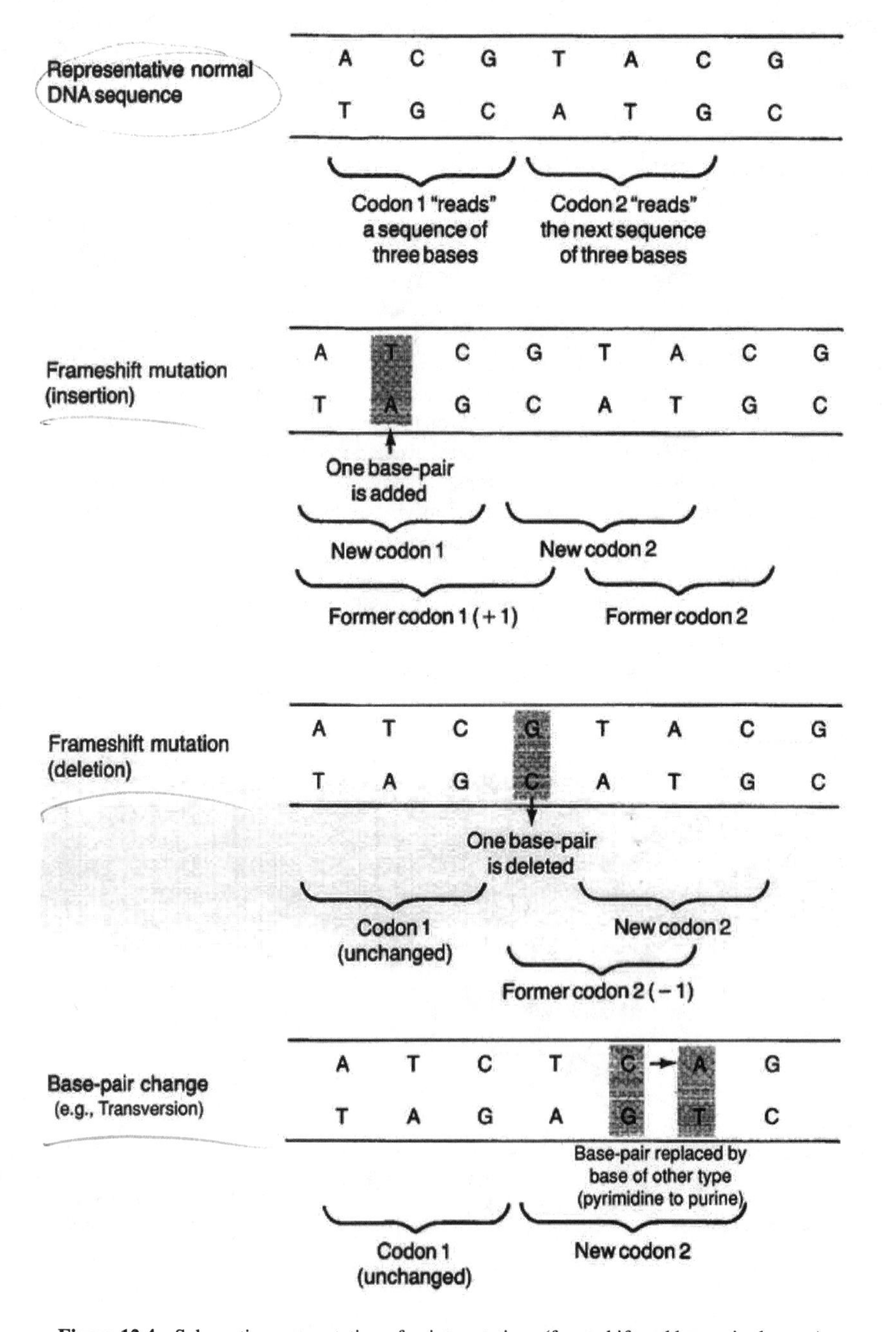

Figure 12.4 Schematic representation of point mutations (frameshift and basepair changes).

the number of bases is unchanged but the sequence is altered. Because the genetic code is "degenerate," this may or may not result in an altered product after transcription and translation. A frameshift mutation, however, results from insertion or deletion of one or more bases from the linear sequence of the DNA. This causes the transcription process to be displaced by the corresponding number of bases and virtually assures an altered genetic product. Proflavine, which has been used as a bacteriostatic agent, is an example of a compound that intercalates within the DNA molecule. It is a flat, planar molecule and inserts itself neatly between the bases. When it intercalates, it forces the DNA strand out of its normal configuration, so that when the replication enzymes or transcription enzymes try to read the bases, the bases are not spatially arranged the normal way, and the enzymes cannot read the base sequence properly. The enzymes may skip over one or several bases, or may put an additional base into the DNA or RNA strand at random. Proflavine does not chemically bind with the bases in DNA. In contrast, many of the environmentally prevalent polynuclear aromatic hydrocarbons (PAHs) may intercalate into the DNA, leading to frameshift, and also may chemically react directly with it, an event that can lead to basepair substitution. An example of this is benzo(*a*)pyrene (BaP), which is found at low concentrations throughout the environment as a product of combustion of fossil fuels, in grilled steaks, tobacco smoke, and many other places. BaP by itself is seldom considered to be mutagenic. However, after metabolism, many highly reactive epoxide intermediate metabolites are formed, one of which (BPDE I) is highly mutagenic. BPDE I combines with guanine to form what is called a *DNA adduct*. These adducts have been found in extremely small quantities by highly specialized and sensitive techniques such as enzyme-linked immunosorbent assay (ELISA) and fluorescence. A scheme of activation and adduct formation for BaP is given in Figure 12.5.

Basepair changes, described earlier, are of two kinds: transitions or transversions. In transitions, one base is replaced by the base of the same chemical class. That is, a purine is replaced by the other purine (e.g., adenine is replaced by guanine); in the case of pyrimidine bases, cytosine would be replaced with thymine or vice versa. An example of a chemical that causes transitions is nitrous acid (see Figure 12.6). Nitrous acid is formed from organic precursors such as nitrosamines, nitrite, and nitrate salts. It reacts with amino (NH_2) groups in nucleotides and converts them to keto (C=O) groups. In transversions, a base pair is replaced in the DNA strand by a base of the other type: a purine is replaced by a pyrimidine or vice versa.

Another group of chemicals that can cause mutations are alkylating agents. Some well-known alkylating agents are the mustard gases, originally developed for chemical warfare. Chemicals in this group add short carbon–hydrogen chains at specific locations on bases. The experimental agent ethyl methanesulfonate (EMS) can alkylate guanine to form 7-ethylguanine (see Figure 12.7), which can cause the bond between the base and deoxyribose in the backbone of the DNA strand to become unstable and break. This leads to a gap in the DNA strand which, if unrepaired at the time of DNA replication, is filled with any of the four available bases.

Not all point mutations are caused by radiation or chemicals; some may occur because of the nature of the bases themselves. The bases have their preferred arrangement of hydrogen atoms, but on rare occasions undergo rearrangements of the hydrogen atoms, called *tautomeric shifts*. The nitrogen atoms attached to the purine and pyrimidine rings are usually in the amino (NH_2) form and only rarely assume the imino (NH) form. Similarly, the oxygen atoms attached to the carbon atoms of guanine and thymine are normally arranged in the keto (C=O) form, but rarely rearrange to the enol (COH) form.

The changes in configuration lead to different hydrogen bonding patterns, and, if a base is in the alternate form during replication, a wrong base can be put into the new growing strand leading to a mutation. A group of chemicals, base analogs, that resemble the normal bases of DNA may lead to mutations by being incorporated into DNA inadvertently during repair or replication. These chemicals go through tautomeric shifts more often and result in inappropriate base pairing during replication so that changes in the base sequence occur. An example of a base analogue is 5-bromouracil, which can replace thymine.

Gene mutation tests measure those alterations of genetic material limited to the gene unit, that are transmissible to progeny unless repaired. Brusick (1980) refers to gene mutations as "microlesions" because the actual genetic lesion is not microscopically visible. Microlesions are classified as either

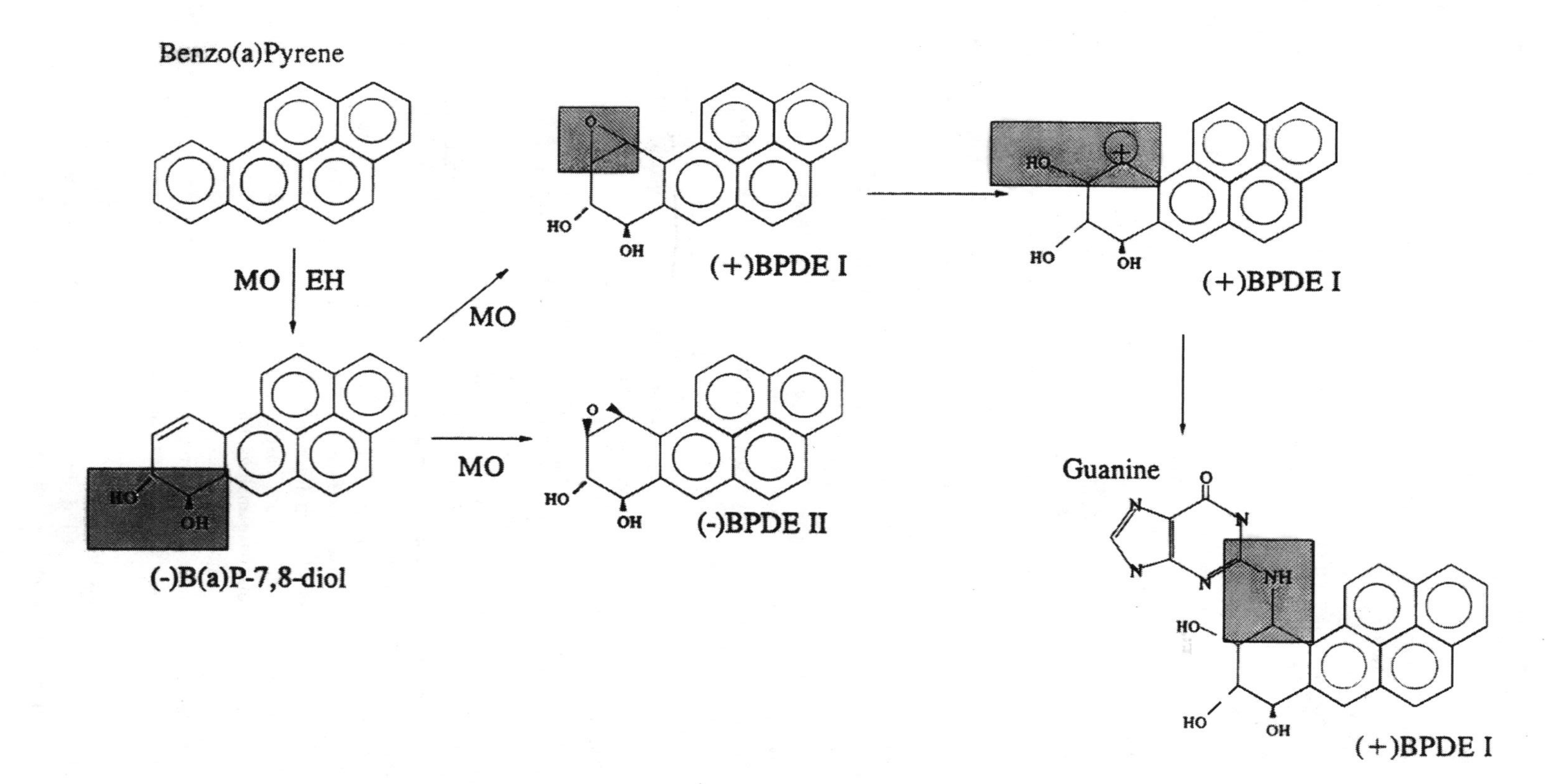

Figure 12.5 Example of DNA adduct formation with benzo-a-pyrene.

Cytosine
NH2
NH
O
N
H2N
O
N
N
N
N
Guanine
O
H2N
NH
N
O
N
N
N
N
Adenine

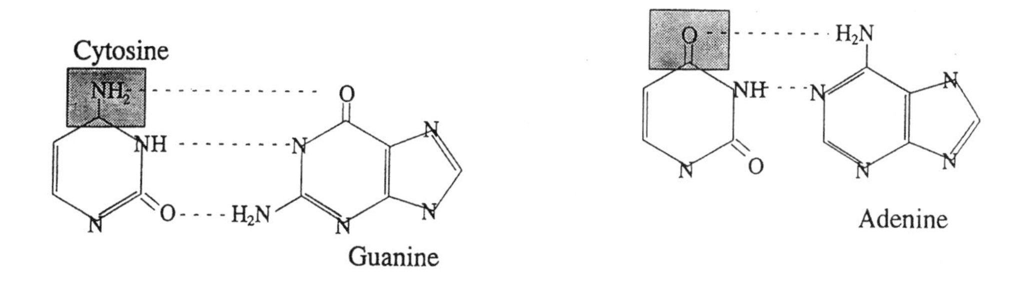

Figure 12.6 Cytosine modified to uracil by nitrous acid.

basepair substitution mutations or frameshift mutations. These two categories of gene mutations are induced by different mechanisms and, often, by distinctly different classes of chemical mutagens. Yet both types of gene mutation are virtually always monitored by measuring some phenotypic change in the test organism. Microlesions occur at a much lower frequency (10^{-5} to 10^{-6}) in comparison to chromosome aberrations or "macrolesions," which may be as frequent as 10^{-2} to 10^{-3}.

As described earlier, the basepair changes induced by point mutations (Figure 12.4) will also alter RNA codon sequences, which, in turn, change the amino acid sequence of the peptide chain being

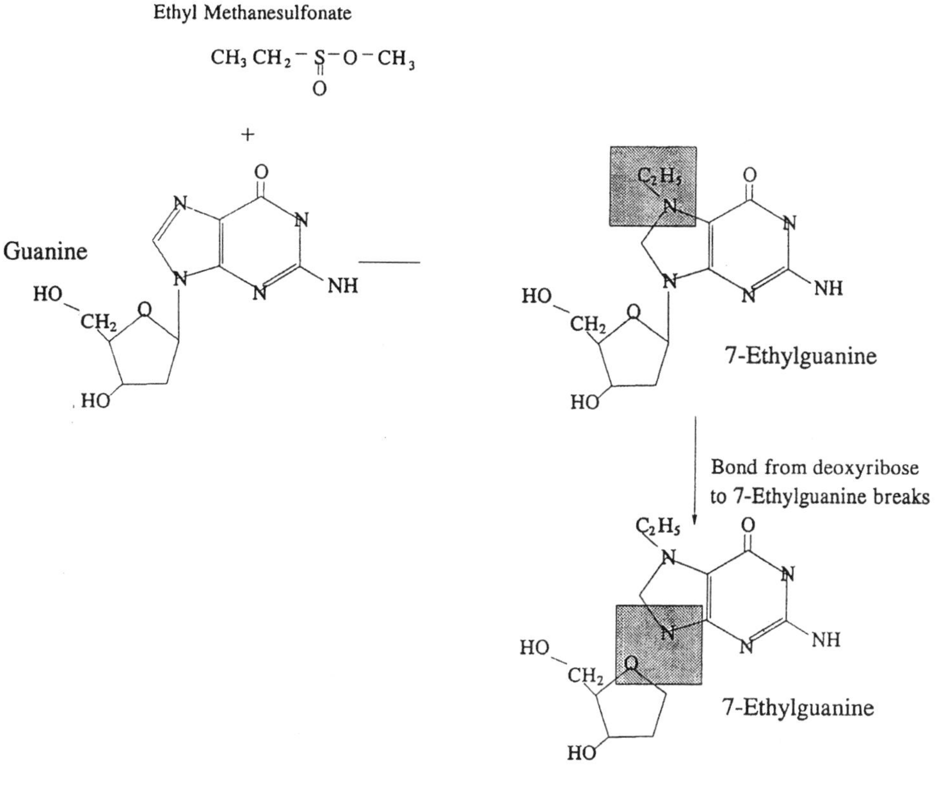

Ethyl Methanesulfonate
CH3 CH2 − S − O − CH3
O
+
Guanine
HO
CH2
HO
NH
C2H5
7-Ethylguanine
Bond from deoxyribose
to 7-Ethylguanine breaks
7-Ethylguanine

Figure 12.7 Alkylating agent (EMS) effects on DNA.

formed, which may result in an alteration of some measurable cellular function. The phenotypic changes that can be monitored by this type of test include auxotrophic changes (i.e., acquired dependence on a formerly endogenously synthesized substance), altered proteins, color differences, and lethality. It is extremely difficult to detect those alterations in mammalian DNA caused by insertions or deletions of one or a few bases, except in rare instances where the specific protein product is known and its formation can be monitored. It is somewhat easier in bacterial or prokaryotic systems, and this has led to the use of bacterial or *in vitro* screening assays to detect potential mutagens. These issues are discussed in greater detail in Brusick (1980, 1994).

Chromosomal aberrations, the third type of genetic change, may be present as chromatid gaps or breaks, symmetrical exchange (exchange of corresponding segments between arms of a chromosome), or asymmetric interchange between chromosomes. Point mutations can result in altered products of gene expression, but chromosomal aberration or alteration in chromosome numbers passed on through germ cells can have disastrous consequences, including embryonic death, teratogenesis, retarded development, behavioral disorders, and infertility. Some naturally occurring abnormalities of human chromosomal structure or number are shown in Table 12.2. The frequency of these events may be increased by mutagenic agents. Because these genetic lesions may be visualized by microscopy, they are referred to as *macrolesions*. One type of macrolesion is caused by an incomplete separation of replicated chromosomes during cell division. This is characterized by the abnormal chromosome numbers that result in the daughter cells and may be recognized as a change in the number of haploid chromosome sets (*ploidy* changes) or in the gain or loss of single chromosomes (*aneuploidy*). A second type of macrolesion caused by damage to chromosome structure (clastogenic effects) is categorized by the abnormal chromosome morphology that results.

Two theories are currently available to explain the mechanism of chromosome aberration. One is the classic "breakage-first" hypothesis. This theory assumes that the initial lesion is a break in the chromosomal backbone that is indicative of a broken DNA strand. Several possibilities exist following such an event: (1) the ends may repair normally and rejoin to form a normal chromosome; (2) the ends

TABLE 12.2 Examples of Human Genetic Disorders

Chromosome Abnormalities
Cri-du-chat syndrome (partial deletion of chromosome 5)
Down's syndrome (triplication of chromosome 21)
Klinefelter's syndrome (XXY sex chromosome constitution; 47 chromosomes)
Turner's syndrome (X0 sex chromosome constitution; 45 chromosomes)
Dominant Mutations
Chondrodystrophy
Hepatic porphyria
Huntington's chorea
Retinoblastoma
Recessive Mutations
Albinism
Cystic fibrosis
Diabetes mellitus
Fanconi's syndrome
Hemophilia
Xeroderma pigmentosum
Complex Inherited Traits
Anencephaly
Club foot
Spina bifida
Other congenital defects

may not be repaired, resulting in a permanent break; or (3) they may be misrepaired or join with another chromosome to cause a translocation of genetic material. A second theory is the "chromatid exchange" hypothesis. If the exchange occurs with a chromatid from another chromosome, an "exchange figure" results. This theory assumes that the initial lesion is not a break and that the lesion can either be repaired directly or may interact with another lesion by a process called *exchange initiation*. Most chromosomal abnormalities result in cell lethality and, if induced in germ cells, generally produce dominant lethal effects that cannot be transmitted to the next generation. The traditional method for determining chromosomal aberrations is the direct visual analysis of chromosomes in cells frozen at the metaphase of their division cycle. Thus, metaphase-spread analysis evaluates both structural and numerical chromosome anomalies directly.

Chemicals inducing changes in chromosomal number or structure also may be identified by the micronucleus test, an assay that assesses genotoxicity by observing micronucleated cells. It is a relatively simple assay because the number of cells with micronuclei are easily identified microscopically. At anaphase, in dividing cells that possess chromatid breaks or exchanges, chromatid and chromosome fragments may lag behind when the chromosome elements move toward the spindle poles. After telophase, the undamaged chromosomes give rise to regular daughter nuclei. The lagging elements are also included in the daughter cells, but a considerable proportion are included in secondary nuclei, which are typically much smaller than the principal nucleus and are therefore called *micronuclei*. Increased numbers of micronuclei represent increased chromosome breakage. Similar events can occur if interference with the spindle apparatus occurs, but the appearance of micronuclei produced in this manner is different, and they are usually larger than typical micronuclei. Historically, lymphocytes and epithelial cells have been the most commonly used cell populations.

Many point mutations are detected by the cell and are deleted by various repair mechanisms. Some, however, remain undetected and are passed to daughter cells. The significance of the mutations varies with the type of cell, and the location within the DNA. If the cell is of somatic lineage, altered gene products can result from gene expression. If the cell is a gonadal cell (or germ cell), the change can be passed on to offspring and may cause problems in future generations. Much of the DNA in organisms is never expressed. If the mutation occurs in that portion of the DNA that is not expressed, no problem occurs. However, if the mutation occurs in the active portion of the DNA, the altered gene products can be expressed. An example of a problematic point mutation is in the gene that causes sickle cell anemia. A change of one basepair (a transversion from thymine to adenine) results in the amino acid glutamate being replaced by another amino acid, valine, in one of the molecules that makes up hemoglobin, the oxygen-carrying molecule in red blood cells. When the blood becomes deoxygenated, such as under heavy exercise conditions, the valine allows the red blood cells to assume a sickle shape instead of the normal circular shape. This leads to clumping of blood cells in capillaries, which in turn may limit blood flow to the tissues. This behavior of the blood cells exacerbates other effects of sickle cell anemia, which result in oxygen deprivation because the hemoglobin content of the blood in persons with sickle cell anemia is about half that of other persons.

12.3 NONMAMMALIAN MUTAGENICITY TESTS

Because results from bacterial or prokaryotic assays often establish priorities for other testing approaches, it is of interest to briefly describe the assays currently used to screen for mutagenic capacity, particularly those done in industrial settings.

Rapid cell division and the relative ease with which large quantities of data can be generated (approximately 10^8 bacteria per test plate) have made bacterial tests the most widely utilized routine means of testing for mutagenicity. These systems are the quickest and most inexpensive procedures. However, bacteria are evolutionarily far removed from the human model. They lack true nuclei as well as the enzymatic pathways by which most promutagens are activated to form mutagenic compounds. Bacterial DNA has a different protein coat than seen in eukaryotes. Nevertheless, bacterial systems have great utility as a preliminary screen for potential mutagens.

In addition to bacteria, fungi have been used in genotoxicity assays. The *Saccharomyces* and *Schizosaccharomyces* yeasts, as well as the molds *Neurospora* and *Aspergillus*, have been utilized in forward mutation tests, which are similar in design to the salmonella histidine revertant assays that will be described in the next section.

Typical Bacterial Test Systems

The most widely utilized bacterial test system for monitoring gene mutations and the most widely utilized short-term mutagenicity test of any type is the *Salmonella typhimurium* microsome test developed by Dr. Bruce Ames and co-workers and commonly called the Ames assay. The phenotypic marker utilized for the detection of gene mutations in all the Ames *Salmonella* strains is the ability of the bacteria to synthesize histidine, an amino acid essential for bacterial division. The tester strains of bacteria have mutations rendering them unable to synthesize histidine; thus, they must depend on histidine included in the culture medium in order to be able to multiply. Bacteria are taken directly from a prepared culture and incorporated with a trace of histidine into soft agar overlay on a dish containing minimal growth factors. The bacteria undergo several divisions, which are necessary for the expression of mutagenicity and, after the available histidine has been used up, a fine bacterial lawn is formed. Bacteria that have back-mutated in their histidine operon sites (and thus have reverted to the ability to synthesize histidine) will keep on dividing to form discrete colonies, while the nonmutated bacteria will die. A chemical that is a positive mutagen will demonstrate a statistically significant dose-related increase in "revertants" (colonies formed) when compared to the spontaneous revertants in control plates.

Five Ames *S. typhimurium* tester strains are recommended for routine mutagenicity testing: TA1535, TA1537, TA1538, TA98, and TA100. The TA1535 tester strain detects basepair substitution mutations. The TA1538 tester strain detects frameshift mutagens that cause basepair deletions. The TA1537 tester strain detects frameshift mutagens that cause basepair additions. The TA100 (basepair substitution) and TA98 (frameshift) strains are sensitive to effects caused by certain compounds, such as nitrofurans, which were not detectable with the previous three strains.

The lack of oxidative metabolism to transform promutagens (those mutagens requiring bioactivation to the active form) is overcome in these bacterial assays by two means. First, a suspension of rat liver homogenate containing appropriate enzymes may be added to the bacterial incubation. The liver preparation is centrifuged at 9000*g* for 20 min at 4°C, and the resultant supernatant (S9) is added to the culture medium. In a slightly more complex procedure, called the *host-mediated assay*, the bacterial tester strains are injected into the body cavity of a test animal such as the mouse. This host is treated with the suspected mutagen and, after a selected period, the bacteria are harvested and assayed for mutation (revertants) as described earlier. Other bacterial species used in mutagenicity screens include *Escherichia coli* and *Bacillus subtilis*.

Assays that measure DNA repair in bacterial systems have also been developed. These tests are based on the premise that a strain deficient in DNA repair enzymes will be more susceptible to mutagenic activity than will a similar strain that possesses repair enzymes that can correct the mutagenic damage. A "spot" test consists of placing the chemical to be tested in a well or on a paper disk on top of the agar in a petri dish. The test chemical will diffuse from the central source, causing a declining concentration gradient as the distance from the source increases. A strain deficient in repair enzymes will exhibit a greater diameter of bacterial kill than the repair-sufficient strain tested with a mutagen. In a "suspension" test, a given number of bacteria are preincubated with and without the test compound. The bacteria are then plated and the colonies counted. The repair-deficient strain will demonstrate a greater percentage kill than will the sister DNA-repair-sufficient strain. A liver S9 activation system can also be incorporated in bacterial DNA repair tests. The most widely used bacterial DNA repair test utilizes the $polA^+$ and $polA^-$ strains of *E. coli*. The $polA^-$ strain is deficient in DNA polymerase I, whereas the $polA^+$ strain is sufficient in this enzyme.

Drosophila Test Systems

The fruitfly (*Drosophila melanogaster*) has received wide use in the sex-linked recessive lethal test. The endpoint phenotypic change monitored in this test is the lethality of males in the F_2 generation. Brusick has gone to the extent of labeling *Drosophila* an "honorary mammalian model" by virtue of its widespread application and correlation with positive mutagens in mammalian testing. *Drosophila melanogaster* has also been utilized to monitor two types of chromosomal aberration endpoints through phenotypic markers: loss and nondisjunction of X or Y sex chromosomes and heritable translocations. The monitoring of translocations has the advantage of a very low background rate, facilitating comparisons between controls and treated groups. Dominant lethal assays are also performed with insects and can theoretically be applied in any organism where early embryonic death can be monitored. The male is treated with the test agent, then mated with one or more females. If early fetal deaths occur, these are demonstrative of a dominant lethal mutation in the germ cells of the treated male.

Plant Assays

A number of assay types are available in plant systems as well, including specific locus tests in corn (*Zea maize*) and multilocus assays in *Arabidopsis*. Cytogenetic tests have been developed for *Tradescantia* (micronucleus test), as have chromosomal aberration assays in the root tips of onions (*Allium sepa*) and beans (*Vicia faba*). Finally, DNA adducts analysis is applicable to somatic and germinal plant cell systems. It is anticipated that one or more plant species may prove to be useful indicators of the potential for genetic damage that may be related to emissions of environmental pollutants.

12.4 MAMMALIAN MUTAGENICITY TESTS

Testing chemicals for mutagenicity *in vivo* in mammalian systems is the most relevant method for learning about mutagenicity in humans. Mammals such as the rat or mouse offer insights into human physiology, metabolism, and reproduction that cannot be duplicated in other tests. Furthermore, the route of administration of a chemical to a test animal can be selected to parallel normal human environmental or occupational conditions of oral, dermal, or inhalation exposures.

Human epidemiologic findings may also be compared with the results of tests done in animals. While the monitoring of human exposures and their effects does not constitute planned, controlled mutagenicity testing, human epidemiology offers the opportunity to monitor and test for correlations suggested by other mutagenicity tests. Thus, these studies are the only opportunity for direct human modeling of a chemical's mutagenic potential. It is worth noting that despite extensive investigation, to date no chemical substances have been positively identified as human mutagens. The advantages and limitations of a wide variety of genetic test systems are presented in Brusick (1994).

One perceived disadvantage of *in vivo* mammalian test systems is the time they require and their cost. A larger commitment of physical resources and personnel is required than is required with *in vitro* testing. Human epidemiology studies are further complicated by the fact that not all of the environmental variables can be controlled. Frequently, the duration and extent of exposure to single or multiple compounds can only be estimated. Nevertheless, progress is being made to lessen the cost and decrease the time required for *in vivo* mammalian testing. Also, new data handling, statistical techniques, and increased cooperation from industry have increased the reliability of human epidemiology studies. More regular sampling of workplace exposures has helped to improve the quality and accessibility of human data.

Mutagenic potential can vary greatly across a class of analytes, as shown for the metals (Costa, 1996). Mutagenicity data for metals can be quite difficult to interpret due to the breadth of mechanisms at work, as illustrated by differences between Cr, Ni, As, and Cd.

Germ Cell Assays

A basic test used to detect specific gene mutations induced in germ cells of mammals is the mouse-specific locus assay. This test involves treatment of wild-type mice, either male or female, with a test compound before mating them to a strain homozygous for a number of recessive genes that are expressed visibly in phenotype. If no mutations occur, then all offspring will be of the wild type. If a mutation has occurred at one of the test loci in the treated mice, then the recessive phenotype will be visibly detectable in the offspring. The mouse-specific locus test is of special significance in human modeling because it is the only standardized assay that directly measures heritable germ cell gene mutations in the mammal. A major drawback of the mouse-specific locus test is that extensive physical plant facilities are required to execute this assay, and it has been estimated that one scientist and three technicians could execute 10 single-dose mouse-specific locus tests in one year, provided there are facilities for 5000 cages.

New and promising test procedures have been described for detecting germ cell mutations by using alterations in selected enzyme activity as the phenotypic endpoint. A large group of somatic cell enzymes can be monitored for changes in activity and kinetics in the F_1 generation. These changes indicate changes in the parental genome.

A similar biochemical approach has been proposed for identifying germ cell mutations in humans through the monitoring of placental cord blood samples. The activity of several erythrocyte enzymes, such as glutathione reductase, can be monitored because the enzyme proteins are the products of a single locus and because heterozygosity of a mutant allele for the chosen enzymes will result in abnormal levels of enzyme activity. Likewise, it has been proposed that gene mutations be directly monitored in mammalian germ cells by searching for phenotypic variants with biochemical markers such as lactic acid dehydrogenase-X (LDH-X), an isozyme of lactic acid dehydrogenase found only in testes and sperm. The test is based on the fact that a monospecific antibody for rabbit LDH-X reacts with rat but not mouse LDH-X in sperm. The rat sperm fluoresce as a result of the reaction but the mouse sperm do not fluoresce unless a phenotypic variant is present. If adapted to humans, this test has potential use as a noninvasive screening test of germ cell mutations in males.

It has been proposed that the induction of behavioral effects in the offspring of male rats exposed to a mutagenic agent may represent a genotoxic endpoint. For example, studies have demonstrated that the mutagen cyclophosphamide can induce genotoxic behavioral effects in the progeny of male rats and that these effects correspond to observed genetic damage caused in the spermatozoa following meiosis. A similar effect has been attributed to vinyl chloride in at least one instance of occupational exposures.

Mammalian germ cells can be monitored for chromosomal aberrations, and normally the testes are used as the cell source. Mammalian male germ cells are protected by a biological barrier comparable in function to the barrier which retards the penetration of chemicals to the brain. The blood–testes barrier is a complex system composed of membranes surrounding the seminiferous tubules and the several layers of spermatogenic cells organized within the tubules. This barrier restricts the permeability of high-molecular-weight compounds to the developing male germ cell. An advantage of *in vivo* mammalian germ cell mutagenicity testing is that the protective contribution of this barrier is automatically taken into account. Conventional procedures for harvesting mammalian male germ cell tissue for metaphase-spread analysis involve mincing or teasing the seminiferous tubules to liberate meiotic germ cells in suspension. This homogenate is centrifuged, the centrifuged pellet is discarded, and the suspended cells are collected and analyzed. However, it was found that the tissue fragments discarded during this conventional procedure contained more spermatogonial cells and meiotic metaphases than did the suspension. Thus, the method has been refined by using tissue fragments and adding collagenase to dissociate them. After collagenase treatment, the tissue fragments are gently homogenized and centrifuged, and the pellet containing meiotic cells is resuspended and prepared for microscopic analysis.

Dominant Lethal Assays

Dominant lethal assays can be performed in any organism where early embryonic death can be monitored. As described earlier, mammals are commonly used in dominant lethal assays, although it is possible to do so with insects as well. The male animals (typically mice or rats) are treated with the suspected mutagen before being mated with one or more females. Each week these females are removed and a new group of females is introduced to the treated male. This process is repeated for a period of 6–10 weeks. The females are sacrificed before parturition, and early fetal deaths are counted in the uterine horns. This test has become standardized, and a large number of compounds have been screened in mouse studies. As with most *in vivo* mammalian assays, costs and commitment of resources can be extensive. However, the applicability of the data is typically quite good. The dominant lethal test in rodents is of significance for human modeling because it gives an indication of heritable chromosomal damage in a mammal. Even though the endpoint of early fetal death may seem of minor significance when considering only its effects on the human gene pool, it does provide a signal that viable heritable chromosomal damage and gene mutations may also be produced.

Heritable Translocation Assay

Results of dominant-lethal assays frequently correlate well with another test used for determining clastogenic effects in mammalian germ cells: the heritable translocation assay (HTA). Translocation represents complete transfer of material between two chromosomes. In HTA procedures, male mice are mated to untreated females after treatment with the test compound, and the pregnant females are allowed to deliver. Male offspring are subsequently mated to groups of females. If translocations are produced through genotoxic action, then the affected first-generation male progeny will be partially or completely sterile; this can be noticed in the litter size produced from those females to which they were mated.

Micronucleus Tests

Application of the micronucleus test to mammalian germ cells recently has been reported. This test procedure is analogous to the bone marrow micronucleus test (somatic cells), but it involves the sampling of early spermatids from the seminiferous tubules of male rats. The number of micronuclei are quantified by using fluorescent stain and counting micronucleated spermatids. To date, the technique has not been widely used in occupational evaluations.

Spermhead Morphology Assay

Some relatively new tests have been developed that evaluate the ability of a test chemical to induce abnormal sperm morphology when compared to controls. It has been proposed that an increase in abnormal sperm morphology is evidence of genotoxicity because there seems to be an association between abnormal sperm morphology and chromosome aberrations. However, recent investigations have reported that induction of morphologically aberrant sperm can be caused by nongenotoxic actions, such as dietary restriction. In addition, some known mutagens, including 1,2-dibromo-3-chloropropane (a pesticide with mutagenic, carcinogenic, and gonadotoxic properties), were reportedly unable to induce production of spermhead abnormalities in mice, when tested. It should be noted that sperm abnormalities are fairly common in humans and may occur at rates of 40–45 percent. Thus, more verification is needed before strong conclusions can be drawn about the mammalian spermhead morphology assay.

Tests for Primary DNA Damage

A historical test thought to monitor primary DNA damage in mammalian germ cells *in vivo* involves the monitoring of sister chromatid exchange. The observation of sister chromatid exchanges through differential staining involves exposing the cells to bromodeoxyuridine for two rounds of replication, so that the chromosomes consist of one chromatid substituted on both arms with 5-bromodeoxyuridine and the other substituted only on a single arm. Differential staining between sister chromatids is due to the differences in bromodeoxyuridine incorporation in the sister chromatids.

Unscheduled DNA repair has been induced by chemical mutagens in mammalian male germ cells from the spermatogonial to midspermatid stages of development. The test is based on the fact that cells not undergoing replication (scheduled DNA synthesis) should not exhibit significant DNA synthesis. Thus, incorporation of radiolabeled tracer molecules into the DNA of these cells should be minimal. However, if a chemical mutagen damages the DNA, the DNA repair system may be activated, causing unscheduled DNA synthesis (UDS). If such is the case, radiolabeled tracers will be incorporated into the DNA; these can be monitored by autoradiography or by direct measurement of radioactivity in the repaired DNA. Male germ cells lose DNA repair capability when they have advanced to the late spermatid and mature spermatozoa stages; unscheduled DNA synthesis cannot then be induced by chemical mutagens. The genotoxic agents methyl methanesulfonate, ethyl methanesulfonate, cyclophosphamide, and Mitomen have been shown to induce unscheduled DNA repair *in vivo* in male mouse germ cells. Similar procedures are available to evaluate UDS in some types of somatic cells as well.

Transgenic Mouse Assays

In the late 1980s and early 1990s the development of a new genotoxicity assay system was reported by Gossen et al. (1989), Kohler et al. (1991), and colleagues. Briefly, the test system involves mutagen dosing to a specific mouse strain (C57BL6) that has been infected with a viral "shuttle vector," isolation of the mouse DNA, recovery of the phage segment (*lac*I or *Lac*Z), and infection of an *E. coli* strain with the recovered phage. The phage will form plaques on a lawn of *E. coli*. The plaques are colorless if no mutation has occurred or blue if a mutation has occurred. The assay may be performed to gather information on mutations in somatic cells or in germ cells. The *lacZ* assay also is known as the "Muta-Mouse" assay, while the *lac*I assay also is known as the "Big Blue" assay.

Advantages of the assay include its *in vivo* treatment regime, the fact that it can be conducted in a few days from the isolation of DNA through plaque formation to mutation scoring. However, it may be difficult to use extremely high dosages (e.g., approaching lethal doses), since the mice must survive for 1–2 weeks in order to fix the mutation in the affected tissues. The performance of the transgenic mouse assays that have been conducted on 26 substances was evaluated by Morrison and Ashby (1994), including a review of the results of the tests that had been performed in the *lac*Z case (14 reports) and in the lacI case (16 reports). These authors concluded that the variability of data reporting formats and the rapid developments and modifications in the assay protocols make it difficult to perform direct comparisons among tests or between this assay type and the results of other historically available methods. Nevertheless, the initial results are generally promising, and there are no examples of internal disagreement between responses for the same chemical in the same tissue.

In Vitro Testing

Test systems have been developed that use mammalian cells in culture (*in vitro*) to detect chemical mutagens. Disadvantages in comparison with *in vivo* mammalian tests are that *in vitro* tests lack organ–system interaction, require a route for administration of the agent that cannot be varied, and lack the normal distributional and metabolic factors present in the whole animal. The obvious advantages are that costs are decreased and that experiments are more easily replicated, which facilitates verification of results. Cases where human cells have been cultured successfully (e.g., lymphocytes) provide the only viable *in vitro* experiments on the human organism. Several endpoints can be used during testing of potential mutagens *in vitro*. One of the most common involves the monitoring of mutations in specific well-characterized gene loci, such as those coding for hypoxanthineguanine phosphoribosyl transferase (HGPRT), thymidine kinase (TK), or ouabain resistance (OVA^r). Mutagenic modification in the segments of the DNA coding for these proteins (enzymes) results in an increased sensitivity of the cell, which can often be evaluated by the cell's heightened susceptibility to other agents (e.g., bromodeoxyuridine or 8-azaguanine).

As described in the section on *in vivo* mammalian testing, evaluations of sister chromatid exchange, DNA repair activity, and chromosomal aberration through interpretation of metaphase spreads may be applied to *in vitro* testing of mutagens. An additional procedure that has been correlated with chemical mutagenicity is examination for cell culture transformation; following treatment with mutagens, some cells in culture lose their normal, characteristic arrangement of monolayered attachment and begin to pile up in a disorganized fashion. Two major drawbacks in looking for this feature are that considerable expertise is necessary to interpret the results accurately and that the criteria for evaluation are more subjective than for other mutagenicity assays.

A comparison of the sensitivity and specificity of selected short-term tests by two recognized systems (National Toxicology Program (NTP) and Gene-Tox) is shown in Table 12.3.

12.5 OCCUPATIONAL SIGNIFICANCE OF MUTAGENS

Areas of Concern: Gene Pool and Oncogenesis

The potential significance of occupationally acquired mutations can be divided into two areas. The first is concern for the protection of the human gene pool. This factor may represent the most significant reason for genetic testing, but it is often underemphasized by nongeneticists involved with safety evaluation because of the inability to demonstrate induced mutation in humans to date. The second area is that of oncogenesis. The intimate relationship between the tumorigenic and genotoxic properties of many chemicals (Figure 12.8) makes genetic testing a potentially powerful screening technique for establishing priorities for future testing of chemicals of unknown cancer-causing potential. This factor has been one of the primary driving forces behind the rapid expansion of genetic toxicology as a discipline. Once again, however, the paucity of proved human carcinogens compared with the number of demonstrated animal carcinogens suggests weaknesses in the process of extrapolating from animal studies to human exposure in the workplace.

At the heart of the present legal and regulatory approach toward environmental and occupational exposure to mutagens is the possibility that they may cause human genetic damage. Two important assumptions underlie this central concept:

- Environmental or occupational mutagens may cause aneuploidy, chromosome breaks, point mutations, or other genetic damage in humans
- Environmental or occupational mutagens that can be controlled by regulatory efforts represent a significant component of total human exposure

Much of the interest in potential environmental and occupational mutagens is related to the prevalent opinion that many cancers are initiated by a mutagenic event. This premise is supported by the strong correlation between some specific occupational chemical exposures and cancer incidence in humans. One good example is the relationship between liver cancer (angiosarcoma) and exposure to vinyl chloride in some manufacturing operations. Another example is the respiratory tract cancers that may be caused by exposure to bis(chloromethyl)ether.

TABLE 12-3 NTP and Gene-Tox Evaluation of Short-Term Test Sensitivities and Specificities

	Sensitivity[a]				Specificity[b]			
	Gene-Tox[c]		NTP[d]		Gene-Tox		NTP	
Assay	(+)/Total	%(+)	(+)/Total	%(+)	(−)/Total	%(−)	(−)/Total	%(−)
Ames/*Salmonella*	175/223	78	20/44	45	29/47	62	25/29	86
Mouse/lymphoma	45/54	87	31/44	70	0/5	0	13/29	45
CHO/HGPRT	40/41	98	—	—	1/1	100	—	—
V79	84/104	81	—	—	3/3	100	—	—
Drosophila SLRL	77/106	73	4/18	22	9/16	60	9/9	100
In vitro cytogenetics	40/54	74	24/44	55	2/6	33	20/29	69
In vivo cytogenetics	8/9	89	9/15	60	0/0	—	11/12	92
In vitro SCE	100/101	99	31/44	70	0/10	0	13/29	45
In vivo SCE	21/21	100	10/15	67	0/0	—	5/12	42
UDS in hepatocytes	19/22	86	6/30	20	0/0	—	13/14	93

Source: Kier (1988).

[a]Sensitivity is the proportion of positive results for carcinogens.

[b]Specificity is the proportion of negative results for noncarcinogens.

[c]Gene-Tox data include combined results for sufficient and limited-evidence carcinogens andnoncarcinogens.

[d]NTP specificity assumes that equivocal evidence compounds are noncarcinogens.

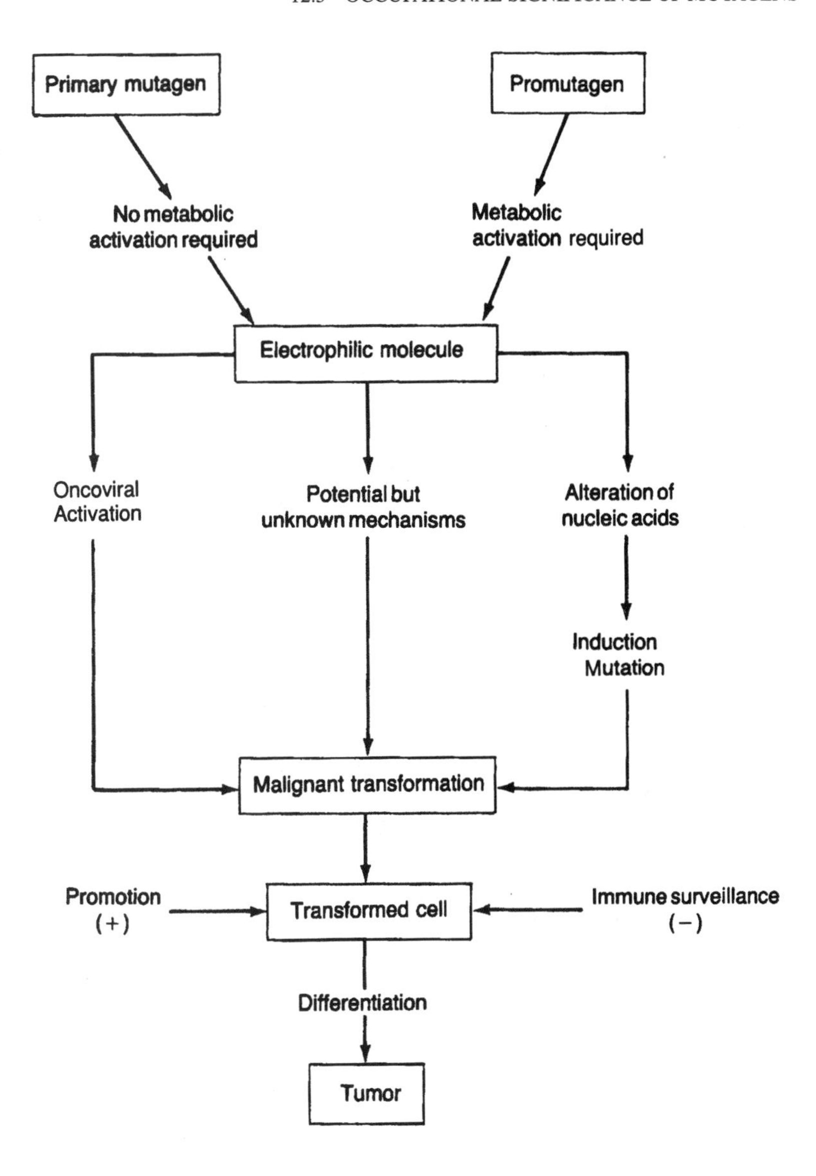

Figure 12.8 Proposed relationships between mutagenicity and carcinogenicity. (Adapted from Fishbein, 1979).

A Multidisciplinary Approach

Although each of the mutagenicity tests described in this chapter has individual strengths, likewise each is weak in some facets of detection capacity. Clearly, therefore, the accurate and efficient testing of chemicals and the protection from potential occupational mutagens require a multidisciplinary approach that integrates toxicology, clinical chemistry, microbiology, pathology, epidemiology, industrial hygiene, and occupational medicine. Testing intact animals has the advantage of increasing the reliability of any extrapolations that must be made from the data; however, cost considerations often limit the application of *in vivo* mammalian assays except when it is expected that they will verify lower tier assays.

There are three areas where the results of mutagenicity testing of a given substance may be applied:

- Extrapolation of test results in order to make a quantitative evaluation of the hazard of exposures for humans
- Prioritization of human hazards caused by specific compounds
- Institution of remedial procedures that should be undertaken to minimize the human hazard

One of the most difficult areas of analysis is the correct application of mutagenicity tests to arrive at a quantifiable human hazard from exposure to a given substance. There is frequently good correlation between the mutagenicity and carcinogenicity of a substance in animal tests (Table 12.4). However, this may be misleading because models for carcinogenicity determination are often characterized by chronic procedures utilizing very high doses in nonprimate species. These may bear little resemblance to aspects of exposure in the human model, such as magnitude and route of exposure, metabolic patterns, and environment (which are qualitative factors), and exposure dose (which is quantitative).

As noted previously, the time and expense that are involved with lifetime carcinogenicity assays have strongly influenced the use of test batteries as predictive measures of carcinogenic potential. Among many others, Ashby and Tennant (1994), Anderson et al. (1994), Benigni and Giuliani (1987), and Blake et al. (1990), have addressed the question of applicability of multiple test systems to the classification of a substance as genotoxic or not, and carcinogenic or not. It is important in these efforts to distinguish among "sensitivity" (ability to identify a known carcinogen), "specificity" (ability to identify a noncarcinogen), and "accuracy" (correct results of either type). Parodi et al. (1990) reported on qualitative correlations associated with studies of up to 300 substances conducted during the period 1976 through 1988. Initial measures of sensitivity, specificity, and accuracy were approximately 90 percent, if the decision is based solely on *Salmonella* assays. As more substances have been tested, this estimate has ranged from 45 to 91 percent. Best results typically are reported for sensitivity, where accuracy generally is on the order of 65 to 75 percent. Consideration of the quantitative correlation between short-term genotoxicity tests and carcinogenic potency has yielded extremely variable estimates, ranging from approximately 30 to over 90 percent. The overwhelming conclusion was that a battery of test systems that addresses differing endpoints is required if the goal is to develop a confident conclusion regarding predictivity.

As is the case with many areas of toxicology, one may choose between *in vivo* and *in vitro* test systems, each with their attendant advantages and disadvantages. The testing of chemicals in experimental animals has all the advantages of any intact *in vivo* system; that is, it has all of the biochemical and physiological requirements to make anthropomorphization more reliable. However, *in vivo* mutagenicity testing may require an investment of many thousands of dollars and a long period of time. These disadvantages often force the tester to use a less expensive, well-established short-term

TABLE 12.4 Comparative Mutagenicity of Various Compounds

Compound	Established human carcinogen	Bacteria	Yeast	*Drosophila*	Mammalian cells	Human cells
Epichlorohydrin	N	N	O	Y	Y	Y
Ethyleneimine	N	O	Y	Y	Y	Y
Trimethyl phosphate	O	Y	O	Y	Y	Y
Tris	N	Y	O	Y	Y	Y
Ethylene dibromide	N	Y	Y	Y	Y	N
Vinyl chloride	Y	Y	Y	Y	Y	Y
Chloroprene	Y	Y	O	O	O	Y
Urethane	N	Y	Y	Y	Y	O

N = no; Y = yes; O = not tested.

bioassay, such as the Ames *Salmonella* bacterial assay, to determine the mutagenicity of a chemical, and then extrapolate these results into the *in vivo* mutagenicity model.

Occupational Monitoring and Biomarkers for Genetic Damage

Cytogenetic analysis (chromosome evaluation) of human lymphocytes has been a standard industrial technique for monitoring human genetic damage. However, several limitations are inherent in the conventional use of human lymphocytes as indicators of exposure to genotoxic chemicals or radiation:

- Individual and population variability in normal levels of chromosomal aberrations may mask small changes in the frequency of mutations. To overcome this obstacle, specific defects that occur with low frequency in normal individuals may also be tested for, but typically several thousand cells must be scored per individual to achieve sufficient sample size.
- Evaluation of chromosomal aberrations is subject to substantial variation between laboratories. Therefore, replicate readings should be obtained; this substantially increases the effort required when thousands of samples are involved.
- Since chromosomal aberrations are considered indicators of relatively gross damage, the techniques may miss many more subtle effects of mutagens.

Evaluation of sister chromatid exchange (SCE) may be potentially valuable in answering some of these difficulties. For example, SCEs are elevated in patients undergoing chemotherapy, which is not unexpected, as many of the chemotherapeutic agents are powerful mutagens. These elevations tend to be dose-related, which supports the usefulness of the technique as a potential screening device. It must be emphasized that SCE may not be a damaging lesion in itself, but may prove a useful marker for other detrimental effects on the DNA induced by the agents in question. This caveat is underscored by the observation that SCE is poorly correlated with radiation exposure and exposure to other agents that break DNA. Agents that alkylate the DNA (bind tightly to the molecule) show a better correlation with mutagenic potential and may be a sensitive indicator for the monitoring of chromosomal aberrations, which are otherwise more difficult and time-consuming to detect.

The Ames-type mutagenicity testing of urine from exposed individuals (e.g., tobacco, chemotherapy patients) has yielded promising results as a simple, rapid, and inexpensive screening technique, although the timing and impact of cumulative versus acute exposures are not yet fully understood. The evaluation of other biomarkers for genetic damage are under development or investigation, particularly with regard to germinal cell populations. These methods include techniques to detect formation of DNA and protein adducts, and changes in sperm morphology or fertility indices. These issues recently were reviewed by the National Research Council (1989).

Areas for Future Activity

Many mutagenicity assays have been proposed, each with a unique attribute and measurable biochemical or visible endpoint. These tests are being incorporated into routine safety assessment programs in all regulatory agencies. Furthermore, these tests have been proposed as part of a regulatory decision-making policy by the Occupational Safety and Health Administration (OSHA) for the classification of chemical carcinogens in the workplace, and by the U.S. Environmental Protection Agency (USEPA) for the regulation of pesticides and for regulating the disposal of toxic wastes. A tremendous amount of information is available through the Environmental Mutagen Information Center (EMIC), housed at the Oak Ridge National Laboratory. The short-term mutagenicity tests actually serve two purposes. They not only assist in the assessment of a chemical's potential for cancer induction but also assess the potential for inducing germ cell mutations in humans. Some of the organizations involved in the development of guidelines for germ cell mutagenicity tests are the International Commission for Protection against Environmental Mutagens and Carcinogens (ICPEMC), the World Health Organization, and the Commission for European Communities. In the past, most estimates of genotoxic risks were more qualitative than quantitative, and the emphasis has rested on somatic effects (e.g., those leading to cancers) rather than on germinal cells (sperm and ovum). On the basis of evidence in animals demonstrating germinal cell effects, it is imperative to develop human screening methods capable of detecting such effects. Therein lies one of the premier challenges to genetic toxicology and occupational medicine.

The uncertainties of accurate extrapolation of mutagenicity test data to a human hazard model have supported the philosophy that if uncertainty is to occur in extrapolation it should favor the side of safety. This concept is particularly important in the consideration of whether or not threshold characteristics may exist. In the case of carcinogens, discussed further in the next chapter, good evidence supports the view that genotoxic (DNA-damaging) carcinogens may be distinct from epigenetic carcinogens (those that induce or potentiate cancer by means other than direct DNA interaction). For the purposes of this discussion, mutagens are assumed to exert nonthreshold effects. That is, even as one approaches zero dose, there is still a calculable risk of DNA effects.

The concern for the potential mutagenic hazards in the workplace from exposure to chemicals should include routine tests of nonpregnant females and males, as well as the more traditional monitoring of pregnant and lactating women. For example, vinyl chloride, mentioned earlier in relation to its suggested role in angiosarcoma of the liver, has been correlated with an increased incidence of nervous system malformations in infants fathered by exposed workers. It has also been demonstrated to cause elevations in chromosomal aberration in the occupationally exposed. 1,2-dibromo-3-chloropropane (DBCP), a pesticide linked to sterility in exposed male workers, causes increases in indices of mutagenic capacity in humans and animals.

Monitoring of male populations may prove particularly important in that the spermatogenic cycle is continuous in adults and therefore poses continuous opportunities for genetic damage to be expressed as damaged chromosomes. Since the female carries the full lifetime complement of ova at birth, susceptibility to propagation of genetic alteration during cell division is reduced except in those periods of division following conception. By the same token, the cessation of exposure in the male should allow for recovery from a mutagenic event in premeiotic spermatocytes, providing that spermatogonia are not affected. If chromosome damage occurs in sperm or ovum, then fetal death frequently occurs. Greater than 50 percent of spontaneous abortions in humans show chromosomal defects.

Once mutagenic potential is established for a compound, the risks posed by exposure under expected conditions must be assessed. As discussed, complications may be encountered in situations where mutagenic effects are due to "multihit" phenomena and therefore reflect threshold-type responses. A more complete discussion on risk assessment is presented in Chapter 18.

12.6 SUMMARY

Modification of genetic material by mutagenic agents poses a serious environmental and occupational threat. Chemical or physical mutagens may induce cancer or lead to germ cell alteration.

- The mutagens that lead to cancer alter the DNA of somatic cells so as to cause modifications in gene expression, which results in tumorigenesis.
- Germ cell (sperm, ovum) mutagens may exert their effects through decreased fertility, birth defects, spontaneous abortion, or through changes that may not become evident for several subsequent generations (such hidden mutagenic effects remain essentially undetectable except when expressed as a gross malformation).

Many screening tests have been developed to investigate the mutagenic potential of chemical agents.

- These assays use bacteria, insects, mammals, and various cells in culture.

- Although *in vitro* tests are less expensive and less complex, *in vivo* mammalian tests give results that can be extrapolated to human circumstances more realistically, but *in vivo* studies are expensive and labor-intensive.

Persons whose occupations expose them to potential mutagens may undergo chemically induced changes at a greater rate than the general population does. Validation of this hypothesis is the subject of extensive ongoing research.

- Epidemiology seeks to identify groups with increased susceptibility to chemical mutagens, or increased incidence of exposure, in order to limit exposures.
- No single method currently stands out as the most comprehensive and thorough screen for identifying mutagenic agents; often, a multidisciplinary approach employing several tests is best suited to the accurate identification of industrial mutagens.

Once mutagenic potential has been demonstrated for a compound, typically an analysis must be made of the risks posed to exposed individuals. Such a determination is essential in the qualitative evaluation of the occupational hazard of mutagens.

REFERENCES CITED AND SUGGESTED READING

Anderson, D., M. Sorsa, and M. D. Waters, "The parallelogram approach in studies of genotox ic effects," *Mutat. Res.* **313:** 101 (1994).

Ashby, J., and R. W. Tennant, "Prediction of rodent carcinogenicity for 44 chemicals: resul ts," *Mutagenesis* **9:** 7 (1994).

Ashby, J., and H. Tinwell, "Use of transgenic mouse lacI/Z mutation assays in genetic toxicol ogy," *Mutagenesis* **9:** 179 (1994).

Auerbach, C., J. M. Robson, and J. G. Carr, "The chemical production of mutations." *Science* **105:** 243 (1947).

Barlow, S. M., and F. M. Sullivan, *Reproductive Hazards of Industrial Chemicals*, Academic Press, New York, 1982.

Benigni, R., "Rodent tumor profiles Salmonella mutagenicity and risk assessment," *Mutat. Res.* **244:** 79 (1990).

Benigni, R., and A. Giuliani, "Which rules for assembling short-term test batteries to predic t carcinogenicity," *Molec. Toxicol.* **1:** 143 (1987).

Berg, K., ed., *Genetic Damage in Man Caused by Environmental Agents*, Academic Press, New York, 1979.

Blake, B. W., K. Enslein, V. K. Gombar, and H. H. Borgstedt, "Salmonella muatgenicity and rod ent carcinogenicity: Quantitative structure-activity relationships," *Mutat. Res.* **241:** 261 (1990).

Brusick, D. J., *Principles of Genetic Toxicology*, Plenum Press, New York, 1980.

Brusick, D. J., ed., *Methods for Genetic Risk Assessment*, Lewis Publishers, New York, 1994.

Calabrese, E. J., *Pollutants and High Risk Groups*, Wiley-Interscience, New York, 1978.

Cohen, B. H., A. M. Lilienfeld, and P. C. Huang, eds., *Genetic Issues in Public Health and Medicine*, Charles C. Thomas, Springfield, IL, 1978.

Costa, M., "Introduction to metal toxicity and carcinogenicity of metals," in L. W. Chang, ed., *Toxicology of metals*, CRC Press, Boca Raton, 1996.

Fishbein, L., *Potential Industrial Carcinogens and Mutagens*, Elsevier Scientific, Amsterdam, 1979.

Gossen, J. A., W. J. F de Leeuw, C. H. T. Tan, E. C. Zwarhoff, F. Berends, P. H. M. Lohman, D. L. Knook, and J. Vijg, "Efficient rescue of integrated shuttle vectors from transgenic mice: A model for stud ying mutations *in vivo*," *Proc. Natl. Acad. Sci.* (USA) **86:** 7971 (1989).

Hoffmann, G. R., "Genetic toxicology," In M. O. Amdur, J. Doull, and C. D. Klaassen, eds., *Casarett and Doull's Toxicology: The Basic Science of Poisons*, 4th ed., Macmillan, New York, 1991.

Hollaender, A., ed., *Chemical Mutagens: Principles and Methods for Their Detection*, Vols. 1–8, Plenum Press, New York, 1971–1984.

ICPEMC (International Commission for Protection against Environmental Mutagens and Carcinogens), "Estimation of genetic risks and increased incidence of genetic disease due to environmental mutagens," *Mutat. Res.* **115:** 255 (1983b).

ICPEMC (International Commission for Protection against Environmental Mutagens and Carcinogens), "Regulatory approaches to the control of environmental mutagens and carcinogens," *Mutat. Res.* **114:** 179 (1983b).

Kier, L. D., "Comments and perspective on the EPA workshop on the relationship between short-term test information and carcinogenicity," *Environ. Molec. Mutag.* **11:** 147–157 (1988).

Kirsch-Volders, M., *Mutagenicity: Carcinogenicitry and Teratogenicity of Industrial Pollutants*, Plenum Press, New York, 1984.

Kohler, S. W., G. S. Provost, A. Fieck, P. L. Kretz, W. O. Bullock, D. L. Putman, J. A. Sorge, and J. M. Short, "Analysis of spontaneous and induced mutations in transgenic mice using a lambda ZAP/lacl shuttle vector," *Environ. Molec. Mutag.* **18:** 316 (1991).

Livingston, G. K., "Environmental mutagenesis," in W. N. Rom, ed., *Environmental and Occupational Medicine*, Little, Brown and Co., Boston, 1992.

Mendelsohn, M. L., and R. J. Albertini, *Mutation and the Environment*, Wiley-Liss, New York, 1990.

Miller, E. C., J. A. Miller, I. Hirono, P. Sugimura, and S. Takayama, *Naturally Occurring Carcinogens, Mutagens, and Modulators of Carcinogenesis*, University Park Press, Baltimore, 1979.

Morrison, V., and J. Ashby, "A preliminary evaluation of the performance of the Muta-Mouse (lacZ) and Big Blue (lacI) transgenic mouse mutation assays," *Mutagenesis* **9:** 367 (1994).

Muller, H. J., "Artificial transmutation of the gene," *Science* **64:** 84 (1927).

NRC (National Research Council), *Biomarkers in Reproductive and Neurodevelopmental Toxicology*, National Academy Press, Washington, DC, 1989.

Osterloh, J. D., and A. B. Tarcher, "Environmental and biological monitoring," in *Principles and Practice of Environmental Medicine*, A. B. Tarcher, ed., Plenum Medical, New York, 1992.

Parodi, S., M. Taningher, P. Romano, S. Grilli, and L. Santi, "Mutagenic and carcinogenic potency indices and their correlation," *Teratogen. Carcinogen. Mutagen.* **10:** 177 (1990).

Rom, W. N., ed., *Environmental and Occupational Medicine*, Little, Brown, Boston, 1983.

Shaw, C. R., ed., *Prevention of Occupational Cancer*, CRC Press, Boca Raton, FL, 1981.

Sorsa, M., and H. Vainio, eds., *Mutagens in Our Environment*, Alan R. Liss, New York, 1982.

Suutari, A., and T. Sjoblom, "The spermatid micronucleus test with the dissection technique detects the germ cell mutagenicity of acrylamide in rat meiotic cells," *Mutat. Res.* **309**(2): 255 (1994).

Tarcher, A. B., ed., *Principles and Practice of Environmental Medicine*, Plenum Medical, New York, 1992.

Teaf, C. M., "Mutagenesis," in *Industrial Toxicology*, P. L. Williams and J. L. Burson, eds., Van Nostrand-Reinhold, New York, 1985.

Thilly, W. G., and R. M. Call, "Genetic toxicology," in *Casarett and Doull's Toxicology: The Basic Science of Poisons*, 3rd ed., C. D. Klaassen, M. O. Amdur, and J. Doull, eds., Macmillan, New York, 1986.

Tweats, D. J., "Mutagenicity," in *General and Applied Toxicology*, B. Ballantyne, T. Marrs, and P. Turner, eds., Stockton Press, New York, 1995.

Waters, M. D., H. F. Stack, J. R. Rabinowitz, and N. E. Garrett, "Genetic activity and pattern recognition in test battery selection," *Mutat. Res.* **205:** 119 (1988).

Weisberger, J. H., and G. M. Williams, "Chemical carcinogenesis," in *Casarett and Doull's Toxicology: The Basic Science of Poisons*, 2nd ed., J. Doull, C. D. Klaassen, and M. O. Amdur, eds., Macmillan, New York, 1980.

13 Chemical Carcinogenesis

ROBERT C. JAMES and CHRISTOPHER J. SARANKO

There are few people living today who have not been affected in some way by cancer, through either personal experience or that of a family member. Current statistics indicate that one out of two men and one out of three women in the United States will develop cancer over the course of their lifetime. Approximately 1.2 million people will be diagnosed with cancer this year alone, and this number excludes common and easily treatable basal and squamous cell skin cancers. While long-term survival rates are improving, cancer is still the second leading cause of death in the United States behind heart disease. In 1999, one of every four deaths or approximately 560,000 will be from cancer. In addition to the price cancer exacts in human lives lost, economic costs are estimated to be a staggering 107 billion dollars per year. This figure includes direct medical expense as well as the cost of lost productivity due to increased morbidity and early death. Clearly, there are many reasons for modern society to be concerned about cancer.

The disease we call *cancer* is actually a family of diseases having the common characteristic of uncontrolled cell growth. In normal tissue, there are a myriad of regulatory signals that instruct cells when to replicate, when to enter a resting state, and even when to die. In a cancer cell these regulatory mechanisms become disabled and the cell is allowed to grow and replicate unchecked. Cancer is largely a disease of aging. The overwhelming majority of cancers are first diagnosed when patients are well over the age of 50. Carcinogenesis, or the sequence of events leading to cancer, is a multistep process involving both intrinsic and extrinsic factors. We know this because certain individuals inherit a genetic predisposition to certain types of cancer. The majority of cancers, however, are not associated with any particular inheritance pattern. Still, many of the same steps have been implicated. These incremental steps typically occur over the span of decades.

At the most fundamental level, cancer is caused by abnormal gene expression. This abnormal gene expression occurs through a number of mechanisms, including direct damage to the DNA and inappropriate transcription and translation of cellular genes. Carcinogenesis has been shown to be induced or at least accelerated by exposure to certain types of chemicals. These chemicals are known as *carcinogens*. In the pages that follow, we will discuss the carcinogenic process and how chemicals can contribute to that process.

This chapter will discuss:

- Tumor classification and nomenclature
- Properties of carcinogenic chemicals
- An overview of the molecular basis of carcinogenesis
- Methods for testing chemicals for carcinogenic activity
- Chemicals identified as human carcinogens
- Risks associated with occupational carcinogens
- Factors that modulate carcinogenic risk

Principles of Toxicology: Environmental and Industrial Applications, Second Edition, Edited by Phillip L. Williams, Robert C. James, and Stephen M. Roberts.
ISBN 0-471-29321-0 © 2000 John Wiley & Sons, Inc.

13.1 THE TERMINOLOGY OF CANCER

Like most scientific disciplines, carcinogenesis has its own language. This section will familiarize you with some of the terms that you will encounter in your study of cancer. Since cancer is a disease that is characterized by uncontrolled or disregulated cell growth, most of the following terms relate to cell growth and differentiation.

Anaplasia. Literally means "without form." Characterized by a marked change from a highly differentiated cell type to one that is less differentiated or more embryonic in nature. Thus, anaplastic tissue is less organized and functional than is the normal tissue. Anaplasia probably occupies the borderline between dysplasia and neoplasia.

Benign. A term applied to neoplasms that are localized and encapsulated. Growth generally occurs via expansion and compression of adjacent tissue. Growth is generally slow, and there may be regression. If the growth is progressive, it is usually orderly and uniform.

Dysplasia. A reversible change in cells, which may include an altered size, shape, and/or organizational relationship. This change usually affects epithelium and often results from chronic irritation.

Hyperplasia. Increased organ or tissue size due to an increase in cell number. Hyperplasia may be physiological (e.g., tissue development and wound healing) or pathological (e.g., nodular liver regeneration in chronic alcoholics). Several hallmarks distinguish neoplasia from hyperplasia:

Neoplasia	Hyperplasia
Growth in excess of needs	Not excessive to needs
Purposeless	Purposeful
Persistent	Ceases when stimulus ceases
Irreversible	Reversible
Autonomous	Regulated

Malignant. A term applied to neoplasms that are locally invasive. Growth may be rapid and is disorderly and progressive. Malignant neoplasms often have areas of necrosis. May spread by extension or metastasis.

Metaplasia. A reversible change in which one differentiated cell type is replaced by another cell type.

Metastasis. Presence of a disease process (usually cancer) at a site distant from the site of origin (the primary tumor). Metastasis (v. *metastasize*, adj. *metastatic*) is the primary hallmark of malignancy.

Neoplasia. Literally means "new growth." Often used synonymously with cancer. The pathologist R. A. Willis offered the following definition, "A neoplasm is an abnormal mass of tissue, the growth of which exceeds and is uncoordinated with that of normal tissues and which persists in the same excessive manner after cessation of the stimuli which evoked the change."

Tumor. A mass or swelling; one of the cardinal signs of inflammation. By common usage, a tumor is specifically a neoplasm.

13.2 CLASSIFICATION OF TUMORS

Neoplasms can be formed in any tissue, and a variety of benign and malignant tumors can occur throughout the body. These tumors are classified using a binomial system based on (1) the tissue or cell type of origin and (2) their actual or predicted behavior (i.e., benign or malignant).

Histogenesis: Tissue Origin

Nearly all tumors arise from either *epithelial* (ectodermal or endodermal) or *mesenchymal* (mesodermal) tissues. Epithelial tissues include lining epithelium (e.g., skin epidermis and the epithelium of the gastrointestinal tract and urinary system) and glands (e.g., pancreas, liver, mammary, prostate, sweat, and sebaceous glands). Mesenchymal or connective tissues include cartilage, bone, muscle, lymphoid, and hematopoietic cells.

Behavior of Neoplasm

The distinction between benign and malignant tumors is extremely important because the malignancy of a tumor is typically what defines human cancer as a disease state. As noted previously, the ability to metastasize is the definitive characteristic of a malignant neoplasm. Not all malignant tumors metastasize (e.g., CNS tumors, intraocular tumors), but no benign tumors do. Metastasis is the major cause of morbidity and mortality associated with cancer. Treatment is much less likely to be successful once metastasis has occurred. In contrast to malignant tumors, benign neoplasms do not grow beyond their boundaries. However, benign tumors can inflict damage by localized obstruction, compression, interference with metabolism, and even secretion of unneeded hormones. Once a benign tumor is removed, so is its harmful influence. Table 13.1 provides further distinction between benign and malignant neoplasms.

Combined Classification

Benign neoplasms, whether arising from epithelial or mesenchymal tissues, are named by adding the suffix *oma* to the tissue type. A benign epithelial neoplasm is an adenoma. This can be made more specific by the addition of modifiers that refer to the tissue of origin (e.g., mammary adenoma). There are many exceptions that are too rooted in common usage to be eliminated. For example, *melanoma* usually refers to a malignancy of melanocytes; the preferred term is *malignant melanoma*. Lymphoma usually refers to a malignancy of lymphoid tissue; the preferred term is lymphosarcoma.

Malignant neoplasms arising from epithelial tissues are named by adding *carcinoma* to the tissue or cell of origin (e.g., thyroid carcinoma). The prefix *-adeno* is added if the tumor is gland forming (e.g., thyroid adenocarcinoma). Malignant neoplasms arising from mesenchymal tissues are named by adding *sarcoma* to the tissue or cell of origin (e.g., fibrosarcoma). Table 13.2 provides several examples of tumors classified using this scheme.

TABLE 13.1 Distinctions between Benign and Malignant Neoplasms

Benign	Malignant
Well differentiated; resembles a cell of origin	Poorly differentiated or anaplastic
Grows by expansion	Grows by expansion and infiltration
Well circumscribed, often encapsulated by a peripheral rim of fibrous tissue	Poorly circumscribed and invades stroma and vessels
Grows at a normal rate	Growth rate usually increased
Few mitotic figures	Frequent mitotic figures
Growth may be limited	Progressive growth
Does not metastasize, seldom dangerous	Metastasis usual; can be fatal
Adequate blood supply	Often outgrows blood supply, becomes necrotic

TABLE 13.2 Some Examples of Tumor Classification and Nomenclature

Tissue (Cell) Origin	Benign	Malignant
Epithelial		
Biliary tract	Cholangioma	Cholangiocarcinoma
Liver cells	Hepatocellular adenoma	Hepatocellular carcinoma
Lung	Pulmonary adenoma	Pulmonary carcinoma
Mammary gland	Mammary adenoma	Mammary adenocarcinoma
Squamous epithelium	Papilloma	Squamous cell carcinoma
Mesenchymal		
Blood vessels	Hemangioma	Hemangiosarcoma
Bone	Osteoma	Osteosarcoma
Fibroblasts	Fibroma	Fibrosarcoma
Hematopoietic cells	No benign form recognized	Leukemia
Fat	Lipoma	Liposarcoma
Striated muscle	Rhabdomyoma	Rhabdomyosarcoma
Smooth muscle	Leiomyoma	Leiomyosarcoma

13.3 CARCINOGENESIS BY CHEMICALS

The following definitions should help clarify subsequent discussions about chemical-induced carcinogenesis.

- A *carcinogen*, as defined in this chapter, is a chemical capable of inducing tumors in animals or humans.
- *Carcinogenesis* is the origin or production of cancer; operationally speaking, this includes any tumor, either benign or malignant.
- A *direct-acting* or *primary carcinogen* is a chemical that is reactive enough to elicit carcinogenic effects in the parent, unmetabolized form. Often these chemicals produce tumors at the site of exposure (e.g., alkylating agents, radiation).
- A *procarcinogen* is a chemical that requires metabolism or bioactivation to another chemical form before it can elicit carcinogenic effects (e.g., polycyclic aromatic hydrocarbons).
- A *cocarcinogen* is a chemical that increases the carcinogenic activity of another carcinogen when coadministered with it. While not carcinogenic itself, the agent may act to increase absorption, increase bioactivation, or inhibit detoxification of the carcinogen administered with it.

Early Epidemiologic Evidence

Some of the earliest evidence that exposure to chemicals could play a role in the development of cancer came from the observations of the British physician Sir Percivall Pott. In 1775, Pott reported on a relationship between scrotal cancer and occupation in men who in their youth, had been employed as chimney sweepers. He suggested that the soot to which these men were exposed on the job played a causal role in the development of their cancer. Over the next century, mounting evidence implicated other chemicals and industrial processes in human cancer. In 1884, Bell and Volkman independently reported on the increased prevalence of skin cancer in workers who were exposed to oils that were distilled from coal and shale. Sporadic reports of other occupational exposures and cancer began appearing in the literature. In 1895, Rehn reported on bladder cancer in aniline dye workers in Germany.

In the early 1900s, radiation was associated with lung tumors in uranium miners and in skin tumors and leukemia in technicians working with the recently discovered X rays.

If it were true that exposure to chemicals could cause cancer in humans, researchers at the beginning of the twentieth century were hopeful that they could be identified and their mechanisms of action studied. Unfortunately, and a bit ironically, early attempts at reproducing cancer in laboratory animals were fruitless. Experimental validation of Pott's original hypothesis finally came in 1915, when the Japanese pathologists Yamagiwa and Ichikawa reported that rabbits developed malignant skin tumors following repeated topical applications of coal tar. The line of research that Pott began was brought full circle when, in the 1930s, a group of investigators led by Cook and Kennaway implicated polycyclic aromatic hydrocarbons as putative carcinogens in coal tar and other industrial oils. They identified benzo[a]pyrene as the first carcinogenic hydrocarbon of known structure isolated from coal tar. Soon afterward, the structures of other chemical carcinogens were identified and their carcinogenic effects replicated in experimental animal models.

The Somatic Mutation Theory

In the early 1900s very little was known about the mechanism of cancer induction by chemicals. Theodor Boveri is often credited with the proposition that cancer involved a permanent alteration of the genetic material in somatic cells. In what became known as the *somatic mutation theory*, he attributed cancer to an "abnormal chromatin complex, no matter how it arises. Every process which brings about this chromatin condition would lead to a malignant tumor." It is important to point out, however, that this theory had as its basis only gross morphological observations of cancer cells. It was only after the pioneering work of Watson and Crick in 1950s, that it became evident that interference with DNA basepairing could be a mechanism by which chemicals induced mutation.

A great deal of evidence has now accumulated in support of the somatic mutation theory (i.e., a genetic mechanism of cancer). Since the early 1970s many carcinogens have been shown to produce permanent, heritable changes in DNA. It has further been shown that these changes are involved in the carcinogenic process. Smart (1994) has described several lines of evidence that support a genetic mechanism for cancer.

- Cancer is a heritable change at the cellular level.
- Tumors are generally clonal in nature.
- Many carcinogens or their activation products can form covalent bonds with DNA and produce mutations.
- The inheritance of certain recessive mutations in genes associated with genomic integrity predisposes affected individuals to cancer.
- Most cancers display chromosomal abnormalities.
- The phenotypic characteristics of a tumor cell can be transferred to a nontumor cell by DNA transfection.

Initiation and Promotion

Since the eighteenth century, investigators have realized that the carcinogenic process involved a period of latency following exposure to chemical carcinogens prior to the appearance of any clinical symptoms. Early attempts at inducing cancer with chemicals in experimental animals met with little success. It was not until the work of Yamigawa and Ichikawa that it was realized that the generation of cancer in experimental animals required long-term repeated exposures to chemicals. With this discovery, researchers quickly developed animal models with which to test the carcinogenic potency of chemicals and mixtures. The most common was the mouse skin model, in which repeated applications of a potential carcinogen were applied to the shaved back of a mouse and the number of tumors generated and the time required for their development was recorded. In the early 1940s, researchers working with rodent skin models discovered that croton oil could stimulate the rapid development of tumors but only if it was applied after treatment with polycyclic aromatic hydrocarbons. Rous and coworkers were the first to use the terms *initiation* and *promotion* to describe two stages of carcinogenesis observed in the experimental induction of skin tumors in rodents. The term *progression* was added later to describe the sequence of events leading to the development of malignant tumors. These three stages are remarkably similar to those described in the development of human skin cancer, induced by soot and paraffin oils.

It is now known that initiation, the first step in this process, involves an irreversible mutation in the DNA of a somatic cell. Chemical initiators are usually electrophiles or metabolically activated to electrophiles (see Chapter 3 for a discussion of metabolic activation). These chemicals bind to nucleophilic centers in DNA, forming DNA adducts. If the DNA is replicated prior to repair of an adduct, a mutation can be "fixed" in the DNA of the daughter cell. This mutation essentially primes the cell for later steps in neoplastic development. Most initiators are mutagens and are thus classified as *genotoxic* carcinogens.

Chemicals that act as tumor promoters may not, by themselves, be carcinogenic. However, if given subsequent to an initiating agent, they increase either the number of tumors or decrease the latency period or both. Tumor promoters typically do not bind DNA; rather, they allow for the clonal expansion of initiated cells by providing a selective growth advantage. For this reason, promoters are considered *epigenetic* (or *nongenotoxic*) carcinogens. For example, the active ingredients of the first tumor promoter, croton oil, are phorbol esters. These compounds mimic endogenous molecules that trigger cell proliferation, thus allowing initiated cells to proliferate. Progression, the third stage of the experimental carcinogenic process, is less well characterized. In general, progression is thought to involve the accumulation of further genetic alterations in a population of initiated cells that have been provided a growth advantage through promotion. These changes ultimately lead to a malignant tumor (Figure 13.1).

A concept that is important in the context of neoplastic progression is that of tumor cell heterogeneity. Investigators studying leukemias and lymphomas have demonstrated that these cancers are almost universally clonal in origin. While there is also evidence of this type of clonal origin in solid tumors (i.e., carcinomas and sarcomas), by the time clinical signs of cancer are evident, the cells that make up tumor have usually developed a certain amount of genotypic and phenotypic diversity. Many researchers believe that the cellular heterogeneity observed in these tumors is the result of genetic instability acquired during tumor progression. Genetic instability suggests that the DNA in tumor cells is mutated at higher rates than the surrounding normal cells rapidly producing subclones. Some of these clones would have adaptations, such as the ability to escape the host's defense mechanisms or invade surrounding tissue that give them a selective advantage. These clones eventually grow to dominate the tumor population. Multiple rounds of this type of selection lead to populations of cells that are increasingly abnormal on a genotypic and phenotypic level and as a result, more aggressive and invasive.

While the initiation-promotion model was first described in a rodent skin model, the process has been experimentally reproduced in other organs such as liver, colon, lung, prostate, and mammary gland. Our experience with both initiators and promoters indicates that many are organ specific. There are, however several features of the initiation–promotion model that remain relatively constant (Figure 13.2).

- Exposure to a sub-threshold dose of an initiator alone results in few, if any tumors.
- Exposure to a sub-threshold dose of an initiator followed by repeated exposure to a promoter results in many tumors.
- Exposure to a promoter will produce tumors even if there has been a latent period following exposure to an initiator. Thus, initiation is irreversible.
- In contrast, if initiation is not followed by promotion of sufficient duration, no tumors are produced. Thus, the effects of tumor promoters are reversible in the early stages.

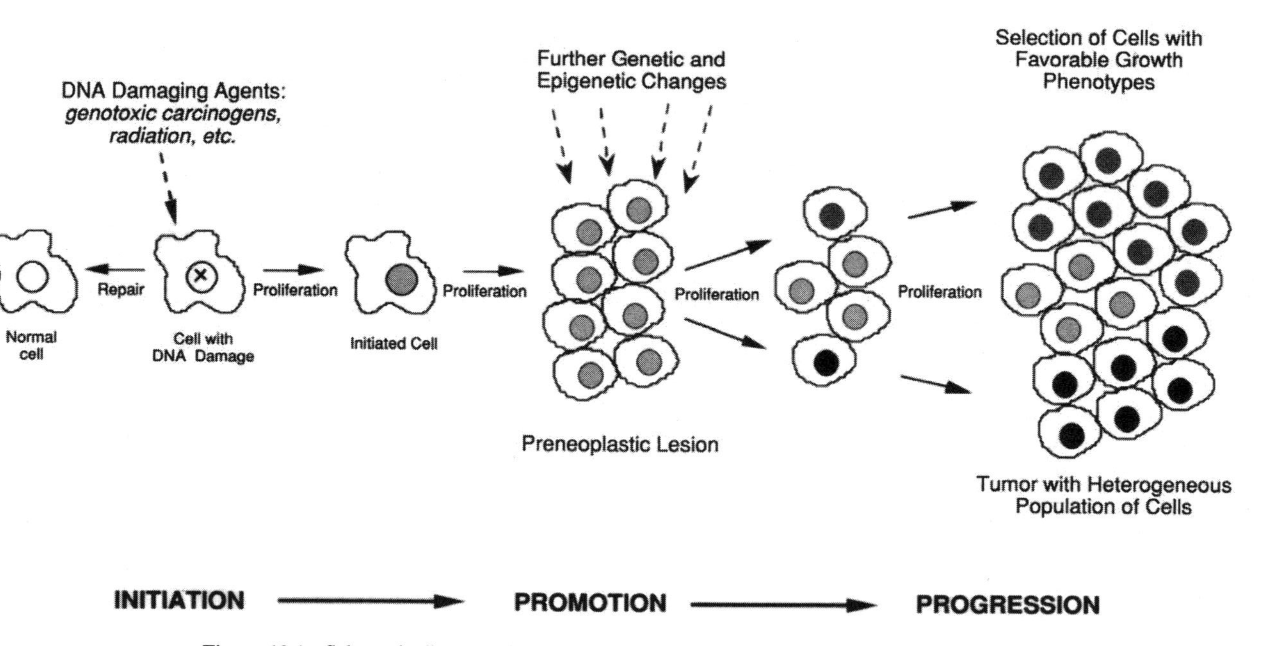

Figure 13.1 Schematic diagram of the development of a malignant tumor from a normal cell.

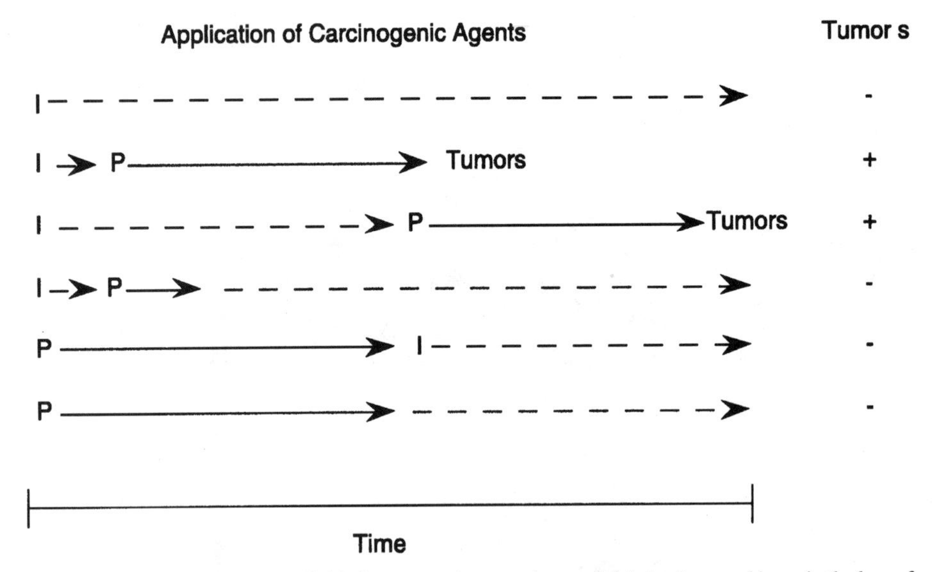

Figure 13.2 Generalized scheme of initiation-promotion experiments. Initiation is caused by a single dose of an initiating agent such as a carcinogenic polycyclic aromatic hydrocarbon; promotion is carried out by repeated application or chronic dosing with a tumor promotor such as TPA. (I, initiator; P, promoter; solid line indicates continual application of agent; dotted lines indicate the duration of time without exposure to an agent.)

- Initiation must occur prior to promotion.
- Repeated exposure to a promoter alone will result in few, if any tumors.

Experiments with initiation–promotion models have demonstrated that there are chemicals that possess both initiator and promoter activity. These chemicals are known as complete carcinogens. By the same token, chemicals that cannot by themselves induce cancer in experimental animal models are called *incomplete carcinogens*. In reality, dosage is a critical factor in determining whether a chemical is a complete carcinogen. At sufficiently low doses, most initiators require subsequent promotion for the development of a tumor, while at very high doses, most carcinogens possess initiating and promoting ability. As you will learn later, this has important implications for the identification and the assessment of risk associated with potential carcinogens.

Electrophilic Theory

The nature of the initiation step in chemical carcinogenesis was the subject of much scientific inquiry and debate for decades. Until 1940, the only known chemical carcinogens were aromatic hydrocarbons and amines. Soon afterward, other aliphatic chemicals were also shown to be carcinogenic and by the 1960s the various chemical carcinogens belonged to over a dozen chemical classes (Figure 13.3). Attempting to explain this structural diversity, in 1969 James and Elizabeth Miller hypothesized that "most, if not all, chemical carcinogens either are, or are converted to, reactive electrophilic derivatives which combine with nucleophilic growth crucial tissue components, such as nucleic acids or proteins." In what became known as the *electrophilic theory of chemical carcinogenesis*, the Millers described the metabolic activation of inactive *procarcinogens* to intermediates they called *proximate carcinogens* and on to *ultimate carcinogens* that covalently bind DNA and cause mutations. Examples of chemical carcinogens that require metabolic activation include benzo[*a*]pyrene and other polycyclic aromatic hydrocarbons, 1,3-butadiene, and 2-acetylaminofluorene. The bioactivation of several chemical carcinogens is illustrated in Figure 13.4.

Carcinogenicity and Mutagenicity

The relationship between mutagenicity and carcinogenicity as it related to the effects of ionizing radiation has been known since the early part of the twentieth century. Sufficient doses of ionizing radiation produce large structural alterations in the genetic material. The first chemical mutagens to be identified were the nitrogen and sulfur mustards (mustard gas). These chemicals were studied because their biological effects were similar to those of ionizing radiation. The nitrogen mustards and other *radiomimetic* compounds have the ability to form crosslinks between strands of DNA or between DNA and proteins. When such lesions are not repaired, large-scale alterations in the DNA can occur. These types of effects are referred to as *clastogenic*, and chemicals that cause them are known as clastogens. The identification of DNA as the genetic material along with the proposition that many chemical carcinogens are, or become, metabolically activated to reactive electrophiles paved the way for the identification of a strong link between chemical mutagenicity and carcinogenicity.

As described in Chapter 12, there are now many short-term tests for the identification of chemical mutagens. The most widely used is the *Salmonella* mutation assay (Ames assay). It is important to note that the combination of this assay with mammalian metabolic activation systems was a critical step in the identification of a large number of chemical carcinogens as mutagens. Analyses conducted by Tennant and Ashby (1991) indicate that there is an approximately 60 percent concordance between carcinogenicity in rodent bioassays and mutagenicity in the *Salmonella* assay. In other words, the *Salmonella* assay predicts whether a chemical will be a carcinogen in rodents approximately 60 percent of the time. These investigators also identified a number of structural "alerts" that are indicative of potential mutagenicity and carcinogenicity. Initially, it had been hoped that such short-term tests would be highly predictive of carcinogenicity in rodents, and by extension, humans. However, much experimental evidence now suggests that there are a large number of chemicals that do not act via a genotoxic mechanism and thus will not be detected by such short-term tests.

Genotoxic and Epigenetic Carcinogens

With the realization that carcinogens elicited their effects via diverse mechanisms, a number of classification schemes based on carcinogenic mechanism were developed. One popular scheme was offered by Williams (Table 13.3). This scheme divides chemical carcinogens into two broad categories based on whether they act in a genotoxic fashion. Accordingly, the two main groups of carcinogenic chemicals are

- Genotoxic carcinogens
- Epigenetic (or nongenotoxic) carcinogens[1]

Genotoxic carcinogens are those chemicals that are capable of modifying the primary sequence of DNA (i.e., initiators). This group includes chemicals that induce mutational and clastogenic changes or changes in the fidelity of DNA replication. Epigenetic carcinogens do not alter the primary sequence of DNA; instead they can affect cell proliferation and differentiation by a number of mechanisms including cytotoxicity and compensatory cell proliferation, receptor mediated events, and by altering the expression or repression of certain genes and cellular events related to cell proliferation and differentiation. It is estimated that at least 40 percent of carcinogens identified by rodent bioassays elicit their affects via an epigenetic mechanism. Many epigenetic agents favor the proliferation of cells with altered genotypes due to an interaction with an initiating carcinogen. While epigenetic carcinogens and tumor promoters share many of the same characteristics, there is some debate regarding whether classic tumor promoters should be considered carcinogens at all. However, since tumor promoters can

[1]The term epigenetic carcinogen may seem confusing since many of these chemicals have the abili ty to alter the regulation of critical genes. This is why some prefer the term *non-genotoxic* carcinogen because it implies that the chemical does not alter the primary sequence of DNA by direct interaction. The two terms are often used synonymously and for our purposes we will use the term epigenetic carcinogen.

Carcinogenic Polycyclic Aromatic Hydrocarbons

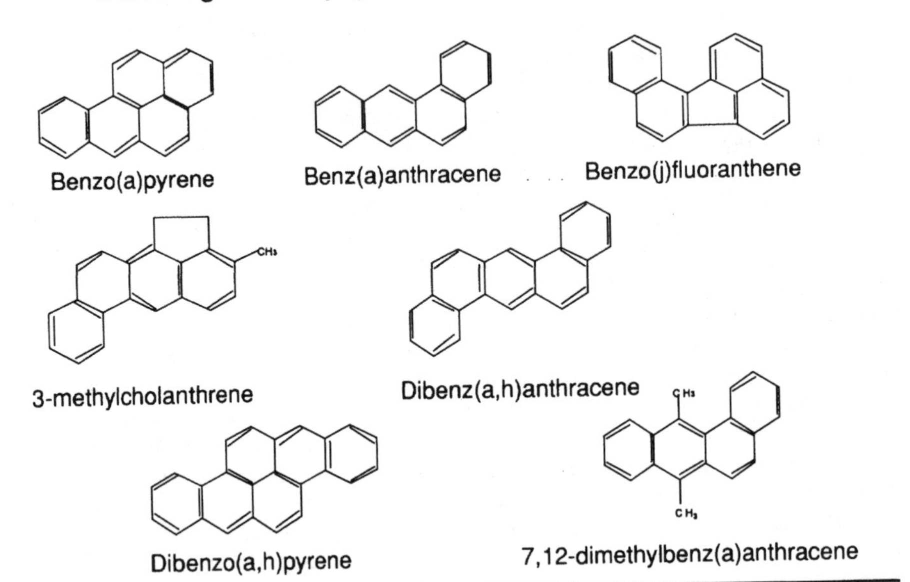

Carcinogenic Aminoazo Dyes

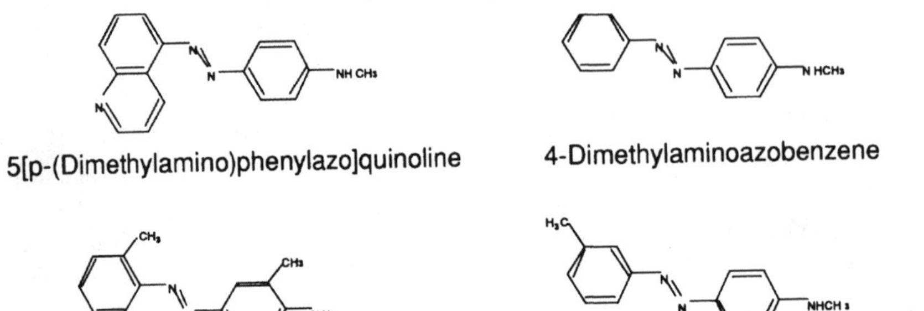

Carcinogenic Alkylating Agents

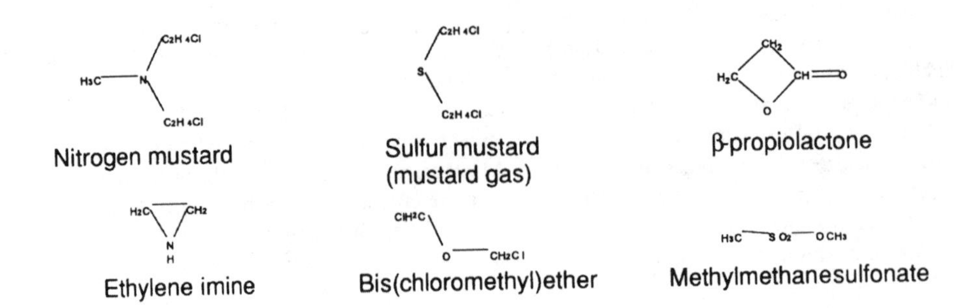

Figure 13.3 Chemical structures of some representative chemical carcinogens. [Adapted from Cancer Biology, Third Edition by Raymond W. Ruddon. copyright © 1981, 1987, 1995 by Oxford University Press, Inc. Used by permission of Oxford University Press, Inc.]

Carcinogenic Aromatic Amines

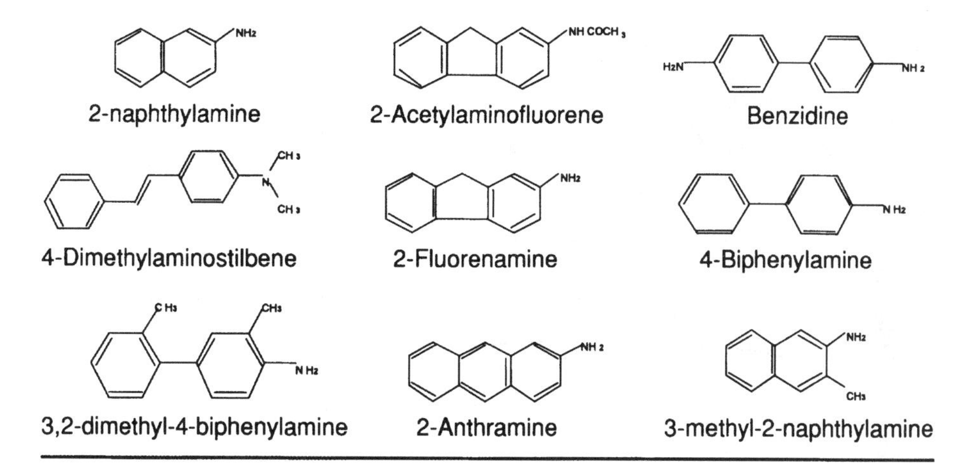

Carcinogenic Nitroso Compounds

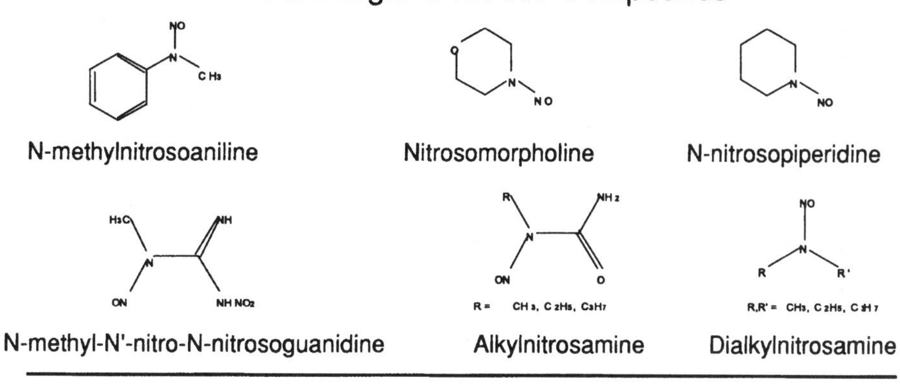

Carcinogenic Natural Products

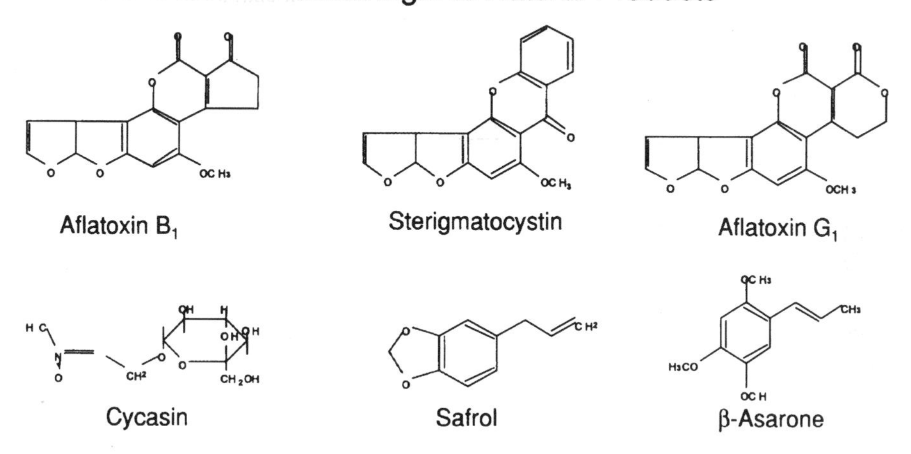

Figure 13.3 (*Continued*)

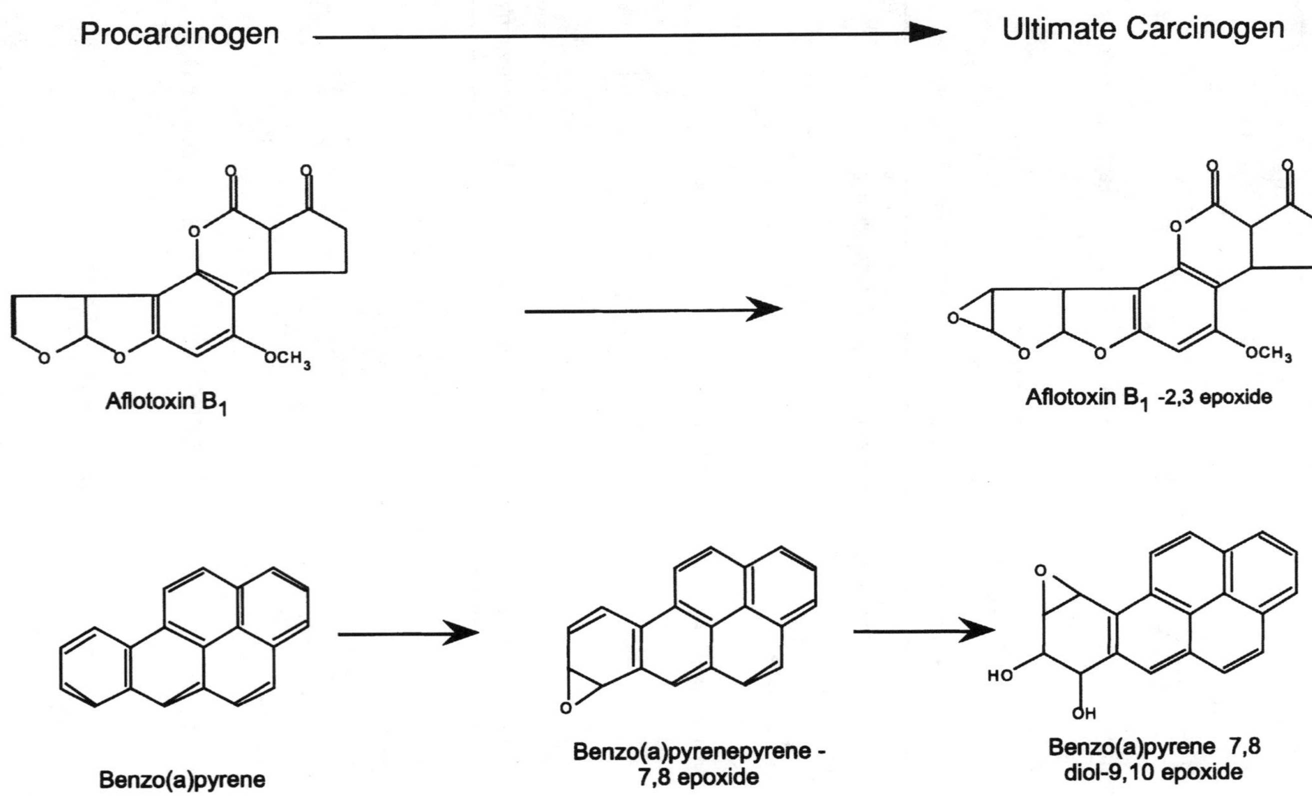

Procarcinogen → Ultimate Carcinogen

Aflotoxin B_1 → Aflotoxin B_1 -2,3 epoxide

OCH_3

Benzo(a)pyrene → Benzo(a)pyrenepyrene - 7,8 epoxide → Benzo(a)pyrene 7,8 diol-9,10 epoxide

HO

OH

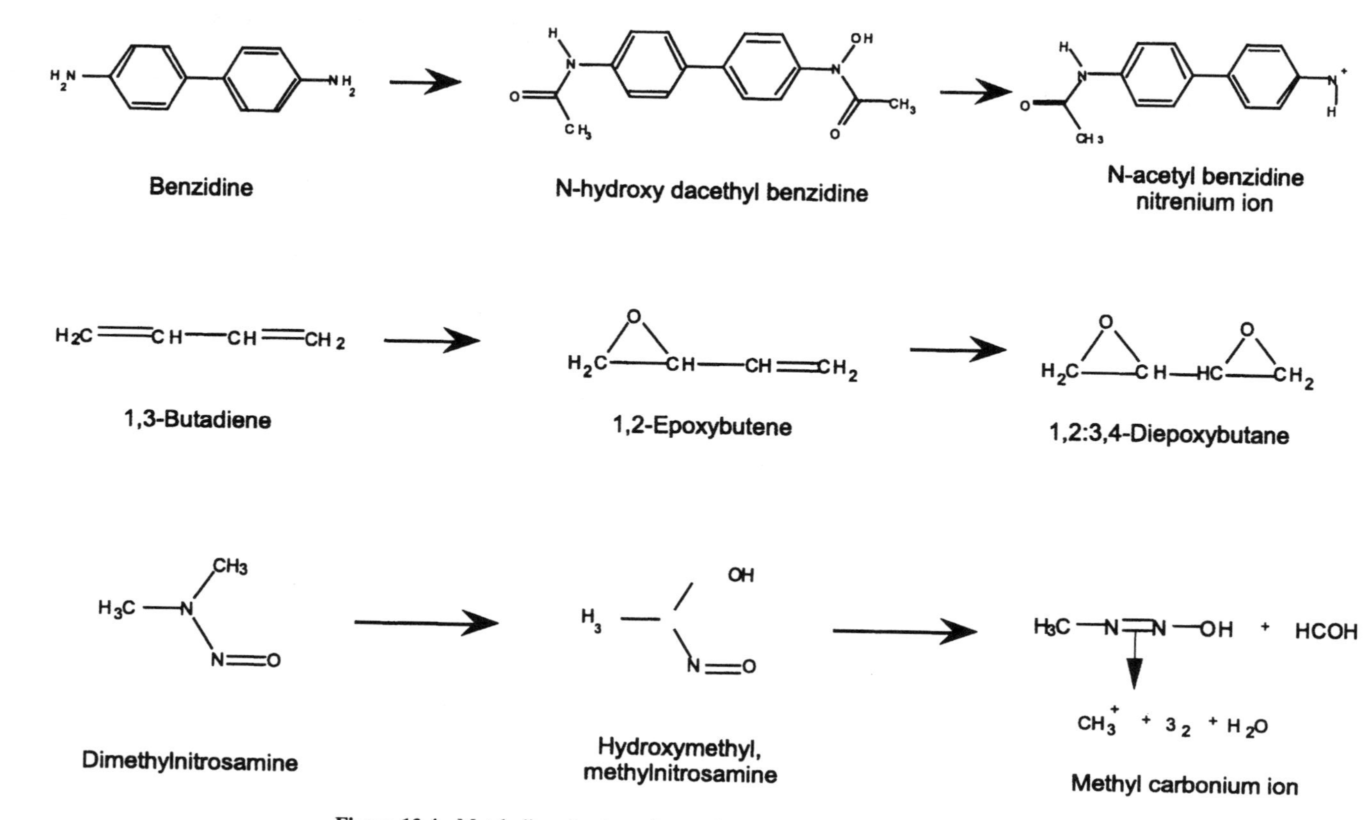

Figure 13.4 Metabolic activation of several representative genotoxic carcinogens.

TABLE 13.3 Classification of Carcinogenic Chemicals

Type	Possible or Probable Mechanism of Action	Examples
Genotoxic carcinogens		
Direct-acting or procarcinogenic	An electrophile, the compound alters the genetic code via mutagenic or clastogenic processes	Bis(chloromethyl) ether, nitrosamines, benzanthracene, epoxides, dimethyl sulfate, nitrosoureas
Inorganic carcinogenic	Alters the fidelity of DNA replication	Cadmium, chromium, nickel
Epigenetic carcinogens		
Solid-state	Mechanical disruption of tissue	Asbestos, metal foils, plastic
Hormonal	Disrupts cellular dedifferentiation, promotes cellular growth	Estrogens, androgens, thyroid hormone, tamoxifen, diethylstilbestrol
Immunosuppressant	Depression of the immune system allows the proliferation of initiated cells or tumors	Azathioprine
Cocarcinogenic	Modifies the response of genotoxic carcinogens when coadministered	Ethanol, solvents, catechol
Promoter	Enhances cell growth, promotes response to initiator or genotoxic carcinogen	Phorbol esters, catechol, ethanol
Cytotoxic	Increases the rate of spontaneous mutation, promotes regenerative cell growth	Trichlorethylene, carbon tetrachloride, chloroform

Source: Adapted from Weisberger and Williams (1981).

affect the proliferation of cells that have spontaneous as well as chemically induced mutations, we will group them with the epigenetic carcinogens.

The mechanism of a chemical's carcinogenic action affects the manner in which it is treated for regulatory purposes. Based on early theories, it was assumed that even a single molecule of a genotoxic chemical could irreversibly damage DNA and that each additional exposure could add to the damage from previous exposures. Thus, regulators initially assumed that there is no safe level of exposure or "threshold" below which harmful effects do not occur. Now, however, many carcinogens, particularly epigenetic carcinogens are thought to elicit their effects in a manner consistent with a threshold.

Epigenetic Mechanisms

Although the precise mechanisms of carcinogenesis by epigenetic chemicals are unknown, progress is being made toward understanding the organ- and species-specific effects of certain classes of epigenetic carcinogens. Some effects of these chemicals appear to be mediated by cytotoxic insult and a compensatory regenerative response while others have been shown to act via receptor-mediated events resulting in the altered transcription of critical cellular genes. Alteration of DNA methylation status such that critical genes are expressed or are inactivated inappropriately is another epigenetic mechanism that is currently receiving increased attention, although as yet, it is uncertain how chemicals might affect this process.

Some chemicals that are not directly genotoxic, but are carcinogenic in chronic rodent bioassays have been shown to exhibit cytotoxic properties. It has been shown that many of these chemicals produce necrosis or cell death due to cytotoxicity at the target organ. This is usually followed by regenerative cell proliferation. The organ- or cell-type-specific effects of carcinogens that induce cytotoxicity may be due to the high concentration of the chemical at that organ or the selective toxicity directed at specific cell populations. Cells in the target tissue could become initiated following an insult with a cytotoxic chemical due to (1) spontaneous mutations produced by defective mitosis or inefficient repair occurring during multiple rounds of DNA replication during regenerative cell proliferation or (2) generation of DNA damage by oxygen free radicals, produced by lipid peroxidation or by recruited inflammatory cells. Such initiated cells would then have a selective growth advantage due to the production of various stimuli (e.g., growth factors) produced by proliferating cells. Examples of chemicals that may work via these types of mechanisms include chloroform, carbon tetrachloride, and saccharine.

There is an increasing amount of evidence that suggests that many epigenetic carcinogens activate receptors and as a result, elicit changes in the expression of critical target genes involved in cellular functions ranging from signal transduction, cell proliferation, and differentiation to apoptosis and cell-to-cell communication. Some of these chemicals and their receptors are shown in Table 13.4. The changes in gene expression induced by certain epigenetic carcinogens may mimic the effects of endogenous growth factors and hormones that similarly affect these cell functions. The effects of natural hormones, growth factors, and dietary constituents on promotion of neoplasia suggest that endogenous tumor promotion and epigenetic carcinogens have common links. The tissue-specific effects of some chemicals may be due to the predominance of particular signaling pathways in a given cell or tissue type. Because of the increased understanding of nongenotoxic mechanisms of action of epigenetic carcinogens, endpoints such as proliferation, differentiation, apoptosis, cell-to-cell communication, and the induction of gene expression have all been used in the assessment of epigenetic carcinogens. Experimental evidence that some of these compounds elicit their effects via receptors provides a solid basis for the contention that there exists a threshold level under which epigenetic carcinogens would not exert carcinogenic activity.

Another epigenetic mechanism that received more attention in the late 1990s is the alteration of DNA methylation patterns in neoplastic cells. Most cells in the body contain the same genetic information. Yet somehow, different cell types express only a subset of that genetic code that is required for the cell to function properly. Cell differentiation is almost always achieved without altering the primary sequence of DNA, yet the phenotypic characteristics of the cell are usually stable and can be passed on to daughter cells during cell division. Much of the control of the gene expression that ultimately determines cell phenotype is maintained by the addition of methyl groups to the 5′ carbon of cytosine residues in cellular DNA, particularly at CpG dinucleotide sequences. The promoter and enhancer elements of many genes have regions high in CpG dinucleotides. There is an inverse correlation between gene expression levels and the degree of methylation in these regions. That is, actively transcribed genes have low levels of methylation (hypomethylation) in their promoter regions while transcriptionally silent genes have heavily methylated (hypermethylation) promoter regions. It has been hypothesized that changes in methylation status play an important role in the neoplastic progression of tumor cells. Once in place, changes in methylation status could be passed to daughter as permanent epigenetic changes. There is evidence for this type of mechanism in the inactivation of the p16 tumor suppressor gene in human tumors by hypermethylation of the promoter region. It is still unclear how epigenetic carcinogens might affect this mechanism of gene regulation, but it has become increasingly apparent that alteration of DNA methylation patterns do play a role in the progression of some tumors.

TABLE 13.4 Some Epigenetic Carcinogens and the Receptors They Activate

Chemical	Receptor
Tetrachloro dibenzo-*p*-dioxin (TCDD)	Ah receptor
12-*o*-Tetradecanoylphorbol-13-acetate (TPA)	Protein kinase C
Peroxisome proliferating compounds	Peroxisome proliferator-activated receptor (PPAR)
Estrogenic compounds	Estrogen receptor
Okadaic acid	Protein phosphatase-2A

13.4 MOLECULAR ASPECTS OF CARCINOGENESIS

To this point we have discussed some of the critical observations that have contributed to our understanding of the roles chemicals play in the carcinogenic process. From these observations we know that a majority of chemical carcinogens produce mutations in DNA, presumably in critical genes. We also know that there are other chemicals that alter cell growth and differentiation through epigenetic mechanisms. Until relatively recently, very little was understood about the nature of genes involved in carcinogenesis. As you will learn, the discovery of oncogenes and tumor suppressor genes as positive and negative regulators of cell growth, respectively, unified many of the earlier observations and theories about the nature of the carcinogenic process. These discoveries have been followed by further refinements in our understanding of the molecular basis of cell growth and differentiation and how these processes are subverted in the development of cancer.

Oncogenes

Thus far, our study of carcinogenesis has focused on information gathered by researchers studying the induction of cancer by chemicals. There was however, another group of cancer researchers who believed that infectious agents, namely viruses were actually the cause of cancer. While we now know that this is not the case for the overwhelming majority of human cancer, the work of these investigators has provided us with some of the most important information on the molecular mechanisms of cancer. Peyton Rous and his coworkers discovered the first known tumor virus in 1909. They demonstrated that virus particles extracted from a sarcoma in one chicken could produce similar tumors when injected into other chickens. The Rous sarcoma virus as it became known was eventually shown to be an *oncogenic* (from the Greek word *onkos*—mass or swelling) retrovirus. In the decades that followed, many other oncogenic retroviruses were discovered. These viruses were active in other avian species, rodents, and even primates.

Retroviruses encode their genetic material in RNA rather than DNA like other organisms. The genome of the retrovirus is limited to several genes that are critical to the production of more virus particles. Following infection, the viral RNA is transcribed into DNA by the enzyme reverse transcriptase. The newly created double stranded DNA integrates into the DNA of the host cell, where the strong promoter sequences of the virus induce the host cell's nuclear machinery to express the viral genes and produce new viral particles. Some oncogenic retroviruses, such as the Rous sarcoma virus, have the ability to rapidly "transform" normal cells into cancer cells. Researchers working with these so called *acute transforming retroviruses* hypothesized that they contained a one or more genes responsible for their rapid transforming ability.

Eventually, a gene responsible for the transforming ability of the Rous sarcoma virus was discovered. It was named *src* for sarcoma, and it was the first *oncogene* or gene capable of inducing cancer identified. In the years that followed, oncogenes from other acute transforming retroviruses were identified, each of these genes was named with a three-letter identifier that corresponded to the virus from which they were first isolated: *ras* from the rat sarcoma virus, *myc* from the avian myelocytomatosis virus, *sis* from the simian sarcoma virus, and *fes* from the feline sarcoma virus. It was ultimately demonstrated that the oncogenes responsible for the transforming ability of the oncogenic retroviruses had normal, highly conserved counterparts in the cells of a wide variety of prokaryotic and eukaryotic organisms. It stood to reason that if these genes had been conserved over so many years of evolutionary history, they must play an absolutely critical role within the cell. These normal cellular genes appear to have been transduced or in essence "stolen" by retroviruses. Following transduction, the cellular genes became "activated," that is, altered in ways that made them oncogenic. The method of activation was not the same for each oncogene. Some like *erb* B were activated by a deletion of several hundred basepairs of DNA from the coding region of the gene, some like *myc* were activated by gene amplification, while others like *ras* were activated by a single-point mutation. The normal cellular counterparts to the oncogenes of the transforming retroviruses were called *protooncogenes*. Subsequently protooncogenes were shown to be activated in a number of human tumors. Many lines of experimental evidence converged when it was shown that oncogenes that had been activated by mutation with a carcinogenic chemical could transform normal cells into cancer cells. We now know that the same oncogenes are often activated in tumors of the same cell type, whether the tumors arose spontaneously or were virally or chemically induced.

A word about nomenclature should be mentioned here. The student may encounter several notations associated with the names of oncogenes (*oncs*). The notation *v-onc* (e.g., *v-sis*) is used to distinguish an oncogene of viral origin from a similar oncogene of cellular origin (*c-onc*, e.g., *c-sis*). It should also be noted that not all oncogenes have been discovered in transforming retroviruses.

Oncogenes and Signal Transduction Pathways

To date, approximately 75 cellular oncogenes and their protooncogene counterparts have been identified. The protein products of nearly all of these genes function in one way or another in cellular signal transduction pathways to precisely regulate cell growth and differentiation. Signal transduction pathways are used by cells to receive and process information and ultimately to effect a biological response. These pathways generally consist of external signaling molecules, receptors on the cell surface, transducer proteins, second messenger proteins, amplifier proteins, and effector proteins such as transcription factors, all of which are involved in the regulation of cellular function or gene expression. A generalized signal transduction pathway is shown in Figure 13.5. The protein products of oncogenes have been grouped according to their function in several different categories (Table 13.5). These categories include growth factors, growth factor receptors, membrane-associated GTP binding proteins (G proteins), nonreceptor tyrosine kinases, cytoplasmic serine/threonine kinases, and nuclear transcription factors. Activation of a protooncogene confers a gain of function to the gene product in the sense that its ability to promote cell proliferation is enhanced. Mechanisms of activation of

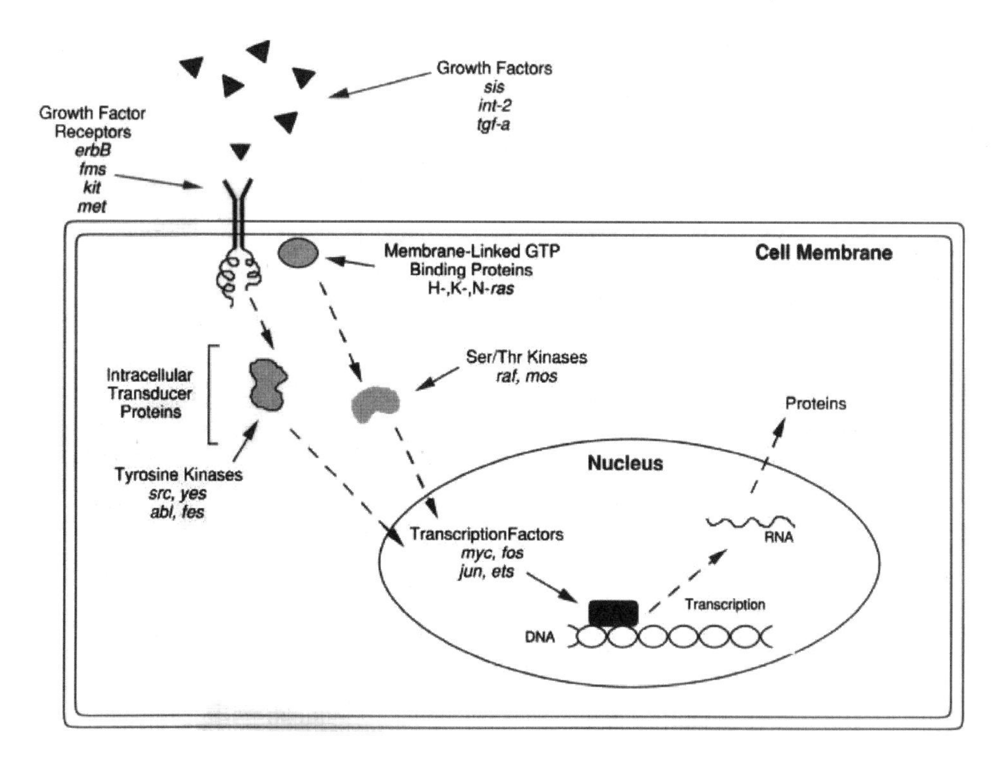

Figure 13.5 Schematic diagram of growth factor-mediated signal transduction pathways.

TABLE 13.5 Selected Protooncogenes and the Functions of Their Encoded Proteins

Oncogene Name	Function
	Growth Factors[a]
sis	Platelet-derived growth factor
int-2	Fibroblast growth factor
tgf-α	Transforming growth factor-α
	Growth Factor Receptors[b]
*erb*B	Epidermal growth factor receptor (tyrosine kinase)
fms	Colony stimulating factor receptor (tyrosine kinase)
kit	Stem cell receptor (tyrosine kinase)
met	Hepatocyte growth factor receptor (tyrosine kinase)
	GTP Binding Proteins (G proteins)[c]
H,N,K-*ras*	Membrane-associated GTP binding/GTPase
	Nonreceptor Tyrosine Kinases[d]
src	Membrane associated—mediates integrin signaling
yes	Membrane associated
abl	Cytoplasmic with nuclear translocation ability—DNA binding and DNA transcription activation
fes	Cytoplasmic
	Cytoplasmic Serine/Threonine Kinases[e]
raf	Phosphorylates MAPKK proteins in cell signaling
mos	Activates and/or stabilizes maturation promoting factor (MPF)
	Nuclear Transcription Factors[f]
myc	Sequence-specific DNA binding protein (transcription factor)
fos	Combines with *jun* to form AP1 transcription factor
jun	Combines with *fos* to form AP1 transcription factor
ets	Transcription factor

[a]These are secreted factors that typically act in an autocrine or paracrine fashion.

[b]Normally these receptors are transiently activated by ligand binding. Mutant forms are peristently activated.

[c]Numerous growth factor receptors normally signal through GTP binding proteins. These protins transiently activated in response to ligand binding at the receptor. Mutant forms are persistently activated.

[d]These are cytoplasmic proteins involved in the relay of signals from growth factor receptors and from the extracellular matrix through cytoskeletal proteins. Activation requires differential, transient phosphorylation of tyrosine residues. Mutant forms are persistently activated.

[e]These are another group of cytoplasmic proteins involved in the relay of signals to the cell nucleus. Activation requires differential, transient phosphorylation at serine and threonine residues. Mutant forms are persistently activated.

[f]These proteins are localized primarily to the cell nucleus; where they function to transcriptionally activate and repress genes associated with cell growth and differentiation.

protooncogenes that commonly occur in human tumors include point mutation, gene rearrangement, gene amplification, chromosomal translocation, and increased transcription (Table 13.6).

Tumor Suppressor Genes

Demonstration of the existence of cellular oncogenes and knowledge of their function as positive regulators of cell growth provided an obvious mechanism by which chemicals could induce the carcinogenic process. The thinking was that an activated oncogene could force the cell and its descendants into unneeded rounds of division ultimately resulting in a tumor. However, there was a problem with such a simplistic view. Researchers soon demonstrated that when tumor cells were fused with normal cells, the resulting hybrid cells were usually nontumorigenic. Thus the transforming ability of oncogenes could be reversed or controlled by some other factor produced by normal cells. It was eventually discovered that normal cells carried genes that coded for proteins that function as negative regulators of cell growth. These genes came to be called *tumor suppressor genes*. There now exists much evidence supporting the existence of tumor suppressor genes and their functions as negative regulators of cell growth. To date, approximately 20 putative tumor suppressor genes have been identified, although, for many of these, a function is not well understood. Like the oncogenes, the products of tumor suppressor genes appear to have diverse functions within the cell. These functions include cell cycle control, transcriptional regulation, regulation of signal transduction, maintenance of cellular structure, and DNA repair. Some tumor suppressor genes and the functions of the proteins they encode are shown in Table 13.7. In contrast to the situation with oncogenes where a mutation in only one allele is often transforming, the inactivation of tumor suppressor genes requires two genetic events, that is, the inactivation of both alleles. The mechanism most commonly invoked in tumorigenesis is a mutation in one allele followed by a subsequent deletion of the second allele or replacement of the second allele with a copy of the mutated allele, resulting in what is commonly known as loss of heterozygosity (LOH).

Tumor suppressor genes are often linked to rare, inherited forms of cancer. In fact, the existence of tumor suppressor genes had been suggested as early as 1971 when Knudson forwarded the "two hit" hypothesis, in which he proposed that the development of retinoblastoma, a rare tumor of the eye in children, required two genetic events. His work eventually led to the cloning of the retinoblastoma

TABLE 13.6 Oncogenes Activated in Human Tumors

Oncogene	Neoplasm(s)	Lesion
abl	Chronic myelogenous leukemia	Translocation
erbB-1	Squamous cell carcinoma; astrocytoma	Amplification
erbB-2 (neu)	Adenocarcinoma of the breast, ovary, and stomach	Amplification
gip	Adenocarcinoma of the ovary and adrenal gland	Point mutations
gsp	Thyroid carcinoma	Point mutations
myc	Burkitt's lymphoma	Translocation
	Carcinoma of lung, breast, and cervix	Amplification
L-*myc*	Carcinoma of lung	Amplification
N-*myc*	Neuroblastoma, small cell carcinoma of lung	Amplification
H-*ras*	Carcinoma of colon, lung, and pancreas; melanoma	Point mutations
K-*ras*	Acute myelogenous and lymphoblastic leukemia; thyroid carcinoma, melanoma	Point mutations
N-*ras*	Carcinoma of the genitourinary tract and thyroid; melanoma	Point mutations
ret	Thyroid carcinoma	Rearrangement
K-*sam*	Carcinoma of stomach	Amplification
trk	Thyroid carcinoma	Rearrangement

TABLE 13.7 Tumor Suppressor Genes in Human Cancer and Genetic Disease

Gene	Consequence of loss	Function of encoded protein
Rb	Retinoblastoma and osteosarcoma	Binds and sequesters the transcription factor E2F to maintain cells in G_0 of cell cycle
p53	Li-Fraumeni syndrome inactivated in >50% of human cancers	Transcription factor with multiple functions, including cell cycle progression, detection of DNA damage, and apoptosis
p16	Familial melanoma, pancreatic cancer	Inhibits CDK4 to block cell cycle progression
Wt1	Wilms' tumor/nephroblastoma	Transcription factor required for renal development
VHL	Von Hippel–Lindau syndrome renal cell carcinoma	Negative regulation of hypoxia-inducible mRNAs
NF1	Neurofibromatosis type 1 schwannoma and glioma	GTPase-activating protein (GAP), which regulates signaling through *ras*
NF2	Neurofibromatosis type 2 acoustic nerve tumors and meningiomas	Connects cell membrane proteins with the cytoskeleton
BRCA1	Familial and sporadic breast and ovarian cancer, also prostate and colon cancers	Secreted growth factor
BRCA2	Breast cancer (female and male) also prostate cancer	Unknown function
DCC	Colon cancer	Cell adhesion molecule
APC	Familial and sporadic adenomatous polyposis colorectal tumors	Interacts with catenins, proteins involved in signaling pathway for tissue differentiation
MMR	Hereditary nonpolyposis colorectal cancer	Mediates DNA mismatch repair

gene (Rb) and the discovery that both copies of the gene are inactivated and/or deleted in retinoblastoma tumors. It is now known that a large proportion of persons with retinoblastoma have inherited a defective copy of the Rb gene. Tumors develop when the second copy is inactivated prior to the terminal differentiation of the retinoblasts. Another group of retinoblastoma patients do not have a defective copy of the Rb gene. In this group, two somatic mutations have occurred sometime after conception. Individuals born with a mutated copy of Rb gene are also at a higher risk of developing other cancers, most notably osteosarcoma, later in life. A number of the known or putative tumor suppressor genes appear to be involved in a relatively small subset of tumors specific to certain tissue types. These include Wt-1 (Wilms' tumor), NF-1 and NF-2 (neurofibromatosis types 1 and 2), APC (adenomatous polyposis coli), and DCC (deleted in colon carcinoma). In contrast to these, the p53 tumor suppressor gene, is inactivated in more than 50 percent of all human tumors. The p53 protein is a remarkable protein that is involved in diverse cell functions including the detection of DNA damage, the regulation of cell cycle progression, and the induction of apoptosis or programmed cell death. Rb and p53 will be discussed briefly below as well as in the context of the cellular functions in which they are involved.

The p53 gene and the protein it encodes has been called the "guardian of the genome" in recognition of the critical role it plays in the life and death of cells. The p53 gene is considered to be the most frequently mutated gene in human tumors. Approximately 40 percent of breast cancers, 70 percent of colon cancers, and 100 percent of small cell lung cancers contain mutations in the p53 gene. The p53 gene encodes a 53-kD nuclear phosoprotein that is active in regulating the transcription of a number of genes relating to cell cycle progression and apoptosis. Levels of p53 are increased in response to several types of cell stress, including DNA damage, hypoxia, and decreases in the levels of nucleotide triphosphates required for DNA replication. The p53 protein has been shown to have a direct role in the detection of DNA damage by some chemical carcinogens and radiation. In the presence of DNA damage p53 has the ability to slow cell cycle progression or bring the cycle to a halt until the damage can be repaired. In the face irreparable damage p53 has been shown to initiate the events leading to apoptosis.

The Rb gene encodes a 107-kD nuclear protein that plays a critical role in the early stage of the cell cycle. When the cell is stimulated by a growth factor, that signal is ultimately relayed to the nucleus. This signal results in the production of proteins that temporarily inactivate Rb. In its active form, the Rb protein is tightly bound to an important transcription factor, E2F. When the cell receives a signal to divide, Rb is hyperphosphorylated, causing a conformational change and the release of E2F. The transcription factor E2F induces the production of other proteins involved in cell cycle progression.

The Cell Cycle and Apoptosis

It is important to discuss some of the processes that govern the life and death of cells to better understand how oncogenes and tumor suppressor genes are involved in these processes. As pointed out previously, proto-oncogenes function in various capacities in the transduction of signals for cell growth and differentiation within and between cells. In normal cells, replication of the DNA and cell division is stimulated by the presence of growth factors that bind receptors at the cytoplasmic membrane and initiate a cascade of intracellular signals. Once these signals reach the nucleus they cause the transcription of a complex array of genes, producing proteins that mediate progression of the cell through the cell cycle culminating in mitosis or cell division.

The cell cycle is divided into five phases (Figure 13.6). The length of each of these phases can vary depending on factors such as cell type and localized conditions within the tissue. After completing mitosis (M), daughter cells enter the Gap 1 (G_1) phase. If conditions are favorable, cells enter the synthesis (S) phase of the cycle, where the entire genome of the cell is replicated during DNA synthesis. Following S phase, cells enter the Gap 2 (G_2) phase before proceeding through mitosis again. There is a critical boundary early in G_1 called the *restriction point.* This is the point at which the cell must

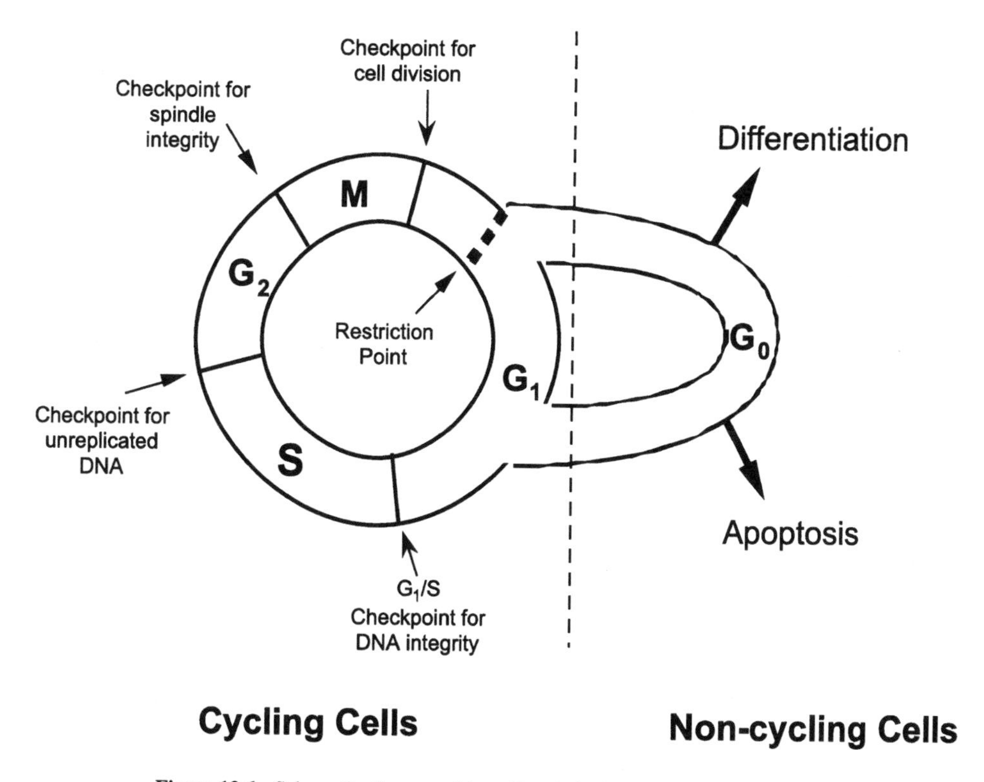

Figure 13.6 Schematic diagram of the cell cycle including primary checkpoints.

make a decision to (1) enter the cell cycle again or (2) move into a state of quiescence also known as G_0 phase. Once in G_0 phase, the cell can either remain in a state of replicative quiescence until it receives a signal to divide again or it can proceed down a path that leads either to terminal differentiation or to apoptosis.

Movement of the cell through the cell cycle is controlled by an enormously complex network of proteins many of which are expressed in a phase-specific fashion. Several major groups of these proteins have been studied to date. These include cyclins, cyclin-dependent kinases (CDKs), cyclin-activating kinases (CAKs), and CDK inhibitory proteins. The cyclins and CDK proteins are categorized by the stage of the cell cycle in which they are the primarily active. The binding of the appropriate growth factor at the cell surface starts a signaling cascade that ultimately leads to the expression of the G_1 phase cyclins. These cyclins combine with appropriate CDKs to form a complex that inactivates the Rb protein. As mentioned previously, in its active form, RB binds the transcription factor E2F. When Rb is inactivated by a cyclin/CDK complex in G_1 phase of the cell cycle E2F is released to transcribe genes necessary for continued progression through the cell cycle. It is important to note that in normal cells, external factors (e.g., growth factors) are absolutely required for the cell to continue past the restriction point. After the restriction point, the cell is committed to DNA replication and cell division. Thus, the interference with normal signal transduction pathways by chemical carcinogens, regardless of mechanism, can force a cell into proliferation that is not governed by normal physiological controls.

Even after passing through the restriction point early in G_1 phase and committing to replication, there are still multiple mechanisms through which the cell regulates progression through the cell cycle. For example, the cell must pass through what is known as a "checkpoint" at the G_1/S boundary. The G_1/S checkpoint serves to insure that DNA has been sufficiently repaired before new DNA is synthesized. The p53 protein plays a critical role at the G_1/S checkpoint. There is evidence that p53 is directly involved in the detection of several types DNA damage. Upon detecting damage, p53 regulates the production of proteins that function to bring a halt to the cell cycle. There is also evidence to suggest that p53 actually mediates the repair of certain genetic lesions by DNA repair enzymes. Once the damaged DNA has been sufficiently repaired, the cell proceeds with the synthesis of new DNA. In this phase of the cell cycle, alterations in the fidelity of DNA synthesis or inefficient repair of replication errors could have detrimental effects on the cell. Following S phase, cells pass through another checkpoint to ensure that the DNA has been fully replicated before moving into the G_2 phase. During this phase, the cells where prepares for mitosis by checking the DNA for replication errors and ensuring that the cellular machinery needed in mitosis is functioning properly. Following G_2, the cells undergo mitosis and a daughter cell is created. Any errors in the made in the replication of the DNA of the original cell are now fixed in the DNA of the daughter cell.

If a cell has sustained an unacceptable level of DNA damage, or in situations where the cell receives irregular growth signals, such as in the overexpression of the transcription factor and protooncogene *myc*, p53 can mediate a process called *apoptosis*. Simply put, apoptosis is cell suicide. Apoptosis is an extremely important component of many physiological processes relating to growth and development. In the developing embryo, for example, apoptosis is responsible for the elimination of superfluous cells that must be eliminated to ensure proper tissue structure and function (e.g., digit formation in developing limbs). Apoptosis is also responsible for the maintenance of the correct number of cells in differentiated tissues and the elimination of cells that have been irreparably damaged. Apoptosis is an orderly process characterized by several morphological stages, including chromatin condensation, cell shrinkage, and the packaging of cellular material into apoptotic bodies (also known as blebing) that can be consumed by phagocytes in the vicinity of the cell. This orderly and well-regulated process is a distinct contrast to cell death by necrosis. As indicated previously, the p53 protein has been implicated in apoptosis resulting from several different types of cell stress, including DNA damage induced by chemical mutagens (Figure 13.7). The mechanisms by which p53 mediates apoptosis are currently a subject of intensive study for cell biologists. Some functions of p53 in the apoptotic pathway are mediated by the transcription of certain genes (e.g., *bax*) that regulate apoptosis, while other effects appear to stem from protein–protein interactions with other intercellular mediators of apoptosis.

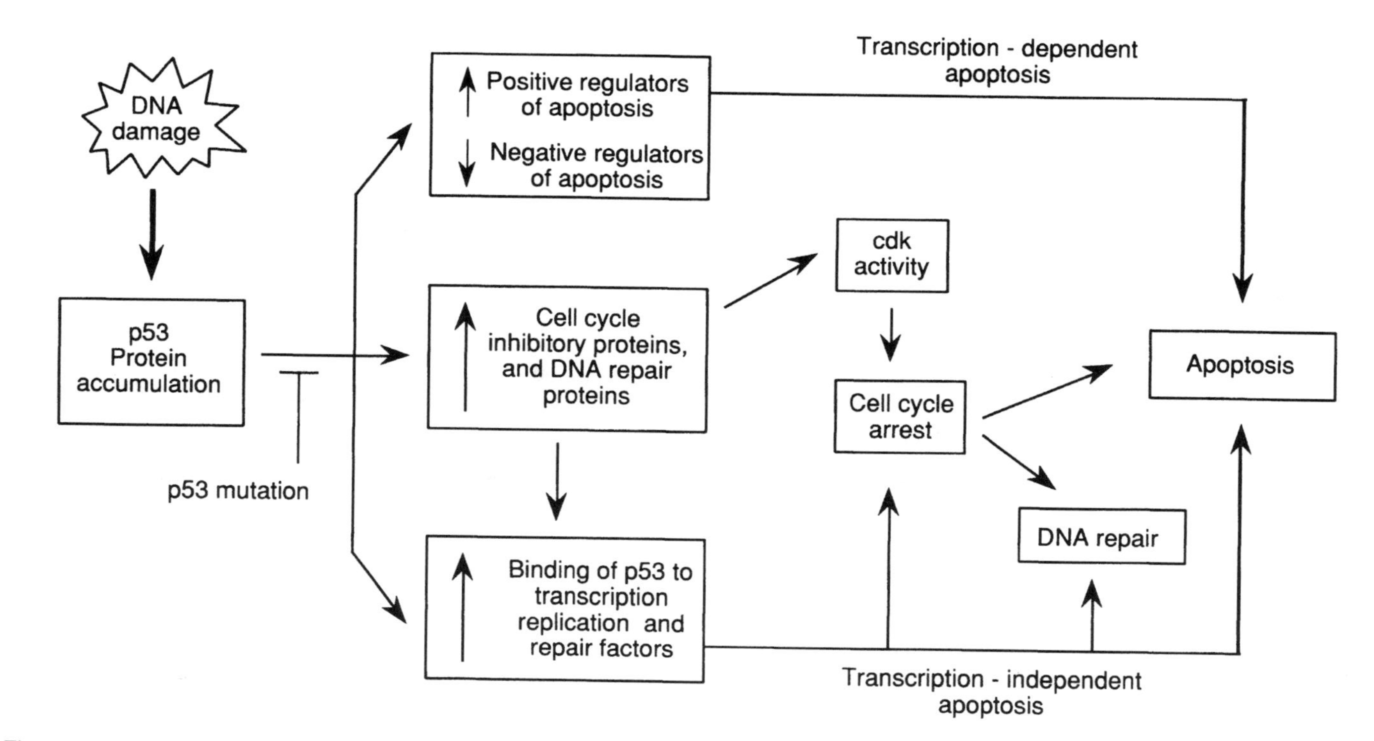

Figure 13.7 DNA damage leads to p53 accumulation and subsequent changes in gene expression and protein-protein interactions [Adapted from Harris (1996).]

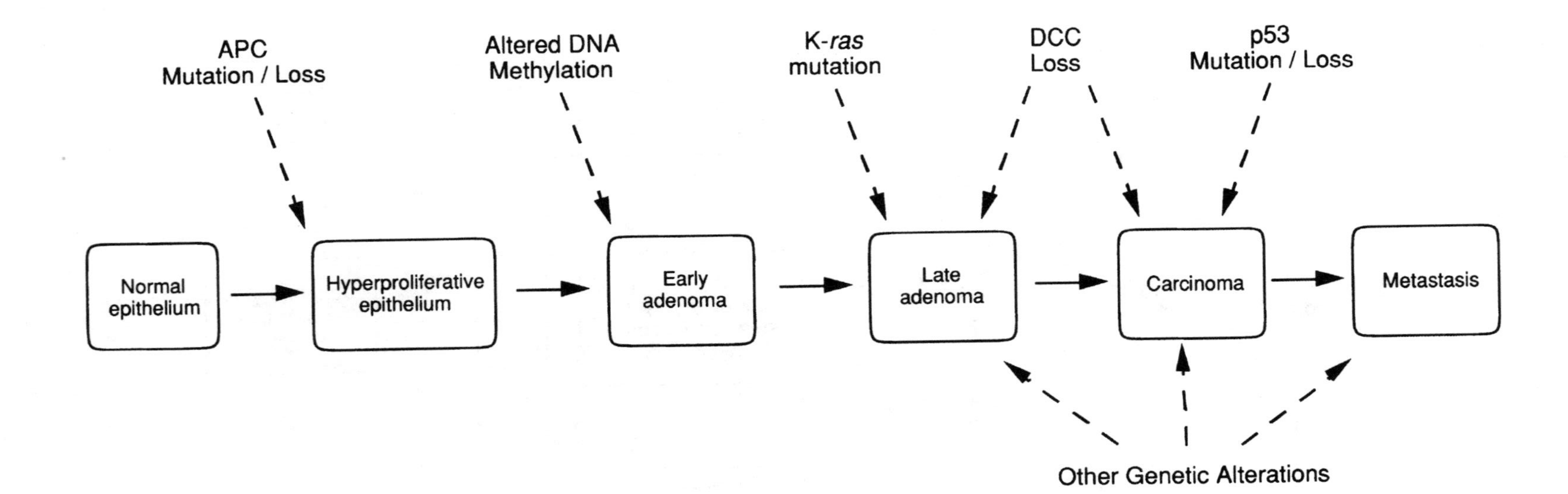

Figure 13.8 A genetic model of the molecular events involved in human colorectal tumor development. [Adapted from Trends in Genetics, Volume 9. Vogelstein and Kinzler, The multistep nature of cancer. p. 140, 1993. with permission from Elsevier Science.]

Clearly apoptosis is a process critical to the balance of cell populations in normal tissue. The loss of the ability for neoplastic cells to undergo apoptosis could tip the scales in favor of cell proliferation and uncontrolled growth. As such, apoptosis is another process that could be detrimentally affected by the loss of p53 function.

This has been a brief and necessarily simplistic overview of some of the cellular processes that can be subverted in the course of the carcinogenic process. It should be evident from this discussion that the formation of a malignant tumor is a multistage process that involves multiple molecular mechanisms, including the activation of oncogenes and the inactivation of tumor suppressor genes (Figure 13.8). What is not often clearly articulated to the student of chemical carcinogenesis is the fact that while different types of chemical carcinogens (e.g., genotoxic and epigenetic) differ mechanistically, these mechanisms have impacts on similar molecular pathways. While there is still much work to be done, the knowledge that has been developed over a relatively short time regarding the mechanisms of action of chemical carcinogens and critical cellular targets for these agents is astounding. This knowledge, has given us valuable insights to the origins of human cancer and will lead to the development of better tools with which to fight it.

13.5 TESTING CHEMICALS FOR CARCINOGENIC ACTIVITY

The General Chronic Animal Bioassay Protocol

The National Toxicology Program (NTP) is the agency currently responsible for testing chemicals for carcinogenic activity in the United States, a responsibility originally held by the National Cancer Institute. But while the responsibility for testing chemicals has changed, the general animal testing protocol currently used to evaluate the carcinogenic potential of a chemical has remained essentially the same for more than two decades. The basic procedure is relatively simple in experimental design, given the complexity of the disease process that constitutes the observational endpoint of this test procedure (cancer) and the consequences and importance accorded any positive findings identified by this test procedure. Furthermore, echoing an early recommendation by the FDA that *testing be done at doses and under experimental conditions likely to yield maximum tumor incidence*, the use of high doses to maximize the sensitivity of the procedure has become an area of considerable controversy. In short, the chemical doses tested, the animal species selected, and the simple, observational nature of this test often later become targets for criticism when the test results are applied in risk or hazard assessments that have a large impact on public health policy. For this reason, this and subsequent sections of this chapter will focus on the basic experimental design of the chronic animal cancer bioassay and the scientific issues commonly raised about these procedures or what interpretation might be given the results.

The commonly recommended requirements for a thorough assessment of carcinogenic potential in a test animal that mold the basic experimental design of a chronic animal bioassay are

- That two species of rodents, both sexes of each species, should be tested as a minimum. This helps ensure that false negative responses are not generated by selecting a non-responsive species for the test.
- That adequate controls are run during the test procedure. Ideally, the tumor incidence in test animals is compared to both historical and concurrent control animal responses. This helps ensure that the observed response is not an aberration of that specific study.
- A sufficient number of animals should be tested so that a positive response is not likely to be missed. The goal is to test enough animals to have a sufficient statistical basis whereby even a weak carcinogenic response should be observed and to be able to determine whether an observed increase in tumors, or lack thereof, was a chance or real observation. Typically, 50–100 animals of each sex and species are considered to be an appropriate-sized test.

Increasing the number of animals tested might increase the sensitivity of the test, but as the number of animals is increased, the cost of the experiment rises and could render the test cost-prohibitive.

- The exposure and observation periods should last a lifetime, if possible, so that the latency of the response does not become an issue.
- At least two doses should be tested. One should be the *maximally tolerated dose* (MTD), the second dose should be some fraction (usually 50% or 25%) of the MTD. The MTD is defined as the highest dose that can be reasonably administered for the lifetime of the animal without producing serious, life-threatening toxicity to the animal that might compromise completion of the study. In the past the MTD has been defined as a dose that causes no more than a 10 percent decrease in body weight gain and does not lead to lethality over time.
- A detailed pathologic examination of all tissues should be held at termination of the experiment (and sometimes at 6-month intervals).

In addition to these recommended guidelines, this test is normally performed following *good laboratory practice* (GLP) procedures. These and other procedures ensure proper animal care during the extended period of the test, that no cross-contamination with other chemicals being tested will occur, and the possibility of having infectious agents or disease affect the outcome of the test is limited.

Using these basic guidelines, any positive result obtained in at least one sex of one species is generally considered sufficient evidence to classify the chemical, for regulatory and public health purposes, as a carcinogen. Four different types of tissue response might be observed in a chronic test and considered positive evidence of carcinogenicity:

1. An increase in the incidence of a tumor type that occurs in control animals but at a significantly lower rate
2. The development of tumors at a significantly earlier period than is observed in the control animals
3. The presence of tumor types that are not seen in control animals
4. An increased multiplicity of tumors (although generally speaking, differences in total tumor load between exposed and unexposed animals is not considered reliable evidence)

Positive results in a test with a more limited power to detect carcinogenicity (e.g., tests of shorter duration or fewer animals), but where the overall test procedures employed are considered adequate, may also become accepted as sufficient evidence of carcinogenicity, particularly where other relevant evidence (e.g., mechanistic data, structural alerts, structure–activity relationships) are also available. In contrast, because it is well recognized that important species differences exist in regard to response, negative results (an observed lack of a tumorigenic response), might not be considered definitive evidence that a chemical is not a carcinogen in other species that were not tested.

The Issue of Generating False-Negative or False-Positive Results

Both false-negative and false-positive results are a potential problem in carcinogen bioassays. Ideally, the number of animals required to provide adequate negative evidence should be great enough that even a false-negative test (a test failing to detect existent carcinogenicity) will not allow an excessive risk to go unnoticed. The likelihood that such a risk will not be detected during the evaluation of bioassay data is dependent on two factors (excluding species differences in response): the number of animals tested and the extent to which the test dose exceeds the usual level of human exposure, therefore increasing either parameter tends to lessen the chance of obtaining a false negative response (with respect to humans).

The probability that a test will generate a false-negative result is also affected by the background tumor rate in the control animals. As the background incidence of tumorigenesis increases, so does the number of animals required to detect a small percent increase in tumor incidence above the animal's background rate. This means that it may be difficult to detect small increases for those tumor types that have large spontaneous background rates in an animal model when the test group contains only 50–100 animals.

To increase the safety of the animal-to-human extrapolations, the number of animals tested in a cancer bioassay may need to be increased if (1) the number of humans to be exposed to the chemical is either expected to be large or (2) a small margin of safety exists between the animal dose tested and the expected human exposure. In general, however, resource limitations are such that only 50 animals of each sex are tested at each dose for both species; this limits the total number of animals tested to about 400 animals, plus 200 animals to serve as controls.

Short-Term Cancer Bioassays and Other Measures of Carcinogenic Potential

Because of the large number of animals, lengthy timeframe, and expense associated with the chronic carcinogenesis bioassay, there has long been a need for reliable shorter-term tests of carcinogenic potential that could be used to complement standard carcinogenicity testing protocols. In the past, such short-term tests were limited to abbreviated initiation–promotion experiments with defined endpoints and the induction of tumors in susceptible animal models (e.g., lung tumors in strain A mice). Recently however, the tools of molecular biology have made it possible to construct genetically altered (transgenic) animals that may prove to be useful models for predicting the carcinogenic potential of chemicals. The U.S. National Toxicology Program is currently in the process of validating two of these transgenic models, Tg.AC mice and $p53^{+/-}$ mice, with chemicals previously tested in the standard 2-year chronic bioassay. These transgenic models are described briefly below.

1. The Tg.AC line was produced in FVB/N mice by the incorporation of a v-H-*ras* transgene into the cellular DNA. Mutations in *ras* oncogenes, which encode a family of GTP binding proteins critical to many growth factor signaling pathways, have been detected in a large proportion of human tumors. Tg.AC mice behave like genetically initiated mice, and rapidly develop epidermal papillomas in response to topical treatment with carcinogens. Researchers have shown that the mutant transgene is overexpressed in the proliferating cells in benign and malignant tumors but is not expressed in normal cells. Interestingly, treatment with initiators *or* tumor promoters induces the development of skin tumors. While treated mice have a dramatic increase in tumor yield with abbreviated time-to-tumor response, untreated mice have a normal skin histology and do not usually develop spontaneous tumors within the testing period. Treatment with carcinogens results in the production of papillomas in less than 6 months, substantially reducing the period of time for a typical initiation-promotion experiment in mouse skin.

2. Heterozygous $p53^{+/-}$ mice possess only a single functional copy of the p53 gene. As discussed previously, p53 function is lost through mutation or deletion in over 50 percent of all human cancers. With only a single functional allele, p53 mice develop normally but are at an increased susceptibility to the induction of tumors. This situation is analogous to an individual who has inherited a defective copy of a tumor suppressor gene. Upon dosing with mutagenic carcinogens, $p53^{+/-}$ mice rapidly develop tumors compared to normal mice, usually within 6 months. Untreated $p53^{+/-}$ mice do not usually develop tumors within the test period.

In the chemicals tested thus far, there has been a high degree of concordance with results from traditional chronic bioassays. Examination of the discordant cases indicates that a number of these are due to the absence of hepatocellular carcinoma in the $p53^{+/-}$ mice. As will be discussed, the $B6C3F_1$ mice typically used in chronic bioassays have a high spontaneous background rate of these tumors, making interpretation of positive carcinogenic responses in this tissue problematic. These and other transgenic animal models hold promise as less expensive and time-consuming adjuncts or replacements for conventional chronic bioassays. In addition, some scientists believe that transgenic animal models may be more relevant to the humans because they possess alterations in genes known to be involved in many human tumors.

13.6 INTERPRETATION ISSUES RAISED BY CONDITIONS OF THE TEST PROCEDURE

Human health hazards and, to allow for some quantitative assessment of the risk, the reliability of the animal-to-human extrapolation of animal cancer data is understandably an important issue. And as is true for any animal test procedure, questions concerning the reliability with which the results of the chronic bioassay can be extrapolated to human exposure conditions are frequently raised. However, in addition to the obvious potential for frank differences to arise in the human response because of species differences, an issue that can be raised with the animal test data for any other toxic endpoint (e.g., liver injury or developmental deficits), a number of interpretation issues have been raised that stem from the experimental conditions of the cancer bioassay test procedure itself. For example, it has long been noted that a number of interpretation problems will simply arise out of the data collected from this procedure because significant species differences may reasonably be anticipated and because of the test's relatively crude and observational approach. That is, after approximately 3 years of test procedure, tissue collection and histopathological examination, we are largely left with a single, simple observation, specifically, the number of tumors in a tissue following lifetime high-dose exposure. Because the test procedure is arguably a screening test for carcinogenic activity, regardless of dose, when a positive result is observed, little else is provided. For example, typically little or no information is provided on dose–response or the mechanism by which the cancer was induced. Thus, it should perhaps not be surprising that the utility of this procedure continues to foster debate in the scientific community, or that much additional research is routinely required to be able to reliably interpret and extrapolate the results obtained by this procedure.

In a seminal article dealing with the problems associated with chronic cancer bioassay tests and their interpretation, Squire listed five experimental design issues that remain relevant today:

1. Use of the MTD as currently defined
2. The number of doses tested
3. The relevance of findings in certain test species
4. Route of exposure and vehicle
5. Extent of the pathological examination

These and related issues raised by this test procedure are discussed in the following paragraphs because they have a considerable impact on the hazard evaluation of the chemical in question, and because they are frequently raised when debates occur over the significance of the observation or the regulation of a particular chemical.

The Doses Used to Test Chemicals Are Too High

This has become perhaps the most frequently raised criticism of chronic animal cancer tests. Because the MTD is often selected as the highest dose the test animal can maintain for a lifetime without shortening the length of the test, it is often a dose where chronic toxicity and biological changes occur. The biological arguments against the applicability of positive results that are seen only at high doses, doses that produce chronic toxicity and substantially exceed the expected human exposure level, are as follows:

1. High doses may alter the metabolism and disposition of the chemical such that the types of reactive, toxic metabolites that are responsible for the critical biochemical changes producing cancer

are not present at lower doses. It has been shown with a number of chemicals that a particular metabolic pathway becomes saturated above a certain dose level. Once saturated either the formation of a specific toxic metabolite begins to increase, or a detoxification–protective pathway now begins to become overwhelmed. This leads to a cellular insult and damage that either does not occur at lower doses, or does so at a significantly lower rate. This phenomenon is often referred to as a dose that "produces zero order kinetics" in an otherwise "first-order reaction process."

2. High doses produce irritation or inflammation. These conditions produce the formation of reactive oxygen species that are capable of inducing DNA damage that simply does not occur at lower doses.

3. High doses may produce changes in immune or endocrine systems, disrupt nutrition, or otherwise produce stressors that induce cancer secondary to changes in the background cancer rate. Because these same organ toxicities have thresholds and so do not occur at lower doses, low doses of the chemical are incapable of inducing cancer secondary to these specific biochemical and molecular changes.

4. High doses can produce damage to important DNA repair enzymes, or the DNA damage will overwhelm the cell's ability to withstand these genetic assaults.

5. High doses produce a recurrent injury, cell death, and cell turnover that are not induced at lower doses. Under these conditions the cytotoxicity that is induced by high doses alters important cellular pools of factors responsible for maintaining genetic integrity within the cell. Thus, high doses may foster conditions within the cell that result in mutations or genetic damage indirectly. In addition, the increase in cell turnover may now cause mutations to become fixed that would normally be repaired. A sustained increase in cell turnover may also increase the rate at which natural errors in DNA replication and spontaneous mutations occur.

In short, the criticism of high-dose testing is that the cancers observed may originate secondarily to other important biochemical changes and toxicities that are induced only at high doses. In this situation, where the chemical induces cancer indirectly and is related to conditions unique to high doses, then low-dose conditions would not be carcinogenic. Whether such chemicals should be viewed as a carcinogenic hazard or as chemicals without carcinogenic activity becomes a function of dose; an issue that may raise considerable controversy when the positive carcinogenicity data are used to regulate the exposures of such chemicals.

The possibility that the carcinogenicity of a particular chemical may be a high-dose phenomenon has been assumed or hypothesized for a number of different chemicals and different mechanisms. For example, a number of different chemicals cause a chronic reduction in the circulating levels of thyroid hormone at high doses. This, in turn, causes a chronic elevation in blood levels of thyroid-stimulating hormone, a normal response to low thyroid levels, that results in a chronic overstimulation of thyroid follicular cells and eventually the development of thyroid follicular cell tumors. This high-dose phenomenon has a threshold (lower doses will not decrease thyroid hormone levels), and no risk of cancer would be associated with lower doses. Similarly, the bladder tumors observed with very high doses of compounds such as vitamin C, saccharin, glycine, melamine, and uracil are believed to be induced only when an excessive dose produces the depositing of insoluble calculi or crystals in the urinary bladder. The occurrence of these physical agents produce a chronic irritation or inflammation, thereby providing a stimulus for the proliferation of the bladder epithelium and ultimately the formation of bladder tumors. Since none of these changes are produced at lower doses, there are clear thresholds for these carcinogens, and their "carcinogenic hazard" can be induced only at unrealistically high exposure levels.

As the evidence accrued that use of the MTD in bioassays frequently produced dose-dependent results, scientists within the NTP assessed the use of the MTD and the long-held view that responses obtained at the MTD could be extrapolated in a linear fashion to lower doses. This assessment concluded that the following implicit assumptions underlie the current use of the MTD, and the associated use by regulatory agencies of a linear extrapolation of the results obtained with it:

- The pharmacokinetics of the chemical are *not* dose-dependent.
- The dose–response relationship *is* linear.
- DNA repair is *not* dependent on dose.
- The response is *not* dependent on the age of the animal.
- The test dose need *not* bear a relationship to human exposure.

Following a review of these assumptions, however, the Board of Scientific Counselors within the NTP concluded that the implicit assumptions underlying linear extrapolation from the MTD do not appear to be valid for many chemicals, and that both the criteria for selecting doses tested in the chronic bioassay, as well as the method for extrapolating these results, should be reevaluated. Regarding the issue of alternative criteria for selecting the highest doses to be tested in a chronic bioassay, the following criteria recommended earlier by Squire seem to address a number of the issues that are raised by implicit assumptions associated with the MTD:

1. The MTD should induce no overt toxicity, that is, no appreciable death, organ pathology, organ dysfunction, or cellular toxicity.
2. The MTD induces no toxic manifestation predicted to shorten lifespan.
3. The MTD does not retard body weight by 10 percent.
4. The MTD is a dose that in two-generational studies is not detrimental to conception, fetal or neonatal development, and postnatal development or survival.
5. Takes into consideration important metabolic and pharmacokinetic data.

There are a number of attractive features of this proposed definition for the MTD. Because the ultimate goal of all toxicity testing is to identify all potential hazards we should be guarding against, and to develop exposure limits that will prevent all toxicities, the criteria listed above allow other toxicological considerations, under specific circumstances, to set a reasonable upper limit on the doses tested for carcinogenicity. If high doses produce biochemical changes not seen at lower doses, and if at these doses the chemical produces other toxicities we have already identified and must prevent by limiting exposure, then these toxic endpoints may set a reasonable upper limit on the dose range we should employ to test for other toxicities (e.g., cancer).

While a number of additional arguments and example compounds can be cited in support of changing the doses tested in chronic animal cancer bioassays, especially if we are going to use the results in the risk assessment and risk management areas, this particular feature of the test protocol has placed regulatory agencies on the horns of a dilemma that is difficult to escape. It seems only logical to attempt to maximize the sensitivity (ability to detect carcinogens) of this test by using the highest dose possible. Testing the maximal dose helps eliminate the chance of producing false negative responses. Testing the maximal dose helps ensure the statistical significance of small but important changes, and helps set a manageable limit on the number of animals that must be tested to be able to statistically identify a positive response. On the other hand, by maximizing the dose that is tested, we seem to be incurring a considerable number of positive responses, the results of which, after further testing for mechanisms at considerable additional expense, do not seem to be relevant, serious human hazards, at least at the doses to which humans are exposed. The possibility that this may be a substantial problem with the current testing scheme is indicated by analyses showing some 44 percent of the positive test results observed in NTP bioassays as positive (carcinogenic) only at the highest dose tested and not at a dose that is 25–50% of the MTD. Thus, it would appear that for almost one-half of the chemicals tested thus far, the carcinogenicity of the chemical is strictly a function of the high doses being tested.

Number of Doses Tested

Once a chemical has been identified as capable of producing cancer in at least one animal species the results of that test are frequently used to develop exposure guidelines or regulatory standards via the development of a cancer slope factor or benchmark dose from these same data. Because in general only two or a very few doses are tested, and as these doses are usually relatively close in magnitude, the chronic bioassay frequently provides a poor database from which the human risk must be modeled. In rare instances both doses are positive at a maximal rate (i.e., 100 cancer incidence is seen at both doses). In this situation no judgement can be made as to the shape of the dose–response curve or how far one must go down in dose before the response begins to decline in a dose-dependent fashion. In other instances, one dose is positive and the second dose is not. In this situation there is again no information concerning the slope of the dose–response curve discernible from the data. Furthermore, in this situation modeling the single positive dose would also appear to inflate the cancer risks associated with low doses as the second dose, which is also a relatively high dose, produced no discernable activity. Both of these problems might be eliminated with the use of more doses, particularly where the doses are selected with the intent of developing usable dose–response data.

A related problem is caused when the doses tested are both positive, and yield some information concerning the shape of the dose–response curve at doses where the increase in the cancer incidence is observable, but both doses are above that point where metabolic processes become saturated and now significant changes are seen in key biochemical pathways responsible for the tumorigenic response (e.g., metabolism, disposition, endocrine, immune or DNA damage–repair responses). In these instances it would be helpful to have tested doses at those points where the biochemical changes believed to be key to the carcinogenic process are either not saturated or do not occur so as to assess their mechanistic significance directly.

The problem, however, with changing the protocol to include more doses is that it will dramatically increase the cost of performing a cancer bioassay (i.e., a 50 percent increase with each additional dose). So, once again regulatory agencies and public health officials are faced with the dilemma of either improving the test results at the expense of having the financial resources to test more compounds, or maximizing the number of tests performed within a specific budget at the possible expense of limiting the interpretation of the data.

The Route of Administration and Vehicle Issues

Because the route of administration of a compound may alter the metabolism and disposition of a chemical, and because local damage at the site of application may induce certain changes necessary for the carcinogenic response, the route of administration tested in animals should mimic that of the intended or most likely route of human exposure. For example, some metals induce sarcomas at the site of injection when injected into the muscles of animals, apparently in response to local inflammatory and other responses, but are not carcinogenic by any other route of exposure. What importance should be attached to these responses? The issue of route specific differences in response has become so well recognized that agencies like the U.S. Environmental Protection Agency now calculate separate cancer slope factors for the inhalation and oral routes of exposure. In so doing they use route-specific animal test data in order to avoid making a route-to-route extrapolation from a single animal experiment.

In some instances a vehicle is used to administer the test compound that is capable or either altering the pharmacokinetics of the compound (absorption, metabolism, etc.) or may cause changes (e.g., inflammation) that potentially influence the tumorigenic response of the chemical being tested. Where the vehicle produces either a qualitative or quantitative change in the response (compared to when no, or another, vehicle is used), the results should be interpreted with the appropriate caution. For example, corn oil has been used as a vehicle to administer chemicals not readily soluble in water. But distinct preneoplastic changes have been observed in organs like the pancreas in the corn-oil-only treatment group. So, tests producing pancreatic tumors when the chemical is administered in corn oil should be evaluated for cocarcinogenic responses rather than attributing all of the activity to the chemical being tested.

Issues Associated with the Histopathological Examination

In some instances, perhaps more so in years past, the histopathological examination of the slides taken from the control animals have not been examined as rigorously as those slides taken from the animals administered the test compound. While at first it might seem that more attention should be paid to those slides where the potential change is anticipated, this can lead to results that are an artifact of the examination. For example, if all animals during the test became infected by a viral organism, and if this infection affected the background cancer incidence in a particular organ of the animal, then placing a greater emphasis on the "exposed" slides might lead one to reach erroneous conclusions. In this situation the pathologist might identify more tumors in exposed animals simply because of the more extensive microscopic search of the exposed tissues even though equivalent numbers of infection-induced tumors might exist in both control and exposed animals.

Other aspects of the histopathological examination may affect the outcome of the study. For example, what organs should be examined? Should we evaluate organs like the Zymbal glands of rats if humans have no anatomic correlate? What relevance should be attached to results where only benign tumors, or tumors that behave benignly, are elicited? What relevance should be attached to a chemical that increases the tumor incidence in one organ while decreasing the tumor incidences in other organs, particularly if the total cancer/tumor risk of the animal group does not increase? Should we attach the same significance to these results? (*Note*: Here the extrapolation to humans would essentially be no net changes in the population's risk of cancer.) As only one chemical example of this phenomenon, PCBs, a chemical of considerable regulatory restriction and interest, has been observed in several studies to produce liver tumors in rats, and relatively low exposure guidelines have been developed for this chemical on the basis of such data. However, two general findings in these studies were that the total tumor incidence in exposed animals was not increased (because the prevalence of other tumor types were decreased) and that these tumors did not behave like malignant masses; in fact, the exposed animals lived on average longer that did the control animals.

One final facet of this issue is the fact that over the years the pathological descriptions (criteria for classifying pathological changes as tumors) have evolved. This means that chemicals using more modern descriptions might be viewed as having lower tumor incidences than they would if their evaluation occurred in years past. While this difference does not affect whether the test was considered to have produced a positive finding for carcinogenicity, it does affect the tumor incidence reported in the test, which, in turn, affects the perceived potency of the chemical as measured by the cancer slope factors derived from the tumor incidence that was reported. Thus, the perceived potency of a chemical carcinogen, as measured by its cancer slope factor, may differ according to which pathological criteria were used.

Dietary and Caloric Restrictions

Over the years we have come to realize that nutritional status and caloric intake during the test can affect the test results. In most test protocols the rats are fed ad libitum; that is, they are given constant access to food in the cage. Given the already restricted activities that can occur within these cages and the propensity of animals to eat as often as allowed, the animals tested under these conditions are generally obese animals during much of their lifetime. Studies with a number of different chemicals have shown that obesity can inflate the final tumor incidence that is observed; that is, there are a number of chemicals that, when administered at the same dose, will result in those animals placed on a normal caloric or restricted caloric intake to have significantly lower tumor rates than observed in animals fed ad libitum. Since ad libitum feeding is the general rule in chronic animal test procedures, the results of many studies have been inflated or possibly made statistically significant by the mere fact that the animals were allowed to ingest more food than their bodies need. Similarly, some chemicals might induce nutritional changes in the animal secondary to organ toxicity, which, if ameliorated, may significantly alter the outcome of the bioassay.

What Animal Species Represents the Most Relevant Animal Model?

While it may be prudent for regulatory purposes to use animal data to predict what the human response might be when human data are unavailable, it should be remembered that when one makes an animal-to-human extrapolation, the basic assumption of that extrapolation is that the animal response is both *qualitatively* and *quantitatively* the same as the human response. However, because two different species may respond differently, either qualitatively and quantitatively, to the same dosage of a particular chemical, any animal-to-human extrapolation should be considered a catch-22 situation. That is, to know whether it is valid to extrapolate between a particular animal species and humans in a sense requires prior knowledge of both outcomes. So, even though toxicologists frequently use animal data to predict possible human outcomes, the potential for significant *qualitative* and *quantitative* differences to exist among species requires that the human response first be known before an appropriate animal model can be selected for testing and extrapolation purposes. But the selection of the appropriate animal model is complicated by the fact that innumerable and vast species differences exist. These differences are related primarily to the anatomical, physiological, and biochemical specificity of each species; these differences may produce significant wide variation in the metabolism, pharmacokinetics, or target organ concentrations of a chemical between species. When these differences are then combined with species-related differences in the physiology or biochemistry of the target organ, it is not surprising that significantly different responses may be achieved when one moves to a different test species. The major point of interest here, however, is that because these differences exist, the extrapolation of animal responses to humans should be viewed as being fraught with considerable difficulty and uncertainty. Important species differences encompass, but are not limited to, the following:

1. Basal metabolic rates
2. Anatomy and organ structure
3. Physiology and cellular biochemistry
4. The distribution of chemicals in tissues (toxicodynamics); pharmacokinetics, absorption, elimination, excretion, and other factors
5. The metabolism, bioactivation, and detoxification of chemicals and their metabolic intermediates

A few well-known examples that illustrate the magnitude of these differences are discussed below.

Anatomic Differences

Laboratory animals possess some anatomic structures that humans lack, and when cancer is observed in one of these structures, the particular relevance to humans is unknown and cannot be assumed with any scientific reliability. For example, the Zymbal gland, or auditory sebaceous gland, is a specialized sebaceous gland associated with the ears in Fischer rats. This gland secretes a product known as sebum. Although there is little information about the specific function of the secretion of the Zymbal gland, there is no known human structural correlate. Thus, the fact that dibromopropanol can cause squamous cell papillomas of the Zymbal gland in Fischer rats might be argued as providing no information relevant to discerning the carcinogenic potential of this chemical in humans.

Another such problem exists with rodent species because they also possess an additional structure with no known human correlate: the forestomach. The esophagus empties into this organ, and it is here that ingested materials are stored before passing to the glandular stomach. The forestomach of rodents has a high pH, as opposed to the low pH of the human stomach, and high digestive enzyme activity.

In rats, hyperplastic and neoplastic changes in the forestomach may result from the chronic administration of compounds like butylated. Once again, however, the relevance to humans of such responses is not known.

Physiologic Differences

Male rats produce a protein known as α-2-microglobulin, which, in combination with certain chemicals or their metabolites, causes a repeated cell injury response in the proximal tubules of the kidney. However, significant levels of α-2-microglobulin are not found in female rats, mice, or humans. Thus, the mechanism believed responsible for the repeated cell injury and tumors formation observed in male rats does not exist in these species. The male rat kidney tumors observed after chronic gasoline exposure, or exposure to certain aliphatic compounds, such as *d*-limonene, are notable examples of this phenomena. The scientific community has concluded that the positive male rat data for such chemicals is not relevant for predicting human cancer risk.

Cellular and Biochemical Differences

The $B6C3F_1$ mouse routinely used in cancer bioassays has a genetically programmed high background incidence of hepatocellular cancer. Approximately 20–30 percent of untreated animals develop this type of cancer. The $B6C3F_1$ mouse is a genetic cross between the C3H mouse, which has almost a 60 percent background rate of liver cancer, and a C57BL mouse, which has a very low incidence rate of liver. Because the $B6C3F_1$ mouse was bred to exhibit a genetic predisposition for developing liver cancer, tests using this animal model have subsequently identified a number of chemicals that are only liver carcinogens in this mouse strain and not the rat. In turn, the relevance of the liver tumors which are so commonly induced in this mouse are frequently questioned when extrapolated to humans, especially in light of the relatively low incidence with which human hepatocellular cancer occurs (3–5 cases per 100,000) in the United States.

The molecular mechanism for the high background cancer incidence in the $B6C3F_1$ mouse appears to be related to its propensity for oncogene activation in the liver. For example, the DNA of the $B6C3F_1$ mouse H-*ras* oncogene is hypomethylated, or deficient in methylation. Methylation of DNA serves to block transcription of a gene. And since the mouse H-*ras* oncogene is not adequately methylated (i.e., not "blocked"), it may be inappropriately expressed more easily, thus providing a mechanistic foundation for the higher background incidence of liver tumors in this mouse strain. Further, certain types of hepatotoxicity may exacerbate the hypomethylation of the H-*ras* gene in this sensitive species, but have no significant effect on the gene methylation rates in less sensitive species. Thus, the relevance to humans of liver tumor development in this test species, or any other animal species which has a propensity for the spontaneous development of the tumor, is questionable.

To summarize, the use of mice and rats is generally a compromise aimed at decreased costs. While primates or dogs might better represent the human response to some chemicals, they cannot be used routinely because of the additional costs incurred and other reasons. In general, the use of rodents as a surrogate animal model for humans might be criticized because rodents typically have a faster rate of metabolism than do humans. So, at high doses the metabolic pattern and percentage of compound ultimately metabolized may be significantly different than that of humans. If the active form of the carcinogen is a metabolite, then the animal surrogate may be more sensitive to the chemical because it generates more of the metabolite per unit of dose. Alternatively, the problem of false negatives also applies in that the selection of an insensitive species may yield a conclusion of noncarcinogenicity whereas further testing would uncover the actual tumorigenic activity. Because significant species differences exist in key aspects of all areas relevant to carcinogenesis (metabolism, DNA repair, etc.), and as these differences are the rule rather than the exception, extrapolating the response in any species to humans without good mechanistic data should be done with caution. In addition, developing mechanistic data that will allow comparisons to be made between humans and both a responsive and nonresponsive species would appear to be the only way to improve our use (extrapolation) of chronic cancer bioassay data.

Are Some Test Species Too Sensitive?

A number of strains or species have a significantly higher tumor incidence in a particular tissue than do humans. The incidence of liver tumors in $B6C3F_1$ mice was discussed earlier. Another example is the strain A mouse, a mouse strain sometimes used to test a chemical's potential to induce lung tumors. In this particular mouse strain the incidence of lung tumors in the control (unexposed) animals will reach 100 percent by the time the animals have reached old age. In fact, because all animals will at some point develop lung tumors, a shortening of the latency (time to tumor) or the number of tumors at an early age are used, rather than the final tumor incidence measured at the end of the animals' lives. The use of positive data from an animal species with a particularly high background tumor incidence poses several problems. For example, are the mechanisms of cancer initiation or promotion the same for this chemical in humans? Can the potency of the chemical be estimated or even ranked when it might not be clear if the enhanced animal response is just a promotional effect of high background rate or the added effect of a complete carcinogen? Where the biology of the test animal clearly differs from that of humans is a positive response meaningful without corroboration in another species?

13.7 EMPIRICAL MEASURES OF RELIABILITY OF THE EXTRAPOLATION

What is the Reliability of the Species Extrapolation?

To test the reliability of making interspecies extrapolations, scientists have analyzed the results of a large number of chronic animal bioassays to ascertain the consistency with which a response in one species is also observed in another species. In one of the largest analyses performed to date, scientists analyzed the results for 266 chemicals tested in both sexes of rats and mice. The data forming this analysis is presented in Table 13.8.

From the findings discussed above, after defining concordance to be species agreement for both positive and negative results, the authors of this analysis concluded the following:

- The intersex correlations are stronger than the interspecies correlations.
- If only the male rat and female mouse had been tested, positive evidence of carcinogenicity would have led to the same conclusions regarding carcinogenicity/noncarcinogenicity in 96 percent of the chemicals tested in both sexes of both species (i.e., 255/266 correct responses).

TABLE 13.8 Correlations in Tumor Response in NCI/NTP Carcinogenicity Studies

Comparison[a]	Observed Outcome					% Concordant (++ or −−) Responses
	+	+−	−+	−−	Total	
Male rats vs. female rats	74	25	12	181	292	87.3
Male rats vs. male mice	46	43	36	145	270	70.7
Male rats vs. female mice	29	33	36	145	273	74.7
Female rats vs. male mice	46	32	37	156	271	74.5
Female rats vs. female mice	57	23	39	156	275	77.5
Male mice vs. female mice	78	10	23	177	288	88.5
Rats vs. mice	67	32	36	131	266	74.4

Source: Adapted from Haseman and Huff (1987).

TABLE 13.9 Correlations across Species of Positive Cancer Bioassays

Comparison	Observed Outcome +	+–	–+	Total	Percent concordance (++ or ––)
	Intraspecies Comparisons				
Male rats vs. female rats	74	25	12	111	67%
Male mice vs. female mice	78	10	23	111	70%
	Interspecies Comparisons				
Male rats vs. male mice	46	43	36	125	37%
Male rats vs. female mice	59	33	36	128	46%
Female rats vs. male mice	46	32	37	115	40%
Female rats vs. female mice	57	23	39	119	48%
Rats vs. mice	208	131	148	487	43%

This, in turn, suggests that the number of animals tested might be reduced (i.e., eliminate the testing of male mice and female rats).

- The high concordance between rats and mice supports the view that extrapolation of carcinogenicity outcomes to other species (humans) is appropriate.

However, the high degree of concordance in this analysis stems from the fact that about half of the studies are negative and the chemical being tested manifested no carcinogenic activity. When a slightly different questions is asked—regarding how reliably positive test results can be extrapolated across species—a much different answer is reached. In Table 13.9 the noncarcinogens have been removed and the comparisons across sexes and species have been reanalyzed. Figure 13.9 contains the same

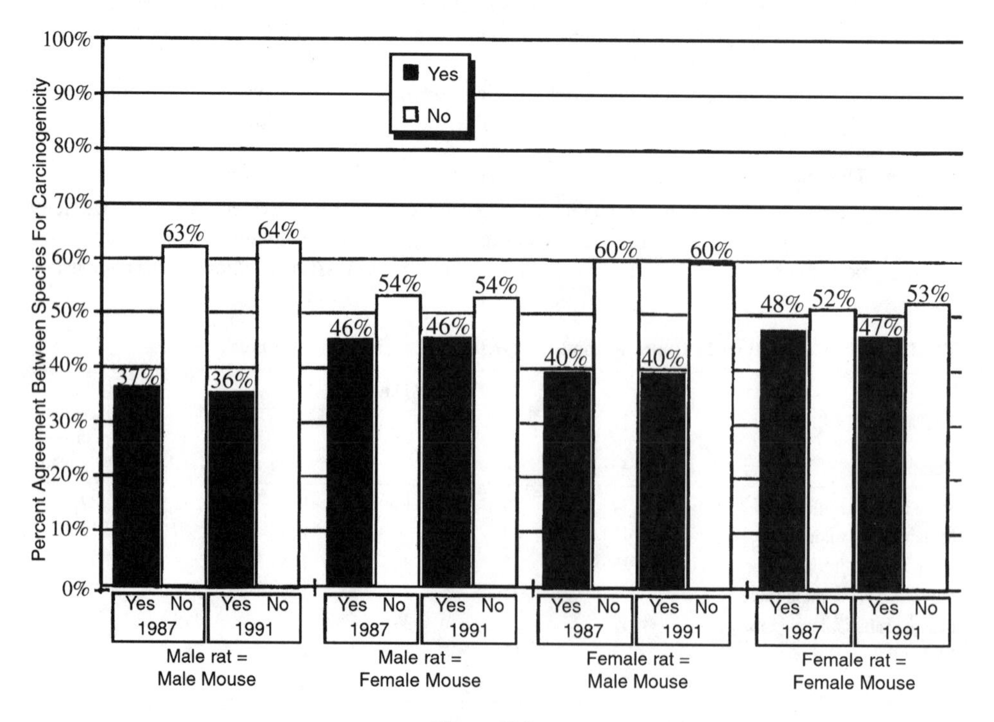

Figure 13.9

TABLE 13.10 The Poor Correlation in Organ Sites among Positive Rodent Tests

Site of Cancer	*N* Rats/Mice	Percent	*N* Mice/Rats	Percent
Liver	25/33	75	25/78	32
Lung	2/7	29	2/18	11
Hematopoietic system	3/14	21	3/11	27
Kidney (tubular cells)	3/21	14	3/4	75
Mammary gland	4/18	22	4/7	57
Forestomach	8/14	57	8/15	53
Thyroid gland	7/16	44	7/9	78
Zymbal gland	2/12	17	2/2	100
Urinary bladder	2/12	17	2/3	67
Skin	3/11	27	3/3	100
Clitoral/Preputial gland	0/7	—	0/3	—
Circulatory system	2/4	50	2/10	20
Adrenal medulla	0/4	—	0/4	—
Total	61/173	35	61/167	37

Source: Adapted from Haseman and Lockhart (1993).

analysis but compares the data from a subsequent update of the original study as well, illustrating that as the number of chemicals tested expands, the agreement in results across species does not seem to be changing.

From this analysis it is evident that when a chemical induces cancer in one of these two rodent species, it is also carcinogenic in the other species less than 50 percent of the time. This lack of concordance between these two phylogenetically similar species raises a concern voiced by many scientists when such data are extrapolated to humans without also considering mechanistic and pharmacokinetic data from both species that might help explain why such large differences exist.

A similar problem arises when the issue of identifying the correct target organ is considered. A recent analysis of the predictivity of the target organ for a carcinogen when extrapolating across two rodent species found one could predict the correct target organ about only about 37 percent of the time (Table 13.10). So, it would appear that not only is the assumption that a positive response in animals can be assumed to predict the human response, but the likelihood that the correct target has been identified would also seem to be of some question.

13.8 OCCUPATIONAL CARCINOGENS

Although the first occupational carcinogen was identified by Sir Percival Pott in 1775, it was not until 1970 with the passage of the Occupational Safety and Health Act and establishment of the Occupational Safety and Health Administration (OSHA) that the United States had enforcement authority granted to an agency to regulate the use of substances that were considered carcinogenic in the workplace. Prior to 1970, the source that was widely considered the most authoritative was the American Conference of Governmental Industrial Hygienists (ACGIH) and industry relied on this organization to regulate worker exposure to chemicals and agents. The other event occurring about this time that has shaped our current view of occupational carcinogens was the emergence of the cancer bioassay. The development and continued use of this bioassay over the years has identified many hundreds of industrial chemicals as having carcinogenic activity, at least in high-dose animal tests, many of which had never before been suspected of human carcinogenic activity. As certain chemicals or groups of chemicals became identified as carcinogens, this, in turn, brought to bear new pressures on industries as lower exposure levels or alternative chemicals were sought to reduce the possible risks associated with exposure to chemicals, many of which, before these new data were developed, were believed to be very safe and industrially useful chemicals.

Since the mid-1970s, several organizations—both private and public—have attempted to identify occupational carcinogens, or possible carcinogens, in an effort to reduce workplace exposure since logically, occupational exposures to carcinogenic chemicals would potentially be their gravest threat to human health because of their duration (a working lifetime) and the magnitude of occupational exposures. For example, the ACGIH ranks the known carcinogenic hazard of the compounds for which it provides TLVs in their annual listing (Table 13.11). Similarly, OSHA has identified its own list of chemical carcinogens that it regulates (Table 13.12), and the National Institute for Occupational Safety and Health (NIOSH), which is often referred to as the "research arm" of OSHA, provides a separate listing of what it considers to be the known or probable carcinogens that might be encountered in the workplace. Additional lists of known human carcinogens and chemicals known to be carcinogenic in animal tests include lists by the National Toxicology Program (Table 13.13) and the International Agency for Research on Cancers (IARC) which publishes a monograph series that evaluates the animal and human data for widely used chemicals and chemical processes (Table 13.14). In reviewing these different lists, it is of interest to note that rather than being identical, as one might expect, there can be significant differences in what is viewed as a possible carcinogen depending upon the agency promulgating the listing.

TABLE 13.11 Known or Suspected Carcinogens Identified by the ACGIH[a]

Confirmed Human Carcinogen (A1)	
4-Aminodiphenyl	Coal tar pitch volatiles
Arsenic	β-Naphthylamine
Asbestos	Nickel, insoluble
Benzene	Nickel subsulfide
Benzidine	Uranium (natural)
Beryllium	Vinyl chloride
Bis(chloromethyl)ether	Wood dust (hard or mixed hard/soft woods)
Chromite ore processing	
Chromium(VI)	Zinc chromates
Suspected Human Carcinogen (A2)	
Acrylonitirile	Diazomethane
Antimony trioxide	1,4-Dichloro-2-butene
Benz[*a*]anthracene	Dimethyl carbamoyl chloride
Benzo[*b*]fluoranthene	Ethylene oxide
Benzo[*a*]pyrene	Formaldehyde
Benzotrichloride	Lead chromate
1,3-Butadiene	4,4′-Methylene bis(2-chloroaniline)
Cadmium	4-Nitrodiphenyl
Calcium chromate	Oil mist, mineral
Carbon tetrachloride	Strontium chromite
Chloromethyl methyl ether	Sulfuric acid
Coal dust	Vinyl bromide
Diesel exhaust	Vinyl fluoride

[a]Including agents identified as carcinogens A1 or A2 in the *Notice of Intended Changes* for the TLVs.

TABLE 13.12 Potential Occupational Carcinogens Listed by NIOSH

Acetaldehyde	Formaldehyde
2-Acetylaminofluorene	Gallium arsenide
Acrylamide	Gasoline
Acrylonitrile	Heptachlor
Aldrin	Hexachlorobutadiene
4-Aminodiphenyl	Hexachloroethane
Amitrole	Hexamethyl phosphoramide
Aniline	Hydrazine
o-Anisidine	Kepone
Arsenic	Malonaldehyde
Arsine	Methoxychlor
Asbestos	Methyl bromide
Benzene	Methyl chloride
Benzidine	4,4′-Methylenebis(2-chloroaniline)
Benzidine dyes	Methylene chloride
Benzo[*a*]pyrene	4,4′-Methylenedianiline
Beryllium	Methyl hydrazine
1,3-Butadiene	Methyl iodide
tert-Butylchromate	α-Naphthylamine
Cadmium (dust and fume)	β-Naphthylamine
Calcium arsenate	Nickel carbonyl
Captafol	Nickel (insoluble, and soluble compounds)
Captan	Nickel subsulfides (and roasting operations)
Carbon black	4-Nitrobiphenyl
Carbon tetrachloride	*p*-Nitrochlorobenzene
Chlordane	2-Nitronaphthalene
Chlorinated camphene	2-Nitropropane
Chloroform	*N*-Nitrosodimethylamine
Bis(chloromethyl) ether	Phenylglycidyl ether
Chloromethyl methyl ether	Phenylhydrazine
β-Chloroprene	*N*-Phenyl-β-naphthylamine
Chromic acid	Polychlorinated biphenyl
Chromates	Propane sultone
Chromyl chloride	β-Propiolactone
Coal tar pitch volatiles	Propylene dichloride
Coke oven emissions	Propylene imine
DDT	Propylene oxide
2,4-Diaminoanisole	Rosin core solder pyrolysis products
o-Dianisidine	Silica, crystalline
o-Dianisidine-based dyes	Silica, Christobolite
1,2-Dibromo-3-chloropropane	Silica, quartz
Dichloroacetylene	Silica, Tridymite
p-Dichlorobenzene	Silica, Tripoli
3,3′-Dichlorobenzidine	Talc, asbestiform
Dichloroethyl ether	2,3,7,8-Tetrachlorodibenzo-*p*-dioxin
1,3-Dichloropropane	1,1,2,2-Tetrachloroethane
Dieldrin	Tetrachloroethylene
Diesel exhaust	Titanium dioxide
Diglycidyl ether	Toluene-2,4-diisocyanate
4-Dimethylaminoazobenzene	Toluenediamine

(*continued*)

TABLE 13.12 Continued

Dimethyl carbamoyl chloride	*o*-Toluidine
1,1-Dimethylhydrazine	*p*-Toluidine
Dimethyl sulfate	1,1,2-Trichloroethane
Dinitrotoluenes	Trichloroethylene
Di-*sec*-octyl phthalate	1,2,3-Trichloropropane
Dioxane	Uranium
Environmental tobacco smoke	Vinyl bromide
Epichlorohydrin	Vinyl chloride
Ethyl acrylate	Vinyl cyclohexene dioxide
Ethylene dibromide	Vinylidene chloride
Ethylene dichloride	Welding fumes
Ethyleneimine	Wood dust
Ethylene oxide	Zince chromates
Ethylene thiourea	

Source: *NIOSH Pocket Guide*, 1999.

13.9 CANCER AND OUR ENVIRONMENT: FACTORS THAT MODULATE OUR RISKS TO OCCUPATIONAL HAZARDS

Increased awareness of the ubiquity of synthetic, industrial chemicals in our environment has led a number of scientists to try to determine what role environmental exposures play in cancer causation. The USEPA devotes a great deal of its resources to this question as do other federal, international and private agencies such as the Agency for Toxic Substances and Disease Registry (ATSDR) of the Centers for Disease Control (CDC), the American Cancer Society (ACS), and the World Health Organization's (WHO) International Agency for Research on Cancer (IARC) (see Table 13.14). While each organization researching the impact of our occupations, lifestyles, diets, and environmental exposures on cancer have differing agendas and views as to the predicted cancer risks associated with environmental exposures or our daily routines, there is widespread agreement that the most substantial risks, and the greatest causes of cancer, are those factors that are controlled by the individual (e.g., diet, smoking, alcohol intake).

The importance of this fact is twofold: (1) it should be recognized that cancer is a phenomenon associated with normal biologic processes, and is therefore impacted by those factors that may affect our normal biologic processes (e.g., diet); and (2) many environmental risk factors exist, and these, in combination with hereditary risk factors, may frequently provide overwhelming influences in epidemiological studies of occupational hazards. Thus, the risk factors not being studied (and so frequently not controlled for) may mask or exacerbate the response being studied and so confound any study that is not normalized in a manner that removes all potential influences from the association being studied.

Estimates of the contribution of various factors to the rate of cancer in humans were perhaps first put forth by Doll and Peto, who produced the results plotted in Figure 13.10. As can easily be seen in Figure 13.10, the vast majority of the cancers were thought to be related to lifestyle factors; tobacco and alcohol use, diet, and sexual behavior accounted for 75 percent of all cancers in this initial analysis. Conversely, industrial products, pollution, and occupation were thought to be related to only 7 percent of all cancers. Currently, the contributions of diet, disease, and viral agents are still being researched as perhaps the most common causes of cancer.

In the years following Doll and Peto's initial assertions, some scientists have questioned whether such a large proportion of the cancers in humans had such clearly defined causal associations. However, the most recent evidence accumulated by researchers in this area indicates that less than 1 percent of today's cancers result from exposure to environmental pollution, and diet has since been identified as a key risk factor for cancer in nearly 200 epidemiologic studies. More importantly, the view that there

TABLE 13.13 Agents Listed in the *Report on Carcinogens* (8th Edition) from the National Toxicology Program, as Known or Suspected Human Carcinogens

Known Human Carcinogens

Aminobiphenyl (4-aminodiphenyl)	Erionite
Analgesic mixtures containing phenacetin	Lead chromate
Arsenic compounds, inorganic	Melphalan
Asbestos	Methoxsalen [with ultraviolet A (UVA) therapy]
Azathioprine	Mineral oils
Benzene	Mustard gas
Benzidine	2-Naphthylamine (β-naphthylamine)
Bis(chloromethyl) ether	Piperazine Estrone Sulfate
1,4-Butanediol dimethylsulfonate (Myleran)	Radon
Chlorambucil	Sodium equilin sulfate
1-(2-Chloroethyl)-3-(4-methylcyclohexyl)-1-nitrosourea	Sodium estrone sulfate
	Soots
Chloromethyl methyl ether	Strontium chromate
Chromium hexavalent	Tars
Coal tar	Thiotepa [tris(1-aziridinyl)phosphine sulfide]
Coke oven emissions	Thorium dioxide
Creosote (coal)	Tris(1-aziridinyl)phosphine sulfide (thiotepa)
Creosote (wood)	Vinyl chloride
Cyclophosphamide	Zinc chromate
Cyclosporin A (cyclosporine A; ciclosporin)	
Diethylstilbestrol	

Agents Reasonably Anticipated to be Human Carcinogens

Acetaldehyde	Beryllium zinc silicate
2-Acetylaminofluorene	Beryl ore
Acrylamide	Bis(chloroethyl) nitrosourea (BCNU)
Acrylonitrile	Bis(dimethylamino)benzophenone
Adriamycin (doxorubicin hydrochloride)	Bromodichloromethane
2-Aminoanthraquinone	1,3-Butadiene
o-Aminoazotoluene	Butylated hydroxyanisole (BHA)
1-Amino-2-methylanthraquinone	Cadmium
Amitrole	Cadmium chloride
o-Anisidine hydrochloride	Cadmium oxide
Azacitidine (5-azacytidine)	Cadmium sulfate
Benz[*a*]anthracene	Cadmium sulfide
Benzo[*b*]fluoranthene	Carbon tetrachloride
Benzo[*j*]fluoranthene	Ceramic fibers
Benzo[*k*]fluoranthene	Chlorendic acid
Benzo[*a*]pyrene	Chlorinated paraffins (C12, 60% chlorine)
Benzotrichloride	1-(2-Chloroethyl)-3-cyclohexyl-1-nitrosourea (CCNU)
Beryllium aluminum alloy	
Beryllium chloride	Chloroform
Beryllium fluoride	3-Chloro-2-methylpropene
Beryllium hydroxide	4-Chloro-*o*-phenylenediamine
Beryllium oxide	*p*-Chloro-*o*-toluidine
Beryllium phosphate	*p*-Chloro-*o*-toluidine hydrochloride
Beryllium sulfate tetrahydrate	Chlorozotocin

(continued)

TABLE 13.13 Continued

CI[a] Basic Red 9 monohydrochloride
Cisplatin
p-Cresidine
Cristobalite [under "Silica, crystalline (respirable size)"]
Cupferron
Dacarbazine
2,4-Diaminoanisole sulfate
2,4-Diaminotoluene
Dibenz[*a,h*]acridine
Dibenz[*a,j*]acridine
Dibenz[*a,h*]anthracene
7*H*-Dibenzo[*c,g*]carbazole
Dibenzo[*a,e*]pyrene
Dibenzo[*a,h*]pyrene
Dibenzo[*a,i*]pyrene
Dibenzo[*a,l*]pyrene
1,2-Dibromo-3-chloropropane
1,2-Dibromoethane [ethylene dibromide (EDB)]
1,4-Dichlorobenzene (*p*-dichlorobenzene)
3,3-Dichlorobenzidine
3,3-Dichlorobenzidine dihydrochloride
Dichlorodiphenyltrichloroethane (DDT)
1,2-Dichloroethane (ethylene dichloride)
1,3-Dichloropropene (technical-grade)
Diepoxybutane
N,N-Diethyldithiocarbamic acid 2-chloroallyl esterDEHP; bis(2-ethylhexyl phthalate)]
Diethylnitrosamine
Diethyl sulfate
Diglycidyl resorcinol ether
1,8-Dihydroxyanthraquinone [Danthron]
3,3-Dimethoxybenzidine
4-Dimethylaminoazobenzene
3,3-Dimethylbenzidine
Dimethylcarbamoyl chloride
1,1-Dimethylhydrazine (UDMH)
Dimethylnitrosamine
Dimethyl sulfate
Dimethylvinyl chloride
1,6-Dinitropyrene
1,8-Dinitropyrene
1,4-Dioxane
Direct Black 38
Direct Blue 6
Disperse Blue 1
Epichlorohydrin
Estradiol-17b
Estrone
Ethinylestradiol
Ethyl acrylate
Ethylene oxide
Ethylene thiourea
Ethyl methanesulfonate
Formaldehyde (gas)
Furan
Glasswool
Glycidol hexachlorobenzene
α-Hexachlorocyclohexane
β-Hexachlorocyclohexane
γ-Hexachlorocyclohexane
Hexachlorocyclohexane
Hexachloroethane
Hexamethylphosphoramide
Hydrazine
Hydrazine sulfate
Hydrazobenzene
Indeno[1,2,3-*cd*]pyrene
Iron dextran complex
Kepone (chlordecone)
Lead acetate
Lead phosphate
Lindane
Mestranol
2-Methylaziridine (propylenimine)
5-Methylchrysene
4,4-Methylenebis(2-chloraniline)
4,4-Methylenebis(*N,N*-dimethylbenzenamine)
Methylene chloride
4,4-Methylenedianiline
4,4-Methylenedianiline dihydrochloride
Methylmethanesulfonate
N-Methyl-*N*-nitro-*N*-nitrosoguanidine
Metronidazole
Mirex
Nickel
Nickel acetate
Nickel carbonate
Nickel carbonyl
Nickel hydroxide
Nickel hydroxide
Nickelocene
Nickel oxide
Nickel subsulfide
Nitrilotriacetic acid
o-Nitroanisole
6-Nitrochrysene
Nitrofen
Nitrogen mustard hydrochloride
2-Nitropropane

(*continued*)

TABLE 13.13 Continued

1-Nitropyrene	1,3-Propane sultone
4-Nitropyrene	β-propiolactone
N-Nitroso-*n*-butyl-*N*-(3-carboxypropyl)amine	Propylene oxide
N-Nitroso-*n*-butyl-*N*-(4-hydroxybutyl)amine	Propylthiouracil
N-Nitrosodi-*n*-butylamine	Quartz [under "silica, crystalline (respirable size)"]
N-Nitrosodiethanolamine	Reserpine
N-Nitrosodi-*n*-propylamine	Saccharin
N-Nitroso-*N*-ethylurea (*N*-ethyl-*N*-nitrosourea (ENU)	Safrole
4-(*N*-Nitrosomethylamino)-1-(3-pyridyl)-1-butanone	Selenium sulfide
N-Nitroso-*N*-methylurea	Silica, crystalline (respirable size)
N-Nitrosomethylvinylamine	Streptozotocin
N-Nitrosomorpholine	2,3,7,8-Tetrachlorodibenzo-*p*-dioxin (TCDD)
N-Nitrosonornicotine	Tetrachloroethylene (perchloroethylene)
N-Nitrosopiperidine	Tetranitromethane
N-Nitrosopyrrolidine	Thioacetamide
N-Nitrososarcosine	Thiourea
Norethisterone	Toluene diisocyanate
Ochratoxin A	*o*-Toluidine
4,4-Oxydianiline	*o*-Toluidine hydrochloride
Oxymetholone	Toxaphene
Phenacetin	2,4,6-Trichlorophenol
Phenazopyridine hydrochloride	1,2,3-Trichloropropane
Phenoxybenzamine hydrochloride	Tridymite
Phenytoin	Tris(2,3-dibromopropyl) phosphate
Polybrominated biphenyls (PBBs)	Urethane (Urethan; ethyl carbamate)
Polychlorinated biphenyls (PCBs)	4-Vinyl-1-cyclohexene diepoxide
Polycyclic aromatic hydrocarbons (PAHs)	
Procarbazine hydrochloride	
Progesterone	

[a]Color Index.

was a "cancer epidemic" in this nation attributable to environmental exposure to pollutants shown to cause cancer in animals has been found to be inaccurate. In the absence of large percentages of cancers attributable to environmental contaminants or occupational exposures, then, we are faced with determining how much of our cancer risk is inevitable (due to aging processes or perhaps genetic predisposition) or could be offset by changes to lifestyle factors such as smoking and diet.

Genetic Makeup of Individuals

The understanding of the role that genetics plays in carcinogenesis increased greatly in the 1990s and the relationship between genetic makeup and carcinogenesis is rapidly becoming a dominant area of cancer research. To date there have been more than 600 genetic traits associated with an increased risk of neoplasia. This relatively recent area of research is focused on how changes in the phenotypic expression of certain enzymes may alter the activation, detoxification, or repair mechanisms and thereby enhance the genetic damage produced by a particular chemical exposure. Genetic predisposition now accounts for perhaps 5–10 percent of all cancers, and it has been identified as a component

TABLE 13.14 IARC Carcinogens

Group 1: Carcinogenic to Humans (75)

Exposure circumstances
- Aluminum production
- Auramine, manufacture of
- Boot and shoe manufacture and repair
- Coal gasification
- Coke production
- Furniture/cabinetmaking
- Hematite mining with exposure to radon
- Iron and steel founding
- Isopropanol manufacture (strong-acid process)
- Magenta, manufacture of
- Painter
- Rubber industry
- Strong-inorganic-acid mists containing sulfuric acid

Agents and groups of agents
- Aflatoxins, naturally occurring
- 4-Aminobiphenyl
- Arsenic and arsenic compounds
- Asbestos
- Azathioprine
- Benzene
- Benzidine
- Beryllium and beryllium compounds
- *N,N*-Bis(2-chloroethyl)-2-naphthylamine (Chlomaphazine)
- Bis(chloromethyl) ether and chloromethyl methyl ether
- 1,4-Butanediol dimethanesulfonate (Busulphan; Myleran)
- Cadmium and cadmium compounds
- Chlorambucil
- 1-(2-Chloroethyl)-3-(4-methylcyclohexyl)-1-nitrosourea (methyl-CCNU; Semustine)
- Chromium VI compounds
- Ciclosporin
- Cyclophosphamide
- Diethylstilboestrol (DES)
- Epstein–Barr virus
- Erionite
- Ethylene oxide
- Estrogen therapy, postmenopausal
- Estrogens, nonsteroidal
- Estrogens, steriodal
- *Helicobacter pylori* (infection with)
- Hepatitis B virus (chronic infection with)
- Hepatitis C virus (chronic infection with)
- Human immunodeficiency virus type 1 (infection with)
- Human papillomavirus type 16
- Human papillomavirus type 18
- Human T-cell lymphotropic virus type I
- Melphalan
- 8-Methoxypsoralen (methoxsalen)
- MOPP and other combined chemotherapy, including alkylating agents
- Mustard gas (sulfur mustard)
- 2-Naphthylamine
- Nickel compounds
- *Opisthorchis viverrini* (infection with)
- Oral contraceptives, combined
- Oral contraceptives, sequential
- Radon and its decay products
- *Schistosoma haematobium* (infection with)
- Silica, crystalline
- Solar radiation
- Talc containing asbestiform fibers
- Tamoxifen
- 2,3,7,8-Tetrachlorodibenzo-*para*-dioxin
- Thiotepa
- Treosulfan
- Vinyl chloride

Mixtures
- Alcoholic beverages
- Analgesic mixtures containing phenacetin
- Betel quid with tobacco
- Coal tar pitches
- Coal tars
- Mineral oils, untreated and mildly treated
- Salted fish (Chinese style)
- Shale oils
- Soots
- Tobacco products, smokeless
- Tobacco smoke
- Wood dust

(*continued*)

TABLE 13.14 Continued

Group 2A: Probably Carcinogenic to Humans (59)

Agents and groups of agents

Acrylamide
Adriamycin
Androgenic (anabolic) steroids
Azacitidine
Benz[*a*]anthracene
Benzidine-based dyes
Benzo[*a*]pyrene
Bischloroethyl nitrosourea (BCNU)
1,3-Butadiene
Captafol
Chloramphenicol
Chlorinated toluenes (benzyl chloride), benzotrichloride, benzyl chloride and benzoyl chloride
1-(2-Chloroethyl)-3-cyclohexyl-1-nitrosourea (CCNU)
para-Chloro-*ortho*-toluidine and its strong-acid salts
Chlorozotocin
Cisplatin
Clonorchis sinensis (infection with)
Dibenz[*a,h*]anthracene
Diethyl sulfate
Dimethylcarbarnoyl chloride
1,2-Dimethylhydrazine
Dimethyl sulfate
Epichlorohydrin
Ethylene dibromide
N-Ethyl-*N*-nitrosourea
Formaldehyde
Human papillomavirus type 31
Human papillomavirus type 33
IQ (2-Amino-3-methylimidazo[4,5-*f*]quinoline)
Kaposi's sarcoma herpesvirus/human herpesvirus 8
5-Methoxypsoralen
4,4′-Methylene bis(2-chloroaniline)
Methyl methanesulfonate
N-Methyl-*N*′-nitro-*N*-nitrosoguanidine (MNNG)
N-Methyl-*N*-nitrosourea (nitrogen mustard)
N-Nitrosodiethylamine
N-Nitrosodimethylamine
Phenacetin
Procarbazine hydrochloride
Styrene-7,8-oxide
Tetrachloroethylene
Trichloroethylene
1,2,3-Trichloropropane
Tris(2,3-dibromopropyl) phosphate
Ultraviolet radiation A
Ultraviolet radiation B
Ultraviolet radiation C
Vinyl bromide
Vinyl fluoride

Mixtures

Creosotes
Diesel engine exhaust
Hot mate
Polychlorinated biphenyls

Exposure circumstances

Art glass, glass containers and pressed ware (manufacture of)
Hairdresser or barber (occupational exposure as a)
Nonarsenical insecticides (occupational exposures in spraying and application of)
Petroleum refining (occupational exposure in)
Sunlamps and sunbeds (use of)

Group 2B: Possibly Carcinogenic to Humans (227)

Agents and groups of agents

A-a-C(2-amino-9*H*-pyrido[2,3-b]indol)
Acetaldehyde
Acetamide
Acrylonitrile
A-F-2[2-(2-Furyl)-3-(5-nitro-2-furyl)acrylamide]
Aflatoxin M1
para-Aminoazobenzene
ortho-Aminoazotoluene
2-Amino-5-(5-nitro-2-furyl)-1,3,4-thiadiazole
Amitrole
ortho-Anisidine
Antimony trioxide
Aramite
Auramine
Azaserine
Aziridine
Benzo[*b*]fluoranthene

(*continued*)

TABLE 13.14 Continued

Benzo[*j*]fluoranthene
Benzo[*k*]fluoranthene
Benzofuran
Benzyl violet 4B
Bleomycins
Bracken fern
Bromodichloromethane
Butylated hydroxyanisole (BHA)
β-Betyrolactone
Caffeic acid
Carbon black
Carbon tetrachloride
Catechol
Ceramic fibres
Chlordane
Chlordecone (Kepone)
Chlorendic acid
para-Chloroaniline
Chloroform
1-Chloro-2-methylpropene
Chlorophenoxy herbicides
4-Chloro-*ortho*-phenylenediamine
Chloroprene
Chlorothalonil
CI Acid Red 114
CI Basic Red 9
CI Direct Blue 15
Citrus Red 2
Cobalt
para-Cresidine
Cycasin
Dacarbazine
Dantron (1,8-Dihydroxyanthraquinone)
Daunomycin
DDT (*p,p*′-DDT)
N,N′-Diacetylbenzidine
2,4-Diaminoanisole
4,4′-Diaminodiphenyl ether
2,4-Diaminotoluene
Dibenz[*a,h*]acridine
Dibenz[*a,j*]acridine
7*H*-Dibenzo[*c,g*]carbazole
Dibenzo[*a,e*]pyrene
Dibenzo[*a,h*]pyrene
Dibenzo[*a,i*]pyrene
Dibenzo[*a,l*]pyrene
1,2-Dibromo-3-chloropropane
para-Dichlorobenzene
3,3′-Dichlorobenzidine
3,3′-Dichloro-4,4′-diaminodiphenyl ether
1,2-Dichloroethane
Dichloromethane (methylene chloride)
1,3-Dichloropropene
Dichlorvos
Di(2-ethylhexyl)phthalate
1,2-Diethylhydrazine
Diglycidyl resorcinol ether
Dihydrosafrole
Diisopropyl sulfate
3,3′-Dimethoxybenzidine (*ortho*-dianisidine)
para-Dimethylaminoazobenzene
trans-2-[(Dimethylarnino)methylimino]-5-[2-(5-nitro-2-furyl)-vinyl]-1,3,4-oxadiazole
2,6-Dimethylaniline (2,6-Xylidine)
3,3′-Dimethylbenzidine (*ortho*-tolidine)
1,1-Dimethylhydrazine
3,7-Dinitrofluoranthene
3,9-Dinitrofluoranthene
1,6-Dinitropyrene
1,8-Dinitropyrene
2,4-Dinitrotoluene
2,6-Dinitrotoluene
1,4-Dioxane
Disperse Blue 1
1,2-Epoxybutane
Estrogen–progestogen therapy, postmenopausal
Ethyl acrylate
Ethylene thiourea
Ethyl methanesulfonate
2-(2-Formylhydrazino)-4-(5-nitro-2-furyl)thiazole
Furan
Glasswool
Glu-P-1 (2-Amino-6-methyldipyrido[1,2-*a*:3′,2′-*d*]imidazole)
Glu-P-2(2-Aminodipyrido[1,2-*a*:3′,2′-*d*]imidazole)
Glycidaldehyde
Griseofulvin
HC Blue No. 1
Heptachlor
Hexachlorobenzene
Hexachloroethane
Hexachlorocyclohexanes
Hexamethylphosphoramide
Human immunodeficiency virus type 2 (infection with)

(*continued*)

TABLE 13.14 Continued

Human papillomaviruses: some types other than 16, 18, 31 and 33
Hydrazine
Indeno[1,2,3-*cd*]pyrene
Iron–dextran complex
Isoprene
Lasiocarpine
Lead
Magenta
MeA-a-C(2-amino-3-methyl-9*H*-pyrido[2,3-b]indol)
Medroxyprogesterone acetate
MeIQ
MeIQx merphalan
2-Methylaziridine (Propyleneimine)
Methylazoxymethanol acetate
5-Methylchrysene
4,4′-Methylene bis(2-methylaniline)
4,4′-Methylenedianiline
Methyl mercury compounds
2-Methyl-1-nitroanthraquinone
N-Methyl-*N*-nitrosourethane
Methylthiouracil
Metronidazole
Mirex
Mitomycin C
Monocrotaline
5-(Morpholinomethyl)-3-[(5-nitrofurfurylidene)amino]-2-oxazolidinone
Nafenopin
Nickel, metallic
Niridazole
Nitrilotriacetic acid
5-Nitroacenaphthene
2-Nitroanisole
Nitrobenzene
6-Nitrochrysene
Nitrofen
2-Nitrofluorene
1-[(5-Nitrofurfurylidene)amino]-2-imidazolidinone
N-[4-(5-Nitro-2-furyl)-2-thiazolyl]acetamide
Nitrogen mustard *N*-oxide
2-Nitropropane
1-Nitropyrene
4-Nitropyrene
N-Nitrosodi-*n*-butylamine
N-Nitrosodiethanolamine
N-Nitrosodi-*n*-propylamine
3-(*N*-Nitrosomethylamino)propionitrile
4-(*N*-Nitrosomethylamino)-1-(3-pyridyl)-1-butanone (NNK)
N-Nitrosomethylethylamine
N-Nitrosomethylvinylamine
N-Nitrosomorpholine
N-Nitrosonornicotine
N-Nitrosopiperidine
N-Nitrosopyrrolidine
N-Nitrososarcosine
Ochratoxin A
Oil Orange SS
Oxazepam
Palygorskite (attapulgite)
Panfuran S
Phenazopyridine hydrochloride
Phenobarbital
Phenoxybenzamine hydrochloride
Phenyl glycidyl ether
Phenytoin
PhIP (2-amino-1-methyl-6-phenylimidazo[4,5-*b*]pyridine)
Polychlorophenols and their sodium salts (mixed exposure)
Ponceau MX
Ponceau 3R
Potassium bromate
Progestins
Progestogen-only contraceptives
1,3-Propane sultone
β-Propiolactone
Propylene oxide
Propylthiouracil
Rockwool
Safrole
Schistosoma japonicum (infection with)
Slagwool
Sodium *ortho*-phenylphenate
Sterigmatocystin
Streptozotocin
Styrene
Sulfallate
Tetrafluoroethylene
Tetranitromethane
Thioacetamide
4,4′-Thiodianiline
Thiourea
Toluene diisocyanates
ortho-Toluidine
Toxins derived from *Fusarium moniliforme*

(continued)

TABLE 13.14 Continued

Trichlormethine (trimustine hydrochloride)	Diesel fuel, marine
Trp-P-1 (3-Amino-1,4-dimethyl-5*H*-pyridol[4,3-*b*]indole)	Engine exhaust, gasoline
	Fuel oils
Trp-P-2 (3-Amino-1-methyl-5*H*-pyrido[4,3-*b*]indole)	Gasoline
Trypan blue	Pickled vegetables
Uracil mustard	Polybrominated biphenyls
Urethane	Toxaphene
Vinyl acetate	Welding fumes
4-Vinylcyclohexene	Exposure circumstances
4-Vinylcyclohexene diepoxide	Carpentry and joinery
Mixtures	Dry cleaning
Bitumens	Printing processes
Carrageenan	Textile manufacturing industry
Chlorinated paraffins (C12 and 60% Cl)	
Coffee	

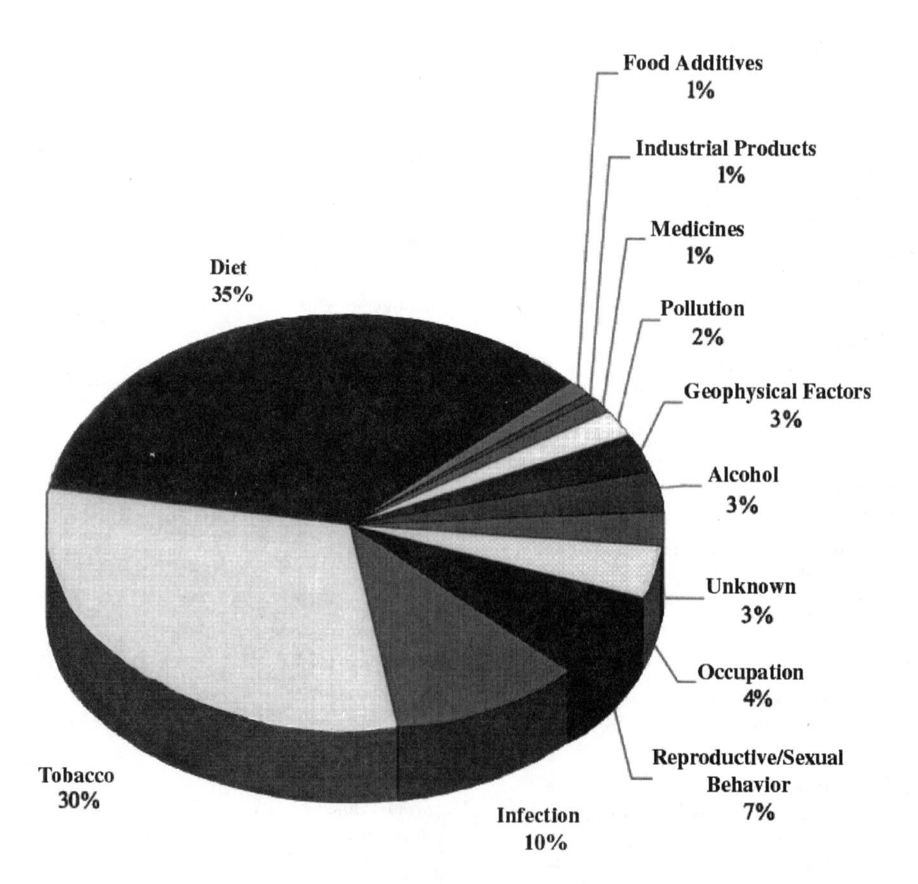

Figure 13.10 Cancer factors: approximate percent contribution (Doll and Peto, 1981)

in lung, colorectal, and breast cancers (among the major cancer types) as well as being a key factor in many rarer forms of cancer such as nevoid basal cell carcinoma. This area of research may well change the way in which we view certain chemical exposures, as the risk of cancer may ultimately be shown to be more a function of an individual's or groups of individuals' unique susceptibility to a given chemical. Such information would not only improve our understanding of the carcinogenic process, but it may alter chemical exposure regulation by allowing screening tests to eliminate potentially susceptible persons from future potentially adverse exposures.

For example, El-Zein et al. report that the inheritance of variant polymorphic genes such as CYP2D6 and CYP2E1 for the activation of certain chemicals, and GSTM1 and GSTT1 for the detoxification of certain chemicals, may predispose smokers with these traits to lung cancer. The importance of identifying the range of phenotypic expression among specific genes is clearly manifest in the impact that such changes may frequently make in the ultimate outcome of chemical exposure. In the future, identifying gene variants have a large impact on epidemiological research, cancer prevention, and the development of more effective intervention and treatment modalities. In addition, the ability to identify those genetic traits that influence certain types of cancer might become useful biomarkers that enable employers to place persons in positions that do not expose them to agents that would otherwise place them at a greater risk than the normal population.

Because of the cell transformation that occurs in carcinogenesis, there is some "genetic" component to every cancer. However, the traits referred to as one's "genetic makeup" are only a portion of the many factors that might occur in the progression from a healthy cell to an immortal, cancerous one. The role of environmental factors, as they might impact or augment hereditary or genetic elements of carcinogenesis are illustrated in Figure 13.11. The "all environmental risks" box in this diagram is intended to represent the sum of all possible environmental insults; these might come from occupational exposures, lower-level environmental chemical exposures (indoor air, drinking water, diet), diets and dietary insufficiency, viruses and other infectious diseases, and important lifestyle factors (e.g., inactivity, smoking, drinking, illicit drug use).

Smoking

The American Cancer Society (ACS) has compiled statistical data for the incidence of cancers in the U.S. population (Figure 13.12). For six major cancer sites in males in the United States, only lung cancers, which are far and away associated with tobacco smoking (perhaps 87 percent of all lung cancer deaths), have shown any demonstrable increase in the last 65+ years. The data for female cancers were similar. Lung cancer in females, driven by smoking, has now outstripped breast cancer as the leading cause of cancer death among U.S. women. The ACS stated:

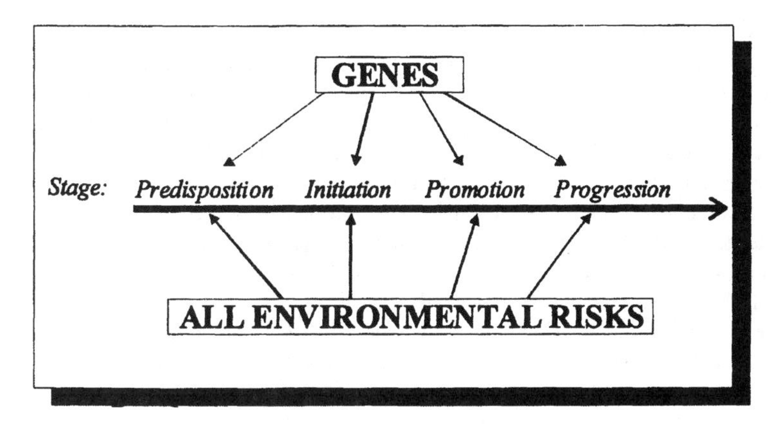

Figure 13.11 Interactions of environmental (lifestyle, diet viral, occupational) exposures and genes.

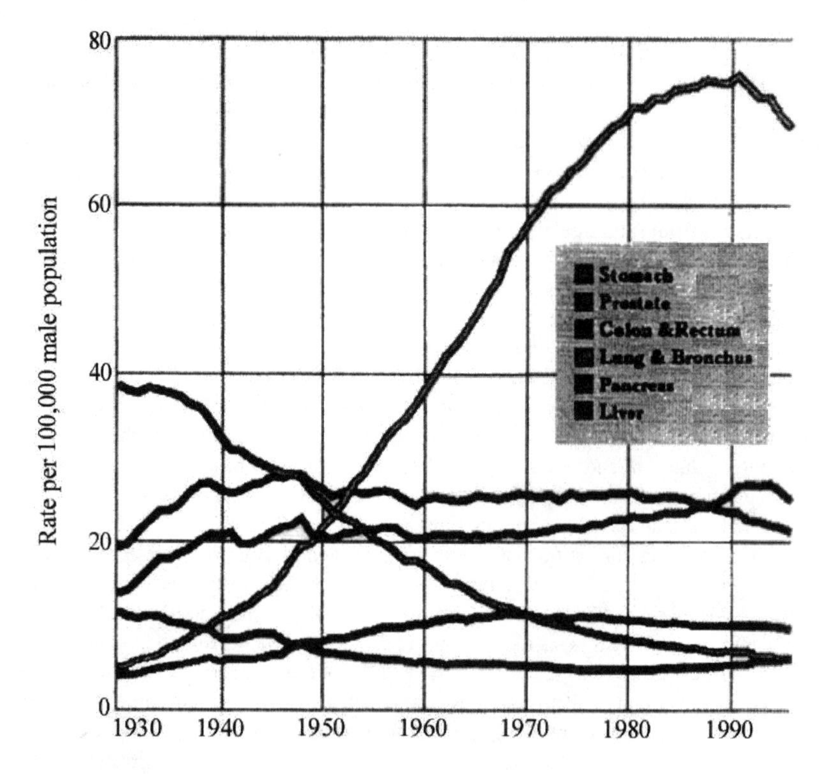

Figure 13.12 Cancer incidence rates for U.S. males, annual trends. (From *Cancer Facts and Figures—1999*, American Cancer Society.)

Lung cancer mortality rates are about 23 times higher for current male smokers and 13 times higher for current female smokers compared to lifelong never-smokers. In addition to being responsible for 87 percent of lung cancers, smoking is also associated with cancers of the mouth, pharynx, larynx, esophagus, pancreas, uterine cervix, kidney, and bladder. Smoking accounts for at least 30 percent of all cancer deaths, is a major cause of heart disease, and is associated with conditions ranging from colds and gastric ulcers to chronic bronchitis, emphysema, and cerebrovascular disease.

The data surrounding smoking is particularly distressing for persons who might be occupationally exposed to other substances as well. Asbestos-exposed workers who smoke reportedly contract lung cancer at a rate that is 60 times that of persons not exposed to either substance. Other risk factors for lung cancer may include exposure to arsenic, some organic chemicals, radon, radiation exposure from occupational, medical, and environmental sources. Smokers who incur such exposures should be aware of the increased risks they face compared to their nonsmoking co-workers.

Research has identified more than 40 carcinogenic substances emitted in tobacco smoke. Many of these substances are initiating agents (genotoxic) and are capable of inducing cancer by themselves at sufficient doses, others are recognized as promoters or cocarcinogens and act to enhance the activity of chemicals initiating the key genetic change. With so many different chemical carcinogens contained in cigarette smoke, it seems logical to ask if cigarette smoking is largely a phenomenon of initiation or promotion. If lung cancer due to cigarette smoking was the result of initiating carcinogens, the observed risk should arguably be proportional to cumulative lifetime exposure, and the cessation of cigarette smoking would not alter the already accumulated pack/year risk (i.e., one's risk of cancer, once achieved, could not be decreased with abstinence). Current data, however, is contradictory to this suggestion, and studies indicate that as the duration of abstinence from smoking increases, a person's lung cancer risk actually becomes lower until it eventually approaches the risk faced by a nonsmoker. For this reason, many have argued that the affect of cigarette smoking is largely one of promotion. Regardless of whether smoking is largely due to promotion or initiation, it is clearly an avoidable health hazard and after factoring in the increased risk from cerebrovascular disease due to smoking is arguably society's greatest contributor to preventable causes of death.

Alcohol

Alcohol is another clearly avoidable cancer risk. Alcohol consumption is causally related to cancers of the oral cavity, pharynx, larynx, esophagus, and liver. The combined use of alcohol and tobacco products also leads to an increased incidence of oral cavity, esophagus, and larynx cancers. Associations between alcohol and breast cancer have also been proposed. Estimates of the contribution of alcohol to cancer in the United States range as high as 5 percent; however, it is estimated that there are some 10 million problem drinkers in the United States, and so, the influence ultimately exerted upon the national cancer incidence by alcohol might not be fully determined at the present time.

There are several theories regarding the carcinogenic activity of alcohol. Alcohol is known to induce specific oxidative enzymes and so is suspected of potentially enhancing the initiation activity of certain carcinogens. It has also been proposed to make tissues more responsive to the action of a carcinogen by increasing cell permeability or by increasing the effective concentration of a carcinogen intracellularly. Ethanol is cytotoxic chemical at high doses, and recurrent cellular injury has been suggested as another possible mechanism for ethanol-induced or enhanced carcinogenesis. The fact that the development of cirrhosis often precedes and frequently ends in primary liver cancer would tend to support this hypothesis. Other possible mechanisms include the generation of free radicals (via lipid peroxidation), and possibly some immunosuppressive effect. Regardless of the mechanism or mechanisms by which chronic alcohol intake induces cancer or enhances the response of other carcinogens, it clearly remains as a clearly important, but avoidable, cancer risk factor.

Diet

When Doll and Peto released their statistical analysis of the causes of cancer, many authors noted the impact that diet had on cancer incidence was as yet unknown, or at best, very much debated. Diet, via the intake of high quantities of animal fats, can have a decidedly negative impact on a person's health and such diets are clearly linked to higher incidences of cancers. However, diet is a double-edged sword in that it can also be an important moderating influence by providing antioxidants, anticarcinogens, and other nutritional benefit that helps the body's detoxification and repair mechanisms to fight off tumorigenic activity. So, with the possible exception of the cessation of smoking, the improvement of our diet can have the greatest impact on our own health and the national cancer rate.

It is now well recognized that the plants we consume as part of our diet contain their own natural pesticides. In fact, certain strains of plants have been cultivated with the purpose of enhancing these natural defense mechanisms and so require less maintenance and care. However, as was seen with the increased use of synthetic chemicals, this can enhance the toxicity of the foods we consume. As with the synthetic chemicals tested in the chronic animal cancer bioassay, the carcinogenic activity of the "natural" pesticides normally contained in vegetables and fruits is running at roughly 50 percent for the chemicals tested. Thus, it has been argued that when chemicals are tested in high-dose animal cancer bioassays one can expect approximately half of the chemicals tested, human-made (synthetic) or natural, to elicit carcinogenic activity. Based on these projections and on the currently available data, it has been estimated that 99.9 percent of our total pesticide intake is via the ingestion of natural, plant-produced pesticides. In fact, it would appear we ingest as much as 1.5 g (1500 mg) of plant-produced, natural pesticides each day.

Recently, the National Research Council's (NRC) Board of Environmental Studies and Toxicology Committee on Comparative Toxicity of Naturally Occurring Carcinogens published a conclusion similar to that of Ames. Although the committee admitted that more research was needed before definitive conclusions could be drawn, it stated that natural components of the diet were likely to be more significant with respect to cancer risk than were synthetic chemicals found in food. The committee's conclusion was based on the amounts of foods consumed by the typical U.S. citizen and the levels of natural or synthetic pesticides present in those foods. The committee refers to various studies, including the National Health and Nutrition Examination Surveys (NHANES, the recent study of pesticides in the diets of infants and children, and the Nationwide Food Consumption survey performed by the US Department of Agriculture (USDA) as sources of data for their analysis. The NRC committee interpreted from these different studies that Americans consume a large number of natural and synthetic carcinogens in their diets. The committee also based its conclusion regarding the potential significance of dietary carcinogens on the fact that the natural dietary substances studied to date have, on average, a greater carcinogenic potency than the synthetic chemicals found in food.

A diet high in animal fats has been implicated in numerous epidemiologic and case-control studies as being a factor in colorectal and possibly prostate cancer. Excess dietary fat is thought to induce cancer by a number of potential mechanisms, including the alteration of hormone levels, a change in the composition of cellular membranes, an increase in fatty acids (which may inhibit immune responses or serve as precursors to prostaglandins, which may then act as promoters), and a stimulation of the production of liver bile acids, some of which can act as promoters. Diet has been linked to numerous other cancers as well (Table 13.15).

Microorganisms normally found in foods, such as fungi, are another potential source of carcinogens. For example, mycotoxins are prominently distributed in the food chain, and the prevalence of *Aspergillus* in the environment, a producer of dietary aflatoxins, appears to contribute significantly to the higher risk of liver cancer that is observed in some third world countries. *Fusarium monilifome* is ubiquitous in corn and produces fumonisins B_1, B_2, and fusarin C, all of which have been implicated in human esophageal cancer.

Cooking is another factor that may alter the dietary carcinogen load. Cooking alters the chemical structure of foods, and cooking has long been known to produce cyclic compounds, a number of which

TABLE 13.15 Cancer Sites and Associated Risks/Benefits of Diets

	Probable		Possible	
Site of Cancer	Increases Risk	Decreases Risk	Increases Risk	Decreases Risk
Colorectum	Red meat, processed meat	Vegetables, nonstarch, polysaccharides	Alcohol, fat	Folate
Breast	Alcohol, red meat, fried meat	Vegetables		Fruit, phytoestrogens
Lung			Alcohol, meat	Fruit and vegetables
Stomach	Salt, pickled and preserved food	Fruit and vegetables, vitamin C		Carotenoids
Prostate		Vitamin E	(Red) meat, fat	Vegetables
Cervix		Fruit and vegetables, vitamin C		Folate, vitamin A
Esophagus	Alcohol	Fruit and vegetables		
Pancreas			Red meat	Fruit and vegetables, vitamin C, nonstarch, polysaccharides
Bladder		Fruit and vegetables		
Liver	Alcohol			

Source: Adapted from Cummings and Bingham (1998).

are mutagens and carcinogens. For example, polycyclic heterocyclic amines (PHAs) are produced when any amino acid is pyrolyzed (e.g., in broiling a beefsteak), and many of these are highly mutagenic. Broiling and charring foods may also increase the presence of polyaromatic hydrocarbons (PAHs).

Other carcinogens are among those chemicals that are frequently found as natural or added constituents of the foods that make up our diet (Tables 13.16 and 13.17) or as synthetic chemical pesticide or other residues (Table 13.18). For example, caffeic acid occurs in higher plants and has produced tumors in both male and female rats. The rodent carcinogen (rabbits, hamsters and mice) *n*-nitrosodimethylamine is found in cheeses, bacon, frankfurters, soybean oil, smoked or cured meats, fish, and some alcoholic beverages, including beer. Nitrates and nitrites occur naturally and are introduced to foods in curing and preserving processes. It has been argued that nitrites may form carcinogenic nitrosamines in the acid environment of the stomach by combining with amines of the aminoacids that form the protein in our diets. Thus, cooking, curing processes, applied chemicals (fertilizers, pesticides, soil or water contamination, etc.), and the selective growth of insect resistant plants are ways in which the carcinogenic load or potential of the foods we ingest may be altered. Typically, these sources outweigh the contributions by the application of synthetic pesticide by perhaps as much as 10,000-fold. So, although it is clear that naturally occurring chemicals outweigh the synthetic chemicals we are exposed to in our diet. However, the relative contribution to the incidence of cancer by these exposures is generally considered to be far less than is caused by the intake of excess calories via animal fat ingestion.

Finally, diets deficient in iron, selenium, and vitamin C have all been associated with increased cancer rates. Vitamin C has been shown to inhibit the formation of certain initiating carcinogens, vitamin E appears to prevent promotion, and vitamin A appears to decrease the susceptibility of epithelial tissue to carcinogens.

Overall, the evidence indicates diet can have a profound effect on the incidence of cancer, and estimates that have diet contributing to as high as 70 percent of the total cancer incidence [perhaps as much as 80 percent of large bowel (colon) and breast cancers] can be found in the scientific literature. In addition, differences in diet may explain some regional geographic differences in the distribution and frequency of the cancer types observed. Like drinking alcohol and smoking, diet can also have a unknown impact on the results of epidemiologic investigations, an impact that is often inadequately investigated.

TABLE 13.16 Natural Pesticides and Metabolites Found in Cabbage

Glucosinolates: 2-propenyl glucosinolate (sinigrin),[a] 3-methylthiopropyl glucosinolate, 3-methylsulfinyl-propyl glucosinolate, 3-butenyl glucosinolate, 2-hydroxy-3-butenyl glucosinolate, 4-met hylsulfinylbutyl glucosinolate, 4-methylsulfonylbutyl glucosinolate, benzyl glucosinolate, 2-phenylethyl glucosinolate, propyl glucosinolate, butyl glucosinolate

Indole glucosinolate and related indoles: 3-indolylmethyl glucosinolate (glucobrassicin), 1-methoxy-3-indolylmethyl glucosinolate (neoglucobrassicin), indole-3-carbinol,[a] indole-3-acetonitrile, bis(3-indolyl)methane

Isothiocyanates and goitrin: allyl isothiocyanate,[a] 3-methylthiopropyl isothiocyanate, 3-methylsulfinylpropyl isothiocyanate, 3-butenyl isothiocyanate, 5-vinyloxazolidine-2-thione (goitrin), 4-methylthio butyl isothiocyanate, 4-methylsulfinylbutyl isothiocyanate, 4-methylsulfonylbutyl isothiocyanate, 4-pent enyl isothiocyanate, benzyl isothiocyanate, phenylethyl isothiocyanate

Cyanides: 1-cyano-2,3-epithiopropane, 1-cyano-3,4-epithiobutane, 1-cyano-3,4-epithiopentane, *threo*--1-cyano-2-hydroxy-3,4-epitiobutane, *erythro*--1 -cyano-2-hydroxy-3,4-epithiobutane, 2-phenylpropionitrile, allyl cyanide,[a] 1-cyano-2-hydroxy-3-butene, 1-cyano-3-methylsulfinylpropane, 1-cyano-4-methylsulfinylbut ane

Terpenes: menthol, neomenthol, isomenthol carvone[a]

[a]Indicates data on mutagenicity or carcinogenicity (see Ames et al. 1990 for discussion of data); others untested.

Source: Adapted from Ames et al. (1990)

TABLE 13.17 Naturally Occurring Carcinogens Potentially Present in U.S. Diets

Constitutive: acetaldehyde, benzene, caffeic acid, cobalt, estradiol 17β, estrone, ethyl acrylate, (with UV light exposure), 8-methoxypsoralen (xanthotoxin) (with UV light exposure), progesterone, safrol e, styrene, testosterone

Derived: A-alpha-C, acetaldehyde, benz(*a*)anthracene, benzene, benzo(*a*)pyrene, benzo(*b*)fluoranthene, benzo(*j*)fluoranthene, benzo(*k*)fluoranthene, dibenz(*a,h*)acridine, dibenz(*a,j*)acridine, dibenz(*a,h*)anthracene, formaldehyde, glu-P1, glu-P2, glycidaldehyde, IQ, Me-A-alpha-C, MEIQ, MeIQx, methyl mercury compounds, *N*-methyl-*N'*-nitro-nitrosoquanidine, *N*-nitroso-*N*-dibutylamine, *N*-nitorosodiethylamine, *N*-nitrosodimethylamine, *N*-nitrosodi-*N*-propylamine, *N*-nitorosomehtylethylamine, *N*-nitrosopiperidine, *N*-nitrosopyrrolidine, *N*-nitrososarcosine, PhIP, Trp-P1, Trp-P2, urethane

Acquired: aflatoxin B_1, aflatoxin M_1, ochratoxin A, sterigmatocystin, toxins derived from *Fusarium moniliforme*

Pass through: arsenic, benz(*a*)anthracene, benzo(*a*)pyrene, beryllium, cadmium, chromium, cobalt, indeno(1,2,3)pyrene, lead, nickel

Added:

Contaminant introduced through tap water: arsenic, asbestos, benzene, beryllium, cadmium, hexavalent chromium, dibenzo(*a,l*)pyrene, indenol(1,2,3,-cd)pyrene, radon

Indirect through use as a drug or in packaging: i) veterinary drugs—estradiol 17β, progesterone, reserpine, testosterone, ii) food packaging material—benzene, cobalt, ethyl acrylate, formaldehyd e, nickel

Direct food additives: acetaldehyde, ethyl acrylate, formaldehyde

Traditional foods and beverages: alcoholic beverages, betel liquid, bracken fern, hot mate, pickled vegetables, salted fish (Chinese style)

Source: Adapted from Table 5-1, NRC (1996). Reprinted with permission from *Carcinogens and Anticarcinogens in the Human Diet.* Copyright 1996 by the National Academy of Sciences. Courtesy of the National Academy Press, Washington, D.C.

Iatrogenic Cancer

The use of drugs that might impact the cancer incidence in a given population, is rarely addressed in the mortality studies of occupational cohorts from which we derive much of our knowledge regarding chemical carcinogenicity. No chemical has only one effect and pharmaceutical medications are no exception to this rule. Pharmaceuticals are known to be capable of producing side effects other than the desired therapeutic effect. A surprising number of drugs are known to have carcinogenic effects. Perhaps the most well-known class of agents with such effects is, of course, the potent chemotherapeu-

TABLE 13.18 Synthetic Carcinogens that Might Be Present in Foods

Pesticide residues: acrylonitrile, amitrole, aramite, atrazine, benzotrichloride, 1,3-butadiene, captafol, carbon tetrachloride, chlordane, Kepone, chloroform, 3-chloro-2-methylpropene, *p*-chloro-*o*-toluidine, chlorophenoxy herbicides, creosotes, DDD, DDE, DDT, Dichlorvos, 1,2-dibromo-3-chloropropane, *p*-dichlorobenzene, 1,2-dichloroethane, 2-dichloroethane, dichloromethane, 1,3-dichloropropene, dimethyl carbamoyl chloride, 1,1-dimethylhydrazine, ethylene dibromide, ethylene thiourea, heptachlor, hexa chlorobenzene, hexachlorocyclohexane, mirex, *N*-nitrosodiethanolamine, pentachlorophenol, *o*-phenylphenate, nitrofen, 1,3-propane sulfone, propylene oxide, styrene oxide, sulfallate, tetrachlorodibenzo-*p*-dioxin, thiourea (past), toxaphene, 2,4,6-trichlorophenol

Potential animal drug residues: diethylstilbesterol (now banned), ethinyl estradiol, medroxyprogesterone acetate, methylthiouracil, *N*-[4-(5-nitro-2-furyl)-2-thiazolyl]acetamide, nortestosterone, propylthiouracil

Packaging or storage container migrants: acrylamide, acrylonitrile, 2-aminoanthraquinone, BHA, 1,3-butadiene, chlorinated paraffins, carbon tetrachloride, chloroform, 2-diaminotoluene, di(2-et hylhexyl)phthalate, dimethylformamide, diethyl sulfate, dimethyl sulfate, 1,4-dioxane, ethyl acrylate, epichloro hydrin, ethylene oxide, ethylene thiourea, 2-methylaziridine, 4,4′-methylenedianiline, 4,4′-methylene bis(2-chloroaniline) (now prohibited), 2-nitropropane, 1-nitropyrene, phenyl glycidyl ether, propylene oxide, s odium phenyl phenate, sodium saccharin, styrene, styrene oxide, tetrachloroethylene, toluene diisocyanate, vinyl chloride

Residues from food processing: dichloromethane, epichlorohydrin, NTA trisodium salt monohydrate

Source: Adapted from Table 5-3, NRC (1996). Reprinted with permission from *Carcinogens and Anticarcinogens in the Human Diet.* Copyright 1996 by the National Academy of Sciences. Courtesy of the National Academy Press, Washington, D.C.

TABLE 13.19 Pharmaceutical Agents[a] with Carcinogenic Effects[b]

Generic Name	Therapeutic Use	Daily Dosage (mg/day)	Tumor Site; Species
Rifampin	Antibiotic: tuberculosis	600	Liver; mice
Isoniazid	Antibiotic: tuberculosis	300	Lung; mice
Clofibrate	Lowers cholesterol	2000	Liver; mice
Disulfiram	Discourages alcohol abuse	125–500	Liver; rats
Phenobarbital	Antiepileptic	100–200	Liver; mice
Acetaminophen	Pain relief (OTC)	2000–4000	Liver; mice
Metronidazole	Antibiotic, antiparasitic	500	Lung; rats/mice
Sulfisoxazole	Antibiotic, urinary tract	8000	
Dapsone	Antibacterial, AIDS, leprosy, etc.	300	Spleen, thyroid, and peritoneum; rats
Methimazole	Hypothyroidism	15	Thyroid and pituitary tumors; rats
Oxazepam	Antianxiety	70	Thyroid, testes, prostate; rats/mice liver; mice
Furosemide	Water retention in disease states	75	

[a]List adapted from Waddell (1996).

[b]Cancer effects as listed in the *Physicians Desk Reference* (PDR), 1996, or Ames and Gold (1991).

tic agents used to treat cancer. Many antineoplastic drugs are potent genotoxic chemicals, and their damage to DNA in rapidly dividing cells like cancer cells is a primary feature of both their therapeutic effects and toxicities. Admittedly, it may be well worth a theoretical risk of developing cancer 20 years after taking medication to cure a current case of cancer, however, a number of drugs whose therapeutic benefits are directed at less serious health conditions are also known to have carcinogenic effects in humans or in animal cancer bioassays. Some potentially carcinogenic pharmaceuticals are listed in Table 13.19.

Not only are many of the drugs listed in Table 13.19 commonly prescribed, but the single daily doses of these chemicals are large relative to the doses of chemicals one is typically concerned with when evaluating environmental pollutants. Thus, the theoretical risks associated with even limited therapy may approach or exceed the theoretical risks posed by the environmental contamination we are often concerned about when remediating sites that contain these contaminants.

13.10 CANCER TRENDS AND THEIR IMPACT ON EVALUATION OF CANCER CAUSATION

Human Cancer Trends in the United States

As mentioned regarding smoking, the incidence of cancer in this nation has remained stable, or declined, for most types of cancer according to the American Cancer Society. The greatest exception is, of course, lung cancer in both males and females. A 1998 report from the National Cancer Institute (NCI) (see Table 13.20) indicated that after increasing 1.2 percent per year from 1973 to 1990, incidence for all cancers combined declined in the United States an average of 0.7 percent from 1990 to 1995. Cancer mortality similarly declined about 0.5 percent per annum for the same period (1990–1995). Cancers of the lung, breast, prostate, and colon–rectum accounted for over half of the new cases. Cancer of the lung, both incidence and mortality, is actually showing a slight decline while in women, such cancers (and the resultant mortality) are still on the increase. Incidence and mortality

TABLE 13.20 Estimated New Cancer Cases and Deaths by Sex for All Sites, United States, 1999[a]

	Estimated New Cases			Estimated Deaths		
Cancer Sites	Both Sexes	Male	Female	Both Sexes	Male	Female
All sites	1,221,800	623,800	598,000	563,100	291,100	272,000
Oral cavity and pharynx	29,800	20,000	9,800	8,100	5,400	2,700
Digestive system	226,300	117,200	109,100	131,000	69,900	61,100
Esophagus	12,500	9,400	3,100	12,200	9,400	2,800
Stomach	21,900	13,700	8,200	13,500	7,900	5,600
Small intestine	4,800	2,500	2,300	1,200	600	600
Colon	94,700	43,000	51,700	47,900	23,000	24,900
Rectum, anus, etc.	38,000	19,400	15,300	8,700	4,800	3,900
Liver and intrahepatic bile duct	14,500	9,600	4,900	13,600	8,400	5,200
Gallbladder and other biliary	7,200	3,000	4,200	3,600	1,300	2,300
Pancreas	28,600	14,000	14,600	28,600	13,900	14,700
Other digestive organs	4,100	1,200	2,900	1,200	400	800
Larynx	10,600	8,600	2,000	4,200	3,300	900
Lung and bronchus	171,600	94,000	77,600	158,900	90,900	68,000
Other respiratory organs	5,400	4,200	1,200	1,100	700	400
Bones and joints	2,600	1,400	1,200	1,400	800	600
Soft tissue (including heart)	7,800	4,200	3,600	4,400	2,100	2,300
Skin (no basal and squamous)	54,000	33,400	20,600	9,200	5,800	3,400
Breast	176,300	1,300	175,000	43,700	400	43,300
Genital system	269,100	188,100	81,000	64,700	37,500	27,200
Uterine corpus	37,400		37,400	6,400		6,400
Ovary	25,200		25,200	14,500		14,500
Vulva	3,300		3,300	900		900
Vagina and other female genitalia	2,300		2,300	600		600
Prostate	179,300	179,300		37,000	37,000	
Testis	7,400	7,400		300	300	
Penis and other genitalia, male	1,400	1,400		200	200	
Urinary bladder	54,200	39,100	15,100	12,100	8,100	4,000
Kidney and renal pelvis	30,000	17,800	12,200	11,900	7,200	4,700
Ureter and other urinary	2,300	1,500	800	500	300	200
Eye and orbit	2,200	1,200	1,000	200	100	100
Brain and other nervous system	16,800	9,500	7,300	13,100	7,200	5,900
Endocrine system	19,800	5,400	14,400	2,000	900	1,100
Hodgkin's disease	7,200	3,800	3,400	1,300	700	600
Non-Hodgkin's lymphoma	56,800	32,600	24,200	25,700	13,400	12,300
Multiple myeloma	13,700	7,300	6,400	11,400	5,800	5,600
All leukemias	30,200	16,800	13,400	22,100	12,400	9,700
Other primary sites	35,100	16,400	18,700	36,100	18,200	17,900

[a]Excludes basal and squamous cell skin cancers and in situ carcinomas except urinary bladder Carcinoma in situ of the breast accounts for about 39,900 new cases annually, and melanoma carcinoma in situ accounts for abut 23,200 new cases annually. Estimates of new cases are based on incidence rates from the NCI SEER program, 1979–1995. Amerian Cancer Society, Surveillance Research, 1999.

from non-Hodgkin's lymphoma and from melanoma are also increasing. These data were confirmed in the 1999 joint release from the CDC, NCI, and ACS.

As awareness of environmental contamination and the ubiquity of synthetic chemicals arose in the 1960s, specifically after the release of Rachel Carson's *Silent Spring* in 1962, speculation persisted that we were awash in a "sea of carcinogens" and that after an appropriate latency interval, a cancer epidemic would hit. As has been shown in Figure 13.1, the hypothesized epidemic of cancer has never arrived, and considering the data indicating decreasing cancers through 1995, it would seem that perhaps our current reductions in smoking, food consumption, and alcohol might be starting to impact the incidence of new cancers in the United States. Considering that the stability of the cancer incidence (aside from lung cancer due primarily to smoking) occurred during a period of great industrialization in the United States, the impact of occupation and environmental pollution on cancer incidence is probably less than what was postulated 30 years ago. That is not to say, however, that exposure reductions are not still warranted in these areas, but merely to point out that the current data indicate that our future, with its concomitant exposure to new synthetic chemicals, is not a dire one.

13.11 SUMMARY

Chemical-induced carcinogenesis represents a unique and complex area within toxicology. The difficulty in assessing the carcinogenic hazards and human risks of chemicals stems from the following characteristics of chemical carcinogenesis:

- It is a multistage process involving at least two distinct stages: initiation, which converts the genetic expression of the cell from a normal to aberrant cell line; and promotion, in which the aberrant cell is stimulated in some fashion to grow, thereby expressing its altered state.
- Since chemicals may increase cancer incidence at various stages and by different mechanisms, the term *carcinogen* by itself is somewhat limiting and a number of descriptive labels are applied to the chemical carcinogens that define or describe these differences, such as cocarcinogens, initiators, promoters, and epigenetic.
- Chemicals may produce or affect only a single stage or a single aspect of carcinogenesis that leads to a number of important differences and considerations about the potential health impacts of chemical carcinogens. Perhaps the most important considerations are the concept of thresholds and that qualitative differences do exist among carcinogens.
- Carcinogenicity testing raises many questions about interpretations of results. Considerations such as mechanism (genotoxic vs. epigenetic), dose, and relevant test species, are important in determining probable human risk; thus, many additional toxicity test data are needed to improve the extrapolation of cancer bioassay data from test species to humans.
- A number of lifestyle-related factors influence carcinogenesis, altering the risks posed by carcinogenic chemicals and acting to confound epidemiological evidence.

Considering the complexities involved in (1) determining the mechanism of cancer causation, (2) using animal and human data to identify carcinogenic substances, and (3) using these data to extrapolate risks with the aim of reducing or eliminating environmental risk factors, it should be clear to the reader that the best approach to occupational carcinogenesis is an interdisciplinary one. As depicted in Figure 13.13, identifying and reducing occupational cancer requires the interfacing of several scientific disciplines and several kinds of health professionals:

- The toxicologist is responsible for testing and identifying chemical carcinogens; through animal testing the toxicologist attempts to provide information about carcino-

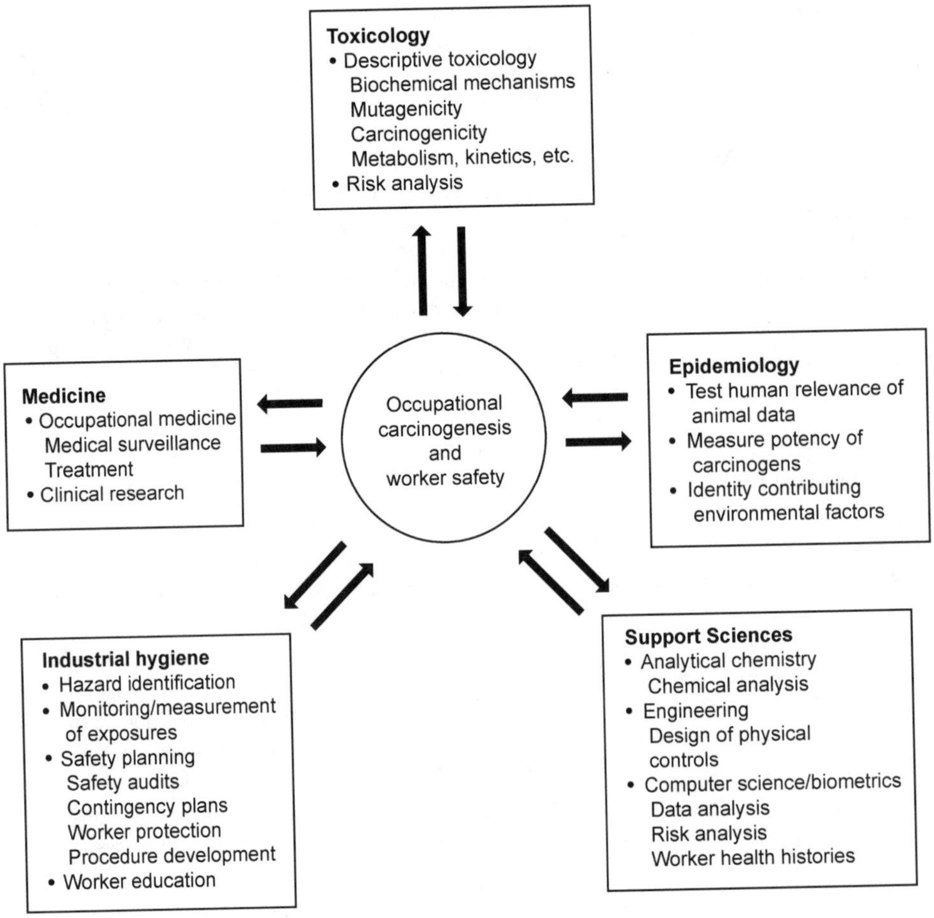

Toxicology
- Descriptive toxicology
  Biochemical mechanisms
  Mutagenicity
  Carcinogenicity
  Metabolism, kinetics, etc.
- Risk analysis

Medicine
- Occupational medicine
  Medical surveillance
  Treatment
- Clinical research

Occupational carcinogenesis and worker safety

Epidemiology
- Test human relevance of animal data
- Measure potency of carcinogens
- Identity contributing environmental factors

Industrial hygiene
- Hazard identification
- Monitoring/measurement of exposures
- Safety planning
  Safety audits
  Contingency plans
  Worker protection
  Procedure development
- Worker education

Support Sciences
- Analytical chemistry
  Chemical analysis
- Engineering
  Design of physical controls
- Computer science/biometrics
  Data analysis
  Risk analysis
  Worker health histories

Figure 13.13 Identifying and reducing chemical carcinogens requires an interdisciplinary approach, in which health professions interface with other scientific disciplines.

genic mechanisms, and about species differences or similarities that can aid in assessing the human risk.

- Epidemiologists add human evidence to risk evaluations or ascertain if a chemical should or should not be considered a human carcinogen for various reasons (it may have weak or undetectable activity).
- Specialists in occupational medicine provide health surveillance programs to protect the health status of the worker and attempt to prevent those exposures that could lead to serious, chronic health problems.
- Industrial hygienists help design better methods for evaluating and preventing worker exposures; and biometrists and computer scientists aid in risk analysis, data storage, and data analysis.

As long as these disciplines are utilized jointly and their relationship to the occupational carcinogenesis problem is understood, occupational health and safety professionals can have good reason to hope for improved success in the prevention of occupational carcinogenesis.

REFERENCES AND SUGGESTED READING

ACS (American Cancer Society), *Cancer Facts and Figures*. New York: American Cancer Society (1999).

Ames, B. N., M. Profet, and L. S. Gold, Dietary pesticides (99.99% all natural). *Proceedings of the National Academy of Science,* **87:** 7777–7781 (1990).

Ames, B. N., L. S. Gold, and M. K. Shigenaga, Cancer prevention, rodent high-dose cancer tests, and risk assessment. *Risk Analysis,* **16:** 613–617 (1996).

Ames, B. N., and L. S. Gold, The causes and prevention of cancer: Gaining perspective. *Environmental Health Perspectives,* **105**(Sup4): 865–873 (1997).

Arcos, J. C., M. F. Argus, and Y. T. Woo, *Chemical Induction Of Cancer: Modulation and Combination Effects,* Birauser, Boston, MA, (1995).

Ashby, J., F. R. Johannsen, G. K. Raabe, N. G. Doerrer, S. C. Lewis, R. C. Reynolds, F. G. Flamm, N. D. Krivanek, J. M. Smith, J. T. Stevens, J. E. Harris, J. F. McCarthy, M. J. Teta, D. H. Hughes, R. J. Moolenar, and J. D. Wilson, *A Scheme for Classifying Carcinogens.* New York: Academic Press, Inc. 0273-2300/90 (1990).

Barrett, J. C., Mechanisms of multistep carcinogenesis and carcinogen risk assessment. *Environmental Health Perspectives,* **100:** 9–20 (1993).

Baylin, S. B., Herman J. G., Graff, J. R., Vertino, P,M., Issa, J. P., Alterations in DNA methylation: a fundamental aspect of neoplasia. *Advances in Cancer Research,* **72:** 141–196 (1998).

Cummings, J. H., and S. A. Bingham, Diet and the prevention of cancer. *British Medical Journal* **317:** 1636–1640 (1998).

Cunningham, M. L., J. Foley, R. R. Maronpot, and H. B. Matthews, Correlation of hepatocellular proliferation with hepatocarcinogenicity induced by the mutagenic noncarcinogen:carcinogen pair 2,6- and 2,4-diaminotoluene. *Toxicology and Applied Pharmacology,* **107:** 562–567 (1991).

D'Amato, R., T. J. Slaga, W. H. Farland, and C. Henry, *Relevance Of Animal Studies To The Human Cancer Risk,* New York: Wiley-Liss (1992).

DeVita, V. T., S. Hellman, and S. A. Rosenberg, *CANCER: Principles & Practice of Oncology, Volume 1,* Philadelphia: Lippincott-Raven (1997).

Doll, R., and R. Peto, The causes of cancer: Quantitative estimates of avoidable risks of cancer in the United States today. *Journal of the National Cancer Institute,* **66:** 1191–1308 (1981).

Ennever, F. K., T. J. Noonan, and H. S. Rosenkranz, The predictivity of animal bioassays and short-term genotoxicity tests for carcinongenicity and non-carcinogenicity to humans. *Mutagenesis,* **2:** 73–78 (1987).

El-Zein, R., N. Conforti-Froes, and W. W. Au, Interactions between genetic predisposition and environmental toxicants for development of lung cancer. Environmental and Molecular Mutagenesis **30:** 196–204 (1997).

Eastin, W. C., Haseman, J. K., Mahler, J. F., and Bucher, J. R., The National Toxicology Program evaluation of genetically altered mice as predictive models for identifying carcinogens. *Toxicologic Pathology,* **26:** 461–473 (1998).

Gold, L. S., T. H. Slone, N. B. Manley, G. B. Garfinkel, E. S. Hudes, L. Rohrbach, and B. N. Ames, The carcinogenic Potency Database: Analyses of 4000 chronic animal cancer experiments published in the general literature and by the U.S. National Cancer Institute/National Toxicology Program. *Environmental Health Perspectives,* **96:** 11–15 (1991).

Gold, L. S., T. H. Slone, and B. N. Ames, What do animal cancer tests tell us about human cancer risk? Overview of analyses of the carcinogenic potency database. *Drug Metabolism Reviews,* **30:** 359–404 (1998).

Harris, C. C., p53 tumor suppressor gene: at the crossroads of molecular carcinogenesis, molecular epidemiology, and cancer risk assessment. *Environmental Health Perspectives,* **104**(suppl. 3): 435–439 (1996).

Haseman, J. K., and J. E. Huff, Species correlation in long-term carcinogenicity studies. *Cancer Letters,* **37:** 125–132 (1987).

Haseman, J. K., and A.-M. Lockhart, Correlations between chemically related site-specific carcinogenic effects in long-term studies in rats and mice. *Environmental Health Perspectives,* **101:** 50–54 (1993).

Huff, J., and J. Haseman, Long-term chemical carcinogenesis experiments for identifying potential human cancer hazards: Collective database of the National Cancer Institute and National Toxicology Program (1976–1991). *Environmental Health Perspectives,* **96:** 23–31 (1991).

IARC (International Agency for Research on Cancer), IARC Monographs Programme on the Evaluation of Carcinogenic Risks to Humans. Lists of IARC Evaluations (1999). *http://193.51.164.11/monoeval/grlist.html.*

Knudson, A. G., Hereditary predisposition to cancer. Annals New York Academy of Sciences, **833:** 58–67 (1997).

Lawley, P. D., Historical origins of current concepts of carcinogenesis. *Advances in Cancer Research*, **65:** 18–112 (1994).

Lyons, S. K., and Clarke, A. R., Apoptosis and carcinogenesis. *British Medical Bulletin*, **52:** 554–569 (1997).

NRC (National Research Council), *Carcinogens and Anticarcinogens in the Human Diet.* Washington: National Academy Press (1996).

NTP (The National Toxicology Program) 1999. Eighth Report on Carcinogens. *http://ntp-server.niehs.nih.gov/NewHomeRoc/CurrentLists.html.*

Pitot, H. C. and Y. P. Dragan, Chemical Carcinogenesis. In: Amdur, M. and Classan, K., eds. *Casarett and Doull's Toxicology*, 5th ed. New York: Pergamon Press (1997).

Ruddon, R. W., *Cancer Biology, Third Edition*. New York: Oxford University Press (1995).

Smart, R. C., and J. K. Akunda, "Carcinogenesis." In Hodgson, E. and Smart, R.C. (Eds.) *Introduction to Biochemical Toxicology*. New York: John Wiley & Sons (2000).

Stanley, L. A. Molecular aspects of chemical carcinogenesis: the roles of oncogenes and tumor suppressor genes. *Toxicology*, **96:** 173–194 (1995).

Squire, R. A., Ranking animal carcinogens: A proposed regulatory approach. Science **214:** 877–880 (1981).

Tennant, R. W., and J. Ashby, Classification according to chemical structure, mutagenicity to Salmonella and level of carcinogenicity of a further 39 chemicals tested for carcinogenicity by the U.S. National Toxicology Program. *Mutation Research*, **257:** 209–227 (1991).

Vogelstein, B., and K. W. Kinzler, The multistep nature of cancer. *Trends in Genetics*, **9:** 138–141 (1993).

Waddell, W. J., Reality versus extrapolation: An academic perspective of cancer risk regulation. *Drug Metabolism Reviews*, **28:** 181–195 (1996).

Weisburger, J. H., and G. M. Williams, Carcinogen testing: Current problems and new approaches. *Science* **214:** 401 (1981).

Weiss, N. S., Ambiguities in the IARC criteria for evaluation of carcinogenic risks to humans, and a recommendation. *Epidemiology*, **7:** 105–106 (1999).

Yoshikawa, K., Anomalous nonidentity between salmonella genotoxicants and rodent carcinogens: Nongenotoxic carcinogens and genotoxic noncarcinognens. *Environmental Health. Perspectives*, **104:** 40–46 (1996).

14 Properties and Effects of Metals

STEVEN G. DONKIN, DANNY L. OHLSON, and CHRISTOPHER M. TEAF

Metals are extensively used in commercial and industrial applications and, as a result, exposure can occur from direct and indirect pathways. These exposures may be associated with such processes as smelting, welding, grinding, soldering, printing, and many other product manufacturing operations. This chapter discusses a number of fundamental characteristics and health effects of metals, including

- Classification of metals
- Chemical and physical properties of metals
- Absorption, distribution, metabolism, and excretion of metals
- Mechanisms of metal-induced toxicity
- Toxicologic information on selected representative metals

14.1 CLASSIFICATION OF METALS

Metals are elements which are naturally occurring, ubiquitous, and resistant to natural degradation. The study of metal toxicity must take into consideration several characteristics unique to this group of toxicants. While all metals are toxic at some level of exposure, many metals are essential nutrients required at some minimum intake level for good health. Therefore, the distinction must be made between necessary minimal exposure and toxic overexposure. Because life has evolved in the constant presence of metals, most organisms, including humans, have various built-in mechanisms for coping with potentially harmful levels of both essential and nonessential metals. It is when the frequency, magnitude, or duration of exposure exceeds the capacity of these detoxifying mechanisms that metal toxicity may become a concern.

Several general physical and chemical properties set metals apart from other elements. Among these are strength, malleability, reflectivity, high electrical and thermal conductivity, and weakly held valence electrons resulting in a tendency to ionize in solution. Some of these properties are of interest from a toxicologic standpoint because they affect the absorption, distribution, metabolism, and resulting biological effects of metals. In addition, the fact that metals are elements and do not degrade in the environment means that they have a very high persistence, resulting in a greater potential for exposure than other, less persistent, toxic chemicals. In addition to their uncomplexed or elemental state, metals may exist in the environment as complexes with other substances. These complexes may differ dramatically in their chemical and toxicological properties (e.g., elemental mercury vs. methyl mercury).

While humans possess some fairly effective means for detoxifying and excreting metals at exposure levels normally encountered in the environment or the workplace, exposure in some occupational settings where metals are routinely used (e.g., smelting, plating) may be substantially higher. These situations require a heightened level of protection usually attained by work practices, protective equipment, or technological innovation. In addition, the increased mobilization of some metals within

Principles of Toxicology: Environmental and Industrial Applications, Second Edition, Edited by Phillip L. Williams, Robert C. James, and Stephen M. Roberts.
ISBN 0-471-29321-0 © 2000 John Wiley & Sons, Inc.

the general environment, brought about by their mining, processing, commerce, and disposal by humans, has resulted in higher background levels of metals in some areas where the general population may be exposed. In this chapter, properties and potential effects of both occupational and environmental exposure to metals are discussed. Table 14.1 lists common uses and toxic effects of some selected metals.

Essential and Nonessential Metals

A number of metals have important biological roles and thus are considered essential for good health. Nevertheless, at sufficient concentrations, a number of these essential metals are potentially toxic. For example, cobalt is a necessary component of vitamin B_{12} and is required for the production of red blood cells and the prevention of pernicious anemia. Copper is an essential component of several enzymes and is necessary for the utilization of iron. Iron, in turn, is necessary for the production of hemoglobin. Magnesium, manganese, and molybdenum are cofactors for a number of enzymatic reactions. Selenium is a component of the enzyme glutathione peroxidase. Zinc is a cofactor for more than 100 metalloenzymes.

Vanadium and tin are also considered essential in some animal species. Arsenic and chromium are regarded as essential at low doses to animals and humans, respectively, but also are considered to be major toxic concerns at higher exposure levels in some specific forms and are discussed in detail in Section 14.6 of this chapter.

Nonessential metals are those metals that have no known beneficial role to play in biological function. These metals include beryllium, cadmium, lead, mercury, thallium, titanium, and uranium.

TABLE 14.1 Common Uses and Principal Toxic Effects of Selected Metals

Metal	Common Industrial Uses	Principal Toxic Effects
Aluminum	Alloys, sheetmetal, appliances, food packaging	Environmental exposures are relatively nontoxic
Arsenic	Pesticides, herbicides, agricultural products	Lung cancer, skin diseases
Beryllium	Electronics, alloys, spacecraft	Lung disease
Cadmium	Batteries, plastics, pigments, plating	Kidney damage, lung cancer, bone disorders
Chromium	Plating, alloys, dyes, tanning	Lung cancer (Cr^{6+}), respiratory effects, allergic dermatitis
Cobalt	Alloys, paints, porcelain	Environmental exposures are relatively nontoxic
Copper	Electrical wiring, water pipes, sheetmetal, alloys	Environmental exposures are relatively nontoxic
Lead	Batteries, wire and cable, alloys	Neurological effects, hematopoietic system damage, reproductive effects
Manganese	Pesticides, ceramics, batteries, steel	Central nervous system effects
Mercury	Chloralkali industry, pesticides, thermometers, batteries	Neurological effects, kidney damage
Nickel	Coins, jewelry, alloys, plating, batteries	Environmental exposures are relatively nontoxic; dermatitis
Thallium	Electronics, alloys	Neurological, heart, lung, kidney, and liver effects
Tin	Plastics, food packaging, pesticides, wood preservatives	Inorganic tin is relatively nontoxic; organic compounds—neurological
Titanium	Paints, alloys, ceramics, plastics	Environmental exposures are relatively nontoxic
Zinc	Batteries, alloys, galvanizing, dyes, pharmaceuticals	Gastrointestinal effects, anemia

Most of the nonessential metals also may be considered toxic, although some (e.g., titanium) appear to be relatively nontoxic, even at high exposure levels.

14.2 SPECIATION OF METALS

Nonbiological Factors Affecting Speciation

Understanding the ways in which environmental parameters determine the speciation of a metal is the first step in assessing its toxicological potential. These conditions, which are essentially physical and chemical parameters, might be broadly categorized as nonbiological factors. A property common to all metals is the tendency to ionize in solution, giving up one or more electrons to become a cation, or species with a net positive charge (e.g., $Cd \rightarrow Cd^{2+} + 2e^-$). The degree to which this process is carried out will, in turn, affect various other behaviors of the metal and ultimately its toxicity. As pH decreases, metals typically become more mobile in the environment and more available to organisms.

One form in which a metal may exist is the uncharged, or elemental, form. This state is often designated by a "0," representing "zero charge" [i.e., Hg(0) or Hg^0 for elemental mercury]. In aqueous solution, metals may exist in a variety of ionization states or valences that differ in the number of missing electrons and therefore exhibit various net positive charges. The standard designation for the first ionization state of many metals is the name of the metal with an *ous* suffix, such as *mercurous* ion for Hg^+, or *cuprous* ion for Cu^+. The next level of ionization is sometimes designated with an *ic* suffix, such as *mercuric* for Hg^{2+} and *cupric* for Cu^{2+}. Other ionization states may exist for some metals, but usually the number of states most commonly encountered under normal conditions is limited to two or three.

Ionization is not a uniform process across all metal types. For instance, tin gives up two electrons in its first ionization state (the *stannous* ion, or Sn^{2+}), while mercury and copper give up only one. In contrast to these examples, the primary valences of interest for chromium are Cr^{3+} (trivalent chromium) and Cr^{6+} (hexavalent chromium), for which the toxicity varies dramatically. Consideration of a metal's ionization state is important because it may substantially affect that metal's toxicity. Trivalent (3^+) chromium compounds, for example, which are noncorrosive and noncarcinogenic, are less of a concern than hexavalent (6+) chromium compounds, which are quite corrosive and may cause cancer on sufficient exposure by the inhalation route only.

These different patterns of metal ionization are influenced by environmental conditions, as well as by the inherent properties of the metals themselves. Since metal cations carry a charge, they will react readily with other charged species, thus making their behavior heavily dependent on parameters such as pH (the concentration of H^+ ions), pOH (the concentration of OH^- ions), the presence of other charged particles or ligands, oxidizing or reducing conditions, temperature, and whether the medium of concern is air, soil, or an aqueous solution. Simply put, metals aren't just metals.

The degree to which these factors may affect the biological hazards posed by metals is illustrated by copper. Copper has an especially high affinity for organic ligands. In aquatic systems, copper readily binds to organic matter and may thus settle out of the water column and be unavailable to most organisms. Likewise, copper binds to organic matter in soil and tends to accumulate in the upper layers of soil if applied at the surface. This decreases the likelihood of oral exposure to copper through contaminated groundwater, but could increase the likelihood of exposure through soil contact.

When a metal becomes ionized, it may form a variety of compounds by combining with available negatively charged groups, such as chlorides, sulfates, nitrates, carbonates, and acetates. These metal complexes may have toxicological properties that differ from those of the uncomplexed metal. Alternatively, metal complexes may behave in different ways that affect their absorption by and distribution within living organisms.

Biological Factors (Biotransformation)

Speciation of metals and conversion from one form to another in the environment may also occur as a result of biological action, termed *biotransformation*. This process may be exploited in engineering applications for remediation purposes, for instance, when contaminated soils are seeded with plants or fungi, which absorb metals and other inorganic substances from the soil and bind the metals in stable forms within their tissues. Removal of the plants or fungi then effectively removes the metals from the site. It is important to remember that since metals are elements, in most instances, they cannot be removed by degradation processes, which often are effective for organic compounds.

The metabolic processes of many microorganisms may be manipulated so as to biotransform metal compounds to less toxicologically important forms. Some yeasts, for instance, can be cultured in conditions whereby they will reduce ionic mercury to elemental mercury, which is a liquid and which typically is less bioavailable. However, vapor forms of elemental mercury may rapidly diffuse across some cellular membranes. There are also many organisms that readily methylate ionic mercury in selected environmental circumstances to the more easily absorbed and much more toxic organic species, methylmercury and dimethyl mercury, thus creating a more hazardous form. Likewise, inorganic lead may be biotransformed by some bacteria to tetraethyl lead, an organic compound that is more easily transported across biological membranes.

14.3 PHARMACOKINETICS OF METALS

Absorption and Distribution

For a metal to exert a toxic effect on a human or other organism, there must be exposure and it must gain entry into the body. There are three main routes of absorption: inhalation, oral, and dermal. The significance of the route depends on the metal being considered. For instance, copper is very soluble in water, so that ingestion (oral exposure) is a common pathway of exposure. Ingestion is also an important exposure pathway of mercury, often through the consumption of fish or other animals in which organic forms of this metal have bioaccumulated to a high level in their tissues. However, the elemental form of mercury has a high vapor pressure so that inhalation exposure also may be significant in certain situations. Dermal exposure to metals may cause local effects, but it is rarely a significant consideration from an absorption perspective.

The route of absorption also may influence the subsequent distribution of the metal within the body and, thus, may affect its metabolism, potential toxic effects, and excretion. While oral exposure to copper may lead to the absorption of high levels through the gastrointestinal mucosa, several efficient excretion mechanisms exist that may minimize the subsequent distribution of absorbed copper to the bloodstream and other tissues. Thus, exposure through inhalation or dermal absorption may ultimately pose the greatest danger of producing copper toxicity, since these routes may deliver the toxicant more directly via the circulation to target tissues.

Inhalation

Inhalation is a primary route of exposure in occupational and, to a lesser extent, environmental settings. A number of metals have the potential for producing toxic effects in the respiratory tract. Some metals volatilize into the atmosphere where they exist for long periods as a vapor. Mercury is one such example, being easily transported into the atmosphere when present in incinerated waste and potentially traveling great distances before settling down back to earth. Mercury may also be present in the breathing space of workers in an area where mercury is used or in the air around hazardous-waste sites where mercury-containing materials are present.

The speciation of mercury has an important influence on its potential for inhalation exposure, and subsequent distribution within the body. For instance, inhaled elemental mercury partitions mostly to the bloodstream, brain, and kidney. When elemental mercury is inhaled by pregnant women, it may also be distributed to the fetus. On the other hand, inorganic mercury salts, such as mercurous chloride and mercuric chloride, are not as easily absorbed via inhalation.

Metals may be found in the air in the form of aerosols (i.e., particulates associated with or formed from fumes and dust). Metal oxides, particularly zinc oxide and cadmium oxide, may cause an acute pulmonary disease called "metal fume fever" when such particles are inhaled in large amounts. The inhaled particles cause inflammation and tissue damage in the alveoli, resulting in respiratory irritation, painful breathing, and impaired pulmonary function.

The site of absorption for very small particles and vapors is usually deep within the lungs in the alveolar space. If the metal is water-soluble, it is then absorbed into the bloodstream. Larger particles may be trapped in the mucosal lining of the upper respiratory tract, where they may either be eliminated by the cough reflex or absorbed through the gastrointestinal tract if swallowed.

Oral Ingestion

Gastrointestinal absorption of metals and related compounds varies widely, depending on factors such as solubility, chemical form, presence of other materials, competition for binding sites, and the physiological state of the gastrointestinal tract. When swallowed, many metals may become bound up with other compounds within the gastrointestinal tract and pass through unabsorbed. Effective elimination in the urine or feces results. Other metals may be absorbed through the stomach or intestinal lining and thus enter the bloodstream. Absorption can occur by passive diffusion, facilitated diffusion, or active transport.

Consumption of drinking water is a common source of metal exposure, from either naturally occurring conditions or from pollutant discharges. The solubility of metals in water is determined in large part by their speciation. Lead and mercury salts, for instance, may dissolve readily in water, as will elemental copper. These solubilized metals are then rather easily absorbed through the gastrointestinal mucosa after ingestion. Tin and elemental mercury, on the other hand, are quite insoluble in water and, if swallowed, will most likely pass through the gastrointestinal tract unabsorbed, to be excreted in the feces.

Exposure to metals in the diet also may occur by eating some foods, such as fruits or vegetables with metal-containing pesticide residues, meat or seafood in which metals have bioaccumulated during the life of the animal, or foods contaminated in some way during processing (e.g., canning foods in lead-soldered cans). If exposure occurs through these pathways, it usually involves small concentrations of limited toxic concern. However, some infamous cases of acute metal poisoning epidemics from contaminated food have occurred. Among these is the severe bone disease "itai-itai," which occurred in Japan in the 1950s among residents living near a mine and smelter operation. Ingestion of rice that was grown in the area, and which was contaminated with cadmium wastes, was implicated. Epidemics of methyl mercury poisoning from eating contaminated fish and shellfish also have appeared in parts of Japan in the 1950s (Minamata Bay) and 1960s (Niigata), while the largest outbreak of methyl mercury poisoning occurred in Iraq in the early 1970s, a result of eating bread contaminated with a mercury-containing fungicide.

The influence of mercury speciation on its absorption following the oral exposure is quite different than that seen with the inhalation route. Orally ingested elemental mercury does not partition to the bloodstream as easily as does inhaled elemental mercury, and thus it is readily excreted. However, inorganic mercury salts, such as mercurous chloride and mercuric chloride, may be more easily absorbed through swallowing than is elemental mercury. Also, these inorganic salts are more efficiently absorbed through the oral route than through the inhalation route.

Dermal Exposure

As noted previously, the skin is typically an effective barrier to metal absorption. Exposure by this route becomes a concern only with selected metal species, or in cases when the integrity of the dermal barrier is compromised, such as with open wounds or burns. Skin sensitization may be a result of exposure to some metals, resulting in irritation, discoloration, or rash.

Dermal exposure is of greatest concern for those few metals in which significant toxicity is combined with the ability to penetrate the skin. Inorganic mercury can enter the bloodstream through the skin in small amounts, where it may then be distributed to other tissues and cause a variety of mercury poisoning symptoms. Nickel can be absorbed when handling nickel alloys, and among the general population through handling coins, although the level of absorption is too low to be of concern in that case.

Contact dermatitis and allergic skin reactions may be seen in individuals exposed dermally to some metals. Nickel and its salts are among the most common causes of metal-induced allergic contact dermatitis, though silver and copper also have been associated with such reactions in highly sensitive individuals. Beryllium also is of concern, with necrotic granulomatous lesions or papulovesicular lesions often resulting from prolonged skin contact. Hexavalent chromium is an oxidizing irritant and corrosive agent that may severely damage the skin on prolonged contact. Cobalt also causes allergic reactions in humans.

Metabolism and Storage

Once metals have entered the bloodstream, they are available for distribution throughout the body. The rate of distribution to organ tissues is determined by blood flow to the particular organ. The eventual distribution of a metal compound is largely dependent on the ability of the compound to pass through cell membranes, coupled with its affinity for binding sites. Metals are often concentrated in a specific tissue or organ (e.g., lead in bone, cadmium in liver) and their metabolism within the body usually involves binding to proteins, such as enzymes, or changes in their speciation. Metals also may bind to other substrates and alter the bioavailability of important cell constituents.

A common means by which the body defends itself against metal poisoning is by producing nonenzymatic proteins, which bind to and inactivate the metal, then transport it out of the body via the excretory system. Lipoproteins located within renal lysosomes appear to function in this way, thus serving as some protection for the kidneys, which are particularly vulnerable to metal damage. Metallothioneins are a class of sulfur-containing, metal-binding proteins, which display a substantially increased expression in response to cadmium, mercury, zinc, and several other metals.

A number of antidotes to metal poisoning have been developed on the basis of knowledge of natural detoxification mechanisms. Chelation therapy involves systematic treatment of the patient with a chelator, defined as a molecule with several electronegative groups able to form coordinate covalent bonds with metal cations, and thus render the metal unavailable and inactive. An example of a chelator is ethylenediaminetetraacetic acid (EDTA), a flexible molecule with four binding sites capable of nonspecifically binding metal ions and escorting them to the kidney for excretion. Calcium EDTA is particularly effective against lead poisoning. British Anti-Lewisite (BAL, or dimercaprol) is another chelator which was developed during World War II as an antidote against a form of arsenic gas (British Lewisite) and has since found application in therapy for chromium, nickel, cobalt, lead, and inorganic mercury poisoning. Dimercaptosuccinic acid (DMSA) is a water-soluble chelating agent similar to BAL, which is used to treat heavy-metal poisoning.

Chelation therapy must be used with discretion. EDTA has the potential to produce kidney toxicity, possibly due to its action in preferentially mobilizing the metals and concentrating them in this organ. BAL may worsen the effects of selenium and methyl mercury intoxication. Since chelators are fairly nonspecific, they may attack essential cations as well.

Excretion

The simplest method of excreting metals is to exclude them before they achieve significant absorption. For instance, inhaled metal vapors may be immediately expired during subsequent breaths, while inhaled particles may be trapped in the mucous and expelled by the cough reflex. Vomiting is a common reflex in response to orally ingested toxicants, including metals. Metals may also be excreted in the sweat and saliva or incorporated into growing hair and fingernails. The body's attempt to excrete excess mercury and lead in the saliva often results in visible "lead lines" along the gums. However, urine and feces remain the primary route of excretion for most ingested metals.

Usually, excretion consists of a combination of these pathways, which may differ according to the route of exposure and the speciation of the metal. Inhaled elemental mercury vapor, for instance, is excreted in the urine, feces, and breath, while ingested elemental mercury is primarily excreted in the feces. Methyl mercury is excreted very slowly, primarily in the feces after transformation to inorganic forms of mercury.

14.4 TOXICITY OF METALS

Acute Exposure

The numerous toxic effects of metals may be broadly divided into two main modes of action. The first is related to the fact that many metals have a strong affinity for common protein moieties, such as the sulfhydryl (–SH) group. By forming covalent bonds with these groups, metals may inhibit the activity of important enzymes or disrupt the integrity of cell membranes. The other way in which metals may exert toxicity is by competing with and displacing essential cations. For instance, lead can displace the essential element zinc in certain zinc-requiring enzymes, thus inhibiting their function.

Acute toxicity is caused by a relatively large dose of a metal over a short period of time. The duration of time from initial exposure to the onset of clinical symptoms is usually short, typically on the order of hours or days. Because of their generally disruptive effects on cell membranes, metals can produce various localized effects at their initial point of contact. Common symptoms of acute metal poisoning by the oral route include nausea, vomiting, and damage to the intestinal mucosa and gastrointestinal tract. Acute inhalation exposure to metals often results in nose and throat irritation, coughing or wheezing, and damage to the lungs and the respiratory lining. Acute dermal exposure can result in localized rash and skin irritation or discoloration.

Certain metals tend to target specific organs or systems. Acute exposure to high lead levels may result in severe neurological symptoms, including convulsions or coma, as well as disruption of the hematopoietic system, specifically heme production. The toxic effects of acute mercury poisoning can vary depending on the form present in the body. Exposure to large amounts of elemental mercury vapors or methyl mercury is more likely to cause central nervous system effects than is exposure to inorganic mercury because of differences in transport to the brain. Methyl mercury, because of its lipophilic nature, can easily cross the blood-brain barrier while inorganic mercury cannot. Elemental mercury vapors can cross membranes with relative ease.

Chronic Exposure

Long-term exposure to low levels of some metals may result in gradual development of symptoms and often, but not always, less severe symptomology than observed in acute events. Carcinogenicity, as discussed in the next section, may be the result of chronic exposure to a few metals. Also, some metals can be stored for long periods within biological tissues. Lead may displace calcium in developing bone, resulting in long-term storage and slow release of this metal, which may cause toxic symptoms that persist long after the exposure has ceased.

Damage to and impaired function of the kidneys and liver are typical effects of chronic exposure to many metals, probably due to the role of these two organs in concentrating, detoxifying, and excreting toxic metals. Mucosal degeneration and inhibition of hepatic and renal enzymes are common results of chronic metal overexposure. Chronic lead, chromium, mercury, and cadmium poisoning are commonly associated with kidney damage, while the role of the liver in copper storage and excretion makes this organ vulnerable to damage from chronic copper exposure.

TABLE 14.2 Target Organ Toxicity of Selected Metals and Metalloids

Metal	Kidney	Nervous System	Liver	Gastrointestinal Tract	Respiratory System	Hematopoietic System	Bone	Endocrine System	Skin	Cardiovascular System
Aluminum		+			+					
Arsenic		+	+	+	+	+		+	+	
Beryllium					+				+	
Bismuth	+		+						+	
Cadmium	+	+		+	+		+			+
Chromium	+	+	+		+				+	
Cobalt		+		+	+			+	+	+
Copper				+		+			+	
Iron		+	+	+	+	+		+		
Lead	+	+		+		+			+	
Manganese		+			+					
Mercury	+	+		+	+					
Nickel		+			+				+	
Selenium	+			+					+	
Strontium										+
Thallium	+	+	+	+	+			+		
Tin (organic)		+		+						
Zinc				+		+	+			

Several metals preferentially target the central nervous system. A neurological condition similar to Parkinson's disease, characterized by strange behavior and a "masklike" facial expression, is often the result of chronic, high level exposure to manganese. Lead intoxication may result in learning problems, behavioral disorders, hyperexcitability, or lethargy and ataxia. Chronic exposure to mercury can produce hyperexcitability and even psychosis. This latter symptom was identified as "mad hatter's syndrome" in the nineteenth century due to its prevalence among hatmakers exposed to mercury compounds used in the hat felting process. Table 14.2 lists principal target organ toxicities for selected metals.

Carcinogenicity

A few metals, such as beryllium, cadmium, hexavalent chromium, and some nickel compounds have been demonstrated to be carcinogenic in humans and experimental animals, primarily by the inhalation route. Arsenic is classified as a human carcinogen by IARC (International Association for Research on Cancer), but evidence for carcinogenicity in animals is not conclusive. The opposite is true for lead, shown to be carcinogenic in animals but not in humans. Other metals, such as copper, zinc, and mercury, have no demonstrated carcinogenic potential. Table 14.3 lists metals that have been demonstrated to be carcinogenic or possibly carcinogenic in humans and animals.

Of the metals that are known or suspected carcinogens, some have shown the effect to be confined to specific forms. For instance, hexavalent chromium is a known human carcinogen by inhalation, while trivalent chromium does not appear to be carcinogenic by any route. Likewise, nickel carcinogenicity has been demonstrated for the sulfide compound, but evidence for the carcinogenic potential of other nickel species is less certain.

Route specificity is also an interesting observation related to metal carcinogenic potential. Cadmium and hexavalent chromium are carcinogenic by inhalation only, while arsenic and beryllium exhibit broader carcinogenic potential by the inhalation, oral, and perhaps the dermal route.

Each carcinogenic metal seems to have a unique mechanism of action. Hexavalent chromium is converted to trivalent chromium in cells and forms tightly bound adducts with DNA and proteins. The biologically active carcinogens are likely the reactive intermediates of chromium (+6) reduction. Nickel ions accumulate within cells, generating oxygen radicals that appear to be the cause of the

TABLE 14.3 Classifications of Selected Metals Known or Suspected to be Carcinogenic to Humans

Metal	USEPA	IARC	ACGIH/OSHA
Arsenic	Human carcinogen[a]	Human carcinogen	Human carcinogen
Beryllium	Probable human carcinogen[b]	Probable human carcinogen	Suspected human carcinogen
Cadmium	Probable human carcinogen	Probable human carcinogen	Suspected human carcinogen
Chromium (VI only)	Human carcinogen[d]	Human carcinogen	Human carcinogen
Cobalt	Not classified	Possible human carcinogen[c]	Animal carcinogen
Lead	Probable human carcinogen	Possible human carcinogen	Animal carcinogen
Nickel	Human carcinogen[d] (refinery dust and nickel subsulfide)	Human carcinogen	Human carcinogen

[a]Human carcinogen = adequate evidence of carcinogenic potential in humans.

[b]Probable or suspected human carcinogen = equivocal evidence of carcinogenic potential inhumans, with adequate evidence of carcinogenicity in animals.

[c]Possible human carcinogen = equivocal evidence of carcinogenic potential in humans.

[d]Inhalation route only.

genotoxic damage associated with nickel carcinogenicity. Arsenic likely acts as a non-genotoxic indirect carcinogen via induction of oncogene expression and inhibition of DNA repair.

14.5 SOURCES OF METAL EXPOSURE

Natural Sources

Metals are naturally occurring elements in the earth's crust, and thus exposure from natural sources is inevitable. Metals are found in various concentrations in soils, sediments, surface and groundwaters, and air. They move through the environment as part of the natural biogeochemical pathways, being deposited on surfaces from the air, taken up from the soil and water by plants, passed up the food chain when ingested and bioaccumulated, released back to the environment through excretion and decay of dead organisms, and often going through various transformations and speciation changes in the process. The importance of metals in these cycles is in large part due to their inherent persistence.

Table 14.4 lists the typical levels of some metals found in the environment. It is important to remember that these levels may vary considerably, depending on whether the site of interest is rural, urban, or near a concentrated source of the metal, such as an ore deposit or hazardous waste facility. Thus, the levels reported in the table are only approximations or ranges for comparative purposes, and should not be seen as reflective of levels to be expected in all areas. A great deal of literature is available concerning the geographic distribution of naturally occurring metals in the United States.

Anthropogenic Sources

In addition to natural biogeochemical cycles, humans play a large role in the mobilization, transformation, and transportation of metals in the environment. Mining, dredging, construction, and manufacturing all remove metals from the locations in which they naturally occur, and may incorporate them into the human economic sphere, thus increasing the potential for human exposure. Metals often find their way back into the environment through the use and disposal of products containing them, through the disposal of manufacturing wastes, or through the discharge of mine tailings or dredge material.

TABLE 14.4 Typical Levels of Selected Metals Observed in the Environment

Metal	Air (ng/m^3)	Drinking Water (μg/L)	Rivers and Lakes (μg/L)	Soil (mg/kg)
Aluminum	<500	<100	<100	700–100,000
Arsenic	20–100	<40	<10	5
Barium	<1500	<1000	<380	<2500
Beryllium	<0.2	<5	10–1000	<5
Cadmium	5–100	<10	<10	<1
Chromium	10–30	<2	<30	<100
Cobalt	<1	<20	<1	1–40
Copper	<200	20–75	<1000	2–250
Lead	<1	1–60	<1	5–25
Manganese	5–30	<50	<50	40–900
Mercury	<10	<0.02	<5	<1
Nickel	<100	2	10	4–80
Thallium	<0.4	0.9	<10	0.3–0.7
Tin	<100	<2	<2	<200
Zinc	<1000	<2000	<50	10–300

Humans often may be exposed to hazardous levels of metals in the environment through the alteration of natural biogeochemical cycles. For instance, the use of lead as a gasoline additive increased the amount of lead to which people in urban areas were exposed to levels far above what would normally be inhaled if unleaded gasoline was used. The discontinuance of lead as an additive has dramatically decreased the impact of this exposure route, as well as the presence of lead in soils. However, the continued presence of lead-based paint and lead pipes (or lead plumbing solder) for transporting drinking water has maintained lead exposure at far above background levels. The presence of lead in the paint of old buildings, where it may be inhaled as paint dust or ingested by children swallowing paint chips, continues to be a major public health problem despite active efforts in the public health field at local, state, and federal levels.

The organomanganese compound, methylcyclopentadienyl manganese tricarbonyl (MMT), was used as an antiknock additive to replace lead in gasoline, but serious concerns regarding its health effects have been raised as well. MMT was banned for this use in the United States in 1977. A court decision in 1995 ordered EPA to lift the ban and allowed the registration of MMT. Testing for health effects of this manganese compound is ongoing.

People living near waste sites, mining operations, or smelters may be exposed to higher-than-background levels of metals in air, drinking water, and soil. Several incidents of massive public poisoning due to accidental or merely ignorant environmental release of metal wastes have underscored the potential dangers. Perhaps the most famous is the *Minamata disease*, named after the area of Japan where many cases of severe neurological impairment and death appeared among the population in the 1950s. It was eventually discovered that mercury wastes discharged into the nearby bay from a chloralkali plant were being bioaccumulated by the fish and shellfish. Since mercury is easily converted into the methyl form under common environmental conditions, and because the form persists for long periods in biological tissues, a magnification of tissue mercury concentration up the aquatic food chain would be expected. Thus, by the time the seafood was harvested for human consumption, it contained extremely toxic levels of methyl mercury, which were reflected in severe effects on adults and children in the area.

Aside from such catastrophic epidemics of metal poisoning among the general public, exposure to high levels of metals is usually of greatest concern for workers in industries where metals are commonly used. These include mining, processing and smelting, manufacturing, and waste disposal operations. Occupational exposure to some metals may be confined to specific industries. For instance, agricultural workers may have exposure to arsenic and mercury, which are ingredients in some herbicides and fungicides. Sheetmetal workers may be exposed to copper and aluminum dust particles. Gold, silver, platinum, and nickel are metals commonly handled by workers involved in manufacturing jewelry, and may exert effects under some conditions of exposure.

Although workers are often exposed to higher metal concentrations than members of the general public, this exposure can be maintained at a safe level through proper enforcement of regulations regarding exposure limits and workplace safety. Because workplace exposure is often confined to a specific site with a specific population at risk, routine monitoring of exposure can be performed, with corrective action taken as necessary to maintain safe limits.

Indicators of Exposure (Biomarkers)

In addition to overt signs of toxicity, exposure to metals may be verified and often quantified by specific biomarkers of exposure. A biomarker of exposure is any measurable biological parameter that indicates exposure to a toxic substance, whether it is an induced protein, enzyme, metabolite, or even the toxic substance itself. In assessing suspected metal exposure, the first step is usually to take blood and/or urine samples for analysis. Since metals cannot be metabolized beyond recognition, and many metals are not normal constituents of biological samples, their detection in the samples discussed above certain defined levels is a reliable indicator of recent metal exposure, and perhaps intoxication. Due to the relatively short half-life of most metals in the blood and urine, sampling is usually required within several days of the last exposure for detection to occur. Some metals, such as lead and cadmium, remain in the blood or other tissues (e.g., bone, kidney) for longer periods of time.

Many metals concentrate in the hair and fingernails, allowing measurements from samples of these keratinous materials to be used as indicators of longer-term exposure. Care must be taken, however, to ensure that the reported levels represent complexed metal in the matrix of the tissue, rather than surface contamination. It has been determined that the level of methyl mercury measured in the hair corresponds to about 250 times that measured in the blood. Both of these measurements in turn may be used to derive a fairly accurate determination of the level of methyl mercury absorbed per kilogram of body weight per day. Past exposure to arsenic may be confirmed through the measurement of *Mee's lines*, which are bands in the fingernails produced by arsenic deposition. Qualitative biomarkers of severe chronic exposure include the presence of gray *lead lines* along the gums in the mouth, resulting from secretion of lead and mercury in the saliva. Similar assessments can be performed with other metals based on their known bioaccumulation behavior.

14.6 TOXICOLOGY OF SELECTED METALS

Arsenic

Arsenic is a gray-colored metal found in the environment in both organic and inorganic compounds. Inorganic arsenic occurs naturally in many kinds of rock. Low levels of arsenic are present in soil, water, air, and food. Arsenic is used in a number of herbicides and insecticides.

The toxicity of arsenic compounds is extremely variable and depends on the animal species tested, the form of arsenic (e.g., As^{3+} vs. As^{5+}), the route of exposure, as well as the rate and duration of exposure. Human exposure may involve inhalation of arsenic dusts; ingestion of arsenic in water, food, or soil; or dermal contact with dust, soil or water.

By the inhalation route, the effect of concern is increased risk of lung cancer, although respiratory irritation, nausea, and skin effects also may occur. Workers exposed to inorganic arsenic dusts often experience irritation to the mucous membranes of the nose and throat, which may lead to laryngitis, bronchitis, or rhinitis. High-level exposures can cause perforation of the nasal septum. Little information is available regarding hepatic, renal, and dermal or ocular effects following inhalation of arsenic in humans or animals. Results of studies in animals suggest that inhalation of inorganic arsenic can affect the immune system and may interfere with its function, though human data are lacking. Inhalation of inorganic arsenic can lead to neurological injury in humans.

Gastrointestinal irritation, peripheral neuropathy, vascular lesions, anemia, and various skin diseases, including skin cancer, may result from high-level oral exposure. There are many case reports of death in humans due to intentional or unintentional ingestion of high doses of arsenic compounds. In nearly all cases, the most immediate effects are vomiting, diarrhea, and gastrointestinal hemorrhage. Death may ensue from fluid loss and circulatory collapse. A number of studies in humans indicate that ingestion of arsenic may lead to serious cardiovascular effects. Anemia and leukopenia, common effects of arsenic poisoning in humans, have been reported following acute, intermediate, and chronic oral exposures. One of the most common and characteristic effects of arsenic ingestion is a pattern of skin changes that include generalized hyperkeratosis and formation of hyperkeratotic warts or corns on the palms and soles, along with areas of hyperpigmentation interspersed with small areas of hypopigmentation on the face, neck, and back.

Relatively little information is available on effects due to dermal contact with inorganic arsenicals, but the primary effect is local irritation and dermatitis.

Epidemiologic studies suggest that inhalation exposure to inorganic arsenic increases the risk of lung cancer. Many of the studies provide only qualitative evidence for an association between duration and/or level or arsenic exposure and risk of lung cancer, but several studies provide sufficient exposure data to permit quantification of cancer risk. When exposure occurs by the oral route, the main carcinogenic effect is increased risk of skin cancer. In addition to the risk of skin cancer, there is mounting evidence that ingestion of arsenic may increase the risk of internal cancers.

The current daily oral reference dose (RfD) for arsenic is 3×10^{-4} mg/kg. The USEPA has placed inorganic arsenic in group A (known human carcinogen) for exposure by both the oral and inhalation route, and similar designations have been assigned by the American Conference of Governmental Industrial Hygienists (ACGIH) and the Occupational Safety and Health Administration (OSHA).

Beryllium

Beryllium is a hard, grayish metal that occurs as a chemical component of certain rocks, coal and oil, soil, and volcanic dust. Most of the beryllium that is mined is converted into alloys, which are used in making electrical or electronic parts and construction materials. Beryllium enters the environment (air, water, and soil) as a result of natural and human activities. In general, exposure to water-soluble beryllium compounds poses a greater threat to human health than does exposure to water-insoluble forms.

Judging from animal studies, the several different forms of beryllium that have been studied are poorly absorbed through both the gastrointestinal tract and the skin. The most important route by which beryllium compounds are taken up by animals and humans is inhalation. There are very few dose–response data regarding the health effects of beryllium and beryllium-containing compounds in humans. Some data exist regarding decreased longevity and pulmonary effects resulting from inhalation exposure, but dose–response relationships are not well defined.

Inhalation exposure of animals and humans to beryllium can result in two types of potentially fatal nonneoplastic respiratory disease: acute pneumonitis and chronic beryllium disease. Lethality and decreased longevity appear to be due to the development of chemical pneumonitis. A 1948 investigation of acute beryllium pneumonitis in three U.S. beryllium plants reported that all of the cases of beryllium pneumonitis studied were associated with inhalation exposures to beryllium concentrations of >0.1 mg/m^3, primarily as beryllium sulfate or beryllium fluoride. For workers who were exposed after 1950, beryllium pneumonitis has been virtually eliminated except in cases of accidental exposure to concentrations above the OSHA standard of 0.002 mg/m^3. For chronic beryllium disease, dose-response relationships are more difficult to determine, due to the lack of an established correlation between exposure histories and the incidence of disease. However, the number of cases of chronic beryllium disease has dramatically decreased in workers who were first exposed after 1950.

Several retrospective cohort studies of workers, who were exposed to beryllium from the 1940s to the 1970s, report significantly higher mortality rates in comparison with the U.S. general mortality rates for the time periods studied. Most of the workers reportedly experienced shortness of breath, general weakness, and weight loss, and autopsies revealed granulomatous lung disease, lung fibrosis, and heart enlargement.

As in humans, animal studies indicate that the respiratory tract is the primary target for inhalation exposure to beryllium and some of its compounds. Pneumonitis, with accompanied thickening of the alveolar walls and inflammation of the lungs, was reported in rats and mice exposed to beryllium for one hour at ≥3.3 and 7.2 mg/m^3 (as beryllium sulfate), respectively, for 12 months or less.

Some animal species exhibit hematological effects from beryllium exposure. Acute exposure had little hematological effect, but intermediate-duration exposure resulted in anemia in several animal species. Weight loss has been reported in some animal species after inhalation exposure to beryllium compounds.

No studies were discovered regarding death, systemic effects (other than some dermatological abnormalities), immunological effects, neurological effects, developmental effects, reproductive effects, genotoxic effects, or cancer in humans after oral or dermal exposure to beryllium or its compounds.

A number of studies have associated inhalation exposure to beryllium with an increased incidence of human lung cancer. In general, these studies have been judged to have limited application due to inadequate controls in the studies related to confounding factors such as smoking, improperly calculated expected deaths from lung cancer, as well as the use of inappropriate comparative control populations. Some beryllium compounds have been shown to be carcinogenic in animals. Rats that were exposed to beryllium concentrations of 0.035 mg/m^3 as beryllium sulfate for 180 days exhibited increased lung cancer rates, compared to controls. In another study, 18 of 19 rats exposed to beryllium concentrations of 0.62 mg/m^3 as beryl ore developed tumors that were classified as bronchial alveolar cell tumors, adenomas, adenocarcinomas, or epidermoid tumors.

The current daily oral RfD for beryllium is 5×10^{-3} mg/kg. USEPA has classified beryllium as a B1 carcinogen (probable human carcinogen) by the oral and inhalation routes of exposure. ACGIH and OSHA consider beryllium to be a potential carcinogen.

Cadmium

Cadmium is usually not found in the environment as a pure metal. It is usually found as a mineral, such as cadmium oxide, cadmium chloride, or cadmium sulfate, or in association with zinc. These solids may dissolve in water and small particles of cadmium may be found in the air. Food and cigarette smoke may be significant sources of cadmium exposure for the general public.

Inhalation exposure to high levels of cadmium oxide fumes or dust can cause severe irritation to respiratory tissue. Symptoms such as tracheobronchitis, pneumonitis, and pulmonary edema can develop within several hours of exposure; however, these symptoms do not occur following low level inhalation exposure. Lung injury following cadmium exposure can be at least partially reversible. Long-term occupational exposure to low levels of cadmium in the air may be associated with emphysema, but cigarette smoking is often a confounder in these studies.

The kidney appears to be the main target organ of chronic cadmium exposure via inhalation. The toxicity of cadmium to proximal renal tubular function is characterized by the presence of low- (but sometimes high- as well) molecular-weight proteins in the urine (proteinuria). Tubular dysfunction develops only after cadmium reaches a minimum threshold level in the renal cortex. Negative effects on calcium metabolism may occur as a result secondary to kidney damage.

Cadmium may enter the blood to a limited extent by absorption from the stomach or intestine after ingestion in food or water. The form of cadmium in food and water is generally the cadmium ion. Oral absorption of cadmium from food and water ranges within 3–5 percent of the ingested dose and is dependent on the iron stores in the body, with low iron levels correlated with increased cadmium absorption. Once cadmium enters the body, it is strongly retained in a number of organs. Cadmium absorbed by the human body is eliminated slowly, with a biological half-life estimated to be 10–30 years, and is accumulated throughout a lifetime with over 30 percent of the body burden stored in the kidneys.

The kidney is the main target organ of cadmium toxicity following chronic oral exposure to cadmium, with effects similar to those seen following inhalation exposure. Oral exposure to high concentrations of cadmium causes severe irritation to the gastrointestinal epithelium, resulting in nausea, vomiting, abdominal pain, and diarrhea. Painful bone disorders have been observed in some humans chronically exposed to cadmium in food. Decreased calcium content of bone and increased urinary calcium excretion are common findings in rats and mice following oral exposure to cadmium. Cadmium compounds have not been observed to cause significant health effects when exposure is by the dermal route.

There is persuasive evidence from studies in rats, but not mice, and equivocal results in hamsters, to conclude that chronic inhalation exposure to cadmium chloride is associated with increased frequency of lung tumors. Ingestion of cadmium compounds has not been associated with cancer in animals. Similarly, epidemiologic studies in humans exposed to elevated levels of cadmium in water or food do not indicate that cadmium is a carcinogen by the oral route. Some epidemiologic studies in humans have suggested that inhaled cadmium is a pulmonary carcinogen. Smoking is a common confounding factor in the determination of the role of cadmium in lung cancer, and cadmium is an identified component of cigarette smoke.

The current daily oral RfD for cadmium is 5×10^{-4} mg/kg for water sources, in comparison to 1×10^{-3} mg/kg daily from food sources. The USEPA has categorized cadmium as a Class B1 carcinogen (probable human carcinogen), by inhalation only.

Chromium

Chromium is a naturally occurring element, which is found in the environment in three major valence states: elemental chromium (0), trivalent chromium (+3), and hexavalent chromium (+6). Chromium (+3) occurs naturally in the environment, while chromium (+6) and chromium (0) typically are generated by industrial processes.

Natural geologic sources represent a component of chromium present in the environment. However, chromium is released to the environment in much larger and more concentrated amounts as a result of human activities. The most stable form of the chromium compounds is the trivalent state, the naturally occurring form. The hexavalent form is uncommon in a natural setting and is easily reduced to the trivalent form by environmental processes.

The three major forms of chromium differ dramatically in their potential for causing effects on human health. Trivalent chromium is an essential nutrient required for normal energy metabolism. Hexavalent chromium is irritating, and short-term high-level exposure can result in adverse effects at the site of contact, such as ulcers of the skin, irritation of the nasal mucosa, perforation of the nasal septum, and irritation of the gastrointestinal tract, as well as adverse effects in the kidney and liver. Exposure to metallic chromium is less common and is not well characterized in terms of levels of exposure or potential health effects.

The respiratory tract in humans is a major target of chromium inhalation exposure. Occupational exposure to chromium (+6) and/or chromium (+3) in a number of industries has been associated with respiratory effects. Irritant effects, decreased pulmonary function, and perforation of the nasal septum have been noted in workers exposed to chromium. Chronic exposure to chromium compounds have also resulted in adverse respiratory effects in animals.

There is little evidence of reproductive effects of chromium in humans, but there is some evidence that chromium has adverse reproductive effects in certain animal species. There is conflicting human and animal evidence of developmental toxicity following inhalation, oral, or dermal exposure to chromium. There is no evidence that exposure to chromium has any developmental effects in humans. However, studies indicate that chromium may be teratogenic in animals. Chromium (+6) compounds have been shown to cross the placenta and to induce neural tube defects and lethality in mice, cleft palates and lethality in hamsters, and a variety of abnormalities in chick embryos. Allergic contact dermatitis in sensitive individuals can result from exposure to chromium compounds.

Chromium (+6) compounds have been tested in a wide range of in vivo, cell culture, and bacterial genotoxicity assay systems and were positive for all endpoints. The genotoxicity data support the hypothesis that chromium (+6) is not genotoxic per se, but that chromium genotoxicity is mediated by the intracellular reduction of chromium (+6) to chromium (+3), which may be the ultimate genotoxic form of chromium. Paradoxically, chromium (+3) does not induce DNA damage, even though it binds to DNA in vitro and in vivo.

Unlike many chemical carcinogens, the majority of information on chromium-induced carcinogenesis comes from human epidemiology studies of occupationally exposed workers rather than from animal studies. Lung cancer is considered to be an occupational hazard for workers exposed to chromium (+6) in a wide variety of industrial and commercial occupations. Studies indicate that workers in industries that use chromium can be exposed to concentrations of chromium two orders of magnitude higher than exposure to the general population. There is a good correlation between the dose of chromium, expressed as a function of concentration and time of exposure, and the relative risk of developing lung cancer, which has been calculated to be as much as 30 times that of appropriate controls. The principal forms of chromium-induced cancer are lung carcinomas and, to a lesser extent, nasal and pharyngeal carcinomas. There also is some evidence for an increased risk of developing gastrointestinal cancer.

The current daily oral RfD for chromium (+3) is 1.0 mg/kg. The current daily oral RfD for chromium (+6) is 5×10^{-3} mg/kg. Chromium (+6) is classified as a Class A (confirmed human carcinogen) by the inhalation route of exposure only. Chromium (+3) is not classified as a carcinogen by any route of exposure.

Lead

Lead is a naturally occurring bluish-gray metal found in the earth's crust. Lead can combine with other chemicals to form what are known as lead salts. These compounds are water-soluble, while elemental lead is not. Lead is used in the production of batteries, ammunition, metal products, as well as scientific and medical equipment. Most of the lead mobilized in the environment is the result of human activities.

Exposure to inorganic lead and inorganic lead compounds or to organic lead compounds can occur in environmental or occupational circumstances. Inorganic lead exposure is most applicable to general and occupational exposure. Human body burdens of lead result from inhalation and oral exposure to inorganic lead. In humans, oral absorption of ingested lead occurs primarily in the gastrointestinal tract; 50 percent of the oral dose is absorbed by children and 15 percent is absorbed by adults. Typically, the lead body burden of an average adult human has been reported to range between 100 and 300 mg. Exposure to lead can affect a number of organs and/or systems in humans and animals. The research on lead contains dose–response data for humans primarily from studies of occupationally exposed groups, as well as the general population. Inhalation contributes a greater proportion of the dose for occupationally exposed groups, and the oral route contributes a greater proportion of the dose for the general population. The effects of lead are the same regardless of the route of entry into the body and are well correlated with blood lead level.

Lead in soil and dust are important sources of exposure in children. The Centers for Disease Control (CDC) in 1985 stated that concentrations of lead in soil or dust at 500–1000 mg/kg are associated with blood levels in children that exceed background levels. Death from lead poisoning occurred in children with blood lead levels >125 μg/dL, including several deaths in children who exhibited severe encephalopathy. The duration of exposure associated with this effect was on the order of a few weeks or more, and, in some cases, it may have been an acute toxic response. For industrial sites, where access to soil may be limited or restricted, it was recommended that the upper end of the range that was identified in the Office of Solid Waste and Emergency Response (OSWER) Directive 9355.4-02 for lead be used in establishing protective concentrations for lead in soil at industrial sites (range of 500–1000 mg/kg). However, subsequent studies in areas of high lead soil concentrations in mining areas have not confirmed those recommendations.

Relatively low exposure levels of lead may cause adverse neurological effects in fetuses and young children. Very low blood lead levels (e.g., 10–20 μg/dL) have been associated with effects on learning ability. The CDC initially defined lead toxicity in a child as a blood lead level in the range of 25–35 μg/dL, coupled with an erythrocyte protoporphyrin level of >35 μg/dL. The threshold has since been lowered to 10 μg/dL by that agency. A blood lead level of 50 μg/dL has been determined to be an approximate threshold for the expression of lead toxicity in exposed adult workers, and the occupational recommended conservative benchmark typically is 30 μg/dL. Neurobehavioral effects have also been documented in animal studies as well as human studies.

Lead affects the hematopoietic system by altering the activity of three enzymes involved in heme biosynthesis. The impairment of heme synthesis has a number of subsequent effects, including decreased hemoglobin levels and anemia. These effects have been observed in lead workers and in children with prolonged lead exposure. A decrease in cytochrome P450 content of hepatic microsomes has been noted in animal studies.

Lead may cause kidney damage as a result of acute or chronic exposure. Reversible proximal tubular damage can result from acute lead exposure. Heavy, chronic exposure can result in nephritis, interstitial fibrosis, and tubular atrophy.

Colic—characterized by abdominal pain, constipation, cramps, nausea, vomiting, anorexia, and weight loss—is a symptom of lead poisoning in occupationally exposed cases and in children.

A review of 65 animal studies concluded that low-level exposure to lead during prenatal or early postnatal life results in retarded growth in the absence of overt signs of toxicity. There is some evidence that exposure to lead also can cause adverse effects on human reproduction and development as well.

There is some evidence that lead acetate and lead phosphate cause renal tumors in laboratory animals. The USEPA has classified lead as a Probable Human Carcinogen (Group B2) based on these results; however, that agency has not developed a *cancer slope factor* (CSF). The USEPA has advised that the risks to children from exposure to lead should be evaluated based on the integrated exposure uptake biokinetic model (IEUBK). The IEUBK model is a method for estimating the total lead uptake (μg Pb/day) in humans that results from diet, inhalation, and ingestion of soil, dust, and paint, to predict a blood lead level (μg Pb/dL), based on the total lead uptake. A number of models also are available for the evaluation of blood lead in occupationally exposed adults.

Mercury

Mercury is found in the environment in a metallic or elemental form, as an inorganic compound, or as organic mercury compounds. Mercury metal is the liquid used in thermometers and some electrical switches. Metallic mercury will evaporate to some extent at room temperature to form mercury vapor. Vaporization increases with higher temperatures. Mercury can combine with other elements to form inorganic and organic compounds. Some inorganic mercury compounds are used as fungicides, antiseptics, and preservatives. Methyl mercury is produced primarily by microorganisms in the environment.

The toxicity of mercury depends on the specific compound in question. Alkyl mercury compounds (e.g., methyl mercury) are extremely toxic in comparison to the inorganic mercury compounds. Absorption of inorganic mercury salts from the gastrointestinal tract typically is less than 10 percent in humans, whereas absorption of methyl mercury exceeds 90 percent. The pattern of distribution in the mammalian body also differs between alkyl and inorganic forms of mercury. The red blood cell/plasma ratio for inorganic forms generally is less than 2, while for organic forms it is approximately 10, indicating a longer body half-life for the latter. Inorganic mercury tends to localize in the kidneys as a result of filtration and reabsorption, while organic mercury exhibits a preference for the brain and, to a lesser extent, the kidneys. Excretion may be in both urine (minor) and feces (major), depending on the form of mercury, the magnitude of the dosage, and the time postexposure.

Blood concentrations of mercury generally represent recent exposure to methyl mercury, while hair concentrations reflect average intake over a long period. The mercury concentrations in successive segments of hair over the period of its formation can indicate the degree of past absorption of mercury compounds.

Metallic (elemental) mercury exposure may result from breathing mercury vapors released from dental fillings. Spills of metallic mercury or release from electrical switches may result in exposure to metallic mercury and vapors released to indoor air. Exposure to metallic mercury may result from breathing contaminated air from various sources. Exposure to mercury compounds can result from contaminated sources, as well as medicinal and household products. Occupational exposure to mercury vapors may occur in various manufacturing and processing industries, as well as medical professions. Ingestion of oral doses of 100–500 g has occurred in humans with little effect other than diarrhea. Inhalation of metallic mercury vapor has been associated with systemic toxicity in humans and animals. At low exposure levels, the kidneys and central nervous system may be affected. At high levels, the respiratory, cardiovascular, and gastrointestinal systems can be affected as well.

The toxicity of the inorganic salts of mercury is related to their comparative absorption rates. Insoluble mercurous salts, such as calomel (Hg_2Cl_2; mercurous chloride), are relatively nontoxic in comparison to the mercuric salts. The immediate effects of acute poisoning with mercuric chloride ($HgCl_2$) are due to primary irritation and superficial corrosion of the exposed tissues. Chronic oral effects for the mercuric salts include kidney damage, intestinal hemorrhage, and ulceration.

The principal problem of mercury toxicity is related to the ingestion of organic mercury compounds, which may accumulate in fish. Ingestion of meat from animals that have been fed grain treated with alkyl mercury compounds, or ingestion of the treated grain, may also result in toxic endpoints. The most sensitive endpoint for oral exposure to alkyl mercury compounds is the developing nervous system. In addition, mercury may adversely affect a wide range of other organ systems after exposure to high levels. Affected systems include the immune, respiratory, cardiovascular, gastrointestinal, hematologic, and reproductive systems.

The *inhalation reference concentration* (RfC) for elemental mercury is 3×10^{-4} mg/m^3. An oral RfD of 3×10^{-4} mg/kg daily for inorganic mercury and 1×10^{-4} mg/kg per day for organic mercury has also been established. These values are under review by USEPA at present. Mercury is not classified as a carcinogen by any route of exposure.

Nickel

Nickel is a hard, silver-white, malleable, ductile metallic element used extensively in alloys and for plating because of its oxidation resistance. Nickel, combined with other elements, occurs naturally in the earth's crust. Nickel released to the atmosphere typically exists in particulate form or adsorbed to particulate matter. Primary removal mechanisms of atmospheric nickel include gravitational settling and precipitation. Nickel released to soil may be adsorbed to soil surfaces depending on the soil conditions. Nickel released to aquatic systems generally exists in particulate forms that settle out in areas of active sedimentation. However, nickel also may exist in soluble form under appropriate conditions.

Nickel salts exhibit significant solubility in water. Nickel occurs naturally in drinking water at an average concentration of about 2 μg/L. Adult daily intake of nickel from water is about 2 μg/day. About 170 μg of nickel is consumed in food per day. Available information indicates that nickel does not pose a toxicity problem following ingestion because the absorption from food or water is low.

The most prevalent effect of nickel exposure is nickel dermatitis in nickel-sensitive individuals. Nickel dermatitis typically exhibits two components: (1) a simple dermatitis localized in the contact area and (2) chronic eczema or neurodermatitis without apparent connection to such contact. Nickel sensitivity, once acquired, may be persistent. Toxicological information of concern to industrially exposed humans is primarily confined to two potential categories of effects: (1) dermatoses, contact and atopic dermatitis, and allergic sensitization; and (2) cancers of the lung and nasal sinuses. Cancers of the lung and nasal sinuses in nickel workers have been described for more than 50 years in association with nickel refining processes (calcination, smelting, roasting, and electrolysis) and from nickel plating and polishing operations (e.g., electrolysis and grinding). Noncarcinogenic respiratory effects, such as bronchitis and emphysema, have also been seen in occupational exposures. These effects occurred at concentrations much higher than those found in the environment.

The current daily oral RfD for nickel is 2×10^{-4} mg/kg. An inhalation RfD has not been established and is currently under review. USEPA has classified nickel refinery dust and nickel subsulfide as class A carcinogens by the inhalation route. Nickel is not presently considered to be a carcinogen by the oral route.

Zinc

Zinc is found in natural form in the air, in soil and water, and occurs in the environment primarily in the +2 oxidation state. It is a bluish-white metal that can combine with a number of other elements. Most zinc ore is zinc sulfide. Zinc is an essential trace element in the diet. In water, zinc partitions to sediments or suspended solids through sorption. Zinc is likely to sorb strongly onto soil, and the mobility of zinc in soil depends on the solubility of the speciated forms of the compound and on soil properties such as pH, redox potential, and cation exchange potential. In air, zinc is present primarily as small dust particles.

Humans are exposed to small amounts of zinc in food and drinking water each day. Levels in air are generally low and fairly constant. Occupational exposure to zinc occurs in a number of mining and industrial activities, such as the manufacture of zinc-containing alloys, paints, and pesticides.

Zinc is required for normal growth and development, reproduction, and immune function. Zinc deficiency can have numerous adverse effects on the normal function of all of these systems. The Recommended Dietary Allowance (RDA) for zinc ranges from 5 mg/day for infants to 19 mg/day for lactating women.

Metal fume fever has been observed in humans exposed to high concentrations of zinc oxide fumes. These exposures have been acute, intermediate, and chronic. Metal fume fever is thought to be an immune response characterized by flulike symptoms and impaired lung function.

Zinc salts of strong mineral acids are astringent, are corrosive to the skin, and are irritating to the gastrointestinal tract. When ingested, they may act as emetics. In these cases, fever, nausea, vomiting, stomach cramps, and diarrhea occurred within 3–13 h following ingestion. The dose associated with such effects is greater than 10 times the RDA. Aside from their irritant action, inorganic zinc compounds are relatively nontoxic by oral exposure. Zinc ion, however, is ordinarily too poorly absorbed to induce acute systemic intoxication.

The USEPA has established a daily oral reference dose (RfD) of 3×10^{-1} mg/kg for zinc; however, no inhalation RfD has been established. Zinc is classified in group D, defined as not classifiable with regard to human carcinogenicity (USEPA, 1998).

14.7 SUMMARY

This chapter has briefly discussed the fundamental concepts of metal toxicity. Because of the large number of metals, their ubiquitous nature, and their chemical and physical diversity, the field of metal toxicology is one of the broadest areas of health effects research.

Metals vary greatly in their physical and chemical properties, and therefore, in their potential for absorption and toxicity. Some metals are considered essential for good health, but these same metals, at sufficient concentrations, can be toxic.

Inhalation and ingestion are the most common routes of metal exposure. Dermal effects may be severe, but typically are limited to the site of application. Some metals can remain in the body for significant periods of time, stored in specific tissues and slowly released over time. Urine and feces are the primary routes of excretion for most ingested metals. Biomarkers of exposure to some metals can thus be detected in these excretory products, as well as in stored forms in hair and fingernails.

Following sufficient acute or chronic exposure to certain metals, a variety of toxic effects can be observed in humans and animals. A review of the toxicology of some selected metals is presented in Section 14.6 of this chapter.

The following bibliography provides some additional sources of information for the toxicity and general characteristics of metals.

REFERENCES AND SUGGESTED READING

ACGIH (American Conference of Governmental Industrial Hygienists), *Documentation of Threshold Limit Values and Biological Exposure Indices,* 6th ed., 1991–1998.

ATSDR (Agency for Toxic Substances and Disease Registry), *Toxicological Profiles*, Atlanta, GA, 1993–1999.

Chang, L. W., L. Magos, and T. Suzuki, eds., *Toxicology of Metals*, CRC Press, Boca Raton, FL, 1996.

Clayton, G. D., and F. E. Clayton, eds., *Patty's Industrial Hygiene and Toxicology*, Vol. II, *Toxicology*, 4th ed., Wiley, New York, 1994.

Ellenhorn, M. J., *Medical Toxicology: Diagnosis and Treatment of Human Poisoning*, 2nd ed., Williams & Wilkins, Baltimore, 1997.

IARC (International Agency for Research on Cancer), *Monographs 1972–present*, World Health Organization. Lyon, France.

Klaassen, C. D., ed)., *Casarett and Doull's Toxicology: The Basic Science of Poisons*, 5th ed., 1996.

Pounds, J. G., "The toxic effects of metals," in *Industrial Toxicology*, P. Williams, and J. Burson, eds., Van Nostrand-Reinhold, New York, 1985.

Rom, W. N., *Environmental and Occupational Disease,* Little, Brown, Boston, 1992.

USEPA (U.S. Environmental Protection Agency), IRIS (*Integrated Risk Information System*) and HSDB (*Hazardous Substance Data Bank*), on line computer databases.

Zenz, C., et al., eds., *Occupational Medicine*, Mosby, St. Louis, MO, 1994.

15 Properties and Effects of Pesticides

JANICE K. BRITT

The term "pesticide" encompasses a group of chemical compounds that are used for the elimination or control of pests. Pesticides are grouped into classes based on their target of action and include such groups as insecticides, fungicides, herbicides, rodenticides, and molluscicides. Pesticides have economic and public health benefits and have been used over the years for the control of vectorborne diseases such as malaria and Rocky Mountain Spotted Fever, for the promotion of agricultural production in the United States as well as in other countries, and by homeowners for the control of domestic pests (e.g., household and garden pests).

Individuals may be exposed to pesticides either occupationally (e.g., from working in a pesticide formulating plant or from commercial pesticide application) or environmentally (e.g., from food products such as fruits and vegetables treated for pests). Individuals may also be exposed to pesticides at their residences (e.g., from use as home or garden insecticide). The most commonly used pesticides have recently been reported by the U.S. Environmental Protection Agency (USEPA) (see Table 15.1).

The registration and regulatory requirements concerning pesticides are governed under the Federal Insecticide, Fungicide, and Rodenticide Act, also known as FIFRA. Rules aimed at the protection of

TABLE 15.1 Most Commonly Used Pesticides in the United States

Most Commonly Used Pesticides in U.S. Agricultural Crop Production
Atrazine
Metolachlor
Metam sodium
Methyl bromide
Dichloropropene
Most Commonly Used Pesticides in Nonagricultural Sectors of the United States
Home and garden market
2,4-Dichlorophenoxyacetic acid
Glyphosate
Dicamba
MCPP
Diazinon
Industrial/commercial/government uses
2,4-Dichlorophenoxyacetic acid
Chlorpyrifos
Glyphosate
Methyl bromide
Copper sulfate

Principles of Toxicology: Environmental and Industrial Applications, Second Edition, Edited by Phillip L. Williams, Robert C. James, and Stephen M. Roberts.
ISBN 0-471-29321-0 © 2000 John Wiley & Sons, Inc.

agricultural and greenhouse workers who used pesticides were passed in 1992 by the USEPA and are located in 40 *Code of Federal Regulations* Parts 156 and 170. The Occupational Safety and Health Administration (OSHA) as well as the American Conference of Governmental Industrial Hygienists (ACGIH) also publishes guidelines for occupational exposures to pesticides in air.

This chapter will discuss the classes of the most commonly used pesticides and will include a discussion of the following with respect to these pesticide classes:

- Uses
- Mechanism of action
- Pharmacokinetics
- Acute and chronic effects from exposure
- Biological monitoring
- Treatment of pesticide overexposure
- Regulatory information

15.1 ORGANOPHOSPHATE AND CARBAMATE INSECTICIDES

Introduction, Use, and History of Organophosphates and Carbamates

Organophosphate compounds have become widely used pesticides as replacements for the more persistent organochlorine insecticides (discussed in Section 15.2). Organophosphate insecticides do not bioaccumulate in tissues and organisms or accumulate in the environment as do the organochlorines. In fact, chlorpyrifos, an organophosphate compound, has become a widely used termiticide, serving as a substitute for the more persistent organochlorine compounds used in the past. However, because of the acute toxicity of some of the organophosphate compounds, another class of pesticide—pyrethrins (discussed in Section 15.3)—are becoming more widely used. Examples of commonly used organophosphates include Dursban (chlorpyrifos), Knox Out 2FM (diazinon), and Vapona (dichlorvos) (see Figure 15.1 for examples of organophosphate insecticides). Examples of carbamate pesticide products commonly used are Sevin (carbaryl) and Temik (aldicarb) (see Figure 15.2 for examples of carbamate insecticides). It should be noted that organophosphate compounds are used not only as pesticides; chemicals in this class are also used as therapeutic agents for the treatment of glaucoma and myasthenia gravis in humans. For example, the organophosphate echothiophate iodide is used to treat glaucoma.

The first organophosphate insecticide developed was tetraethyl pyrophosphate (TEPP), developed in Germany during World War II as a substitute for nicotine. Because TEPP, although an effective compound, was unstable in the environment, there was an effort to develop more stable compounds. This effort resulted in the development of the organophosphate insecticide parathion in 1944.

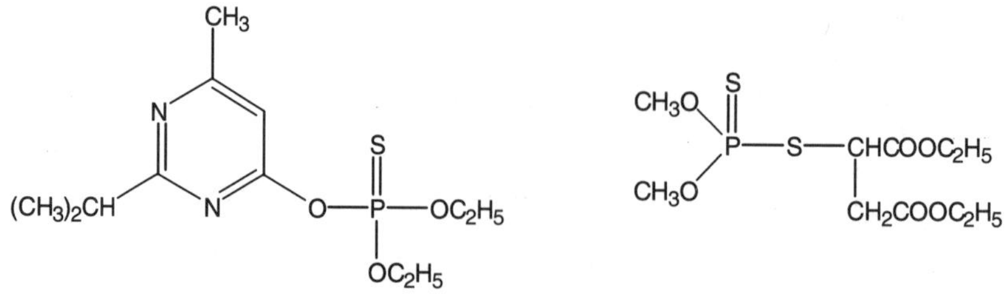

CH3
N
(CH3)2CH
N
O—P—OC2H5
S
OC2H5
Diazinon

CH3O
S
P—S—CHCOOC2H5
CH3O
CH2COOC2H5
Malathion

Figure 15.1 Examples of organophosphate insecticides.

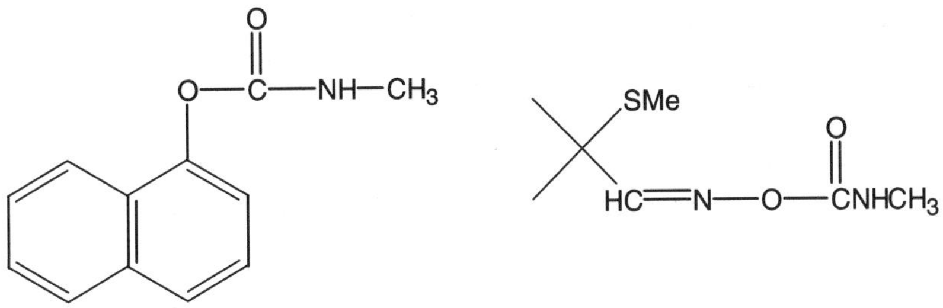

Carbaryl

Aldicarb

Figure 15.2 Examples of carbamate insecticides.

Mechanism of Action

Both organophosphate and carbamate classes of compounds have the same mechanism of action in insects as well as in mammals (including humans): the inhibition of the enzyme acetylcholinesterase. The inhibition of acetylcholinesterase by organophosphates and carbamates is the mechanism of action that is responsible for the acute symptomatology associated with these compounds.

Acetylcholinesterase is an enzyme located in the synaptic cleft and its function is the breakdown of acetylcholine, the neurotransmitter present at the following cites: postganglionic parasympathetic nerves, somatic motor nerves endings in skeletal muscle, preganglionic fibers in the parasympathetic and sympathetic nerves, and in some synapses in the central nervous system. Organophosphate and carbamate insecticides act by inhibiting the enzyme acetylcholinesterase at its esteratic site, resulting in an accumulation of the neurotransmitter acetylcholine in nerve tissue and at the effector organ. This accumulation then results in the continued stimulation of cholinergic synapses and at sufficient levels leads to the signs and symptoms associated with overexposure to these compounds (discussed later in this section).

Absorption and Metabolism

Organophosphates and carbamates can be readily absorbed via ingestion, dermal, and inhalation routes because of their lipophilic nature. For organophosphate insecticides not requiring metabolic activation (discussed below), also called direct inhibitors, these can produce local toxic effects at the site of exposure, including sweating (dermal exposure), miosis or pinpoint pupils (eye contact), and/or bronchospasms (inhalation exposure). Both the organophosphate and carbamate insecticides have relatively short biological half-lives and are fairly rapidly metabolized and excreted.

Within the class of organophosphate insecticides, there are direct organophosphate inhibitors (those containing =O) and organophosphate indirect inhibitors (those containing =S), depending on whether or not they require metabolic activation before they can inhibit acetylcholinesterase. In other words, the indirect organophosphate compounds (containing =S) must undergo bioactivation to become biologically active (containing =O). The indirect inhibiting compounds, including organophosphates such as parathion, diazinon, malathion, and chlorpyrifos, become more toxic than the parent compound on metabolism. In the case of these indirect inhibitors, oxidative desulfuration (replacement of the sulfur atom with an oxygen atom as described above) results in the formation of the oxon of the parent compound (e.g., parathion → paraoxon, diazinon → diazoxon, malathion → maloxon, and chlorpyrifos → chlorpyrifos–oxon). This metabolism occurs via the mixed function oxidase system of the liver.

Once cholinesterase activity has been inhibited in the body by an organophosphate compound, the recovery of that compound is dependent on the reversal of inhibition, aging, and the rate of regeneration of a new enzyme. A chemical reaction that organophosphate insecticides can undergo in the body once they are bound to the cholinesterase enzyme is called "aging." Aging involves the dealkylation of the compound once it is bound to the cholinesterase enzyme. In this "aged" form, the organophosphate compound is tightly bound to the enzyme and will not release itself from the enzyme. Once the aging reaction has occurred, treatment with medications (such as pralidoxime, that is discussed later) is not effective on these aged complexes. Once a cholinesterase molecule has been irreversibly inhibited (via the aging process), the only manner in which the enzyme activity may be restored is through synthesis of new enzyme.

In addition to the aging reaction, organophosphates can also undergo various phase I and II biotransformation pathways, including oxidative, hydrolytic, GSH-mediated transfer, and conjugation reactions.

Carbamate compounds do not require metabolic activation in order to inhibit cholinesterase. Further, carbamate insecticides are not considered irreversible inhibitors like some organophosphate insecticides. Cholinesterase inhibition by carbamate compounds is readily reversible, with reversal of inhibition occurring typically within a few hours after exposure. This rapid reversal of the cholinesterase enzyme activity leads to a much shorter duration of action and thus shorter period of intoxication than is seen in cases of organophosphate overexposure. Also, carbamates do not undergo "aging" as do the organophosphates. As with the organophosphates, carbamates also can undergo various phase I and phase II metabolism reactions.

Acute Effects of Organophosphate and Carbamate Insecticides

The effects of organophosphate and carbamate insecticides can be either local (e.g., sweating from localized dermal exposure) or systemic. Signs and symptoms of overexposure to organophosphate and carbamate compounds occur fairly rapidly after exposure, with effects typically seen beginning from 5 min to 12 h after exposure. A diagnosis of organophosphate intoxication typically is based on an exposure history of 6 h or less before the onset of signs and symptoms. It has been suggested that if symptoms appear more than 12 h after the exposure, then another etiology should be considered, and if the symptoms begin 24 h after the exposure, then organophosphate intoxication should be considered to be equivocal.

Symptoms of carbamate overexposure generally develop within 15 min to 2 h of exposure and typically last only several hours, a duration much shorter than that of the typical overexposure to organophosphate pesticide. Symptoms that are present 24 h following exposure are likely not a result of overexposure to carbamate insecticides.

The acute signs and symptoms seen in cases of over-exposure to both organophosphate and carbamate pesticides are related to the degree of inhibition of acetylcholinesterase in the individual. The clinical manifestation of overexposure to organophosphate and carbamate compounds is a result of muscarinic, nicotinic, and CNS symptoms. In systemic intoxications with organophosphate and carbamate compounds, the muscarinic effects are generally the first effects to develop.

Muscarinic symptoms (parasympathetic nervous system) include sweating, increased salivation, increased lacrimation, bronchospasm, dyspnea, gastrointestinal effects (nausea, vomiting, abdominal cramps, and diarrhea), miosis (pinpoint pupils), blurred vision, urinary frequency and incontinence, wheezing, and bradycardia (decreased heart rate). Nicotinic effects (sympathetic and motor nervous system) include pallor, hypertension, muscle fasciculations, muscle cramps, motor weakness, tachycardia (increased heart rate), and paralysis. Central nervous system signs include giddiness, tension, anxiety, restlessness, insomnia, nightmares, headache, tremors, drowsiness, confusion, slurred speech, ataxia, coma, Cheyne–Stokes respiration, and convulsions.

Organophosphate intoxication is diagnosed on the basis of an opportunity for exposure to the compound, signs and symptoms consistent with organophosphate overexposure, and significant inhibition (i.e., 50 percent inhibition) of cholinesterase enzyme as measured in the plasma and in red blood cells (this will be discussed later). Signs and symptoms resulting from overexposure to these organophosphate and carbamate compounds can be best described by the mnemonic DUMBELS: Diarrhea, Urination, Miosis (pinpoint pupils), Bronchospasm, Emesis (vomiting), Lacrimation (tearing), and Salivation. Signs and symptoms associated with overexposure to organophosphates and carbamate compounds generally do not occur unless acetylcholinesterase activity is approximately 50 percent or less of normal activity.

Signs and symptoms in cases of mild to moderate organophosphate intoxication typically resolve within days to weeks following exposure. In cases of severe organophosphate intoxication, it can be 3 months or so before cholinesterase red blood cell levels return to normal. Death from organophosphate intoxication is usually due to respiratory failure from depression of the respiratory center in the brain, paralysis of the respiratory muscles, and excessive bronchial secretions, pulmonary edema, and bronchoconstriction. Death in individuals with acute organophosphate intoxication that are untreated typically occur within the first 24 h, and within 10 days in treated individuals. If there is no anoxia, complete recovery will occur, in general, within about 10 days after the exposure incident.

Carbamate intoxication presents similar to that of organophosphate intoxication. Cases of carbamate intoxication resolve much more quickly than cases of organophosphate overexposure, due to the rapid reversal of acetylcholinesterase enzyme as well as to the rapid biotransformation in vivo.

Chronic Effects of Organophosphate and Carbamate Insecticides

In general, the main reported chronic effect that may result from exposure to organophosphate insecticides is delayed neuropathy. Organophosphate-induced delayed neuropathy has been associated with exposure to only a few organophosphate compounds, with cases occurring almost exclusively at near-lethal exposure levels. Studies of individuals involved with the handling or formulation of organophosphate compounds (e.g., chlorpyrifos) have not shown permanent adverse health effects. No permanent effects generally result from carbamate intoxication; delayed neuropathy does not occur as a result of carbamate poisoning (see discussion below).

Organophosphate-Induced Delayed Neuropathy

A few of the organophosphates have been associated with the development of a delayed predominantly motor peripheral neuropathy, termed organophosphate-induced delayed neuropathy (OPIDN). In the United States in the 1930s, individuals developed OPIDN, also called "ginger jake" paralysis after consuming ginger liquor contaminated with triorthyl cresyl phosphate (TOCP). Other outbreaks of OPIDN have occurred in relation to the consumption of cooking oil contaminated with TOCP. Organophosphates that have been associated with OPIDN include TOCP, mipafox, trichlorphon, leptophos, and methamidophos. It should be pointed out that only a few of the organophosphate compounds actually are capable of causing OPIDN.

The development of OPIDN is not physiologically related to cholinesterase inhibition. The nerve lesion in OPIDN is that of a distal symmetric predominantly motor polyneuropathy of the long, large-diameter axons (the short, small diameter nerves appear to be spared) in the peripheral nerves. OPIDN, as its name suggests, has a delayed onset of approximately 1–3 weeks following an acute life-threatening exposure to an organophosphate capable of causing delayed neuropathy. The initial complaints of OPIDN include cramping of the calves with numbness and tingling in the feet and then later in the hands. Next, weakness develops in the lower limbs. Bilateral foot drop and wrist drop may develop, and there are usually absent or normal reflexes. A high-stepping gait has also been described in individuals with OPIDN. There also may be motor weakness involving the limbs and motor nerve conduction studies may show abnormalities.

The current theory as to the cause of OPIDN involves a two-step process that occurs in the nervous system. First, it is thought that phosphorylation of a target protein in the nervous system is required. This enzyme is known as *neuropathy target esterase* (NTE), formerly known as *neurotoxic esterase*. The biological action of NTE in the body is not known. The second, and essential, step leading to OPIDN is thought to be the transformation, or "aging," of the enzyme. This "aging" process involves cleavage of an R group from phosphorous, resulting in a negatively charged residue attached to the active site of the enzyme. It appears that compounds that are capable of inhibiting NTE and aging can only cause OPIDN if a threshold of inhibition is reached. A high level of inhibition—70 percent to 80 percent inhibition of NTE in the brain, spinal cord, or peripheral nerve of the experimental animal—soon after dosing with an organophosphate capable of causing OPIDN is necessary before this condition can develop. Thus, the determining factor in the development of OPIDN is the formation of a critical mass of aged-inhibited NTE.

A term used to express the concentration of a substrate (such as an organophosphate compound) that is needed to inhibit 50 percent of an enzyme is IC_{50}. One way of predicting whether a compound will produce OPIDN compared to the levels that cause acute cholinergic signs, is to compare the AChE and NTE IC_{50}s for a specific compound. The in vitro and in vivo IC_{50}s for NTE and AChE in humans and hens (the test species used to evaluate the delayed neuropathic potential of organophosphate insecticides) for several organophosphate compounds have been compared, and it was found that for organophosphate compounds with a IC_{50} AChE/IC_{50} NTE ratio of less than one, OPIDN can occur only after recovery and treatment from acute, otherwise fatal, cholinergic crisis.

The scientific literature indicates that for at least the few compounds known to cause OPIDN, that have been analyzed, the AChE IC_{50}/NTE IC_{50} typically is less than unity. This means that the concentration of a chemical that will inhibit 50 percent of the AChE molecules is less than the concentration of the same chemical that is required to inhibit 50 percent of the NTE molecules. Therefore, at a given concentration, AChE will be inhibited to a greater degree than NTE. As previously stated, it is currently theorized that more than 50 percent of NTE (i.e., 70 percent to 80 percent) must be inhibited in order to develop OPIDN. Likewise, a 50–80 percent inhibition of AChE results in clinical manifestations. In fact, human case reports indicate that virtually all patients who develop OPIDN were managed for a cholinergic crisis first.

It is important to note that carbamates do not cause delayed neuropathy. While some carbamates are capable of inhibiting NTE, aging does not occur. In fact, experimental evidence has showed that some carbamates that inhibit NTE actually protect hens against developing OPIDN.

Neurobehavioral Sequelae

Several studies have been conducted in persons exposed to organophosphates, either occupationally or by accidental or intentional poisoning, to examined the delayed sequelae of organophosphate poisoning. A majority of these studies did not detect any change in permanent memory impairment or other psychological problems in individuals exposed or poisoned by organophosphate insecticides. A majority of the papers in which neuropsychological changes in persons exposed to organophosphates have been reported to contain serious methodological flaws, including failure to control for exposure level, age, education, and alcohol consumption. One study examined 117 individuals who had experienced acute organophosphate poisoning and found no neuropsychiatric symptoms attributable to the organophosphate intoxication. While certain neurobehavioral symptoms (e.g., headache, tension, giddiness, confusion, insomnia) may occur during the acute phase of organophosphate poisoning, there is no objective evidence of permanent neurobehavioral sequelae associated with organophosphate intoxication.

Biological Monitoring for Organophosphates and Carbamates

The organophosphates and carbamates have the ability to inhibit pseudocholinesterase, red blood cell cholinesterase, and nervous system cholinesterase, with biological effects due to the actual inhibition of nervous system cholinesterase only. The levels of cholinesterase present in the blood, especially in the red blood cells, can be used to estimate the degree the nervous system is being affected by anticholinesterases. As mentioned earlier, a 50 percent depression of plasma and red blood cell cholinesterase levels is typically necessary before clinical manifestations are seen. Acute overexposure to organophosphates can be classified as mild (20–50 percent of baseline cholinesterase levels), moderate (10–20 percent of baseline cholinesterase levels), or severe (10 percent or less of baseline cholinesterase level).

Plasma cholinesterase, while susceptible to the inhibitory actions of organophosphate insecticides, has no known biological use in the body. Plasma cholinesterase can vary in an individual based on a number of disease states or conditions (e.g., decreased plasma cholinesterase levels in liver disease such as cirrhosis and hepatitis, multiple metastases, during pregnancy). Plasma cholinesterase is produced by the liver, and this enzyme is found in the nervous tissue, heart, pancreas, and white matter of the brain. Plasma cholinesterase levels typically decline and regenerate more rapidly than red blood cell cholinesterase levels. Plasma cholinesterase levels typically regenerate at the rate of 25 percent in the first 7–10 days. Following organophosphate intoxication, plasma cholinesterase levels may remain depressed for a period of 1–3 weeks.

The enzyme in red blood cell cholinesterase is the same enzyme that is present in the nervous system. Red blood cell cholinesterase regenerates at the rate of approximately 1 percent per day in the body and is dependent on the synthesis of new red blood cells in the body. As mentioned earlier, in severe intoxications from organophosphate pesticide exposure, red blood cell cholinesterase could take as long as 3 months to regenerate.

The measurement of cholinesterase activity in cases of carbamate intoxication are not useful, due to quick reactivation of cholinesterase following carbamate overexposures.

Treatment for Organophosphate and Carbamate Symptomatology

There are two effective treatments for organophosphate intoxication: atropine and pralidoxime. Atropine competes with muscarinic sites, and treatment ameliorates symptoms of nausea, vomiting, abdominal cramps, sweating, salivation, and miosis. Atropine treatment has no effect on the nicotinic signs, such as muscle fasciculations and muscle weakness. Atropine does not affect muscle weakness of respiratory failure. Additionally, atropine does not reactivate cholinesterase.

The second therapeutic agent, pralidoxime (also called 2-PAM), is a medication that reactivates the organophosphate-inhibited cholinesterase enzyme by the removal of the phosphate group that is bound to the esteratic site. However, 2-PAM should be given fairly soon after exposure because the aged enzyme cannot be reactivated. 2-PAM is effective in improving the symptoms of respiratory depression and muscle weakness. Individuals suffering from carbamate intoxication should not be treated with 2-PAM possibly because the reversal of the carbamate inhibitor could add insult to injury.

Decontamination of the individual should include the removal of any contaminated clothing (rubber gloves should be worn to avoid contact with contaminated clothing and materials) and thorough washing of the contaminated skin with soap and water.

Regulatory Information on Organophosphates and Carbamates

OSHA PELs and ACGIH TLVs exist for some of the organophosphate and carbamate insecticides. Biological exposure indices (BEIs) also exists for exposure to parathion (see Table 15.2). There is an ACGIH BEI *p*-nitrophenol levels (metabolite of parathion) in urine as well as a BEI for cholinesterase activity for workers exposed to organophosphate cholinesterase inhibitors.

TABLE 15.2 1999 ACGIH Biological Exposure Indices (BEI)

Pesticide Determinant	Sampling Time	BEI
Organophosphorus cholinestease inhibitors		
Cholinesterase activity in red cells	Discretionary	70% of individual's baseline
Parathion		
Total *p*-nitrophenol in urine	End of shift	0.5 mg/g creatinine
Cholinesterase activity in red cells	Discretionary	70% of individual's baseline

15.2 ORGANOCHLORINE INSECTICIDES

Introduction, Use, and History of Organophosphates and Carbamates

While organochlorine insecticides had widespread use in the 1940s through the mid-1960s in agricultural and malarial control programs, their use has become almost completely discontinued because of their environment effects. Examples of organochlorine insecticides that were commonly used in the past include toxaphene (Toxakil), endrin (Hexadrin), aldrin (Aldrite), endosulfan (Thiodan), BHC (hexachlorocyclohexane), dienoclor (Pentac), heptachlor (Heptagran), dicofol, mirex (Declorane), chlordane, and DDT. One organochlorine compound that is still in use today is lindane, which is used in the medicinal product Kwell for human ectoparasite disease (as well as in products for use in the home and garden and on animals, e.g., Acitox, Gammex). In additional to their use as insecticides in agricultural and forestry settings, organochlorine compounds were also widely used as structural protection against termites in the past. Some of these compounds are still commonly used in some developing countries.

Physical and Chemical Properties

The physical and chemical properties of the organochlorine compounds—their lipophilicity, low vapor pressures, and slow rate of degradation—not only made them effective pesticides, but these same qualities also resulted in their persistence in the environment and their bioaccumulation in the food chain leading to the eventual discontinuance of their use.

Mechanisms of Action

Organochlorine compound chemicals act on the nervous system to produce adverse effects. This class of chemicals is thought to act by the interference with cation exchange across the nerve cell membranes resulting in hyperactivity of the nerves.

Benzene hexachloride compounds (BHCs) (lindane and related compounds) are examples of isomers that produce different effects on the nervous system. The γ isomer, also referred to as *lindane*, causes severe convulsions with rapid onset, while other isomers of BHC, generally cause central nervous system depression. The relative contribution of each of the isomers may explain toxicological differences between formulations of these products. The effect of the cyclodiene organochlorine compounds (e.g., dieldrin) is on the central nervous system.

Pharmacokinetics

Organochlorine compounds are lipophilic and can be absorbed not only through the intestines but also across the lung and skin. Some of these compounds (e.g., lindane, endrin, and chlordane) are more readily absorbed dermally than other compounds in this class, such as DDT or toxaphene.

This class of insecticides (e.g., DDT) can also be stored in fatty tissues in the body. Organochlorine compounds can be detected in adipose tissue, serum, and in milk. Some compounds (e.g., DDT) are mainly stored unchanged in adipose tissue (some DDE is stored in adipose tissue), while others (e.g., endrin) are stored in a metabolized form, aldrin metabolized to endrin (this transformation also occurs in the environment) or heptachlor metabolized to heptachlor epoxide. Organochlorine insecticides are eliminated primarily via the feces.

Acute and Chronic Health Effects of Organochlorine Insecticides

The principal adverse effect associated with over-exposure to organochlorine insecticides is nervous system hyperactivity (e.g., headache, dizziness, paresthesias, tremor, incoordination, or convulsions). Early symptoms seen in chlorinated insecticide intoxications, such as with DDT, include hyperesthesias and paresthesias of the face and limbs, dizziness, nausea and vomiting, headache, tremor, and mental disturbances. Myoclonic movement and convulsions are sometimes seen in severe cases of poisoning. It should be noted that with overexposure to the toxaphene and cyclodiene compounds (e.g., aldrin, endrin, chlordane, and heptachlor) the first sign seen is convulsions, in the absence of the early symptoms just mentioned. Convulsions seen in these cases may not first occur until 2 days after exposure. A group of factory workers overexposed to chlordecone (Kepone) manifested signs and symptoms including gait disturbances, opsoclonus, headache, tremors, hepatomegaly/splenomegaly, and neurobehavioral changes.

While studies of the carcinogenicity of organochlorine compounds have demonstrated positive effects in mice, but generally not in rats, at high doses, there is generally no evidence of cancer in humans, even in the most highly exposed individuals (e.g., workers involved in the manufacture and formulation of organochlorine insecticides).[1]

Biological Monitoring for Organochlorine Insecticides

Levels of organochlorine insecticides can be detected at background levels in biological tissues of individuals not occupationally exposed to these compounds. Serum and adipose tissue testing for the presence of organochlorine pesticides in the general population has been conducted. The NHAT (National Human Adipose Tissue) survey was conducted in 1982. In this study, 763 individual adipose tissue specimens collected from the general population were tested for various compounds, including several organochlorine compounds (β-BHC, *p,p′*-DDE, dieldrin, heptachlor epoxide, and DDT). These results are presented in the NHATS Broad Scan Analysis. Results for organochlorine insecticides detected in the serum of the general population are reported in Health and Nutrition Examination Survey II (HANES II).

Treatment of Organochlorine Intoxication

Treatment of organochlorine intoxication is supportive (e.g., control of convulsions with benzodiazepines or barbiturates). One chlorinated insecticide that can be effectively removed from the body is chlordecone (Kepone, the compound involved in the poisoning of factory workers discussed above). In this case, chloestyramine was used therapeutically to treat the workers who had been poisoned with chlordecone. In nine patients, administration of 24 g of chloestyramine per day resulted in a 3.3–17.8-fold increase in fecal elimination of chlordecone. Treatment also resulted in a reduction of chlordecone half-life in blood from 165 to 80 days.

15.3 INSECTICIDES OF BIOLOGICAL ORIGIN

Many compounds are present in nature that have insecticidal qualities, including extracts from the chrysanthemum flower and from the *Legumionocae* genera (e.g., rotenone). Trade names of insecticides in this classification include Pyrocide (pyrethrum) and Prentox (rotenone).

Pyrethrum and Pyrethrins

Pyrethrum is an extract from the chrysanthemum flower, *Pyrethrum cinerariaefollium* ("Dalmatian insect flowers") and other species. This extract contains approximately 50 percent natural pyrethrins—

[1]In addition, in a recent review of the relationship between environmental estrogens and breast cancer, Safe (1997) concluded, from a review of epidemiologic studies, that there is no scientific evidence that o rganochlorine xenoestogens are causally associated with the development of breast cancer. Safe (1997) poi nts out that analysis of the data available to date do not demonstrate that levels of organochlorines are significant ly higher in individuals with breast cancer compared to controls and that there is no evidence of increased risk of brea st cancer in women exposed occupationally to relatively high levels of PCBs or DDT/DDE.

the insecticidal component of the extract. The pyrethrins jasmolins I and II, cinerins I and II, and pyrethrins I and II are extracted from the powder for formulation into commercial aerosols and spray products. These compounds are often formulated with a synergist such as piperonylbutoxide or *n*-octyl bicycloheptene dicarboximide. These synergists are incorporated in order to slow down the degradation of the pyrethrin compounds. This class of compounds are commonly used in household insecticides and in pet products (e.g., flea and tick dips and sprays).

Pyrethrins and pyrethrum are very rapidly metabolized and excreted from humans and have very low mammalian toxicity. Crude pyrethrums have been associated with allergic responses in individuals, although this action is most likely due to the noninsecticidal components of this compound. A study of 59 workers who had been employed in a pyrethrum factory ranging from 1 to 25 years showed essentially no adverse health effects, with the exception of inflammatory pleural lesions in some individuals exposed to high air levels of pyrethrums. Treatment of pyrethrin and pyrethrum exposure is primarily symptomatic.

Synthetic Pyrethroids

Pyrethrins and pyrethrum insecticides are unstable in light and heat. Because of this instability, synthetic pyrethroids, which have better stability to light and heat, have been developed and are used in agricultural settings as well as for home pest control. Over 1000 synthetic pyrethroids have been developed over the years and include compounds such as cyfluthrin, cypermethrin (Cymbush), deltamethrin, fenpropathrin (Danitol), fluvalinate (Mavrik), permethrin (Ambush), resmethrin (Chryson), and tralomethrin (Scout) (Figure 15.3).

The action sites of the pyrethroids are the voltage-dependent sodium channels in nerves. The general basis for nerve impulse generation and conduction is the ionic permeability of the membrane combined with the sodium (high levels outside the cell) and potassium (high levels inside in the cell) concentration gradients. The resting cell membrane is maintained by the sodium–potassium pump, and the inside of the cell is negatively charged with respect to the outside of the cell. A normal nerve impulse is caused by a quick transient increase in the permeability of the membrane to sodium ions, causing an inward

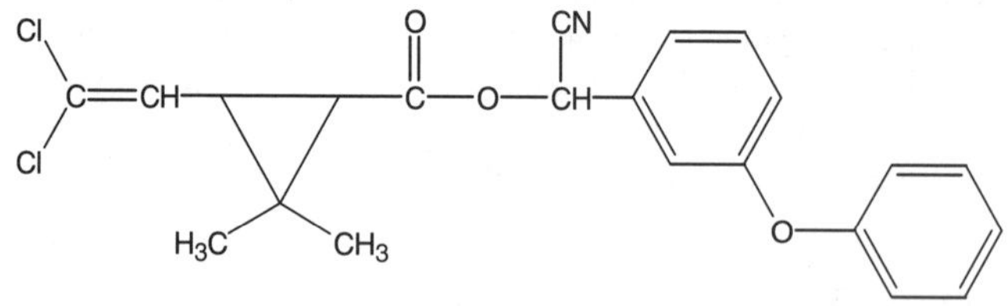

Cypermethrin

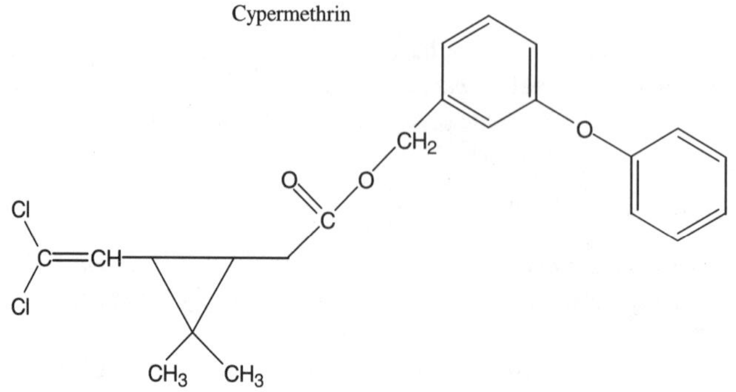

Permethrin

Figure 15.3 Synthetic pyrethroids.

influx of sodium, followed by an increase in the potassium permeability, causing an outward flow of potassium. The ionic currents cause a temporary reversal of the membrane potential from negative to positive resulting in nerve impulse conduction along the nerve fiber. Pyrethroids exert their effect by slowing the closing of the sodium activation gate. Type I pyrethroids prolong individual channel currents causing whole cell sodium influx to be prolonged, elevating the after-potential until the threshold potential is reached and repetitive discharges occur. Examples of type I pyrethroids include allethrin, cismethrin, permethrin, and resmethrin. Type I pyrethroids, at high levels in animals, have been reported to cause increased sensitivity to external stimuli, tremors, increased body temperature, and rigor immediately preceding death. Type II pyrethroids cause an extremely prolonged sodium current, leading to depolarization of the nerve and impulse conduction block. Type II pyrethroids include cyfluthrin, cyhalothrin, cypermethrin, deltamethrin, and fenvalerate. Type II pyrethroids cause behavioral problems early in the intoxication, leading to salivation, miosis, bradycardia, tremor, decreased startle response to sound, and ataxia.

Like pyrethrins and pyrethrum, the synthetic pyrethroids are rapidly metabolized and excreted in humans and do not bioaccumulate. In fact, the relatively resistant nature of mammals, including humans, stems from the ability to metabolize these compounds quickly and efficiently. Synthetic pyrethroids have greater insecticidal activity and lower mammalian toxicity than the organophosphate, carbamate, and organochlorine insecticides. Experimental animals that have been treated with high doses of pyrethroids experience symptoms such as tremors, salivation, and/or convulsions. In general, animals surviving an acute intoxication to pyrethroids recover within several hours of exposure.

The primary reported reaction to exposure to synthetic pyrethroid insecticides in humans occurs with exposure to those pyrethroids containing cyano groups (e.g., fenvalerate and cypermethrin). This reaction consists of paresthesia, typically occurring around the mouth region in workers exposed to these compounds. This paresthesia is reversible and dissipates usually within 24 h of cessation of exposure. An occupational study of 199 workers who were involved in dividing and packaging pyrethroids (fenvalerate, deltamethrin, and cypermethrin) showed that aside from transient paresthesias occurring in the facial area and sneezing and increased nasal secretions, there were essentially no adverse health effects attributable to the pyrethroid exposure. Treatment of pyrethrin overexposure consists of decontamination and supportive treatment.

Rotenone

Rotenone (Noxfish) occurs naturally in several plants species (e.g., the *Leguminocae* genera) and is used mainly as an insecticide as well as to eliminate fish in lakes and ponds. The mechanism of action for rotenone is as a respiratory toxin, blocking electron transport at ubiquinone, preventing oxidation of NADH. Rotenone seems to have low toxicity in man, and few reports of serious injury appear to have been reported. Occupational exposure to the powder of the plant that contains rotenone has reportedly caused dermal and respiratory tract irritation and numbness in mouths of workers. Treatment of rotenone overexposure consists mainly of decontamination and supportive therapy.

Bacillus thuringiensis

Microbial insecticides, such as several strains *Bacillus thuringiensis* (e.g., Dipel, variety *kurstaki*), have been developed as effective insecticides. The endotoxin of *Bacillus thuringiensis* is insecticidal in certain sensitive species. *Bacillus thuringiensis* has not generally been associated with mammalian or human toxicity; only a few instances of adverse effects in humans have been reported. A group of 18 human volunteers ingesting 1 gram of a *B. thuringiensis* formulation for 5 days, with 5 of these 18 subjects also inhaling 100 mg of the powder for 5 days, reported no adverse effects. Furthermore, a group of workers exposed to various processes involved in the formulation of a commercial product containing the biological insecticide showed no adverse health effects.

15.4 HERBICIDES

Chlorophenoxy Herbicides

The chlorophenoxy herbicides 2,4-dichlorophenoxy acetic acid (2,4-D) and 2,4,5-trichlorophenoxy acetic acid (2,4,5-T) are probably the most commonly recognized of the chlorophenoxy herbicides. These compounds exert their action in plants by acting as growth hormones, but have no such hormonal action in animals or humans. Some of the commonly used chlorophenoxy herbicides include Banvel (dicamba), Weedone (2,4-D), and Basagran M (MCPA) (see Figure 15.4).

Acute Toxicity

2,4-Dichlorophenoxyacetic acid (2,4-D) is prepared commercially by the reaction of 2,4-dichlorophenol and monochloroacetic acid. Other chlorophenoxy herbicide analogs include 2,4-DB, 2,4-DP, MCPA [(4-chloro-2-methylphenoxy) acetic acid], MCPP, and the herbicides 2,4,5-trichlorophenoxyacetic acid (2,4,5-T) and 2-(2,4,5-trichlorophenoxy) propionic acid (2,4,5-TP; Silvex) that are no longer used.

Dioxin (2,3,7,8-TCDD) has not been identified in 2,4-D formulations (WHO, 1984). While in the past, 2,3,7,8-TCDD contamination may have occurred in 2,4-D, this was due to contamination from the production of 2,4,5-T. The synthesis of 2,4-D does not produce 2,3,7,8-TCDD.

The primary routes of exposure to chlorophenoxy herbicides are dermal and inhalation. Chlorophenoxy compounds act by uncoupling oxidative phosphorylation and decreasing oxygen consumption in tissue. These compounds are fairly rapidly excreted and do not accumulate in the body. These compounds are excreted via the urine primarily, and apart from conjugation of acids, little biotransformation occurs in the body.

Following ingestion, the acute toxicity of chlorophenoxy herbicides includes irritation of the mucous membranes and gastrointestinal lining. Large intentional overdoses with chlorophenoxy acids have resulted in symptoms of coma, metabolic acidosis, myotonia, mucous membrane irritation, and myalgias. While cases of peripheral neuropathy following exposure to 2,4-D have been reported sporadically throughout the literature, no causal association between this compound and neuropathy has been proved.

Treatment of cases of overexposure with chlorophenoxy herbicides is symptomatic and also involves decontamination.

Carcinogenicity

2,4-D is currently classified as a "D" carcinogen (not classifiable) by the USEPA. A recent mortality study of chemical workers exposed to 2,4-D and its derivatives found no evidence of increased mortality from cancer, including non-Hodgkin's lymphoma.[2] A recent review of the available animal and human data for the chlorophenoxy herbicides 4-chloro-2-methyl phenoxyacetic acid (MCPA), 2-(4-chloro-2 methylphenoxy) propionic acid (MCPP), and 2-(2,4-dichlorophenoxy) propionic acid (2,4-DP) concluded that there was no evidence to indicate that these compounds were carcinogenic to humans.

[2] 2,4-D has been classified as a group D carcinogen by the USEPA Office of Pesticide Programs. In their recent review of 2,4-D (USEPA OPP, 1996) entitled "Carcinogenicity Peer Review (4th) of 2,4-Dichlor ophenoxyacetic Acid," the Office of Pesticide Programs concluded: "The Health Effects Division Carcinogen icity Peer Review Committee (CPRC) met on July 17, 1996 to discuss and evaluate the weight-of-the-evidence on 2,4 ,-D with particulate reference to its carcinogenic potential. The CPRC concluded that 2,4-D should r emain classified as a Group D—Not Classifiable as to Human Carcinogenicity" and "The CPRC agree that 2,4-D should r emain classified as a Group D . . . . In two new adequate studies in rodents, which were conducted a t doses high enough to assess the carcinogenic potential of 2,4-D, there were no compound related statisticall y significant increases in tumors in either rats or mice."

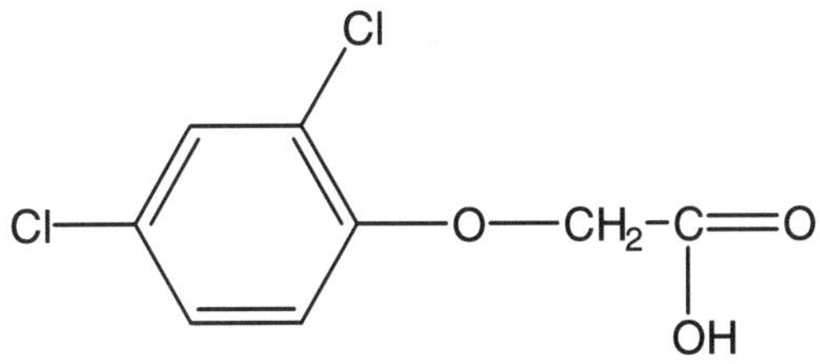

Figure 15.4 2,4-Dichlorophenoxyacetic acid (2,4-D).

Bipyridyl Compounds—Paraquat and Diquat

Paraquat (1,1′-dimethyl-4,4′-dipyridylium) (see Fig. 15.5) and diquat (1,1′-ethylene-2,2′-bipyridylium) are bipyridylium herbicides, with common trade names including Gramoxone (paraquat) and Aquacide (diquat). A majority of reported cases of toxicity associated with both paraquat and diquat are seen in cases of accidental or intentional (suicidal) ingestion, with paraquat having greater toxicity than diquat. An emetic and stenching agent, valeric acid, is added to paraquat solutions.

Paraquat poisoning (e.g., from suicide attempts) can lead to multiorgan toxicity (e.g., gastrointestinal tract, kidney, heart, and liver) including pulmonary fibrosis. Early deaths occurring after intoxication with paraquat result from acute pulmonary edema, oliguric renal failure, and hepatic failure. Deaths occurring one to three weeks following an intoxication episode are typically the result of pulmonary fibrosis. Paraquat is not typically readily dermally absorbed, but reports of toxicity following sufficient dermal absorption have been seen in individuals with skin abrasions or individuals with continued dermal exposure to paraquat. Sufficient dermal exposure to paraquat can also cause dermal irritation, blistering, and ulceration. Similar irritant effects are seen in the esophagus and stomach of individuals swallowing paraquat. Paraquat concentrates in the lung, where its proposed mechanism of action leading to pulmonary fibrosis is that by which free radicals are generated leading to lipid peroxidation. Pulmonary fibrosis, which can be fatal in cases with sufficient exposure, begins within 2 days to 2 weeks following paraquat exposure.

Inhalation is not believed to be a toxic route of exposure. Aerosol paraquat droplets have been measured as having diameters exceeding 5 μm, indicating that they do not reach the alveolar membrane to cause either direct or systemic toxicity via inhalation. In two field trials in which absorption of paraquat was measured by urinary paraquat levels, systemic absorption was apparently not significant. The authors of that study concluded that "ordinary care in personal hygiene is sufficient to prevent any hazard from surface injury or from systemic absorption." Also, a recent study conducted on a group of 85 paraquat spraymen revealed no adverse health effects (aside from irritant-type effects), including no lung effects, attributable to long-term occupational use of this herbicide.

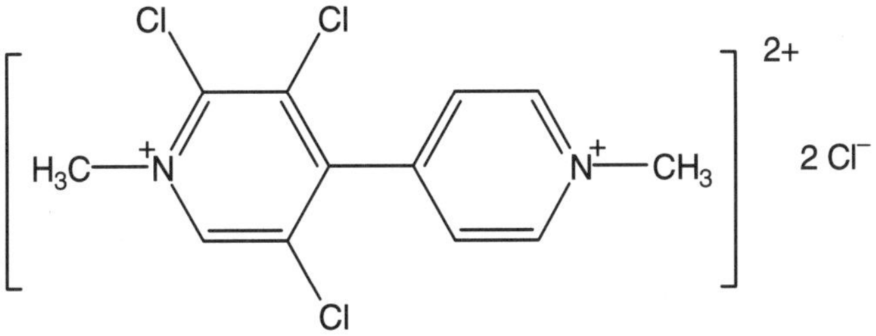

Figure 15.5 Paraquat.

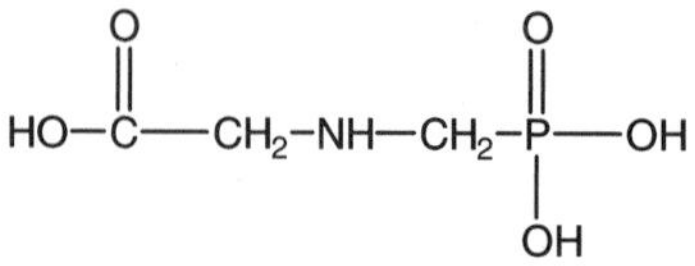

Figure 15.6 Glyphosate.

Diquat causes less dermal irritation and injury than does paraquat, and diquat is not selectively concentrated in pulmonary tissue like paraquat. Diquat, in contrast to paraquat, causes little to no injury to the lungs; however, diquat has an effect on the central nervous system, whereas paraquat does not. The mechanism of action of diquat is thought to be similar to that of paraquat, involving the production of superoxide radicals that cause lipid membrane destruction. Dermal exposure to sufficient levels of diquat can cause fingernail damage and irritation of the eyes and mucous membranes. Intoxication by diquat via the oral route has reportedly caused signs and symptoms including gastrointestinal irritation, nausea, vomiting, and diarrhea. Both paraquat and diquat are reportedly associated with renal toxicity. There is no known specific antidote for either paraquat or diquat poisoning.

Glyphosate (Round-Up) [*N*-(phosphonomethyl) glycine] (see Figure 15.6) is a widely used herbicide that interferes with amino acid metabolism in plants. In animals it is thought to act as a weak uncoupler of oxidative phosphorylation. Glyphosate is moderately absorbed through the gastrointestinal tract, undergoes minimal biotransformation, and is excreted via the kidneys. There have been several reports in the literature of intoxications, typically resulting from accidental or suicidal ingestion, following overexposure to the glyphosate-containing product Round-Up. Various signs and symptoms include gastro-intestinal irritation and damage, as well as dysfunction in several organ systems (e.g., lung, liver, kidney, CNS, and cardiovascular system). It has been proposed that the toxicity seen following intoxication with Round-Up is due to the surfactant agent in the commercial product. One study conducted determined that the irritative potential of the commercial preparation of Round-Up is similar to that of baby shampoo.

Triazines

Examples of triazine and triazole herbicides include atrazine (2-chloro-4-ethylamino-6-isopropylamine-*s*-triazine), propazine, simazine [2-chloro-4,6-bis(ethylamino)-*s*-triazine], and cyanazine [2-chloro-4-(1-cyano-1-methylethylamino)-6-ethylamino-*s*-triazine]. Triazine herbicides have relatively low toxicity, and no cases of systemic poisoning have appeared to have been reported. Occasional reports of dermal irritation from exposure to triazine herbicides has been reported in the literature.

15.5 FUNGICIDES

Fungicides are compounds that are used to control the growth of fungi and have found uses in many different products, from their use to protect grains after harvesting while they are in storage to their use in paint products.

Pentachlorophenol, also known as penta, is used as a wood preservative for fungus decay or against termites, as well as a molluscicide. Trade names of pentachlorophenol include Pentacon, Penwar, and Penchlorol (Figure 15.7).

Pentachlorophenol is readily absorbed via the skin, lung, and gastrointestinal tract. Pentachlorophenol and its biotransformation products are excreted primarily via the kidneys. The biochemical mechanism of action of pentachlorophenol is through an increase in oxidative metabolism from the uncoupling of oxidative phosphorylation. This increase in oxidative metabolism in poisonings can lead to an increase in body temperature. In fatal cases of poisoning from pentachlorophenol, body temperatures as high as (almost) 41.8 °C (107.4 °F) have been reported. Severe overexposure to pentachlorophenol can cause signs and symptoms such as delirium, flushing, pyrexia, diaphoresis, tachypnea, abdominal pain, nausea, and tachycardia.

Because pentachlorophenol volatilizes from treated wood and fabric, excessively treated indoor surfaces can lead to irritation of the skin, eyes, and upper respiratory tract. Contact dermatitis has been reported in workers exposed dermally to pentachlorophenol. Treatment of pentachlorophenol poisoning consists mainly of decontamination of clothing and skin and/or gastrointestinal tract as well as supportive treatment for symptoms associated with the exposure (e.g., temperature control).

Pentachlorophenol can be assayed for in blood, urine, and adipose tissue. The ACGIH biological exposure index for pentachlorophenol is 2 mg/g creatinine total pentachlorophenol in urine prior to the last shift of the workweek or 5 mg/L free pentachlorophenol in plasma at the end of the workshift.

Dithiocarbamates/Thiocarbamates

The dithiocarbamates and the thiocarbamates are used as fungicidal compounds and have little insecticidal toxicity, unlike the *N*-methyl carbamates (e.g., the acetylcholinesterase-inhibiting carbamate, carbaryl) discussed earlier. Examples of thiocarbamate fungicides include thiram (AAtack), metam-sodium (Vapam), ziram (Ziram 76), ferbam, and the ethylene bis dithiocarbamate (EBDC) compounds—maneb, zineb, and mancozeb.

In general, the thiocarbamate class of fungicides has low acute toxicity. Thiram dust has been reported to cause eye, skin, and mucous membrane irritation, with contact dermatitis and sensitization reportedly occurring in a few workers. Systemic intoxications that have been associated with exposure to thiram have resulted in symptomatology similar to that cause by reactions to disulfiram (Antabuse), a dithiocarbamate medication used to treat alcoholism. Thiram, like disulfiram, is not a cholinesterase inhibitor, but does cause inhibition of the enzyme acetaldehyde dehydrogenase (responsible for the conversion of acetaldehyde to acetic acid), and reportedly, in rare cases, workers who have been exposed to thiram have complained of "Antabuse" reactions after ingestion of alcoholic beverages.

Exposure to ziram, ferbam, and the EBDC compounds have been associated with skin, eye, and respiratory tract irritation in humans. Maneb and zineb have been associated with cases of chronic dermatological disease, possibly due to dermal sensitization to these compounds in workers.

Chlorothalonil

Chlorothalonil (Bravo, Daconil) (2,4,5,6-tetrachloro-1,3-benzenedicarbonitrile) has been reported to cause dermal and mucous membrane irritant effects in humans exposed to this compound. Chlorothalonil appears to have low potential for toxicity in humans.

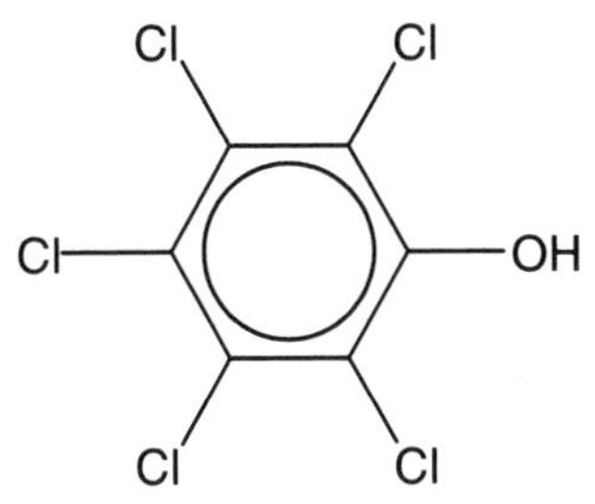

Figure 15.7 Pentachlorophenol.

Copper Compounds

Exposure to dust and powder formulations of copper-based fungicides has been reported to cause irritation of the skin, eyes, and respiratory tract. Systemic intoxication in humans by copper fungicides has been rarely reported. Ingestion of the compound has reportedly caused gastrointestinal irritation, nausea and vomiting, diarrhea, headache, sweating, weakness, liver enlargement, hemolysis and methemoglobinemia, albuminuria, hemoglobinurina, and occasionally renal failure. Treatment of copper intoxication can include an effort to prevent absorption (e.g., lavage) followed by chelation therapy.

15.6 RODENTICIDES

The rodenticides, as the name indicates, are a class of compounds designed to specifically target rodents. These compounds have, in some cases, taken advantage of physiological differences between rodents and other mammals (viz., humans) that make rodents more susceptible to their toxic effects. The most efficient route of exposure of these compounds is via ingestion.

This class of rodenticides works by depression of the vitamin K synthesis of the blood clotting factors II (prothrombin), VII, IX, and X. This anti-coagulant property manifests as diffuse internal hemorrhaging occurring typically after several days of rodenticide bait ingestion. Warfarin (see Figure 15.8) is a commonly used coumarin rodenticide that causes its toxic effects by inhibiting the formation of prothrombin and the inhibition of vitamin K–dependent factors in the body. Other anticoagulant rodenticides include coumafuryl, brodifacoum, difenacoum, and prolin. Warfarin is known to be absorbed both dermally and from ingestion. Signs and symptoms of intoxication with warfarin include epistaxis, hemoptysis, bleeding gums, gastrointestinal tract and genitourinary tract hemorrhage, and ecchymoses.

The indandiones, unlike the coumarins, cause nervous system, cardiac, and pulmonary effects in laboratory animals preceding the death from the anticoagulant effects. These types of adverse effects have not been reported in cases of human exposure. Examples of indandione rodenticides include diphacinone, diphacin, and chlorphacinone.

The most prominent clinical laboratory sign from the administration of these classes of compounds is an increased prothrombin time and a decrease in plasma prothrombin concentration. Treatment of toxicity from coumarins and indandions consists of the administration of vitamin K_1.

Thallium Sulfate

Thallium sulfate is readily absorbed via ingestion and dermally, as well as via inhalation. The target organs of thallium sulfate include the gastrointestinal tract (hemorrhagic gastroenteritis), heart and blood vessels, kidneys, liver, skin, and the hair. Symptoms such as headache, lethargy, muscle weakness, numbness, tremor, ataxia, myoclonia, convulsions, delirium, and coma are seen in cases of

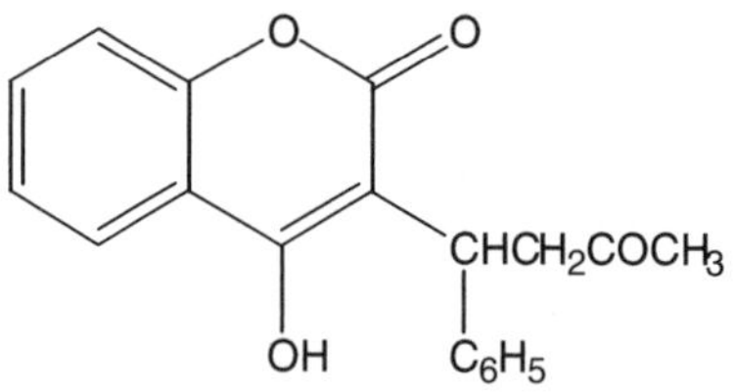

Figure 15.8 Warfarin.

thallium sulfate–induced encephalopathy. Death from thallium sulfate intoxication is due to respiratory paralysis or cardiovascular failure.

Serum, urine, and hair thallium levels can be used to assess exposure to this compound. There is no specific treatment for thallium sulfate poisoning, and treatment is supportive. Syrup of ipecac and activated charcoal can be used to decrease gastrointestinal absorption.

Sodium Fluoroacetate

Sodium fluoroacetate is also known as 1080 (registered trademark). This compound is easily absorbed via ingestion as well as through inhalation and dermal routes. The toxicity of sodium fluoroacetate is due to the reaction of three molecules of fluoroacetate which form fluorocitrate in the liver. Fluorocitrate adversely affects cellular respiration through disruption of the tricarboxylic acid cycle (inhibiting the enzyme *cis*-aconitase). It is thought that the accumulation of citrate in tissues also accounts for some of the acute toxicity associated with this compound. The target organs of sodium fluoroacetate are the heart (seen as arrhythmias leading to ventricular fibrillation) and the brain (manifested as convulsions and spasms), following intoxication (typically following suicidal or accidental ingestion). A specific antidote to sodium fluoroacetate intoxication does not exist. Treatment consists of decontamination and supportive therapy, including gastric lavage and catharsis.

15.7 FUMIGANTS

The fumigants (e.g., see Figure 15.9) are a group of compounds that are volatile in nature. Some of the fumigants exist in a gas phase at room temperature while others are liquids or solids.

Fumigants are in general readily absorbed via dermal, respiratory, and ingestion routes. Treatment for overexposure to fumigants typically includes irrigation of the contaminated areas (skin, eyes). Following irrigation of eyes, medical treatment should be sought because some of these compounds are severely corrosive to the cornea. Sufficient dermal absorption may occur as to produce systemic effects. Patients with inhalation exposure should be monitored for pulmonary edema and treated accordingly if edema develops. Contaminated clothing should be removed and discarded. It should be

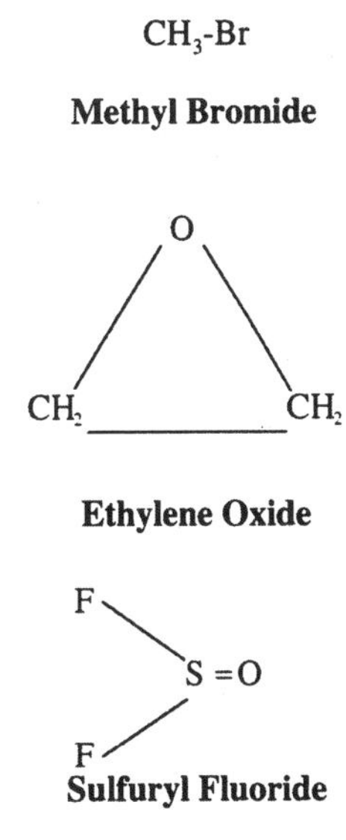

Figure 15.9 Chemical structures of selected fumigants.

noted that certain fumigants have the ability to penetrate rubber and neoprene (often used for personnel protective equipment).

Methyl Bromide

Methyl bromide (Brom-O-Sol, Terr-O-Gas) has been in use as a fumigant since 1932 and is a colorless and practically odorless compound (at low levels), with its low warning potential contributing to its toxicity. At higher concentrations, the odor of methyl bromide is similar to chloroform. Fatalities have been reported during application and from early reentry into treated areas. Methyl bromide has been used to treat dry packaged foods in mills and warehouses as well as used as a soil fumigant to control nematodes and fungi.

Methyl bromide is very irritating to the lower respiratory tract. It is thought that the parent compound is responsible for the toxicity of the methyl bromide, with the mechanism of toxicity possibly having to do with its ability to bind with sulfhydryl enzymes. Exposure to high concentrations of methyl bromide can lead to pulmonary edema or hemorrhage, and those exposed typically experience delayed onset (several hours after exposure). Symptoms of acute intoxication include those consistent with central nervous system depression such as headache, dizziness, nausea, visual disturbances, vomiting, and ataxia. Exposure to very high concentrations can lead to unconsciousness. In cases of exposure to fatal levels of methyl bromide, death typically occurs within 4–6 h to 1–2 days postexposure; the cause of death is respiratory or cardiovascular failure resulting from pulmonary edema. Dermal exposure to liquid methyl bromide can cause skin damage in the form of burning, itching, and blistering. Treatment of methyl bromide poisoning is symptomatic.

Ethylene Oxide

Ethylene oxide, also known as epoxyethane (ETO), is a sterilant and fumigant that exists as a colorless gas and which has a high odor threshold. Ethylene oxide also is a severe mucous membrane and skin irritant. Dermal exposure at sufficient levels can result in edema, burns, blisters, and frostbite. Acute intoxications can result in CNS depression characterized by headache, nausea, vomiting, drowsiness, weakness, and cough. Exposure to extreme concentrations of ethylene oxide can cause the development of pulmonary edema and cardiac arrhythmias.

Sulfuryl Fluoride

Sulfuryl fluoride (Vikane) (SO_2F_2), a colorless and odorless gas, is used as a structural fumigation. Fatalities have been reported from individuals entering buildings recently fumigated with sulfuryl fluoride before reentry was allowed. The acute toxic effects from sulfuryl poisoning include mucous membrane irritation, nausea, vomiting, dyspnea, cough, severe weakness, restlessness, and seizures.

15.8 SUMMARY

This chapter has discussed the toxicology of some of the most commonly used groups of pesticides:

- Organophosphate and carbamate insecticides
- Organochlorine insecticides
- Insecticides of biological origin
- Herbicides
- Fungicides
- Rodenticides
- Fumigants

From the discussion included in this chapter, the following are the main points to be gained:

- Pesticides are used for a variety of different reasons, including control or eradication of pests from homes, pets, or crops. Pesticides are also important in the control of vector-borne diseases (e.g., malaria).
- Individuals may be exposed to a variety of pesticides via inhalation, ingestion, or dermal routes. Exposure can be either occupational, dietary, accidental, or intentional (e.g., suicide).
- Pesticides work via numerous mechanisms in pest species as well as in humans and animals.
- The persistent organochlorine insecticides have been replaced by organophosphate compounds. These organophosphate insecticides are now being replaced by pesticides such as pyrethrins which are even of lower toxicity and are not very persistent.
- Industrial hygiene standards, such as OSHA PELs and ACGIH TLVs and BEIs, exist for a number of pesticides.

REFERENCES AND SUGGESTED READING

American Conference of Governmental Industrial Hygienists (ACGIH), *1999 Threshold Limit Values (TLVs) for Chemical Substances and Physical Agents and Biological Exposure Indices (BEIs)*, ACGIH, Cincinnati, 1999.

Austin, H., J. E. Keil, and P. Cole, "A prospective follow-up study of cancer mortality in rel ation to serum DDT," *Am. J. Publ. Health* **79:** 43–46 (1989).

Baselt, R. C., and R. H. Cravey, *Disposition of Toxic Drugs and Chemicals in Man*, 4th ed., Chemical Toxicology Institute, Foster City, CA, 1995.

Bolt, H. M., "Quantification of endogenous carcinogens. The ethylene oxide paradox," *Biochem. Pharmacol.* **52:** 1–5 (1996).

Bond, G. G., and R. Rossbacher, "A review of potential human carcinogenicity of the chlorop henoxy herbicides MCPA, MCPP, and 2,4-DP," *Br. J. Ind. Med.* **50:** 340–348 (1993).

Burns, C. J., "Update of the morbidity experience of employees potentially exposed to chl orpyrifos," *Occup. Environ. Med.* **55:** 65–70 (1998).

Cannon, S. B., J. M. Veazey, R. S. Jackson, V. W. Burse, C. Hayes, W. E. Straub, P. J. Landrigan , and J. A. Liddle, "Epidemic Kepone poisoning in chemical workers," *Am. J. Epidemiol.* **107**(6): 529–537 (1978).

Cohn, W. J., J. J. Boylan, R. V. Blanke, M. W. Farriss, J. R. Howell, and P. S. Guzelian, "Treatm ent of chlordecone (Kepone) toxicity with chloestyramine. Results of a controlled clinical trial," *NEJM* **298:** 243–248 (1978).

Costa, L. G., "Basic toxicology of pesticides," *Occup. Med. State of the Art Rev.* **12**(2): 251–268 (1997).

Coye, M. J., P. G. Barnett, J. E. Midtling, A. R. Velasco, P. Romero, C. L. Clements, and T. G. Ro se, "Clinical confirmation of organophosphate poisoning by serial cholinesterase analyses," *Arch. Intern. Med.* **147:** 438–442 (1987).

Daniell, W. S. Barnhart, P. Demers, L. G. Costa, D. L. Eaton, M. Miller, and L. Rosenstock, "Neu ropsychological performance among agricultural pesticide applicators," *Environ. Res.* **59:** 217–228 (1992).

Dannaker, C. J., H. I. Maibach, and M. O'Malley, "Contact urticaria and anaphylaxis to the fu ngicide chlorothalonil," *Cutis* **52:** 312–315 (1993).

de Jong, G., G. M. H. Swaen, and J. J. M. Slangen, "Mortality of workers exposed to dieldrin and aldrin: A retrospective cohort study," *Occup. Environ. Med.* **54:** 702–707 (1997).

Ditraglia, D., D. P. Brown, T. Namekata, and N. Iverson, "Mortality study of workers employed a t organochlorine pesticide manufacturing plants," *Scand. J. Work Environ. Health.* **7:** 140–146 (1981).

Durham, W. F., and W. J. Hayes, "Organic phosphorous poisoning and its therapy with special reference to modes of action and compounds that reactivate inhibited cholinesterase," *Arch. Environ. Health* **5:** 21–47 (1962).

Ellenhorn, M. J., *Ellenhorn's Medical Toxicology: Diagnosis and Treatment of Human Poisonings*, 2nd ed., Williams & Wilkins, Baltimore, 1997.

Farm Chemicals Handbook, Meister Publishing, Willoughby, OH, 1997.

Fisher, R., and L. Rosner, "Insecticide safety. Toxicology of the microbial insecticide, Thuricide," *J. Agric. Food Chem.* **7:** 686–688 (1959).

Gadoth, N., and A. Fisher, "Late onset of neuromuscular block in organophosphate poisoning," *Ann. Int. Med.* **88:** 654–655 (1978).

Gombe, S., and T. A. Ogada, "Health of men on long term exposure to pyrethrins," *East Afr. Med. J.* **65:** 734–743 (1988).

Grob, D., and A. M. Harvey, "The effects and treatment of nerve gas poisoning," *Am. J. Med.* **14:** 52–63 (1953).

Guzelian, P. S., "Therapeutic approaches for chlordecone poisoning in humans," *J. Toxicol. Environ. Health* **8:** 757–766 (1981).

Guzelian, P. S., "Comparative toxicology of chlordecone (Kepone) in humans and experimental animals," *Ann. Rev. Pharmacol. Toxicol.* **22:** 89–113 (1982).

Hall, S. W., and B. B. Baker, "Intermediate syndrome from organophosphate poisoning," Abstract 103, *Vet. Hum. Toxicol.* **31:** 35 (1989).

Hayes, W. J., and E. R. Laws, *Handbook of Pesticide Toxicology*, Vols. 1–3, Academic Press, New York, 1991.

He, F., J. Sun, K. Han, Y. Wu, P. Yao, S. Wang, and L. Liu, "Effects of pyrethroid insecticides on subjects engaged in packaging pyrethroids," *Br. J. Ind. Med.* **45:** 548–551 (1988).

Higginson, J., "DDT: Epidemiological evidence," in *Interpretation of Negative Epidemiological Evidence for Carcinogenicity*, N. J. Wald and R. Doll, eds., IARC Scientific Publication 65, 1985, pp. 107–117.

Howard, J. K., "A clinical survey of paraquat formulation workers," *Br. J. Ind. Med.* **36:** 220–223 (1976).

Howard, J. K., N. N. Sabapathy, and P. A. Whitehead, "A study of the health of Malaysian plantation workers with particular reference to paraquat spraymen," *Br. J. Ind. Med.* **38:** 110–116 (1981).

Hunter, D. J., S. E. Hankinson, F. Laden, G. A. Colditz, J. E. Manson, W. C. Willett, F. E. Speizer, and M. S. Wolff, "Plasma organochlorine levels and the risk of breast cancer," *NEJM* **337**(18): 1253–1258 (1997).

Hustinx, W. N. M., R. T. H. van de Laar, A. C. van Huffelen, J. C. Verwey, J. Meulenbelt, and T. J. F. Savelkoul, "Systemic effects of inhalation methyl bromide poisoning: A study of nine cases occupationally exposed due to inadvertent spread during fumigation," *Br. J. Ind. Med.* **50:** 155–159 (1993).

Ibrahim, M. A., G. G. Bond, T. A. Burke, P. Cole, F. N. Dost, P. E. Enterline, M. Gough, R. S. Greenberg, W. E. Halperin, E. McConnell, I. C. Munro, J. A. Swenberg, S. H. Zahm, and J. D. Graham, "Weight of the evidence on human carcinogenicity of 2,4-D," *Environ. Health Persp.* **96:** 213–222 (1991).

Johnson, M. K., "Initiation of organophosphate-induced delayed neuropathy," *Neurobehav. Toxicol. Teratol.* **4:** 759–765 (1982).

Johnson, M. K., and M. Lotti, "Delayed neurotoxicity caused by chronic feeding of organophosphates requires a high-point of inhibition of neurotoxic esterase," *Toxicol. Lett.* **5:** 99 (1980).

Karademir, M., F. Erturk, and R. Kocak, "Two cases of organophosphate poisoning with development of intermediate syndrome," *Hum. Exp. Toxicol.* **9:** 187–189 (1990).

Kiesselbach, N., K. Ulm, H.-J. Lange, and U. Korallus, "A multicentre mortality study of workers exposed to ethylene oxide," *Br. J. Ind. Med.* **47:** 182–188 (1990).

Kogevinas, M., H. Becher, T. Benn, et al., "Cancer mortality in workers exposed to phenoxy herbicides, chlorophenols, and dioxins—an expanded and updated international cohort study,*Am. J. Epidemiol.* **145**(12): 1061–1075 (1997).

Lilienfeld, D. E., and M. A. Gallo, "2,4-D, 2,4,5-T, and 2,3,7,8-TCDD: An overview," *Epidemiol. Rev.* **11:** 28–58 (1989).

Lotti, M., A. Moretto, R. Zoppellari, R. Dainese, N. Rizzuto, and G. Barusco, "Inhibition of lymphocytic neuropathy target esterase predicts the development of organophosphate-induced delayed polyneuropathy," *Arch. Toxicol.* **59:** 176–179 (1986).

Lotti, M., C. E. Becker, and M. J. Aminoff, "Organophosphate polyneuropathy: Pathogenesis and prevention," *Neurology* **34:** 658–662 (1984).

Maibach, H. I., "Irritation, sensitization, photoirritation and photosensitization assays with a glyphosate herbicide," *Contact Dermatitis* **15:** 152–156 (1986).

Mattsson, J. L., and D. L. Eisenbrandt, "The improbable association between the herbicide 2,4-D and polyneuropathy," *Biomed. Environ. Sci.* **3:** 43–51 (1990).

Milby, T. H., "Prevention and management of organophosphate poisoning," *JAMA* **216:** 2131–2133 (1971).

MMWR (*Morbidity and Mortality Weekly Report*) "Fatalities resulting from sulfuryl fluoride exposure after home fumigation—Virginia," *Morbid. Mortal. Weekly Rep.* **36:** 602–611 (1987).

Reigart, J., and J. Roberts, *Recognition and Management of Pesticide Poisoning*, 5th ed., EPA 735-R-98-003, Washington, DC, 1999.

Ribbens, P. H., "Mortality study of industrial workers exposed to aldrin, dieldrin, and endrin," *Int. Arch. Occup. Environ. Health* **56:** 75–79 (1985).

Richardson, R. J., "Interactions of organophosphorous compounds with neurotoxic esterase," in *Organophosphates: Chemistry, Fate and Effects,* Academic Press, New York, 1992, pp. 299–323.

Safe, S. H., "Is there an association between exposure to environmental estrogens and breast cancer?" *Environ. Health Persp.* **105**(Suppl. 3): 675–678 (1997).

Samal, K. K., and C. S. Sahu, "Organophosphorous poisoning and intermediate neurotoxic syndrome," Assoc. Physicians, India, **38:** 181–182 (1990).

Sawada, Y., Y. Nagai, M. Ueyama, and I. Yamamato, "Probable toxicity of surface-active agent in commercial herbicide contained glyphosate," *Lancet* **1:** 299 (1988).

Senanayake, N., and L. Karalliedde, "Neurotoxic effects of organophosphate insecticides. An intermediate syndrome," *NEJM* **316:** 761–763 (1987).

Senanayake, N., G. Gurunathan, T. B. Hart, P. Amerasinghe, M. Babapulle, S. B. Ellapola, M. Udupihille, and V. Basanayake, "An epidemiological study of the health of Sri Lankan tea plantation workers associated with long term exposure to paraquat," *Br. J. Ind. Med.* **50:** 257–263 (1993).

Sharp, D. S., B. Eskenazi, R. Harrison, P. Callas, and A. H. Smith, "Delayed health hazards of pesticide exposure," *Ann. Rev. Public Health* **7:** 441–471 (1986).

Shindell, S., and S. Ulrich, "Mortality of workers employed in the manufacture of chlordane: An update," *J. Occup. Med.* **28:** 497–501 (1986).

Shore, R. E., M. J. Gardner, and B. Pannett, "Ethylene oxide: An assessment of the epidemiological evidence on carcinogenicity," *Br. J. Ind. Med.* **50:** 971–997 (1993).

Swan, A. A. B., "Exposure of spray operators to paraquat," *Br. J. Ind. Med.* **26:** 322–329 (1969).

Tabershaw, I. R., and W. C. Cooper, "Sequelae of acute organic phosphate poisoning," *J. Occup. Med.* **8:** 5–20 (1966).

Tafuri, J., and J. Roberts, "Organophosphate poisoning," *Ann. Emerg. Med.* **16**(2): 193–202 (1987).

Talbot, A. R., M. Shiaw, J. Huang, S. Yang, T. Goo, S. Wang, C. Chen, and T. R. Sanford, "Acute poisoning with a glyphosate-surfactant herbicide (Round-Up): A review of 93 cases," *Hum. Exp. Toxicol.* **10:** 1–8 (1991).

Taxay, E. P., "Vikane inhalation," *J. Occup. Med.* **8**(8): 425–426 (1966).

Taylor, J. R., J. N. Selhorst, S. A. Houff, and A. J. Martinez, "Chlordane intoxication in man. I. Clinical observations," *Neurology* **28:** 626–630 (1978).

Temple, W. A., and N. A. Smith, "Glyphosate herbicide poisoning experienced in New Zealand," *NZ Med. J.* **105**(933): 173–174 (1997).

United States Environmental Protection Agency (USEPA), *Health and Nutrition Examination Survey II: Laboratory Findings of Pesticide Residues*, National Survey. USEPA, Washington, DC, 1980.

United States Environmental Protection Agency (USEPA), *NHATS Broad Scan Analysis: Population Estimates from Fiscal Year 1982 Specimens*, Office of Toxic Substances, Washington, DC, 1989.

United States Environmental Protection Agency (USEPA), "Environmental Protection Agency. Worker Protection Standard, Hazard Information, Hand Labor Tasks on Cut Flowers and Fern; Final Rule, and Propose Rules," 57 *Federal Register* 38101–38176 (Aug. 21, 1992).

United States Environmental Protection Agency (USEPA), "Worker Protection Standard," 40 *Code of Federal Regulations*, Parts 156 and 170, 1992.

United States Environmental Protection Agency (USEPA), *An SAB Report: Assessment of Potential 2,4-D Carcinogenicity: Review of the Epidemiological and Other Data on Potential Carcinogenicity of 2,4-D by the SAB/SAP Joint Committee*, Science Advisory, March 1994, EPA-SAB-EHC-94-005, Environmental Protection Agency, Washington, DC.

United States Environmental Protection Agency (USEPA), *Office of Pesticide Programs Reference Dose Tracking Report*, 1997.

United States Environmental Protection Agency (USEPA), *Pesticides Industry Sales and Usage. 1994 and 1995 Market Estimates*, Office of Prevention, Pesticides and Toxic Substances, Aug. 1997, 733-R-97-002.

USEPA OPP, *Carcinogenicity Peer Review (4th) of 2,4-Dichlorophenoxyacetic acid*, July 17, 1996, memorandum from Jess Rowland, M.S. and Ester Rinde, Ph.D. to Joanne Miller and Walter Waldrop, Office of Prevention, Pesticides, and Toxic Substances.

Vanholder, R., F. Colardyn, J. De Reuck, M. Praet, N. Lameire, and S. Ringoir, "Diquat intoxication: Report of two cases and review of the literature," *Am. J. Ind. Med.* **70:** 1267–1271 (1981).

Wadia, R. S., C. Sadagopan, R. B. Amin, and H. V. Sardesia, "Neurological manifestations of organophosphorous insecticide poisoning," *J. Neurol. Neurosurg. Psych.* **37:** 841–847 (1974).

Wang, H. H., and B. MacMahon, "Mortality of workers employed in the manufacture of chlordane and heptachlor," *J. Occup. Med.* **21**(11): 745–748 (1979).

Williams, P. L., "Pentachlorophenol, an assessment of the occupational hazard," *Am. Ind. Hyg. Assoc.* **43**(11): 799–810 (1982).

Wong, O., and L. S. Trent, "An epidemiological study of workers potentially exposed to ethylene oxide," *Br. J. Ind. Med.* **50:** 308–316 (1993).

Wood, S., W. N. Rom, G. L. White, and D. C. Logan, "Pentachlorophenol poisoning," *J. Occup. Med.* **25:** 527–530 (1983).

World Health Organization (WHO), Environmental Health Criteria 29, 2,4-Dichlorophenoxyacetic acid (2,4-D), Geneva, Switzerland, 1984.

16 Properties and Effects of Organic Solvents

CHRISTOPHER M. TEAF

The organic solvents comprise a large and diverse group of industrially important chemical compounds, and a detailed individual discussion for the hundreds or thousands of such agents is beyond the scope of this text. However, the chapter provides information concerning the following areas of solvent toxicology and potential health effects:

- Chemical properties of selected classes and individual organic solvents
- Relationships between solvent chemical structures and toxicological effects
- Toxicology of selected solvent examples, including some substances that have not traditionally been considered as solvents, though they are used as such. The chapter also examines selected compounds which may be present as constituents of commercial solvents
- Potential health hazards that may result from industrial use of organic solvents

16.1 EXPOSURE POTENTIAL

The potential for solvent exposure is common in the home and in many industrial applications. Despite advances in worker protection standards, such exposures remain a health concern to millions of workers throughout the world. In some countries, 10–15 percent of the occupational population may be exposed to solvents of one type or another on a regular basis. In the United States, the National Institute of Occupational Safety and Health (NIOSH) estimated that in the late-1980s about 100,000 workers were likely to have some degree of toluene exposure, and about 140,000 individuals have potential exposure to xylene in their work. In some professions (e.g., painters) nearly all workers may have some degree of exposure, although education and protective measures, coupled with the introduction of water-based paints and adhesives, have reduced such exposures. In addition to what may be considered more conventional industrial exposure, potential exposure in household products and handling of petroleum fuels remains a significant source of exposure to hydrocarbon solvent chemicals of various types. Not only is it important to address potential exposure to individual solvent agents; there is also a need to consider the possible interactive effects of multiple incidents of exposure, since these are the rule, rather than the exception.

Solvent exposure typically varies among individuals in an occupational population and clearly will vary over time for a specific individual, based on consideration of job type, specific duties, and work schedule. Thus, assessment of the magnitude of exposure is often complicated and may require detailed evaluation of worker populations concerning airborne concentrations and/or dermal contact, as well as estimates of the frequency and duration of exposure. For example, industrial practices which result in the controlled or uncontrolled evaporation of volatile solvents (e.g., metal degreasing, application of surface coatings) are of particular interest in an exposure context. Appropriate protective equipment,

Principles of Toxicology: Environmental and Industrial Applications, Second Edition, Edited by Phillip L. Williams, Robert C. James, and Stephen M. Roberts.
ISBN 0-471-29321-0 © 2000 John Wiley & Sons, Inc.

engineering controls, and adequate work practices can be instrumental in limiting exposures, but careless or inexperienced handling of solvents may still occur not only in small facilities (e.g., automobile paint and body shops, metal fabricators) but also may be a problem during short-term activities in large and otherwise well-run factories and service industries. Methods that may be used for the characterization and quantification of occupational exposure history are discussed in greater detail in Chapters 18,19, and 21.

16.2 BASIC PRINCIPLES

The breadth of structural variability and the range of physicochemical properties exhibited by organic solvents limits the number of generalized observations that can be made regarding physiological effects and exposure hazards. However, because of their common industrial, commercial, and household use, often in large quantities, it is useful to discuss some fundamental characteristics that are common to at least the principal classes of organic solvents. Table 16.1 summarizes selected important physicochemical properties for a number of the solvents that are discussed in subsequent sections of this chapter. Of particular interest are the properties of volatility (vapor pressure) and water solubility, as well as organic carbon partition coefficients, since these attributes greatly influence exposure potential and environmental behavior.

Occupational guidelines, which are designed to control exposures to solvents and other materials in the workplace, may be expressed in units of volume:volume [e.g., parts per million (ppm)], or in units of mass:volume [e.g., milligrams per cubic meter (mg/m^3)]. For vapors and gases, these data if expressed in either form may be interconverted according to the following expression:

$$X\,\text{ppm} = \frac{Y\,\text{mg/m}^3}{\text{MW}}\,24.45$$

where X ppm = concentration in units of volume:volume
Y mg/m^3 = concentration in units of mass:volume
MW = molecular weight of the chemical
24.45 = molar volume of an ideal gas at standard temperature and pressure.

Rearranging this expression provides the opportunity to convert in the other direction as well.

$$Y\,\text{mg/m}^3 = \frac{(X\,\text{ppm})(\text{MW})}{24.45}$$

For purposes of dose estimation, the units of mg/m^3 are more useful since they may be used in conjunction with inhalation rates (in units of m^3/h or m^3/day) to calculate chemical intake. These unit conversion relationships do not apply for dusts, aerosols, or other chemical forms that may be airborne.

Table 16.2 presents the occupational guidelines for selected solvents and solvent constituents. These guidelines include those developed by the American Conference of Governmental Industrial Hygienists (ACGIH), termed *threshold limit values* (TLV®), as well as those developed by the Occupational Safety and Health Administration (OSHA) employed as legally enforceable standards permissible exposure limits (PELs). These guidelines and standards may be viewed as a long-term protective concentration, represented by a time-weighted average (TWA), or a protective value for a more limited time frame, represented by a short-term exposure limit (STEL) or a ceiling concentration. To the extent that they are available, carcinogen classifications have been included as well. Table 16.3 provides the definitions and differences among the available occupational guidelines. The U.S. Environmental Protection Agency (USEPA) also has established acceptable exposure limits for many of the substances discussed in this chapter [e.g., reference dose (RfD), cancer slope factor (CSF), and reference concentration (RfC) for air]; however, these values are not discussed in detail as they generally do not

TABLE 16.1 Physicochemical Properties of Representative Solvents and Related Materials

Chemical	CAS No.	Molecular Formula	Molecular Weight (g/mol)	Melting Point (°F)	Boiling Point (°F)	Vapor Pressure (mm Hg)	Water Solubility (mg/L)	Vapor Density	Specific Gravity
Halogenated									
Carbon tetrachloride	56-23-5	CCl_4	153.8	−9	170	91	793	5.32	1.59
Chloroform	67-66-3	$CHCl_3$	119.4	−82	143	160	7,920	4.12	1.48
Methyl bromide	74-83-9	CH_3Br	95.0	−137	38	1,444	900	3.36	1.73
Methyl chloride	74-87-3	CH_3Cl	50.5	−144	−12	3,800	15,200	2.15	ND[a]
Methylene chloride	75-09-2	CH_2Cl_2	62.5	−139	104	350	13,000	ND	1.33
Trichloroethene	79-01-6	C_2HCl_3	131.4	−99	189	58	1,100	4.53	1.5
Vinyl chloride	75-01-4	CH_2CHCl	84.9	−256	7	3.3	2,760	1.78	0.91
Nonhalogenated									
Acetaldehyde	75-07-0	CH_3CHO	44.1	−190	69	740	Miscible	1.52	0.79
Acetone	67-64-1	$(CH_3)_2CO$	58.1	−140	133	180	Miscible	ND	0.79
Acrolein	107-02-8	C_2H_3CHO	56.1	−126	127	210	208,000	1.94	0.84
Aniline	62-53-3	$C_6H_5NH_2$	93.1	21	363	0.6	34,000	3.22	1.02
Benzene	71-43-2	C_6H_6	78.1	42	176	75	1,750	2.7	0.88
Benzidine	92-87-5	$NH_2(C_6H_4)_2NH_2$	184.3	239	752	low	400	6.36	1.25
Benzo(a)pyrene	50-32-8	$C_{20}H_{12}$	252	179	>360	>1	0.00162	8.7	1.35
Carbon disulfide	75-15-0	CS_2	76.1	−169	116	297	1,190	2.67	1.26
Dimethylaniline	121-69-7	$C_6H_5N(CH_3)_2$	121.2	36	378	1	1,450	4.17	0.96
1,4-Dioxane	123-91-1	$C_4H_8O_2$	88.2	50	213	30	Miscible	3.03	1.03
Ethanol	64-17-5	C_2H_5OH	46.1	−173	173	44	Miscible	1.59	0.79
Ethyl acetate	141-78-6	$C_4H_8O_2$	88.1	−118	171	100	64,000	3	0.89
Ethyl ether	60-29-7	$(C_2H_5)_2O$	74.1	−177	94	440	69,000	2.55	0.71
Ethylene glycol	107-21-1	$C_2H_4(OH)_2$	62.1	9	388	0.06	1,000	2.14	1.11
Formaldehyde	50-00-0	$HCHO$	30.0	−134	−6	<760	550,000	1.08	ND
n-Hexane	110-54-3	C_6H_{14}	86.2	−139	156	150	140	3	0.89
Hydrazine	302-01-2	N_2H_4	32.1	36	236	10	Miscible	ND	1.01
Isopropanol	67-63-0	C_3H_8O	60.1	−127	181	32	Miscible	2.1	0.785
Isopropyl ether	108-20-3	$((CH_3)_2\ CH)_2O$	102.2	−76	154	119	2	3.5	0.73

(continued)

TABLE 16.1 (*Continued*)

Chemical	CAS No.	Molecular Formula	Molecular Weight (g/mol)	Melting Point (°F)	Boiling Point (°F)	Vapor Pressure (mm Hg)	Water Solubility (mg/L)	Vapor Density	Specific Gravity
Methanol	67-56-1	CH_3OH	32.1	−144	147	96	Miscible	ND	0.79
Methyl ethyl ketone	78-93-3	$CH_3OC_2H_5$	72.1	−86.3	79.6	77.5	353,000 (10°C)	2.41	0.81
Naphthalene	91-20-3	$C_{10}H_8$	128.2	176	424	0.08	31	4.42	1.15
Nitrobenzene	98-95-3	$C_6H_5NO_2$	123.1	42	411	0.3 (77°F)	2,090	4.3	1.2
Nitromethane	75-52-5	CH_3NO_2	61.0	−20	214	28	110,000	2.11	1.14
Phenol	108-95-2	C_6H_5OH	94.1	109	359	0.4	82,800	3.24	1.06
Pyridine	110-86-1	C_5H_5N	79.1	−44	240	16	Miscible	2.72	0.98
Styrene	100-42-5	$C_6H_5(C_2H_4)$	104.2	−23	293	5	310	3.6	0.91
Tetrahydrofuran	109-99-9	C_4H_8O	72.1	−163	150	131	Miscible	2.5	0.89
Toluene	108-88-3	$C_6H_5CH_3$	92.1	−139	232	21	526	3.2	0.87

[a] No data.

TABLE 16.2 Occupational Exposure Limits for Selected Solvents and Related Materials

Compound	CAS No.	1999 ACGIH TLV[a] (ppm)	1999 ACGIH STEL[b] (ppm)	1999 ACGIH Carcinogen Class	1999 OSHA-PEL (ppm)	1999 OSHA-STEL (ppm)	1999 USEPA Carcinogen Class[d]
Acetaldehyde	75-07-0	NE	25	A3	200	NE	B2
Acetone	67-64-1	500	750	A4	1,000	NE	D
Acrolein	107-02-8	0.1	0.23	A4	0.1	NE	C
Aniline	62-53-3	2	NE	A3	5	NE	B2
Benzene	71-43-2	0.5	2.5	A1	1	5	A
Benzidine	92-87-5	NE	NE	A1	NE	NE	A
Carbon disulfide	75-15-0	10	NE	NE	20	30	ND
Carbon tetrachloride	56-23-5	5	10	A2	10	25	B2
Chloroform	67-66-3	10	NE	A3	NE	50	B2
Dimethylaniline	121-69-7	5	10	A4	5	NE	ND
1,4-Dioxane	123-91-1	20	NE	A3	100	NE	B2
Ethanol	64-17-5	1,000	NE	A4	1,000	NE	ND
Ethyl acetate	141-78-6	400	NE	NE	400	NE	ND
Ethyl ether	60-29-7	400	500	NE	400	NE	ND
Ethylene glycol, aerosol	107-21-1	NE	100	A4	NE	NE	ND
Formaldehyde	50-00-0	NE	0.3	A2	0.75	2	B1
n-Hexane	110-54-3	50	NE	NE	500	NE	ND
Hexane isomers (other)	ND	500	1,000	NE	NE	NE	ND
Hydrazine	302-01-2	0.01	NE	A3	1	NE	B2
Isopropanol	67-63-0	400[c]	500	A4	400	NE	ND
Isopropyl ether	108-20-3	250	310	NE	500	NE	ND
Methanol	67-56-1	200	250	NE	200	NE	ND
Methyl bromide	74-83-9	1	NE	NE	NE	20	D
Methyl chloride	74-87-3	50	100	A4	100	200	ND
Methylene chloride	75-09-2	50	NE	A3	25	125	B2
Methyl ethyl ketone	78-93-3	200	300	NE	200	NE	D
Naphthalene	91-20-3	10	15	A4	10	NE	C
Nitrobenzene	98-95-3	1	NE	A3	1	NE	D
Nitromethane	75-52-5	20	NE	NE[c]	100	NE	ND
Phenol	108-95-2	5	NE	A4	5	NE	D
Pyridine	110-86-1	5	NE	NE	5	NE	ND
Styrene	100-42-5	20	40	A4	100	200	ND
Tetrahydrofuran	109-99-9	200	250	NE	200	NE	ND
Toluene	108-88-3	50	NE	A4	200	300	D
Trichloroethene	79-01-6	50	100	A5	100	200	ND
Vinyl chloride	75-01-4	1	NE	A1	1	5	A

NE Not Established

ND No Data, or not classified with regard to carcinogen status by U.S. EPA.

[a]ACGIH Time-Weighted Average Threshold Limit Value (TWA-TLV).
[b]ACGIH Short Term Exposure Limit (STEL) or Ceiling Value.
[c]Changes are pending.
[d]U.S. EPA Carcinogen Classification system presently undergoing reveiw.

TABLE 16.3 Occupational Exposure Guideline Definitions

ACGIH: American Conference of Governmental Industrial Hygienists

TLV-TWA: threshold limit value—time-weighted average—time-weighted average concentration for a normal 8-h workday and a 40-h workweek, to which nearly all workers may be repeatedly exposed, da y after day, without adverse effects

STEL: short-term exposure limit—defined as 15 min TWA exposure that should not be exceeded dur ing a workday; concentration to which workers can be exposed continuously for a short period with out suffering irritation, chronic or irreversible tissue damage, or narcosis sufficient to increase li kelihood of injury, impair self-rescue or materially reduce work efficiency, provided the TLV-TWA is not exceeded

Categories for carcinogenic potential:

- A1 Confirmed Human Carcinogen
- A2 Suspected Human Carcinogen
- A3 Animal Carcinogen
- A4 Not Classifiable as a Human Carcinogen
- A5 Not Suspected as a Human Carcinogen

OSHA: Occupational Safety and Health Administration

PEL-TWA: permissible exposure limit–time-weighted average—concentration not to be exceede d during any 8-h workshift of a 40-h workweek

C: ceiling limit—ceiling concentrations must not be exceeded during any part of the workday; if instantaneous monitoring is not feasible, the ceiling must be assessed as a 15-min TWA exposure

USEPA *Integrated Risk Information System* (IRIS database)

Categories for carcinogenic potential:

- A Known Human Carcinogen
- B1 Probable Human Carcinogen (based on human data)
- B2 Probable Human Carcinogen (based on animal data)
- C Possible Human Carcinogen
- D Not Classifiable as to Human Carcinogenicity (based on lack of data concerning carcino genicity in humans or animals)

have direct applicability in industrial settings. They can be acquired directly from USEPA on databases such as the *Integrated Risk Information System* (IRIS).

Absorption, Distribution, and Excretion

Most commonly, due to the characteristic of volatility, solvent exposure occurs via the inhalation route, but there also may be absorption through the skin following exposures to vapors or through direct contact with the liquid form. While penetration of solvent vapors through the skin typically is considered to be negligible at low air concentrations, ACGIH and OSHA specifically note for a number of substances that this route may be significant, hence the "skin" designation in occupational guidelines. This is particularly true in cases where high concentrations exist in confined spaces and where respiratory protection (e.g., use of air-purifying or air-supplied respirators) limits the potential for inhalation exposure. As an example, it has been demonstrated that exposure to vapor of 2-butoxyethanol, a glycol ether, under some conditions may result in uptake through the skin which exceeds uptake via inhalation.

Characteristic of all volatile materials, the quantity of solvent that is absorbed by the lungs is dependent on several factors, including pulmonary ventilation rate, depth of respirations, and pulmonary circulation rate, all of which are influenced by workload. The partition coefficients that are representative of solvent behavior in various tissues (i.e., for air:blood, fat:blood, brain:blood) are specific to the chemical structure and properties of the individual solvent. Toluene, styrene, and acetone are examples of rapidly absorbed solvents.

Once absorbed, solvents may be transported to other areas of the body by the blood, to organs where biotransformation may occur, resulting in the formation of metabolites that can be excreted. Significant differences exist between the uptake and potential for adverse effects from solvents, based on the route of exposure. Absorption following ingestion or dermal exposure results in absorption into the venous circulation, from which materials are rapidly transported to the liver where they may be metabolized. Following inhalation exposure, however, much of the absorbed chemical is introduced into the arterial circulation via the alveoli. This means that the absorbed solvent may be distributed widely in the body prior to reaching the liver for metabolism, degradation, and subsequent excretion.

Since solvents constitute a heterogeneous group of chemicals, there are many potential metabolic breakdown pathways. However, in many instances there is involvement of the P450 enzyme system and the glutathione pathways, which catalyze oxidative reactions and conjugation reactions to form substances that are water-soluble and can be excreted in the urine and, perhaps, the bile. Several pathways may exist for the biotransformation of a specific solvent and some of the excreted metabolites form the basis for biological monitoring programs that can be used to characterize exposure (e.g., phenols from benzene metabolism, trichloroacetic acid obtained from trichloroethene, and mandelic acid from styrene). These metabolic processes are discussed in greater detail in Chapter 3.

Although it is well recognized that the metabolism of most solvents occurs primarily in the liver, other organs also exhibit significant capacity for biotransformation (e.g., kidney, lung). Some organs may be capable of only some of the steps in the process, potentially leading to accumulation of toxic metabolites if the first steps of the biotransformation pathway are present, but not the subsequent steps. For example, whereas an aldehyde metabolite may be metabolized readily in the liver, the same aldehyde may accumulate in the lung and cause pulmonary damage due to a lack of aldehyde dehydrogenase enzyme in that organ. In addition to the generally beneficial aspects of biotransformation and excretion, metabolism may generate products that are more toxic than the parent compound. This process is termed *metabolic activation* or *bioactivation*, and the resultant reactive metabolic intermediates (e.g., epoxides and radicals) are considered to be responsible for many of the toxic effects of solvents, especially those of chronic character (see Chapter 3).

Enzymes that are critical to the metabolic processes may be increased in activity, or "induced," by various types of previous or concomitant exposures to chemicals, such as those from therapeutic drugs, foods, alcohol, cigarette smoke, and other industrial exposures, including other solvents. Competitive interactions between solvents in industrial contexts also may influence the toxic potential, complicating the question of whether exposure to multiple chemicals always should be considered to be worse than individual exposures. A well-described example of interactive effects relates to methanol and ethanol, both of which are substrates that compete for the alcohol dehydrogenase pathway. This observation of biochemical competition led to the use of ethanol as an early treatment for acute methanol intoxication. As another example, induction of the enzyme that is active in the biotransformation of trichloroethene (TCE), as a result of chronic ethanol consumption, may influence sensitivity to the adverse effects of TCE. Interactions between alcohols (e.g., ethanol, 2-propanol) and other solvents (e.g., carbon tetrachloride, trichloroethene) have been described.

Saturation of the typical metabolic pathways that are responsible for biological breakdown may cause a qualitative shift in metabolism to different pathways. Whereas the normal pathway may be one of detoxification, saturation of that pathway may result in "shunting" to another pathway, resulting in bioactivation. Examples in which this phenomenon has been demonstrated include 1,1,1-trichloroethane, *n*-hexane, tetrachloroethene, and 1,1-dichloroethene.

In addition to the process of biotransformation and subsequent urinary excretion described above, many solvents may be eliminated in changed or unchanged form by exhalation, an action that varies with workload. This observation forms the basis for the practice of sampling expired air as a measure of possible occupational exposure in some industrial medical surveillance programs.

Depression of Central Nervous System Activity

One of the common physiological effects which is associated with high levels of exposure to some organic chemicals, including volatile solvents, is depression of central nervous system (CNS) activity. Chemicals that act as CNS depressants have the capacity to cause general anesthetic effects, inhibit activity in the brain and the spinal cord, and lower functional capacity, render the individual less sensitive to external stimuli, and ultimately may result in unconsciousness or death as the most severe consequence. A general feature of many solvents is their highly lipophilic ("fat-loving") character. As discussed in Chapter 2, lipophilic chemicals exhibit a high affinity for fats (lipids), coupled with a low affinity for water (hydrophobic). Thus, these compounds tend to accumulate in lipid-rich areas of the body, including lipids in the blood, lipid zones of the nervous system, and depot fats. Neurotoxic chemicals have been shown to accumulate in the lipid membranes of nerve cells after repeated high-level acute exposure or lower-level, chronic exposure, in some cases disrupting normal excitability of the nerve tissues and adversely effecting normal nerve impulse conduction.

While organic solvents with few or no functional groups are lipophilic and exhibit some limited degree of CNS-depressant activity, this property increases with the carbon chain length, to a point. This increased toxicity is most evident when larger functional groups are added to small organic compounds, since the increase in molecular size generally disproportionately decreases the water solubility and increases the lipophilicity. As a practical consideration, this observation is relevant only to industrial exposures for chemicals up to a five- or six-carbon chain length. As molecular size increases beyond this point for any of the functional classes (amines, alcohols, ethers), the vapor pressure is decreased and the exposure considerations, particularly with regard to inhalation, change dramatically.

The unsaturated chemical analogs (organic structures where hydrogens have been deleted, forming one or more double or triple bonds between carbon atoms; see Section 16.3) typically are more potent CNS-depressant chemicals than their saturated (single-bond) counterparts. In a similar fashion, the CNS-depressant properties of an organic compound are generally enhanced by increasing the degree of halogenation [e.g., chlorine (Cl), bromine (Br)] and, to a lesser extent, by addition of alcoholic (–OH) functional groups. For example, while methane and ethane have no significant anesthetic properties and act as simple asphyxiants at high concentrations, both of the corresponding alcohol analogs (methanol and ethanol) are potent CNS depressants. Likewise, while methylene chloride (i.e., dichloromethane, CH_2Cl_2) has appreciable anesthetic properties, chloroform ($CHCl_3$) is more potent than methylene chloride, and carbon tetrachloride (CCl_4) is the most potent in terms of anesthetic considerations.

Several solvents have been associated in the literature with behavioral toxicity, including carbon disulfide, styrene, toluene, trichloroethene, and jet fuel, though reports are often difficult to corroborate.

Peripheral Nervous System

A selected group of organic solvents are capable of causing a syndrome known as *distal axonal peripheral neuropathy*. Among these solvents are *n*-hexane, methyl *n*-butyl ketone, and carbon disulfide. The occupation development of the disease condition is slow, but may be accelerated in cases of those guilty of solvent abuse (e.g., inhalation). In at least some cases, the disease state may progress for 3–4 months after the cessation of exposure.

Membrane and Tissue Irritation

Another adverse response of common interest for organic solvents is the potential for membrane and tissue irritation. Because cell membranes are composed principally of a protein–lipid matrix, organic solvents at sufficient concentrations may act to dissolve that matrix, or extract the fat or lipid portion out of the membrane. This "defatting" process, when applied to skin, may cause irritation and cell damage and, by similar processes, may seriously injure the lungs, or eyes. As described previously, the addition of classes of functional groups to organic molecules predictably influences the toxicological properties of the molecule. For example, amines and organic acids confer irritative or corrosive properties when added as functional groups, while alcohol, aldehyde, and ketone groups tend to increase the potential for damage to cell membranes by precipitating and denaturing membrane proteins at high exposure concentrations or durations.

As noted previously for CNS-depressant actions, unsaturated compounds generally are stronger irritants than are corresponding saturated analogs. As the size of the molecule increases, the irritant properties typically decrease and the solvent defatting action of the hydrocarbon portion becomes more important.

Table 16.4 presents the relative potency of selected functional groups with regard to general CNS-depressant and irritant properties. These approximate rankings rely on basic comparisons among the unsubstituted chemical analogs and become less applicable in broader comparisons among the larger, more complex and multisubstituted compounds.

Carcinogenicity

As with toxicological evaluation of the other potential adverse effects of solvents, the often complex nature of industrial exposure situations complicates most objective evaluations of malignancy with regard to a specific solvent. Thus, many occupational studies end up considering solvent exposure as a general "risk factor" for neoplasia, but are unable to establish "cause and effect." Some exposure circumstances, however, more specifically may indicate a relevant human cancer risk for industrial activities (e.g., vinyl chloride production workers, high-level benzene exposure).

With regard to nonchlorinated hydrocarbons, there is historical documentation for benzene as a human carcinogen under some intense exposure circumstances. Multiple factors may be responsible for the observed effects, but the prevailing conclusion is that the metabolism of benzene to a number of reactive metabolites (e.g., epoxides) is responsible for the myelotoxicity. An alternative or complementary hypothesis suggests that a depressant effect by benzene or its metabolites on cell-mediated immunity may influence basic carcinogenesis. The substituted benzene analog styrene (or vinyl benzene) also forms reactive metabolites, notably styrene oxide. Styrene, like the other substituted benzenes, toluene and xylene, undergoes ring hydroxylation, suggesting at first glance a common pathway through reactive and potentially cancer-causing intermediates. Although the latter two substances generally are not considered to be carcinogenic, a limited carcinogenic potential for styrene

TABLE 16.4 Relative CNS Depressant and Irritant Potency of Selected Organic Solvent Classes

Decreasing CNS depressant potential	
Most:	halogen-substituted compounds
	ethers
	esters
	organic acids
	alcohols
	alkenes
Least:	alkanes
Membrane and tissue irritant potential	
Most:	amines
	organic acids
	aldehydes = ketones
	alcohols
Least:	alkanes

has been suggested by some researchers on the basis of results from genotoxicity assays, animal experiments, and sporadic reports of excess human leukemias and lymphomas. These reports have been difficult to substantiate.

Several chlorinated solvents (e.g., carbon tetrachloride, chloroform, tetrachloroethene, trichloroethene, vinyl chloride) exhibit varying degrees of carcinogenic potential, notably hepatic tumors in animals. The carcinogenic potential associated with trichloroethene (TCE) exposure has been of interest since the mid-1970s, when the National Cancer Institute reported increases in liver cancer in male mice that received TCE by gastric intubation. TCE, like some other chlorinated hydrocarbons, exhibits limited and controversial mutagenic activity in bacterial test systems after microsomal activation, so the mutagenic effect is probably dependent on the products of metabolism of this compound. This has influenced recent concern about actual TCE potency. A similar conclusion applies to the carcinogenic potential of tetrachloroethene, or perchloroethene (PERC). Although USEPA still is reviewing the classification of TCE, other groups such as the American Conference of Governmental Industrial Hygienists (ACGIH) no longer consider the substance to be a significant human carcinogenic risk under occupational circumstances. A third group of agencies, including the Occupational Safety and Health Administration (OSHA), and the International Agency for Research on Cancer (IARC), have not released final positions with respect to carcinogenic potential of TCE.

Except for leukemogenic effects from extreme benzene exposure and hepatic angiosarcoma in vinyl chloride workers, no unequivocal human reports are available that document cancer hazards from exposure to the organic solvents. However, there are a number of epidemiologic observations that have been published regarding cancer and exposure to chlorinated solvents. For example, both Hodgkin's and non-Hodgkin's lymphomas have been linked to occupational exposure to some organic solvents of the aliphatic, aromatic, and chlorinated types. In a cohort of laundry and dry-cleaning workers (with putative TCE and PERC exposure), there was a slight excess of liver cancers (approximately 2.5-fold), and in a case-referent study, a similar elevated incidence was reported for laundering, cleaning, and other garment service workers. Additional data have suggested an association between exposure to a variety of solvents and liver cancer, one of them showing an association for females only, whereas the other study was restricted to males and found about a twofold risk. In a study of nearly 1700 dry-cleaner workers with potential exposure to PERC, an increased incidence of urinary tract cancer was reported. The conclusions from this and other studies are complicated by the fact that exposure to petroleum solvents was likely as well. Another study of over 5300 dry-cleaner workers reported a slight excess of cancer, with an overall ratio of only 1.2.

Recently, a cohort of nearly 15,000 aircraft maintenance workers with exposure to trichloroethene and other solvents reportedly showed a decreased overall cancer mortality, but a calculated excess in non-Hodgkin's lymphoma, multiple myeloma, and bile duct cancers.

In addition to benzene and vinyl chloride, both of which are classified as Group A (Known Human Carcinogen), several of the chlorinated solvents or their relatives still are classified by USEPA in the B2 (Probable Human Carcinogen) or C (Possible Human Carcinogen) categories, based on historical information. That information presently is under review by that agency. This approach generally is consistent with both ACGIH and OSHA, as discussed elsewhere in this chapter.

Other Selected Acute Toxic Properties

As noted previously, the CNS-depressant and irritant properties are common to the chemicals usually referred to as "solvents." These two properties, as well as carcinogenic potential, are the focus of this chapter because one or more of the properties are consistently observed in each chemical class discussed. These classes of chemicals also may produce a number of other acute toxic effects upon prolonged or high intensity exposure. After systemic absorption, acute effects may include hepatotoxicity, nephrotoxicity, and cardiac arrhythmias that have been reported as a result of sensitization of the heart to catecholamines (i.e., adrenaline). Although these effects are seldom reported in occupational circumstances, they may occur for certain classes such as halogenated hydrocarbons, particularly in chronic, high-level exposure. As noted previously, many of these substances were historically found or now are found in common household products and, thus, poisonings may occur in children. In addition to ingestion in those cases, aspiration into the lungs may occur, causing chemical pneumonitis that may complicate treatment.

Many of the adverse effects attributed to solvents are rather nonspecific, and the symptomology of any particular unknown solvent poisoning often provides few clues to the specific solvent in question. Acute overexposure to organic solvents initially may produce a generalized "chemical malaise" with a wide range of subjective complaints. It also may produce temporal changes in effects that appear contradictory (e.g., euphoria, narcosis). Therefore, initial treatment often is symptomatic with regard to the systemic toxicity, coupled with measures designed to limit further systemic absorption.

To illustrate the generally nonspecific nature of solvent intoxication and the related problems that may be faced by the health specialist in attempting to diagnose uncharacterized exposure situations, acute symptoms are described below for a few common agents. It should be noted that, in contrast to the acute effects, the effects of chronic exposure to these agents may differ dramatically, as discussed in other sections of this chapter.

- *Benzene*—euphoria, excitement, headache, vertigo, dizziness, nausea, vomiting, irritability, narcosis, coma, death
- *Carbon tetrachloride*—conjunctivitis, headache, dizziness, nausea, vomiting, abdominal cramps, nervousness, narcosis, coma, death
- *Methanol* (wood alcohol)—euphoria, conjunctivitis, decreased visual acuity, headache, dizziness, nausea, vomiting, abdominal cramps, sweating, weakness, bronchitis, narcosis, delirium, blindness, coma, death

16.3 TOXIC PROPERTIES OF REPRESENTATIVE ALIPHATIC ORGANIC SOLVENTS

Saturated Aliphatic Solvents: C_nH_{2n+2}

Alkanes

The chemical class known as the *saturated aliphatic hydrocarbons*, or *alkanes* (also termed *paraffins*) have many members and generally rank among the least potentially toxic solvents when acute effects are considered. This group represents the straight-chain or branched hydrocarbons with no multiple bonds. The vapors of these solvents are mildly irritating to mucous membranes at the high concentrations that are required to induce their relatively weak anesthetic properties. The four chemicals in this series with the lowest-molecular-weight (methane, ethane, propane, butane) are gases with negligible toxicity and their hazardous nature is limited almost entirely to flammability, explosivity, and basic asphyxiant potential.

The higher molecular weight members of this class are liquids and have some CNS-depressant, neurotoxic, and irritant properties, but this is primarily a concern of the lighter, more volatile fluid compounds in this series (i.e., pentane, hexane, heptane, octane, nonane). The liquid paraffins, beginning with the 10-carbon compound decane, are fat solvents and primary irritants capable of dermal irritation and dermatitis following repeated, prolonged or intense contact.

The symptoms of acute poisoning by this group are similar to those previously described as generally present in solvent intoxication (i.e., nausea, vomiting, cough, pulmonary irritation, vertigo or dizziness, slow and shallow respiration, narcosis, coma, convulsions, and death) with the severity of the symptoms dependent upon the magnitude and duration of exposure. Accidental ingestion of large amounts (exceeding several ounces, or about 1–2 mL/kg body weight) may produce systemic toxicity. If less than 1–2 mL/kg is ingested, a cathartic, used in conjunction with activated charcoal to limit absorption, is the therapeutic approach. In either situation, aspiration of the solvent into the lungs is the initial primary concern from a medical perspective. Low-viscosity hydrocarbons attract particular attention in this context because their low surface tension allows them to spread over a large surface area, thereby having the potential to produce damage to the lungs after exposure to relatively small quantities. These chemicals may sensitize the heart to epinephrine (adrenaline), but that feature is rarely a practical consideration since a narrow dose range typically separates cardiac sensitization from fatal narcosis.

The chronic exposure to some aliphatics, notably hexane and heptane, reportedly has the capacity to produce polyneuropathy in humans and animals, characterized by a lowered nerve conduction velocity and a "dying back" type of degenerative change in distal neurons. Symptoms of this condition may include muscle pain and spasms, muscular weakness, and paresthesias (tingling or numbness). Normal metabolites have been implicated as the causative agents in this case, with 2,5-hexanedione and 2,6-heptanedione as the respective toxic metabolites of hexane and heptane. Since these metabolites represent oxidative breakdown products, first to the alcohol and then to the respective diketone, it has been suggested and observed that structurally similar alcohols and ketones at sufficient concentration may produce similar neuropathies compared to the parent aliphatic hydrocarbons.

The alkanes generally are not considered to have carcinogenic potential.

Unsaturated Aliphatic Solvents: C_nH_{2n}

Olefins (Alkenes)

Alkenes, which are the double-bonded structured analogs of the alkanes, also are referred to as *olefins*, and generally exhibit qualitative toxicological properties similar to those of the alkanes.

The double bond typically enhance(s) the irritant and CNS-depressant properties in comparison to the alkanes, but this enhancement often is of limited practical significance. For example, ethylene is a more potent anesthetic than its corresponding alkane (ethane), which acts as a simple asphyxiant. However, since a concentration greater than 50 percent ethylene is required to induce anesthesia, the potential for hypoxia and the explosive hazard are major drawbacks that preclude its clinical use as an anesthetic. Such an ethylene concentration in an industrial setting would sufficiently displace the oxygen present so that asphyxiation (as is the case with ethane) would be the major concern, rather than narcosis and respiratory arrest. Of greater toxicological interest is the observation that the unsaturated nature of the hexene and heptene series apparently largely abolishes the neurotoxic effects that have been reported following chronic hexane or heptane exposure. This change may be related to substantive metabolic differences between the groups.

16.4 TOXIC PROPERTIES OF REPRESENTATIVE ALICYCLIC SOLVENTS

Alicyclic hydrocarbons functionally may be viewed as alkane chains of which the ends have been joined to form a cyclic, or ring, structure (see, e.g., structures in Figure 16.1). Their toxicological properties resemble those of their open-chain relatives and they generally exhibit anesthetic or CNS-depressant properties at high exposure concentrations. Industrial experience indicates that negligible chronic effects typically are associated with long term exposure to these compounds. The lower-molecular-weight alicyclics (e.g., cyclopropane) received some limited attention as surgical anesthetics, but the larger compounds (e.g., cyclohexane) are not as useful because the incremental

Figure 16.1 Cyclopropane and cyclobutane.

difference between narcosis and a lethal concentration is small. While there are qualitative similarities between the groups, the irritant qualities of cycloalkenes (cycloolefins) tend to be of greater concern than those of the unsaturated analogs.

16.5 TOXIC PROPERTIES OF REPRESENTATIVE AROMATIC HYDROCARBON SOLVENTS

The class of organic solvents that commonly are referred to as "aromatics" are composed of one or more six-carbon (phenyl) rings. The simplest member of the class (defined by lowest molecular weight) is the single-ringed analog termed benzene, followed by the aliphatic-substituted phenyl compounds (alkylbenzenes) and then the aryl- and alicyclic-substituted, multiring benzenes. Diphenyl and polyphenyl compounds are represented in this class, which includes the polynuclear aromatic hydrocarbons (PNAs or PAHs), such as naphthalene, which are common as constituents of petroleum fuels, as well as other commercial products. Benzene and its alkyl relatives are important industrial compounds, with over 1.5 billion gallons of benzene annually produced or imported in the United States. Even larger quantities of several of the alkylbenzenes (e.g., toluene, xylenes) are produced. Benzene and the alkylbenzenes are common as raw materials and solvents in the ink, dye, oil, paint, plastics, rubber, adhesives, chemical, drug, and petroleum industries. Most commercial motor gasolines contain at least 1 percent benzene, a value which may range up to several percent, and alkylbenzenes may be present in or may be added to unleaded fuels to concentrations reaching 25–35 percent of the total commercial product.

Aromatic hydrocarbons typically cause more tissue irritation than the corresponding molecular weight aliphatics or alicyclics. These phenyl compounds may cause primary dermatitis and defatting of the skin, resulting in tissue injury or chemical burns if dermal contact is repetitive or prolonged. Conjunctivitis and corneal burns have been reported when benzene or its alkyl derivatives are splashed into the eyes, and naphthalene has been reported to cause cataracts in animals at high dosages. If the aromatics are reaspirated into the lungs after ingestion (e.g., following vomiting), they are capable of causing pulmonary edema, chemical pneumonitis, and hemorrhage. Inhalation of high concentrations can result in conditions ranging from bronchial irritation, cough, and hoarseness to pulmonary edema. Once absorbed and in systemic circulation, these hydrocarbons are demonstrably more toxic than aliphatics and alicyclics of comparable molecular weight. While CNS depression is a major acute effect of this class of compounds, its severe form differs fundamentally from that observed following exposure to the aliphatics. The aliphatic-induced anesthesia and coma is characterized by an inhibition of deep tendon reflexes. In comparison, aromatic-induced unconsciousness and coma is characterized by motor restlessness, tremors, and hyperactive reflexes, sometimes preceded by convulsions.

Representative members of the aromatic hydrocarbon family are profiled in the following section (see, e.g., benzene structure in Figure 16.2).

Benzene is a colorless liquid with a characteristic odor that generally is described as pleasant or balsamic. The term benzene should not be confused with *benzine,* as the latter historically refers to a mixed-component, low-boiling-range, petroleum fraction composed primarily of aliphatic hydrocarbons. Because of its extensive use for many years, this compound has been studied perhaps more extensively than any other. Benzene can be toxic by all routes of administration at sufficient dosage; however, the acute inhalation LC_{50} in animals begins at about 10,000 ppm. This may be compared

Figure 16.2 Benzene.

with observations in humans where lethal effects are observed at about 20,000 ppm within 5–10 min of exposure. Air concentrations on the order of 250 ppm often produce vertigo, drowsiness, headache, nausea, and mucous membrane irritation. Ingested benzene exhibits comparatively greater systemic toxicity than the corresponding aliphatic homologs, and the fatal adult human dose usually is reported to be on the order of 0.2 ml/kg (about 10–15 mL). Although CNS effects generally dominate over other systemic toxic effects in acute exposure circumstances, cardiac sensitization and cardiac arrhythmias also may be observed, particularly in severe intoxication cases. Pathology observed in acutely poisoned benzene victims includes severe respiratory irritation, pulmonary edema and hemorrhage, renal congestion, and cerebral edema.

Benzene in pure liquid form is an irritating liquid that is capable of causing dermal erythema, vesiculation, and a dry, scaly dermatitis. Prolonged dermal contact with benzene (or analogous alkylbenzenes) may result in lesions that resemble first- or second-degree thermal burns, and skin sensitization has been reported, though rarely. If splashed into the eyes, it may produce a transient corneal injury.

Benzene differs from most other organic solvents in that it is a myelotoxin, with effects on the blood-forming organs (e.g., marrow). The hematological findings following chronic exposure are variable, but effects have been noted in red cell count (which may be 50 percent of normal), decreased hemoglobin levels, reduced platelet counts, and altered leukocyte counts. The most commonly reported effect at significant, acute, repeated exposure is a fall in white blood cell count. In fact, in an example of what later was recognized to be misguided therapeutics, benzene actually was used in the early 1900s to decrease numbers of circulating leukocytes in leukemia patients.

Three separate stages or degrees of severity usually can be identified in the benzene-induced change in blood-forming tissues. Initially, there may be reversible blood-clotting defects, as well as a decrease of all blood components (mild pancytopenia or aplastic anemia). With continued exposure, the bone marrow may first become hyperplastic and a stimulation of leukocyte formation may be the earliest clinical observation. While chronic benzene exposure probably is best known for its link to specific types of leukemia, aplastic anemia actually is a more likely chronic observation. Several metabolites of benzene have been implicated as the putative causative agents in these effects. Leukopenia and anemia in animals have been reported following chronic hydroquinone and pyrocatechol administration, both of which are benzene metabolites. However, the benzene syndrome has not been observed in humans exposed to phenol, hydroquinone, or catechol.

Urinary phenol, expressed in conjunction with urinary creatinine, represents an acceptable measure of industrial exposure.

Selected Substituted Aromatic Compounds

The group of aliphatic substituted benzenes, also described by the term *alkylbenzenes*, includes toluene (or methyl benzene) (see Figure 16.3), ethylbenzene, xylenes or dimethylbenzenes, styrene (or vinyl benzene), cumene (or isopropylbenzene) and many others. Unlike benzene, these substances are seldom considered as carcinogens and rarely cause effects in genotoxicity assays. However, toluene exerts a more powerful CNS-depressant effect than benzene, and human exposures at 200 ppm for periods of 8 h generally will produce such symptoms as fatigue lasting for several hours, weakness, headache, and dermal paresthesia. At 400 ppm, mental confusion becomes a symptom and at 600 ppm,

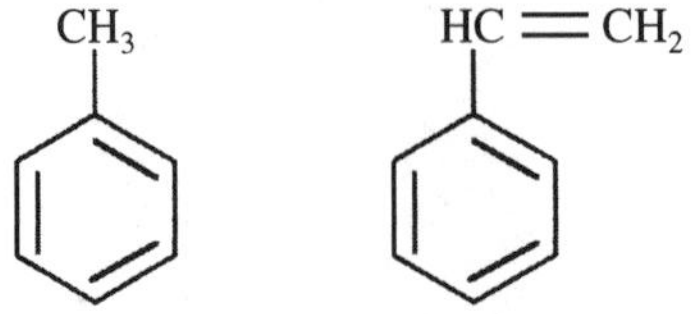

CH_3 $HC{=}CH_2$

Figure 16.3 Toluene and styrene.

extreme fatigue, confusion, exhilaration, nausea, headache, and dizziness may result within a short time. In comparison, the acute toxicity of the xylene isomers is qualitatively similar to that of toluene, although they are less potent. In addition to considerations of occupational exposure, concern recently has been directed toward the reports of intentional inhalant abuse of alkylbenzenes and alkylbenzene-containing products.

A number of indicators of industrially important exposure have been developed for the alkylbenzenes, including urinary hippuric acid (toluene, xylenes), mandelic acid (ethylbenzene, styrene), and phenylglyoxylic acid (styrene).

Polycyclic Aromatic Hydrocarbon (PAH) Compounds

This chemical class includes many members, all of which are cyclic-substituted benzenes. While this group often is not commonly classed with solvents, many of the PAHs are common components of petroleum fuels and some solvent mixtures, and are presented here for comparative purposes (see also Figure 16.4).

The PAHs are nonpolar, lipid-soluble compounds that may be absorbed via the skin, lungs, or digestive tract. Once absorbed, they can be concentrated in organs with a high lipid content. They are metabolized by a subpopulation of cytochrome P450 enzymes, which they also induce. These cytochromes are commonly referred to generically as aryl hydrocarbon hydroxylase (AHH), or cytochrome P448. Since PAHs are composed of aromatic rings with limited available sites for metabolism, hydroxylation is the prevalent physiological means to initiate metabolism of PAHs to more water-soluble forms that facilitate excretion. In this process, potentially toxic and carcinogenic epoxide metabolites may be formed. While the ubiquitous environmental presence of the PAHs suggests that regular exposure would more commonly lead to adverse effects, other routes of metabolism have been identified that appear to act as protective mechanisms by degrading these reactive PAH metabolites. Similarly, natural or added constituents of foods such as flavenoids; selenium; vitamins A, C, and E; phenolic antioxidants; and food additives (e.g., BHT, BHA) all can exert protective effects against these metabolites. Recent evidence indicates that the simple, initial epoxide metabolites of PAHs are not the ultimate carcinogens because secondary metabolites of PAHs have been shown to be more potent mutagenic and carcinogenic agents, and because they form DNA adducts, which are more resistant to DNA-repair processes. However, a detailed discussion of these processes is beyond the scope of this chapter. The reader is referred to the bibliography at the end of this chapter for further references in this area, such as the ATSDR Toxicological Profiles.

Naphthalene is the simplest member of the PAHs (two phenyl rings) and is a common fuel component, as well as a commercial moth repellent. Naphthalene inhalation at sufficient concentration may cause headache, confusion, nausea, and profuse perspiration. Severe exposures may cause optic neuritis and hematuria. Cataracts have been produced experimentally following naphthalene exposure in rabbits and at least one case has been reported in humans. Naphthalene is an irritant and hypersensitivity has been reported, though rarely. The teratogenic and embryotoxic effects of PAHs have only been documented for a few of the more potent, carcinogenic PAH compounds, and then only in extreme exposure regimes in animal studies.

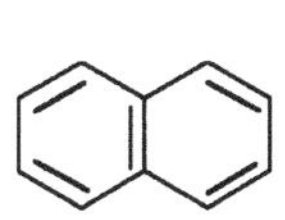

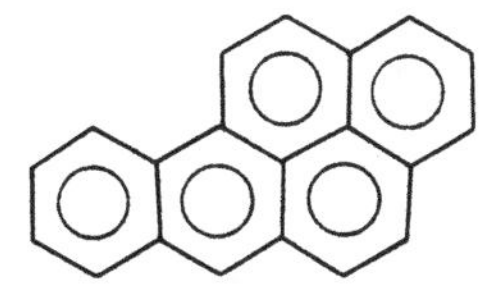

Figure 16.4 Naphthalene and benzo[*a*]pyrene.

While PAHs can be acutely toxic, this characteristic generally is relevant only at doses sufficiently great that they are not of interest in an industrial or environmental setting. At high, acute doses, PAHs are toxic to many tissues and degenerative changes may ultimately be observed in the kidney and liver, but the thymus and spleen are particularly sensitive to acute effects. For example, the noncarcinogen PAH acenaphthene, given in doses as high as 2000 mg/kg, produces only minor changes in the liver or kidney and is relatively nontoxic when compared to the hematoxicity produced by 100 mg/kg of dimethylbenzanthracene, a much more potent PAH.

Several of the PAHs with four, five, or more rings (e.g., benzo-*a*-pyrene, benzo-*a*-anthracene, benzo-*b*,*k*-fluoranthene) have been classified as possible carcinogens by a number of environmental regulatory agencies. Occupational guidelines have been established for a chemical category known as "coal tar pitch volatiles," which includes some PAHs.

16.6 TOXIC PROPERTIES OF REPRESENTATIVE ALCOHOLS

Alcohol Compounds: R–OH

As a general observation, alcohols are more powerful CNS depressants than their aliphatic analogs. In sequence of decreasing depressant potential, tertiary alcohols with multiple substituent OH groups are more potent than secondary alcohols, which, in turn, are more potent than primary alcohols. The alcohols also exhibit irritant potential and generally are stronger irritants than similar organic structures that lack functional groups (e.g., alkanes) but are much less irritating than the corresponding amines, aldehydes, or ketones. The irritant properties of the alcohol class decrease with increasing molecular size. Conversely, the potential for overall systemic toxicity increases with greater molecular weight, principally because the water solubility is diminished and the lipophilicity is increased. Alcohols and glycols (dialcohols) rarely represent serious hazards in the workplace, because their vapor concentrations are usually less than the required irritant levels, which, in turn, prevents significant CNS effects as well.

Methanol (see Figure 16.5), also known as methyl alcohol or wood alcohol, is the simplest structural member of the alcohols and is widely employed as an industrial solvent and raw material for manufacturing processes. It also is used as one of several possible adulterants to "denature" ethyl alcohol, which then is used for cleaning, paint removal, and other applications. The denaturing process in theory prevents its ingestion.

Methanol is of toxicological interest and industrial significance because of its unique toxicity to the eye, and it has received considerable attention from the medical community over the years due to misuse, as well as accidental or intentional human consumption. It has been estimated that methanol ingestion may have been responsible for 5–10 percent of all blindness in the U.S. military forces during World War II. Methanol intoxication typically exhibits one or more of the following features:

- CNS depression, similar to or greater than that produced by ethyl alcohol (ethanol)
- Metabolic acidosis, caused by degradation of methanol to formic acid and other organic acids
- Ototoxicity, expressed as specific toxicity to retinal cells caused by formaldehyde, an oxidation product of methanol

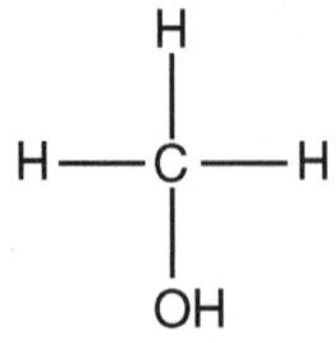

```
      H
      |
  H — C — H
      |
      OH
```

Figure 16.5 Methanol.

Despite its occasional consumption for that purpose, inebriation is not a prominent symptom of methanol intoxication, unless an extreme quantity is consumed. It is significant that if ethanol is simultaneously ingested in sufficient quantity, methanol poisoning may be considerably delayed, or even averted completely. The subsequent administration of ethanol after methanol intake forms the basis for treatment of methanol poisoning, based on competition for the same metabolic enzyme system (alcohol dehydrogenase).

Acute methanol poisoning is characterized by headache, vertigo, vomiting, upper abdominal pain, back pain, dyspnea, restlessness, cold or clammy extremities, blurred vision, ocular hyperemia, and diarrhea. Visual disturbance can proceed to blindness. The pulse may slow in severely ill patients, and coma can develop rapidly. Death may be sudden, with accompanying inspiratory apnea and convulsions, or may occur only after long coma, depending on circumstances of exposure.

The rate of methanol oxidation is only about 10–15 percent that of ethanol; therefore, complete oxidation and excretion may require several days. An asymptomatic latent period of up to 36 hours may precede the onset of adverse effects. As little as 15 mL of methanol reportedly has caused blindness, and several ounces (70–100 mL) may be fatal. Aside from the well-described ocular effects, neurologic damage of various types may follow methanol poisoning.

Ethyl alcohol (ethanol) (see Figure 16.6) in high concentrations acts as a mild to moderate local irritant, having the ability to injure cells by precipitation and dehydration. The CNS typically is affected more markedly than other systems. The initial apparent stimulation that accompanies ethanol ingestion results from altered activity in areas of the brain that have been freed of inhibition through the depression of control mechanisms. Ethanol increases the pain threshold considerably in most individuals even at moderate doses.

Vasodilation of cutaneous blood vessels, resulting in flushing, may accompany ethanol ingestion. Because of this, and despite folklore to the contrary, its use is contraindicated during hypothermia or exposure to cold. Another principal acute effect is cardiovascular depression of CNS origin. It may directly damage tissues at high chronic doses, producing skeletal myopathy and cardiomyopathy. Because ethanol increases gastric secretion at high doses, it has been linked to erosive gastritis, which can, in turn, increase the severity of ulcers. It promotes fat accumulation in the liver in some circumstances, and chronic intake may lead to cirrhosis, liver cancer, or lethality. It promotes urine flow by inhibiting the release of steroids and adrenaline from the adrenal glands. Ethanol may exert a direct depressant action on bone marrow and may lead to a depression of leukocyte levels in inflamed areas, which may explain in part the poor resistance to infection that often is reported in alcoholic individuals.

Other Simple Alcohols

Aside from the common examples of methanol and ethanol, high concentrations of propanols (propyl alcohols; structural variants including isopropanol, *n*-propanol) may cause intoxication and CNS depression. They also have bactericidal properties.

Isopropanol [$(CH_3)_2CHOH$] generally is less toxic than *n*-propanol [$CH_3(CH_2)_2OH$], but both substances are more acutely toxic than ethanol. In humans, brief exposure to several hundred parts per million of isopropyl alcohol in air generally causes mild irritation of eyes, nose, and throat. *n*-Butanol (C_4H_9OH) is potentially more toxic than the lower-molecular-weight homologs, but it also is less volatile, a fact that limits airborne exposure in most circumstances. Its symptoms may include eye,

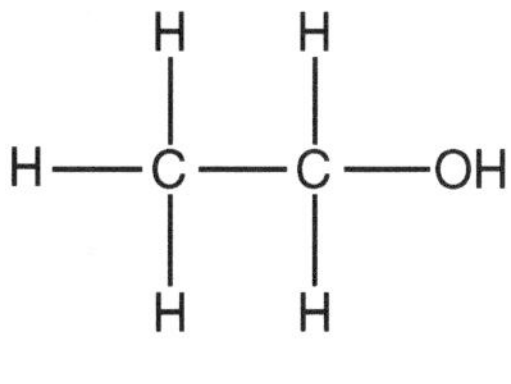

$$H-\overset{\displaystyle H}{\underset{\displaystyle H}{\overset{|}{\underset{|}{C}}}}-\overset{\displaystyle H}{\underset{\displaystyle H}{\overset{|}{\underset{|}{C}}}}-OH$$

Figure 16.6 Ethanol.

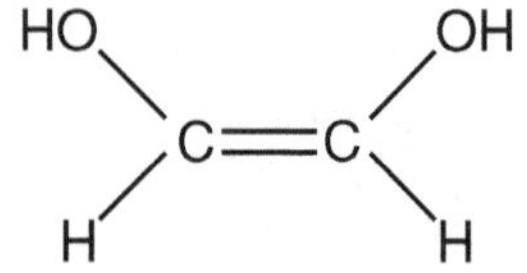

Figure 16.7 Ethylene glycol.

nose, and throat irritation; vertigo; headache; drowsiness; contact dermatitis; and corneal inflammation. No systemic effects from *n*-butanol typically occur at exposures less than 100 ppm. Skin irritation is common from allyl alcohol and absorption through the skin can lead to deep pain. It may cause severe burns of the eye and other ocular symptoms include lacrimation, photophobia, and blurring vision. Allyl alcohol is metabolized by the liver to allyl aldehyde, a potent hepatotoxin.

The alcohols may interact in industrial circumstances with chlorinated solvents to enhance toxicity that would occur from either group (potentiation and synergism).

Glycols

The larger alkyl-chain glycols (e.g., some of the dihydroxy alcohols) typically exhibit a lower degree of acute oral toxicity in comparison to the monohydroxy alcohols. They are not significantly irritating to eyes or skin, and have vapor pressures that are sufficiently low so that toxic air concentrations are not usually observed at ambient temperature (e.g., 60–80 °F). Ethylene glycol (see Figure 16.7) is a common example that may be used to represent the glycol family. A single oral dose on the order of 100 mL is lethal in humans, because of its metabolism to oxalate (or oxalic acid) which is toxic to the kidneys and may cause obstructive renal failure from formation of oxalate crystals. As in the case of methanol, ethanol can be used as a competitive inhibitor of ethylene glycol toxicity by blocking the aldehyde dehydrogenase-mediated metabolism.

16.7 TOXIC PROPERTIES OF REPRESENTATIVE PHENOLS

The aromatic alcohols (also termed phenols), in which the hydroxyl group is attached to a benzene ring, have the ability to denature and to precipitate proteins in a manner similar to their that of aliphatic counterparts. This property makes phenol useful as a bacteriostatic agent at concentrations exceeding 0.2 percent and an effective bactericide at concentrations in excess of 1.0 percent. However, it also renders these compounds quite corrosive and severe burns may result from direct contact. Fatalities have resulted in individuals inadvertently splashed with liquid phenol. Phenolic compounds also exhibit limited local anesthetic properties (hence their use in over the counter throat lozenges) and, in general, are CNS depressants at high concentrations.

Dihydroxy aromatics act like simple phenols but their effects are largely limited to local irritation. The trihydroxy compounds may reduce the oxygen content of blood at sufficient exposure levels. Methyl phenols (or cresols), while widely used in industrial applications, typically do not pose a significant inhalation hazard due to their relatively low vapor pressure and objectionable odor. Their physiological effects are similar to those of phenol, and dermal exposure, if prolonged, may result in significant absorption, even to the extent that fatalities have been reported from such exposures. Chlorinated phenols are strong irritants and exhibit significant oral toxicity because of their direct inhibition of cellular respiration. They also may produce muscle tremors, weakness, and, in overdose, convulsions, coma, and death.

Phenol (see Figure 16.8) can be cytotoxic to cells and tissues on sufficient exposure, given the ability to complex with and denature proteins. Because it is easily absorbed and it forms a loose complex with proteins, phenol may quickly penetrate the skin and underlying tissue, causing deep burns and tissue necrosis. This penetrating capacity, coupled with its nonspecific toxicity, renders it a serious handling hazard and all routes of exposure should be controlled carefully. If splashed on the

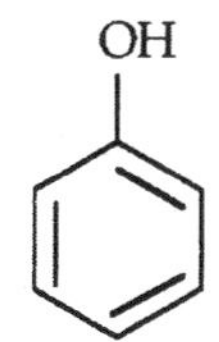

Figure 16.8 Phenol.

skin, it produces redness and irritation with dermal injury ranging from eczema or discoloration and inflammation to frank necrosis or gangrene, depending on the degree and duration of the exposure.

Following ingestion in concentrated solution, the extensive local necrosis produced by phenol in the mucous membranes of the throat, esophagus, and stomach can cause severe pain, vomiting, and tissue corrosion, which may lead rapidly to shock and death. If inhaled at sufficient concentrations, phenol may cause chemical pneumonitis. Like many other solvents, phenol may cause tissue damage and necrosis in the liver and kidneys following absorption and systemic distribution. It is more potent in this respect than most organics, and phenol may produce degenerative or necrotic changes in the urinary tract and the heart as well. A rapid fall in blood pressure may result from the CNS-depressant properties of phenol on vasomotor control, and because it exerts direct effects on the myocardium and small blood vessels. If acute poisoning occurs, death is usually the result of respiratory depression. However, a brief period of CNS stimulation and convulsions may be initially observed.

Estimated lethal oral doses of phenol have been reported as low as 1–2 g in some individuals. On a ppm basis, the TLV of phenol is equivalent to that established for vinyl chloride (i.e., 5 ppm). Phenol's dermal hazard is underscored by anecdotal reports involving individuals on whom phenol accidentally was splashed. In one case, accidentally spraying over the thighs was fatal within 10 min of exposure (illustrating its rapid dermal absorption) despite attempts to remove the clothing and to rinse the phenol off with water.

Other Phenolic Compounds

Substituted phenolics include catechol (*ortho*-dihydroxybenzene), resorcinol (*meta*-dihydroxybenzene), hydroquinone (*para*-dihydroxybenzene), and the cresols (or methylphenols), all of which have toxicological properties that are qualitatively similar to phenol. The alkyl substitutions tend to increase the toxicity and all of the substituted phenols may be considered to be more toxic than the parent phenol. Catechol may induce methemoglobinemia in addition to those toxic effects previously described for phenol.

16.8 TOXIC PROPERTIES OF REPRESENTATIVE ALDEHYDES

The aldehydes (see Figure 16.9) may cause primary irritation of the skin, eyes, and mucosa of the respiratory tract. These phenomena are most evident in aldehydes with lower molecular weights and those with unsaturated aliphatic chains. Although a number of the aldehydes can produce narcosis, this effect rarely is observed because the irritation that accompanies exposure generally serves as a sufficient warning property. Some of the aldehydes, such as fluoroacetaldehyde, can be converted to the corresponding fluorinated acids, giving them an extraordinarily high degree of systemic toxicity. The irritant properties of the dialdehydes have not been intensively studied. However, in some instances, concentrated solutions can be severe ocular and dermal irritants. The acetals and the aromatic aldehydes generally exhibit a greatly reduced potential for direct irritation.

An endpoint of toxicity that is common to aldehydes, but is not common to most other organic solvent constituents, is sensitization. Formaldehyde is the most common agent among the aldehydes with respect to this problem, and sensitization reactions have been reported in persons

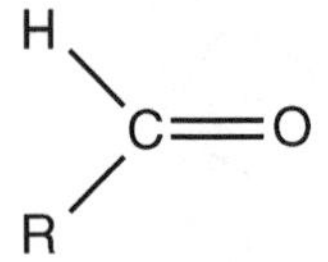

Figure 16.9 Aldehyde compounds.

who have been exposed to formaldehyde merely by wearing "permanent press" fabrics containing melamine-formaldehyde resins. Because their irritant effects limit inhalation exposure, the industrial use of aldehydes is relatively free of problems associated with other systemic or organ toxicities of a serious nature. However, at sufficient concentration, damage to the respiratory tract is possible.

Unsubstituted aldehydes (e.g., acrolein and ketene) are particularly toxic. The double bond in close association with the aldehyde functional group renders these compounds more reactive, and therefore more toxic, than the unsubstituted analogs. For example, ketene and acrolein are on the order of 100 times more potent when measuring acute lethality by inhalation than either acetaldehyde or proprionaldehyde. Reflecting this increase in potency is the fact that potential damage to the respiratory system is also more severe, resembling the deep lung damage of phosgene. Once absorbed, the systemic toxicity of the unsaturated aldehydes is also more severe than that of the saturated members of the class.

Mutagenicity of some aldehydes (e.g., acrolein, formaldehyde, and acetaldehyde) suggests carcinogenic potential of these compounds. Some of these substances are regulated as possible carcinogens by various occupational and environmental agencies.

Formaldehyde (see Figure 16.10) is the simplest structural member of the aldehyde family. It also is the most important aldehyde in both commerce from an economic perspective, and the environment from a regulatory perspective, and over 7 billion pounds are produced in the United States annually. Because of its reactivity and instability in the pure form, it is generally marketed in aqueous solution ranging from 37 to 50 percent formaldehyde (known as formalin). This product generally is diluted 1:10 for laboratory use, resulting in a typical usable concentration of approximately 4 percent. Formaldehyde also is readily available in two other forms, trioxymethylene (the cyclic trimer), and a low-molecular-weight homopolymer, paraformaldehyde. The latter is used primarily in the plastic and resins industries, in the synthesis of chemical intermediates and, less frequently, in sealants, cosmetics, disinfectants, foot-care creams, mouthwashes, embalming fluids, corrosion inhibitors, film hardeners, wood preservatives, and biocides.

Formaldehyde is a fairly strong dermal irritant, and its local actions dominate the adverse effects that are observed following excessive exposure, in comparison to the systemic effects that might otherwise occur. Table 16.5 describes the continuous nature of the dose-response relationship for irritant effects from formaldehyde exposures. It also illustrates the range of effective concentrations reported for the human population. Formaldehyde can produce dermal sensitization reactions in approximately 4 percent of the population, making it the tenth most common cause of dermatitis. With repeated application of high concentration, irritating solutions of formaldehyde induces sensitization in about 8 percent of the male subjects tested, but with lower concentrations (<2 percent),

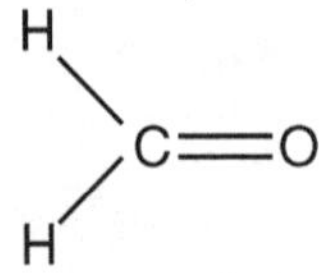

Figure 16.10 Formaldehyde.

TABLE 16.5 Dose–Response Relationship for Formaldehyde in Humans

Health Effects Reported	Formaldehyde Air Concentration (ppm)
None	0–0.5
Odor detection	0.05–1.50
Eye sensation and irritation	0.05–2.0
Upper airway irritation	0.10–25
Lower airway/pulmonary effects	5–30
Pulmonary inflammation/edema	50–100
Death	>100

the incidence is 5 percent or less. The primary irritant effects of formaldehyde are considered the most significant problem.

The fatal oral dose of formaldehyde is estimated to be 60–90 mL of formalin (37 percent). Depending on the dose, ingestion may cause headaches, corrosion of the gastrointestinal tract, pulmonary edema, fatty degeneration of the liver, renal tubular necrosis, unconsciousness, and vascular collapse.

Formaldehyde has been tested in a variety of animal species and found not to cause reproductive toxicities or teratogenicity. However, concerns have been raised with the recent findings of its mutagenic activity in certain test systems and with its corresponding local carcinogenicity in rodents. While the exact mechanism of its mutagenic actions remains to be resolved, a recurrent finding in several test systems was that formaldehyde produces crosslinks within DNA, which generally are recognized and repaired by the DNA repair enzyme system. The rodent carcinogenicity tests reveal a steep dose–response relationship, which suggests that a threshold may exist for this toxicity. The following points are consistent with this suggestion:

- Formaldehyde is a common metabolite of normal cellular processes and serves as a cofactor in the synthesis of several essential biochemical substances. Tissue concentrations of formaldehyde may reach several ppm in normal physiological circumstances.
- The mechanism for formaldehyde carcinogenicity appears to be a recurrent tissue injury and resultant hyperplasia caused by the high, irritating, and necrotizing exposures in the test systems.
- Epidemiological studies to date have not linked formaldehyde exposure to human cancer, nor have they indicated that persons chronically exposed to formaldehyde are at increased risk.

Acetaldehyde (see Figure 16.11) is the next larger molecular weight aldehyde beyond formaldehyde in the series and is also a common industrial chemical. Acetaldehyde is a normal metabolite of mammalian ethanol metabolism and has been implicated by some authors as the "hangover" associated with expressive ethanol consumption. It is less reactive than formaldehyde and, therefore, is generally less irritating and toxic than formaldehyde. It has not been found to be carcinogenic in animal tests,

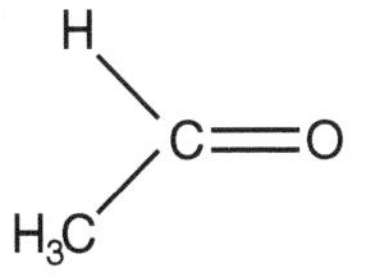

Figure 16.11 Acetaldehyde.

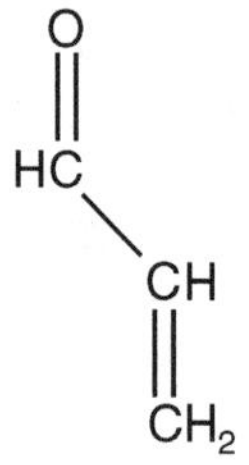

O
||
HC
CH
||
CH_2

Figure 16.12 Acrolein.

but its carcinogenic potential has not been investigated by extensive long-term tests. Acetaldehyde is embryotoxic and teratogenic in animal tests and is the presumptive proximal toxicant, inducing these effects when ethanol has been tested.

At concentrations in excess of 200 ppm in air, it is capable of eye irritation marked by conjunctivitis, and even at 25–50 ppm, some individuals may experience its irritant effects. The odor threshold is less than 0.1 ppm, thus affording a considerable margin of safety if used as a warning property, since the OSHA PEL based on health effects is 200 ppm. While it is a minimal industrial health hazard, it may represent an explosion hazard in some cases.

Acrolein (see Figure 16.12) is the unsaturated analog of propionaldehyde and the double bond greatly enhances its toxicity in comparison. Acrolein is toxic by all routes of administration and is capable of severe eye and pulmonary irritation. Since contact with the skin may produce necrosis, its direct contact with the eyes must be carefully avoided. Even though it is a highly reactive chemical, no carcinogenicity has been observed in tests thus far. The occupational guideline for acrolein is 0.1 ppm, not much less than the level that is considered to be moderately irritating (0.25 ppm).

16.9 TOXIC PROPERTIES OF REPRESENTATIVE KETONES

The paucity of reports in the literature regarding serious injury suggests that ketone compounds (see Figure 16.13) do not present serious hazards to health under most circumstances, probably at least in part because they have fairly effective warning properties (e.g., easily identifiable odor). Ketones are recognized CNS depressants, but the vapors present at concentrations that are great enough to cause sedation also are strongly irritating to the eyes and respiratory passages and, thus, typically are avoidable. Lower concentrations, however, may be inhaled easily and may accumulate to levels that impair judgment. Death, owing to overdose, typically results from respiratory failure. In general, the toxic properties of the ketones increase with increasing molecular weight, and the unsaturated compounds are more toxic than the saturated analogs.

Acetone (see Figure 16.14) is an extremely common industrial solvent and raw material in industrial manufacturing processes. Several daily ingestions of acetone in doses as high as 15–20 g produced limited adverse effects, including drowsiness. Skin irritation typically occurs only after repeated prolonged contact. Persons unaccustomed to acetone may experience eye irritation at 500 ppm, while workers who are used to daily exposure can easily tolerate several thousand parts per million in air. At 9000–10,000 ppm, unambiguous irritation of the throat and lungs occurs. The odor threshold rises after initial contact in acclimated people. Studies of exposed employees with average exposure to acetone concentrations of 2000 ppm typically reveal no serious injury.

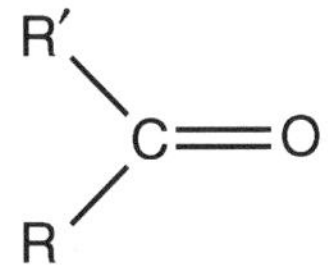

R′
C=O
R

Figure 16.13 Ketone compounds.

Other Ketones

Methyl ethyl ketone (2-butanone; MEK) is a very common industrial raw material and solvent. Uses include production of 1,3-butanedione and MEK peroxide. It is detectable by odor at a few parts per million in air, which represents a reasonable warning property several hundred times less than its occupational exposure guideline. High concentrations are irritating to the eyes, nose, and throat, and dermal or ocular irritation will accompany exposure to liquid splashed on the skin or in the eyes. CNS depression may result from prolonged exposure. The toxicological literature concerning MEK is extensive both in the form of animal studies and human exposure data.

Acetophenone (phenylethylketone) saw historical use as an anesthetic, but it is used today as a component of perfumes because it has a persistent odor that is not unlike orange blossoms or jasmine. It is a strong skin irritant, but it is not a potent CNS depressant. Eye contact may cause irritation and transient corneal burns.

Methyl-*n*-butyl ketone is a potent neurotoxin under some exposure circumstances and is metabolized to 2,5-hexanedione, which also is the neurotoxic metabolite that has been attributed to hexane. Therefore, it may induce a polyneuropathy like that described previously for hexane.

Industrially important exposures to ketones may be evaluated by urinary measurement of the relevant chemical (e.g., acetone, MEK, methyl isobutyl ketone).

16.10 TOXIC PROPERTIES OF REPRESENTATIVE CARBOXYLIC ACIDS

As with aldehydes and ketones, the irritant properties of these compounds dominate the observed effects and may mask CNS-depressant potential. The acidity (low pH), and therefore irritancy, decreases with increasing molecular size. Halogenation of carboxylic acids (see Figure 16.15) increases the strength of the acid and makes a stronger irritant. Dicarboxylic acids and unsaturated carboxylic acids are comparatively more corrosive. Hydroxyl or halogen substitution at the α-carbon enhances irritant potential. For example, acetic acid (CH_3COOH) is moderately irritating, but the unsaturated acrylic acid ($CH_2{=}CHCOOH$) and crotonic acid ($CH_3CH{=}CHCOOH$), or trichloroacetic acid (CCl_3COOH; a mammalian metabolite of trichloroethene), may produce burns and tissue damage.

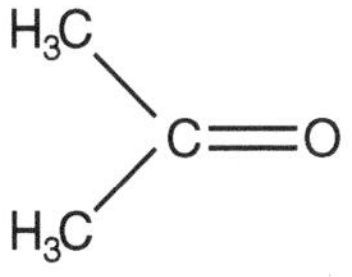

H_3C
C=O
H_3C

Figure 16.14 Acetone.

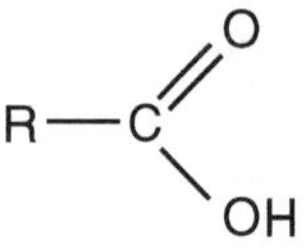

$$R-C(=O)-OH$$

Figure 16.15 Carboxylic acid.

16.11 TOXIC PROPERTIES OF REPRESENTATIVE ESTERS

Esters of similar structure are more potent anesthetics than the corresponding alcohols, aldehydes, or ketones, but are weaker than the ethers or halogenated hydrocarbons. Esters (see Figure 16.16) typically are degraded in the bloodstream by plasma esterases to the corresponding carboxylic acids and alcohols. Similarly, the lower molecular weight esters are more potent irritants than the alcohols and are known to cause eye irritation and lacrimation. Halogen substitution serves to increase the irritant effects, and unsaturated double bonds in the side chain may increase the toxicity by a large margin. Unsaturated esters have increased irritancy, and some may act to cause CNS stimulation, rather than CNS depression. Additional functional groups other than halogens or double bonds between carbons in the alkyl chain tend to reduce the vapor pressure and the systemic toxicity. Therefore, these esters are usually of limited interest as irritants to the skin and eyes.

Phthalate esters, used widely as *plasticizers*, may produce CNS damage at very high concentrations, may be irritating and have the capacity to act as either convulsants or depressants. Because of the wide variety of ester structures and the dramatic effect that additional functional groups may have regarding toxicity, the exact consequences of exposure to each compound should be sought in the literature, and few generalizations other than the ones indicated here are possible concerning these compounds.

Di(2-ethylhexyl)phthalate (DEHP; BEHP) is classified as a potential carcinogen by some regulatory agencies. However, the proposed mechanism of action (repetitive tissue injury) renders this classification of limited practical significance.

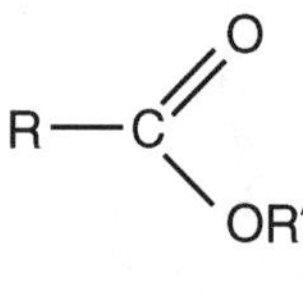

$$R-C(=O)-OR'$$

Figure 16.16 Ester compounds.

16.12 TOXIC PROPERTIES OF REPRESENTATIVE ETHERS

The ethers (see Figure 16.17) are recognized to have a broad variety of industrial and commercial uses. They are used in the production of rubber, plastics, cosmetics, paints and coatings, refrigeration, and foods. They also are widely used in medicine. Ethers as a class are effective anesthetics, and this property increases with the molecular size. The utility of this property is, however, limited by the irritant effects of ethers, and the fact that they are easily oxidized or photodegraded to peroxides which may be quite explosive.

Generally speaking, as chain length increases for the ethers, the oral and inhalation toxic potential decreases. In contrast, however, dermal penetrability and skin irritation increases with increasing chain length.

Diethyl ether [or ethyl ether (see Figure 16.18)] is absorbed rapidly through the lungs and, subsequently, is rapidly excreted through the lungs. It has potent anesthetic properties, and was widely

$$R\text{-}O\text{-}R'$$

Figure 16.17 Ether compounds.

used as an anesthetic at one time, but it is slightly irritating to the skin and contact with the eyes should be avoided. It produces anesthesia in humans in a concentration range of 3.6–6.5 percent in air, but respiratory arrest occurs at 7–10 percent, providing only a small margin of safety. Ether can produce profound muscular relaxation by means of corticospinal and neuromuscular blockade. However, nausea and vomiting are common subsequent side effects and were limiting factors in its use.

Isopropyl ether (see Figure 16.19) (or diisopropyl ether) is comparatively more toxic than ethyl ether, and causes irritation at much lower concentrations than those which are required to produce anesthetic effects. This effectively limits its use as an anesthetic. In humans, 500 ppm for 15 min causes no irritation, but odor is noticeably unpleasant at 300 ppm. At 800 ppm, disagreeable irritation of the eyes and nose is noticeable.

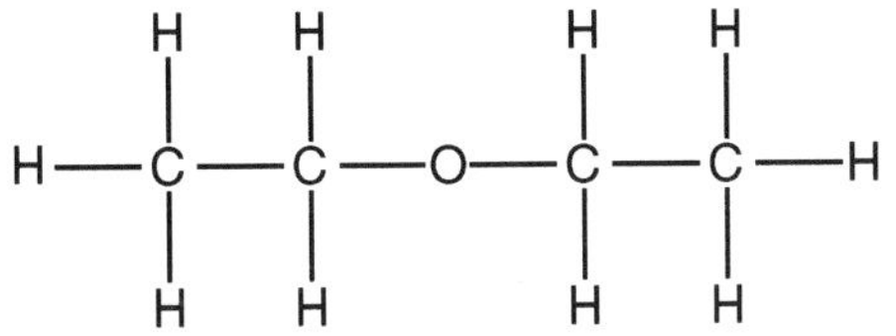

Figure 16.18 Ethyl ether.

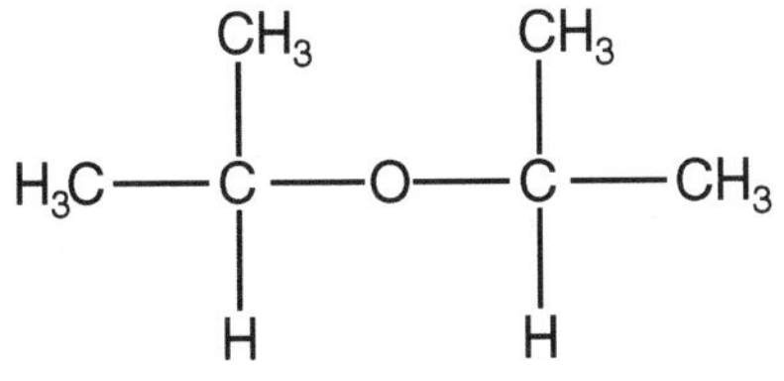

Figure 16.19 Isopropyl ether.

Other Ethers

The unsaturated ethers are, in general, more toxic than the saturated ethers, produce anesthesia faster, and possibly cause liver damage. Divinyl ether is more potent than ethyl ether. Halogenated ethers can cause very severe irritation to the skin, eyes, and lungs. For example, the vapors of chloromethyl ethers are painful at 100 ppm. The chlorinated ethers may also be potent alkylating agents, and compounds such as bis(dichloromethyl) ether are classified as carcinogens by some regulatory agencies (see also Chapter 13). Aromatic ethers, on the other hand, generally are less volatile, less irritating, and less toxic than the alkyl ethers.

The solvent 1,4-dioxane (or diethylene dioxide) is a member of the glycol ethers and is used in a wide range of lacquers, paints, dyes, cosmetics, deodorants, stains, and detergent products. It should be carefully distinguished from the dioxins (e.g., 2,3,7,8-tetrachlorodibenzodioxin; or 2,3,7,8-TCDD) in any discussions of toxicity and environmental. Liquid dioxane is a painful irritant to the eyes and skin, and can be absorbed dermally in significant quantities. Renal and hepatic damage may be observed following prolonged exposure. While dioxane has been reported to be carcinogenic based on animal data, the mutagenicity data have been generally negative, suggesting an epigenetic mechanism (see Chapter 13).

16.13 TOXIC PROPERTIES OF REPRESENTATIVE HALOGENATED SOLVENTS

Many halogenated (e.g., chlorinated, brominated, fluorinated) solvents are widely used in industry for metal degreasing extraction processes, refrigerants or aerosol propellants, paint removers, fumigants, as precursors in the manufacture of fluorocarbons, and as chemical intermediates in numerous chemical syntheses. They are largely nonflammable, and they generally exhibit potent anesthetic properties; several halogenated alkanes are the systemic anesthetic agents of choice in contemporary surgery (e.g., enflurane, halothane). Halogenated hydrocarbon anesthetics have negative effects on muscular rhythmicity and contractility as well as nerve conduction velocity (negatively chronotropic, inotropic and dromotropic, respectively) at concentrations that typically are effective for anesthesia.

Halogenated compounds may exhibit strong dermal irritant effects. Brominated compounds are more toxic systemically and locally than chlorinated compounds, while fluorine replacement of the chlorine may decrease the observed toxicity. A drawback to the widespread use of halogenated alkanes is that some of these compounds (e.g., 1,2-dichloroethane) have been shown to induce liver cancer in rodent bioassays, although this rarely has been confirmed in human epidemiology studies. Also, many highly substituted halogenated alkanes are environmentally persistent.

Chronic exposure to certain haloalkanes has been implicated in human cases of degenerative cardiac disease. This speculation has resulted in investigations of the cardiodepressant mechanisms of haloalkanes, their capacity for interference with energy production, and their effect on intracellular calcium transport between subcellular compartments. Halogenated alkanes may sensitize the heart to endogenous epinephrine or to β-adrenergic agonist drugs. The cardiotoxicity of low-molecular-weight halogenated hydrocarbons is considerably greater than that of low-molecular-weight unsubstituted hydrocarbons. Systemic toxicity of chlorinated haloalkanes to humans typically increases with increasing molecular size, the degree of halogen substitution, and the number of unsaturated bonds (e.g., progression from ethanes to ethenes). Conversely, halogenated substitution of the aromatic ring may significantly decrease the systemic responses in humans.

Halogenated aliphatics, particularly those with short alkyl chains and one or more chlorine or bromine atoms, constitute a class of chemicals that are acutely nephrotoxic and hepatotoxic in experimental animals. Experimental evidence suggests that the nephrotoxicity is related to metabolic products, rather than the parent haloalkane.

Representative Halogenated Methane Compounds

Structures of methyl chloride and bromide are shown in Figure 16.20. Methyl chloride (or chloromethane) is used as a refrigerant, aerosol propellant, "blowing" agent for plastic foams, solvent, and as a chemical intermediate in methylating reactions. Industrial exposure typically has involved operations in which foamed plastic is cut or shaped. While methyl chloride use as a refrigerant is rare, occasional exposure reports related to leaking equipment still occur. Since methyl chloride is an odorless gas at room temperature, its use should be restricted to well-ventilated areas. Inhalation is considered the only significant route of toxic exposure under typical industrial environments. Most cases of intoxication by chloromethane have involved air concentrations above 500 ppm and the major problem encountered in mild exposure is a state similar to drunkenness or inebriation. The onset of symptoms following exposure to methyl chloride may be confused with mild viral illness; in cases of more severe intoxication the symptoms can be delayed for many hours and may be mistaken for viral encephalitis or heavy-metal poisoning. Toxicity of selected halogenated methanes increases in the following order: methyl iodine, methyl bromide, methyl chloride.

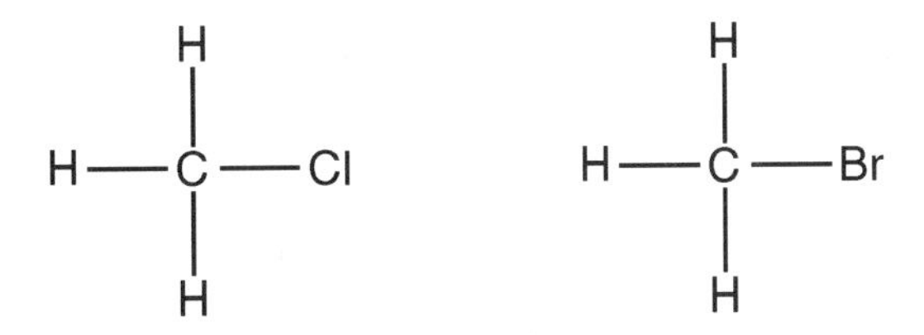

Figure 16.20 Methyl chloride and methyl bromide.

Chloromethane is metabolized to methanol and hydrochloric acid, and the methanol subsequently is oxidized to formaldehyde. There is a remarkable difference in individual human response to methyl chloride exposure, with some subjects consistently showing several times slower metabolism and hence much greater blood and expired air concentrations compared with other subjects. Chronic and subacute exposure to methyl chloride predominantly affects the CNS, producing ataxia, staggering gait, weakness, tremors, vertigo, speech difficulties, drowsiness, blurred vision, diplopia, personality changes, visual or auditory hallucinations, nausea, and EEG abnormalities. In addition, cardiac degeneration and depression of bone marrow activity have been reported. Incidents of acute exposure to methyl chloride concentrations below lethal levels has resulted in pulmonary congestion and edema in laboratory animals. Methyl chloride is considered a weak to moderate irritant to the eyes and skin. Liquid methyl chloride splashing in the eyes or on the skin may result in defatting of the affected tissues, anesthesia through freezing of the tissues, erythema, and blistering.

PHYSIOLOGIC RESPONSES TO METHYL CHLORIDE

Concentration (ppm)	Response
10	Odor threshold in humans
50	TLV
510	Irritation threshold
3000	Lethal concentration in animal studies

Testicular degeneration, renal adenocarcinoma, and male infertility have been produced in experimental animals following chronic exposure to methyl chloride. Teratogenic effects have been produced in mice and rats following high-level exposure to methyl chloride, but its role as a human reproductive hazard is not proven. Renal tumors have been induced in mice, but there is insufficient evidence to classify methyl chloride as an animal or human carcinogen. Methyl chloride induced unscheduled DNA synthesis, mutations, sister chromatid exchanges, and oncogenic transformation in a variety of short-term test systems, and is a weak, direct-acting mutagen for bacteria and human cells in culture.

Methyl bromide, the brominated analog, primarily is used as a soil fumigant, as well as in a wide range of grains, mills, warehouses, and homes. It also is used as a chemical intermediate, principally as a methylating agent. It has found use as a fire-extinguishing agent, particularly in automatic equipment for the control of engine fires on aircraft, but due to the inhalation toxicity of this material, use as a fire extinguisher was limited to specialized applications. Methyl bromide is a dangerous cumulative poison with delayed symptoms of central nervous system intoxication that may appear as long as several months after exposure.

PHYSIOLOGIC RESPONSE TO METHYL BROMIDE IN HUMANS

Concentration (ppm)	Response
1	TLV®
21	Odor threshold
35–100	Reported initial acute effects range
300	Lethal inhalation concentration

Elimination of methyl bromide is predominantly via the lungs as unchanged methyl bromide. A significant amount of methyl bromide, however, is metabolized in the body and may appear as inorganic bromide in the urine. "Normal" bromide ion concentration is below 1 mg/100 mL of blood serum, in the absence of the dietary sources (e.g., grains, beverages, medications). Five milligrams per 100 mL of blood may be considered evidence that exposure to bromide has occurred. If blood bromide is to be used in determining the exposure to methyl bromide, consideration must be given to other possible sources of bromide. The most common sources of inorganic bromide are medications, food, or water.

Methylene chloride (see Figure 16.21) (or dichloromethane) is used as a common "blowing" agent for foams and as a solvent for many applications, including the coating of photographic films, in aerosol formulations, and to a large extent in paint processes, where its high degree of volatility is desirable. Owing to this volatility, high concentrations may be rapidly attained in poorly ventilated areas. It is a powerful solvent that is effective in dissolving cellulose esters, fats, oils, resins, and rubber. Methylene chloride is more water-soluble than most other chlorinated solvents.

Methylene chloride is probably the least toxic of the four chlorinated methanes; the predominant toxic effect is CNS depression (expressed as narcosis). Other systemic effects in humans following nonlethal exposures to methylene chloride may include headache, giddiness, stupor, irritability, numbness, psychomotor disturbances, and increased carboxyhemoglobin levels. It is mildly irritating to the skin and dermal absorption is not considered a significant threat to human health unless massive concentrations are encountered. Eye contact may be painful but is not likely to cause serious injury. Adaptation to methylene chloride vapors occurs with repeated exposure, decreasing the ability to detect exposure.

PHYSIOLOGIC RESPONSE TO METHYLENE CHLORIDE IN HUMANS

Concentration (ppm)	Response
200–300	Odor threshold
1000	Unpleasant odor level
2300	Dizziness, fainting

The symptoms of excessive exposure may be dizziness, nausea, tingling or numbness of the extremities, sense of fullness in the head, sense of heat, stupor, or dullness, lethargy, and drunkenness. Exposure to sufficiently high concentrations of methylene chloride may lead to rapid unconsciousness and death. Prompt removal from exposure area typically results in complete recovery.

Historical industrial experience with methylene chloride has been remarkably free of serious adverse effects. Nephrotoxic and hepatotoxic potential of methylene chloride is considered low under typical industrial environments. Reports of systemic injury are rare and, although dermatitis has been reported because of common usage in paint remover formulations, only a few anesthetic deaths have occurred, all at extreme concentrations. Methylene chloride is known to be metabolized to carbon monoxide, but symptoms of carbon monoxide poisoning, such as headaches, have not been a common feature of methylene chloride reports. This indicates that carboxyhemoglobin levels alone are not a good measure of the toxic effect of methylene chloride. Acute exposures have resulted in liver or kidney

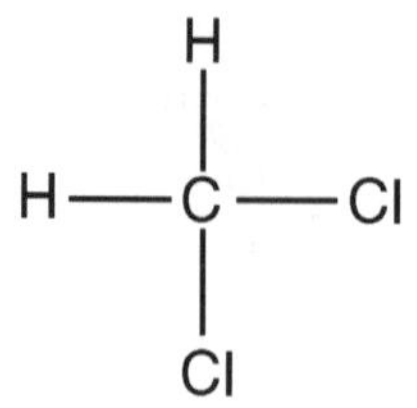

H
H—C—Cl
Cl

Figure 16.21 Methylene chloride.

damage and cardiac sensitization in animal studies, although the latter has been demonstrated in animals only when adrenaline is injected and the practical significance of these findings to industry is considered to be low.

Although studies with methylene chloride in an Ames test with *Salmonella typhimurium* TA98 and TA100 resulted in increased reversions in both strains of bacteria, observed effects were limited. Other tests of genotoxic potential are typically negative.

Methylene chloride was not positive in a pulmonary tumor assay in strain A mice. Although it is regulated by USEPA and some occupational agencies as a potential carcinogen, methylene chloride does not appear to present a practical risk of carcinogenesis in humans at currently acceptable levels of exposure.

Chloroform (see Figure 16.22), or trichloromethane, was used for many years as an anesthetic, solvent, insecticide, and chemical intermediate. Because of delayed liver injury and reports of cardiac sensitization, its use as an anesthetic is obsolete. Chloroform is presently considered a carcinogen by USEPA and some other agencies, based on animal studies. Chloroform may be produced at low levels during the chlorination/disinfection of water.

As with other volatile solvents, inhalation is considered the main exposure route for chloroform. Much of the toxicological information available has been developed following the interest in chloroform as a CNS depressant or anesthetic, and high concentrations of chloroform may result in narcosis, anesthesia, and death. Rapidly absorbed and distributed in all organs, chloroform is exhaled for the most part unchanged, or as carbon dioxide. Chloroform appears to be the most cardiotoxic of the anesthetics, with effects that are the least reversible. It is known to cause permanent hepatic and renal injury following sufficient exposures, and hepatic necrosis has been reported in humans following ingestion of pharmaceutical preparations containing 16.5 percent chloroform.

Signs of severe chloroform exposure in humans include a characteristic sweetish odor on the breath, dilated pupils, cold and clammy skin, initial excitation alternating with apathy, loss of sensation, abolition of motor functions, prostration, respiratory depression, cardiac sensitization to adrenaline, unconsciousness, coma, and death. Lethal exposures to chloroform may be delayed and may result from associated kidney or liver damage. Responses associated with exposure to chloroform concentrations below the anesthetic level, including occupational exposures, are typically inebriation and excitation passing into CNS depression.

PHYSIOLOGIC RESPONSE TO CHLOROFORM IN HUMANS

Concentration (ppm)	Response
200–300	Odor threshold
4100	Vomiting, sensation of fainting
14,000–16,000	Narcotic limiting concentration

Chloroform appears to be unique among the smaller chlorinated aliphatics in that it is the only one that has been reported to be teratogenic and highly embryotoxic in animals. It has been found in

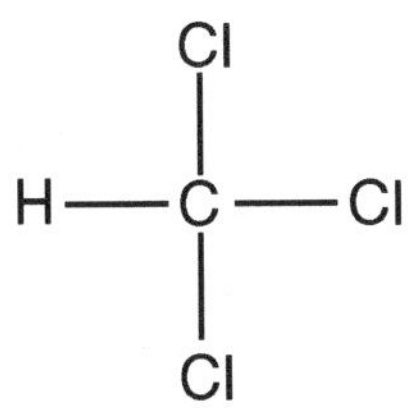

Figure 16.22 Chloroform.

umbilical cord blood at quantities greater or equal to maternal blood levels, indicating transplacental acquisition.

Although the toxicity of chloroform and carbon tetrachloride has been ascribed to their solubility in cellular lipid, metabolism appears necessary to explain their toxicity. Since chloroform does not form the free radical CCl_3, it would not be expected to be as toxic as carbon tetrachloride. However, the mechanisms and paths of metabolism are still not certain and both enzymatic and nonenzymatic processes may be important. No metabolite has been identified in the blood or urine that can be considered as a useful guide for evaluating occupational chloroform exposure. Chloroform does not appear to induce chromosome breakage or sister–chromatid exchanges in human lymphocytes and failed to produce mutagenic changes in cultures of Chinese hamster lung fibroblast cells.

Carbon tetrachloride (see Figure 16.23) (or tetrachloromethane) has seen widespread industrial uses, including: fire extinguishers, refrigerants, metal degreasing, semiconductor production, and as a chemical raw material. Significant amounts were used in grain fumigant mixtures. Carbon tetrachloride is an active insecticide and is effective in suppressing the flammability of more flammable fumigants. The odor is one to which individuals often becomes adapted, and odor is not a satisfactory warning of excessive exposure. Carbon tetrachloride is not significantly teratogenic, but can be shown to be variably embryotoxic and fetotoxic to animals at high exposure levels. Carbon tetrachloride is not considered to be strongly mutagenic. Hepatocellular carcinomas and adrenal tumors developed in mice that received gavaged doses of 2500 or 1250 mg/kg per day for 78 weeks of carbon tetrachloride. The substance is regulated as a potential carcinogen by a number of occupational and environmental regulatory agencies.

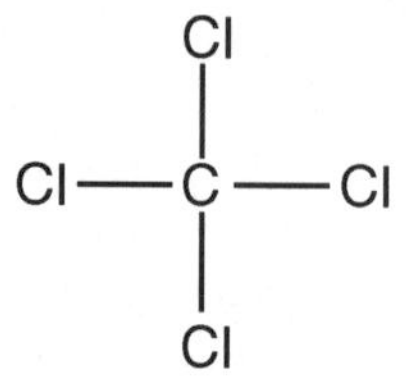

Cl
Cl—C—Cl
Cl

Figure 16.23 Carbon tetrachloride.

Acute exposure to carbon tetrachloride may result in systemic effects including CNS depression, loss of consciousness, dizziness, dyspnea, cyanosis, proteinuria, optic neuritis, vertigo, headache, mental confusion, incoordination, nausea, vomiting, abdominal pain, diarrhea, visual disturbances, ventricular fibrillation, kidney and/or liver injury, oliguria, albuminuria, edema, and anorexia. From the available clinical evidence carbon tetrachloride is suspected of causing retrobulbar neuritis, optic neuritis, and optic atrophy, although no experimental demonstration of retinal or optic nerve injury has been made.

Hepatic and renal injury may occur from a single acute exposure, but is more likely following repeated exposures. The lower the exposure level, the greater the likelihood that the injury will occur predominantly in the liver. The concurrent intake of significant amounts of alcohol with exposure to carbon tetrachloride may greatly increase the probably of liver injury. In nonfatal poisoning, recovery of renal function occurs in three phases. In the first, after 1–3 days, oliguria stops, but creatinine and urea plasma concentrations remain elevated. The second phase begins with a decline in these concentrations. In the third phase, about 1 month after the initial injury, renal blood flow and glomerular filtration begin to improve and renal function is recovered after 100–200 days.

PHYSIOLOGIC RESPONSE TO CARBON TETRACHLORIDE IN HUMANS

Concentration (ppm)	Response
21–79	Odor threshold
200	Severe toxic effects with strong odor
1000–2000	Reported lethal concentration

Local application of carbon tetrachloride to human skin produces distinct pain with erythema, hyperemia, and wheal formation followed by vesiculation, which may facilitate secondary infection. Studies have indicated that absorption of carbon tetrachloride through the skin may present a potential problem based on a limited study on human subjects. In view of the potential hepatotoxicity of carbon tetrachloride, repetitive contact of the skin with liquid carbon tetrachloride should be prevented.

Vinyl chloride (see Figure 16.24) is a gas under ambient conditions, and is a potent skin irritant. Contact with the liquid form may cause frostbite. The eyes may be immediately and severely irritated. Vinyl chloride depresses the central nervous system, causing symptoms that resemble mild alcohol intoxication. Lightheadedness, nausea, and dulling of visual and auditory responses may develop in acute exposures. Severe vinyl chloride exposure has been reported to result in death. Workers entering polyvinyl chloride reactor vessels for cleaning have exhibited a triad combining arthro-osteolysis, Raynaud's phenomenon, and scleroderma. Chronic exposure may damage the liver and induce a highly specific liver cancer (angiosarcoma), which is an established risk for chronic exposures to vinyl chloride at the old TLV® of several hundred ppm. Increased rates of cancer of the lung, lymphatic, and nervous systems have been reported. Experimental evidence links vinyl chloride to tumor induction in a variety of organs, including liver, lung, brain, and kidney, and to nonmalignant alterations (fibrosis, connective tissue deterioration).

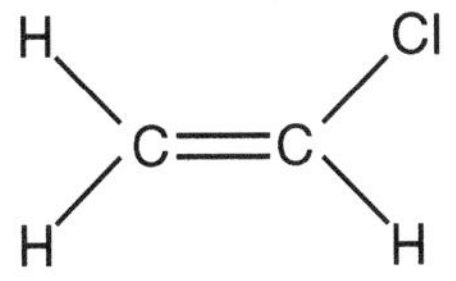

Figure 16.24 Vinyl chloride.

The mutagenic, carcinogenic, and reproductive hazard of vinyl chloride is further discussed in Chapters 11, 12, and 13, respectively.

Trichloroethene (TCE) and perchloroethene (PERC) (see structures in Figure 16.25) are among the most widely used chlorinated solvents, on an historical basis. The toxicological literature is extremely large for these substances, and the reader is referred to the ATSDR Toxicological Profiles for a more detailed treatment of available information. These substances exhibit a generally low degree of acute toxic potential, and industrial use experience has been relatively good in cases where appropriate exposure controls were in place. Although historically both have been regulated as potential carcinogens by some occupational and environmental regulatory agencies, recently the ACGIH reclassified TCE in Group A5 (Not Suspected as a Human Carcinogen, and placed PERC in Class A3 (Animal Carcinogen). The classification for both of these agents by U.S. EPA is under review at this time. It is of interest to note that in some environmental conditions both TCE and PERC may be degraded by sequential dechlorination to vinyl chloride, which was described previously in this section.

Figure 16.25 TCE and PERC.

16.14 TOXIC PROPERTIES OF REPRESENTATIVE NITROGEN-SUBSTITUTED SOLVENTS

The nitrogen substituted compounds have been used in a wide range of industrial applications. Aromatic amino compounds are important as intermediates in the manufacture of dyestuffs and pigments, and are also used in the chemical, textile, rubber, and paper industries. Some aromatic nitro compounds are used as dyestuffs while others, such as trinitrotoluene and pyridine, are used in manufacture of chemical explosives. Pyrrolidine, morpholine, aniline, pyridine, and similar compounds are used in crop protection as herbicides, fungicides, or insecticides. Still other nitrogen substituted compounds are used as drugs and research tools. Finally, both β-chloroethylamine and trichloronitromethane are strong irritants, and have been used for this purpose as effective chemical warfare agents.

The aliphatic amines (see Figure 16.26) are among the most toxic classes of organic chemicals. Most members are potent irritants and sensitizers, a property related to the base strength of the amine substituent. The corrosive properties generally are unrelated to the molecular size, in contrast to most other organic chemical groups (e.g., esters, ethers, alcohols), which exhibit decreasing irritancy as molecular size increases. The aliphatic amines are well absorbed by all routes of exposure. Thus, in addition to physical damage to the skin, they may cause systemic toxicity, including methemoglobinemia, pulmonary hemorrhage, hepatic necrosis, nephrotoxicity, and cardiac degeneration. In addition, amines may mimic the physiologic effect of adrenaline (epinephrine), which has a similar structure.

The tumorigenic properties of the aliphatic amines generally are limited to the nitrosamines. However, it is thought that alkyl amines may be converted to the nitrosamine form in the gastrointestinal tract. The following list summarizes general toxicological properties of the alkyl amines:

- Irritancy increases up to six carbons, then decreases with further increasing molecular weight.
- Unsaturated congeners are well absorbed following dermal exposure, while saturated congeners are not.
- Direct irritant potential is not affected by other functional groups, although sensitization potential may be altered.
- Salts of aliphatic amines are typically weaker irritants than the parent molecule.

As noted previously, they are strong irritants and therefore probably represent a greater potential handling hazard than most of the other chemical classes discussed in this chapter. Their strong irritant properties stem from the fact that the amine portion of the molecule is a very corrosive functional constituent. The skin has some resistance to changes in pH and can withstand chemical attack in the pH range of 1–10 for short periods without significant damage. However, if the base strength of the chemical is much above 10 or the acid has a pH lower than 1, significant skin injury may occur very quickly after initial contact with the chemical. The base strength of most of the simple amines is in the immediately injurious range of pH greater than 10, as shown in Table 16.6.

The molecular size and the degree of substitution (i.e., primary, secondary, or tertiary) have little effect on the corrosiveness of the amine group. Thus, while the irritant nature of the other functional groups (alcohols, ethers, carboxylic acids, etc.) is decreased as the size of the organic portion of the chemical increases, the irritation of the amines is not affected.

$$R\text{-}NH_2$$

Figure 16.26 Amine compounds.

TABLE 16.6 Relative Base Strength of the Amines

Amine Compound	Base Strength (p*K*)
Methylamine	10.6
Dimethylamine	10.6
Trimethylamine	10.7
Ethylamine	10.8
Diethylamine	11.0
Triethylamine	10.7
Propylamine	10.6
Butylamine	10.6
Allylamine	9.5
Cyclohexylamine	10.5

A characteristic of the amines that influences their acute toxicity and handling hazards is that they are well absorbed by all routes. Thus, they represent a significant dermal hazard not only because of the skin injuries they produce, but also because they tend to penetrate skin easily enough that the lethal or acutely toxic dose whether the dose is taken in via the oral or the dermal route. Therefore, contact with the skin should be avoided. Accompanying Table 16.7 summarizes the dose response. Note that LD_{50} is the amount lethal to 50 percent of the animals tested. Necrosis means cell or tissue death.

Because of their tissue-penetrative and tissue-corrosive characteristics, amines are toxic to all tissues in which they are absorbed in measurable amounts and adversely affect a number of organs. What organs will be most affected probably depends to a large extent on the distribution of the chemical within the body. Some of the systemic effects observed for lethal exposures are edema and hemorrhage of the lungs, necrosis of the liver, necrosis and nephritis in the kidneys, and muscular degeneration of the heart.

Two other common characteristics observed in amine compounds are methemoglobin formation in the red blood cells (see Chapter 4), and sensitization to the chemical itself. Sensitization to amines probably happens because the amine group is fairly reactive. The chemicals may bind to cellular proteins to form haptens, or molecules that go unrecognized by the body's immune defenses. The body produces antibodies against these haptens, and on subsequent chemical exposure an allergic reaction may ensue. During this antibody-hapten response, the body releases histamine, which, in turn, induces arterial vasoconstriction, capillary dilation, a fall in blood pressure, itching, and a bronchoconstriction. These effects explain many of the effects seen in a severe allergic response: labored or difficult breathing, fainting or possibly anaphylactic shock, and a reddening or irritation-like response where contact with the skin has occurred. Some of the amine compounds are potent sensitizers, and dermal exposure should be avoided.

A somewhat uncommon feature of the alkyl amines is their ability to simulate the actions of epinephrine (or adrenaline) within the body. Epinephrine is an important neurohormonal transmitter

TABLE 16.7 Relative Toxicity of the Amines[a]

Amine	Oral LD_{50} (mg/kg)	Skin LD_{50} (mg/kg)	Dermal Effects
Methylamine	0.02	0.04	Necrosis
Ethylamine	0.4	0.4	Necrosis
Propylamine	0.4	0.4	Necrosis
Butylamine	0.5	0.5	Necrosis
Hexylamine	0.7	0.4	Slight necrosis

[a]Note that LD_{50} is the amount lethal to 50 percent of the animals tested. Necrosis means cell or tissue death.

within the body, and it consists of a propylamine side chain attached to a catechol ring. Thus, it is easy to understand why some of the alkyl amines may be able to mimic some epinephrine physiologic responses. Following is a list of some of the general conclusions about alkyl amine-induced sympathomimetic activity:

- Activity increases with the size of the alkyl chain, up to six carbons
- For alkylamines over six carbons, heart rate decrease and blood vessel dilation is more common
- A branching of the alkyl carbon chain decreases activity of the chemical
- Pressor activity (blood vessel constriction) varies with structure
- Repeated exposure causes cardiac depression and vasodilation
- Convulsions may cause mortality at high, acute, repeated exposure

Human exposure to amine compounds should be avoided where possible, since a number of these chemicals are regulated as demonstrated carcinogens. In particular, this includes the aromatic or diphenylamine compounds such as the dye components benzidine, 2-naphthylamine, or 4-aminobiphenyl, all of which induce bladder tumors. Other amine compounds may exhibit carcinogenic potential as well, including the alkyl amine group of nitrosamines. The nitrosamines, potential liver carcinogens, represent an interesting human risk issue, since any alkyl amine (including those generated during the digestion of food) absorbed orally theoretically may be converted to a nitrosamine by the acid and nitrite found in the gut. However, animal studies have failed to demonstrate that this is a significant problem in humans.

In summary, the amines are very toxic substances presenting diverse dermal and handling hazards. The following general characteristics apply:

- Irritation increases to six carbons, then decreases with loss in volatility
- For smaller alkylamines, irritantcy is not related to degree of substitution
- Unsaturated amines have greater systemic and dermal toxicity
- Irritation by these chemicals is not usually affected by functional groups
- Salts of amine chemicals are usually weaker irritants
- Additional functional groups may increase sensitization potential
- Sensitization is greater in aromatic amines
- Aromatic rings do not decrease the various toxicities
- Aromatic rings add methemoglobinemia and cancer as potential hazards

Aromatic Amino Compounds

The aromatic amino compounds are aromatic hydrocarbons in which at least one hydrogen atom has been replaced by an amino group. The aromatic amines are synthesized by nitration of the aromatic hydrocarbon with subsequent reduction to the amine or by the reaction of ammonia and a chloro- or hydroxyhydrocarbon. Most of the aromatic nitro compounds are fat-soluble and water-insoluble such that they readily penetrate the skin and can quickly be absorbed through the lungs and into the blood, immediately becoming systemic in action. The most dominant toxic effects of the aromatic amino compounds are methemoglobin formation and cancer of the urinary tract. Other toxic effects include hematuria, cystitis, anemia, and skin sensitization. It is thought that the aromatic amines act indirectly on methemoglobin formation through their metabolites. Most of the compounds are absorbed through the skin and through respiration.

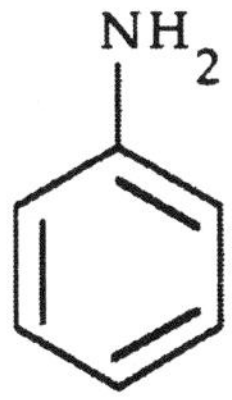

NH2

Figure 16.27 Aniline.

In the chemical industry, aniline (see Figure 16.27) is a parent substance for the synthesis of many compounds, including dyes, rubber accelerators, antioxidants, drugs, photographic chemicals, isocyanates, herbicides, and fungicides.

The most characteristic effect associated with aniline exposure is methemoglobinemia. Because it is fat-soluble, aniline readily penetrates intact skin and is easily absorbed by inhalation. Moderate exposure to aniline by any route may cause cyanosis, but as oxygen deficiency increases, associated symptoms of headache, weakness, irritability, drowsiness, dyspnea, and unconsciousness may occur. Ingestion of aniline causes splenic enlargement, hemosiderosis, as well as hyperplasia of the bone marrow. The existence of chronic aniline poisoning is controversial; however, some CNS symptoms have been reported in association with low dose chronic exposure to aniline. Severe acute toxic effects may occur at 80 ppm, with symptoms typically appearing at approximately 20 ppm. The most common symptoms of acute aniline intoxication are cyanosis, lacrimation, tremors, tachypnea, lethargy, methemoglobinemia, sulfhemoglobinemia, and Heinz body formation. The carcinogenicity of aniline to humans is not clear, since it was found to be mildly carcinogenic to rats, but was not carcinogenic to mice. This may be due to interspecies metabolic differences.

Industrially important exposures to aniline may be evaluated by measurements of urinary *p*-aminophenol, expressed in the context of urinary creatinine.

Dimethylaniline (see Figure 16.28) is used in the synthesis of dyestuffs, as a solvent, and as an analytical reagent. This compound can enter a system by inhalation, skin absorption, ingestion, and eye and skin contact. Signs of intoxication in man are headaches, cyanosis, dizziness, labored breathing, paralysis, and convulsions. Like aniline, dimethylaniline is readily absorbed through the skin, with the potential methemoglobinemia. Few reports of industrial exposure are available with which to estimate the hazards of dimethylaniline. However, cases of severe exposure are accompanied by visual disturbances and severe abdominal pain. No evidence of reproductive or genetic toxicity is available for this compound. The TLV® and PEL have been set at 5 ppm (25 mg/m^3) by ACGIH and OSHA, respectively.

Benzidine (see Figure 16.29) principally is used in the synthesis of dyes, but may also be used as a hardener for rubber, and as a laboratory reagent. A high incidence of bladder cancer has been reported among workers who have been exposed to benzidine. Inhalation and skin absorption are significant routes for benzidine exposure which may result in contact dermatitis. Benzidine is a known human

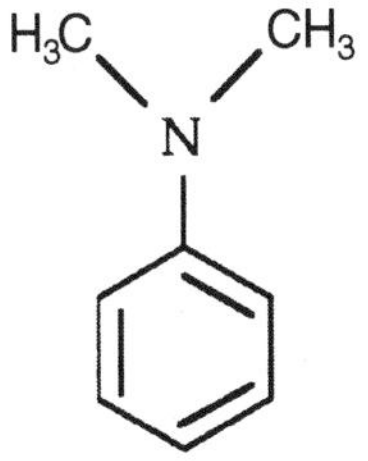

H3C CH3
N

Figure 16.28 Dimethylaniline.

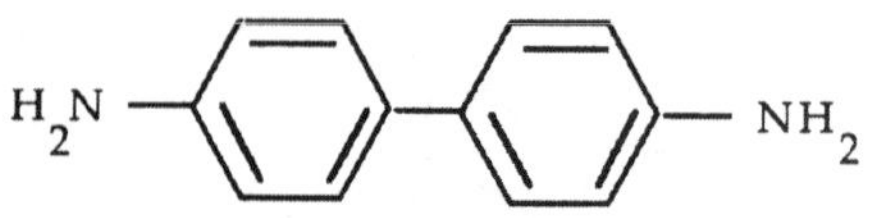

$H_2N-C_6H_4-C_6H_4-NH_2$

Figure 16.29 Benzidine.

H_2N-NH_2

Figure 16.30 Hydrazine.

carcinogen and exhibits mutagenic potential in the bacterial tests. The USEPA has established a carcinogenic potency factor (CPF) of 0.067 $(mg/kg{\cdot}day)^{-1}$ for inhalation exposure and 0.0067 $(mg/kg{\cdot}day)^{-1}$ for oral exposure. Although benzidine is one of the compounds regulated by OSHA as a human carcinogen, no exposure standards or recommendations have been set by OSHA or ACGIH.

Other Nitrogen Compounds

Because of its strong reducing capabilities, hydrazine (see Figure 16.30) is used in a wide range of industrial chemical syntheses, photographic processes, and metallurgy. At sufficient concentration, hydrazine is a strong skin and mucus membrane irritant, convulsant, hepatotoxin, and a moderate hemolytic agent. It is readily absorbed through the lungs, the gastrointestinal tract, and through intact skin, as evidenced by the fact that equivalent median lethal doses are observed in animal studies following oral, intravenous, and intraperitoneal administration. The effects noted after absorption by all routes include anorexia, weight loss, weakness, vomiting, excitement, and convulsions. Airborne exposure may produce eye and respiratory tract irritation, lung congestion, bronchitis, and pulmonary edema.

Hydrazine has been shown to be tumorigenic in mice, where lungs are the primary target organs. A carcinogenic potency factor (CPF) or slope factor (SF) of 17.1 (mg/kg·day) for inhalation exposure, and 3.0 (mg/kg·day) for oral exposure, has been established by USEPA. This compound has also shown to be mutagenic in phage, bacteria, plants, *Drosophila*, and mammalian test systems. The ACGIH TLV® and the OSHA PEL for hydrazine have been set at 0.01 ppm and 1.0 mg/m^3, based on observation of damage to lung, liver, and kidney tissue at higher exposure levels.

16.15 TOXIC PROPERTIES OF REPRESENTATIVE ALIPHATIC AND AROMATIC NITRO COMPOUNDS

These compounds have little industrial use and in general are oily liquids of low solubility and volatility. Toxicologically, they can be classified as moderate irritants, because their anesthetic symptoms are mild. Unsaturated compounds may be absorbed via the skin to a significant extent, while the saturated chemicals are not. Halogenation produces definite skin irritation, some systemic absorption and, therefore, toxicity. Aromatic nitro compounds are a class of chemicals, several members of which may produce methemoglobin formation and/or sensitization to the compound.

Nitro derivatives of benzene and toluene also have prominent toxic effects other than sensitization, CNS depression, or methemoglobinemia. The trinitrotoluene or dinitrobenzene compounds are well absorbed by all routes of exposure. These compounds can uncouple oxidative phosphorylation, and liver injury is often seen along with the toxicities previously mentioned. Other problems that have been observed are dermatitis, anemia, heart irregularities, and peripheral neuritis. Persons deficient in glucose 6-phosphate dehydrogenase are sensitive to hemolytic anemia. Some of these compounds cause bladder tumors as well. Again, the specific toxicities of any individual member of this class of compounds should be reviewed individually if it is to be used.

Aliphatic and Alicyclic Nitro Compounds and Related Substances

The aliphatic/alicyclic nitro compounds generally are oily liquids that exhibit low solubility and low volatility. They are of limited industrial significance, but several members have important uses as specialized fuels and as strong solvents. The nitro paraffins are acidic compounds and thus are rapidly neutralized with strong bases and readily titrated. Its tautomerism forms the basis for its ability to be easily chlorinated and form chloronitroparaffins such as trichloronitromethane. These compounds are moderate skin and mucous membrane irritants, particularly the halogenated derivatives. Saturated members of this class are not well absorbed, while unsaturated members exhibit significant dermal absorption.

Nitromethane (see Figure 16.31) and other nitroparaffins are synthesized by the vapor-phase nitration of propane. Nitromethane may be used as rocket fuel, a gasoline additive, a solvent for cellulosic compounds, polymers and waxes, and in chemical synthesis.

Nitromethane, like other nitroparaffins, may produce narcosis, methemoglobinemia, mucus membrane irritation, CNS excitation, and liver damage. However, based on periodic physical examinations of plant operators engaged in the manufacture and handling of nitromethane for several years, no chronic effects have been attributed to this compound. Nitromethane has not exhibited significant mutagenic capacity in tests performed with Salmonella typhimurium. The TLV® and PEL values which have been established by ACGIH and OSHA for nitromethane are 20 and 100 ppm, respectively.

Aromatic Nitro Compounds

The aromatic nitro compounds are formed by substituting a nitro group directly on the benzene ring. In general, aromatic nitro compounds produce dermal sensitization, CNS depression, and methemoglobinemia. These compounds can have the capacity to uncouple oxidative phosphorylation and to produce liver injury. Other reported adverse effects include dermatitis, anemia, cardiac irregularities, peripheral neuritis, and bladder tumors. Persons deficient in glucose-6-phosphate dehydrogenase are demonstrably sensitive to hemolytic anemia as well.

Nitrobenzene (see Figure 16.32) is used as a solvent intermediate in the preparation of aniline, benzidine, and other chemicals, in shoe and metal polishes, and as a solvent. The vapor is readily absorbed through the skin and lungs. Following absorption nitrobenzene undergoes transformations into its metabolites, which consist of *p*-aminophenol, *p*-hydrooxacetanilide sulfate, *p*-nitrophenol sulfate, and *m*-nitrophenol sulfate. Ring hydroxylation and reduction are important steps in the biotransformation of nitrobenzene, and the resulting metabolites are excreted mainly in the urine.

The primary toxic effect of nitrobenzene involves methemoglobin formation, although the method by which nitrobenzene causes this condition is not well understood. The primary opinion is that heme iron is oxidized by the oxidation-reduction cycling of the metabolites. In subacute and chronic forms of poisoning, hemolysis causes anemia as the main symptom, but other symptoms such as headache,

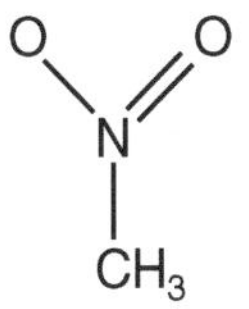

O O N CH_3

Figure 16.31 Nitromethane.

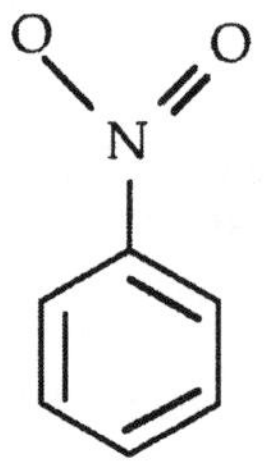

Figure 16.32 Nitrobenzene.

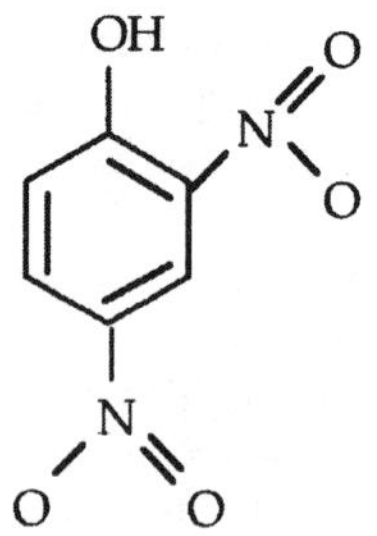

Figure 16.33 Dinitrophenol.

confusion, vertigo, nausea, loss of cognition, hyperalgesia, paresthesia, and polyneuritis have been reported as well as spleen and liver damage. Both the cyanogenic and anemiagenic potential of nitrobenzene were listed as considerably greater than those of aniline, and the overall potential for producing the blood effects was second only to that of dinitrobenzene. There is some risk of reproductive toxicity shown by a decrease in rat fertility following exposure. No genetic toxicity has been noted. Industrially important exposure to nitrobenzene may be evaluated by measurements of *p*-nitrophenol, expressed in conjunction with urinary creatinine.

Among the six isomers of dinitrophenol (DNP) (see Figure 16.33), the one most commonly used for industrial purposes is 2,4-dinitrophenol. The isomers often are kept as a mixture and are involved in the synthesis of dyestuffs, picric acid, picramic acid, as herbicides, and in the manufacture of the photographic developers.

Local application of DNP causes yellow staining of skin and may cause dermatitis due to either primary irritation or to allergic sensitivity. In general, DNP disrupts oxidative phosphorylation, resulting in increased metabolism, oxygen consumption, and heat production. Acute poisoning is characterized by the onset of fatigue, sudden thirst, sweating, and oppression of the chest. There may be rapid respiration, tachycardia, and a rise in body temperature. In cases of less severe poisoning, the symptoms may include nausea, vomiting, anorexia, weakness, dizziness, vertigo, headache, and sweating. The onset of effects is rapid, and death or recovery may occur within 1 or 2 days following massive exposure. Chronic exposure may result in kidney and liver damage and in cataract formation. While no federal standards have been set for dinitrophenol, an exposure limit of 0.2 mg/m^3 has been suggested on the basis of data for dinitro cresol.

16.16 TOXIC PROPERTIES OF REPRESENTATIVE NITRILES (ALKYL CYANIDES)

The nitriles (e.g., acrylonitrile, acetonitrile) are organic cyanide compounds (see structures in Figure 16.34). They are nonpolar and are readily absorbed by all routes. Because some of these compounds dissociate to produce free cyanide, the adverse effects they produce are comparable to those of cyanide

$R-C\equiv N$

Figure 16.34 Nitrile compounds.

poisoning. However, many of these compounds do not readily release cyanide once absorbed and their toxicity cannot simply be characterized as that of cyanide itself. Systemic toxic effects among the unsaturated nitriles are similar but, as noted previously for other series of organic solvents, the unsaturated forms are more irritating than the corresponding saturated homolog.

The most commonly used of the nitriles, acrylonitrile, is regulated as a suspected carcinogen by a number of occupational and environmental regulatory agencies, based primarily on the data from animal studies.

16.17 TOXIC PROPERTIES OF THE PYRIDINE SERIES

Pyridine (see Figure 16.35) is the parent compound for the pyridine series of substituted analogs. It is a flammable, unsaturated six-membered ring resembling benzene, but consisting of five carbons and one nitrogen, as opposed to six carbons (see Section 16.5). The compound exhibits an extremely objectionable, nauseating odor. For most substituted benzene compounds there is an analogous compound in the pyridine series. Pyridine and its derivatives are used as solvents and raw materials in the manufacture of chemicals, explosives, paints, disinfectants, herbicides, insecticides, antihistamines, and vitamins. Use of pyridine as a therapeutic agent in epilepsy treatment has been reported.

The alkyl pyridine derivatives, as well as the parent molecule, are well absorbed from the gastrointestinal tract, peritoneal cavity, lungs, and from intact skin. The metabolic fate is not completely known, but hydroxylation, *N*-methylation, oxidation, and conjugation reactions have been identified. Reported elimination is rapid, limiting the potential for accumulation in tissues. Despite its wide industrial application and limited medicinal use, reports of human poisoning are uncommon.

Pyridine principally exerts its adverse effects on the central nervous system, gastrointestinal tract, liver, and kidneys. Local skin irritation also has been reported and pyridine has been reported to be a photosensitizer. Inhalation exposure to pyridine at 125 ppm, 4 h per day for 2 weeks caused anorexia, nausea, vomiting, gastric distress, headache, fatigue, faintness, and depression. Hepatotoxicity, kidney damage, and death were reported in cases where the dose was in excess of 2 mL/day for 2 months (approximately 0.029 mg/kg·day). Inhalation of vapor irritates the mucus membranes.

16.18 SULFUR-SUBSTITUTED SOLVENTS

Dimethyl sulfoxide (DMSO) (see Figure 16.36) is an industrial solvent that also has a wide applicability in the pharmaceutical area to solubilize water-insoluble medication. It has the ability to carry solutes into the skin's stratum corneum from which they are slowly released into the blood and lymph system.

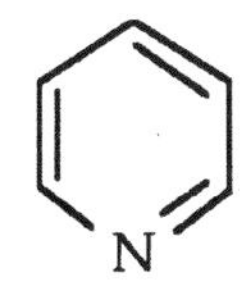

Figure 16.35 Pyridine.

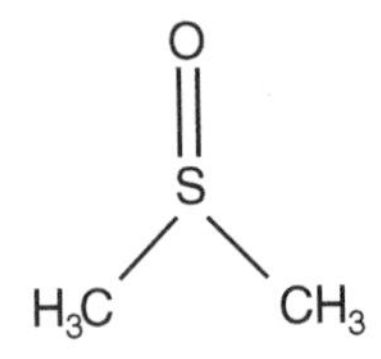

Figure 16.36 Dimethyl sulfoxide.

Signs and symptoms have pointed to the fact that dimethyl sulfoxide has systemic effects on the hepatic and renal system. It can inhibit enzyme reactions, but does not affect thyroid function. It has been suggested that inhaled or otherwise administered DMSO is metabolically reduced to the dimethyl sulfide and subsequently respired. DMSO in rats lowered the body temperature and enhanced the taurine excretion and the toxic effects of aromatic hydrocarbons. It was found to influence the preservation of leukemia cells and proved to be a potent inducer of erythroid differentiation in cultured erythroleukemic cells. DMSO has teratogenic potential as well as the ability to increase activity and tumor-inducing effects of materials with carcinogenic potential.

Carbon disulfide (see Figure 16.37) is a highly flammable, highly toxic solvent that historically has been used in production of carbon tetrachloride, as well as the manufacture of rayon, cellulose fibers, rubber vulcanizers, and pesticides. Its previous use as a grain fumigant has been discontinued. Although the pure substance is odorless, impurities may impart an objectionable sulfurous odor.

The principal toxicological effect at high air concentrations is narcosis, perhaps accompanied by headache, visual disturbances, respiratory disturbances, and gastrointestinal effects. Ingestion of as little as 15 mL may be fatal. Aside from the general neurological changes associated with acute exposures, the greatest toxicological concern regarding carbon disulfide relates to the demonstrated ability to induce peripheral polyneuropathy and psychoses in some chronically exposed individuals. The latter effects reportedly resolve following exposure cessation, but the characteristic neuropathies may persist. These include reflex decrements in the extremities, glove/stocking sensory loss, and decreased nerve conduction velocity.

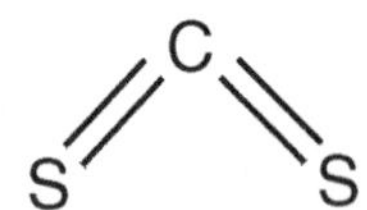

Figure 16.37 Carbon disulfide.

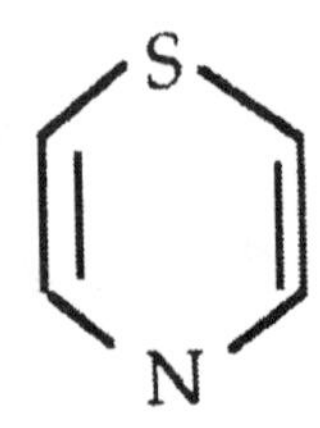

Figure 16.38 Thiazoles basic structure.

Carbon disulfide is not regulated as a carcinogen in an industrial or environmental context. Industrially important exposure to carbon disulfide may be evaluated by measurements of 2-thiothiazolidine-4-carboxylic acid, expressed in conjunction with urinary creatinine.

The thiazoles (see Figure 16.38), in particular benzothiazole and mercaptobenzothiazole, are used in the rubber vulcanizing process and as fungicides. Benzothiazole occurs in such a small quantity that it is not considered to be a major health threat. It is, however, considered moderately toxic. Mercaptobenzothiazole, when heated, may react with oxidizing material and emit toxic decomposition products. The main consequence of exposure is allergic contact dermatitis. This compound is considered to be a potent allergen. Some subcutaneous tests on mice showed a possible carcinogenic potential as well, although it is not regulated as such.

16.19 SUMMARY

As discussed in this chapter, the common toxicological effects attributed to individual solvents and related materials include

- CNS depression and other neurotoxic effects
- Respiratory irritation
- Dermal effects, including irritation
- Nephrotoxicity
- Carcinogenicity

The emphasis of this chapter reflects chemical properties, behavior, and effects, citing examples as appropriate. The chapter summarizes the range of chronic toxic effects that may be expected from selected chemical classes, which should serve as a good introduction to other sources of more detailed information. The following section provides valuable supplementary information sources related to solvents.

REFERENCES AND SUGGESTED READING

ACGIH (American Conference of Governmental Industrial Hygienists), *Documentation of Threshold Limit Values and Biological Exposure Indices*, 5th ed., 1986–1999.

ATSDR (Agency for Toxic Substances and Disease Registry), *Toxicological Profiles*, Atlanta, 1988–1999.

Axelson, O., and C. Hogstedt, "The health effects of solvents," in *Occupational Medicine*, C. Zenz et al., eds., Mosby, St. Louis, MO, 1994.

Baselt, R. C., and R. H. Cravey, *Disposition of Toxic Drugs and Chemicals in Man*, Chemical Toxicology Institute. Foster City, CA, 1995.

Browning, E., *Toxicology and Metabolism of Industrial Solvents*, Elsevier, Amsterdam, 1965.

Calabrese, E. J., and E. M. Kenyon, *Air Toxics and Risk Assessment*, Lewis Publishers, Chelsea, MI, 1991.

Commission of the European Communities (CEC), *Solvents in Common Use: Health Risks to Workers*, Royal Society of Chemistry. London, 1988.

Clayton, G. D., and F. E. Clayton, eds., *Patty's Industrial Hygiene and Toxicology: Volume II, Toxicology*, 4th ed., Wiley, New York, 1994.

Ekberg, K., M. Hane, and T. Berggren, "Psychologic effects of exposure to solvents and other n eurotoxic agents in the work environment," in *Occupational Medicine*. C. Zenz et al., eds., Mosby, St. Louis, MO, 1994.

Ellenhorn, M. J., *Medical Toxicology: Diagnosis and Treatment of Human Poisoning*, 2nd ed., Williams & Wilkins, Baltimore, 1997.

Gerr, F., and R. Letz, "Organic solvents," in *Environmental and Occupational Medicine*, 3rd ed., W. N. Rom, ed., Lippincott-Raven, New York, 1992.

Hardman, J. G., and L. E. Limbird, eds., *The Pharmacological Basis of Therapeutics*, 9th ed., McGraw-Hill, New York, 1996.

HSDB (*Hazardous Substance Data Bank*), on-line computer database. National Library of Medicine, 1999.

IARC (International Agency for Research on Cancer), *Monographs 1972–present*, World Health Organization, Lyon, France, 1972–1999.

James, R. C., "Organic solvents," in *Industrial Toxicology*, P. Williams and J. Burson, Van Nostrand-Reinhold, New York, 1985.

NIOSH (National Institute of Occupational Safety and Health), *Pocket Health Guide to Chemical Hazards*, U.S. Department of Health and Human Services, Washington, DC, 1997.

Parmegianni, L., *Encyclopedia of Occupational Health and Safety*, 3rd ed., International Labour Office. Geneva, Switzerland, 1983.

Plog, B. A., J. Niland, and P. Quinlan, eds., *Fundamentals of Industrial Hygiene*. National Safety Council. Itasca, IL, 1996.

Sax, N. I., and R. J. Lewis, *Dangerous Properties of Industrial Materials*, 7th ed., Van Nostrand-Reinhold, New York, 1989.

Sittig, *Handbook of Toxic and Hazardous Chemicals and Carcinogens*, 2nd ed., 1985, p. 590.

Snyder, R., and L. S. Andrews, "Toxic effects of solvents and vapors," in *Casarett and Doull's Toxicology: The Basic Science of Poisons*, 5th ed., C. D. Klaassen, ed., McGraw-Hill, New York, 1996.

17 Properties and Effects of Natural Toxins and Venoms

WILLIAM R. KEM

This chapter will discuss

- Differences between poisons, toxins, and venoms
- Major sites and mechanisms of toxin action
- Important microbial, plant, and animal toxins
- Animal venoms and their active constituents
- Plants and animals causing contact dermatitis
- Strategies for treating intoxications and envenomations

We live in a world containing a wide variety of organisms—animal, plant, and microbial—possessing substances that are potentially harmful to our health. Fortunately for most urban inhabitants, the chances of developing morbid or fatal reactions to naturally occurring toxins are relatively small. Still, even in an urban setting we are vulnerable to at least some natural toxins, such as those occurring in foods, our ornamental plants, or our places of habitation. Furthermore, as human populations expand into rural regions, they inevitably become more vulnerable to poisonous creatures.

In this chapter we shall discuss some of the most common natural toxins, their mechanisms of action, and some modern principles of their treatment.

17.1 POISONS, TOXINS, AND VENOMS

First we need to understand what is meant by the terms: poison, toxin, venom. A poison is any substance or mixture of substances which can be life-threatening. Poisonous organisms either secrete or contain one or more chemicals (toxins) that seriously interfere with normal physiological functions. A toxin is a single substance with definable molecular properties that interferes with normal function. Most toxins are exogenous substances made by an organism to adversely affect another organism. However, even humans produce endogenous toxins (complement, defensins) to resist attack by foreign organisms such as bacteria and viruses. Venoms are secretions containing a mixture of biologically active substances, including enzymes, toxins, neurotransmitters, and other compounds. They are generally used both for prey capture and as a chemical defense against other predators. Some toxins are used solely as chemical defenses against predators, and in these cases, the toxins are often released from relatively simple integumentary glands, and may even be stored within visceral organs. One example of such a toxin is pufferfish toxin, which will be discussed below.

Principles of Toxicology: Environmental and Industrial Applications, Second Edition, Edited by Phillip L. Williams, Robert C. James, and Stephen M. Roberts.
ISBN 0-471-29321-0 © 2000 John Wiley & Sons, Inc.

17.2 MOLECULAR AND FUNCTIONAL DIVERSITY OF NATURAL TOXINS AND VENOMS

Every major class of molecules synthesized by living organisms—protein, lipid, carbohydrate, nucleoside, alkaloid—has been exploited by some species to produce a toxin. Some of the most important natural toxins that will be discussed in this chapter are listed in Table 17.1.

The most potent toxins are usually proteins, probably because their larger molecular surfaces allow more bonding contact with the receptors on which they act. Besides this high potency, there is another possible reason that many toxins are peptides or proteins. Biosynthesis of protein toxins does not require unusual substrates or catalysts, just a messenger RNA template that specifies the amino acid sequence of the toxin; the rest of the required biosynthetic machinery (ribosomes, messenger RNA, transfer RNA, nucleotides, RNA polymerase, etc.) is already present. It is not yet clear how the protein toxins originated during the course of evolution. However, some snake polypeptide toxins have amino acid sequences which are very similar to endogenous polypeptides that act as proteolytic enzyme inhibitors, and it is suspected that these toxins evolved from duplicated (extra) genes for these enzyme inhibitors.

It is relatively common for chemically similar toxins to be manufactured by creatures that are taxonomically unrelated. Thus, certain echinoderms (starfish and sea cucumbers) synthesize sterol glycosides, which are chemically and pharmacologically very similar to the saponins found in some plants. Anabaseine, an alkaloid toxin occurring in certain marine worms, is almost the same as the tobacco alkaloid anabasine. This evolutionary convergence at the molecular level is perhaps to be expected because many toxins are synthesized by enzymes that serve as catalysts for metabolic pathways which are of general occurrence in living organisms. Plants have long been known to produce an amazing variety of "secondary" metabolism products containing nitrogen, usually referred to as alkaloids. Many of these metabolites serve as a defense against herbivores. Animals and protozoans also produce such compounds, and some of these will be discussed below.

TABLE 17.1 Mouse Lethality of Skeletal Natural Toxins (modified from Middlebrook, 1989)

Toxin	Molecular Weight	MLD[a] μg/kg Mouse	Relative Number of Molecules Causing Death[b]
Botulinum	150,000	0.0003	1
Tetanus	150,000	0.001	4
Diphtheria	60,000	0.03	3×10^2
Ricin	60,000	3	3×10^4
α-Latrotoxin	130,000	10	5×10^4
Pseudomonas exotoxin A	60,000	5	5×10^4
β-Bungarotoxin	20,000	14	4×10^5
Conotoxin M	1,500	5	2×10^6
Cholera	84,000	250	2×10^6
Batrachotoxin	538	2	3×10^6
α-Bungarotoxin	8,500	300	3×10^6
Tetrodotoxin	319	8	2×10^7
Saxitoxin	354	9	2×10^7
Tubocurarine	334	500	1×10^9
Diisopropylfluorophosphate	184	1,000	4×10^9
Sodium cyanide	49	10,000	1×10^{11}

[a]MLD, minimum lethal dose.

[b]Relative to botulinum toxin.

Toxic organisms store their toxic substances in specialized organs (plant vacuoles, animal venom glands) for several reasons. First, the toxic organism otherwise could be exposed to its own poison. By sequestering the toxin within a membranous sac that is impermeable to the toxin, the other tissues of the organism can be protected from exposure to the substance or collection of substances (venom). Second, it is usually advantageous to store the chemical in a concentrated form, which can be efficiently injected into the victim, with the assistance of a barb or fang. Finally, the venom apparatus must be connected with an effector system, which senses the presence of the intended victim. In most venomous animals, the venom is released in response to instructions from the central nervous system, but in some venomous invertebrates like jellyfishes, the entire sensory and motor apparatus for activating venom release is built into each venom-emitting cell.

17.3 NATURAL ROLES OF TOXINS AND VENOMS

The functional value of a venom for the procurement of prey or as a defence against predators in most cases is rather obvious. A venomous predator can immobilize relatively large prey animals, and consume them at a more leisurely pace. A suitably toxic, but not venomous, plant or animal similarly avoids consumption. Even if the toxicity of a single individual is not sufficient to protect its own life, a herbivore or predator will be forced to eat fewer individuals than otherwise, in order to avoid lethal intoxication. In this manner, survival of the unpalatable species will be enhanced.

In toxic prokaryotic organisms, the biological function of a toxin may not be at all obvious. Examples that come readily to mind are the dinoflagellate or red tide organisms that occasionally reach such high population densities in aquatic communities that toxin concentrations in seawater are sufficient to cause massive fish kills, for instance. It has been suggested that these toxins usually serve as regulators of cell growth or metabolism and only rarely act as toxins, but these postulated endogenous functions are yet to be found.

17.4 MAJOR SITES AND MECHANISMS OF TOXIC ACTION

Neurotoxic Actions

Since the nervous system functions primarily as a master communication network that quickly coordinates the operation of practically all cells, tissues, and organs of the body, it is a prime target for toxins, which are intended to rapidly alter the functioning of the target organism. Rapid communication within the nervous system relies on the generation of two types of electrical signal. Initially, small processes (dendrites) emanating from the neuron's cell body respond to neurotransmitters released from adjacent neurons by generating a relatively slow depolarizing junctional potential; this elicits an action potential, which then rapidly travels to the end of the axon where neurotransmitter is again released to activate or inhibit some effector cell (Figure 17.1*a*).

A wide variety of toxins act on electrically active tissues—muscle and neuronal cells—that use neurotransmitter- and voltage-gated ion channels for generating their electrical signals. The peripheral nervous and muscular systems are particularly vulnerable cellular targets for rapidly acting toxins, since no blood–brain barrier protects them from exposure to toxins.

Around 1920 a physiologist named Langley, by locally applying nicotine at only places along the length of the muscle, first showed that the tobacco alkaloid nicotine acted at a few discrete sites, which he called *receptors*, along the length of a muscle cell. Little was known about the molecular properties of these nicotinic receptors until 1971, when they were purified from a particularly rich source, electric fishes. Each muscle-like nicotinic receptor is a pentameric complex containing five polypeptide subunits, which are held together only by noncovalent bonds (Figure 17.1*b*). Two of the five subunits are the same. These so-called alpha subunits actually contain the acetylcholine (ACh) binding sites. Substances such as ACh and nicotine that activate the receptor are called *agonists*, whereas substances

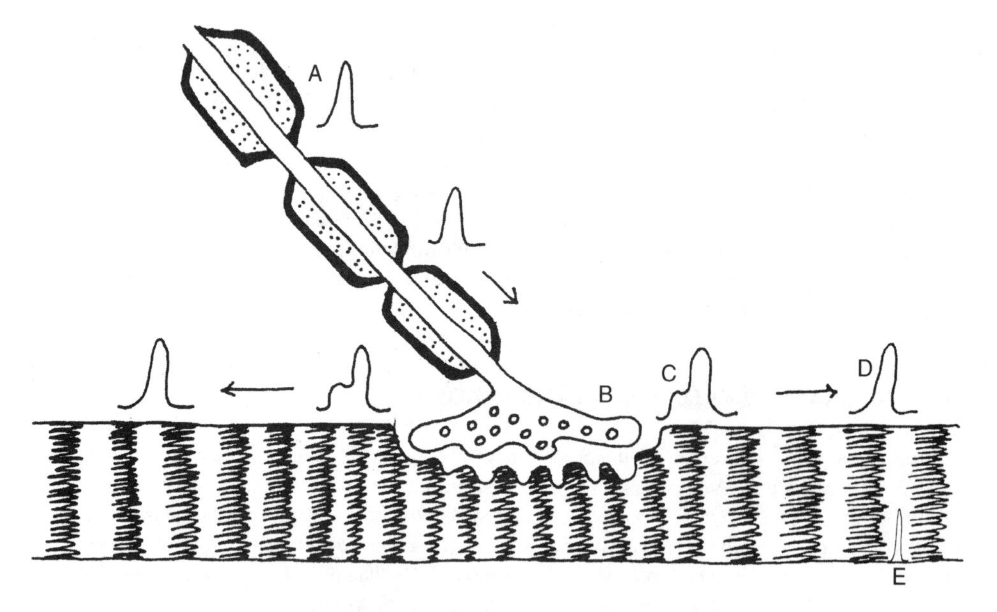

Figure 17.1a Common sites of neurotoxin action at the mammalian neuromuscular junction. Toxins can affect the (A) nerve action potential, (B) release of neurotransmitter from the nerve terminal, (C) the membrane depolarizing response of neurotransmitter receptors to neurotransmitter, (D) muscle action potential, and (E) coupling of muscle membrane depolarization with calcium release from the sarcoplasmic reticulum, mediated by transverse invaginations of the cell membrane called t-tubules.

Figure 17.1b A molecular view of the nicotinic acetylcholine receptor found in the skeletal muscle membrane. The receptor is a pentamer consisting of four different polypeptide subunits; the two α subunits must be occupied by an agonist like ACh to open the ion channel in the center. Three potential sites of toxin action on the nicotinic receptor are shown: circular molecule (A), toxin binding to the neurotransmitter (acetylcholine) recognition site; triangular molecule (B), toxin binding to the edge of the receptor protein, interfering with its interaction with the lipid bilayer; square molecule (C), toxin entering and directly blocking the ion channel.

(including the South American Indian arrow poison tubocurarine and certain snake venom toxins) that reversibly bind to the same site without activating it are competitive *antagonists*. Besides the ACh recognition site, there are other places on this large membrane protein complex at which toxins can act. For instance, many small alkaloid toxins can enter the ion channel and physically plug it! Other toxins probably bind at the interface between the polypeptide subunits and the adjacent lipid bilayer (Figure 17.1*b*).

The muscle cell membrane, continuous with the postsynaptic membrane of the muscle cell, is electrically excitable and must be able to generate an action potential in response to the ACh-induced depolarization. An action potential involves the sequential opening and closing of two different ion channels in response to membrane depolarization. First, the sodium-selective channel opens; sodium ions flow into the cell, reversing its electrical potential so that the inside of the cell momentarily becomes electropositive. This causes the adjacent membrane to be depolarized, which, in turn, activates the opening of more sodium channels. In this manner, a wave of electrical change rapidly passes along the length of the muscle fiber. This depolarizing wave is quickly followed by a repolarizing wave due to the opening of potassium ion selective channels, which allows potassium to flow out of the cell, down its concentration gradient. It is counterproductive for the sodium channels to remain open while potassium channels are opening, so an additional process called *sodium channel inactivation* takes place during the opening of the potassium channels. In some muscle cells, a calcium-selective channel either substitutes for the sodium channel (in smooth muscle) or supplements its ability to depolarize the cell (in cardiac muscle).

In smooth muscles and many neurons, voltage-gated calcium channels substitute for sodium channels in causing at least part of the initial cationic influx, which generates action potentials. At nerve terminals, calcium ions flowing through these calcium channels also mediate the exocytotic release of neurotransmitter. There are several extremely potent toxins that activate or inhibit these calcium ion-passing channels, which we will discuss later. There are even intracellular calcium channels within muscle cell sarcoplasmic reticulum membranes, which must be able to quickly release calcium ions for muscle contraction; these calcium channels can be blocked by a plant toxin called *ryanodine*.

Every ion channel seems to be a potential target for some natural toxin, which usually acts at a much lower concentration than do drugs acting on the same site. For instance, one of the most potent local anesthetics, tetracaine, blocks sodium channels of nerve at concentrations above 10^{-6} M, whereas the pufferfish toxin tetrodotoxin achieves the same blockade at a thousand-fold lower concentration! There are at least seven different sites on sodium channels where toxins act; some of these are listed in Table 17.2.

Cardiovascular Toxins

The cardiovascular system is also quite vulnerable to many natural toxins that act on ion channels in cardiac or smooth muscles or on autonomic nerve terminals. Many lethal actions of venoms probably are due to rapid action on these excitable cells. Once the victim is envenomated, the active constituents spread locally according to their molecular size and other chemical properties. Their entry into the systemic circulation will be greatly enhanced if they rapidly spread into tissues surrounding the bite; this can be enhanced by a venom enzyme, hyaluronidase, which breaks down the hyaluronic acid in connective tissue. Some venoms also contain hemorrhage-inducing, anticoagulant, and hemolytic proteins, which together can cause much loss of blood volume, tissue oedema, and cytolysis. Thus the cardiovascular system can be affected in many different ways by venoms and their toxic constituents.

Toxins Affecting the Liver and Kidneys

Two other organs that are especially vulnerable to toxins are the liver and the kidney. The hepatic portal venous system first delivers substances absorbed from the gastrointestinal tract to the liver. This organ contains many catabolic enzymes and is thus capable of metabolizing practically any type of exogenous compound, usually to a less active or toxic form. However, metabolites may be even more toxic than their molecular precursors. Certain cyclic peptide toxins from poisonous mushrooms (amatoxins) and freshwater algae (microcystins) are relatively selective hepatotoxins because they are able to gain easy entry into the hepatic cells by means of a special solute transport system normally used for reabsorption of bile salts. The kidney is relatively susceptible to certain toxins, particularly those that enter the renal tubule by glomerular filtration but are not readily reabsorbed. This causes them to be concentrated in the nephron and urine, enhancing their ability to damage renal cells.

Cytotoxins

This is probably the most common group of toxins. Cytotoxins generally affect life-requiring processes such as protein synthesis, DNA replication, RNA synthesis, oxidative phosphorylation (metabolism), or cell electrolyte balance. Cytolysins are cytotoxins that create an osmotic imbalance, causing cell swelling and subsequently cell lysis. The most potent cytolysins create large holes in the cell membrane permitting the egress of many proteins as well as low-molecular-weight substances. Others act like detergents, disrupting the lipid bilayer organization of cell membranes. While most cytotoxins are able to attack a variety of cells, at sublytic concentrations they are often cardiotoxic through their ability to directly contract muscle cells by depolarizing the resting membranes of excitable cells.

Toxins Affecting Second Messengers

Signal Transduction Most slow receptor-mediated responses are indirectly coupled to an effector, like an ion channel, through a "second messenger" signal, usually a cyclic nucleotide or some phosphoinositide, which must find another effector molecule and modify its activity. Cyclic AMP

TABLE 17.2 Toxins Affecting Voltage-Gated Sodium Channels

		Effect	
Site Toxin (Source)	Sodium Channel	Action Potential	Systemic
Guanidinium toxins Tetrodotoxin (pufferfish) Saxitoxin (shellfish)	Pore block	Decreased amplitude	Flaccid muscular paralysis
Steroidal toxins Batrachotoxin (frog) Veratridine (false hellebore) Grayanotoxin (rhododendron)	Enhanced activation, inhibited inactivation	Prolonged AP Spontaneous AP	Hyperexcitability Convulsions Cardiac arrhythmias
Peptide toxins α-Scorpion Sea Anemone	Delayed inactivation	Prolonged AP	Hyperexcitability Convulsions Cardiac arrhythmias
Peptide toxins β-Scorpions	Enhanced activation	Repetitive APs	Hyperexcitability Convulsions
Polyether toxins Brevetoxin (dinoflagellate) Cignatoxin (fish)	Enhanced activation	Repetitive ASPs	Hyperexcitability Convulsions Diarrhea
Alkaloid and protein toxins Pyrethrum (chrysanthemum) Goniopora (coral)	Enhanced activation, delayed inactivation	Prolonged AP Repetitive APs	Hyperexcitability Convulsions

stimulates various phosphoryl kinase enzymes, which catalyze the phosphorylation of ion channels and other signaling systems, thereby modulating their function. Several toxins have been found to specifically alter the cAMP-generating system. Some act indirectly by affecting guanosine nucleotide-binding (so-called G) proteins, which modulate adenylate cyclase. For instance, cholera toxin stimulates G_s (the stimulatory G protein subunit) formation and therefore enhances cAMP synthesis, while pertussis toxin inhibits binding of the inhibitory G protein subunit G_i to the cyclase and thereby also stimulates cAMP synthesis (Figure 17.2). The sponge toxin okadaic acid acts in an entirely different fashion, inhibiting certain phophatases that normally reverse the cAMP-catalyzed phosphorylation, and this leads to an enhancement in cAMP concentration.

Inflammatory and Carcinogenic Toxins

These types of toxins are usually meant to discourage consumption or even contact with the toxic organism. Many sedentary organisms like plants and some marine animals synthesize inflammatory substances. These may be similar to endogenous chemical mediators, such as histamine, prostaglandins, or phospholipids, or may liberate the endogenous mediators from basophils and other cells mediating inflammatory processes.

Some of the most potent carcinogens are natural substances, like the ochratoxins. In many cases their mechanisms of action are not yet known. Ames has presented the provocative hypothesis that the dangers of exposure to some industrial carcinogens may not be any greater than the risks associated with daily consumption of small amounts of natural carcinogens occurring in some food plants.

17.5 TOXINS IN UNICELLULAR ORGANISMS

Bacterial Toxins

There are so many bacterial toxins that we are here forced to consider only a few of the most common and interesting ones. In fact, the most potent natural toxins are bacterial protein neurotoxins (Table 17.1).

Botulinum poisoning is primarily a foodborne disease, which can develop when food is improperly canned, allowing anaerobic *Clostridium botulinum* bacterial spores to survive and multiply. There are several strains of this anaerobe that synthesize related toxins. All botulinum toxins act by inhibiting neurotransmitter release at the skeletal muscle neuromuscular junction (Figure 17.1*a*). This peripheral action is dominant with botulinum toxins and leads to flaccid paralysis and eventually death if unabated. Treatment is difficult. After binding, toxin is internalized at the motor nerve terminal, and then acts internally. Botulinum antiserum can neutralize toxin that has not yet been internalized. Neuromuscular transmission may be enhanced by treating the patient with an acetylcholinesterase inhibitor such as neostigmine. Artificial respiration may be necessary until the patient regains new transmitter release sites.

Tetanus poisoning is due to another anaerobe, *Clostridium tetani*. Again, several strains form related toxins. Although they all act to inhibit neurotransmitter release in a manner superficially similar to botulinum toxin, the tetanus toxins in mammals predominantly inhibit the release of an inhibitory neurotransmitter, glycine, within the central nervous system. This inhibition of an inhibitory influence (called "disinhibition") on central motor neurons permits full expression of the excitatory synaptic input to these neurons, causing peripheral excitation of all skeletal muscles. Since extensor muscles are usually most powerful, victims may become immobilized in a contorted contraction that is life-threatening. "Lockjaw" is only symptom of this condition. Most of us are vaccinated with tetanus toxoid as children, so the likelihood of developing tetanus poisoning is greatly reduced; a "booster" vaccination should also be taken about every 10 years, particularly for various outdoors persons who are more likely to be exposed to this bacterium (gardeners, farmers, trash handlers, etc.).

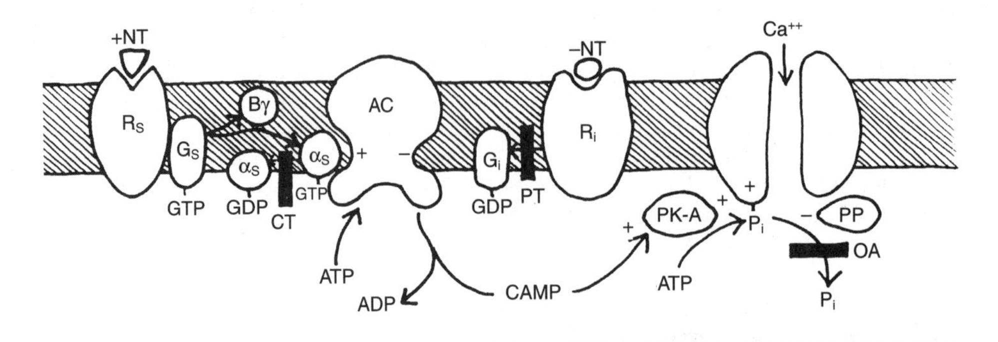

Figure 17.2 Control of adenylate cyclase (AC) by toxins affecting G proteins. Initially, stimulatory (+NT) or inhibitory neurotransmitters interact with their respective receptor proteins, which are indirectly coupled to AC through different G proteins. Cholera toxin (CT) enhances cyclic AMP production by enhancing the interaction of stimulatory Gs protein with the adenylate cyclase, while pertussis toxin (PT) enhances cyclase activity by inhibiting the interaction of the inhibitory GI protein with another site on the cyclase. A sponge toxin, okadaic acid (OA), enhances cyclic AMP action by an entirely different mechanism, namely, by inhibiting certain phosphatase

Cytotoxic proteins produced by infectious bacteria frequently contribute to the fever, vasodilation, and tissue damage associated with infection. Staphylococcal alpha toxin is one of the best understood representatives of this group. Several molecules of this toxin aggregate together on the target cell membrane and form a large pore allowing release of various cellular constituents, even large proteins. Cytolytic toxins are rarely lethal, although they contribute to the symptoms of bacterial infection.

Cholera toxin (Fig. 17.2) activates G_S protein coupling to adenylate cyclase. This stimulates intestinal ion and water secretion and results in dehydration and death if not properly treated.

Other bacterial cytotoxins act in an entirely different manner by inhibiting protein synthesis in the target cell. Diphtheria and *Pseudomonas* toxins are good examples. It has been calculated that a single molecule of one of these toxins is sufficient to inhibit enough protein synthesis that the cell cannot replenish the proteins that are also being continuously broken down. Fortunately, however, the process by which the cytotoxin is internalized and gains entry into the cytoplasmic compartment is not so efficient; it actually takes thousands of cell membrane-bound toxin molecules to result in one molecule lethally reaching its ribosomal destination.

Many bacteria are practically ubiquitous because their spores are widely distributed by air and water movements around the earth. However, some species have a much more localized distribution. One example is the marine bacterium *Vibrio vulnificus*, which can cause life-threatening infections in persons who consume raw marine shellfish or swim in the sea with skin abrasions that are vulnerable to infection. In a recent study many patients developed necrotic tissue or liver disease; about 20 percent died. A 56 kilodalton cytolytic protein called vibriolysin is thought to mediate the tissue damage.

Although the alkaloidal neurotoxin tetrodotoxin was originally thought to be synthesized only by pufferfish and certain newts (amphibians), in the past decade it has been found in a wide variety of marine animals including worms, crabs, and an Australian octopus. Since tetrodotoxin was recently found to be synthesized by several species of marine bacteria, it is likely that the animals obtain the toxin from microbial symbionts.

A Toxin Implicated in Amebic Dysentery

Infection by a freshwater amoeba, *Entamoebae histolytica*, can cause life-threatening meningitis in addition to inflammation of the colon and intestinal abscesses. Tissue damage only occurs on direct physical contact of the protozoan with its target cell. Then a pore-forming peptide called *amobapore*, which causes osmotic swelling and lysis of the target cell, is secreted next to the target cell membrane.

Dinoflagellate (Shellfish) Toxins

Several toxic marine dinoflagellate species, under particularly favorable conditions for population growth, cause toxic algal blooms. These "red tides," so named because water containing high concentrations of these dinoflagellates sometimes is reddish-colored, can also cause massive mortality of fish and other marine animals. Algal blooms occur more frequently along coasts that are polluted by agricultural and human waste. Filter-feeding molluscs are able to concentrate many of these toxins without being intoxicated. Different kinds of symptoms are produced by eating poisonous clams, mussels, and other organisms, their nature depends on the toxins involved. Intoxications include paralytic shellfish poisoning (PSP), neurotoxic shellfish poisoning (NSP), and diarrheic shellfish poisoning (DSP), which will be discussed below.

One of the most common red tides in the northern hemisphere is due to a dinoflagellate called *Gonyaulax catenella*, which secretes a family of over 30 related toxins called *saxitoxins*, which block sodium channels. The saxitoxins are classified as paralytic shellfish poisons (PSPs) because their deleterious actions on the nervous system are reversible, if death is avoided. They only become dangerous to man when shellfish containing high concentrations of these toxins are consumed. Shellfish are relatively resistant to the saxitoxins, and their tissues retain the toxins as they filter feed on the poisonous dinoflagellates. It has been shown that these toxins, like the pufferfish toxin tetrodotoxin, interact with a pore-forming segment on the sodium channel protein alpha-subunit; this guanidinium toxin binding site has been previously called "site 1" on the sodium channel (Table 17.2). The mammalian heart sodium channel is pharmacologically different from nerve and skeletal muscle sodium channels by being about 500 times less sensitive to these toxins, which is probably fortunate for us!

In 1987, an usual form of neurotoxic shellfish poisoning (NSP) occurred off the coast of Nova Scotia. The victims experienced amnesia, a loss of ability to recall information. The toxin was found to be domoic acid, an active analog of the excitatory neurotransmitter glutamic acid. Persistent activation of glutamate ion channels by this toxin causes neuron degeneration, probably by causing an excessive influx of calcium ions. DSP is caused by okadaic acid and by ciguatoxin.

Brevetoxins, produced by the dinoflagellate *Gymnodinium breve*, open, rather than block, sodium channels (Table 17.2). This organism is found in warmer waters such as are found along the Florida coastline, and is responsible for massive fish kills every few years. Brevetoxins, like ciguatoxin, are complicated polycyclic ether molecules that cause the sodium channel to open even under resting conditions. This causes nerve and muscle cells to spontaneously generate action potentials in the absence of stimulation, which, of course, is potentially lethal. Since fish are killed by relatively small amounts of these toxins, humans are not apt to be poisoned by eating exposed fish. However, during a bloom some of the *Gymnodinium* become airborne in ocean spray, and people can experience respiratory distress after inhaling these toxic droplets.

17.6 TOXINS OF HIGHER PLANTS

Mushrooms and Other Fungi

Fewer than 1 percent of the mushroom species are poisonous to humans, but these can be extremely dangerous. Interest in mushroom hunting is increasing, so it is expected that intoxications will also increase. Mushrooms of the genus *Amanita* (Figure 17.3) are the most dangerous. These contain about equal amounts of two relatively small (seven amino acids) cyclic peptide toxins called amatoxins and phallotoxins. Unfortunately these cyclic peptides are quite stable at high temperatures, so they survive cooking. Consumption of a single *Amanita phalloides* mushroom may be lethal. The amatoxins are

Figure 17.3 The death cap mushroom, *Amanita phalloides*. This is the most poisonous mushroom in the world, and occurs in Asia, Europe, and North America. Mushrooms belonging to this genus account for >95 percent of reported human fatalities to mushrooms.

about 20 times more toxic than the phallotoxins, so they are the toxins that must be reckoned with the most. They are particularly hepatotoxic because of their ability to be taken up through the bile acid transport mechanism. Once they enter the hepatic parenchymal cell, they inhibit the key transcriptional enzyme RNA polymerase within the nucleus, thus shutting down the ability of the cell to replace cellular proteins, which are continually being broken down. This results in hepatic necrosis and death in 10–30 percent of intoxicated persons. A major problem in diagnosing and treating amanita poisoning is that the characteristic symptoms due to amatoxin appear only about 15 h after ingestion, regardless of the dose. This represents the minimum time for uptake and enzyme inhibition by the toxin, and for the hepatic protein depletion to begin affecting hepatic function. About 6–10 h after ingestion one experiences gastric distress and diarrhea caused by the phallotoxins. They bind to the actin filaments in the inner surface of the cell membrane, preventing them from dissociating into monomeric actin, which is required for normal cell functioning.

Amatoxins can be detected in the blood and urine with various techniques, in order to verify the cause of the poisoning. Ingestion of oral activated charcoal is effective in absorbing much of the toxins if it is done within 4 hours after ingestion. Of course, the victim may not yet have experienced many symptoms at that critical time.

Other species of mushrooms produce alkaloidal toxins, which are much less life-threatening. Muscarine, isolated over a century ago and used in classifying cholinergic receptors, is one example. Its actions are quite predictable as well as swift. Fortunately, specific antidotes such as the muscarinic antagonist atropine exist. Other mushrooms produce biogenic amines such as bufotenin (originally isolated from venom glands of the toad *Bufo*) and psilocybin. These are hallucinogenic compounds which primarily stimulate certain serotonin receptors in the brain. Finally, several mushrooms synthesize ibotenic acid, a potentially neurotoxic glutamate receptor agonist similar to domoic acid.

In addition to the mushrooms, there are other toxic fungi. Ergot is a fungus that grows upon certain grains in damp climates. This fungus produces a variety of biogenic amines which act as agonists on alpha-type adrenergic receptors including ergotamine, which is used therapeutically to treat migraine headaches. Methysergide, a serotonin antagonist, is probably the major hallucinogenic component of ergot. Some molds have been found to produce carcinogenic substances called *aflatoxins* and *ochratoxins*; proper storage of vegetable crops susceptible to these molds eliminates conditions favorable for their growth.

Flowering Plants

Cardiac Glycosides and Saponins Cardiac glycosides are animal as well as plant products. The traditional source of these compounds for medicinal use in the West has been the foxglove, a beautiful flowering plant (Figure 17.4) now extensively cultivated in many countries. The major glycosides of the foxglove are called digitoxin and digoxin. In the Orient, toad venom glands were used as a major source of very similar medicinal compounds (bufotoxins). The primary therapeutic use of digitalis glycosides is the treatment of congestive heart failure, a condition characterized by a loss of myocardial contractility. For various reasons (including long-term hypertension, atherosclerosis, kidney failure, etc.), the heart is unable to pump the blood sufficiently to avoid its pooling in the lungs and extremities.

Over 300 years ago, Withering found that the leaf of the foxglove was very effective in treating this condition, then known as *dropsy*. Unfortunately, digitalis glycosides are also amongst the most toxic of drugs, frequently causing cardiac arrhythmias at concentrations required to significantly enhance the cardiac output. The site of their action is the sodium, potassium pump (also known as the Na,K-activated Mg-ATPase) in the cell membrane. This active transport system is responsible for maintaining the high potassium, low-sodium intracellular environment of all cells. However, in the heart it appears that blockade of a fraction of the pumping sites with digitalis allows the intracellular sodium concentration to transiently rise above normal during each myocardial action potential, and this elevated sodium then is exchanged with calcium from outside the cell by a membrane carrier called the *sodium–calcium exchanger*. This causes elevation in the intracellular calcium during the heart beat, which stimulates the actomyosin system to contract more forcefully. It is quite remarkable that these glycosides can be used as inotropic drugs at all, considering that all cells possess sodium, potassium pumps which are inhibited by digitalis.

Other plants (Table 17.3) that produce dangerous quantities of digitalis compounds are the oleander bush (*Nerium*), which is an extremely common ornamental shrub in the southeastern United States, the lily-of-the-valley (*Convallaria*) ornamental flower, and a wildflower, the butterfly weed (*Asclepias*). A single oleander leaf contains enough cardiac glycoside to be lethal to an adult human. The danger with foxglove is that during the nonflowering season its leaves are confused with those of the *common comfrey* plant, whose leaves are popularly used in the preparation of herbal teas. This has led to several deaths due to inadvertent use of foxglove leaves.

Toxic saponins are found in potato spuds, green tomatoes (major saponin, α-tomatine), and other members of the family Solanaceae. They are also produced by sea cucumbers and starfish. Many saponins are capable of disrupting the normal bilayer packing of phospholipids in cell membranes, and this may cause the affected cells to become abnormally leaky to ions, ultimately bringing about lysis (cell death). The major saponin present in foxglove is called digitonin; it is an extremely active detergent.

Ginseng (*Panax*) is a traditional herbal medicine supposedly useful for a wide variety of ailments, including fatigue, sexual impotency, heart disease, and even cancer. The ginseng root contains large amounts of saponins called *glycyrrhizins*. These natural products are apparently safe when adminis-

Figure 17.4 The common foxglove, *Digitalis purpurea*. The leaves of this beautiful flowering perennial contain several cardiac glycosides that are used in the medical treatment of congestive heart failure. Unfortunately, foxglove leaves are easily confused with the leaves of the common comfrey, whose leaves are commonly used to prepare herbal teas, and there have been several medical reports of foxglove poisoning due to this error in plant identification.

tered orally at recommeded doses, usually as a tea or a tablet. However, individuals who chronically consume excessive amounts of ginseng may experience deleterious side effects including insomnia, skin eruptions, diarrhea, and hypertension.

Fortunately for us, most saponins are not readily absorbed from the gastrointestinal tract as glycosides. Instead, intestinal glycosidase enzymes cleave away the sugar groups attached to the 3-B–OH group on the sterol skeleton, and this practically abolishes their toxicity. The non-polar aglycones are readily absorbed and probably are pharmacologically active components. The saponins are a large, chemically diverse group. Despite a vast effort by chemists to decipher their complex structures, very little is yet known about their pharmacological mechanisms of action. They probably exert a variety of actions through multiple cell receptors. In spite of their popularity

TABLE 17.3 Some Common Flowering Plants, their Alkaloid or Peptide Toxins, and Major Symptoms Associated with their Ingestion

Toxin	Type of Compound	Plant (Toxic Parts)	Symptoms
Solanine	Saponin	Potato (Spuds, Stressed tuber)	Headache, fever, abdominal pain, hemorrhagic vomiting, diarrhea
Oleandrin	Cardiac glycoside	Oleander (all parts)	Headache, nausea, vomiting, diarrhea, bradycardia, irregular pulse, coma, respiratory depression
Grayanotoxin	Diterpene	Rhododendron, Azalea (all parts)	Salivation, vomiting, hypotension, convulsions, weakness
Coniine	Piperidine	Poison hemlock (all parts)	Tremor, motor weakness, vomiting, diarrhea, dilated pupils, bradycardia, coma
Lupinine	Quinolizidine	Lupine (all parts, esp. seeds)	Vomiting, salivation, nausea, dizziness, headache, abdominal pain
Cicutoxin	Complex alcohol	Water hemlock (all parts, esp. roots)	Tremors, dilated pupils, convulsions, respiratory depression
Ricin	Peptide	Castor bean (chewed seed)	Pain in mouth; delayed onset: abdominal pain, vomiting, severe diarrhea, hemolysis, renal failure
Viscotoxin	Peptide	Mistletoe (all parts, esp. berries)	Vomiting, diarrhea, hypotension, bradycardia

in traditional herbal medicine, their clinical efficacy in the treatment of most of these disorders has not yet been demonstrated.

Alkaloid Toxins Thousands of compounds of this type have been isolated and investigated, in many cases quite superficially. Most of these substances can also be called heterocyclic compounds, as they generally possess a ring structure containing at least one non-carbon atom, usually N or O. Flowering plants have been a particularly rich source of alkaloids, and apart from the antimicrobial drugs, which are mostly derived from bacteria, most drugs have originated directly or indirectly from alkaloids found in the flowering plants. Some flowering plant alkaloid toxins are listed in Table 17.3.

One of the most commonly used alkaloids is nicotine, the substance that stimulates "nicotinic" cholinergic receptors. In addition to its self-administration as tobacco, nicotine and related compounds are useful toxins for controlling certain insect pests. Because the free base form of nicotine rapidly diffuses across the skin, this substance can be quite toxic to farm workers applying it as an insecticide or to laboratory scientists who are handling the free base. Another heterocyclic compound, reputedly taken by the Greek philosopher Socrates, is coniine, a major alkaloid in poison hemlock potion. Two thousand years later, the mechanism of action of this infamous toxin is still unknown! A South American arrow poison alkaloid, tubocurarine, acts as a competitive antagonist of ACh and nicotine at the skeletal muscle neuromuscular junction.

In recent years a significant number of alkaloids were also isolated from less traditional sources such as marine organisms, and some of these are also toxins.

Flowering Plants Containing Peptide and Protein Toxins Several plants contain protein toxins that are lethal when orally ingested or parenterally administered. *Rosary bean* seeds are quite attractive red seeds with a black spot, and as the name indicates, are often used to make necklaces. These seeds contain a 70-kD protein called abrin, which is a ribosomal protein synthesis inhibitor. The *castor bean*, which is now naturalized in southern California, is similar and is also used for making decorative necklaces. It contains ricin, a homologous protein with the same mechanism of action and potential lethality. These toxins, like diphtheria toxin, are composed of two polypeptide chains: the A chain is the active inhibitor of protein synthesis, while the B chain is needed to bind to the cell membrane and stimulate internalization of the toxin. The symptoms of poisoning by these two toxins develop rather slowly during the first 24 h after ingestion, but if the victim has ingested several seeds, he or she may suffer much during the ensuing couple days and then succumb to an awful death (Table 17.3). The toxins are embedded within the fibrous seed pit; if it is not broken up by chewing, the person may not receive much toxin. Induced vomiting by ipecac syrup followed by gastric lavage is recommended as soon as possible during the first few hours after ingestion; otherwise, symptomatic treatment is all that can be done, since the toxins are internalized within the cell.

As herbal medicines, mistletoe leaves and berries have been used to prepare orally administered extracts and teas for the treatment of a variety of conditions including high blood pressure, tachycardia, insomnia, depression, sterility, ulcers, and cancer, to name only a few. While a few of these conditions, such as hypertension and tachycardia, might ostensibly be ameliorated, based upon present knowledge of the contents of mistletoe, at present, there are no medical reports supporting the therapeutic use of mistletoe extracts. Ingestion of mistletoe extracts is likely to be injurious to one's health, due to the presence of a toxin called viscumin whose action is similar to ricin and abrin, as well as smaller peptide toxins called viscotoxins (Table 17.3), which depolarize muscle cell membranes and can cause hypotension, bradycardia, and other problems.

Plants Causing Contact Dermatitis

A wide variety of plants and animals are known to trigger inflammatory reactions. At the beginning of the twentieth century the Nobel-prize winning French physiologist Edward Richet initiated a study

Figure 17.5 Poison ivy, *Toxicodendron radicans*. Contact with this vine releases several chemically related compounds called *urushiols*, which cause contact dermatitis on repeated contact. Virginia creeper, lower right, is commonly mistaken for poison ivy. Its leaves and stems are harmless, although its berries are poisonous.

of natural inflammatory substances. While investigating the toxicity of the Portuguese *man-o'war* jellyfish he discovered anaphylaxis, an acute life-threatening immune inflammatory response. Some venoms can trigger large inflammatory responses of similar magnitude without an immune component. Other natural compounds, because of their allergenic nature, cause a delayed hypersensitivity response called contact dermatitis. One of the best known cases is the response to poison ivy (Figure 17.5), poison oak, or poison sumac. This is a major hazard to most inhabitants of certain countries like the United States and Canada where these plants abound in cities as well as in rural environments. Contact with these plants causes exudation of a mixture of similar compounds called *urushiols*, which are 4-alkyl-substituted dihydroxyphenyl compounds (catechols). These substances are seldom inflammatory during the first exposure, but subsequently trigger a delayed immune response. The mechanism involves initial oxidation to the quinone, which then reacts with skin proteins and becomes an immunogen. The stimulated Langerhans cells of the skin migrate to the thymus, where they, in turn, stimulate the production of thymic lymphocytes capable of responding to urushiol. These thymus lymphocytes then migrate to the skin and participate in the inflammatory response to subsequent exposures to the urushiol compounds. It is interesting that the lacquer used to provide a glossy surface for Japanese pottery is made from a plant related to poison ivy, which also contains urushiols. As the lacquered surface is allowed to dry in the heat, the urushiols are inactivated. Workers cannot entirely avoid exposure to the urushiols in the fluid they initially apply. Fortunately, many become hyposensitized or resistant after chronic exposure.

17.7 ANIMAL VENOMS AND TOXINS

Reptiles and Amphibians

Snake venoms are complex mixtures of active components, which make their scientific investigation and envenomation treatment quite a challenge. The vast literature on the folklore, natural history, scientific investigation, and medical treatment of poisonous snake bites has attracted the interest of most "toxinologists." Many presentations at meetings of the International Society of Toxinology (announced in the Society journal, *Toxicon*) are on snake venoms.

There are four families of poisonous snakes. The similar venoms of the pit vipers (family Crotalidae) and vipers (Viperidae) will be considered first. Then, we shall examine the cobra (Elapidae) and sea snake (Hydrophiidae) venoms, which also share common biochemical and pharmacological properties.

The pit vipers (Figure 17.6) possess a heat-sensitive sensory organ within a pit next to each eye that is used to sense the presence of warm-blooded prey; rattlesnakes, water mocassins, and copperheads belong to this group. Many pit vipers occur in North and South America, whereas vipers occur only in Africa and Europe. In general (and there are some exceptions), pit viper and viper venoms have greater local effects on the tissues where the bite occurs and on the cardiovascular system. Localized tissue swelling (edema) results from protein hemorrhagic toxins, which attack the capillary endothelium, making it leaky to blood cells as well as plasma proteins. Protein myotoxins cause a pathological release of intracellular calcium stores in skeletal muscle, which may produce muscle necrosis. Hyaluronidase and collagenase enzymes break down the connective tissue elements, promoting the spread of the venom from the original site of the bite. Motor paralysis rarely occurs in the absence of cardiovascular crisis, with one notable exception. The venom of the Brazilian rattlesnake, *Crotalus durissus terrificus*, possesses a potent neurotoxin called crotoxin, which paralyzes peripheral nerve terminals, causing loss of neuromuscular transmission and flaccid paralysis.

Since crotalid venoms for the most part contain similar toxins and enzymes, and species identification is often impossible, most immunotherapeutic treatments of pit viper bites utilize a polyvalent horse antivenin originally prepared with an antigenic mixture of several crotalid venoms. This approach has been quite successful.

Figure 17.6 The Eastern diamondback rattlesnake (*Crotalus adamanteus*) is one of the most dangerous pit vipers. On a weight basis its venom is not nearly as powerful as cobra or coral snake venom, but it compensates for this by injecting a much larger quantity of venom with an efficient venom delivery apparatus.

Cobra or sea snake envenomation often causes respiratory arrest before any signs of local tissue or systemic cardiovascular damage are apparent. The major neurotoxin occurring in elapid and hydrophiid venoms is α-neurotoxin. This is a basic polypeptide of 65–80 amino acid residues that is crosslinked with four or five disulfide bonds. The toxin acts as a competitive antagonist of the neurotransmitter acetylcholine (ACh) at the skeletal muscle neuromuscular junction. Unlike the nondepolarizing muscle relaxants used in surgery, which act at the same site, α-neurotoxin binds very tightly because its greater molecular size permits many contacts with the nicotinic receptor. In fact, a toxin found in the Taiwanese krait (*Bungarus multicinctus*), alpha-bungarotoxin, binds essentially irreversibly to the skeletal muscle nicotinic receptor, preventing ACh from interacting with its postsynaptic receptor. As if this potent neurotoxin were not sufficient to paralyze the skeletal muscle, this snake also makes a larger protein toxin called beta-bungarotoxin (Table 17.1), which inhibits the release of ACh from the motor nerve terminal; these two toxins, working together in a synergistic fashion, can reduce the probability of neuromuscular transmission to zero. Besides the postsynaptic alpha-neurotoxic peptides, elapid venoms also generally contain phospholipase A and a peptide called *cardiotoxin*, which is a cytolysin that tends to attack cardiac myocardial cells. Cardiotoxin disrupts the bilayer structure of membrane lipids, and thereby makes these lipids more accessible substrates for the phospholipase A.

Coral snakes are the only new-world elapids. About 50 species have been described. In the United States there are only two species, but in central America and the northern parts of South America there are many species. Coral snake bites are rarely as life-threatening as cobra bites because the volume of venom injected is usually quite small. Elapid snakes lack the fangs observed in the pit vipers, and therefore, they must resort to a more lengthy chewing method of envenomation, which is not nearly as efficient. The major danger for elapid snake envenomation victims is respiratory arrest due to blockade of neuromuscular transmission, and secondarily, cardiac systolic arrest due to the synergistic action of the cardiotoxin and phospholipase A. Generally there is little or no localized edema soon after the bite, as in crotalid envenomations, which sometimes leads to an incorrect initial perception that the life of the victim is not endangered.

Although antivenin therapy remains the most powerful approach towards treating snake envenomations, in many situations the antivenin is not immediately available, so a rational therapeutic approach based on knowledge of the actions of the toxic constituents is required.

Most amphibians possess skin toxins serving as some chemical defense against predators, but only a few species present a danger to humans. Some of the brightly colored tropical South American frogs possess extremely potent toxins, and touching these may be enough to become intoxicated! Apparently, some of these frogs are collected for the exotic pet market and kept in vivariums as pets; fortunately for the owner, these frogs soon lose their toxicity in captivity, which suggests that they make their toxins from precursor molecules in their natural diet. Batrachotoxin (Tables 17.1 and 17.2), which comes from one of these frogs, is a lipophilic sodium channel activator, making it popular in the preparation of poison darts by Indian hunters. Human symptoms of intoxication, although they have not been reported, should be similar to those caused by the grayanotoxins or the veratrum alkaloids found in the false hellebore (Table 17.1 and 17.3). Another frog alkaloidal toxin, histrionicotoxin, causes neuromuscular paralysis by binding to the open channel of the skeletal muscle nicotinic receptor.

Toads of the genus *Bufo* possess a very potent venom in their skin and parotid glands behind their eyes. The major toxic constituents are cardiac glycosides called *bufotoxins*, but there also are biogenic amines, including epinephrine and bufotenin, a methylated form of the neurotransmitter serotonin. Because bufotenin is hallucinogenic, some enthusiasts have taken up "toad licking." This is a dangerous way to get high, as the white milky venom is rich in bufotoxins!

Fish Venoms and Toxins

Only a relatively small proportion of fish species are venomous, and in all cases the venoms are used defensively to deter predators. Probably the most commonly encountered venomous fishes are the catfishes and sting rays. Experienced fisherman are aware of the irritating stings caused by marine catfish venom, but novices often learn the hard way. Little is known about the active constituents, although a recent paper reports smooth muscle stimulating and hemolytic activity of a large protein toxin. Sting rays contain a dorsal spine near the base of their tail; when the ray is stepped on in shallow water, the tail is thrust upward so that the spine can penetrate the skin of the intruder. When waders shuffle their feet along the surface of the bottom, the sting rays almost always are frightened away, so this is the best way of avoiding this fish. In contrast, the tropical Pacific stonefish (*Synancega* sp.) is not easily frightened, and simply raises its spine when it senses the presence of an intruder. Like catfish venom, stonefish and sting ray spine venoms probably contain several protein toxins that cause smooth muscles to contract and cause inflammation. The stonefish toxin has recently been isolated and shown to be a large protein that enhances neurotransmitter release from nerve terminals. While these stings are quite unpleasant, they are rarely life-threatening, and can usually be treated with antiinflammatory drugs such as antihistamines and corticosteroids.

Tetrodotoxin is certainly one of the most potent fish toxins. Pufferfish are considered a dangerous delicacy in Japan, and consequently cooks must be carefully trained in the removal of poisonous viscera and skin when preparing "fugu" flesh for consumption. In the United States pufferfish are rarely consumed, but several cases of poisoning have been reported over the years. A person intoxicated while consuming pufferfish will generally experience tingling and numb sensations in the mouth area within an hour after ingestion. Muscular weakness also develops, and the victim can be completely paralyzed. Endoscopic removal of the consumed fish is recommended if it can be done without delay. Treatment is otherwise supportive; bradycardia and hypotension can be countered with atropine, intravenous fluids, and oxygen. Anticholinesterases may restore neuromuscular function if it is not entirely blocked.

While tetrodotoxin is usually present in puffers, regardless of the place or season, some other toxins like ciguatoxin are less predictable in their occurrence, as they are slowly passed up the food-chain from algae or bacteria to herbivores, then predatory fish and marine mammals.

Ciguatoxin (Table 17.2), which activates voltage-gated sodium channels in nerve and muscle cells, is a prime example. Ciguatera poisoning is quite unpredictable; the predatory fish is edible most of the time in a particular place. It causes a variety of symptoms such a lethargy, tingling and numbness of the lips, hand and/or feet weakness, itching, joint pains, and gastrointestinal symptoms including diarrhea. These problems may last up to several months because this lipophilic toxin is eliminated very slowly. It is active in such minute concentrations that research on its structure was hampered for over a decade because insufficient amounts were available for analysis. Ciguatera infestations occur in the Carribean Sea as well as in the tropical Pacific. The symptoms differ in these sites, suggesting that the toxins are not exactly the same. Administration of hyperosmotic mannitol seems to be an effective symptomatic therapy for controlling the Schwann cell edema caused by this complicated molecule.

Arthropod Toxins and Venoms

This animal phylum consists of such different animals as scorpions, spiders, and insects. Many arthropods use neuroactive substances as repellents, alarm pheromones, or as toxins. While the insects are the largest group in terms of biodiversity, only a small proportion of species seem to possess toxins, whereas almost all scorpions and spiders routinely use venomous secretions to capture their prey and deter predators. Fortunately for us, most arthropod toxins have evolved in the direction of immobilizing animals other than mammals. Only a relatively small group of spider species are known to be poisonous to humans.

Scorpion venom is one of the richest sources of peptide toxins known; it is comparable in diversity to the cone shell venoms, which will be described in the next section. Scorpions quickly immobilize their prey, generally insects, by injecting a complex mixture of peptides that act on the voltage-gated sodium and potassium channels, which then produce action potentials. There are two kinds (called *alpha* and *beta*) of toxins, that bind at sites 3 and 4, respectively, on the external surface of the sodium channel (Table 17.2). Both enhance electrical excitability by modulating the probability that the sodium channel will remain open, even when the electrical potential of the membrane is nearly the same as in the resting state (about 60–90 mV negative on the inside surface of the membrane).

The α-scorpion toxins specifically slow a process, referred to as *inactivation*, by which the open sodium channel turns off in the presence of membrane depolarization. A normally brief (duration about one millisecond) action potential is turned into an abnormally long signal whose duration may be several hundred milliseconds. This causes a massive release of neurotransmitters at peripheral nerve terminals on skeletal and other muscles. The consequences for the victim are disastrous, namely hyperexcitability, convulsions, paralysis, and sometimes death. The β-scorpion toxins by a different mechanism also cause peripheral nervous system hyperexcitability by stimulating the nerves and muscles to generate trains of multiple action potentials in response to each depolarizing stimulus. The β-scorpion toxins reduce the rate at which the opened sodium channel returns to its resting state, a process often referred to as "deactivation." Old-world scorpions generally contain only the alpha-type sodium channel toxins, whereas the new-world species often contain both α- and β-neurotoxins. Antivenins are available for the most dangerous scorpions and offer the most effective means of treatment.

Since the late 1980s, another group of smaller peptide toxins, which block various potassium channels, has been discovered in scorpion venoms. Since the electrical excitability of a nerve or muscle cell at any instant depends on the relative permeability of the membrane to sodium and potassium ions, it makes good sense for a scorpion venom to also contain toxins that block potassium channels. Charybdotoxin, the first of these toxins to be characterized, primarily blocks calcium-activated potassium channels found in smooth and skeletal muscles. This channel protects the cell against excessive membrane depolarization and internal calcium loading. Charybdotoxin also blocks some voltage-activated potassium channels in the brain.

Because of this multiplicity of toxins in scorpion venom that enhance electrical excitability, an alternative approach for treating scorpion envenomation would be to reduce excitability, particularly in the peripheral nervous system (these peptides do not readily cross the blood–brain barrier). This could be at least partially achieved by reducing postsynaptic membrane responsiveness to ACh with nicotinic and muscarinic receptor antagonists. This potential method of treatment could supplement the use of antivenins.

Spiders generally poison their insect prey. Fortunately, vertebrate nervous system receptors are pharmacologically different enough from those of insects that most spider toxins are not very active on humans. It also helps that we are so big and their normal prey and predators are so small! Nevertheless, several spiders are exceedingly dangerous. Black widow spiders (*Latrodectus* sp.) occur throughout the world, so we shall consider them first. Their venom is primarily neurotoxic due to the presence of a powerful protein toxin called alpha-latrotoxin (Table 17.1). This large protein enhances neurotransmitter release from nerve terminals, and can even cause nerve terminal secretory vesicle depletion. Victims concurrently suffer from skeletal muscle spasms and autonomic overstimulation (causing sweating, salivation, nausea, and hypertension). Again, treatment is primarily based upon administration of *Latrodectus* antivenin. Some relief from these symptoms can be achieved with centrally acting muscle relaxants like diazepam, and autonomic overstimulation can be ameliorated with muscarinic and/or adrenergic antagonists, depending on the symptoms.

Brown recluse spider venom (*Loxoceles* sp.) acts in an entirely different way because its venom primarily contains an enzyme, sphingomyelinase, which causes tissue damage. While this venom is less dangerous than black widow venom, it can cause significant tissue necrosis at the site of the bite.

Although the bees, hornets, and wasps all belong to the order Hymenoptera, their venoms are different. The most serious reactions to hymenopteran stings are of the immediate hypersensitivity type and are due to an immune response from previous stings mediated by immunoglobulin E. Bee venom has been found to be an exceedingly rich mixture of enzymes and toxins. The primary enzyme of importance is phospholipase A, which acts synergistically with a peptide detergent called *mellitin* (named after the common honeybee *Apis mellifera*) to break down phospholipids in the plasma membrane, thereby liberating prolytic fatty acids and lysolecithin. While mellitin can act alone to disrupt the cell membrane, its action is greatly facilitated by the presence of these phospholipid breakdown products. Like many snake venoms, bee venom also contains the enzyme hyaluronidase, which breaks down connective tissue and thus facilitates the spreading of the venom from its site of injection. Bee venom also contains two peptide toxins, apamin and mast cell degranulating peptide, which respectively block calcium-activated and voltage-activated potassium channels.

In contrast to bee venom, the wasp and hornet venoms primarily contain small peptides called kinins which, like our endogenous bradykinin, have a triple action: stimulation of sensory nerve endings resulting in neurogenic inflammation, increased capillary permeability, and relaxation of vascular smooth muscle.

Fire ants (*Solenopsis*) are quite abundant in the southeastern United States, and many people are stung each year. The venom contains piperidine alkaloids, which have been found to block the nicotinic receptor ion channel. Protein constituents are thought to be at least partly responsible for the painful sensation associated with the sting. Irritating pustules and some minor tissue necrosis may result at the sting, extending the period of discomfort to several days. The role that the alkaloids (called *solenopsins*) play in the inflammatory responses associated with fire ant stings is not entirely clear, but solenopsins are known to cause histamine release from basophils.

Mollusc Venoms and Toxins

The molluscan exoskeleton provides considerable protection against predators but also limits mobility. This poses a problem for predatory snails. However, one group of gastropods called "cones" possesses a formidable harpoon-like venom apparatus for paralyzing its prey. *Conus* venom was extensively investigated in the 1990s. Almost all *Conus* toxins are peptides or small proteins. The venom is a virtual cocktail of ion channel modulators including nicotinic receptor antagonists (α-conotoxins), sodium channel blockers (μ-conotoxins), calcium channel blockers (ω-conotoxins), and glutamate channel blockers (conantokins). Only a relatively small fraction of the 300 known species of *Conus* are dangerous to humans, and these mainly occur in the tropical Pacific. Inexperienced divers should avoid handling cone shells.

The octopus envenomates its prey with a posterior salivary gland secretion. The only octopus that is toxic to man is the tiny Australian blue-ringed octopus, which appeared in the James Bond movie "Octopussy." Bathers have been known to play with this pretty little animal, often found among beach rocks, without realizing how dangerous it is! While all other octopus venoms contain protein toxins that are not dangerous to humans, this species instead secretes tetrodotoxin, the same toxin used by pufferfish.

The ability of bivalve molluscs to concentrate dangerous quantities of dinoflagellate toxins such as saxitoxin and domoic acid has already been discussed above.

Coelenterate (Cnidarian) Venoms

Cnidaria is a more recent name for this phylum, which indicates that all species contain small stinging capsules called *cnidae* (nematocysts). A wide variety of cnidae exist, even within a single animal. The largest, most formidable cnidae, capable of discharging venom deep within the victim's skin, are found in the classes Scyphozoa (jellyfish) and Hydrozoa (*man-o'-war*, etc.), so it is not surprising that most cnidarian human envenomations result from jellyfish (Figure 17.7) or Portuguese *man-o'-war* stings. However, all species (10,000) belonging to this phylum are potentially toxic, if not venomous. The world's most dangerous species of jellyfish, *Chironex fleckeri*, is found along the Australian coast. Swimmers have been know to collapse within seconds after multiple stings by this species, which precludes swimming at certain times of the year. Barriers are used to keep these jellyfish out of swimming areas, and lifeguards must undergo extensive training in order to assist the unfortunate victims. Most other jellyfish can also cause very unpleasant stings, but these are rarely life-threatening. The fire corals occurring in tropical waters, like the *man-o'-war*, are actually hydrozoans rather than true corals. Their inflammatory sting is probably due to the presence of toxins similar to that of the *man-o'-war*.

Nematocysts discharge when the nematocyte cell in which they are contained is mechanically and chemically stimulated. The tubule within the nematocyst is explosively evaginated, causing a proteinaceous venom to be injected into the skin of the victim. Only recently have a few of the major jellyfish toxins been isolated, since they are large, unstable proteins that are difficult to purify. Most of the limited data on these toxins suggest that they primarily act as pore-formers, causing the depolarization of nerve, muscle, and inflammatory (basophil, etc.) cells.

Most symptoms observed in envenomated persons and experimental animals can be predicted assuming massive release of numerous chemical mediators of inflammation and transient stimulation of nerve terminals in various kinds of muscle including cardiac and vascular. While antihistamines provide considerable relief for the purely inflammatory symptoms, they are not sufficient to counteract all actions of the most active venoms, such as that of *Chironex*. Many treatments have been suggested for limiting the further discharge of nematocysts on the victims skin, including alcohol, acetic acid, and protease mixtures like meat tenderizer. Topically applied vinegar (acetic acid) is probably the best common means of initial treatment. Development of a more rational therapy for these envenomations awaits further analyses of the pharmacological actions of individual toxic components of jellyfish venoms.

One of the most potent marine toxins, palytoxin, is found in zoanthids, which are small colonial sea anemones found in tropical reefs. This toxin, which acts by converting the sodium-potassium pump into an ion channel, actually is synthesized by a marine bacterium that lives in the zoanthid. Like ciguatoxin, palytoxin occasionally causes human food-born intoxications because it can also be passed up the food chain into edible fishes.

Sea anemones possess a variety of peptide and protein toxins that affect ion channels in electrically excitable cells in a manner similar to scorpions. In fact, the anemone toxins bind to the same site on sodium channels as the scorpion α-toxins, and slow down the process of sodium inactivation in essentially the same fashion. Some anemones also contain smaller peptide toxins that selectively block

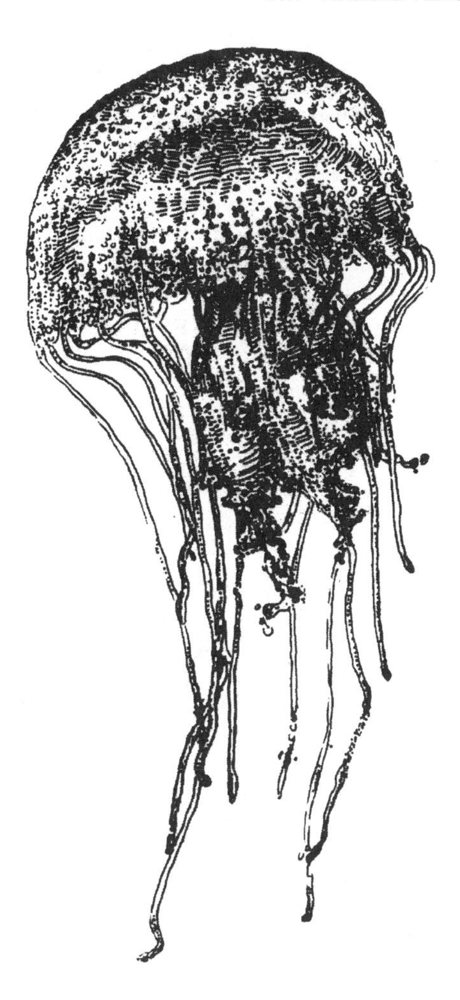

Figure 17.7 The sea nettle jellyfish, *Chrysaora quinquecirrha*. This jellyfish occurs along the eastern coast of North America, south of Cape Cod, Massachusetts, and in the Gulf of Mexico. Its venom contains protein toxins that can cause severe skin inflammation.

certain potassium channels. Most sea anemones also contain potent cytolytic proteins called actinoporins, which permeabilize cell membranes and ultimately cause cell death. It is fortunate that only a small proportion of sea anemone species sting humans when they are handled, perhaps because sea anemone nematocysts are often too small to penetrate far into the skin. Nevertheless, it is best not to handle these organisms with bare hands, as a few species can cause quite a sting!

Other Toxic Marine Invertebrates

While many sedentary or slow moving invertebrates possess defensive toxins to deter predators, it is fortunate that only a relatively small number of species are harmful to humans. Some sea urchins (mostly in the IndoPacific) possess either venomous spines or flower-like venomous organs called *pedicellaria*, which can be observed to frequently rotate about their base, apparently guarding the surface of the sea urchin from small predators or fouling organisms. Recently, several inflammatory protein toxins that contract smooth muscle have been isolated from the pedicellaria.

Some sponges, such as the Caribbean fire sponge (*Tedania ignis*) and the nolitangere sponge (French for do not touch!), cause chronic contact dermatitis in addition to a more immediate inflammatory action; the active constituents of these sponges have not yet been identified.

Certain marine worms are quite poisonous. These include several species of segmented worms, belonging to the phylum Annelida. The toxin of a species occurring in Japanese waters, nereistoxin, has become an agriculturally useful insecticide because it blocks nicotinic cholinergic receptors. Bristle worms commonly found in coral reefs cause quite an irritating sting, which is probably due to release of inflammatory substances in addition to the bothersome irritation caused by the fine bristles lodged in the skin. Another stinging annelid, used as fishing bait in the New England area, is the blood worm *Glycera dibranchiata*. The proboscis "fangs" of this rather large worm inject a protein toxin that stimulates neurotransmitter release from nerve terminals and also commonly causes some tissue necrosis.

17.8 TOXIN AND VENOM THERAPY

Identification of the Toxic Organism

Since the toxins of closely related species often have a different chemical structure and even mechanism of action, it is usually imperative to identify the toxic organism in order to select the appropriate therapy. An immunologic method for identifying the species involved in a snake bite is now available in Australia. Such kits are likely to become available in other parts of the world.

Several excellent guides for the identification of poisonous plants and animals of North America are available to readers wishing to learn about their identification. Many food-related intoxications result when persons sample natural or ornamental plants without proper identification, consuming the wrong plant, sometimes with tragic consequences.

Immediate Therapeutic Procedures to Counteract Ingestion of Poisons

A highly effective initial response to the ingestion of a poisonous substance or plant is to induce vomiting with ipecac syrup, a mixture of plant alkaloids called *emetines*. It is only to be used relatively soon after poison ingestion has occurred. Under certain conditions when the patient is extremely drowsy or unconscious it should not be used, since failure to vomit (usually 15–30 min later) might lead to additional difficulties due to the emetine alkaloids. If the patient fails to vomit within a reasonable period of time an additional dose is swallowed to finally cause vomiting and removal of unabsorbed poisons.

If vomiting is unsuccessful or incomplete, gastric lavage may be carried out by trained medical personnel; this involves placing a tube in the stomach and applying suction to remove the harmful contents and is usually done after the patient has been sedated. After this procedure the patient is often administered an oral dose of activated charcoal, which is effective in sequestering many poisons from the gastrointestinal tract and is ultimately eliminated in the feces. A cathartic saline solution is often administered to speed up elimination of the poisonous substance, charcoal adsorbed or not, from the intestines.

This description is only meant to inform the reader of treatment options. A medical person trained to treat intoxications should be contacted immediately for assistance of the patient!

Prospects for Improved Immunotherapy of Venoms and Toxins

Polyvalent antivenins are commonly used to treat bites from related species of snakes, since they are often difficult to distinguish from each other. The polyvalent serum used to treat crotalid bites is prepared by immunizing horses with a mixture of several *Crotalus* species; it has been shown to be effective in the treatment of bites caused by over 65 crotalid species.

One traditional disadvantage of immune therapy of venomous animal bites and stings is the lack of effectiveness of the IgG antibody in neutralizing venom constituents once they have entered the interstitial fluid space. It has recently been found that preparation of truncated antibody molecules, such as the F(AB)2 fragments, is one way to enhance neutralization of tissue bound toxins.

One of the major problems associated with the use of antivenins is the high incidence of serum sickness, which is a human immune response to the intravenous administration of horse serum antivenin. One method of reducing the problem would be to enrich the antigenic mixture used to produce the antivenin with the major toxins or other proteins that are most dangerous. This may permit a reduction in the total amount of freeze-dried horse serum required for therapy. Monoclonal antibodies have not yet gained much acceptance for immunotherapy of intoxications or envenomations, probably for a combination of reasons: (1) monoclonals are so specific they would only work on single components of a venom, (2) they would be less likely to be of use against envenomation by a different species, and (3) they are considerably more expensive to produce. Expense is a factor; most pharmaceutical firms view the production of antivenins more as a responsibility, due the rather small market for antivenins. In the future it is likely that better antivenins will become available, as the costs of preparing "humanized" monoclonal antibodies decreases.

In Australia, it is routinely recommended that intramuscular epinephrine be administered prior to injection of the antivenin, in order to reduce the intensity of any immediate hypersensitivity reaction to the antivenin, and then oral corticosteroids be taken for several days afterwards in order to reduce the delayed hypersensitivity response to administration of antivenin. With these precautions, the incidence of serum sickness has been less than 10 percent, compared to the almost 30 percent estimated for U.S. victims.

Toward a Rational Pharmacotherapy Based on Knowledge of the Toxic Constituents of a Venom

Probably most envenomation victims are not treated rapidly enough with antivenins to fully respond to immunotherapy, particularly when envenomation occurs outside a geographic area in which envenomations are frequent. Other patients cannot tolerate the antivenom because of allergic sensitivity. While recognizing the therapeutic power of the immune approach, it seems prudent to develop rational pharmacotherapies based on a scientific knowledge of the chemistry and biological actions of the toxins involved. Small toxins that act on receptors may be antagonized by using antagonists if the toxin is an agonist and vice versa. This is most common when receptors to neurotransmitters are involved (e.g., atropine can counteract the actions of the mushroom toxin muscarine). When a competitive antidote is unavailable, it may be possible to physiologically antagonize the intoxication, based on a knowledge of the opposing system. Of course, the very basis of rational therapy is the biochemical and pharmacological understanding of the most active constituents.

For a rational therapy to succeed, it must be based not only upon scientific knowledge of the separate actions of venom constituents, but must also take into account the synergistic actions of many of the constituents. After all, a venom has usually evolved rather than just a single toxic substance! That is why toxicological studies must also be carried out with whole venoms as well as their purified constituents, in order to detect such interactions between venom constituents.

Toxins as Drugs

Besides serving as chemical defenses and offenses for the organisms that create them, some naturally occurring toxins are also being used as chemical tools for investigating biomedical problems and as models for designing novel new drugs. The use of toxins and venoms as therapeutic agents is not a new phenomenon, but rather, an activity that is probably as old as the most primitive humanoid species. In the nineteenth century, drug development based on natural products was made possible by the emergence of organic chemistry. In recent years, the availability of radioligand binding and molecular biological techniques for investigating drug receptors *in vitro* has further accelerated drug development. Some of the toxins mentioned in this chapter are serving as molecular models for designing drugs with novel mechanisms of action. For instance, the worm toxin anabaseine has been modified to eliminate its peripheral nicotinic agonist activity, and the resulting compound, DMXB-anabaseine (also known as GTS-21), is now undergoing human clinical tests as a possible Alzheimer's drug (Kem, 1995).

The goal of drug development is to sever the connection between toxicity and therapeutic activity of compounds intended as drugs, but this ideal is rarely completely attainable. It is useful to keep in mind that the difference between toxin and drug is often a seemingly minor alteration of chemical structure, or at the least, proper selection of dosage. The sixteenth-century physician and chemist Paracelsus understood the dual nature of *Materia Medica* when he stated that drugs are also poisons, and it is often a matter of dose whether the therapeutic or toxic effect predominates. For all the problems that natural toxins and venoms cause, our collective ability to use them as tools in biomedical research and drug design makes them valuable reagents in medical research. Ultimately, these substances can benefit, more than damage, human existence.

17.9 SUMMARY

A *toxin* is a single substance that adversely affects some biological process or organism, whereas a *venom* is a heterogeneous mixture of many substances, some of which are toxic. A poison is either a single injurious substance or a mixture of substances and can be human-made (synthetic) or natural.

Knowledge of the mechanism by which a toxin acts on some biological process provides the ultimate basis for rational treatment of intoxication. While many protein intoxications are successfully treated by immunotherapy, treatment of smaller nonpeptide toxins must be based upon pharmacologic antagonism as well as symptomatic treatment. It is extremely important to identify the toxin or venom involved in an intoxication in order to select the appropriate treatment.

Initial treatments, such as induction of vomiting, and gastric lavage, for orally ingested toxins and immobilization of individuals bitten by poisonous snakes or other animals can reduce entry of the toxin(s) into the systemic circulation, and thereby delay the onset and reduce the intensity of the response. Success often depends on the training of personnel responsible for initial care of the victim.

While few human intoxications due to natural toxins or venoms are lethal when properly treated, delayed or inadequate treatment can be life-threatening.

Toxins and venoms are not only potentially injurious to health but can also be beneficial in providing new research tools for biomedical research and unique molecular models for designing new drugs.

ACKNOWLEDGMENTS

The author thanks Barbara Seymour for artistic renderings of the poisonous organisms and Judy Adams for word-processing the manuscript.

REFERENCES AND SUGGESTED READING

Ames, B., M. Profet, and L. S. Gold, "Nature's chemicals and synthetic chemicals: Comparative toxicology," *Proc. Natl. Acad. Sci.* (USA) **87**: 7782–7786 (1990).

Anderson, D. M., "Red tides," *Sci. Am.* **271**: 62–70 (1994).

Auddy, B., M. I. Alam, and A. Gomes, "Pharmacological actions of the venom of the Indian catfi sh (*Plotosus canius* Hamilton)," *Ind. J. Med. Res.* **99**: 47–51 (1994).

Daly, J. W., "The chemistry of poisons in amphibian skin," *Proc. Natl. Acad. Sci.* (USA) **92**: 9–13 (1995).

Dickstein, E. S., and F. W. Kunkel, "Foxglove tea poisoning," *Am. J. Med.* **69**: 167–169 (1980).

Epstein, W., "Occupational poison ivy and oak dermatitis," *Occup. Derm.* **12**: 511–516 (1994).

Foster, S., and R. A. Caras, *A Field Guide to Venomous Animals and Poisonous Plants*, Houghton-Mifflin, Boston, 1994.

Florsheim, G. L., "Treatment of human amatoxin mushroom poisoning. Myths and advances in therapy," *Med. Toxicol.* **2**: 1–9 (1987).

Hall, A. H., D. G. Spoerke, and B. H. Rumack, "Assessing mistletoe toxicity," *Ann. Emerg. Med.* **15**: 1320–1323 (1986).

Halstead, B. W., *Poisonous and Venomous Marine Animals of the World*, Darwin Press, Princeton, NJ, 1988.

Howard, R. J., and N. T. Bennett, "Infections caused by halophilic marine vibrio bacteria,"*Ann. Surg.* **217**: 525–531 (1993).

Kalish, R. S., "Recent developments in the pathogenesis of allergic contact dermatitis," *Arch. Dermatol.* **127**: 1558–1663 (1991).

Kawai, K., M. Nakagawa, K. Kawai, F. M. Liew, and Yasuno, "Hyposensitization to urushiol among Japanese lacquer craftsmen: Results of patch tests on students learning the art of lacquerware," *Contact Dermatitis* **25**: 290–295 (1991).

Kem, W. R., "Worm toxins," in *Handbook of Natural Toxins*, Vol. 3, *Marine Toxins and Venoms*, A. T. Tu, ed., Marcel Dekker, New York, 1985, pp. 353–378.

Kem, W. R., "Peptide chain toxins of marine animals," in *Biomedical Importance of Marine Organisms*, D. Fautine, ed., Calif. Acad. Sci. Memoirs, San Francisco, 1988, pp. 69–83.

Kem, W. R., "Alzheimers's drug design based upon an invertebrate toxin (anabaseine) which is a potent nicotinic receptor agonist," *Invert. Neurosci.* **3**: 251–259 (1997).

Knight, B., "Ricin—a potent homicidal poison," *Br. Med. J.* 350–351 (1979).

Middlebrook, J. L., "Cell surface receptors for protein toxins," in *Botulinum Neurotoxin and Tetanus Neurotoxin*, L. L. Simpson, ed., Academic Press, San Diego, 1989, pp. 95–119.

Moffett, M. W., "Poison-dart frogs: lurid and lethal," *Natl. Geogr. Mag.* **187**: 98–111 (1995).

Palafox, N. A., L. G. Jain, A. Z. Pinano, T. M. Gulick, R. K. Williams, and I. J. Schantz, "Successful treatment of ciguatera fish poisoning with intravenous mannitol," *J. Am. Med. Assoc.* **259**: 2740–2742 (1988).

Rauber, A., and J. Heard, "Castor bean toxicity re-examined: A new perspective," *Vet. Hum. Toxicol.* **27**: 498–502 (1985).

Russell, F. E., "Snake venom poisoning in the United States," *Ann. Rev. Med.* **31**: 247–259 (1980).

Russell, F. E., "Snake venom immunology: Historical and practical considerations," *J. Toxicol. Toxin Rev.* **7**: 1–82 (1988).

Simpson, L. L., "The actions of clostridial toxins on storage and release of neurotransmitters," in *Natural and Synthetic Neurotoxins*, A. L. Harvey, ed., Academic Press, New York, 1993, pp. 277–318.

Spoerke, D. G., and B. H. Rumack, *Handbook of Mushroom Poisoning. Diagnosis and Treatment*, CRC Press, Boca Raton, FL., 1995, p. 464.

Sutherland, S. K., "Antivenom use in Australia. Premedication, adverse reactions and the uses of venom detection kits," *Med. J. Austral.* **157**: 734–739 (1992).

Tu, A. T., *Venoms: Chemistry and Molecular Biology*, Wiley, New York, 1977.

Turner, N. J., and A. F. Szczawinski, *Common Poisonous Plants and Mushrooms*, Timber Press, Portland, OR, 1991.

Van Mierop, L. H. S., "Poisonous snakebite: A review. 1. Snakes and their venom, 2. Symptomatology and treatment," *J. Fla. Med. Assoc.* **63**: 191–210 (1976).

Yaffee, H. S., and F. Stargardter, "Erythema multiforme from *Tedania ignis*," *Arch. Dermatol.* **87**: 601–604 (1963).

Yamamoto, K., A. C. Wright, J. B. Kaper, and J. G. Morris, Jr., "The cytolysin gene of *Vibrio vulnificus*: Sequence and relationship to the Vibrio cholerae El Tor hemolysin gene," *Infect. Immun.* **58**: 2706–2709 (1990).

Yasumoto, T., and M. Yotsu, "Biogenetic origin and natural analogs of tetrodotoxin," in *Natural Toxins: Toxicology, Chemistry and Safety*, R. F. Keeler, N. B. Mandava, and A. T. Tu, eds., *Amererican Chemical Society Press*, Washington, DC, 1992, pp. 226–233.

PART III
Applications

18 Risk Assessment

ROBERT C. JAMES, D. ALAN WARREN, N. CHRISTINE HALMES, and
STEPHEN M. ROBERTS

Risk assessment is an ever-evolving process whereby scientific information on the hazardous properties of chemicals and the extent of exposure results in a statement as to the probability that exposed populations will be harmed. The probability of harm can be expressed either qualitatively or quantitatively, depending on the nature of the scientific information available and the intent of the risk assessment. Risk assessment is not research per se, but rather a process of collecting and evaluating existing data. As such, risk assessment draws heavily on the disciplines of toxicology, epidemiology, pathology, molecular biology, biochemistry, mathematical modeling, industrial hygiene, analytical chemistry, and biostatistics. The certainty with which risks can be accurately assessed, therefore, depends on the conduct and publication of basic and applied research relevant to risk issues. While firmly based on scientific considerations, risk assessment is often an uncertain process requiring considerable judgment and assumptions on the part of the risk assessor. Ultimately, the results of risk assessments are integrated with information on the consequences of various regulatory options in order to make decisions about the need for, method of, and extent of risk reduction.

It is clear that society is willing to accept some risks in exchange for the benefits and conveniences afforded by chemical use. After all, we knowingly apply pesticides to increase food yield, drive pollutant-emitting automobiles, and generate radioactive wastes in the maintenance of our national defense. We legally discharge the byproducts of manufacturing into the air we breathe, the water we drink, and the land on which our children play. In addition, we have a history of improper waste disposal, the legacy of which is thousands of uncontrolled hazardous-waste sites. To ensure that the risks posed by such activities are not unacceptably large, it is necessary to determine safe exposure levels in the workplace and environment. Decisions must also be made on where to locate industrial complexes, on remediation options for hazardous-waste sites, tolerance levels for pesticides in foods, safe drinking-water standards, air pollution limits, and the use of one chemical in favor of another. Risk assessment provides the tools to make such determinations.

This chapter provides an overview of the risk assessment process, and discusses

- the basic steps of risk assessment
- how risk assessments are performed in a regulatory context
- differences between human health and ecological risk assessments
- differences in the estimation of cancer and non-cancer risks
- differences between deterministic and probabilistic risk assessments
- issues associated with estimating risks from chemical mixtures
- comparisons of risks from chemical exposure with other health risks
- risk communication from chemical exposure with other health risks

Principles of Toxicology: Environmental and Industrial Applications, Second Edition, Edited by Phillip L. Williams, Robert C. James, and Stephen M. Roberts.
ISBN 0-471-29321-0 © 2000 John Wiley & Sons, Inc.

18.1 RISK ASSESSMENT BASICS

A Basic Risk Assessment Paradigm

In 1983, the National Research Council described risk assessment as a four-step analytical process consisting of hazard identification, dose-response assessment, exposure assessment, and risk characterization. These fundamental steps have achieved a measure of universal acceptance and provide a logical framework to assemble information on the situation of potential concern and provide risk information to inform decision making (see Figure 18.1). The process is rigid enough to provide some methodological consistency that promotes the reliability, utility, and credibility of risk assessment outcomes, while at the same time allowing for flexibility and judgment by the risk assessor to address an endless variety of risk scenarios. Each step in the four-step process known as *risk assessment* is briefly discussed below.

Step 1: Hazard Identification. The process of determining whether exposure to a chemical agent, under any exposure condition, can cause an increase in the incidence or severity of an adverse health effect (cancer, birth defect, neurotoxicity, etc.). Although the matter of whether a chemical can, under any exposure condition, cause cancer or other adverse health effect is theoretically a yes/no question, there are few chemicals for which the human data are definitive. Therefore, not only epidemiological studies but also laboratory animal studies, in vitro tests, and structural and mechanistic comparability to other known chemical hazards are considered. This step is common to qualitative and quantitative risk assessment.

Step 2: Dose–Response Assessment. The process of characterizing the relationship between the dose of a chemical and the incidence or severity of an adverse health effect in the exposed population. A dose–response assessment factors in not only the magnitude, duration, and frequency of exposure but also other potential response-modifying variables such as age, sex, and certain lifestyle factors. A dose–response assessment frequently requires extrapolation from high to low doses and from animals to humans.

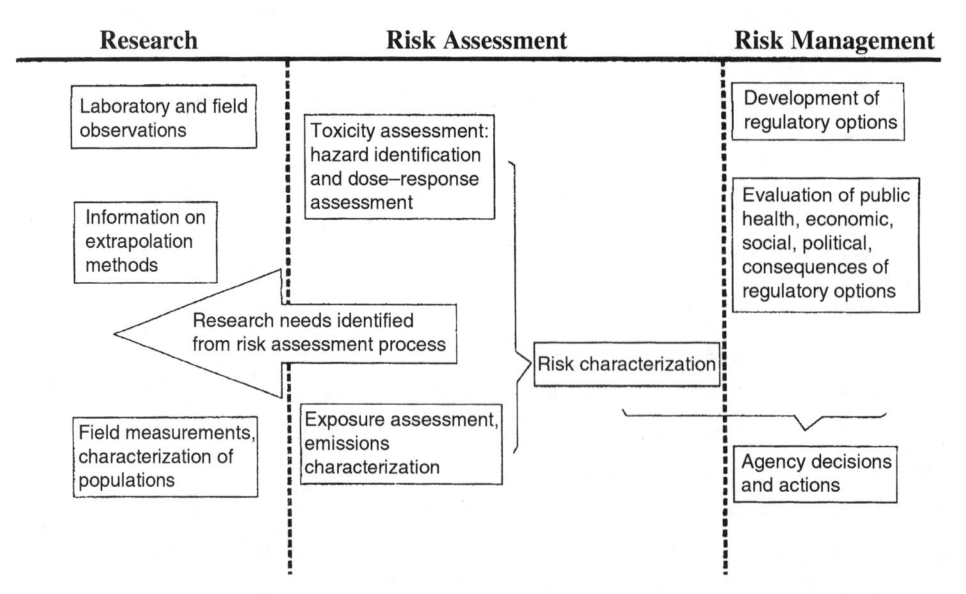

Figure 18.1 Elements of risk assessment and risk management. Risk assessment provides a means to organize and interpret research data in order to inform decisions regarding human and environmental health. Through the risk assessment process, important data gaps and research needs are often identified, assisting in the prioritization of basic and applied toxicological research. [Adapted from NRC (1983).]

Step 3: Exposure Assessment. The process of specifying the exposed population, identifying potential exposure routes, and measuring or estimating the magnitude, duration, and frequency of exposure. Exposures can be assessed by direct measurement or estimated with a variety of exposure models. Exposure assessment can be quite complex since exposure frequently occurs to a mixture of chemicals from a variety of sources (air, water, soil, food, etc.).

Step 4: Risk Characterization. The integration of information from steps 1–3 to develop a qualitative or quantitative estimate of the likelihood that any of the hazards associated with the chemical(s) of concern will be realized. The characterization of risk must often encompass multiple populations having varying exposures and sensitivities. This step is particularly challenging as a variety of data must be assimilated and communicated in such a way as to be useful to everyone with an interest in the outcome of the risk assessment. This may include not only governmental and industry risk managers but also the public as well. This step includes a descriptive characterization of the nature, severity, and route dependency of any potential health effects, as well as variation within the population(s) of concern. Any uncertainties and limitations in the analysis are described in the risk characterization, so that the strengths, weaknesses, and overall confidence in the risk estimates can be understood.

Having defined the four classical steps in risk assessment, it is important to note that the hazard identification step alone may sometimes support a conclusion that a chemical presents little or no risk to health. Also, circumstances may exist in which no risk can be inferred from an exposure assessment that reveals no opportunity for individuals to receive a dose of the chemical. Therefore, situations sometimes exist where a comprehensive risk assessment is unnecessary. In such instances, it may be more practical to communicate findings in a qualitative manner, that is, to state simply that it is highly unlikely chemical X will pose any significant health risk. At other times, quantitative expressions of risk might be more appropriate, as in the case of a population chronically exposed to a known human carcinogen in drinking water. An expression of such risk might be that the lifetime excess cancer risk from exposure is 3 in 1,000,000. Often, such numerical expressions of risk convey an unwarranted sense of precision by failing to convey the uncertainty inherent in their derivation. They may also prove difficult for nontechnical audiences to comprehend. On the other hand, qualitative risk estimates may appear more subjective and not invoke the same degree of confidence in the risk assessment findings as a numerical expression of risk. Also, qualitative expressions of risk do not readily allow for comparative risk analyses, a useful exercise for putting added risk into context. Although addressed later in this chapter, it is worth mentioning here that effective risk communication plays a key role in utilizing risk assessment findings for the protection of public health.

Risk Assessment in a Regulatory Context: The Issue of Conservatism

Regulatory agencies charged with protecting public health and the environment are constantly faced with the challenge of setting permissible levels of chemicals in the home, workplace, and natural environment. For example, the Occupational Safety and Health Administration (OSHA) is responsible for setting limits on chemical exposure in the workplace, the Food and Drug Administration (FDA) has permissible limits on chemicals such as pesticides in the food supply, and the Environmental Protection Agency (USEPA) regulates chemical levels in air, water, and sometimes soils. Ideally, the level of chemical contamination or residues in many of these media (food, water, air, etc.) would be zero, but this simply is not feasible in a modern industrial society. Although it may not be possible to completely eliminate the presence of unwanted chemicals from the environment, there is almost universal agreement that we should limit exposures to these chemicals to levels that do not cause illness or environmental destruction. The process by which regulatory agencies set limits with this goal in mind is a combination of risk assessment and risk management.

The risks associated with chemical exposure are not easily measured. While studies of worker health have been extremely valuable in assessing risks and setting standards for occupational chemical exposure, determining risks from lower doses typically associated with environmental exposures has been difficult. Epidemiologic studies of environmental chemical exposure can provide some estimate of increased risk of specific diseases associated with a particular chemical exposure compared with a control population, but there are several problems in attempting to generalize the results of such studies. Exposure of a population is often difficult to quantify, and the extrapolation of observations from one situation to another (e.g., different populations, different manners of exposure, different exposure levels, different exposure durations) is challenging. For the most part, risk assessments for environmental chemical exposures must rely on modeling and assumptions to generate estimates of potential risks. Because these risk estimates usually cannot be verified, they represent hypothetical or theoretical risks. This is an important facet of risk assessment that is often misunderstood by those who erroneously assume that risk estimates for environmental chemical exposure have a strong empirical basis.

As discussed in subsequent sections, there are many sources of uncertainty in making risk estimates. Good data regarding chemical exposure and uptake are seldom available, forcing reliance on models and assumptions that may or may not be valid. Toxicity information often must be extrapolated from one species to another (e.g., use of data from laboratory mice or rats for human health risk assessment), from one route of exposure to another (e.g., use of toxicity data following ingestion to evaluate risks from dermal exposure), and from high doses to the lower doses more commonly encountered with environmental exposure. In view of all of these uncertainties, it is impossible to develop precise estimates of risks from chemical exposures. Choices made by the risk assessor, such as which exposure model to use or how to scale doses when extrapolating from rodents to humans, can have a profound impact on the risk estimate.

Regulatory agencies address uncertainty in risk assessments by using conservative approaches and assumptions; that is, in the face of scientific uncertainty, they will select models and assumptions that tend to overestimate, rather than underestimate, risk so as to be health protective. Since most risk assessments are by, or for, regulatory agencies, this conservatism is a dominant theme in risk assessments and a continuous source of controversy. Some view the conservatism employed by regulatory agencies as excessive, resulting in gross overestimation of risks and unwarranted regulations that waste billions of dollars. Others question whether regulatory agencies are conservative enough, and suggest that the public (particularly more sensitive individuals such as children) may not be adequately protected by contemporary risk assessment approaches.

Defining Risk Assessment Problems

A coherent risk assessment requires a clear statement of the risk problem to be addressed. This should be developed very early in the risk assessment process and is shaped by the question(s) the risk assessment is expected to answer. Ideally, both the risk assessor(s) and the individuals or organizations that will ultimately use the risk assessment will have input. This helps ensure that the analysis will be technically sound and serve its intended purpose.

One of the first issues to address is which chemicals or agents should be included in the analysis. In some situations this may be straightforward, such as a risk assessment focused specifically on occupational exposure to a particular chemical. In other circumstances, the chemicals of concern may not be obvious. An example of this would be risk assessment for a chemical disposal site where the chemicals present and their amounts are initially unknown. A related issue is which health effects should the risk assessment address. While it is tempting to answer "all of them," it must be recognized that each chemical in a risk assessment is capable of producing a variety of adverse health effects, and the dose–response relationships for these effects can vary substantially. Developing estimates of risks for each of the possible adverse effects of each chemical of interest is usually impractical. A simpler approach is to estimate risks for the health effect to which individuals are most sensitive, specifically, the one that occurs at the lowest dose. If individuals can be protected from this effect, whatever it might be, they will logically be protected from all other effects. Of course, this approach presumes that the most sensitive effect has been identified and dose–response relationship information for this effect exists. Obviously, for this to be effective, the toxicology of each chemical of interest must be reasonably well characterized.

In defining the risk problem, populations potentially at risk must be identified. These populations would be groups of individuals with distinct differences in exposure, sensitivity to toxicity, or both.

For example, a risk assessment for a contaminated site might include consideration of workers at the site, occasional trespassers or visitors to the site, or individuals who live at the site if the land is (or might become) used for residential purposes. If residential land use is contemplated, risks are often calculated separately for children and adults, since they may be exposed to different extents and therefore have different risks. Depending on the goals of the risk assessment, risks may be calculated for one or several populations of interest.

Many chemicals move readily in the environment, from one medium to another. Thus, a chemical spilled on the ground can volatilize into the air, migrate to groundwater and contaminate a drinking water supply, or be carried with surface water runoff to a nearby stream or lake. Risk assessments have to be cognizant of environmental movement of chemicals, and the fact that an individual can be exposed to chemicals by a variety of pathways. In formulating the risk problem, the risk assessor must determine which of many possible pathways are complete; that is, which pathways will result in movement of chemicals to a point where contact with an individual will occur. Each complete pathway provides the opportunity for the individual to receive a dose of the chemical, and should be considered in some fashion in the risk assessment. Incomplete exposure pathways—those that do not result in an individual coming in contact with contaminated environmental media (e.g., air, water, soil)—can be ignored, since they offer no possibility of receiving a dose of chemical and therefore pose no risk.

Risk assessments can vary considerably in the extent to which information on environmental fate of contaminants is included in the analysis. Some risk assessments, for example, have attempted to address risks posed by chemicals released to the air in incinerator emissions, and subsequently deposited on the ground where they are taken up by forage crops that are consumed by dairy cattle. Consumption of meat or milk from these cattle was regarded as a complete exposure pathway from the incinerator to a human receptor. As the thoroughness of the risk assessment increases, so does the complexity. As a practical matter, complete exposure pathways that are thought to be minor contributors to total exposure and risk are often acknowledged but not included in the calculation of risk to make the analysis more manageable.

Often, exposure can lead to uptake of a chemical by more than one route. For example, contaminants in soil can enter the body through dermal absorption, accidental ingestion of small amounts of soil, or inhalation of contaminants volatilized from soil or adherent to small dust particles. Consequently, the manner of anticipated exposure is important to consider, as it will dictate the routes of exposure (i.e., inhalation, dermal contact, or ingestion) that need to be included in the risk assessment for each exposure scenario.

As discussed in the following section contrasting human health and ecological risk assessment, problem formulation is more challenging when conducting ecological risk assessments. Instead of one species, there are several to consider. Also, the exposure pathway analysis is more complicated, at least in part because some of the species of interest consume other species of interest, thereby acquiring their body burden of chemical. Unlike human health risk assessments, where protection of individuals against any serious health impact is nearly always the objective, goals for ecological risk assessments are often at the population, or even ecosystem, level rather than focusing on individual plants and animals. Consequently, development of assessment and measurement endpoints consistent with the goals of the ecological risk assessment is essential in problem formulation for these kinds of analyses.

Human Health versus Ecological Risk Assessments: Fundamental Differences

Ecological risk assessments are defined as those that address species other than humans, namely, plant and wildlife populations. Historically, the risk assessment process has focused primarily on addressing potential adverse effects to exposed human populations, and the development of well-defined methods for human health risk assessment preceded those for ecological risk assessment. However, increasing concern for ecological impacts of chemical contamination has led to a "catching up" in risk assessment methodology. While detailed methods for both human health and ecological risk assessment are now in place, they aren't identical. The conceptual basis may be similar, including some form of hazard identification, exposure assessment, dose–response assessment, and risk characterization. However, there are some important differences in approaches, reflecting the reality that there are some important differences in evaluating potential chemical effects in humans versus plants and wildlife.

The most obvious difference between human health and ecological risk assessments is that the ecological risk assessments are inherently more complicated. Human health risk assessments, of course, deal with only one species. Ecological risk assessments can involve numerous species, many of which may be interdependent. Given the nearly endless array of species of plants and animals that might conceivably be affected by chemical exposure, there must be some process to focus on species that are of greatest interest to keep the analysis to a manageable size. A species may warrant inclusion in the analysis because it is threatened or endangered, because it is a species on which many others depend (e.g., as a food source), or because it is especially sensitive to toxic effects of the chemical and can therefore serve as a sentinel for effects on other species.

The increased complexity of analysis for ecological risk assessments extends to evaluation of exposure. In human health risk assessment, the potential pathways by which the chemical(s) of interest can reach individuals must be assessed and, if possible, the doses of chemicals received by these pathways estimated. In an ecological risk assessment, the same process must be undertaken, but for several species instead of just one. Also, an ecological risk assessment typically must evaluate food-chain exposure. This is particularly important when chemicals of interest tend to bioaccumulate, resulting in very high body burdens in predator species at the top of the food chain. Not only must the potential for bioaccumulation be assessed, but the escalating doses for species of interest must be estimated according to their position in the food chain. This type of analysis is rarely included in human health risk assessments.

A third distinction between human health and ecological risk assessments lies in the assessment objectives. Human health risk assessments characteristically focus on the most sensitive potential adverse health effect, specifically, that which occurs at the lowest dose. In this way, they are directed to evaluating the potential for *any* health effect to occur. For ecological risk assessments, the analyses generally address only relatively severe toxic endpoints such as mortality or reproductive failure. Thus, the goal of an ecological risk assessment might be to determine whether the presence of a chemical in the environment at a particular concentration would result in declining populations of specific species (e.g., due to mortality or reproductive failure), disappearance of a species in a particular area, or loss of an entire ecosystem, depending on risk management objectives. It is entirely possible that chemical exposure could result in the deaths of many animals, but as long as the populations were stable, the risk would be considered acceptable. This reflects philosophical and risk management differences in terms of what constitutes an unacceptable chemical impact on humans versus plants or wildlife.

Because of the greater potential complexity of an ecological risk assessment, more attention must be given to ensuring that an analysis of appropriate scope and manageable size is achieved. For this reason, ecological risk assessments are more iterative in nature than their human health counterparts. An ecological risk assessment begins with a screening-level assessment, which is a form of preliminary investigation to determine whether unacceptable risks to ecological receptors may exist. It includes a review of data regarding chemicals present and their concentrations, species present, and potential pathways of exposure. It is a rather simplified analysis that uses conservative or worst-case assumptions regarding exposure and toxicity. If, using very conservative models and assumptions, the screening analysis finds no indication of significant risks, the analysis is concluded. If the results of the screening analysis suggest possible ecological impacts, a more thorough analysis is conducted that might include additional samples of environmental media, taking samples of wildlife to test for body burdens of chemicals, carefully assessing the health status of populations exposed to the chemical, and conducting toxicity tests, more sophisticated fate and transport analysis of the chemicals of potential concern, and a more detailed and accurate exposure assessment.

18.2 HAZARD IDENTIFICATION

Hazard identification involves an assessment of the intrinsic toxicity of the chemical(s) of potential concern. This assessment attempts to identify health effects characteristically produced by the chemical(s) that may be relevant to the risk assessment. While this may appear to be a straightforward exercise, in reality it requires a good deal of careful analysis and scientific judgment. The reason for this is that the risk assessor rarely has the luxury of information that adequately describes the toxicity of a chemical under the precise set of circumstances to be addressed in the risk assessment. Instead, the risk assessor typically must rely on incomplete data derived from species other than the one of interest, under exposure circumstances very different from those being evaluated in the risk assessment. The existence in the scientific literature of poorly designed studies with misleading results and conclusions, as well as conflicting data from seemingly sound studies, further complicates the task.

This section of the chapter discusses some of the considerations when reviewing and evaluating the toxicological literature for assessment of intrinsic toxicity. Many of these considerations address suitability of data for extrapolation from one set of circumstances to another, while others pertain to the fundamental reliability of the information. Much of the discussion regarding extrapolation deals with assessing the value of animal data in predicting responses in humans, since human health risk assessments are forced to rely predominantly on animal studies for toxicity data. Keep in mind that most of the same extrapolation issues are equally relevant for ecological risk assessments, where toxicity in wildlife species has to be inferred from data available only from laboratory animal species.

Information from Epidemiologic Studies and Case Reports

Observations of toxicity in humans can be extremely valuable in hazard identification. They offer the opportunity to test the applicability of observations made in animal studies to humans and may even provide an indication of the relative potency of the chemical in humans versus laboratory animal models. If the human studies are of sufficient size and quality, they may stand alone as the basis for hazard identification in human health risk assessment.

Despite the attractiveness of human studies, it is important to keep in mind that they often have significant limitations. For example, it may be impossible to eliminate all of the confounding variables in any epidemiological study (see Chapter 21). A less-than-rigorous effort to properly match exposed and control populations makes it difficult or impossible to attribute with confidence any observed differences in health effects to chemical exposure. Even in well-designed epidemiologic studies, there is always the possibility that an unknown critical factor causally related to the health effect of interest has been missed. For this reason, a consistent association between chemical exposure and a particular effect in several studies is important in establishing whether the chemical produces that effect in humans.

Other criteria in evaluating epidemiologic studies include the following:

- The positive association (correlation) between exposure and effect must be seen in individuals with definitive exposure.
- The positive association cannot be explained by bias in recording, detection, or experimental design.
- The positive association must be statistically significant.
- The positive association should show both dose and exposure–duration dependence.

Information from Animal Studies

Typically, data from studies using laboratory animals must be used for some or all of the intrinsic toxicity evaluation of a chemical in humans. There are several aspects that need to be considered when interpreting the animal data, as discussed below.

Breadth and Variety of Toxic Effects The toxicological literature should be reviewed in terms of the types of effects observed in various test species. This is an important first step in chemical toxicity evaluation because:

- It identifies potential effects that might be produced in humans. To some extent, the consistency with which an effect is observed among different species provides greater confidence that this effect will occur in humans as well. An effect that occurs in some species but not others, or one sex but not the other, signals that great care will be needed in extrapolating findings in animals to humans without some form of corroborating human data.
- A comparison of effects within species (e.g., sedation vs. hepatotoxicity vs. lethality) helps establish a rank order of the toxic effects manifested as the dose increases. This aids in identifying the most sensitive effect. Often this effect becomes the focus of a risk assessment, since protecting against this effect will protect against all effects. Also, comparisons of dose–response relationships within species can provide an estimation of the likelihood that one toxic effect will be seen given the appearance of another.

Mechanism of Toxicity Understanding the mechanism of action of a particular chemical helps establish the right animal species to use in assessing risk, and to determine whether the toxicity is likely to be caused in humans. For example, certain halogenated compounds are mutagenic and/or carcinogenic in some test species but not others. Differences in carcinogenicity appear to be related to differences in metabolism of these chemicals, because their metabolism is an integral part of their mechanism of carcinogenesis. For these chemicals, then, a key issue in selecting animal data for extrapolation to humans is the extent to which metabolism in the animal model resembles that in humans. A second example is renal carcinogenicity from certain chemicals and mixtures, including gasoline. Gasoline produces renal tumors in male rats, but not female rats or mice of either sex. The peculiar susceptibility of male rats to renal carcinogenicity of gasoline can be explained by its mechanism of carcinogenesis. Metabolites of gasoline constituents combine with a specific protein, α-2μ-globulin, to produce recurring injury in the proximal tubules of the kidney. This recurring injury leads to renal tumors. Female rats and mice do not produce this protein, explaining why they do not develop renal tumors from gasoline exposure. Humans do not produce the protein either, making the male rat a poor predictor of human carcinogenic response in this situation.

In a sense, choosing the best animal model for extrapolation is always a catch-22 situation. Selection of the best model requires knowledge of how the chemical behaves in both animals and humans, including its mechanism of toxicity. In the situations in which an animal model is most needed—when we have little data in humans—we are in the worst position to select a valid model. The choice of an appropriate animal model becomes much clearer when we have a very good understanding of the toxicity in humans and animals, but in this situation there is, of course, much less need for an animal model.

In addition to helping identify the best species for extrapolation, knowledge of the mechanism of toxicity can assist in defining the conditions required to produce toxicity. This is an important aspect of understanding the hazard posed by a chemical. For example, acetaminophen, an analgesic drug used in many over-the-counter pain relief medications, can produce fatal liver injury in both animals and humans. By determining that the mechanism of toxicity involves the production of a toxic metabolite during the metabolism of high doses, it is possible to predict and establish its safe use in humans, determine the consequences of various doses, and develop and provide antidotal therapy.

Dosages Tested Typically, animal studies utilize relatively high doses of chemicals so that unequivocal observations of effect can be obtained. These doses are usually much greater than those received by humans, except under unusual circumstances such as accidental or intentional poisonings. Thus, while animal studies might suggest the possibility of a particular effect in humans, that effect may be unlikely or impossible at lower dosages associated with actual human exposures. The qualitative information provided by animal studies must be viewed in the context of dose–response relationships. Simply indicating that an effect might occur is not enough; the animal data should indicate at what dosage the effect occurs, and equally importantly, at what dosage the effect does *not* occur.

Validity of Information in the Literature Any assessment of the intrinsic toxicity of a chemical begins with a comprehensive search of the scientific literature for relevant studies. While all of the studies in the literature share the goal of providing new information, the reality of the situation is that all are not equally valuable. Studies may be limited by virtue of their size, experimental design, methods employed, or the interpretations of results by the authors. These limitations are sometimes not readily apparent, requiring that each study be evaluated carefully and critically. The following are some guidelines to consider when evaluating studies:

- Has the test used an unusual, new, or unproved procedure?
- Does the test measure a toxicity directly, or is it a measure of a response purported to indicate an eventual change (a pretoxic manifestation, etc.)?
- Have the experiments been performed in a scientifically valid manner?
- Are the observed effects statistically significant against an appropriate control group?
- Has the test been reproduced by other researchers?
- Is the test considered more or less reliable than other types of tests in which the chemical has also been tested but has yielded different results?
- Is the species a relevant or reliable human surrogate, or does this test conflict with other test data in species phylogenetically closer to humans?
- Are the conclusions drawn from the experiment justified by the data, and are they consistent with the current scientific understanding of the test or area of toxicology?
- Is the outcome of the reported experiment dependent on the test conditions, or is it influenced by competing toxicities?
- Does the study indicate causality or merely suggest a correlation that could be due to chance?

Other Considerations Numerous confounders can affect the validity of information derived from animal studies and its application or relevance to human exposure to the same chemical. Issues regarding selection of the appropriate species for extrapolation are discussed above. Even if the selection of species is sound, certain other characteristics of the experimental animals can influence toxic responses, and therefore the extrapolation of these responses to humans. Examples include the age of the animal (e.g., whether studies in adult animals are an appropriate basis for extrapolation to human children), the sex of the animal (obviously, studies limited to just male or female animals cannot address all of the potential toxicities for both sexes of humans), disease status (e.g., whether results obtained in healthy animals are relevant to humans with preexisting disease, and vice versa), nutritional status (e.g., whether studies in fasted animals accurately reflect what occurs in fed humans), and environmental conditions.

Other confounders go beyond the animal models themselves and pertain to the type of study conducted. For example, studies involving acute exposure to a chemical are usually of limited value in understanding the consequences of chronic exposure, and chronic studies generally offer little insight into consequences of acute exposure. This is because chronic toxicities are often produced by mechanisms different from those associated with acute toxicities. For this reason, good characterization of the intrinsic toxicity of a chemical requires information from treatments of varying duration, ranging from a single dose to exposure for a substantial portion of the animal's lifetime.

18.3 EXPOSURE ASSESSMENT: EXPOSURE PATHWAYS AND RESULTING DOSAGES

Exposure assessment can be defined as the measurement or estimation of the amount or concentration of a chemical(s) coming into contact with the body at potential sites of entry (e.g., skin, lung, GI tract). Not only are the amount and route of exposure concerns, but so too are the exposure duration, exposure frequency, and any factors that modify the ability of the chemical to traverse the portals of entry into the body. In cases where a potential chemical hazard exists, exposure assessment is an obligatory part of the risk assessment process. Without exposure, even the most hazardous chemical poses no risk. Conversely, excessive exposure to minimally hazardous chemicals may pose an unacceptable risk. Therefore, risk assessment requires that toxicity and exposure assessments be coupled. Initially, exposure assessments should identify all potential exposure pathways and assess their completeness, after which the quantification of exposure via each relevant pathway should be determined.

There are two basic methods for quantifying exposure: exposure measurement and exposure modeling. Measurement results in the most accurate and realistic exposure data, but fully characterizing variable exposures that might occur to multiple receptors via multiple pathways for an extended period of time is seldom feasible. In general, the measurement of occupational exposure is easier than environmental exposure, since the former usually occurs in a confined facility whereas the latter involves more complex time–activity patterns. Also, occupational exposure limits are typically based on 8-h time-weighted averages and 15-min short-term exposure limits, making monitoring for compliance purposes manageable from a time standpoint. Examples of personal exposure measures include analyzing a person's intake of food and water and the contaminants therein, collecting and analyzing a urine sample at the end of a workshift, and measuring airborne exposure with a portable sampling device suspended in a person's breathing zone. Where environmental exposure to a large population is at issue, personal exposure monitoring is not a realistic approach. Rather, environmental media suspected of being contaminated are sampled and population-based assumptions about intake rates are made. Frequently, personal questionnaires and time–activity logs are helpful in making accurate exposure estimates within a large population.

In those cases where monitoring data are unavailable or inadequate for exposure assessment, models are used to simulate the behavior of chemicals and predict their concentrations in the environment. Hundreds of such exposure models exist, including atmospheric models, surface-water models, groundwater models, and food-chain models. All of these models are limited by uncertainty in the data input, as well as uncertainty as to the predictive capability of a generic model for a specific exposure scenario. In recognition of this uncertainty, models used for regulatory purposes tend to provide liberal estimates of exposure that may overstate risk. Whenever models are used, an attempt should always be made to collect site- or situation-specific data for the purpose of model validation. Despite their limitations, exposure models are of value in that they can make predictions for an unlimited number of exposure scenarios and predict past and future exposures. Exposure measurement, on the other hand, is limited to the present.

Exposure or concentration is often expressed in units of $\mu g/m^3$ (air), $\mu g/L$ (water), or $\mu g/cm^2$ (skin). Air and water concentrations are also frequently reported in parts per million (ppm) or parts per billion (ppb), units that reflect the weight or volume of chemical per unit volume of the carrier medium. For some chemicals, risk can be directly calculated from these concentration terms using unit risk factors that are expressed as risk per $\mu g/L$ water or risk per $\mu g/m^3$ air. In such cases, risk is simply the product of the chemical concentration and the unit risk factor. A word of caution is in order, however. Since unit risks are based on exposure factors typical of an average adult (e.g., 70 kg body weight, consumption of 2 L/day water, breathing rate of 20 m^3/day air), they must be adjusted accordingly when used to calculate the risk to receptors having different exposure factor values (e.g., children). Such adjustment is necessary since exposure factor values determine dose, which in turn determines the nature of the toxic response. In other words, exposure of an adult and child to the same concentration of a chemical is likely to result in different doses that may translate into different toxic responses. Exposure or concentration data can also be directly compared to many occupational [e.g., OSHA permissible exposure limits (PELs), and American Conference of Governmental Industrial Hygienists (ACGIH) threshold limit values (TLVs)] and environmental [USEPA maximum contaminant levels (MCLs), national ambient air-quality standards, and reference concentrations (RfCs)] exposure standards that have risk considerations inherent in their derivation.

While some exposure to a hazardous chemical is required in order to have risk, it is dose that relates more closely to the toxic response. Dose, often expressed in units of mg/kg·day, is the amount of chemical that is either absorbed, or available to be absorbed, into the body where it can interact with the target tissue (liver, thyroid, red blood cells, etc.). Knowledge of the exposure or concentration of a chemical is essential to determine the magnitude of the dose received. So, too, is knowledge of certain exposure factors such as the volume of contaminated air inhaled or food and water ingested per unit time. In fact, in many cases, it is quite simple to calculate dose when exposure concentration and exposure rate are known. To assist the risk assessor in making dose calculations, the USEPA has published equations that are applicable to a variety of the most frequently encountered exposure scenarios. These equations, including the one used below, are found in the document entitled, *Risk Assessment Guidance for Superfund* (*RAGS*), *Human Health Evaluation Manual Part A*.

To illustrate dose calculation, assume that a 16-kg child ingests 200 mg soil/day containing 400 mg/kg of chemical X, a volatile solvent. As shown below, the child's dose of chemical X from the ingestion of soil is 1.25×10^{-3} mg/kg·day. This figure may not represent total dose, however, since dermal and/or inhalation exposure to the volatile chemical is likely. This illustrates the importance of considering all exposure pathways when assessing exposure for the purpose of risk assessment. Failing to do so may result in the underestimation of risk.

The following formula can be used to determine the residential exposure from ingestion of a chemical in soil by a 5-year-old child receptor:

$$\text{Dose (mg/kg·day)} = \frac{\text{CS} \times \text{IR} \times \text{CF} \times \text{FI} \times \text{EF} \times \text{ED}}{\text{BW} \times \text{AT}}$$

$$= \frac{400 \text{ mg/kg} \times 200 \text{ mg/day} \times 10^{-6} \text{ kg/mg} \times 0.25 \times 365 \text{ days/year} \times 6 \text{ years}}{16 \text{ kg} \times 2{,}190 \text{ days}}$$

$$= 1.25 \times 10^{-3} \text{ mg/kg·day}$$

where: CS = chemical concentration in soil (mg/kg) = 400 mg/kg (site-specific value)

IR = ingestion rate (mg soil/day) = 200 mg soil/day (default value for children 1–6 years old)

CF = conversion factor (10^{-6} kg/mg)

FI = fraction ingested from contaminated source (unitless) = 0.25 (site-specific value)

EF = exposure frequency (days/year) = 365 days/year (site-specific value)

ED = exposure duration (years) = 6 years (site-specific value)

BW = body weight (kg) = 16 kg (default value for children 1–6 years old)

AT = averaging time (period over which exposure is averaged in days) = 6 years × 365 days/year = 2,190 days for non-cancer effects

As shown in this example, where the intake of chemical X from soil ingestion was calculated for a 5-year-old child, default values for input variables can be used where site-specific data are lacking. The most complete collection of default values has been compiled and published by the USEPA as the *Exposure Factors Handbook*, Volumes I–III. Numerous distributions (versus discrete values) for input variables used in exposure calculations have also been reported that are of value to probabilistic risk assessment. In addition, the USEPA has published several guidance documents that address many of the issues related to characterizing exposures for selected pathways:

- *Guidelines for Exposure Assessment* (USEPA 1992a)
- *Dermal Exposure Assessment: Principles and Applications* (USEPA 1992b)
- *Methodology for Assessing Health Risks Associated with Indirect Exposure to Combustor Emissions* (USEPA 1990)
- *Risk Assessment Guidance for Superfund* (USEPA 1989)

- *Estimating Exposures to Dioxin-like Compounds* (USEPA 1994)
- *Superfund Exposure Assessment Manual* (USEPA 1988a)
- *Selection Criteria for Mathematical Models Used in Exposure Assessments* (USEPA 1988b)
- Standard Scenarios for Estimating Exposure to Chemical Substances During Use of Consumer Products (USEPA 1986a)
- Pesticide Assessment Guidelines, Subdivisions K and U (USEPA 1984, 1986b)
- Methods for Assessing Exposure to Chemical Substances, Volumes 1–13 (USEPA 1983–1989)

Once doses have been estimated for all exposure pathways, they can be directly compared to toxicity constants such as USEPA reference doses (RfDs) and Agency for Toxic Substances and Disease Registry minimal risk levels (MRLs) to assess noncancer risks, or alternatively, multiplied by cancer slope factors to obtain an estimate of cancer risk. Another word of caution is in order, however. Toxicity constants, including cancer slope factors, may be specific for particular exposure routes (MRLs vary by exposure route and exposure duration) since target organ dose, and for some chemicals the target organ itself, can be exposure route-dependent. As mechanistic data play a greater role in the risk assessment process, it is also likely that multiorgan carcinogens will have organ-specific cancer slope factors in recognition of organ-specific cancer mechanisms and dose–response curves.

While the administered dose and absorbed dose are common dose measures, they do not reflect the amount of the chemical or its metabolite(s) that ultimately produce the toxic response (except in cases where chemicals exert their action locally, as in the case of strong acids or bases that produce dermatotoxicity on contact). The toxic response is more closely linked to the dose in the target tissue of interest (see Figure 18.2 for a schematic showing the relationships between exposure and various dose measures). For example, solvent-induced neurobehavioral toxicity may be a function of peak brain concentration of the parent compound, whereas liver toxicity from the same chemical may be related to the hepatic tissue concentration of one or more metabolites over time [called the *area under the tissue concentration–time curve* (AUC)]. The identification of such internal dose measures that are mechanistically linked to various toxicities holds promise for improving the risk assessment process.

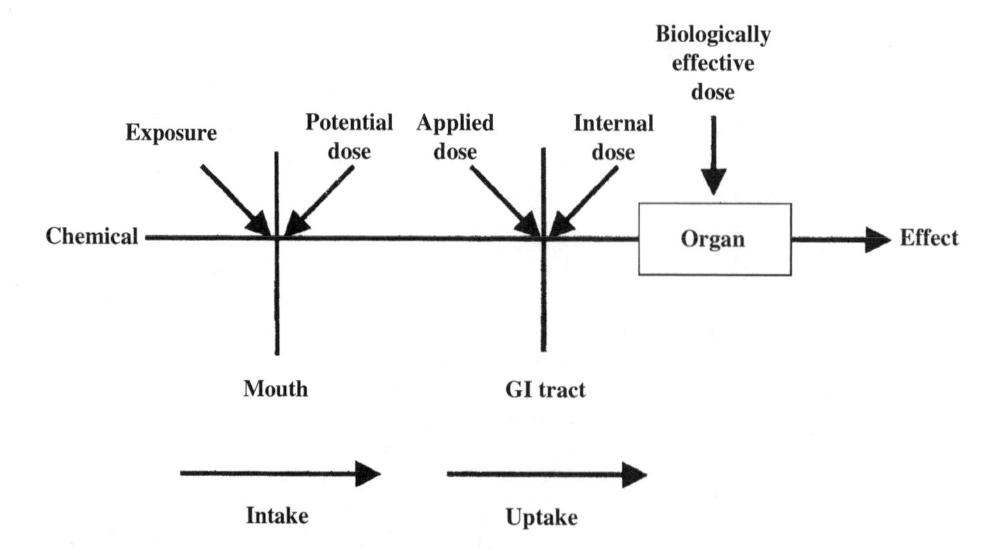

Figure 18.2 The relationship between exposure and various measures of dose. [From *Exposure Factors Handbook*, Vol. 1, *General Factors*, USEPA (1997).]

After all, it is commonly assumed in interspecies extrapolation that the target tissue dose required to produce a biological effect of a given intensity is quantitatively equivalent across species. Therefore, dose–response curves generated with measures of target tissue dose should be readily extrapolatable across species since they obviate the need for consideration of interspecies differences in the toxicokinetics of an administered dose. Unfortunately, dose–response curves of this nature are rare primarily because of the technical difficulties inherent in internal dose measurement. This is likely to change, however, as advancements are made in analytical chemistry and physiologically based pharmacokinetic (PBPK) models find their way into the mainstream. Such models are powerful tools with which to estimate internal dose measures from an endless variety of exposure scenarios to physiologically diverse receptors. As such, they are particularly valuable for the purpose of interspecies extrapolation.

Biological monitoring is another means of exposure assessment. When it is conducted to measure a chemical or its metabolites in the urine, blood, or tissue (including hair and fingernails) of an exposed individual, the chemical and its metabolites are referred to as *biomarkers of exposure*. Other potential biomarkers include DNA and protein adducts, mutations, chromosomal aberrations, genes that have undergone induction, and a host of other "early" cellular or subcellular events thought to link exposure and effect. The characterization and quantification of these latter biomarkers is known as *molecular dosimetry*. If found to be correlated with susceptibility, exposure, and effect, these biomarkers could considerably alter conventional approaches to risk assessment. Perhaps the best known example of such a correlation is urinary aflatoxin-DNA adducts and liver cancer. While molecular dosimetry holds promise for risk assessment, it is yet to be developed well enough for routine application.

Despite advancements in analytical chemistry, mathematical modeling, and biomonitoring, exposure assessment remains a challenge. It is important to realize that most exposure assessments result in estimates rather than definitive values. This stems in part from the fact that site- or situation-specific values are rarely available for all of the input variables necessary to calculate exposure. Despite the challenge, efforts should continue toward conducting exposure assessments that reflect realistic exposures, rather than worst-case scenarios to which no one is exposed. The identification of the dose metric that best correlates with various toxicities should also be a priority. Since this depends on a thorough knowledge of a chemical's mode of action, advancement in exposure assessment is inextricably linked to advances in toxicology.

18.4 DOSE–RESPONSE ASSESSMENT

In this portion of the risk assessment, the dose–response relationships for the toxicities of concern must be measured, modeled, or assumed, in order to predict responses to doses estimated in the exposure assessment. While dose–response relationships could theoretically be obtained for a variety of effects from each chemical of potential concern, in practice attention is usually centered on the most sensitive effect of the chemical.

In risk assessment, two fundamentally different types of dose–response relationships are thought to exist. One is the threshold model, in which all doses below some threshold produce no effect, while doses above the threshold produce effects that increase in incidence or severity as a function of dose. The second model has no threshold—any finite, nonzero dose is thought to possess some potential for producing an adverse effect. The derivation of these two types of dose-response relationships and their use to provide estimates of risk are very different, as described in the following sections.

Threshold Models

It has long been held that for all toxicities other than cancer, there is some dose below which no observable or statistically measurable response exists. This dose, called the *threshold dose*, was graphically depicted in Chapter 1 (see also Figure 18.3). Conceptually, a threshold makes sense for most toxic effects. The body possesses a variety of detoxification and cell defense and repair

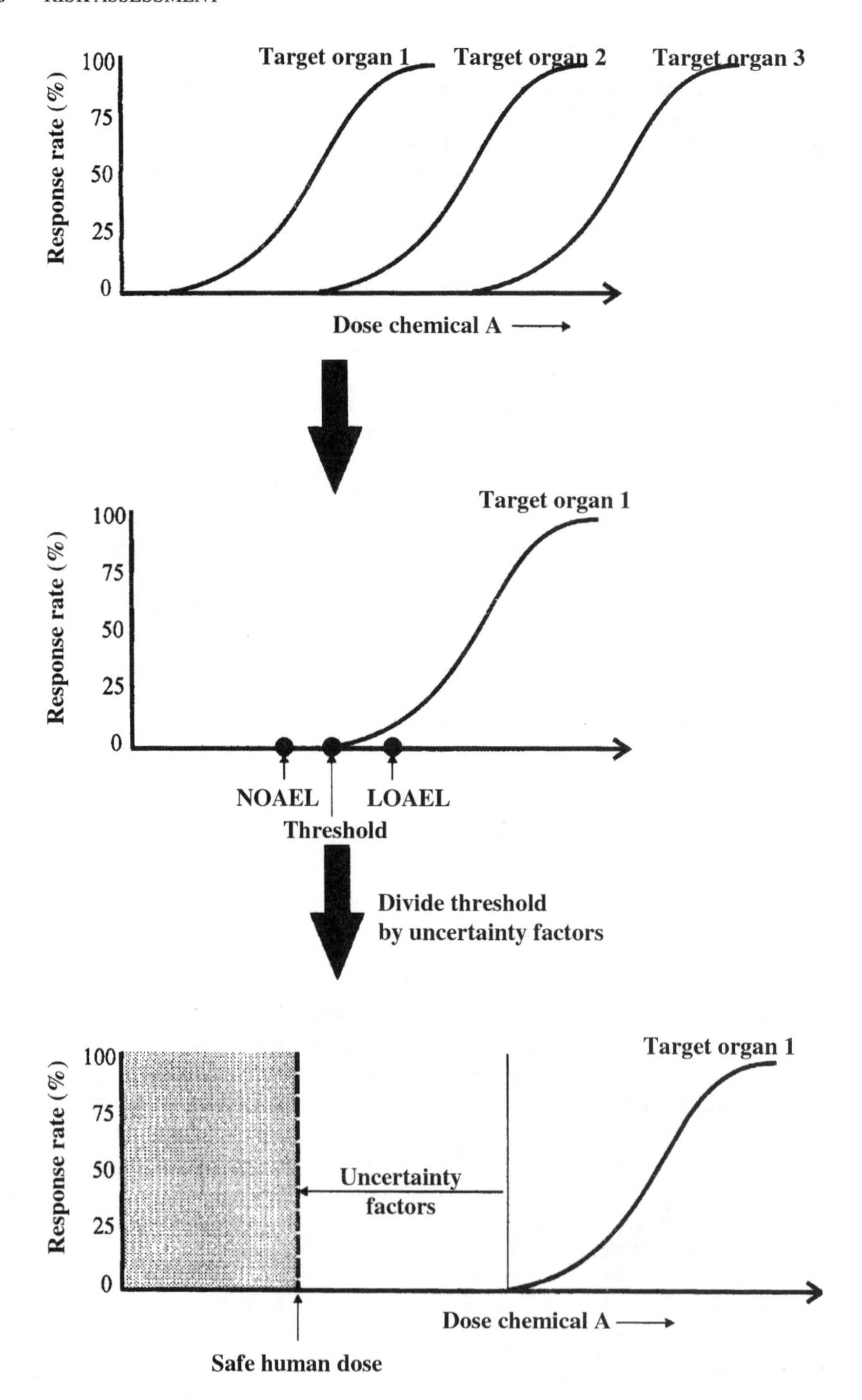

Figure 18.3 Estimation of a safe human dose (SHD). The first step is identification of the target organ or effect most responsive to the chemical (in this case, target organ 1 in the upper panel). Dose–response data for this effect are used to identify no-observable-adverse-effect-level (NOAEL) and/or lowest-observable-adverse-effect-level (LOAEL) doses in order to approximate the threshold dose. Either the NOAEL or LOAEL is divided by a series of uncertainty factors to generate the SHD.

mechanisms, and below some dose (i.e., the threshold dose), the magnitude of effect of the chemical is so small that these detoxification and defense/repair mechanisms render it undetectable.

In the most common form of threshold dose–response modeling, the threshold dose becomes the basis for establishing a "safe human dose" (SHD). Because we rarely, if ever, are able to define the true threshold point on the dose–response curve, the threshold dose must usually be approximated. Two methods of estimating the threshold are used. The more desirable method uses the highest reported dose or exposure level for which no toxicity was observed. This dose, known as the "no observable adverse effect level" (NOAEL), is considered for practical purposes to represent the threshold dose. Sometimes, the available data do not include a NOAEL; that is, all of the doses tested produced some measurable toxic effect. In this situation, the lowest dose producing an adverse effect, termed the "lowest observable adverse effect level" (LOAEL), is identified from the dose–response data. The threshold dose will lie below, and hopefully near, this dose. A threshold dose is then projected from the LOAEL, usually by dividing the LOAEL by a factor of 10 (see Figure 18.3).

From the estimates of the threshold dose, a SHD can be calculated. Different agencies have different terminology they apply to the SHD; the USEPA refers to this dosage as a "reference dose" (RfD), or if it is in the form of a concentration of chemical in air, as a "reference concentration" (RfC). Other agencies have adopted different terminology; for example, the U.S. Food and Drug Administration (FDA) uses the term "allowable daily intake" (ADI). The basic concept is the same, and the approach to development of a SHD is relatively simple, as illustrated in the flow diagram in Figure 18.3. Because a chemical may produce more than one toxic effect, the first step is to identify from the available data the adverse effect that occurs at the lowest dose. By basing the risk assessment on the most sensitive toxic endpoint or target organ, one can be reasonably certain that toxicities requiring higher doses will also be prevented. Second, the threshold dose or some surrogate measure of the threshold dose (e.g., the NOAEL, or LOAEL reduced by some amount) is identified for the most sensitive toxic endpoint. The threshold dose (or its surrogate measure) is then divided by an uncertainty factor to derive the SHD, and this dose can then be converted into an acceptably safe exposure guideline for that chemical (see Chapter 1 for examples of such conversions).

The uncertainty factor is really a composite of several uncertainty factors intended to address weaknesses in the data or uncertainties in extrapolation from animals to humans. These uncertainties arise because of our inability to directly measure the actual human threshold dose. The weaker the data set available for evaluation (few studies, limited doses tested, etc.) and the more assumptions required, the greater the uncertainty that the NOAEL or LOAEL from the literature actually represents the threshold dose in humans. The purpose of dividing the NOAEL or LOAEL by uncertainty factors is to ensure that the SHD used in the risk assessment is below the actual human threshold dose for toxicity for all individuals in the exposed population, thereby avoiding any underestimation of risk. The greater the uncertainty associated with the data, the larger the uncertainty factor required to insure protection.

The general rationale for selecting the size of the uncertainty factor for a particular area of uncertainty is as follows:

- An uncertainty factor of up to 10 is used to account for variability in sensitivity to toxicity among subjects. In practice, an uncertainty factor of 10 is almost universally applied to ensure that the final toxicity value is protective for sensitive individuals within a population.
- An uncertainty factor of up to 10 is applied in extrapolating toxicity data from one species to another. It is used to account for the possibility that humans are more sensitive to toxicity than the test species.
- An uncertainty factor of up to 10 might be applied if only acute or subchronic data are available. It is possible under these circumstances that the threshold dose for longer exposures might be lower, and this uncertainty factor is intended to protect against this possibility.

- As discussed above, an uncertainty factor of up to 10 may be applied if the only value with which to estimate the threshold dose is a LOAEL value. Division by this uncertainty factor is meant to accomplish a reduction in the LOAEL to a level at or below the threshold dose.
- An additional uncertainty factor (or in some terminology, a "modifying factor") is applied if the overall quality of the database is poor—the number of animal species tested is few, the number of toxic endpoints evaluated is small, or the available studies are found to be deficient in quality.

These uncertainty factors are compounded; that is, an uncertainty factor of 10 for sensitive individuals combined with an uncertainty factor of 10 for extrapolation of data from animals to humans results in a total uncertainty factor of 100 (10×10). Total uncertainty factors applied to develop a SHD commonly range between 300 and 1000, and values up to 10,000 or more are possible, although regulatory agencies may place a cap on the size of compounded uncertainty factors (e.g., a limit of 3000).

There are two major criticisms of the NOAEL approach for developing a SHD. One is that the ability of the NOAEL to approximate the threshold dose is dependent on dose selection and spacing in available studies, and in many cases these are not well suited to determining the threshold. A second criticism is that the approach fails to consider the shape or slope of the dose–response curve, focusing instead on results from one or two low doses exclusively. The benchmark dose (BMD) approach to estimating a SHD avoids both criticisms by making full use of the dose–response data. Dose–response data for the toxic effect of concern are fit to a mathematical model, and the model is used to determine the dose corresponding to a predetermined benchmark response. As an example, dose–response data might be used to determine the dose required to produce a 10% incidence of malformed fetuses from pregnant mice treated with a chemical. This dose would be referred to as the ED_{10}, or dose effective in producing a 10% incidence of effect. Often, for regulatory purposes, statistical treatment of the data is used to derive upper and lower confidence limit estimates of this dose. The more conservative of these is the lower confidence limit estimate of the dose, which in this case would be designated as the LED_{10} (see Figure 18.4). In order to develop a SHD from the ED_{10} or LED_{10}, a series of uncertainty factors would be applied, analogous to the NOAEL approach. In a sense, the BMD approach is like extrapolating a SHD from a LOAEL, except the LOAEL is much more rigorously defined. The BMD approach works best if there are response data available for a variety of doses and the data are in a form such that the percentage of animals responding can be readily ascertained. Other types of data presentation (e.g., continuous data) make using the BMD approach more difficult, and advantages of the BMD approach are lost if data from only one or two doses are available.

Although the term "risk" often implies probability of an adverse event, the threshold approach to assessing chemical risk does not result in risk expression in probability terms. This approach is instead directed to deriving a safe limit for exposure, and then determining whether the measured or anticipated exposure exceeds this limit. All doses or exposures below this "safe level" should carry the same chance that toxicity will occur—namely, zero. With this model the acceptability of the exposure is basically judged in a "yes/no" manner. The most common quantitative means of expressing hazard for noncancer health effects is through a hazard quotient (HQ). Agencies such as the USEPA calculate a HQ as the estimated dose from exposure divided by their form of the SHD, the RfD:

$$\mathrm{HQ} = \frac{\mathrm{D}}{\mathrm{RfD}} \quad \left(\text{or } \mathrm{HQ} = \frac{\mathrm{D}}{\mathrm{SHD}}\right)$$

where HQ = hazard quotient

D = dosage (mg/kg·day) estimated to result from exposure via the relevant route

RfD = reference dose (mg/kg·day)

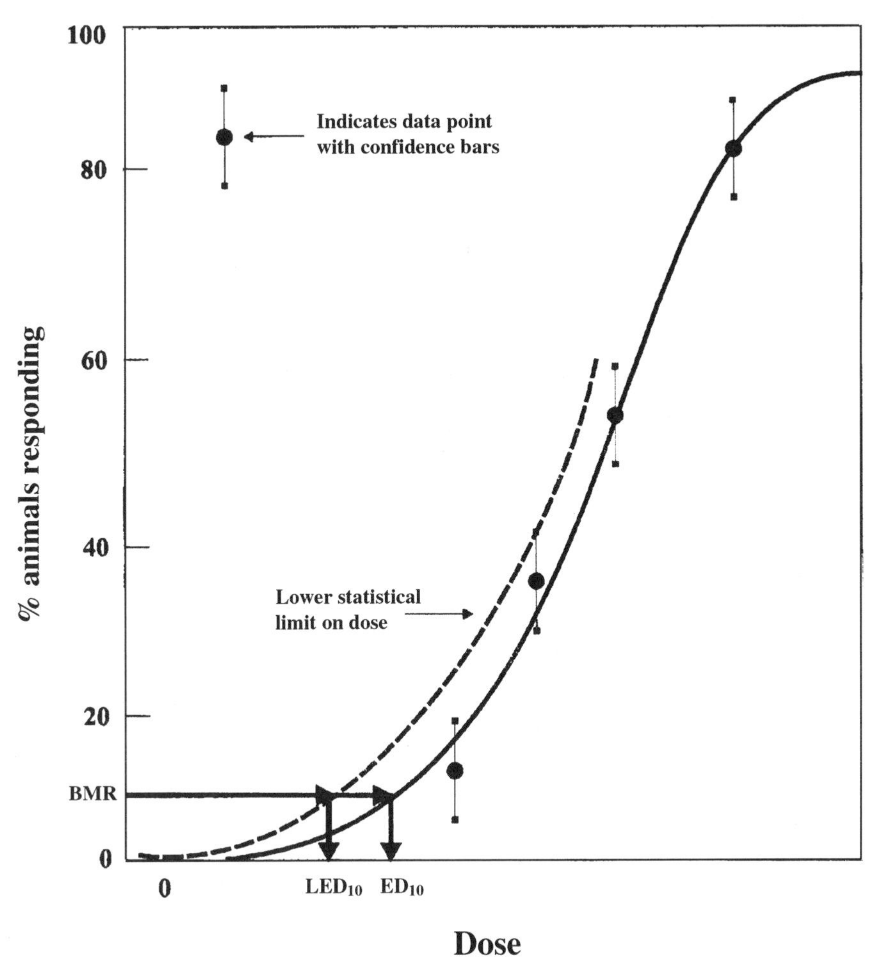

Figure 18.4 Derivation of the benchmark dose. Dose–response data for the toxic effect of concern are fit to a mathematical model, depicted by the solid line. This model is used to determine the effective dose (ED) corresponding to a predetermined benchmark response (BMR) (e.g., 10% of the animals responding). Statistical treatment of the data can be used to derive the lower confidence limit estimate of the ED, termed the LED (see dashed line). Either the ED or LED may be used to represent the benchmark dose (BMD), although the more conservative LED is typically used for regulatory purposes. [Adapted from USEPA (1995).]

Interpretation of the HQ is relatively straightforward if the value is less than one. This means that the estimated exposure is less than the SHD, and no adverse effects would be expected under these circumstances. Interpretation of HQ values greater than one is more complicated. A value greater than one indicates that the estimated exposure exceeds the SHD, but recall that the SHD includes a number of uncertainty factors that impart a substantial margin of safety. Therefore, exposures that exceed the SHD, but lie well within this margin of safety, may warrant further analysis but are unlikely to produce adverse health effects.

Dose–response relationships can vary from one route of exposure to another. A safe dose for inhalation of a chemical may be different from a safe dose for its ingestion, for example. As a result, a given chemical may have different SHDs for different routes of exposure. Since individuals are often exposed to a chemical by more than one route, separate route-specific HQ values are calculated. For example, the dose estimated to be received by inhalation would be divided by the SHD for inhalation to calculate a HQ for inhalation, while the estimated dose received from dermal contact would be divided by a dermal SHD to derive the HQ for this route of exposure. Typically, the HQ values for each relevant route of exposure are summed to derive a hazard index (HI) for that chemical. Interpretation of the HI is analogous to the HQ—values less than one indicate that the safe dose (in this case, in the aggregate from all routes of exposure) has not been exceeded. A value greater than one suggests that effects are possible, although not necessarily likely. The HI is also a means by which effects of different chemicals with similar toxicities can be combined to provide an estimate of total risk to the individual. This is discussed in more detail in Section 18.7.

Another means to convey the relationship between estimated and safe levels of exposure is through calculation of a margin of exposure. This is most often used in the context of the BMD approach. The margin of exposure is the BMD divided by the estimated dose. An acceptable margin of exposure is usually defined by the uncertainty factors applied to the BMD. If, for example, available data suggest that a total uncertainty factor of 1000 should be applied to the BMD for a specific chemical and effect, and the margin of exposure for that chemical is greater than 1000 (i.e., the estimated dose is less than the BMD divided by 1000), the exposure would be regarded as safe.

The above-described methods are almost universally applied in assessing the potential for noncancer health effects. There is, however, one exception for which a radically different approach is used: the evaluation of noncancer effects from lead in children. The Public Health Service has determined that blood lead concentrations in children should not exceed 10 μg/dL in order to avoid intellectual impairment. Thus, the main objective in lead risk assessment is to determine whether childhood lead exposure is sufficient to result in an unacceptable blood lead level. For this purpose, the USEPA has developed a PBPK model known as the "integrated exposure uptake biokinetic model for lead in children" (IEUBK). The IEUBK model has four basic components (i.e., exposure, uptake, biokinetics; and probability distribution) and uses complex mathematics to describe age-dependent anatomical and physiological functions that influence lead kinetics. The model predicts the blood concentration (the dose metric most closely related to the health effect of interest) that results from an endless variety of exposure scenarios that can be constructed by the risk assessor (i.e., exposure to various concentrations of lead in soil, dust, water, food, and/or ambient air). The model also predicts the probability that children exposed to lead in environmental media will have a blood lead concentration exceeding a health-based level of concern (e.g., 10 μg/dL) (see Figure 18.5). The IEUBK approach is rather unique because it is among the few approaches that rely on an internal dose metric (i.e., blood lead level) and PBPK modeling for risk assessment purposes.

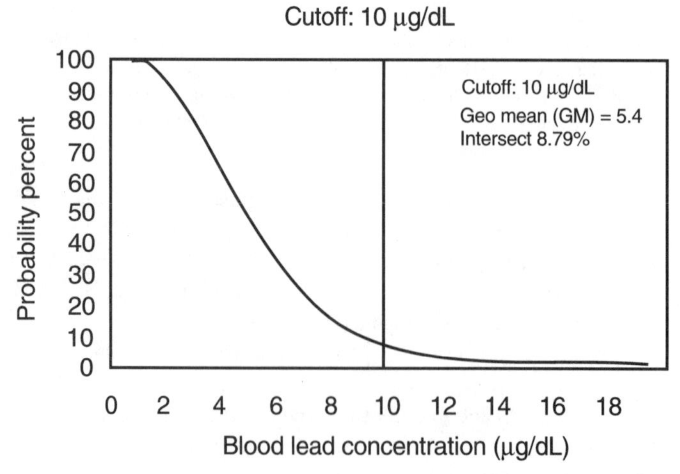

Figure 18.5 Example of output from the IEUBK model. The curve displays the cumulative probability of developing a blood lead concentration at varying levels as a result of the specified exposure. In this example, there is a probability of virtually 100% that the modeled exposure will result in a blood lead concentration greater than 1 μg/dL, but only about a 9% probability that the blood lead concentration will exceed 10 μg/dL.

Nonthreshold Models for Assessing Cancer Risks

Conceptual Issues The nonthreshold dose–response model is typically reserved for cancer risk assessment. The assumption by regulatory agencies that chemical carcinogenesis has no dose threshold began several decades ago. This assumption was initially based largely on empirical evidence that radiation-induced cancer had no threshold and on the theory that some finite amount of DNA damage was induced by all doses of radiation. Smaller doses simply carried smaller risks, but all doses were assumed to carry some mathematical chance of inducing cancer. Following this lead, theories of chemical-induced carcinogenesis began to evolve along the same lines, centering around effects of highly reactive, DNA damaging carcinogens. It was presumed that, like radiation, chemical carcinogens induced cancer via mutations or genetic damage and therefore had no thresholds. So, like radiation before it, chemical-induced carcinogenesis was assumed to carry some quantifiable risk of cancer at any dose.

If viewed somewhat simplistically, a biologic basis for the absence of a practical threshold for carcinogens can be hypothesized. If one ignores the DNA repair processes of cells, or assumes that these protective processes become saturated or overwhelmed by "background" mutational events, it can be postulated that some unrepaired genetic damage occurs with each and every exposure to a carcinogenic substance. As this genetic damage is presumed to be permanent and carry the potential to alter the phenotypic expression of the cell, any amount of damage, no matter how small, might carry with it some chance that the affected cell will ultimately evolve to become cancerous.

With this viewpoint, scientists and regulatory agencies initially proposed that the extrapolation of a cancer hazard must be fundamentally different from that used to extrapolate noncancer hazards, and cancer risk assessment models became probability-based. In contrast to assessing the risk of noncancer health effects, where the dose at which no toxic effect will occur is determined, cancer risk assessment is a matter of assigning probabilities of cancer to different doses. The determination of safety or a safe dose is then a matter of deciding what cancer risks are so small that they can be regarded as *de minimis* or inconsequential.

Determining the relationship between carcinogen dose and cancer risk is very difficult for a number of reasons. One reason is that the concept of latency complicates the interpretation of dose–response relationships for carcinogens. *Latency* is the interval of time between the critical exposure and the ultimate development of disease. While noncancer effects tend to develop almost immediately or very soon after a toxic dose is received, cancer may not develop until an interval of 20 years or more has elapsed. For some carcinogens, increasing the dose shortens the latency period, causing tumors to develop more quickly. A positive carcinogenic response can then be thought of in two ways: as increased numbers of tumors or subjects with tumors, or as a decrease in the time to appearance of tumors. The latter is important, because a dose capable of producing tumors has no consequence if the time required for the tumors to develop exceeds the remaining lifespan of the human or animal.

Another problem is that the critical portion of the dose–response curve for most risk assessments, the low-dose region applicable to most environmental and occupational exposures, is one for which empirical data are not available. Chronic cancer bioassays in animals are expensive, and seldom test more than two or three doses. Also, cost limits the number of animals tested to about 50 or less per dose group. With this group size, only tumor responses of about 10% or more can be detected with statistical significance. Detection of the kinds of cancer responses that might be of interest to the risk assessor, e.g. a response of 0.1%, 0.001%, or 0.00001% (10^{-3}, 10^{-5}, or 10^{-6}, respectively) are therefore beyond the capabilities of these experiments. Consequently, the doses needed to produce these cancer responses are not determined. Expanding the number of animals routinely tested is not economically feasible, and even very large studies may not eliminate this problem. One attempt to test the utility of using larger dose groups, the so-called "megamouse" experiment, was still unable to increase the sensitivity of measurement beyond about 1%, even though almost 25,000 animals were used in this experiment. In short, animal cancer bioassays will typically provide only one or two dose–response points, and these points are always several orders of magnitude above the range of small risks/doses in which we are ultimately interested.

Because low-dose responses cannot be measured, they must be modeled. There are three types of models:

1. The first category of models consists of the "mechanistic" models. These are dose–response models that attempt to base risk on a general theory of the biological steps that might be involved in the development of carcinogenesis. Examples of mechanistic models include the early "one-hit" and the subsequent "multihit" models for carcinogenesis. These models were based on assumptions concerning the number of "hits" or events of significant genetic damage that were necessary to induce cancer. A related model, the "linearized multistage" (LMS) model of carcinogenesis, is based on the theory that cancer cells develop through a series of different stages, evolving from normal cells to cancer cells that then multiply.

2. The second category of cancer extrapolation models includes the "threshold distribution" models. Rather than attempting to mimic a particular theory of carcinogenesis, these models are based upon the assumption that different individuals within a population of exposed persons will have different risk tolerances. This variation in tolerance in the exposed population is described with different probability distributions of the risk per unit of dose. Models that fall within this category include the probit, the logit, and the Weibull.

3. The third category of model is the "time-to-tumor" model. This type of model bases the risk or probability of getting cancer on the relationship between dose and latency. With this model, the risk of cancer is expressed temporally (in units of time), and a safe dose is selected as one where the interval between exposure and cancer is so long that the risk of other diseases becomes of greater concern.

Each of these models can accommodate the assumption that any finite dose poses a risk of cancer, the essential tenet of a nonthreshold model. However, the shape of the dose–response curve in the low-dose region can vary substantially among models (see Figure 18.6). Because the shape of the dose–response curve in the low-dose region cannot be verified by measurement, there is no means to determine which shape is correct. A simple example of the impact of choosing one cancer extrapolation model over another is given in Table 18.1, which compares the results of dose–response modeling using three different models where it was assumed in each model that a relative dose of 1.0 produced a 50% cancer incidence. The results generated by all three models are essentially indistinguishable at high doses where the animal cancer incidence might be observable, and so one would conclude they all "fit" the experimental data equally well. However, when modeling the risks associated with lower doses, the dose/risk range in which regulatory agencies and risk assessors are most frequently interested, there is a wide divergence in the risk projected by each model for a given low dose. In fact, at 1/10,000th of the dose causing a 50% cancer incidence in animals, the risks predicted by these three models produce a 70,000-fold variation in the predicted response.

Regulatory agencies utilize cancer risk estimates in regulating carcinogens, but they are faced with many models that yield a wide range of risk estimates. In the absence of any scientific basis to determine which is most correct, they must make a science policy decision in selecting the model to use. Generally, in the face of this uncertainty, they have selected models that tend to provide higher estimates of risk, particularly when combined with conservative exposure assumptions (see Table 18.2). This is consistent with their mission to protect public health, and consequently the need to avoid underestimating risks. For example, the USEPA has historically used conservative models such as the one-hit or LMS model in calculating cancer risks from exposure to all carcinogens. These models assume linearity in the low-dose range, and as shown in Table 18.1, tend to require a larger reduction in dose to attain a certain low level of risk relative to other models.

Extensive research in the area of chemical carcinogenesis indicates that many chemical carcinogens act via epigenetic or promotional mechanisms that, like noncancer toxicities, do not involve or require genetic damage. It has been proposed that these mechanisms and carcinogenic responses should have thresholds. Similarly, numerous enzyme systems have been identified as responsible for maintaining the integrity of the genetic code. These repair enzymes and pathways could provide an effective dose

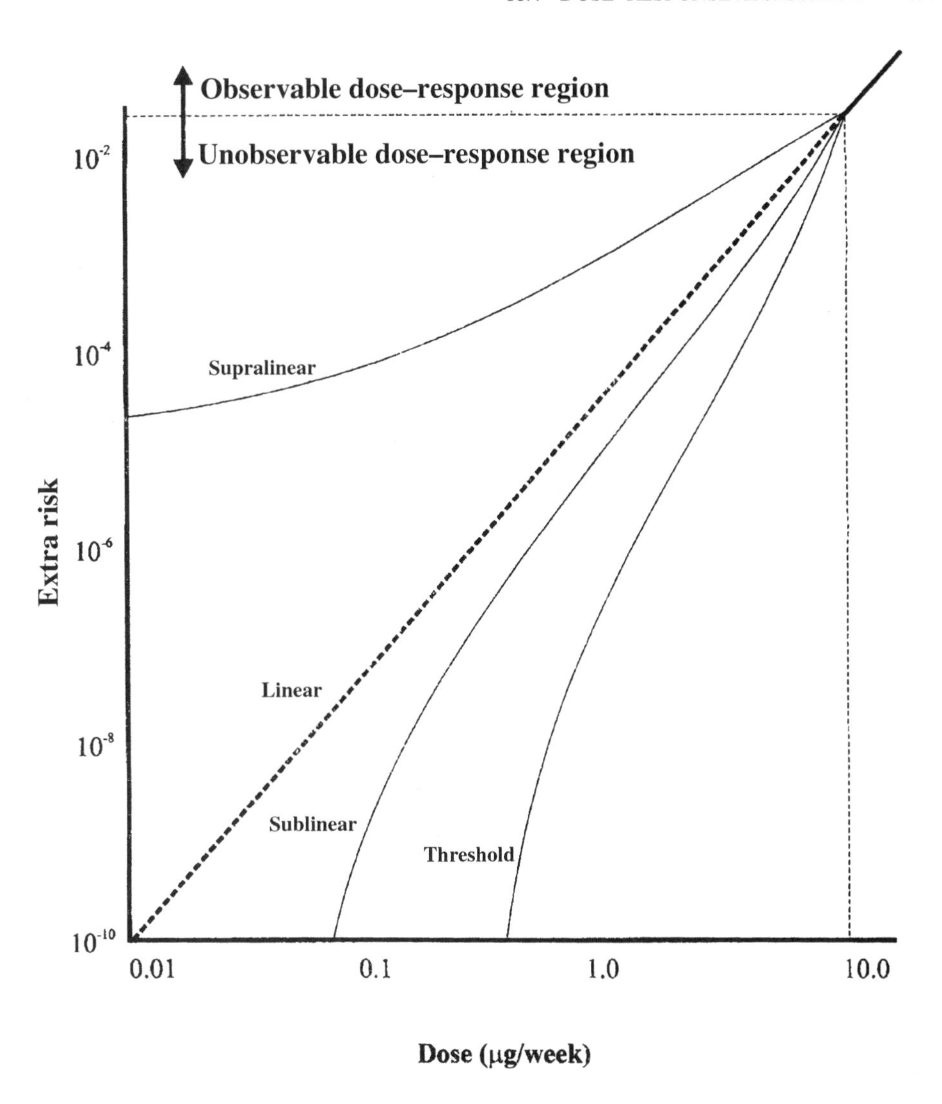

Figure 18.6 Four different extrapolation models applied to the same experimental data. All fit the data equally well in the observable range, but each yields substantially different risk estimates in the low-dose region most applicable to occupational and environmental exposures. [Adapted from NRC (1983).]

threshold for even those carcinogens whose mechanism is believed to involve some mutational event or other form of genetic damage. Thus, it is widely held today that thresholds exist for many carcinogens, and that cancer risk assessment for these chemicals should be conducted using threshold dose–response models similar to those used for noncancer effects.

In recent years, debates involving the actual shape of the dose–response curve for carcinogens in the low-dose region, and the issue of thresholds for carcinogens, have caused scientists and regulatory agencies to reevaluate earlier cancer risk assessment methodologies. Out of this reevaluation has come a movement to adopt two major policy changes in the cancer risk assessment methodologies employed by regulatory agencies. One proposed change is to use risk extrapolation models that make fewer assumptions about the shape of the dose–response curve (e.g., the benchmark dose and margin-of-exposure method). Data within the observation range can be used to develop a "point of departure,"

TABLE 18.1 Expected Risk (Cancer Incidence) Calculated by Three Models When a Relative Dose of 1.0 is Assumed to Cause a 50% Tumor Incidence in Test Animals

	Modeled Tumor Incidence (%)		
Relative Dose	Lognormal	Loglogistic	One-Hit
16	98	96	100
4	84	84	94
1[a]	**50**[a]	**50**[a]	**50**[a]
$\frac{1}{4}$	16	16	16
$\frac{1}{16}$	2	4	4
$\frac{1}{100}$th	0.05	0.4	0.7
$\frac{1}{1000}$th	0.00035	0.026	0.07
$\frac{1}{10,000}$th	0.0000001	0.0016	0.007

Source: Adapted from Office of Technology Assessment (1981).

[a]Boldface numbers represent the only data assumed for each model; all other tumor incidences were calculated from this single dose–response point.

the critical point for extrapolating responses in the low-dose range. For dose–response relationships assumed to have no threshold, the simplest extrapolation model is used: a straight line drawn between the point of departure and zero. The second proposed change is to allow for the consideration and use of nonlinear and threshold models for carcinogens where empirical and mechanistic evidence argues strongly that this type of dose–response model is appropriate for a particular chemical. In this situation, risk of cancer would be evaluated in a manner entirely analogous to noncancer health effects, such as through calculation of a margin of exposure.

TABLE 18.2 The Estimated Impact of Six Conservative USEPA Assumptions on Agency Risk Assessments

Factor	Range of Possible Overstatement in Estimated Cancer Risk[a]
Body weight vs. surface area as a scalar for an interspecies extrapolation	2–12
MLE[b] vs. 95% UCL[c] for the cancer slope factor	2–3
Malignant tumors only vs. malignant + benign	1–2
Average species sensitivity vs. most sensitive species	2–5
Pharmacodynamics vs. effective dose	1–6
Risks at shorter than equilibrium buildup time	2–5
Total risk exaggeration	15–10,800

Source: Adapted from Barnard (1994) and based on information supplied by Dr. E. Anderson.

[a]*Note*: Instead of presenting ranges of possible overstatement in cancer risk, the Barnard paper presents ranges of possible reduction in estimated cancer risk if the alternative factors to the current default factors are applied.

[b]Maximum likelihood exposure.

[c]Upper confidence limit.

Generating Cancer Risk Estimates Estimating the lifetime cancer risk associated with a particular dose is a relatively simple mathematical process. Because most regulatory agencies such as the USEPA use the conservative assumption that cancer risk should be modeled via a linear, nonthreshold model like the LMS model, the risk associated with a particular dose is calculated by the following formula:

$$\mathrm{R} = \mathrm{D} \times \mathrm{CSF} \qquad \text{or} \qquad R = \mathrm{LADD} \times \mathrm{CSF}$$

where R = risk

D = dose [normally expressed as the lifetime average daily dose (LADD) (mg/kg·day)]

CSF = cancer slope factor [the slope of the dose–response curve in units of $(\text{mg/kg·day})^{-1}$]

With this equation, the total dose the individual or population has accumulated during their entire exposure interval is first converted into a lifetime average daily dose (LADD), a dose that if received everyday for a lifetime would be equivalent to the total dose accumulated during the actual exposure period. For example, if the exposure assessment projected a daily dosage of 70 mg/kg·day for a 30-year exposure interval, then the LADD assuming a 70-year lifespan, would be 30 mg/kg·day (i.e., 70 mg/kg·day × 30 years ÷ 70 years = 30 mg/kg·day). The dose is expressed in units of mg/kg·day and the CSF is in units of reciprocal mg/kg·day or $(\text{mg/kg·day})^{-1}$. Thus, the product of dose × CSF is unitless, and is intended to represent the excess probability of cancer associated with the dose. For example, assume that the CSF for a chemical is 0.001 $(\text{mg/kg·day})^{-1}$ (in scientific notation a value of 1.0×10^{-3} $(\text{mg/kg·day})^{-1}$) and that the LADD derived during the exposure assessment was 0.03 mg/kg·day (3.0×10^{-2} mg/kg·day). The risk would be as follows:

$$R = 0.001 \times 0.03 = 0.00003$$

which can also be written as

$$R = \frac{3}{100{,}000} \qquad \text{or} \qquad R = 3.0 \times 10^{-5}$$

In this example the risk estimate represents a 3/100,000 chance or mathematical probability that a cancer will develop from exposure. It should also be noted, however, that because regulatory agencies strive for conservative, health protective risk calculations, the CSF used is statistically an upper-bound estimate of the dose–cancer relationship. The true cancer risk of the chemical at this dose may be much less than that calculated, and in fact, could be as low as zero.

Dose Metrics

A common issue for both threshold and nonthreshold dose–response relationships is the metric used to express dose. The dose metric is important because animal data must often be used as a surrogate for dose–response information in humans. Humans are, of course, much different in size than most laboratory animals. How then should doses be scaled between one animal species and another, and between animals and humans?

One can improve the accuracy of SHD calculations by starting with the best "dose metric" (measure of the dose) for the actual amount of chemical required to induce toxicity in the most sensitive target organ. Most dose information in animal studies is reported in terms of the "applied dose" (the dose administered to the whole animal). Remember, however, that it is only the "absorbed dose" (the amount of the chemical actually absorbed into the body) that is eligible for inducing toxicity. Further, from the dose that is absorbed, it is the dose that reaches the target tissue that is most important in determining the extent of response. The relationship between applied dose and target organ dose can be different among species, due to differences in metabolism and/or distribution of the chemical within the body, leading to important differences in apparent dose–response relationships (i.e., those based strictly on applied dose). One approach used to enhance extrapolation among species is PBPK modeling. Using PBPK models, scientists are able to predict target organ doses of a chemical (or a critical metabolite, if that is important for toxicity) in test species and humans. With this information, corrections can be made for pharmacokinetic differences among species, leading to better extrapolation of dose–response relationships. The principal limitation of PBPK analyses is that they are very data-intensive, and PBPK models have been constructed and validated for only a few chemicals.

Since PBPK models are rarely available, simpler approaches to extrapolating doses must be used in most situations. One of the simplest approaches is to convey doses per unit body weight. Larger animals (or humans) are assumed to require larger doses to produce the same toxic effect in direct proportion to their body weight. This is the dose metric most commonly used when extrapolating information on noncancer health effects among species. Interestingly, a different dose metric has been used traditionally for extrapolating dose–carcinogenicity relationships. In this situation, doses have usually been scaled according to the surface area of the animal. Since surface area for most species is a function of body weight, and can be approximated by body weight raised to the 0.67 power, this function has been used to convert doses used in animal studies (e.g., rodent cancer bioassays) to equivalent human doses. In biology, empirical observations suggest that many biochemical and physiological processes seem to scale among species according to surface area, while differences in others seem to correspond more closely to changes in weight. The correct scaling for doses is not entirely obvious, and could conceivably be different for different chemical classes or toxicological effects. Recently, there have been recommendations that scaling for carcinogen doses use a factor intermediate between body weight (or body weight raised to the power of 1) and surface area (body weight raised to the power of 0.67); that is, body weight raised to the power of 0.75.

Does the choice of scaling factor really make a difference? To illustrate the answer, consider the extrapolation of dose information between a mouse and a human. If a dose for a noncancer effect in a mouse were converted to a human dose based on surface area, rather than on body weight as is customary, the SHD dose would be reduced by a factor of 12–14. On the other hand, switching from surface area scaling to body weight scaling for carcinogenicity data would result in a 12–14-fold decrease in cancer risks estimated from the same dose–response information. The difference in use of body weight versus surface area for extrapolating between rats and humans is not as large—about 6-fold—but still might be considered significant.

18.5 RISK CHARACTERIZATION

The purpose of the risk characterization step is to integrate information provided by the hazard identification, dose–response assessment, and exposure assessment in order to develop risk estimates. Risk information may be conveyed in a qualitative manner, quantitative manner, or both. A qualitative assessment may describe the hazard posed by chemicals of concern, discuss opportunities for exposure, and reach some general conclusions that the risks are likely to be high or low, but would not provide numerical estimates of risk. A quantitative risk characterization, on the other hand, includes numerical risk values. Theoretically, there are many ways that numerical risks could be calculated depending on the specific questions being addressed in the risk assessment. For example, risks could be expressed as an individual's excess lifetime risk of developing a particular health effect as a result of chemical exposure. Risks could also be expressed on a population basis (e.g., estimated number of extra cases of a disease per year attributable to chemical exposure), or as the relative risk of an exposed population versus an unexposed population. Other ways of expressing risks, or health impacts, could be in terms of loss of life expectancy or lost days of work.

As discussed in Section 18.4, the most common means of expressing cancer risk associated with chemical exposure is in the form of individual excess lifetime cancer risk. When calculated for regulatory purposes, these values are intended to represent upper-bound estimates. That is, a cancer risk of one in one million means that an individual chosen at random from the exposed population is likely to have a probability no greater than one in one million of developing cancer as a result of that exposure. Note that this is an excess probability of developing cancer associated specifically with the chemical exposure addressed in the risk assessment, not the overall probability of developing cancer. The risk assessment provides an estimate of excess cancer risks, but does not determine whether the excess cancer risks are acceptable or unacceptable. That determination lies in the province of risk management, which must balance the risk estimate with other considerations (e.g., likelihood of actual exposure, uncertainties in the risk estimate, costs and feasibility of risk reduction strategies) to make decisions regarding steps, if any, to be taken to address chemical exposures. In some situations, an excess cancer risk of 1×10^{-3} (one in one thousand) from chemical exposure has been acceptable to regulatory agencies, while in others any excess risk above 1×10^{-6} (one in one million) has been deemed too high. Travis et al. (1987) reviewed the risks associated with 132 federal regulatory decisions involving environmental carcinogens to determine the level of risk that led to regulatory action. Their analysis revealed that with large populations an action was always taken when the risk exceeded 10^{-4}. For small populations, historically the *de manifestis level*, that is, the level at which action is always taken, was a risk of 10^{-3}. The *de minimis risk level* for these 132 regulatory actions, namely, the level of risk where no action or consideration is deemed necessary, was 10^{-5} to 10^{-4} for small populations and 10^{-6} to 10^{-7} for large populations. Others have suggested that the risk to smaller populations (e.g., a specific workforce) may be justifiably higher as long as the projected risk does not result in the expectation of an additional cancer. For example, if 100 persons were exposed to a 10^{-3} lifetime risk, the total population risk would be only 0.1, and no additional cancers would be expected. The range of "acceptable risks" that has been applied by the USEPA across its various regulatory programs seems to support the conclusions of the regulatory analysis performed by Travis and co-workers. The acceptable risk ranges of several USEPA programs are

- 10^{-4} to 10^{-6}: the cleanup policy under the USEPA Superfund Cleanup Program of the National Oil and Hazardous Substances Pollution Contingency Plan
- 10^{-4} to 10^{-6}: USEPA drinking-water standards [maximum contaminant levels (MCLs)] under the Safe Drinking Water Act
- 10^{-2} to 10^{-6}: National Emission Standards for Hazardous Air Pollutants (NESHAPs) under the Clean Air Act
- 10^{-4} to 10^{-6}: for corrective actions under the Resource Conservation and Recovery Act

For carcinogens with thresholds, risk is portrayed as a margin of exposure (*Note*: The concept of margin of exposure is discussed in Section 18.4.) Just as the definition of an acceptable cancer risk in probability terms is outside the scope of the risk assessment, an acceptable margin of exposure is essentially a policy and risk management issue. The margin-of-exposure concept is also applicable to noncancer effects, and is used to convey the difference between the estimated exposure to a chemical and the benchmark dose, usually for the most sensitive effect. The other, more common, means of expressing hazard for noncancer effects is the hazard index (HI). By convention, a HI greater than 1 signals concern for the possibility of adverse effects. The likelihood that health effects will actually occur with a HI greater than 1 depends in part on the chemicals in question and the margin of safety inherent in the toxicity values used to calculate the HI. These issues bear discussion in the risk characterization, so as to better inform the risk management decisions.

During the risk characterization step, risks from various chemicals, reaching individuals by various pathways and conceivably entering the body by various routes, must be combined in some way such that the total risk to individuals from chemical exposure can be assessed. Methods for combining risks are discussed in Section 18.7. It is not uncommon for risk estimates to be presented for a number of different populations. This may include groups of individuals exposed in different ways (e.g., workers at a contaminated site vs. visitors to the site vs. residents living nearby the site), or individuals that may differ in their sensitivity to hazards posed by the chemicals of concern (e.g., children, pregnant women). Development of these various risk estimates is important, not only in providing a complete characterization of potential risks posed by the chemicals but also in developing effective strategies for managing the risks.

A particularly important aspect of the risk characterization is a discussion of the uncertainties associated with the risk estimates. Each individual step in the risk assessment process is a potential source of uncertainty. Many of these are discussed throughout the chapter, and include uncertainty associated with estimating exposure (e.g., measurement errors, uncertainty in selecting the best exposure models, uncertainty regarding exposure conditions that will exist in the future) as well as determining safe levels of exposure (e.g., uncertainty regarding the shape of the dose–response relationship in the low-dose region and extrapolating results from animals to humans, uncertainty that the most sensitive health effect has been identified, uncertainty regarding ways that multiple chemicals might interact). These need to be articulated in the risk characterization so that an appreciation of the level of confidence and conservatism in the risk estimate can be gained. Without a discussion of uncertainty, risk assessment results are often perceived as being more precise than they really are, which could lead to misuse. Minimally, uncertainties should be discussed qualitatively, identifying the source or nature of each uncertainty and how, in a general way, it could affect the risk estimation (i.e., whether the approach taken, in view of the uncertainty, is likely to contribute to an over- or underestimation of risk). A semiquantitative perspective is helpful, in which the implications of model and assumption choices on the risk estimate are described in rough, order-of-magnitude terms. As discussed in Section 18.6, probabilistic techniques can be used to provide more precise quantitative expression of the uncertainties associated with risk estimates, as well as a description of variability in risks encountered in exposed populations. This approach is attractive in that it offers a much richer characterization of the risks and uncertainties than do the more traditional risk estimation techniques. It is technically demanding, however, requiring much greater time, resources, data, and technical expertise.

18.6 PROBABILISTIC VERSUS DETERMINISTIC RISK ASSESSMENTS

Risk assessments are rarely performed for a single individual. Instead, risk assessments are designed primarily to characterize risks to populations of individuals. Many—perhaps most—of the factors that affect risk can vary from one person to another. Differences in body weight, inhalation rate, frequency and duration of contact with contaminated media, and even sensitivity to toxicity, are examples of factors that can lead to different risks among individuals, even if the concentration of chemical to which they are all exposed is the same. Theoretically, there is no single risk for a particular exposure circumstance, but rather as many different risk values as there are individuals in the exposed population. In confronting the issue of variability in risk assessment, the traditional approach used for regulatory purposes has been to simply characterize the risk to individuals within the population with the greatest exposure. A *deterministic approach* to risk calculation is used, where a single value is selected for each exposure variable and a single risk estimate is produced. The exposure assumptions are chosen to represent the plausible upper bound of exposure, and the risk estimate is said to be associated with reasonable maximal exposure (RME) or high-end exposure. The development of a single, high-end risk estimate for regulatory use is consistent with the goal of regulatory agencies to develop risk management strategies protective of the entire population. For perspective, a deterministic approach may also be used to develop an estimate of risk for the average individual in the population, that is, a central tendency estimate of the risk. The problem with this approach is that it provides little information on the extent to which risk varies within the population. For example, while exposure values may be selected to develop a high-end estimate of risk, it is difficult to know whether this value is merely conservative or extreme. Does it represent an exposure that might be exceeded by 1 or 2 out of every 10 individuals, or does it represent an exposure circumstance so extreme that it is unlikely ever to take place? This vagueness regarding the degree of conservatism in deterministic risk estimates undermines their value and creates controversy regarding their use in regulatory decision making.

A second problem confronting the risk assessor is management of uncertainty in the risk assessment process. As described elsewhere in this chapter, there are numerous sources of uncertainty in risk calculations, including uncertainty in the selection of models and assumptions, and in measurements of risk related parameters. As part of a deterministic calculation of risk, a choice must be made for each of these so that a risk estimate can be made. For regulatory purposes, conservative choices are usually made; models and assumptions that tend to provide higher estimates of risk are selected from among the range of plausible alternatives. The reason for conservative choices by regulatory agencies in the face of uncertainty is well understood, but the extent of conservatism imparted by the various choices is usually unclear. As with the issue of variability, this makes it difficult or impossible for the risk assessor to effectively convey the inherent conservatism associated with the risk estimate.

Probabilistic risk assessment is an alternative approach that can address the shortcomings of deterministic calculations in terms of variability and uncertainty. In probabilistic risk assessment, input variables are entered as probability density functions (PDFs) instead of single values. For example, instead of using a single body weight of 70 kg in the risk calculation, a distribution of body weights would be entered that reflects the variability in body weight of the exposed population. PDFs might also be entered for other variables such as inhalation rate, skin surface area, and frequency of contact with contaminated media—anything that would be expected to vary from one individual to another. These PDFs are then combined in such a way as to yield a risk distribution, representing the range and frequency of risks anticipated to exist in the exposed population. Although there are several ways to combine PDFs, one of the most commonly used techniques is Monte Carlo simulation. With Monte Carlo simulation, a computer program in essence creates a simulated population designed to resemble the exposed population in every key respect. For each risk calculation, it takes a value from each input PDF and calculates a numerical risk. This process is repeated, usually thousands of times, and the resulting range of risk values is tallied in the form of a distribution. This distribution represents the risk distribution for the population. From this distribution, the variability in risk among individuals can be visualized and the risk level at various percentiles of the population determined (see Figure 18.7).

Probabilistic risk assessment can also provide quantitative representation of the uncertainties in the risk calculation. For each input or model, some estimate of the uncertainty is entered. For example, the concentration of chemical X for which a risk estimate is desired is assumed to be 100, but could be as low as 50 or as high as 200. In this case, the chemical concentration could be entered as a distribution of values, with 100 as the most likely estimate, but with a range extending from 50 to 200. As with variability, the uncertainty associated with various inputs can be combined to produce a PDF showing boundaries of uncertainty associated with a risk estimate. An additional benefit of this approach is that a sensitivity analysis can be used to rank the various sources of uncertainty in terms of their relative contribution to overall uncertainty. If the uncertainty is unacceptably large, this can be used to identify the best areas for further analysis or research to reduce uncertainty.

It is possible for a probabilistic risk assessment to address both variability and uncertainty simultaneously. This requires the development of PDFs for both uncertainty and variability. For example, a PDF might be used to portray variability in body weight in the exposed population, and a separate PDF would be used to deal with any uncertainty that the body weight distribution selected accurately reflects the actual body weight distribution of the population in question. (*Note*: This is not an unreasonable uncertainty, since risk assessors almost never have the time and resources to actually weigh everyone in an exposed population, and therefore must rely on published body weight data for the general population to create their body weight PDF.) The variability and uncertainty PDFs are then combined separately to generate a risk distribution with confidence boundaries provided by the uncertainty distributions. This is called a *two-dimensional probabilistic risk assessment*.

The principal advantage of a probabilistic risk assessment is that it provides much greater information on variability and uncertainty associated with risk estimates. The manner in which risk is distributed within the exposed population is transparent, and the magnitude of uncertainty associated with the risk estimate is conveyed in quantitative terms. There are, however, a number of disadvantages to probabilistic risk assessment, including the facts that (1) it is technically demanding, requiring much

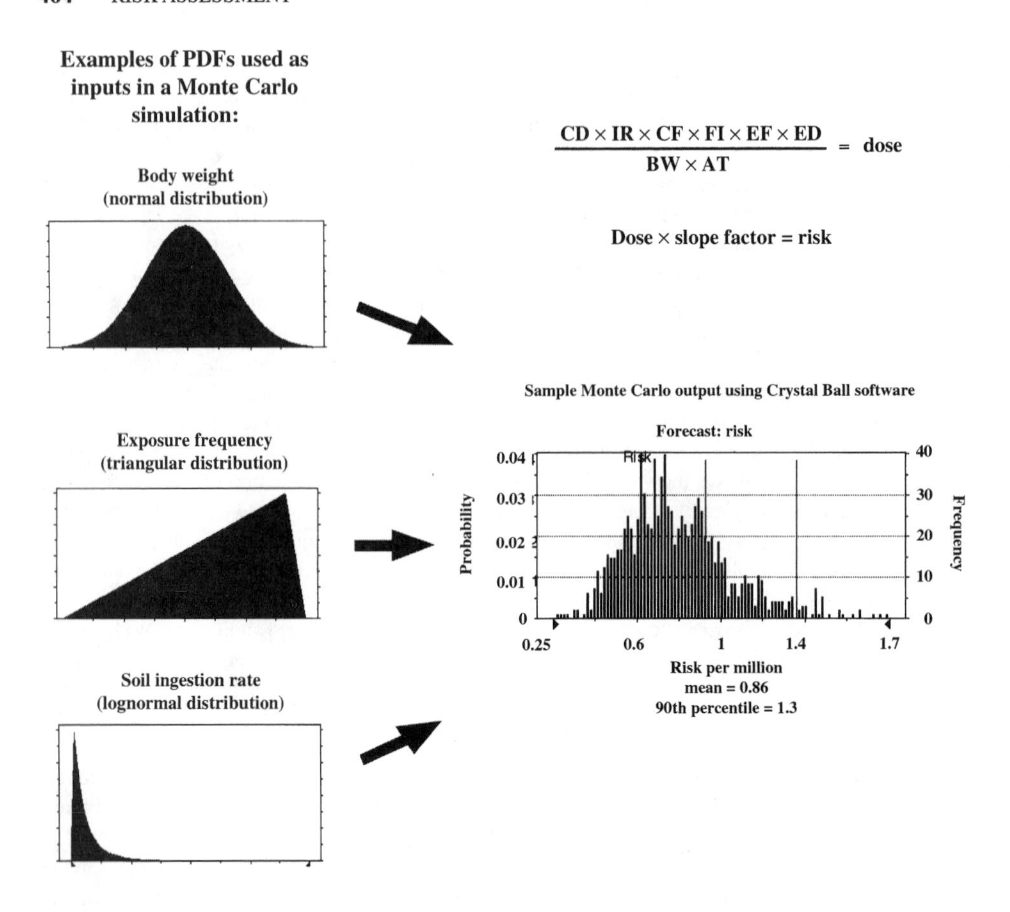

Figure 18.7 Simplified diagram of combining probability density functions (PDFs) to yield a risk distribution in probabilistic risk analysis. In this example, dose is calculated as described in Section 18.3, except each of the inputs for which variability exists is entered as a PDF. Examples of PDFs for body weight, exposure frequency, and soil ingestion rate are shown. Using Monte Carlo simulation, the computer takes values from these PDFs to calculate risk values for individuals in a simulated population. The distribution of these risk values is provided in the output of the simulation, as shown in the figure.

greater expertise than deterministic risk calculations; (2) it is information-intensive, requiring data on exposure characteristics within populations that may not exist; (3) because of the two previous points, it is much more time consuming and expensive than deterministic risk assessments; (4) although they provide much more information, the outputs can be complex and difficult for nontechnical audiences to understand; and (5) it requires a different set of policy assumptions regarding acceptable risk. Unlike the situation with a single risk value, acceptable risk must be defined in terms of an acceptable risk distribution.

18.7 EVALUATING RISK FROM CHEMICAL MIXTURES

The simplest form of risk assessment deals with health risks posed by exposure to a single chemical from a single source. Unfortunately, in reality things are seldom this simple. In most situations, exposure occurs not to a single chemical in high doses, as in toxicology studies in animals, but rather to multiple chemicals in lower doses. Often, more than one of these chemicals is capable of affecting the same target organ or tissue. In this situation, evaluating the risk for individual chemicals one by one may not accurately portray the risk associated with these chemicals in combination. In developing credible risk assessments, it is important not only to consider the cumulative impact of different chemicals affecting the same target organ but also to recognize the potential for these chemicals to interact. Effects of chemicals in combination may not be simply additive. Biological and chemical interactions among the chemicals can lead them to antagonize the effects of one another or produce effects greater than the sum of their individual effects. The ability to account for such interactions and develop meaningful estimates of risks of chemicals in combination is one of the most significant challenges in risk assessment.

There are three basic approaches to evaluating the toxic potential of chemicals in combination. If toxicity data are available for the specific chemical mixture of interest, a preferred approach is to treat the mixture as a single toxicological entity. That is, toxicological data from animals treated with the mixture can be used to identify a reference dose, benchmark dose, or slope factor for use in the risk assessment. For this approach to be valid, the chemical mixture to which individuals are exposed must be the same as the mixture used in the toxicity studies, not only in terms of the specific chemicals present but also their proportions. A second approach involves using toxicity data from a "sufficiently similar" mixture, if available, to develop a risk estimate. A similar mixture might, for example, have the same constituents but slightly different proportions, it might have several common components but lack one or two, or it might have one or more additional components. Similar mixtures would be expected to act by the same mechanism of action or produce the same type of toxicity. Beyond these general expectations, there are no firm criteria as to what constitutes "sufficiently similar," leaving this decision up to the judgment of the toxicologist or risk assessor.

If inadequate toxicity data are available for an identical or similar mixture, a third approach is to assess the toxicity of the mixture based on toxicity of its components. This last approach invariably requires assumptions regarding the presence and nature of chemical interactions. Interaction in this context means that one chemical alters the toxicity of one or more other chemicals in the mixture. The default assumption is usually no interaction among the chemicals; that is, in the absence of evidence to the contrary, the chemicals are assumed to act independently—each neither enhancing nor reducing the effect of the others. Chemicals that produce the same toxic effects are considered to act in an additive fashion in this situation, and the total risk is the sum of the risks posed by the individual component chemicals.

Although seemingly simple in concept, in practice there are several ways to add the effects of chemicals. One way is to use *dose addition*, where the chemicals are considered functional clones of each other. This means that they produce the same toxic effects (or at least the same toxic effect of interest) through the same mode of action and have similar pharmacokinetic properties. These chemicals don't necessarily have identical dose–response curves, and in fact there can be substantial differences in toxic potency. However, the relationships between the dose–response curves are such that differences in potency between chemicals can be represented by some constant proportion (e.g., one chemical might produce the same toxic response as another, but always at 1% of the dose). Experimentally, dose–response curves of such agents are parallel.

For groups of chemicals that fit this description, combined risks can be calculated using the *relative potency factor* (RPF) approach. One chemical in the mixture (usually the best characterized toxicologically) is designated as the *index chemical* and assigned an arbitrary potency factor of 1. Dose–response information for other chemicals is used to assign each a potency factor relative to the index chemical. For example, a chemical with a potency of 1/100th the index chemical would be assigned an RPF of 0.01, while a chemical 10 times as potent as the index chemical would have an RPF of 10. In the risk assessment, these RPF values are used to convert doses of the various chemicals in the group to toxicologically equivalent doses of the index chemical. These doses are then summed and used, along with a toxicity value for the index chemical (e.g., reference dose, slope factor, as appropriate) to derive a risk estimate for the group as a whole.

The RPF values for chemicals in the group may vary depending on the toxic effect of concern and perhaps the exposure circumstances. There are a few examples where all of the toxic effects of concern share a common mode of action and a single scaling factor is applicable for all effects and exposure conditions. This represents a special case of the RPF method termed the *toxic equivalency factor* (TEF) approach. An example of the use of the TEF approach is the risk assessment of polyhalogenated aromatic compounds. Most of the adverse health effects of concern for these compounds are thought to arise from a common mode of toxicity: Ah receptor activation. In this example, the index chemical is 2,3,7,8-tetrachlorodibenzo-*p*-dioxin (2,3,7,8-TCDD), which is assigned a relative potency of 1. Based on studies of comparative potency in terms of Ah-receptor-mediated toxicity, comparative potency factors (termed TEFs) have been determined for other polyhalogenated aromatics (e.g., PCDDs, PCDFs, PCBs). TEFs for PCB congeners relative to 2,3,7,8-TCDD are listed in Table 18.3. In assessing risks from exposure to a mixture of PCB congeners using the TEF approach, the TEF equivalents for all congeners present in the environmental sample are summed to derive a risk estimate for the PCB mixture.

Another means of adding chemical effects is the *hazard index* approach. This approach does not require the assumption of a common mode of toxicity, only that the chemicals share the same target organ or effect. In this approach, the dose of each chemical is compared with some representation of a threshold dose for toxicity. In practice, that may be a reference dose or a benchmark dose (see Section 18.4 for a discussion of reference and benchmark doses and their derivations). The dose for which a risk estimate is sought is divided by the threshold dose for that chemical in the target organ of interest and the result is termed the *hazard quotient*. For example, if exposure to a chemical is predicted to result in a dose of 1 mg/kg·day, and the threshold dose for the toxicity of concern is 10 mg/kg·day, the hazard quotient is 1/10 or 0.1. Hazard quotients for each chemical affecting the target organ are then summed to obtain the hazard index (HI). The interpretation of the magnitude of the HI is similar to that already discussed.

Yet another way in which effects can be added is through *response addition*. This differs from dose addition methods in that the chemicals and their effects are assumed to be completely independent. For this approach, the percent of animals or humans expected to develop toxicity from each of the individual chemicals at their respective doses is estimated. These percentages are termed the "responses." The probability that a toxic event will result from a combination of two chemicals can be expressed as follows:

TABLE 18.3 Proposed Toxic Equivalency Factors (TEFs) for PCB Congeners Relative to Dioxin

IUPAC No.	Congener	TEF
	2,3,7,8-TCDD	1
77	3,3′,4,4′-TCB	0.0005
105	2,3,3′,4,4′-PeCB	0.0001
114	2,3,4,4′,5-PeCB	0.0005
118	2,3′,4,4′,5-PeCB	0.0001
123	2′,3,4,4′,5-PeCB	0.0001
126	3,3′,4,4′,5-PeCB	0.1
156	2,3,3′,4,4′,5-HxCB	0.0005
157	2,3,3′,4,4′,5′-HxCB	0.0005
167	2,3′,4,4′,5,5′-HxCB	0.00001
169	3,3′,4,4′,5,5′-HxCB	0.01
170	2,2′,3,3′,4,4′,5-HpCB	0.0001
180	2,2′,3,4,4′,5,5′-HpCB	0.00001
189	2,3,3′,4,4′,5,5′-HpCB	0.0001

Source: Ahlborg et al. (1994).

$$R_{\text{both}} = 1 - (1 - R_{\text{chemical A}}) \times (1 - R_{\text{chemical B}})$$

When the probabilities are small, this reduces to simply

$$R_{\text{both}} = R_{\text{chemical A}} + R_{\text{chemical B}}$$

This approach is considered to be useful in summing a series of small component risks, but does not work well when one or more of the risks is large. In practice, response addition is used primarily in developing estimates of total cancer risks from more than one chemical or from chemical exposure by more than one route.

Each of the above mentioned approaches to combining risks assumes no interaction among chemicals. This is not always the case. It is possible that in some instances one chemical might antagonize or inhibit the toxicity of another. In this situation, the combination of chemicals would produce less-than-additive toxicity. This could conceivably occur through a variety of means depending on the mechanism(s) of toxicity of the chemicals and their toxicokinetics. Examples include effects to decrease toxicant absorption, increase its elimination or decrease its bioactivation, competition for receptor binding, or production of an opposing biochemical or physiological effect. Chemicals in combination can also produce greater-than-additive effects. When both chemicals are capable of

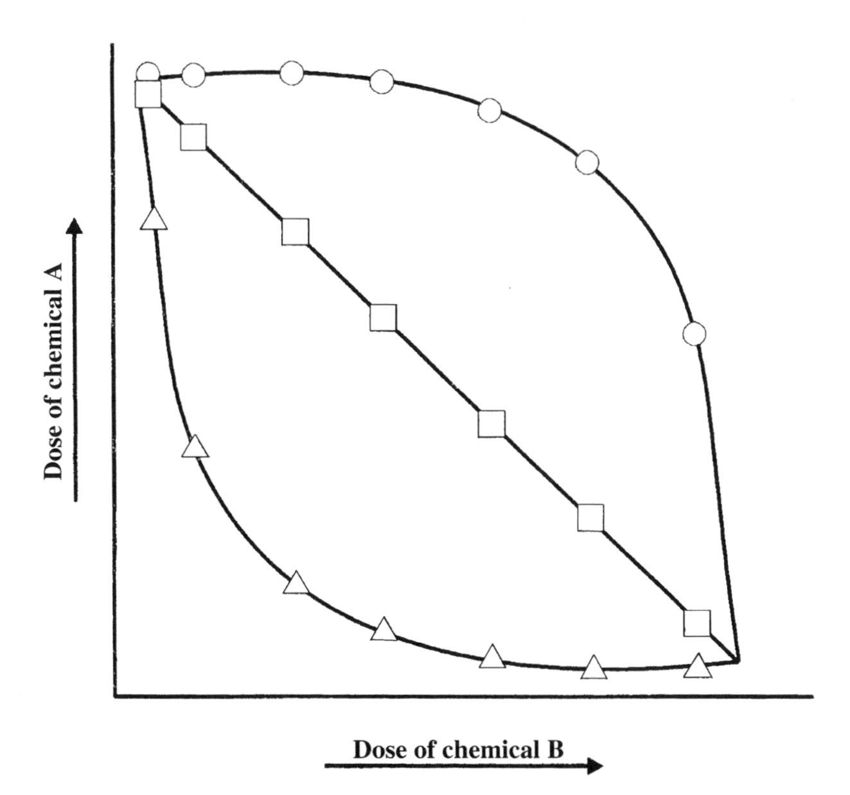

Figure 18.8 Isobologram of effects of two chemicals administered in varying dose combinations. The response obtained from chemical A alone is on the *y* axis, and the response from chemical B alone is plotted on the *x* axis. When there is no interaction between the chemicals, the responses from doses comprised of varying proportions of chemicals A and B will fall on a straight line connecting the response for 100% chemical A to 100% chemical B (squares in the figure). If there is antagonism between the chemicals, responses to combinations of the chemicals will lie below and to the left of this line (triangles), and synergistic responses will lie above and to the right of the line (circles).

producing the effect, this is termed *synergism*. The special case in which one of the two chemicals has no effect on its own, but nonetheless increases the toxicity of another, is termed *potentiation*.

There are a number of tests available to determine whether two chemicals interact in an additive, subadditive (i.e., antagonistic), or supraadditive (i.e., synergistic) fashion. One of the most straightforward is the construction of an isobologram (see Figure 18.8). Two chemicals are administered in varying proportions, ranging from 100% chemical A to 100% chemical B. If the interaction between the two chemicals follows dose addition, their responses will lie along a line that connects the response for 100% chemical A to the response for 100% chemical B. If the responses to chemical combinations are greater than would be predicted by dose addition, that is, if they lie above and to the right of the dose-addition line, a synergistic effect can be inferred. On the other hand, responses below the line and to the left indicate antagonism.

From a practical standpoint, interactions among chemicals are very difficult to deal with quantitatively in a risk assessment. These interactions are seldom well characterized and can be dose-dependent—synergism or antagonism that occurs at one dose combination of the chemicals may not occur at other dose combinations. Also, while tests exist to examine the nature of interactions between two chemicals, as described in the paragraph above, interactions among multiple chemicals are much more difficult to assess and characterize. Although the problem of addressing chemical interactions has been recognized for some time, research to solve this problem is still in a relatively early stage of development. Scientists are still struggling to identify circumstances where important interactions might take place, and rigorous techniques for adjusting risk estimates to account for interactions do not yet exist.

18.8 COMPARATIVE RISK ANALYSIS

For the purposes of this chapter, comparative risk analysis is a means of placing estimates of risk into a larger context in order to provide risk managers and stakeholders with a better perspective for decision making. Comparative risk analysis can also help nontechnical audiences understand the implications of a risk assessment, particularly when findings are reported in unfamiliar quantitative jargon. Furthermore, risk comparisons may be of value in setting priorities and allocating resources within regulatory agencies. In response to many risk problems posed by chemical exposure, the following questions might be asked, all of which should prompt the conduct of a comparative risk analysis: (1) whether the receptors are exposed to the same chemical from other sources, (2) whether exposure to the chemical also occurs from other environmental media, and (3) whether other chemicals from the same sources pose additional risks to receptors. Several types of risk comparisons are listed below.

1. Comparisons of magnitude such as equating a "one in one million" risk to the length of 1 inch in 16 miles, 30 seconds in a year, or 1 drop in 16 gallons
2. Comparisons of risk posed by the same chemical from different sources
3. Comparisons of risk posed by different chemicals from the same source
4. Comparisons of risk posed by different chemicals for the same target organ
5. Comparisons of familiar versus less familiar risks
6. Comparisons of voluntary versus involuntary risks
7. Comparisons of natural versus anthropogenic or technologic risks
8. Comparisons of risks of the same magnitude posed by different risk factors

Just as risk comparisons can be of value, they can also hinder risk communication. For example, inappropriate comparisons can be confusing and may serve to minimize risks that, in reality, deserve serious consideration. To maximize the benefits of risk comparison and avoid its pitfalls, it is recommended that substantially dissimilar risks (e.g., risk of cancer versus risk of losing money in the stock market) not be compared since the relative magnitudes of such risks are difficult to comprehend. Also, research on risk perception has suggested that directly comparing voluntary and involuntary risks or natural and technologic risks does not always improve a lay person's understanding of an environmental risk. However, the risk comparisons described in list items 2, 3, and 4 above are thought to be of considerable communicative value.

There are no shortages of data available for risk comparisons, since we all incur risks by virtue of our continuous exposure to chemicals at work and at home. Indeed, the potentially hazardous chemicals in the food we eat, the water we drink, and the air we breathe are numerous and the list continues to growth as new studies are published. In addition, some medications carry a risk of cancer, and because the dosages of these chemicals are high relative to those chemicals found in the environment, over-the-counter medications and prescription drugs may carry significant theoretical risks even when used as intended. The following tables of risk comparisons have been provided to illustrate some different types of risk comparisons that can be made. Tables 18.4*a* and 18.4*b* illustrate the risks projected for volatile organic chemicals and pesticides measured in homes during the USEPA study of residential environments (a type 3 risk comparison). Table 18.5 illustrates the theoretical risks associated with taking a daily dose of 12 different drugs (again, a type 3 risk comparison). Table 18.6 shows risk in a slightly different manner. In this table, risks for various activities, diseases, or lifestyle choices are compared by the number of days each is believed to decreases one's life expectancy (a comparison mixing categories 5, 6, and 7). Table 18.7 compares many different activities, all of which carry the same one-in-a-million level of risk (a type 8 risk comparison that combines aspects of types 5, 6, and 7).

TABLE 18.4*a* Cancer Risks for Indoor Air Exposures to VOCs (TEAM Studies)

Chemical	Indoor Exposure Levels ($\mu g/m^3$)	Potency, $(\mu g/m^3)^{-1} \times 10^{-6}$	Lifetime Cancer Risk ($\times 10^{-6}$)
Benzene			
Air	15	8	120
Smokers	90	8	720
Vinylidene chloride	6.5	50	320
Chloroform			
Air	3	23	70
Showers (inhalation)	2	23	50
Water	30	2.3	70
Food and beverages	30	2.3	70
p-Dichlorobenzene	22	4	90
1,2-Dibromoethane	0.05	510	25
Methylene chloride	6	4	24
Carbon tetrachloride	1	15	15
Tetrachloroethylene	15	0.6	9
Trichloroethylene	7	1.3	9
Styrene			
Air	1	0.3	0.3
Smokers	6	0.3	2
1,2-Dichloroethane	0.5	7	4
1,1,1-Trichloroethane	30	0.003	0.1

Source: Adapted from Wallace (1991).

TABLE 18.4*b* Cancer Risks for Household Exposures to Pesticides (TEAM Studies)

Pesticide	Exposure (ng/m^3)	Potency (mg/kg·day)$^{-1}$	Lifetime Cancer Risk (× 10^{-6})
Banned termiticides			
Heptachlor	71	4.5	90 (19)
Chlordane	198	1.3	70 (15)
Aldrin	13	17	60 (13)
Dieldrin	3	16	14 (3)
Heptachlor epoxide	0.4	9.1	1 (0.2)
DDE	2.2	0.34	0.2 (0.4)
DDT	0.7	0.34	0.1 (0.02)
Other pesticides			
Dichlorvos	33	0.29	2.7
g-BHC (lindane)	6.6	1.3	2.5
a-BHC	0.5	6.3	1
Propoxur	100	0.0079	0.2
Hexachlorobenzene	0.3	1.67	0.1
Dicofol	2.6	0.34	0.05
o-Phenylphenol	58	0.0016	0.02
2,4-D	0.6	0.019	0.003
Atrazine	0.05	0.22	0.003
cis-Permethrin	0.4	0.022	0.003
trans-Permethrin	0.1	0.022	0.001

Source: Adapted from Wallace (1991).

TABLE 18.5 The Therapeutic and Virtually Safe Dosages (VSD) of a Few Medications

Drug	Cancer Slope Factor (mg/kg·day)$^{-1}$	VSD (mg/kg·day)	Dose Ratio (daily dose/VSD)	Incremental Lifetime Cancer Risk per Daily Dose of Drug (10^{-6})[a]
Rifampin	2.1	4.8×10^{-7}	18,000,000	704.5
Isoniazid	4.9×10^{-1}	2.1×10^{-6}	2,400,000	93.9
Clofibrate	5.4×10^{-2}	1.9×10^{-5}	1,500,000	58.7
Disulfiram	3.6×10^{-1}	2.8×10^{-6}	1,300,000	50.9
Phenobarbital	2.9×10^{-1}	3.5×10^{-6}	825,000	32.3
Acetaminophen	1.3×10^{-2}	7.6×10^{-5}	747,000	29.2
Metronidazole	4.4×10^{-2}	2.3×10^{-5}	467,000	18.3
Sulfisoxazole	1.9×10^{-3}	5.3×10^{-4}	215,000	8.4
Dapsone	1.7×10^{-1}	5.8×10^{-6}	123,000	4.8
Methimazole	4.8×10^{-1}	2.1×10^{-6}	102,000	4.0
Oxazepam	1.0×10^{-1}	9.8×10^{-6}	87,700	3.4
Furosemide	6.0×10^{-2}	1.7×10^{-5}	68,723	2.7

Source: Adapted from Waddell (1996).

[a]Calculated by dividing Waddell's dose ratio by the 25,550 days in a 70-year lifetime to get the incremental lifetime risk per daily dose of drug above the 10^{-6} risk representing the VSD.

TABLE 18.6 Estimated Average Loss of Life Expectancy from Various Risks, Activities, & Diseases

	Days Lost
Being an unmarried male	3500
Smoking cigarettes and being male	2250
Heart disease	2100
Being an unmarried female	1600
Being 30% overweight	1300
Being a coal miner	1100
Cancer	980
Being 20% overweight	900
Having less than an 8th grade education	850
Smoking cigarettes and being female	800
Poverty	700
Stroke	520
Smoking cigars	330
Having a dangerous job	300
Smoking a pipe	220
Increasing your daily food intake 100 calories	210
Driving motor vehicle	207
Pneumonia, influenza	141
Alcohol addiction	130
Accidents in the home	95
Suicide	95
Diabetes	95
Homicide	90
Misusing legal drugs	90
Having an average-risk job	74
Drowning	41
Employment that entails radiation exposure	40
Falls	39
Walking down the street	37
Having a safer-than-average job	30
Fires and burns	27
Generation of energy	24
Using illegal drugs	18
Solid and liquid poisons	17
Suffocation	13
Firearm accidents	11
Natural radiation	8
Poisonous gases	7
Medical X-ray exposure	6
Drinking coffee	6
Oral contraceptives	5
Riding a bicycle	5
Drinking diet sodas	2
Nuclear reactor accidents	2
Radiation from the nuclear industry	0.02

Source: Allman (Oct. 1985).

TABLE 18.7 Activities Estimated to Increase Your Chances of Dying in Any Year by One in a Million

Activity	Cause of Death
Smoking 1.4 cigarettes	Cancer, heart disease
Drinking 0.5 L of wine	Liver cirrhosis
Spending 1 h in a coal mine	Black lung disease
Spending 3 h in a coal mine	Accident
Living 2 days in New York or Boston	Air pollution
Traveling 6 min by canoe	Accident
Traveling 10 miles by bicycle	Accident
Traveling 150 miles by car	Accident
Flying 1000 miles by jet	Accident
Flying 6000 miles by jet	Cancer from cosmic radiation
Living 2 months in Denver on vacation from New York	Cancer from cosmic radiation
Living 2 months in average stone or brick building	Cancer from natural radioactivity
One chest X-ray taken in a good hospital	Cancer from radiation
Living 2 months with a cigarette smoker	Cancer, heart disease
Eating 40 tablespoons of peanut butter	Liver cancer from aflatoxin B
Drinking Miami drinking water for 1 year	Cancer from chloroform
Living 5 years at site boundary of a nuclear power plant	Cancer from radiation
Eating 100 charcoal-broiled steaks	Cancer from benzopyrene

Source: Allman (Oct. 1985).

18.9 RISK COMMUNICATION

In order to be useful, risk assessment results must be effectively communicated to nontechnical audiences. This can include risk managers, legislators, the public, industry, and environmental groups. If risk managers don't understand the results, it can lead to bad regulatory and policy decisions. Public understanding of risk assessment results is also essential if they are to participate in, or at least accept the results of, risk-based decision-making.

Effectively communicating the results of risk assessments is an enormous challenge. Problems lie in virtually all aspects of the risk communication process, including (1) the individual, agency, or company that conducts and presents the risk assessment; (2) the risk assessment itself; (3) the means to convey risk information; and (4) the audience. Examples of these problems are listed in Table 18.8. One of the biggest hurdles is the fact that risk analyses are often very complex, technical exercises. Making the process and outcome of the risk analysis transparent to laypersons is next to impossible unless there is some opportunity to provide background education to "bring them up to speed" on the subject. In most situations, this opportunity doesn't exist. The public is arguably one of the most important recipients of risk information, yet one of the most difficult audiences for risk assessors to communicate with. One problem is that the most common channel for communicating risk information to the public is through the news media. This presents at least three difficulties in trying to communicate a clear and accurate message: (1) reporting of the information may be biased, incomplete, or inaccurate; (2) news accounts may tend to sensationalize or focus on ancillary issues, such as disagreements between parties or human interest stories; and (3) news media have generally shown little interest in providing the background information needed to educate the public on risk analysis and to help them interpret findings for themselves.

No doubt one reason why the media have not invested much effort in educating the public about risk assessment is that the public itself, for the most part, has shown little interest in the technical complexities and nuances of risk analysis. In most situations for which a risk assessment is needed, they just want a straight answer to the simple question, "Is it safe?" Anything other than a clear "yes" answer to this question signals cause for concern. Herein lies a second major problem for risk

TABLE 18.8 Examples of Risk Communication Problems

Source of Problem	Examples
Source of the message	The source of the risk information, usually a governmental or industrial entity or representative, is not trusted
	Any disagreements among scientific experts make the information appear to be guesswork
	There is often a reluctance to disclose limitations and uncertainties in the risk estimates
	The risk assessment may not address issues of greatest concern to individuals and communities
The message	Risk estimates may have large uncertainties due to limitations in models, methods, and data used in the risk assessment
	The inherent technical nature of risk assessments makes them difficult for laypersons to understand
	Use of jargon and bureaucratic and legal language make risk assessments even more incomprehensible
Channel for conveying the message	Media interpretation may result in presentation of oversimplified, distorted, or erroneous information
	Media emphasis on drama, wrongdoing, or conflicts clouds presentation of risk information
	Eagerness by media to report may result in premature disclosures of scientific information
Receiver of the message	Public perceptions of risk are often inaccurate
	There may be unrealistic demands for scientific certainty in risk estimates
	There is usually a lack of interest in the technical complexities of the risk assessment, and therefore a poor understanding of what risk estimates represent
	Not everyone will be open-minded; some individuals with strong opinions and beliefs will not be receptive to new information
	There is often an unwillingness or inability to view risks in context, understand risk tradeoffs, or view risk problems from a perspective other than that of their own perceived immediate interests

Source: Adapted from Cohrssen and Covello (1989).

communication. Unfortunately, all too often, the answers conveyed by the risk assessment can seem ambiguous. Scientists are trained to be circumspect in their conclusions and carefully point out any caveats in their analysis. This certainly applies to risk assessments, where responsible presentation of risk estimates is always accompanied by a discussion of the many areas of uncertainty and limitations in the analysis. When all of the caveats and uncertainties are presented along with the risk estimate, the uncertainty looms large and it is easy for the public to conclude that "They don't really know what the risk is." When this happens, regardless of whether the risk estimates themselves are large or small, they have little credibility. Thus, the dilemma for the risk communicator is how to adequately convey the underlying uncertainties in the risk estimates without losing the essential message that the risks are large or small, as the case may be.

Deciding whether a risk is acceptable requires, in part, placing that risk in context. Thus, the risk from a particular chemical or set of exposure circumstances must be compared with other risks to the individual or population in order to place that risk in perspective. While this is straightforward in concept, it is difficult in practice, particularly when communicating risk to the general public. One reason is that the public, unaccustomed to seeing typical risk assessment outputs, may have little basis for comparison. Unless someone has experience with, or is shown, comparative risk data for a variety of hazards, it is difficult for them to know whether a 1×10^{-5} risk is significant. For noncancer health effects, the meaning of outputs in terms of hazard index or margin of exposure is even more obscure. How, for example, would you help citizens place a hazard index of 3 for a chemical exposure in the context of risk from events in their everyday lives?

A second reason that placing risks in context for the public is difficult is that the public often has distorted views of the risks posed by common and uncommon events in their lives. Comparing risks from chemical exposure to risks the public is more familiar with is valuable only if their point of reference is accurate and, unfortunately, it seldom is. This has been demonstrated repeatedly in studies in which survey respondents' estimates of risks or comparative risk rankings for various hazards were compared with the actual, measured risks. Presenting the public with accurate risk comparisons can be helpful, but doesn't necessarily solve the problem. There are at least two reasons for this. One is that the meaning of the term "risk" itself is often different for the risk assessor and the public. The risk assessor tends to define risk as a probability of an adverse health effect, and thinks of risk in purely probability terms. It is not surprising, then, that risk assessors once thought that a comparison of probabilities is all the public needs to place risks in perspective. The public, however, does not view risk simply in probability terms. The perception of the risk can be shaped powerfully by the nature of the risk (e.g., what health effect is at risk, such as cancer), whether the risk is voluntary or involuntary, and whether the risk is accompanied by any perceived benefits.

Several strategies have evolved for improving risk communication. The first is to pay very careful attention to the language that is used in risk communication. Of course, jargon and acronyms unfamiliar to the public should be avoided. It is also important to understand that terms and expressions in common use in risk assessment have very different meanings to the public. For example, a "conservative approach" is understood in risk assessment to mean one protective of health, while the public might mistakenly interpret this as a risk assessment approach endorsed by one end of the political spectrum (e.g., as opposed to a "liberal approach"). In order to be more protective, an agency might "lower the standards" for a chemical, meaning to decrease permissible concentrations. To the public, however, lowering standards might be misinterpreted as allowing some sort of deterioration in their protectiveness. To avoid awkward and sometimes disastrous misunderstandings, it is important to carefully scrutinize the risk communication message and remove terms and phrases that will be unclear or have a different meaning for the public.

It is an unfortunate fact that there are few sources that the public explicitly trusts for risk information. Risk information provided by industry is often met with skepticism. In particular, risk messages that indicate no harm or basis for concern for chemical exposure are seen as self-serving. Credibility of governmental agencies charged with protecting public health and the environment is better, but not much. In dealing with the public, particularly when engaging them directly (e.g., through public meetings), it is extremely important to be open and honest. An individual seen as not forthcoming with information, or who provides information solely as "technical gibberish," will be regarded as either completely out of touch or hiding something. From a risk communication standpoint, one is just as bad as the other. It is also important to listen to the public and gain an appreciation for their concerns and fears. Engaging in dialog early in the risk assessment process has several benefits, including the following:

1. It helps ensure that the risk assessment will be able to answer questions of greatest interest to the public.
2. Individuals in the public may be able to offer knowledge useful to the risk assessment, such as historical perspective and information regarding the manner in which individuals are (or have been) exposed to the chemicals in question.
3. It affords the opportunity to establish trust with the public. Of course, demeanor is important; a condescending manner is a sure way to cut the lines of risk communication.

18.10 SUMMARY

Conceptually, the basic components of any risk assessment are (1) hazard identification (what health effects may be produced by specific chemicals), (2) dose–response assessment (what dose of chemical is required to produce these effects), (3) exposure assessment (whether persons are actually exposed to chemicals and what doses they receive), and (4) risk characterization (how likely is it that adverse effects will occur, and what are the potential limitations of the risk assessment as performed).

In order to fulfill their goal of ensuring protection of public health, regulatory agencies usually choose conservative exposure and modeling assumptions, namely, those that tend to overestimate rather than underestimate risk. Because the impact of each conservative assumption is frequently multiplicative and cumulative, the final risk estimate may overstate the true population risk substantially. Nonetheless, it is difficult to deviate from this approach given that considerable uncertainty exists for many components of the risk assessment.

While risk assessment has traditionally focused on human health, ecological risk assessments, which address potential impacts to plants and wildlife, are now more commonly performed. Ecological risk assessments differ from human health risk assessments in that they are inherently more complex—there are many more species to consider, including interspecies relationships and more complicated exposure modeling—and they tend to focus more on population-, species-, and ecosystem-level effects.

Traditionally, cancer risks have been expressed in probability terms using linear, nonthreshold dose–response relationships. These relationships assume that any dose of a carcinogen poses some risk of developing cancer. The potential for noncancer health effects is evaluated using threshold models, where a dose is assumed to exist below which no health effects will occur. There has been increasing recognition that the dose–response relationship for some carcinogens may also involve a threshold, and methods to take this threshold into consideration in evaluating cancer risk from these chemicals have been proposed.

Deterministic risk assessments develop a single estimate of risk for a population, usually derived in such a way as to represent an upper-bound estimate. Probabilistic risk assessments can provide a description of the variability of risks within the population and quantitative estimates of uncertainty associated with those risks. While probabilistic risk assessments potentially offer more risk information, deterministic risk assessments are easier to perform and less expensive, and there exists a greater consensus as to how risk outputs should be conveyed and interpreted. At present, deterministic risk assessments are more routinely used because of their simplicity and ease of application.

The risk assessment should be performed in a *transparent* manner; that is, the steps performed should be easy to identify, understand and evaluate. Also, the outcome of the risk assessment must be communicated in a way that can be understood by those without technical backgrounds, including the public. This is very challenging because risk assessors and the public may view risks and risk issues very differently.

A criticism of quantitative risk assessments, specifically, risk assessments that produce a numerical estimate of risk, is that they often convey the impression of greater precision than actually exists. It is vitally important that risk assessments include qualitative information as well, such as a discussion of the uncertainties associated with the risk estimate and the extent to which evidence of a true human hazard is weak or controversial.

It must be recognized that risk assessment is just one aspect of the larger process of risk management. In the development of strategies and procedures to address health concerns for chemical exposures, risk estimates undoubtedly play an important role. They are often not the sole consideration, however, and economic, social, and political factors, as well as technical feasibility, may also influence the management of chemical exposures in modern society.

REFERENCES AND SUGGESTED READING

Ahlborg, U. G., G. C. Becking, L. S. Birnbaum, A. Brouwer, H. J. G. M. Derks, M. Feeley, G. Golor, A. Hanberg, J. C. Larsen, A. K. D. Liem, S. H. Safe, C. Schlatter, F. Waern, M. Younes, and E. Yrjanheikki, "Toxic equivalency factors for dioxin-like PCBs," *Chemosphere* **28**, 1049–1067, 1994.

Allman, W. F., "Staying alive in the 20th century," *Science* **6**(8), 31–41, 1985.

Andersen, M. E., "Physiologically based pharmacokinetic (PB-PK) models in the study of the disposition and biological effects of xenobiotics and drugs," *Toxicol. Lett.* **82/83**, 341–348 (1995).

Barnard, R. C., "Scientific method and risk assessment," *Regul. Toxicol. Pharmacol.* **19**, 211–218 (1994).

Barnes, D. G., G. P. Daston, J. S. Evans, A. M. Jarabek, R. J. Kavlock, C. A. Kimmel, C. Park, and H. L. Spitzer, "Benchmark dose workshop: Criteria for use of a benchmark dose to estimate a reference dose," *Regul. Toxicol. Pharmacol.* **21**, 296–306 (1995).

Center for Risk Analysis, *A Historical Perspective on Risk Assessment in the Federal Government*, Harvard School of Public Health, Boston, 1994.

Cohrssen, J. J., and V. T. Covell, *Risk Analysis: A Guide to Principles and Methods for Analyzing Health and Environmental Risks*, Council on Environmental Quality, Office of the President, Washington DC, NTIS PB 89-137772, 1989.

Crump, K. S., "The linearized multistage model and the future of quantitative risk assessment," *Hum. Exp. Toxicol.* **15**, 787–798 (1996).

Cullen, A. C., and H. C. Frey, *Probabilistic Techniques in Exposure Assessment. A Handbook for Dealing with Variability and Uncertainty in Models and Inputs*, Plenum Press, New York, 1999.

Fischhoff, B., "Risk perception and communication unplugged: Twenty years of process," *Risk Anal.* **15**, 137–145 (1995).

Gaylor, D. W., R. L. Kodell, J. J. Chen, and D. Krewski, "A unified approach to risk assessment for cancer and noncancer endpoints based on benchmark doses and uncertainty/safety factors," *Regul. Toxicol. Pharmacol.* **29**, 151–157 (1999).

Lioy, P. J., "Analysis of total human exposure for exposure assessment: A multidisciplinary science for examining human contact with contaminants," *Environ. Sci. Technol.* **24**, 938–945 (1990).

McMahan, S., and J. Meyer, "Communication of risk information to workers and managers: Do industrial hygienists differ in their communication techniques?" *Am. Ind. Hyg. Assoc. J.* **57**(2), 186–190 (1996).

Molak, V., in *Fundamentals of Risk Analysis and Risk Management*, V. Molek, ed., CRC Press, Boca Raton, FL, 1997.

National Academy of Sciences (NAS), *Understanding Risk. Informing Decisions in a Democratic Society*, National Academy Press, Washington, DC, 1996.

National Research Council (NRC), *Drinking Water and Health*, Vols. 1–9, National Academy Press, Washington, DC, 1977.

National Research Council (NRC), *Risk Assessment in the Federal Government: Managing the Process*, Committee on the Institutional Means for Assessment of Risks to Public Health. Commission on Life Sciences, National Academy Press, Washington, DC, 1983.

National Research Council (NRC), *Issues in Risk Assessment*, Committee on Risk Assessment Methodology, Board on Environmental Studies and Toxicology, Commission on Life Sciences, National Academy Press, Washington, DC, 1993.

National Research Council (NRC), *Science and Judgement in Risk Assessment*, Committee on Risk Assessment of Hazardous Air Pollutants, Board on Environmental Studies and Toxicology, Commission on Life Sciences, National Academy Press, Washington, DC, 1994.

National Research Council (NRC), *Carcinogens and Anticarcinogens in the Human Diet*, National Academy Press, Washington, DC, 1996.

Needham, L. L., J. L. Pirkle, V. W. Burse, D. G. Patterso Jr., and J. S. Holler, "Case studies of relationship between external dose and internal dose," *J. Exposure Analysis Envir. Epidemiol.* **1**(Suppl), 209–221 (1992).

Neil, N., T. Malmfors, and P. Slovic, "Intuitive toxicology: Expert and lay judgments of chemical risks," *Toxicol. Pathol.* **22**(2), 198–201 (1994).

Office of Technology Assessment, *Assessment of Technologies for Determining Cancer Risks from the Environment*, Congress of the United States, Washington, D.C., June, 1981.

Omenn, G. S., "Genetic variation as a key parameter in risk assessment and risk communication: Policy aspects," *Prog. Clin. Biol. Res.* **395**, 235–247 (1996).

Poirier, M. C., "DNA adducts as exposure biomarkers and indicators of cancer risks," *Environ. Health Persp.* **105**(Suppl. 4), 907–912 (1997).

Presidential/Congressional Commission on Risk Assessment and Risk Management, *Framework for Environmental Health Risk Management, Final Report*, Vol. 1, U.S. Government Printing Office, 1997.

Presidential/Congressional Commission on Risk Assessment and Risk Management, *Risk Assessment and Risk Management in Regulatory Decision-Making, Final Report*, Vol. 2, U.S. Government Printing Office, 1997.

Purchase, I. F. H., and P. Slovic, "Quantitative risk assessment breeds fear," *Hum. Ecol. Risk Assess.* **5**(3), 445–453 (1999).

Sielken, R. L., R. S. Bretzlaff, and D. E. Stevenson, "Challenges to default assumptions stimulate comprehensive realism as a new tier in quantitative cancer risk assessment," *Regul. Toxicol. Pharmacol.* **21**, 270–280 (1995).

Swenberg, J. A., D. K. La, N. A. Scheller, and K. Y. Wu, "Dose–response relationships for carcinogens," *Toxicol. Lett.* **82**, 751–756 (1995).

Travis, C. C., S. A. Richter, E. A. C. Crouch, R. Wilson, and E. D. Klema, "Cancer risk management. A review of 132 federal regulatory decisions," *Environ. Sci. Technol.* **21**(5) (1987).

Travis, C. C., and S. T. Hester, "Background exposure to chemicals: What is the risk?" *Risk Anal.* **10**, 463–466 (1990).

United States Environmental Protection Agency (USEPA), *Risk Assessment Guidance for Superfund*, Vol. I; *Human Health Evaluation Manual*, Part A, *Interim Final*, Office of Emergency and Remedial Response, Washington, DC, 1989.

United States Environmental Protection Agency (USEPA), *Seminar Publication. Risk Assessment, Management and Communication of Drinking Water Contamination*, Office of Research and Development, USEPA, Washington, DC, EPA/625/4-89/024, 1990.

United States Environmental Protection Agency (USEPA), *The Use of Benchmark Dose Approach in Health Risk Assessment*, Office of Research and Development, Risk Assessment Forum, Washington, D.C., EPA/630/R-94/007, February, 1995.

United States Environmental Protection Agency (USEPA), *Guidance Manual for the Integrated Exposure Uptake Biokinetic Model for Lead in Children*, OSWER #9285.7-15-1, Feb. 1994.

United States Environmental Protection Agency (USEPA), *Proposed Guidelines for Carcinogen Risk Assessment*, Office of Research and Development, Washington, DC EPA/600/P-92/003C, 1996.

United States Environmental Protection Agency (USEPA), *Guiding Principles for Monte Carlo Analysis*, EPA/630/R-97/001, March 1997.

United States Environmental Protection Agency (USEPA), *Exposure Factors Handbook*, Vol. I, *General Factors*, Office of Research and Development, Washington, DC, 1997.

United States Environmental Protection Agency (USEPA), *Guidelines for Ecological Risk Assessment*, Risk Assessment Forum, USEPA, Washington, DC, EPA/630/R-95/002F, Jan. 1998.

Vose, D., *Quantitative Risk Analysis: A Guide to Monte Carlo Simulation Modelling*, Wiley, New York, 1996.

Waddell, W. J., "Reality versus extrapolation: An academic perspective of cancer risk regulation," *Drug Metab. Rev.* **28**(1–2), 181–195 (1996).

Wallace, L. A., "Comparison of risks from outdoor and indoor exposure to toxic chemicals," *Environ. Health Persp.* **95**, 7–13 (1991).

Wilson, R., "Risk/benefit for toxic chemicals," *Ecotoxicol. Environ. Safety* **4**, 37–383 (1980).

Yang, R. S. H., H. A. El-Masri, R. S. Thomas, A. A. Constan, and J. D. Tessari, "The application of physiologically based pharmacokinetic/pharmacodynamic (PBPK/PD) modeling for exploring risk assessment approaches of chemical mixtures," *Toxicol. Lett.* **79**, 193–200 (1995).

Young, A. L., "A White House perspective on risk assessment and risk communication," *Sci. Total Environ.* **99**(3), 223–229 (1990).

19 Example of Risk Assessment Applications

ALAN C. NYE, GLENN C. MILLNER, JAY GANDY, and PHILLIP T. GOAD

As described in the preceding chapter, human health risk assessment is a flexible, occasionally complex, often controversial process used to characterize the probability and types of adverse health effects that may result from chemical exposure. Historically, risk assessments have been criticized for many reasons, such as failing to quantitatively account for the effects of variability and uncertainty in the characterization of human health risk. Despite these and other shortcomings, risk assessment is an accepted decision-making tool for evaluating the adverse health effects resulting from environmental and occupational chemical exposure.

19.1 TIERED APPROACH TO RISK ASSESSMENT

The risk assessor is often confronted with practical concerns in assessing risks from chemical exposures. These include, but are certainly not limited to

- The lack of toxicity and dose-response information in humans or inadequate data in animals
- Lack of identification of the most sensitive individual
- Extrapolation of toxicity data from one route of exposure to another
- Extrapolation of toxicity data from high doses in animals to much lower doses in humans
- Quantifying uncertainty and variability in the risk assessment
- Accounting for all sources of chemical exposure and not just the source of exposure of immediate concern
- Consideration of the varying physicochemical properties of the chemical and how these may effect exposure and toxicity
- The toxic effects resulting from exposure to more than one chemical
- The use of varying risk assessment methods by different regulatory agencies

One way of dealing with several of these problems is a tiered, or iterative, approach to risk assessment. In every risk assessment, the risk assessor is expected to assess these problems in a manner that conservatively protects human health. The manner in which this is done is governed by the assessor's ability to obtain exposure and toxicity information. In discussing the tiered approach to risk assessment, the NRC (National Research Council) indicates that a risk assessment includes a conservative, first level of analysis. Use of higher, more complex tiers of risk assessment is more costly. The decision to use more complex and costly risk assessment practices will depend on whether the results of simple conservative screening risk assessment indicates the need for further study, whether additional study or data will provide more accurate estimates of risk, and whether increased accuracy is worth the additional cost.

Principles of Toxicology: Environmental and Industrial Applications, Second Edition, Edited by Phillip L. Williams, Robert C. James, and Stephen M. Roberts.
ISBN 0-471-29321-0 © 2000 John Wiley & Sons, Inc.

Use of the USEPA soil screening levels is one example of a first tier risk assessment. The USEPA has calculated soil concentrations of chemicals called *soil screening levels* (SSLs) as a preliminary means of assessing human health risks from exposure to chemicals in soil. The exposure assumptions used are quite conservative in that they assume that an individual ingests soil on a daily basis for 30 years and that the chemical concentration remains constant over the 30-year exposure period. Concentrations of a chemical less that the SSL are generally acknowledged to be associated with acceptably low levels of human health risk. Thus, if the concentration of a chemical in soil is lower than the SSL, the risk assessment process is often concluded at this initial step.

The risk assessor should be cautious in applying risk-based screening levels for soil, water, or air, since exposure pathways used in the calculation of the screening level may not address all pathways of exposure relevant to a site or exposure scenario. For example, the USEPA soil screening levels consider possible residential exposure to soil via incidental soil ingestion. Before using the SSLs, the risk assessor should examine whether the exposure assumptions used in calculating the SSLs are applicable to the specific site of interest.

The decision to use a higher, more complex risk assessment approach is often governed by the need for more accurate estimates of risk from environmental or occupational exposure. More complex risk assessments are inevitably more costly. Higher tiers of the risk assessment process generally include the collection of more detailed and refined exposure data. For example, it may be important to monitor exposure to airborne contaminants in the workplace using personal monitoring devices rather than use air samples collected near a point of release. This allows estimates of chemical exposure to be individualized to workers with specific tasks or work habits rather than assume that all workers are exposed to levels of airborne chemicals near the source. While collection and analysis of this additional data would likely be more costly, it nonetheless allows for better estimates of worker exposure.

Higher tiers of the risk assessment process may also require further investigation of the toxicity of a chemical. This is particularly true of potential carcinogens, since extrapolation of cancer data from high dose animal studies to low levels of exposure in humans is an area of great uncertainty in the cancer risk assessment process. Further animal studies regarding the mechanism of carcinogenic action and the applicability of the mechanism to humans may provide much needed information to increase the accuracy of the risk assessment. For example, elucidation of a receptor-mediated mechanism of action for a potential carcinogen may indicate the existence of a threshold for the carcinogenic response. Because current risk assessment methods default to the position that there is no threshold for the carcinogenic response, this type of information would significantly affect the determination of cancer risk at low, environmentally relevant exposures. Few would argue that more complex risk assessment methods and additional basic research will provide more accurate characterization of human health risks. However, the added cost of this improved accuracy may not be justifiable except in situations where the health or economic impact of the regulatory decision is great.

19.2 RISK ASSESSMENT EXAMPLES

This chapter describes short examples of human health risk assessments of chemical exposure. In its broadest sense, risk assessment may address the effects of any hazardous agent on living things. We have restricted our few examples to characterization of risks posed by chemical exposure in humans. Thus, for the purpose of this chapter, we use the term "risk assessment" to describe human health risks posed by chemical exposure.

Brief summaries of risk assessments for persons exposed to lead, petroleum hydrocarbons, arsenic, and antimony are included as diverse examples of risk assessments for persons exposed to chemicals in occupational or residential settings. These examples illustrate some of the more complex risk assessment problems and how they may be addressed. The lead study presents an example of a novel biokinetic approach in risk assessment where human data regarding the absorption and distribution of lead in the body are integrated into the risk assessment process. The examples of arsenic and petroleum hydrocarbons closely parallel the risk assessment steps described above. The case of antimony trioxide is used to illustrate the toxicity assessment step of the risk assessment process and emphasizes the need for incorporation of new, mechanistic data regarding the carcinogenic effects of chemicals.

In each example, conservative default risk assessment procedures are used to familiarize the reader with these assumptions and methods. Use of these procedures is not necessarily intended to be an endorsement of their use and no systematic critique of the technical or scientific validity of each assumption or method was performed. Rather, the reader is encouraged to critically evaluate the scientific basis for the procedures and, where applicable, adopt alternate assumptions or methods when justified by site-specific information or other data.

19.3 LEAD EXPOSURE AND WOMEN OF CHILD-BEARING AGE

The human health effects of lead are better known than nearly all industrial or environmentally important chemicals. The extensive database of human information regarding the toxicity of lead allows exposure and risks to be characterized with greater certainty than other chemicals. Human lead exposure is most often evaluated by measuring the blood lead concentration. The blood lead concentration provides information regarding the absorbed dose of lead from environmental sources. In contrast, risk assessments for nearly all other chemicals calculate exposures rather than absorbed doses and provide no information regarding the absorbed dose of the chemical, the time course of the chemical in the body, or the concentration of the chemical in the target organ or tissue.

Lead is different in that good information exists to predict human blood lead concentrations resulting from inhaled or ingested lead. Studies of the absorption, distribution, metabolism, and excretion of lead in humans allow the determination of constants that relate ingested or inhaled blood lead to the amount of lead in blood. For example, the USEPA uses a kinetic factor of 0.4 μg/dL per μg/day to relate the amount of lead absorbed from the gastrointestinal tract to the amount of lead in the blood in adults. Thus, for every microgram of lead absorbed from the gastrointestinal the blood lead concentration will increase 0.4 μg/dL.

In addition, there exists a large human toxicological database that allows the blood lead concentration to be related to lead's toxic effects. At blood lead concentrations less than 20 μg/dL, lead may cause neurobehavioral and developmental effects in children and affect vitamin D metabolism. There is an obvious difference in the sensitivity of children and adults to the effects of lead. For example, young children absorb more lead from the gastrointestinal tract. The immature nervous system of young children is also more sensitive to the adverse effects of lead. In the case of a pregnant worker exposed to lead, elements of adult lead exposure and a child's greater sensitivity to lead must be considered. In a pregnant worker, lead absorption, distribution, and rate of excretion from the body is that of an adult. In this way, the lead exposure of the fetus is nearly the same as that of the adult.

The Occupational Safety and Health Administration requires medical monitoring of workers with blood lead levels of 40 μg/dL and higher. However, for a pregnant worker, the rapidly growing fetus is the sensitive individual of greatest concern. OSHA regulations are not specifically designed to protect the fetus. The Centers for Disease Control (CDC) in Atlanta has established a blood lead level of concern for children of 10 μg/dL. Thus, blood lead levels tolerated under OSHA regulations may be potentially harmful to the fetus. It is particularly important to assess the risks posed by lead exposure for female workers of child-bearing age.

Given information concerning the kinetic behavior of lead in the human body and the relationship of blood lead concentration to lead toxicity, the toxic effects of lead exposure can be assessed with a greater degree of certainty than nearly every other environmental chemical. The USEPA interim lead exposure model may be used to evaluate fetal lead exposure that may result from the mother's exposure in the workplace. A brief description of the equations used to predict fetal blood lead concentrations resulting from maternal ingestion of lead in soil or dust is presented below.

The equation for adult lead exposure is

$$PbB_{fetal,0.95} = GSD_i^{1.645} \times \frac{(PbS \times BKSF \times IR_S \times EF_S)}{AT} + PbB_{adult,0} \times R_{fetal/maternal}$$

where $PbB_{fetal,0.95}$ = The 95th percentile blood lead concentration for fetuses born to women similarly exposed to lead in dust. This lead concentration indicates that the likelihood of a greater blood lead concentration is 5 percent. The 95th percentile is often used as a regulatory target by USEPA.

$GSD_i^{1.645}$ = Individual geometric standard deviation for the variability in lead exposure and absorption. The value of 1.645 is the *t* value used to calculate the 95th percentile from a lognormal blood lead distribution. The USEPA recommends a value of 1.8 for the GSD in a fairly homogenous population.

PbS = Soil or dust lead concentration in μg/g.

BKSF = Biokinetic slope factor increased in typical adult blood lead concentration to average daily lead uptake (μg/dL blood lead increase per μg/day lead uptake). The BKSF obtained from reliable human studies is 0.4 μg/dL per μg/day.

IR_S = Daily intake of soil or dust. The USEPA recommends 0.05 g/day as a typical value.

AF_S = Absorbed fraction of lead in dust from the gastrointestinal tract. The USEPA recommended value for the fraction of lead in soil or dust released from the gastrointestinal tract and absorbed is 0.12.

EF_S = Exposure frequency for contact with lead in dust. The USEPA recommends a typical value of 219 days/year.

AT = Total period over which dust exposure occurs. This value is typically assumed to be 365 days/year.

$PbB_{adult,0}$ = Background blood lead concentration for women of child-bearing age. A representative value is 2.0 μg/dL.

$R_{fetal/maternal}$ = Ratio of fetal blood lead concentration to the maternal blood lead concentration at birth. The value typically assumed by USEPA is 0.9.

Assume that workers employed as furniture refinishers may ingest lead in dust during a particular phase of the operation for 3 days a week (approximately 150 days/year). Sampling of the lead-contaminated dust indicates an average lead concentration of 1300 μg/g. What is the calculated 95th percentile blood lead concentration of a fetus of an exposed female worker? This calculation is illustrated below.

$$PbB_{fetal,0.95} = 1.8^{1.645} \times \frac{(1300\ \mu g/g \times 0.4\ \mu g/dL \cdot g/day \times 0.05\ g/day \times 0.12 \times 150\ days/year)}{365\ days/year}$$

$$+ 2.0\ \mu g/dL \times 0.9$$

$$PbB_{fetal,0.95} = 7.8\ \mu g/dL$$

Interpretation of Risk Assessment Results and Comment

For pregnant workers, this indicates that there is a 5 percent likelihood that the fetal blood lead concentration may exceed 7.8 μg/dL in similarly exposed pregnant women. The calculated lead concentration is below the CDC and USEPA level of concern of 10 μg/dL. A greater exposure frequency, a higher dust lead concentration, or exposure to a highly soluble form of lead (such as lead chloride or lead acetate) may result in a calculated $PbB_{fetal,0.95}$ that could potentially exceed 10 μg/dL. In practice, blood lead concentrations could also be measured in women of child-bearing age to provide reassurance that they were not being overexposed.

Although the preceding equation does not evaluate inhalation exposures to lead, it could easily be modified to do so. The Agency for Toxic Substances and Diseases Registry (ATSDR) has summarized human inhalation studies of lead and determined biokinetic slope factors relating the air lead concentration to increases in blood lead. For example, individuals exposed to lead concentrations in air ranging from 3.2 to 11 $\mu g/m^3$ had average blood lead increases of 1.75 μg/dL for every $\mu g/m^3$ lead in air. Individuals in the study reviewed by ATSDR were exposed for 23 h/day for 18 weeks. Given that workers would not be exposed to the workplace atmosphere for 23 h/day, it would be reasonable to assume that only half the air breathed in a day was from the affected workplace (i.e., a correction factor of 0.5). If the above equation is modified to reflect exposure to lead in air at a concentration of 0.5 $\mu g/m^3$, the equation would be revised as follows:

$$PbB_{fetal,0.95} = 1.8^{1.645} \times \frac{(1300\ \mu g/g \times 0.4\ \mu g/dL \cdot \mu g/day \cdot 0.05\ \mu g/day \cdot 0.12) \cdot 150\ days/year}{365\ days/year}$$

$$+ \frac{(0.5\ \mu g/m^3 \times 1.75\ \mu g/dL \cdot \mu g/m^3 \cdot 0.5) \cdot 150\ days/year}{365\ days/year} + 2.0\ \mu g/dL \cdot 0.9$$

$PbB_{fetal,0.95} = 8.2$ μg/dL when inhalation exposure to lead is added to ingestion of lead

19.4 PETROLEUM HYDROCARBONS: ASSESSING EXPOSURE AND RISK TO MIXTURES

Chemical mixtures present special problems to risk assessors. Mixtures may be made up of hundreds of individual chemicals that are inadequately characterized with regard to their toxicity. Further, it is often difficult or impractical to completely characterize the composition of the mixture. Such is the case with petroleum fuels such as gasoline and diesel fuel that contain hundreds of organic compounds.

The USEPA indicates that when adequate information is available, it is preferable to use mixture-specific toxicity information to evaluate the risks of complex chemical mixtures. Mixture-specific toxicity information is preferred since the risk assessor does not have to make assumptions regarding the toxicological interaction of the chemicals of the mixture. However, use of mixture-specific toxicity information is only useful when the mixture in question is the same as the toxicologically characterized mixture. This is an important caveat for risk assessments of petroleum hydrocarbon mixtures. After being released to the environment, petroleum mixtures "weather" with time. Weathering causes the loss of more volatile, water-soluble, and degradable petroleum hydrocarbons. As a result, weathered petroleum fuel mixtures may no longer be chemically or toxicologically similar to the unweathered fuel. Until toxicological data are available for weathered petroleum mixtures, risk assessments of weathered petroleum mixtures are typically performed using either an "indicator chemical" or a "surrogate" chemical approach.

The indicator chemical approach to petroleum hydrocarbon risk assessment assumes that certain compounds in a petroleum hydrocarbon mixture can be used to represent the environmental mobility, exposure potential, and toxicological properties of the entire petroleum mixture. For example, indicator chemicals typically used in risk assessments of unleaded gasoline include benzene, ethylbenzene, toluene, xylenes, and hexane. The Amerian Society for Testing and Materials (ASTM) has prepared a thorough guidance document for conducting risk assessments of petroleum mixtures using the indicator chemical approach.

Examples of the surrogate chemical risk assessment approach for petroleum hydrocarbons include the Massachusetts Department of Environmental Protection and Total Petroleum Hydrocarbon Committee Working Group methods. These methods identify specific carbon ranges for both aliphatic and aromatic hydrocarbons and assign a reference dose to each fraction. The primary difference between the two methods is the number of separate petroleum fractions identified (MADEP method, 6; TPHCWG method, 13) and the manner in which toxicological surrogates are assigned. The MADEP method uses single chemicals to represent the toxicity of a petroleum fraction whereas the TPHCWG method uses petroleum-fraction-specific toxicological data as available.

We illustrate the use of the TPHCWG method to assess the risks posed by weathered diesel fuel in an industrial exposure scenario. In this example, a railyard worker is assumed to be exposed to diesel fuel in soil via incidental ingestion of dust and absorption of petroleum hydrocarbons from soil into the skin. Air monitoring did not detect the presence of petroleum hydrocarbons that could be attributed to site sources.

Table 19.1 presents the soil concentrations of diesel fuel constituents by petroleum fraction, the reference doses (RfDs) used to assess the toxicity that may result from exposure to these fractions, and the target organ or critical effect associated with exposure to each fraction. Animal toxicity data is the basis for the RfD for each petroleum fraction.

The USEPA defines the RfD to be an estimate of the daily exposure that is likely to be without adverse health effects. The exposure (in milligrams of chemical intake per kilogram of body weight per day) divided by the RfD is termed the "hazard quotient" or HQ. The sum of the HQ values for different routes of exposure or chemicals is termed the "hazard index" (HI) (see also Chapter 18 for a discussion of HQ and HI). If the

TABLE 19.1 Example—Petroleum Hydrocarbon Risk Assessment Concentrations of Petroleum Hydrocarbon Fractions in Soil, Reference Doses, Critical Effects

Petroleum Hydrocarbon Fraction	Concentration Detected in Soil (mg/kg)	Oral Reference Dose (mg/kg-day)	Critical Effect
	Aliphatics		
C_5–C_6	ND[a]	5	Neurotoxicity
$C_{>6}$–C_8	ND	5	Neurotoxicity
$C_{>8}$–C_{10}	ND	5	Liver and hematologic changes
$C_{>10}$–C_{12}	ND	0.1	Liver and hematologic changes
$C_{>12}$–C_{16}	2,200	0.1	Liver and hematologic changes
$C_{>16}$–C_{21}	18,000	2	Liver granuloma
$C_{>21}$–C_{35}	6,600	2	Liver granuloma
	Aromatics		
$C_{>7}$–C_8	ND	0.04	Decreased body weight
$C_{>8}$–C_{10}	ND	0.04	Decreased body weight
$C_{>10}$–C_{12}	ND	0.04	Decreased body weight
$C_{>12}$–C_{16}	1,500	0.04	Decreased body weight
$C_{>16}$–C_{21}	9,300	0.03	Kidney toxicity
$C_{>21}$–C_{35}	9,100	0.03	Kidney toxicity

[a]Not detected.

TABLE 19.2 Example—Worker Exposure to Diesel Fuel Hydrocarbons in Soil, Typical Reasonable Maximum Exposure Soil Exposure Parameters

Exposure Parameter	Value	Reference
ABS_{gi}	1	Default
ABS_{sk}	0.05	Professional judgment
AF	0.2 mg/cm^2	USEPA (1997)
AT_{nc}	9125 days	USEPA (1991)
BW	70 kg	USEPA (1991)
ED	25 years	USEPA (1991)
EF	250 days/year	USEPA (1991)
IR	50 mg/day	USEPA (1991)
SA	2000 cm^2	USEPA (1992)

HQ or HI exceeds one, there may be a concern for adverse effects. Exposure assumptions used to calculate exposure to petroleum hydrocarbons in soil are presented in Table 19.2.

The average daily intakes (ADIs) of the six petroleum hydrocarbon fractions are presented in Table 19.3. These ADIs were calculated using the soil concentrations in Table 19.1, the exposure assumptions presented in Table 19.2, and the equations presented later in this chapter (see Table 19.10).

HQs associated with the calculated levels of exposure to petroleum hydrocarbons in soil are calculated by dividing the calculated ingestion and dermal intake by RfD for the appropriate petroleum hydrocarbon fraction. The calculated HQs for the six petroleum hydrocarbon fractions are presented in Table 19.4.

Several petroleum fractions may affect the same target organ or have similar critical effects. In the absence of strong evidence indicating another type of interaction (an antagonistic effect or a synergistic effect), the USEPA assumes that the effects of chemicals affecting the same target organ are additive. Thus, the hazard quotients for chemicals affecting the same target organ are summed. The sum of the HQs for a particular target organ is termed the HI. The calculated HIs for liver toxicity, decreased body weight, and kidney toxicity are presented below.

HI for liver toxicity = sum of the oral and dermal HQs for aliphatic petroleum fractions $C_{>12}$–C_{16}, $C_{>16}$–C_{21}, and $C_{>21}$–C_{35} = 0.024

TABLE 19.3 Example—Worker Exposure to Diesel Fuel Hydrocarbons in Soil, Calculated Average Daily Intakes of Diesel Fuel Hydrocarbons

	Average Daily Intake	
Petroleum Hydrocarbon Fraction	Ingestion (mg/kg)	Dermal (mg/kg)
	Aliphatic	
$C_{>12}$–C_{16}	1.08×10^{-3}	4.31×10^{-4}
$C_{>16}$–C_{21}	8.81×10^{-3}	3.52×10^{-3}
$C_{>21}$–C_{35}	3.23×10^{-3}	1.29×10^{-3}
	Aromatic	
$C_{>12}$–C_{16}	7.34×10^{-4}	2.94×10^{-4}
$C_{>16}$–C_{21}	4.55×10^{-3}	1.82×10^{-3}
$C_{>21}$–C_{35}	4.45×10^{-3}	1.78×10^{-3}

TABLE 19.4 Example—Worker Exposure to Diesel Fuel Hydrocarbons in Soil, Calculated Hazard Quotients for Ingestion and Dermal Exposure

	Hazard Quotient	
Chemical	Ingestion	Dermal
	Aliphatic	
$C_{>12}$–C_{16}	1.08×10^{-2}	4.31×10^{-3}
$C_{>16}$–C_{21}	4.40×10^{-3}	1.76×10^{-3}
$C_{>21}$–C_{35}	1.61×10^{-3}	6.46×10^{-4}
	Aromatic	
$C_{>12}$–C_{16}	1.83×10^{-2}	7.34×10^{-3}
$C_{>16}$–C_{21}	1.52×10^{-1}	6.07×10^{-2}
$C_{>21}$–C_{35}	1.48×10^{-1}	5.94×10^{-2}

HI for decreased body weight = sum of the oral and dermal HQs for aromatic petroleum fraction $C_{>12}$–C_{16} = 0.026

HI for kidney toxicity = sum of the oral and dermal HQs for aromatic petroleum fractions $C_{>16}$–C_{21} and $C_{>21}$–C_{35} = 0.42

Interpretation of Risk Assessment Results and Comment

As calculated above, concurrent exposure to relatively high concentrations of diesel fuel–related petroleum hydrocarbons in soil resulted in calculated hazard indices that are less than one for the liver toxicity, decreased body weight, and kidney toxicity endpoints. These calculations indicate that workers exposed to concentrations of these petroleum hydrocarbons in soil would be unlikely to experience adverse health effects as a result of direct exposure to weathered diesel fuel in soil.

19.5 RISK ASSESSMENT FOR ARSENIC

Risk assessors must consider several important factors when assessing the risks posed by arsenic exposure. First, the chemical form of arsenic must be considered since toxicity varies with the chemical species. Inorganic arsenic occurs in either the trivalent [arsenite (As^{3+})] or the pentavalent [arsenate (As^{5+})] state. Arsenite is more toxic than arsenate and these inorganic forms are more toxic than organic arsenic compounds. Arsenobetaine is an organic form of arsenic that is also called "fish arsenic" since it occurs naturally in fish. Arsenobetaine is rapidly excreted in the urine and does not accumulate in the tissues.

Arsenic in the environment may cycle from one form to another based on the chemical conditions in soil or water and the activity of microbes. Arsenic may be reduced, oxidized, and methylated or demethylated under certain environmental conditions, potentially resulting in a mixture of arsenite, arsenate, and organic forms of arsenic in the environment.

The environmental medium in which arsenic occurs will also affect its absorption from the gastrointestinal tract. Dissolved arsenic in drinking water is well absorbed from the gastrointestinal tract. In comparison, as a result of tight binding, arsenic absorption from a mineral or soil matrix will be decreased relative to absorption from food or water.

Arsenic occurs naturally in air, water, soil, and food in low concentrations. Thus, daily exposure to very low amounts of arsenic is unavoidable. Thus, risk assessments of arsenic must often deal with "background" exposure from everyday living in addition to exposures resulting from occupational or environmental sources.

Inorganic forms of arsenic are known to be carcinogenic to humans. Since 1888, elevated arsenic exposure has been associated with an increased incidence of skin cancer. Arsenic exposure has also been linked to lung, bladder, and liver cancer. Although high levels of arsenic exposure are indisputably carcinogenic to humans, there is growing evidence of an apparent threshold for arsenic carcinogenicity. A number of epidemiologic studies indicate that arsenic may cause cancer by a nonlinear or a threshold mode of action. In large part, this nonlinear action may explain the lack of association between relatively low levels of arsenic exposure and the development of skin, bladder, or other cancers. A nonlinear carcinogenic relationship to dose indicates that the carcinogenic response induced by the chemical decreases more than a linear relationship to dose. In other words, dose-response is sublinear at low doses.

A risk assessment for arsenic using USEPA default exposure factors is presented below. However, the impact of the bioavailability of arsenic in soil is included as an important modifying factor in the USEPA risk assessment process. The impact of these default factors and the adjustment for soil bioavailability is evaluated in this arsenic risk assessment example.

Consider the case of a medium density residential development being built on top of fill partly composed of mining waste containing elevated concentrations of arsenic. Investigation of the site soil indicated surface soil arsenic concentrations ranging from 12 to 140 mg/kg with a mean concentration of 90 mg/kg. The family living in the residence includes both adults and young children. Possible pathways of exposure to arsenic in soil include incidental ingestion of arsenic in soil, absorption of arsenic into the skin from soil adhering to the skin, inhalation of arsenic-containing dust, and ingestion of arsenic taken up from the soil by home-grown produce. Since a residential housing development offers very limited space to plant a garden, ingestion of home-grown produce is not considered relevant for this site.

The USEPA soil screening level (SSL) for arsenic is 0.4 mg/kg. The arsenic SSL is based on ingestion of soil and an added lifetime cancer risk of 1×10^{-6}. As a first tier risk-based screening level, use of the USEPA SSL is problematic since the average background concentration of arsenic in soil in the United States is about 5 mg/kg. Nonetheless, the mean arsenic concentration exceeds the SSL and the typical background concentrations, indicating that a higher tier of risk assessment is needed to address potential health risk at the site due to arsenic.

With the exception of arsenic bioavailability in soil, default USEPA assumptions used to evaluate arsenic exposure due to ingestion, skin contact, and inhalation of soil particles are presented in Table 19.5. The bioavailable fraction of arsenic from soil was assumed to be 0.28 based on studies in monkeys. This is below the typical USEPA default bioavailability of 0.8–1. The exposure equations used to perform these calculations are presented in Table 19.10, later in this chapter.

The following average daily intakes (ADIs) were calculated for a child and adult resident exposed to arsenic in soil. Lifetime ADIs are also calculated to assess the added lifetime cancer risk associated with exposure to arsenic in soil. These calculations are presented in Table 19.6.

The noncarcinogenic risks associated with exposure to arsenic in soil are assessed using the hazard quotient (HQ) method. As discussed earlier in this chapter, the hazard quotient (HQ) is calculated by dividing the ADI by the reference dose (RfD). For arsenic, only an oral RfD is available. However, because skin absorption and inhalation may add to overall exposure, hazard quotients may also be calculated for these routes of exposure using the oral RfD (0.0003 mg/kg/day). The sum of the HQs is known as the hazard index (HI). The HI for the ingestion, skin absorption, and inhalation soil exposure pathways for the child is thus calculated as

$$\frac{3.22 \times 10^{-4}\ \text{mg/kg·day}}{3 \times 10^{-4}\ \text{mg/kg·day}} + \frac{1.15 \times 10^{-6}\ \text{mg/kg·day}}{3 \times 10^{-4}\ \text{mg/kg·day}} + \frac{1.07 \times 10^{-7}\ \text{mg/kg·day}}{3 \times 10^{-4}\ \text{mg/kg·day}}$$

TABLE 19.5 Arsenic Risk Assessment Example: Typical USEPA Reasonable Maximum Exposure Soil Exposure Parameters[a]

Exposure Parameter	Value	Reference
ABS_{gi}	0.28	Freeman et al. (1995)
ABS_{sk}	0.001	USEPA (1995)
AF	0.2 mg/cm^2	USEPA (1997)
AT_{nc}	8760 days (adult); 2190 days (child)	USEPA (1991)
AT_c	25,550 days	USEPA (1991)
BW	70 kg (adult); 15 kg (child)	USEPA (1991)
CA	1.8×10^{-7} mg/m^3	Modeled air concentration
CS	90 mg/kg	Site-specific average arsenic concentration in soil
ED	24 years (adult); 6 years (child)	USEPA (1991)
EF	350 days/year	USEPA (1991)
IR	100 mg/day (adult) 200 mg/day (child)	USEPA (1991)
SA	2900 (adult) 1000 (child)	USEPA (1997)
VR	20 (adult) 10 (child)	USEPA (1997)

[a]*Note:* USEPA typically assumes 80–100 percent bioavailability for arsenic in soil. Therefore, the USEPA default value for ABS_{gi} is 0.8–1.

The calculated HI is rounded to one significant figure. Because the HI does not exceed one, arsenic exposure would be unlikely to cause noncancer effects. However, even if the HI value slightly exceeded one, this is would be unlikely to be of significant health consequence. This is particularly the case since the oral RfD for arsenic is based on a no-observed-adverse-effect level (NOAEL) in humans of 8×10^{-4} mg/kg·day. As stated by the USEPA, a case can be made for setting the oral RfD as high as the NOAEL. The USEPA adjusted the NOAEL downward using an uncertainty factor of 3 to account for uncertainty associated with an incomplete database regarding the noncarcinogenic effects of arsenic.

Note that if calculated for the adult, the HI for exposure to arsenic in soil would be lower because a child is exposed to more soil than an adult when dose is calculated on the basis of body weight.

TABLE 19.6 Arsenic Risk Assessment Example: Calculated Daily Exposure (in mg/kg) to Arsenic in Residential Soil

	Child Resident		Adult Resident	
Exposure Pathway	ADI[a]	LADI[b]	ADI	LADI
Ingestion	3.22×10^{-4}	2.76×10^{-5}	3.45×10^{-5}	1.18×10^{-5}
Skin absorption	$*1.15 \times 10^{-6}$	$*9.86 \times 10^{-8}$	$*7.15 \times 10^{-7}$	$*2.45 \times 10^{-7}$
Inhalation	1.06×10^{-7}	9.05×10^{-9}	3.46×10^{-8}	1.19×10^{-8}

[a]Average daily intake.

[b]Lifetime average daily intake.

[c]Expressed as an absorbed dose rather than a daily intake.

Cancer risks posed by exposure to soil are calculated using the lifetime average daily intake (LADI) and the oral or inhalation slope factor. The oral slope factor for arsenic is 1.5 kg·mg/day. Thus, the lifetime cancer risk for the child's ingestion of arsenic in soil is calculated as 2.76×10^{-5} mg/kg·day $\times$ 1.5 kg·mg/day = 4×10^{-5} (*Note*: Lifetime cancer risk estimates are expressed to only one significant digit.) Lifetime cancer risks posed by dermal exposure are estimated by multiplying the dermal LADI by the oral slope factor.

Inhalation lifetime cancer risks may be calculated using a unit risk factor (expressed in units of m^3/µg) or an inhalation slope factor (kg·day/mg). Since inhalation exposure is expressed in terms of body weight (mg/kg·day), the inhalation slope factor should be used. If only an inhalation unit risk factor is available, it can be converted to an inhalation slope factor by multiplying the unit risk factor by (70 kg/20 m^3) $\times$ 1000 µg/mg. The inhalation slope factor for arsenic is 15 kg·day/mg. Multiplication of the child's inhalation LADI by this slope factor yields an estimated lifetime cancer risk of 1×10^{-7} (9.05×10^{-9} mg/kg·day $\times$ 15 kg·day/mg).

According to default USEPA policy, the cancer risks for adult and child residents are summed together using the assumption that an individual will live at the affected residence from infancy until 30 years of age. The overall sum of calculated lifetime cancer risks from childhood and adult exposure is 6×10^{-5}.

The lifetime cancer risk associated with exposure to arsenic in soil is 6×10^{-5}. This risk is within the range of additional lifetime cancer risks considered acceptable by the USEPA (i.e., 1×10^{-6} to 1×10^{-4}). However, many states have set the acceptable level of allowable added lifetime cancer risk at 1×10^{-5} or even 1×10^{-6}. In these cases the calculated lifetime cancer risk exceeds these targets by 6- or 60-fold, respectively.

It is important to put the risks of site-related arsenic exposure and risk in perspective with unavoidable arsenic exposures. For example, the USEPA estimated that daily inorganic arsenic intake from food and water is approximately 0.018 mg/day. For a 70-kg individual, this amounts to 2.6×10^{-4} mg/kg per day. Using the USEPA oral slope factor for arsenic (1.5 kg·day/mg), the lifetime cancer risk for unavoidable ingestion of arsenic in food and water is 4×10^{-4}, greater than the USEPAs upper bound acceptable lifetime cancer risk level of 1×10^{-4}. By placing site-related arsenic risk into context with the higher risk from unavoidable sources of exposure, it may not be necessary to undertake action to decrease site-related risks by limiting the residents exposure to arsenic in soil.

Furthermore, at the arsenic intakes from soil described in this example, default USEPA cancer risk assessment methods may cause risk to be overestimated at low exposure levels. The default method assumes that the carcinogenic response to arsenic intake is linear at low doses. However, according to recent reviews of the possible carcinogenic mechanism of action in humans, a cancer threshold or sublinear carcinogenic response may exist at lower doses such as those calculated in the residential exposure scenario above.

The form of arsenic considered in this example is important consideration to the risk assessment. Default risk assessment policy often assumes that organic chemicals in soil are absorbed to the same extent as the form of the chemical studied in developing the oral RfD. Typically, these studies involve exposure to the chemical in food or water. Studies in monkeys indicate that the oral bioavailability of arsenic in soil or dust resulting from mining or smelting activities is only 10–28 percent that of sodium arsenate in water. Mineralogic factors appear to control the solubility and therefore, the release of arsenic from the soil impacted by smelting. Only soluble arsenic is available for absorption from the gastrointestinal tract. This example stresses the need to consider the form the chemical in the environment and the impact that chemical form may have on the bioavailability of the chemical. Use of the default assumption that arsenic in soil is as bioavailable as arsenic in water would result in the calculation of a hazard index above 1 and lifetime cancer risks in excess of 1×10^{-4} in the preceding example. Thus, even a change in one USEPA default exposure assumption (the bioavailability of arsenic in soil) may greatly affect the degree to which regulatory action is taken.

Human exposure monitoring can be used as a check on calculated estimates of exposure to arsenic in soil. Human arsenic exposure may be monitored by determining arsenic concentrations in urine, hair, and nails. Although human exposure monitoring is not routinely conducted at most

TABLE 19.7 Comparison of Arsenic Concentrations in Surface Soil to Urinary Arsenic Concentrations in Children 0–6 Years of Age

Reference and Site	Number of Children	Mean Concentration of Arsenic in Surface Soil (mg/kg)	Mean Urinary Arsenic Concentration (μg/L)[a]
Binder et al. (1987)			
Mill Creek, MT	10	648	66.1
Anaconda, MT	92	127	14.4
Opportunity, MT	25	113	10.6
Livingston, MT	105	44	10.6
Kalman et al. (1990)			
Ruston, WA	108	353	50.6
Tacoma/Bellingham, WA	87	7–57	11.7
Fort Valley, GA	15	14–140	< 10

[a]Binder et al. (1987) based on total urinary [As]; Kalman et al. (1990), based on speciated urinary [As].

sites, the USEPA encourages the inclusion of site-specific human exposure studies to strengthen the overall conclusions of the risk assessment. For arsenic, there have been a number of studies relating human exposure to arsenic (measured by excretion of arsenic in the urine) to concentrations of arsenic in soil.

As discussed above, children 6 years of age or younger are generally considered the age group at most risk of exposure to chemicals in soil because of their higher assumed soil ingestion rates. If it is assumed that a 15-kg child ingests 200 mg of soil per day that contains 90 mg/kg of arsenic and that 80 percent of the arsenic in soil is absorbed, a child's intake of arsenic is 14 μg/day. If it is further assumed that the average daily urinary for a 3-year-old child is 355 mL, the urinary arsenic concentration for a young child would be 41 μg/L.

Studies that have examined the relationship between surface soil arsenic concentration and urinary arsenic concentration in this age group are summarized in Table 19.7. Note that the 41 μg/L urinary arsenic concentration calculated for a young child is well above mean urinary arsenic concentrations calculated for children exposed to similar arsenic concentrations in soil in the Binder et al. (1987) and Hewitt et al. (1995) studies. This comparison suggests that exposure factors used in calculating soil arsenic exposure may substantially overestimate actual exposure. These factors may include the assumption of high bioavailability of arsenic in soil (80 percent) as well as upper end estimates of a child's daily soil ingestion.

19.6 REEVALUATION OF THE CARCINOGENIC RISKS OF INHALED ANTIMONY TRIOXIDE

We examine the animal carcinogenicity data for antimony trioxide and possible mechanisms to explain the carcinogenic action of antimony trioxide as an example of the hazard identification step of the human health risk assessment process. The hazard identification step evaluates whether a chemical causes a particular toxic effect in humans (i.e., cancer), the strength of human, animal, or other evidence for making this determination, and the overall quality of the toxicological data for predicting human toxicity. The hazard identification step also considers the possible mechanism of toxicity to humans and the relevance of animal data in predicting human toxicity.

The case of antimony trioxide also emphasizes the need for inclusion of up-to-date toxicological information in risk assessment. The National Research Council emphasized the iterative nature of risk assessment and encouraged inclusion of new, in-depth, toxicological data and the investigation of toxic mechanisms other than the default regulatory position. For example, California's Proposition 65 defaults to the position that there is no threshold for the carcinogenic effect of a chemical "known to the State to cause cancer." This "no threshold" default policy assumes that at low levels of exposure, the cancer risk associated with exposure to a carcinogen is linear to an exposure level at zero. Simply stated, calculated cancer risk is zero only when there is zero exposure to the chemical.

In contrast to the "no threshold" default policy of chemical carcinogenesis, a review of recent evidence suggests that some agents that are carcinogenic to the rat lung at very high levels of exposure may not be carcinogenic at lower, more environmentally relevant levels of exposure in humans. These studies suggest that the response of the rat lung to accumulated particles is different from the mouse and human. Even in the rat, exposure to lower concentrations of particles that do not overwhelm lungs' ability to clear the particles do not appear to be carcinogenic. Importantly, these observations suggest that the rat may not be the best model for assessing the carcinogenicity of particular chemicals in humans. However, even if the rat is considered to be a relevant model for humans, studies in the rat suggest that the response in the rat lung at high levels of exposure is different that that seen at environmentally relevant levels of exposure. The response of the rat lung to antimony trioxide particles appears to fit the pattern of a threshold response—lung tumors develop at very high concentrations of particle exposure but do not occur at lower levels of exposure. For this reason, the default regulatory position of no carcinogenic threshold does not appear applicable to antimony trioxide.

Antimony trioxide is used as a flame retardant in a diverse array of products. As a result of the International Agency for Research on Cancer (IARC) ranking of antimony trioxide as "possibly carcinogenic to humans (Group 2B)" in 1989, antimony trioxide was listed as a chemical "known to the State to cause cancer" on October 1, 1990 under the State of California's Proposition 65. The IARC classification of antimony trioxide as "possibly carcinogenic to humans" is based on two studies of inhaled antimony trioxide in rats conducted in the 1980s. Unlike IARC and State of California, the USEPA does not consider antimony trioxide to be a potential human carcinogen. In this way, antimony trioxide is an example of inconsistencies that may exist between regulatory agencies regarding the risks resulting from chemical exposure.

A review of information published before and after the 1990 listing of antimony trioxide as "Possibly Carcinogenic to Humans" is presented below. This information is particularly important to the hazard identification step in assessing the human health risks from inhaled antimony trioxide. As such, inclusion of this updated information is a new iteration in the assessment of health risks resulting from inhalation of antimony trioxide.

Human Studies of Antimony Carcinogenicity

In cancer risk assessment, the results of well-conducted human epidemiology studies are generally preferable to animal studies since interspecies extrapolation is not required. In the case of antimony trioxide, two studies of antimony exposed workers were available for review (Jones, et al., 1994; Schnorr et al., 1995) (see Table 19.7). However, neither of these studies was considered to provide conclusive evidence for or against a carcinogenic effect of antimony trioxide in humans.

Carcinogenicity Studies of Antimony Trioxide in Rodents

The results of three carcinogenicity studies of inhaled antimony trioxide in rats are summarized in Tables 19.8 and 19.9.

On initial review, the rodent studies of Watt (1983) and Groth et al. (1986) appear to indicate that antimony trioxide is a rat lung carcinogen. However, in-depth examination of the mechanism of antimony trioxide toxicity to the rat lung and the technical problems with these studies suggest that such a conclusion is uncertain. In addition, the results of the most recent and well-designed study find no evidence that antimony trioxide is a potential lung carcinogen in rats (Newton, et al., 1994).

TABLE 19.8 Summary of Rodent Inhalation Studies of Antimony Trioxide

Species	Exposure	Animals with Lung Tumors	Reference
Rat (female; Fischer)	0, 1.6, 4.2 mg/m^3 6 h/day, 5 days/week for 13 months; 1 year postexposure observation	0 mg/m^3—0/13 1.6 mg/m^3—1/17 4.2 mg/m^3—14/18	Watt (1983)
Rat (male and female; Wistar)	45 mg/m^3 7 h/day, 5 days/week for 52 weeks; 20 weeks postexposure observation	Male rats—no lung tumors; Female rats—19/70	Groth et al. (1986)
Rat (male and female; Fischer 344)	0, 0.06, 0.51, and 4.50 mg/m^3 6 h/day, 5 days/week for 52 weeks; 12-month postexposure observation	Male rats—no lung tumors; Female rats—no lung tumors	Newton et al. (1994)

The Watt study is limited by the use of only one sex for carcinogenicity testing. In addition, the precision of dose measurements in this study has been questioned, suggesting that antimony trioxide exposures may have actually been higher than reported (Newton et al., 1994).

Groth et al. (1986) treated male and female Wistar rats with 0 or 45 mg/m^3 (time-weighted average) antimony trioxide for 7 h/day, 5 days/week for 52 weeks followed by a 18–20 observation period before terminal sacrifice (71–73 weeks after initiation of the study). Groth et al. (1986) also reported significant fluctuations in the antimony exposure concentrations generated in the exposure chambers. During the latter 6 months of exposure, air concentrations occasionally exceeded the calculated time-weighted average concentration by 50–100 percent. Lung changes in treated rats included interstitial fibrosis, alveolar-wall hypertrophy and hyperplasia, and cuboidal and columnar cell metaplasia. These changes were more severe with increasing duration of exposure. The extent of interstitial fibrosis continued to progress even after exposure ceased. Overall, 27% of treated females (19/70) were observed with lung tumors. It is unusual that no tumors were observed in treated males.

Interpretation of the results of the Groth et al. study is limited by the use of only one very high dose level, so no dose-response information can be derived from the study. Chronic tissue injury appears likely as the mechanism for the eventual neoplasms, yet no insight can be gained from this study regarding possible no-effect levels. Also, there is considerable uncertainty in the actual exposure levels experienced by the test animals. Taken together, there are significant limitations in relying on this study to extrapolate any potential human carcinogenic potential of antimony.

Newton et al. reported the effects of subchronic and chronic inhalation toxicity of antimony trioxide in Fischer 344 rats. Male and female rats were exposed to air concentrations of 0, 0.06, 0.51 or 4.5

TABLE 19.9 Toxicity of Antimony Trioxide versus Carcinogenicity Potentials for Carbon Black and Talcum Powder

Test Material	Duration (months)	Exposure Rate (h/week)	Exposure Period (h)	Concentration (mg/m^3)	Cumulative Exposure [(mg/m^3) (h)]	Tumor Incidence (percent)
Antimony trioxide[a]	12	35	1820	38	69,160	27
Carbon black	20	85	7395	6.0	44,370	25
	24	80	8400	2.5	21,000	11
	24	80	8400	6.5	54,600	67
Talc[a]	28	30	3660	6	21,960	0
	28	30	3660	18	65,880	54

Source: Adapted from Hext (1994).

[a]Female rats only.

mg/m^3 for 6 h/day, 5 days/week for 12 months. In addition to clinical observations and microscopic pathology assessments, the authors measured antimony tissue levels in the lung at different time during the exposure period and during the observation period. Although inflammatory lung changes were observed at the 4.5 mg/m^3 exposure level, no increase in lung tumors was observed in either sex at any of the exposure levels. The authors concluded that the lung burden resulting from the highest exposure level decreased pulmonary clearance approximately 80%, with an increase in clearance half-time of 2–10 months.

The differences in carcinogenic outcome in the positive Watt (1983) and Groth et al. (1986) studies and the negative Newton et al. (1994) study may be the result of differences in the amount of antimony deposited in the lung. Newton et al. suggested that the different results may be due to higher exposure levels in the Watt study than were actually reported. The increased lung burden of particles in the Watt and Groth reports and the lung damage resulting from antimony trioxide may explain the positive lung tumor results in contrast to the negative results of Newton. Increasing lung burdens result in impaired clearance of particles from the lung, leading to prolonged and more severe chronic lung damage (Strom et al., 1989; Pritchard, 1989; Morrow, 1992).

Short-Term Genetic Toxicity Studies

Short term genetic toxicity (genotoxicity) studies are believed to provide important information regarding the potential carcinogenicity of a chemical. These studies evaluate the potential for chemicals to cause genetic damage such as gene mutations, damage to chromosomes, and changes in the number of chromosomes (aneuploidy). Chemically-induced genetic damage is believed to be an important event in chemical carcinogenesis.

The results of genotoxicity studies of antimony trioxide are mixed and provide no clear indication that inhaled antimony trioxide is genotoxic. Studies of antimony trioxide mutagenicity in bacteria are largely negative, (CalEPA, 1997) although antimony trioxide is reported to cause DNA damage in the bacterium *B. subtilis.* Antimony trioxide was not mutagenic in the mouse lymphoma cell assay but caused chromosomal aberrations in human lymphocytes and leukocytes (CalEPA, 1997). Both positive and negative results have been obtained from whole animal tests of the ability of antimony trioxide to cause chromosomal damage. These whole animal studies used orally administered antimony trioxide. The applicability of these oral studies to the genotoxic potential of inhaled antimony trioxide is unknown.

Putative Carcinogenic Mechanism of Antimony Trioxide in the Rat Lung

As discussed by Newton et al. (1994), the high lung burden of antimony trioxide resulting from exposures used in the Watt and Groth et al. studies may explain the positive carcinogenic effect. At the high concentrations used in the Watt and Groth et al. studies, clearance of antimony trioxide particles from the lung is reduced. The result of reduced lung clearance is increased retention of particles in the lung. Even particles of relatively innocuous materials such as titanium dioxide may cause lung tumors in the rat. These tumors appear to result as a secondary effect of impaired lung clearance, leading to inflammation and hyperplasia of the surrounding lung tissue. The putative mechanism of carcinogenity of these chemically inert particles appears to result from the inflammatory response of the rat lung to foreign particles rather than from a chemical-specific response. The impairment of lung clearance and subsequent response of the lung to retained foreign bodies is believed to explain the carcinogenicity of relatively nontoxic and insoluble particles including talc, carbon black, and titanium dioxide in the rat (Nikula et al., 1997).

The results of a recent study by Nikula (Nikula et al., 1997) support the doubts of the relevance of inhalation studies in rats to humans. As reviewed by Nikula et al., the lung of the cynomolgus monkey is anatomically much more like the human lung. Furthermore, particle clearance rates from the lung of the cynomolgus monkey are similar to humans and unlike the rat. Nikula et al. evaluated the effect of coal dust, diesel soot, and a mixture of coal dust and diesel soot on the lungs of Fisher 344 rats and the cynomolgus monkey at a concentration of 2 mg/m^3, 7 h/day, 5 days per week for 24 months. Importantly, rats, but not monkeys, developed significant alveolar epithelial hyperplastic, inflammatory, and septal fibrotic responses to the retained particles. These data indicate that if human lungs respond more like the monkey than the rat, the pulmonary response of the rat to particles may not be predictive of the response in humans at particle concentrations representing high occupational exposures.

While "particle overload" alone does not necessarily account for the lung toxicity of antimony trioxide in the Newton et al. study, it is possible that decreased clearance of particulates from the lung may be the cause of lung tumors seen in the Groth et al. and Watt studies. Hext (1994) compared the results of studies demonstrating particle-related pulmonary tumors by agents such as antimony trioxide, diesel exhaust, coal, carbon black, titanium dioxide, and others. To compare particle exposure between the studies, Hext calculated cumulative particle exposure in mg/m^3-hr. This comparison is presented for selected agents below.

Test Material	Duration (months)	Exposure rate (hrs/wk)	Exposure period (hrs)	Concentration (mg/m^3)	Cumulative exposure (mg/m^3-hr)	Tumor incidence (percent)
*Antimony trioxide	12	35	1820	38	69,160	27
Carbon black	20	85	7395	6.0	44,370	25
	24	80	8400	2.5	21,000	11
	24	80	8400	6.5	54,600	67
Talc	28	30	3660	6	21,960	0
	28	30	3660	18	65,880	54

*Female rats only

Adapted from Hext, 1994

At similar cumulative particle exposures, antimony trioxide caused fewer tumors than did carbon black or talc, two substances generally regarded as relatively nontoxic. Although the differences in cancer incidence between antimony trioxide-treated rats and carbon black and talc-treated rats may partly result from differences in experimental design, the size of particles tested, and other factors, it nonetheless suggests that tumors observed by Groth et al. may result from reduced lung clearance caused by "particle overload."

Of the available antimony trioxide inhalation studies, only the Newton et al. study used an experimental design that permits a dose–response assessment of the effects of inhaled antimony trioxide at concentrations above and below the concentrations that affect particle clearance from the lung. The technical deficiencies of the Watt and Groth et al. studies limit interpretation of the study results.

Weight of Evidence Characterization of the Potential Carcinogenicity of Inhaled Antimony Trioxide to Humans

According to all weight-of-evidence schemes, the greatest emphasis is placed on the results of well-conducted human epidemiology studies. In the case of antimony trioxide, human evidence is inadequate to establish a link between antimony trioxide exposure and cancer.

According to NRC and USEPA criteria, weight of evidence for the carcinogenicity of inhaled antimony trioxide in animals must also be regarded as equivocal. Although two studies in rats indicate that high concentrations of inhaled antimony trioxide cause lung tumors in female rats (Watt, 1983; Groth et al., 1986), male rats did not develop lung tumors in two studies that males were tested (Groth et al., 1986 and Newton et al., 1994). Neither female nor male rats developed lung tumors in the Newton et al. study. Watt observed lung tumors in rats at only one of two antimony trioxide concentrations tested. Groth et al. tested only one concentration of antimony trioxide. Thus, there is little dose–response data available from these studies. Further reducing the weight-of-evidence for a carcinogenic effect of inhaled antimony trioxide in humans is the fact that positive results have only been obtained from a single species (rat), single site (lung), and a single sex (females).

Other data may also be considered in weight-of-evidence determinations. Genotoxicity is an important component in determining weight of evidence for the potential carcinogenicity of a chemical. In the case of antimony trioxide, genotoxicity test results are mixed. This data is inconclusive regarding the potential for antimony trioxide to cause genetic damage in humans.

TABLE 19.10 Air and Soil Exposure Equations

Air

Inhalation of vapor-phase or particulate-phase chemicals in air:

$$\text{Daily intake in mg/kg} = \frac{CA \times VR \times EF \times ED}{BW \times AT}$$

where CA = modeled or actual concentration of chemical in air (mg/m^3)
VR = inhalation rate (m^3/day or event)
EF = exposure frequency (days/year)
ED = exposure duration (years)
BW = body weight (kg)
AT = averaging time [period over which exposure is averaged (AT_{nc} for noncarcinogens: ED × 365 days/year; AT_c for carcinogens: 70 years × 365 days/year)]

Soil

Ingestion of chemicals in soil:

$$\text{Daily intake in mg/kg} = \frac{CS \times IR \times ABS_{gi} \times EF \times ED \times CF}{BW \times AT}$$

where CS = chemical concentration in soil (mg/kg)
IR = ingestion rate (mg soil/day)
ABS_{gi} = fraction of chemical absorbed from soil relative to fraction absorbed from food or water
EF = ingestion exposure frequency (days/year)
ED = exposure duration (years)
CF = conversion factor (1×10^{-6} kg/mg)
BW = body weight (kg)
AT = averaging time [period over which exposure is averaged (AT_{nc} for noncarcinogens: ED × 365 days/year; AT_c for carcinogens: 70 years × 365 days/year)]

Dermal absorption of chemicals in soil:

$$\text{Absorbed dose in mg/kg/day} = \frac{CS \times SA \times AF \times ABS_{sk} \times EF \times ED \times CF}{BW \times AT}$$

where CS = chemical concentration in soil (mg/kg)
SA = skin surface area available for contact (cm^2)
AF = adherence of soil to skin (mg/cm^2)
ABS_{sk} = fraction of chemical absorbed though the skin
EF = exposure frequency (days/year)
ED = exposure duration (years)
CF = conversion factor (1×10^{-6} kg/mg)
BW = body weight (kg)
AT = averaging time [period over which exposure is averaged (AT_{nc} for noncarcinogens: ED × 365 days/year; AT_c for carcinogens: 70 years × 365 days/year)]

Other important weight of evidence factors include the potential carcinogenic mechanism or mechanisms of the chemical of concern. As considered by the USEPA, if the metabolism, toxicokinetics, and carcinogenic mechanism of action of a chemical are similar in rodents and humans, the weight of evidence for a carcinogenic effect of the chemical in humans is strengthened. Alternatively, if data show that the results of animal studies are not relevant to humans, the weight of evidence for a carcinogenic effect of the chemical in humans is weakened. As discussed above, recent data provide an indication that rodent inhalation studies may not predict the carcinogenic potential of low-level antimony trioxide exposure in humans.

The relevance of inhalation tests of high concentrations of particulate chemicals (such as antimony trioxide) in rats to human exposures has been questioned in recent years. Recent data (Nikula et al., 1997) indicates that the pattern of accumulation of particles in the rat lung is different from the same particles in the lung of monkeys. Furthermore, the rat lung shows greater inflammatory response to the particles than does the lung of the monkey. Because the lung of the monkey is structurally and functionally much more like the human lung than the rat lung, the recent information suggests that the relevance of high concentration inhalation studies in rats to humans should be reexamined.

Considered in total, the available evidence does not support a conclusion that inhaled antimony trioxide is carcinogenic to humans. This conclusion is different from the weight-of-evidence conclusions reached by IARC in 1989 and the State of California in 1990. However, these agencies did not have the benefit of important and more recent studies that cast doubt on the carcinogenicity of antimony trioxide and the relevance of rat inhalation studies of particulates in predicting carcinogenicity in humans.

Comments

The reassessment of carcinogenicity data demonstrates the need for iteration in risk assessment and its impact on antimony trioxide. The update of the toxicity assessment of antimony trioxide presented above suggests that low levels of inhaled antimony trioxide are not carcinogenic to humans.

While current evidence indicates that low-level antimony trioxide exposure may not be carcinogenic to humans, conservative public health policy may nonetheless require a risk assessor to assume that antimony trioxide is a potential human carcinogen. Thus, the use of a threshold or a nonlinear method to assess the possible carcinogenic effects of antimony trioxide may be a more reasonable alternative to the "no threshold" linearized multistage model used in Proposition 65. While the term "nonlinear" does not necessarily imply a threshold for the carcinogenic effect, it indicates that the carcinogenic response declines much more quickly than linearly with dose. A nonlinear model is also appropriate when the carcinogenic mode of action may theoretically have a threshold, for example, the carcinogenicity may be a secondary effect of toxicity or of an induced physiological change. Thus, if antimony trioxide must be considered a potential human carcinogen on the basis of conservative public health policy, the risk of cancer should be quantified using a nonlinear cancer response model.

19.7 SUMMARY

This chapter illustrates several of the practical problems that often face a risk assessor. Each example blends the use of default risk assessment procedures with higher tiers of risk evaluation. The example of lead identifies and addresses the effect of lead on a sensitive individual—the developing fetus. The antimony example highlights how inconsistency among regulatory agencies may affect the risk assessment process. The antimony example also addresses the uncertainty associated with extrapolation of high exposure animal studies to low exposures in humans and the relevance of these studies in predicting the human carcinogenic response to antimony. Risk assessment of chemical mixtures (see equations for air and soil exposure in Table 19.10) is evaluated in the petroleum hydrocarbon example. The example of arsenic evaluates the importance of considering the physical/chemical form of the chemical and how this may affect the bioavailability, human exposure, and risk.

REFERENCES AND SUGGESTED READING

ATSDR (Agency for Toxic Substances and Disease Registry), *Toxicological Profile for Lead. Draft for Public Comment* (update), U.S. Department of Health and Human Services, Aug. 1997.

ASTM (American Society for Testing and Materials), *Standard Guide for Risk-Based Corrective Action Applied at Petroleum Release Sites*, E1739-95, 1995.

Bhumbla, D. K., and R. F. Keefer, "Arsenic mobilization and bioavailability in soils," in *Arsenic in the Environment, Part I: Cycling and Characterization*, J. O. Nriagu, ed., Wiley, New York, 1994, p. 51.

Binder, S., D. Forney, W. Kaye, D. Paschal, "Arsenic exposure in children living near a former copper smelter," *Bull. Environ. Contam. Toxicol.* **39:** 114–121 (1987).

Binder, S., "The case for the NEDEL (the no epidemiologically detectable exposure level," *Am. J. Publ. Health* **78:** 589–590 (1988).

CalEPA, *Public Health Goal for Antimony in Drinking Water*, prepared by the Pesticide and Environmental Toxicology Section, Office of Environmental Health Hazard Assessment, California Environmental Protection Agency; draft for public comment and scientific review, Nov. 1997.

Dourson, M. L., and J. F. Stara, "Regulatory history and experimental support of uncertainty (safety) factors," *Reg. Toxicol. Pharmacol.* **3:** 224–238 (1983).

Eastern Research Group, *Report on the Expert Panel on Arsenic Carcinogenicity: Review and Workshop*, National Center for Environmental Assessment. U.S. Environmental Protection Agency, Aug. 1997.

Freeman, G. B., R. A. Schoof, M. V. Ruby, A. O. Davis, J. A. Dill, S. C. Liao, C. A. Lapin, and P. D. Bergstrom, "Bioavailability of arsenic in soil and house dust impacted by smelter activities following oral administration in cynomolgus monkeys," *Fund. Appl. Toxicol.* **28:** 215–222 (1995).

Groth, D. H., L. E. Stettler, J. R. Burg, W. M. Busey, G. C. Grant, and L. Wong, "Carcinogenic effects of antimony trioxide and antimony ore concentrate in rats," *J. Toxicol. Environ. Health* **18**(4): 607–626 (1986).

Hewitt, D. J., G. C. Millner, A. C. Nye, M. Webb, R. G. Huss, "Evaluation of residential exposure to arsenic in soil near a Superfund site," *Hum. Ecol. Risk Assess.* **1:** 323–335 (1995).

Hext, P. M., "Current perspectives on particulate induced pulmonary tumors," *Hum. Exp. Toxicol.* **13:** 700–715 (1994).

IRIS *(Integrated Risk Information Service)*, USEPA database accessed on Dec. 12, 1998.

Kalman, D. A., J. Hughes, G. van Belle, T. Burbacher, D. Bolgiano, K. Coble, N. K. Mottet, and L. Polissar. "The effect of variable environmental arsenic contamination on urinary concentrations of arsenic species," *Environ. Health Persp.* **89:** 145–151 (1990).

MADEP (Massachusetts Department of Environmental Protection), *Characterizing Risks Posed by Petroleum Contaminated Sites: Implementation of MADEP VPH/EPH Approach*, Public Comment Draft, Sept. 23, 1997.

Morrow, P. E., "Dust overloading of the lungs: Update and reappraisal," *Toxicol. Appl. Pharmacol.* **113:** 1–12 (1992).

Newton, P. E., H. F. Bolte, I. W. Daly, B. D. Pillsbury, J. B. Terrill, R. T. Drew, R. Ben-Dyke, A. W. Sheldon, and L. F. Rubin, "Subchronic and chronic inhalation toxicity of antimony trioxide in the rat," *Fund. Appl. Toxicol.* **22:** 561–576 (1994).

Nikula, K. J., K. J. Avila, W. C. Griffith, and J. L. Mauderly, Lung tissue responses and sites of particle retention differ between rats and cynomolgus monkeys exposed chronically to diesel exhaust and coal dust, *Fund. Appl. Toxicol.* **37:** 37–53 (1997).

NRC (National Research Council), *Science and Judgment in Risk Assessment*, National Academy Press, Washington, 1994.

Presidential/Congressional Commission on Risk Assessment and Risk Management, *Risk Assessment and Risk Management in Regulatory Decision-Making*, Final Report, Vol. 2, 1997.

Pritchard, J. N., "Dust overloading causes impairment of pulmonary clearance: Evidence from rats and humans," *Exp. Pathol.* **37**(1–4): 39–42 (1989).

Rudel, R., T. M. Slayton, and B. D. Beck, "Implications of arsenic genotoxicity for dose response of carcinogenic effects," *Reg. Toxicol. Pharmacol.* **23:** 87–105 (1996).

Schultz, M., "Comparative pathology of dust-induced pulmonary lesions: Significance of animal studies to humans," *Inhalation Toxicol.* **8:** 433–456 (1996).

Shacklette, H. T., and J. G. Boerngen, "Element concentrations in soil and other surficial materials of the conterminous United States," U.S. Geological Survey Professional Paper 1270, United States Government Printing Office, Washington, DC 1984.

Snipes, M. B., "Current information on lung overload in nonrodent mammals: Contrast with rats," *Inhalation Toxicol.* **8**(Suppl.): 91–109 (1996).

Strom, K. A., J. T. Johnson, and T. L. Chan, "Retention and clearance of inhaled submicron carbon black particles," *J. Toxicol. Environ. Health* **26**(2): 183–202 (1989).

TPHCWG (Total Petroleum Hydrocarbon Criteria Working Group), *Development of Fraction Specific Reference Doses (RfDs) and Reference Concentrations (RfCs) for Total Petroleum Hydrocarbons*, Vol. 4, Amherst Scientific Publishers, Amherst, MA, 1997.

USEPA, "Guidelines for the Health Risk Assessment of Chemical Mixtures," *Federal Register* **51**, (185), 34014–34025 (Sept. 24, 1986).

USEPA, *Special Report on Ingested Inorganic Arsenic. Skin Cancer and Nutritional Essentiality*, EPA/625/3-87/013, July 1988.

USEPA, *Risk Assessment Guidance for Superfund.* Vol. I, *Human Health Evaluation Manual* (Part A), USEPA/540/1-89/002, 1989.

USEPA, *Risk Assessment Guidance for Superfund.* Vol. 1, *Human Health Evaluation Manual. Supplemental Guidance "Standard Default Exposure Factors,"* PB91-921314, March 25, 1991.

USEPA, *Methods of Derivation of Inhalation Reference Concentrations and Application of Inhalation Dosimetry,* EPA/600/8-90/066F, Oct. 1994.

USEPA, *Supplemental Guidance to RAGS,* Region 4 Bulletins, *Exposure Assessment. Human Health Risk Assessment*, Bulletin 3. Nov. 1995.

USEPA, *Region 9 Preliminary Remediation Goals (PRGs) 1997*, from Stanford J. Smucker, Ph.D. to PRG Table Mailing List.

Walker, S., and S. Griffin, "Site-specific data confirm arsenic exposure predicted by the U.S. Environmental Protection Agency," *Environ. Health Persp.* **106:** 133–139 (1998).

Watt, W. D., *Chronic Inhalation Toxicity of Antimony Trioxide: Validation of the Threshold Limit Value*, Wayne State Univ., Detroit, 1983.

20 Occupational and Environmental Health

FREDRIC GERR, EDWARD GALAID, and HOWARD FRUMKIN

The objectives of this chapter are to introduce the medical specialty called *occupational and environmental medicine*, its goals and methods. This chapter

- Defines, categorizes, and quantifies occupational and environmental diseases
- Describes the professions that work in occupational health care
- Describes the activities of occupational health care, including diagnosis and treatment, screening and surveillance, evaluation for attribution, and training and education
- Describes the settings in which occupational and environmental medicine is practiced
- Introduces ethical issues that arise in delivering occupational and environmental health care

20.1 DEFINITION AND SCOPE OF THE PROBLEM

Hazards can be found in the workplace and the non-work environment that increase the risk of both illness and injury. Illness tends to develop over time following repeated exposure to a hazard whereas injury usually occurs instantly. Because this textbook focuses on toxicology, the main focus of this chapter will be on occupational illness resulting from chemical exposure. Some chemical exposures, however, such as organic solvents can increase the risk of injury by impairing coordination and judgment.

Occupational illness and environmental illness are adverse health conditions, the occurrence or severity of which is related to exposure to factors on the job or in the nonwork environment. Such factors can be chemical (solvents, pesticides, heavy metals), physical (heat, noise, radiation), biological (tuberculosis, hepatitis B virus, HIV) or psychosocial/organizational stressors (machine pacing, piecework, lack of control over work, inadequate personal support). Examples of occupational illness include

1. Scarring of the lungs following inhalation of airborne asbestos dust fibers among insulation workers
2. Loss of memory following long-term exposure to organic solvents among spray painters
3. Headache, low blood counts (anemia), and abdominal pain following exposure to lead among battery workers
4. Hearing loss among noise-exposed textile plant workers
5. Hepatitis B infection following needlestick accidents among health care workers in a hospital
6. Neck and shoulder pain among journalists with intense deadline pressures

The leading categories of work-related diseases are presented in Table 1.

Principles of Toxicology: Environmental and Industrial Applications, Second Edition, Edited by Phillip L. Williams, Robert C. James, and Stephen M. Roberts.
ISBN 0-471-29321-0 © 2000 John Wiley & Sons, Inc.

Illnesses associated with hazardous exposures both in the workplace and in the general environment have been recognized for thousands of years. For example, the toxic effects of lead, including abdominal pain, pallor (anemia), and paralysis, appear to have been described by several observers among the ancient Greeks and Romans. In the first known textbook of occupational medicine, *De Morbis Artificum Diatriba*, the Italian physician Bernardino Ramizzini (1633–1717), often called the father of occupational medicine, described diseases of the occupations and instructed physicians of the time: "and to the questions recommended by Hippocrates, the physician should add one more—what is your occupation?" In the United States, Dr. Alice Hamilton (1869–1970) had a major role in establishing occupational medicine as a legitimate clinical discipline. Dr. Hamilton, the first woman appointed to the faculty of the Harvard Medical School, wrote in her autobiography: "American medical authorities had never taken industrial diseases seriously . . . employers could, if they wished, shut their eyes to the dangers their workmen faced, for nobody held them responsible, while the workers accepted the risks with fatalistic submissiveness." Among her many legacies, Dr. Hamilton fought, without success, the introduction of tetraethyl lead into gasoline, correctly predicting that it would result in widespread lead contamination of the environment and adverse health effects in the exposed population.

How big a problem is occupational diseases? Two kinds of numbers are informative: counts and rates. Suppose there are two industries, one employing 1,000 workers nationally, the other employing 50,000 workers nationally. Suppose that the incidence of work-related asthma is 12 per 100 workers per year in the first industry, and only 4 per 100 workers per year in the second industry. By this measure, the first industry is more hazardous. But 120 workers in the first industry develop asthma each year, compared to 2,000 workers in the second industry. From a public health point of view, the larger burden of illness in the second industry might merit more attention. Counts and rates both provide useful information, but they can yield different conclusions.

There are two principal sources of data that help answer this question: employer reports, and insurance records. Employers are required by OSHA to record all work-related injuries and illnesses, and each year, a sample of employers provide information to the Bureau of Labor Statistics. This serves as the national data source on occupational illnesses. As for insurance, the Workers Compensation system acts as the health insurer for workers with occupational illnesses, and the records of claims made or claims paid also serves as a potential data source. In both cases, there is considerable under-reporting. Employers and workers may not recognize that an illness is work-related, or employers may deny a worker's claim of work-relatedness. Employers may in some cases fail to report recognized cases. Sometimes, occupational illnesses arise long after the exposure, perhaps after employment has ended, making data recording difficult.

Other sources of information on occupational illnesses exist. Examples include clinical laboratories, which can yield data on cases of elevated blood lead, and physician reporting of specific diseases. While such sources are important in specific settings, none has gained widespread use.

TABLE 20.1 Leading Categories of Work-Related Diseases

Occupational lung diseases: asbestosis, byssinosis, silicosis, coal worker's pneumoconiosi s, lung cancer, occupational asthma
Musculoskeletal injuries: disorders of the back, trunk, upper extremity, neck, lower extr emity, trauma-induced Raynaud's phenomenon
Occupational cancers (other than lung cancer): leukemia, mesothelioma, cancers of the bladd er, nose, and liver
Occupational cardiovascular diseases: hypertension, coronary artery disease, acute myo cardial infraction
Disorders of reproduction: infertility, spontaneous abortion, teratogenesis
Neurotoxic disorders: peripheral neuropathy, toxic encephalitis, psychoses, extreme perso nality change (exposure-related)
Noise-induced hearing loss
Dermatologic conditions: dermatoses, burns (scaldings), chemical burns, contusions (abras ions)
Psychological disorders: neuroses, personality disorders, alcoholism, drug dependen cy

TABLE 20.2. Number of Reported Occupational Illnesses by Category of Illness, Private Industry, 1998 (in thousands)

Industry	Total Cases	Skin Diseases	Dust Diseases of the Lungs	Toxic Respiratory Conditions	Poisoning	Disorders Due to Physical Agents	Disorders Associated with Repeated Trauma	All Other
All private industry	391.9	53.1	2.1	17.5	4.0	16.6	253.3	45.4
Agriculture, forestry and fishing	4.3	2.4	<0.1	0.5	0.1	0.1	0.6	0.5
Construction	7.7	1.8	0.2	0.8	0.3	1.2	2.0	1.5
Manufacturing	236.3	24.4	0.8	6.6	2.2	9.0	180.9	12.5
Transportation and public utilities	16.6	1.7	0.3	1.2	0.3	1.2	9.2	2.7
Wholesale and retail trade	38.8	4.3	0.2	2.6	0.3	2.2	20.9	8.4
Finance, insurance, and real estate	15.2	0.8	<0.1	0.6	<0.1	0.1	12.0	1.6
Services	71.7	17.7	0.4	5.1	0.8	2.7	27.0	18.0

Ideally, data on occupational illnesses are linked directly to prevention efforts. For example, if data show that cases of asbestosis are occurring in a particular location, public health authorities could investigate the source of exposure and take steps to control it. However, with rare exceptions, occupational disease data in the United States are not directly linked to prevention efforts. In many European countries, and is certain states, this linkage has been successfully implemented, and control efforts are guided by health data.

To provide some indication of the magnitude of the problem, the number of occupational illness cases by industry type and illness category reported in the United States during 1998 is presented in Table 2. Just under three hundred ninety-two thousand cases of occupational illness were reported during 1998, with the largest number of cases come from the manufacturing sector. The single largest category of occupational illness was "disorders associated with repeated trauma", which includes tendinitis, carpal tunnel syndrome, and noise-induced hearing loss. The next most prevalent illness was skin diseases, the most common being rashes from chemical irritation or skin allergy.

Patterns of occupational illness change over time. For example, in 1982, skin diseases or disorders accounted for approximately 40 percent of all reported occupational illness in the United States. In 1998 it accounted for only 14 percent of all reported occupational illness. In contrast, in 1982, disorders associated with repeated trauma accounted for 21 percent of all reported occupational illness in the

TABLE 20.3. Percent Distribution of Reported Occupational Illnesses by Category of Illness, Private Industry, 1982–1998

Category	1982	1986	1990	1994	1998
Total illness cases	100	100	100	100	100
Skin diseases or disorders	40	30	18	13	14
Dust diseases of the lungs	2	2	1	1	1
Respiratory conditions due to toxic agents	8	9	6	5	4
Poisoning	3	3	2	1	1
Disorders due to physical agents	8	7	6	4	4
Disorders associated with repeated trauma	21	33	56	65	65
All other occupational illness	18	17	15	12	12

United States whereas in 1998 they accounted for 65 percent of all reported occupational illness. The percent distribution of reported occupational illnesses by category of illness for private industry in the United States is presented for years 1982–1998 in Table 3.

20.2 CHARACTERISTICS OF OCCUPATIONAL ILLNESS

Health care providers often overlook the occupational cause of human illness. This is due to several special characteristics of occupational disease that may obscure its occupational origin.

1. The clinical and pathological presentation of occupational disease is often identical to that of nonoccupational disease. For example, asthma (excessive airways narrowing in the lungs) due to airborne exposure to toluene diisocyanate is clinically indistinguishable from asthma due to other causes.
2. Occupational disease may occur after the termination of exposure. An extreme example would be asbestos-related mesothelioma (a cancer affecting the lining of the lung and abdomen) that can occur 30–40 years after the exposure. Even relatively acute illness can occur after the exposure episode. Some forms of occupational asthma manifest at night, several hours after the end of the exposure.
3. The clinical manifestations of occupational disease can vary with the dose and timing of exposure. For example, at very high airborne concentrations, elemental mercury is acutely toxic to the lungs and can cause pulmonary failure. At lower levels of exposure, elemental mercury has no pathologic effect on the lungs but can have chronic adverse effects on the central and peripheral nervous systems.
4. Occupational factors can act in combination with nonoccupational factors to produce disease. A classic example is the interaction between exposure to asbestos and exposure to tobacco smoke. Long-term exposure to asbestos alone increases the risk of lung cancer about fivefold. Long-term smoking of cigarettes increases the risk of lung cancer about 10–20-fold. When exposed to both, however, the risk of lung cancer is increased about 50–70-fold.

20.3 GOALS OF OCCUPATIONAL AND ENVIRONMENTAL MEDICINE

Occupational and environmental medicine is both a preventive and a clinical specialty. Prevention activities are often divided into three categories, primary, secondary, and tertiary. Primary prevention is accomplished by reducing the risk of disease. In the occupational setting, this is most commonly done by reducing or eliminating exposure to hazardous substances. As exposure is reduced, so is the risk of adverse health consequences. Such reductions are typically managed by industrial hygiene personnel and are best accomplished by changes in production process or associated infrastructure. Such changes might include substitution of a safer substance for a more hazardous one, enclosure or special ventilation of equipment, as well as rotation of workers through areas in which hazards are present to reduce the dose to each worker. (Note that this method does increase the number of workers exposed to the hazard.)

Secondary prevention is accomplished by identifying health problems before they become clinically apparent (i.e., before workers report feeling ill) and making interventions to limit the resulting disease. This is a major goal of occupational health surveillance, which is discussed in greater detail below. The underlying assumption is that such early identification will result in a more favorable outcome. An example of secondary prevention in occupational health is the measurement of blood lead levels in workers exposed to lead. An elevated blood lead level indicates a failure of primary prevention but can allow for corrective action before clinically apparent lead poisoning occurs. Corrective action would be to improve the primary prevention activities listed above.

Tertiary prevention is accomplished by minimizing the adverse clinical effects on health of an illness or exposure. Treatment of lead poisoning (headache, muscle and joint pain, abdominal pain, anemia, kidney dysfunction) by administration of chelating medication is an example of tertiary prevention.

The goal is to limit symptoms or discomfort, minimize injury to the body, and maximize functional capacity.

20.4 HUMAN RESOURCES IMPORTANT TO OCCUPATIONAL HEALTH PRACTICE

Occupational health is a multidisciplinary effort, and professionals of diverse backgrounds are part of the successful occupational health team. Industrial hygienists recognize and assess hazards through process analysis, visual inspection, direct measurement, and other methods. Because the goal is to prevent the occurrence of adverse health effects due to toxic exposure before they occur, these professionals collaborate with health care providers in identifying potential hazards. As described above, primary prevention is most often accomplished by designing new workplaces and work processes that are free from exposure to hazardous substances or reengineering existing workplaces and work processes to reduce occupational exposures to acceptable levels. Industrial engineers, ventilation engineers, and industrial hygienists accomplish these design tasks. Secondary prevention is typically accomplished by a multidisciplinary group that includes physicians, nurses, epidemiologists, industrial hygiene and other exposure control experts, and members of management and labor. Tertiary prevention is typically accomplished by traditional clinical specialists including nurses, doctors, and other specialized therapists such as occupational and physical therapists.

20.5 ACTIVITIES OF THE OCCUPATIONAL HEALTH PROVIDER

Diagnosis and Treatment of Occupational Illness

Diagnosis and treatment are the activities most commonly associated with the clinical practice of medicine in almost any setting. Diagnosis is the process of determining the specific health problem affecting a person and treatment is the application of therapies intended to restore function to that person. Many occupational and environmental medicine specialists diagnose and treat both acute and chronic occupational illnesses. An example of an acute occupational illness is respiratory difficulty immediately following airborne exposure to chlorine gas. Diagnosis is based on the presence of characteristic symptoms, such as shortness of breath, signs such as the sound of wheezing in the chest, and test results such as abnormalities on a chest X ray. Treatment of the respiratory difficulty might include hospitalization, administration of supplemental oxygen, use of medicine to promote air exchange, and, in severe cases, mechanical assistance for breathing. An example of a chronic occupational illness is lead poisoning after 20 years of occupational exposure to airborne lead vapor at a secondary smelter. Diagnosis is based on symptoms of depression and memory loss, signs such as elevated blood pressure, and test results such as low blood counts, kidney dysfunction, and poor performance on tests of mental ability. Treatment of lead toxicity might include administration of medication to promote excretion of lead, as well as enrollment of the worker in a memory rehabilitation program to provide skills that reduce the impact of the impairment on daily activities.

Routine Clinical Examinations

Diagnosis and treatment, as described above, are usually triggered when a patient or clinician suspects a health problem. In contrast, some clinical examinations in occupational health are conducted routinely. Often these are required by applicable government regulations, but occupational health professionals may recommend them in the absence of regulatory requirements. The objectives of routine clinical examinations are to (1) assess an individual's fitness to carry out certain job functions, such as wearing a respirator, (2) protect the health and safety of the public who may be affected by an individual's illness, and (3) protect the individual from illnesses associated with workplace exposures.

Routine clinical examinations occur in three settings:

1. A preplacement examination, as part of the hiring process, to determine the applicant's ability to perform the job.
2. A periodic examination, at regular intervals during employment, to assess fitness to perform the job, evidence of toxic exposure, and/or evidence of disease. Periodic examinations are usually part of surveillance programs, which are discussed in the next section.
3. A return-to-work evaluation after recovering from an injury or illness (either work- or non-work-related), to determine the employee's ability to perform the job.

In some cases, routine examinations are highly standardized. Examples include Department of Energy regulations covering nuclear power plant operators and Department of Transportation regulations covering truck drivers, commercial airplane flight crews, air traffic controllers, aircraft mechanics, and the merchant marines. Similarly, many employers now require testing for evidence of illegal drug use.

In other cases, routine clinical examinations are tailored to specific workplace situations, based on the job demands and risks associated with particular jobs. Clinicians use their knowledge of the workplace environment and the job demands to focus examinations on specific origin systems, such as the musculoskeletal system. Information about the demands and risks of a job may be supplied by the industrial hygienist or safety professional. To supplement information collected in the physical examination, the clinician may request the applicant to participate in a work capacity evaluation (WCE) that simulates the demands of the job. Using these data, the clinician determines whether the applicant can safely perform the essential functions of the job without or with workplace modifications, or whether the applicant should be disqualified because there are no reasonable accommodations that could enable the applicant to perform the essential functions of the job. Use of medical information in this manner is delineated in the federal law, The Americans with Disabilities Act of 1990.

Occupational Health Surveillance

Most clinical examinations focus on the evaluation of individual patients. In occupational health, routine clinical examinations can focus instead on the health of an entire population, such as a workforce. When the health of a workforce is systematically and continuously assessed, this is known as *occupational health surveillance*. A standard definition is

> The ongoing systematic collection, analysis, and interpretation of health data essential to the planning, implementation, and evaluation of public health practice, closely integrated with the timely dissemination of these data to those who need to know. The final link in the surveillance chain is the application of these data to prevention and control.

In general medical practice, surveillance programs aim to detect cases of disease early, so that they can be treated promptly to improve the patients' long-term outcome. Familiar examples include mammograms to detect breast cancer, Pap (Papanicolaou) smears to detect cervical cancer, and blood pressure screening to detect hypertension. In occupational health, surveillance also aims to detect cases of disease early. However, there are additional objectives: (1) to identify and characterize worker exposure to health hazards, (2) to assess the success of preventive interventions, (3) to monitor trends over time, and ultimately (4) to prevent disease associated with exposures.

The components of a medical surveillance program may include (1) collection of health history information from individual workers, (2) collection of exposure information from personnel records, (3) performance of physical examinations with emphasis on organ systems known to be affected by the exposure, (4) tests that check for evidence of exposure (such as a blood lead test), and (5) tests that check for disease of dysfunction (such as urine tests for proteins, lung function tests for decreased airflow, and chest X rays). Medical surveillance examination results may be analyzed in several ways. Of course, they may be used to assess an individual's health and may lead to further evaluations, treatment, or medical removal from the workplace. The results may be scrutinized for the occurrence of sentinel health events, "red flags" such as asbestosis or mercury poisoning that indicate the presence of a preventable exposure. Systematic epidemiologic analysis is extremely useful. For example, two groups of workers, one with potential exposure to a toxin and one without exposure, might be compared to determine whether the exposed group has any excess of disease. Similarly, the disease rates of one group over time might be followed, to verify that a preventive intervention has been successful. The necessary skills in data collection, management, and analysis are an increasingly important part of the occupational health toolbox.

An example of a medical surveillance program is the Cadmium Medical Surveillance Program, mandated by the United States Occupational Safety and Health Administration in the Cadmium Standard, 29 CFR 1910.1027. Studies of human populations have suggested that excessive cadmium exposure is associated with an increased risk of lung cancer, kidney damage, and prostate cancer. Therefore, the Cadmium Medical Surveillance Program focuses on evaluating the respiratory, renal, and genitourinary systems of exposed workers. For example, elements of the mandatory medical surveillance program for cadmium are presented in Table 20.4. One limitation of most medical surveillance programs, including the cadmium program, is that tests and methods traditionally used in clinical medicine to detect and diagnose disease among individuals with symptoms who come forward for medical care cannot always be relied on for detection and diagnosis of the health effects of occupational exposures among those who are free of symptoms but may be in an early stage of disease.

Evaluations for Attribution

The occupational and environmental medicine specialist is frequently asked to make a determination of attribution. The specific question is whether an exposure at work caused or contributed to an illness in an individual. The results of this evaluation may be used to help diagnose and treat the disease, to compensate the employee monetarily for lost wages due to the injury or illness, and to implement prevention programs. This often difficult and sometimes controversial task must be based upon the similarity of the exposure–disease relationship in the individual to those reported in the medical literature in systematic studies of large groups. Several main characteristics of occupational illness, as described above, can make the occupational origins of illness obscure to all except the most committed observers. Critical issues include the fact that occupational disease is often clinically indistinguishable from nonoccupational disease, that occupational disease can occur a long time after the end of exposure, and that occupational exposures often have synergy with nonoccupational exposure.

An example of a case involving attribution is a 25-year-old male who experienced shortness of breath and wheezing of three months duration. Although he had a history of seasonal allergies that caused nasal congestion, he had no problems with wheezing prior to the past 3 months. The occupational history revealed that 6 months prior to the onset of his respiratory symptoms, he began to work on the production line of a company that repackages bulk quantities of isocyanate-based paint into smaller containers. He stated that hoses leading from the bulk tanks to the filling machine would periodically leak. He did not use personal protective equipment. Examination of the worker was positive for wheezing and objective lung function testing revealed a pattern diagnostic of asthma. Because exposure to isocyanates has been associated with asthma in large studies, the physician determined that there was a causal link between the workplace exposure and the new onset of disease.

TABLE 20.4 Specific Elements of the Mandatory Medical Surveillance Program for Workers Exposed to Cadmium

Questionnaire, completed by the employee, pertaining to health effects associated with cad mium exposure
Directed physical examination, with emphasis on the respiratory, genitourinary, and ren al systems
Chest X ray and pulmonary function tests
Physiologic monitoring of kidney function (blood urea nitrogen, creatinine, B2-microglobul in, and urinalysis)
Biologic monitoring (blood and urine cadmium levels)

The patient was restricted from any further exposure to the paint packaging department or other areas where exposure to isocyanates might occur and his symptoms improved.

Training and Education

Another critical function of occupational health professionals is training. They are responsible for communicating with management, government, and workers about the hazards of workplace exposure, and about proper remedial actions. According to OSHA's Hazard Communication Standard, workers have a "right to know" about chemicals to which they are exposed, through information sheets (Material Safety Data Sheets), labels on chemical containers, and training programs. Important information includes the identity of chemicals, their acute and chronic health effects, how to respond to emergency situations, and how to prevent toxicity. Not only workers, but also supervisors and managers must be thoroughly familiar with chemical hazards. Available changes rapidly as more research results are reported, so keeping abreast of new developments is essential. Increasingly, occupational health professionals must not only recognize and control hazards, but also communicate this information to those they serve.

Setting of Occupational Medicine Service Delivery

Occupational medicine services are delivered in a variety of settings. Over time, with changes in business practices and the health care system, these settings have evolved.

In the past, the prototype setting for occupational medicine service delivery was the workplace itself, usually in a medium- to large-sized manufacturing facility. Plant physicians and nurses, based in dispensaries close to the work process, would look after workers with injuries, conduct preplacement and return-to-work physical examinations, and in some cases evaluate injury and illness trends in the workforce and initiate prevention programs. Some industries still maintain on-site physicians and nurses, especially in very large and/or remote plants. The physicians may be community practitioners who spend only part of their time at the plant. But increasingly, this work is being "outsourced" to private practices outside the plant.

The private practice of occupational medicine is growing rapidly. Occupational medicine practices may be based at community hospitals, multispecialty group practices, managed care organizations such as health maintenance organizations (HMOs), or freestanding specialty practices. Typically an occupational medicine practice will serve dozens or even hundreds of client companies, treating acute injuries, conducting routine examinations, and providing other services, including unnecessary ones, to client companies. Critics argue that company physicians and nurses become thoroughly familiar with their companies' facilities, enabling them to provide in-depth expertise that multiclient practices cannot match. On the other hand, multiclient occupational medicine practices offer important advantages. Providers in multiclient practices can amass broad, diverse experience in program development, data management and analysis, regulatory compliance, and other occupational health activities, which can, in turn, enable them to deliver a high level of service. Providers in multiclient practices can remain independent of individual employers, which may help avoid some ethical dilemmas (see discussion below). Small and medium-sized firms, which are unable to afford in-house occupational medicine services, can better afford the services of multiclient practices. Even larger firms often find it more economical to outsource their occupational medicine. Finally, occupational health providers in managed care organizations can potentially integrate their services with primary medical care, leading to more continuous, less fragmented care.

A third setting for occupational medicine service delivery is the academic setting. Many major medical centers, with links to medical schools and/or schools of public health, now have occupational medicine units. These may be located in departments of medicine, family practice, or preventive medicine. Academic occupational medicine units provide many of the clinical services noted above. However, they differ in important ways from community-based practices. Typically their staffs are highly trained, with board certification in several medical specialties including occupational medicine.

Academic practices welcome complicated referral cases that pose medical or medicolegal diagnostic challenges and require much time to evaluate and treat; such cases are often used in training physicians and/or nurses who hope to specialize in occupational health. Most academic units have active programs of research and service and blend clinical care with study, collaboration with local employers, unions, and government agencies, and similar activities.

Other occupational medicine providers work in the insurance industry, in consulting firms, and in government agencies. All of these settings provide opportunities for treating and diagnosing patients with work-related ailments, and perhaps more importantly, for recognizing, assessing, and controlling workplace hazards.

20.6 ETHICAL CONSIDERATIONS

Occupational medicine sits astride several kinds of competing interests, most notably labor-management disputes. Sometimes practitioners find themselves caught "between medicine and management." The ethical issues that arise are interesting and challenging.

Confidentiality is one issue. An accepted principle of medical ethics is that medical information about a patient is private and should be released only with the patient's consent. However, employers sometimes have access to medical information about their employees obtained through occupational medical evaluations. In some situations, this information is not protected; it is accessible to personnel managers, supervisors, and others. Clinicians who collect the information may feel that they owe it to the employer, since the employer paid for the examination and is in some sense the "customer." Occupational health professionals must strive to maintain medical information confidential. A standard approach is to maintain medical information in locked files, accessible only to medical personnel and to provide employers only with statements of fitness to work and necessary accommodations.

A second issue has to do with notification of hazards. Physicians and other health care workers are usually considered to have some ethical responsibility to public health. This implies an obligation to inform health authorities, and people at risk, of a hazard that is uncovered. However, history records an unfortunate number of instances in which occupational health professionals were prevented from disclosing hazards, usually by companies that would be financially threatened by such disclosure. For example, suppose that a physician contracts with a paint manufacturer to conduct medical examinations of the workers. The physician finds an elevated prevalence of asthma and dermatitis and localizes these problems to one area of the plant where chemical exposure levels are high and then reports this finding to management and plans to notify the workers of their diagnoses. However, management is concerned that this might trigger workers' compensation claims and informs the physician that her contract will be terminated if she informs the patients of their findings.

A related dilemma arises when disclosure would violate the confidentiality of an individual. For example, suppose that a worker is diagnosed with severe occupational asthma, and the physician determines that the cause is excessive exposure to epoxy resins. Other workers are potentially exposed and are at risk of developing asthma. The physician plans to notify the employer, to recommend hazard abatement, and to inform OSHA of the problem. The patient pleads with the physician not to do so, claiming that he or she would be identified as the complainant and be fired. In these cases, the physician's duty to inform is challenged by competing considerations. A standard approach is to define, in advance, the occupational health professional's ethical obligations, including the duty to inform and to build this into any contract.

A third ethical issue involved employment discrimination. A famous case arose in the 1980s when a manufacturing facility that used lead prohibited women from working in certain jobs (incidentally, those with the best pay) unless they had been sterilized. The employer reasoned that if women became pregnant, their fetuses would be especially susceptible to the toxic effects of lead and that a ban would prevent this undesirable outcome. However, employees argued that the ban amounted to blatant gender discrimination and took their claim all the way to the U.S. Supreme Court case, where they prevailed.

Occupational health policies can have a major effect on people's employment and require careful consideration of fairness and equity.

Informed consent is generally accepted as a fundamental element of medical ethics. In general medical care, patients cannot be coerced into accepting tests and treatments. The same is true in the workplace setting, but this principle sometimes clashes with job requirements. For example, employers can compel employees to submit to drug testing, within certain guidelines. Occupational health illustrates the difficulty of balancing individual autonomy with the requirements of employers and government.

Several professional groups have issued codes of ethics for occupational health practice. The most widely accepted is the International Code of Ethics of the International Commission of Occupational Health (ICOH), issued of 1992. Selections from this Code are presented here:

- Duties and obligations of occupational health professionals
  - a. *Knowledge and expertise.* Occupational health professionals must continuously strive to be familiar with the work and the working environment as well as to improve their competence and to remain well informed in scientific and technical knowledge, occupational hazards and the most efficient means to eliminate or to reduce the relevant risks. Occupational health professionals must regularly and routinely, whenever possible, visit the workplaces and consult the workers, the technicians, and the management on the work that is performed.
  - *Commercial secrets.* Occupational health professionals must not reveal industrial or commercial secrets of which they may become aware in exercising their activities. However, they cannot conceal information that is necessary to protect the safety and health of workers or of the community. When necessary, the occupational health professionals must consult the competent authority in charge of supervising the implementation of the relevant legislation.
  - *Information to the worker.* The results of the examinations, carried out within the framework of health surveillance, must be explained to the worker concerned. The determination of fitness for a given job should be based on the assessment of the health of the worker and on a good knowledge of the job demands and the worksite. The workers must be informed of the opportunity to challenge the conclusions concerning their fitness for work that they feel are contrary to their interests. A procedure of appeal must be established in this respect.
- Condition of execution of the function of occupational health professionals
  - *Professional independence.* Occupational health professionals must maintain full professional independence and observe the rules of confidentiality in the execution of their functions. Occupational health professionals must under no circumstances allow their judgment and statements to be influenced by any conflict of interest, in particular when advising the employer, the workers, or their representatives in undertaking on occupational hazards and situations that present evidence of danger to health or safety.

20.7 SUMMARY AND CONCLUSION

Occupational and environmental illnesses include a wide range of health conditions.

- These are common, with an estimated 300,000 new cases annually in the United States.
- These include pulmonary, dermatologic, muscoskeletal, neurologic, and reproductive conditions, as well as cancers and others.
- Cases of occupational and environmental illness are usually clinically indistinguishable from cases of the same illness that are not exposure-related.
- Occupational and environmental illnesses are highly preventable.

Occupational and environmental medicine is a medical specialty that

- Diagnoses and treats occupational and environmental illnesses
- Provides medical assessments of fitness, risk, and attribution to cause
- Collaborates with other professionals, such as industrial hygienists, to achieve primary prevention
- Offers screening and surveillance programs to achieve secondary prevention
- Provides training and education regarding workplace hazards
- Confronts challenging ethical dilemmas and functions in accordance with widely accepted codes of ethics

REFERENCES AND SUGGESTED READING

Fischbein, A., "Occupational and environmental lead exposure," in *Environmental and Occupational Medicine*, 3rd Ed., W. Rom, ed., Lippincott-Raven, Philadelphia, 1998. Ch. 68.

Hamilton, A., *Exploring the Dangerous Trades*, Little Brown, Boston, 1943.

Rom, W., "The discipline of environmental and occupational medicine," in *Environmental and Occupational Medicine*, 3rd ed., W. Rom, ed., Lippincott-Raven, Philadelphia, 1998.

Thacker, S. B., and R. L. Berkelman, "History of public health surveillance," in *Public Health Surveillance*, W. Halperin and E. Baker, Jr., eds., Van-Nostrand-Reinhold, New York, 1992.

US DOL, BLS, Bulletin USDL 99-358, *Workplace Injuries and Illnesses in 1998*. December, 1999.

Walsh, D. C., *Corporate Physicians: Between Medicine and Management*, Yale Univ. Press, New Haven, CT, 1987.

21 Epidemiologic Issues in Occupational and Environmental Health

LORA E. FLEMING and JUDY A. BEAN

Epidemiology is the study of the distribution and determinants of disease or death in human populations. In the case of the environment or the workplace, epidemiology attempts to determine associations between a chemical exposure and particular human health effects.

This chapter will discuss:

- What epidemiologists study, and describe the scientific discipline of epidemiology
- Epidemiologic causation, its implications for other scientific disciplines, and the interpretation of the results of epidemiologic studies
- Types of epidemiologic studies, and advantages and disadvantages
- Definitions of exposure, disease, population, and their measures in epidemiologic studies
- Types of measures of risk in epidemiologic studies
- Bias and other issues, and how to approach these issues in epidemiology
- Occupational and environmental epidemiologic issues

21.1 A BRIEF HISTORY OF EPIDEMIOLOGY

Over 2000 years ago, the famous Greek physician Hippocrates noted that environmental factors can influence the occurrence of disease. However, until the nineteenth century no one measured the distribution and determinants of disease or death in human populations in a formal way. In particular, John Snow in 1855 noted a possible association between drinking water and deaths from cholera in London. Using epidemiologic principles (not defined as such at the time), Snow showed that cholera was spread by contaminated water, long before the bacterial organism for cholera had even been discovered. His work lead to public health interventions to prevent the spread of cholera.

The data, which Snow (1855) used to perform this investigation, form the basic building blocks of an epidemiologic study. He collected information based on a case definition of the disease (i.e., death due to cholera), a definition of exposure (i.e., drinking water source), and a definition of the denominator population (i.e., the total number of at risk people living in the particular district). Snow used this information to construct a standard rate or risk for comparison: the number of cholera deaths associated with a particular type of drinking water divided by the number of at risk people living in that particular district. Thus, he was able to compare the rates (or risk) of cholera deaths by the different water supplies (Table 21.1).

Snow's investigation also illustrates some common sources of epidemiologic data. These include vital records (deaths, births, etc.), Census data and questionnaires (source of drinking water). Other

Principles of Toxicology: Environmental and Industrial Applications, Second Edition, Edited by Phillip L. Williams, Robert C. James, and Stephen M. Roberts.
ISBN 0-471-29321-0 © 2000 John Wiley & Sons, Inc.

TABLE 21.1 Cholera Deaths in London (1984) by Water Supply

Water Company	Population (1851)	Cholera Deaths	Rate per 1000 Population
Southwark	167,654	844	5.0
Lambeth	19,133	18	0.9

Sources: Snow (1855); Beaglehole et al. (1993).

record sources commonly used by epidemiologists include employment records, trade union files, hospital records, motor vehicle registrations, and disease registries. All of these data sources have their own individual advantages and limitations.

Since Snow, epidemiology has expanded from a method for the investigation of acute infectious disease epidemics to a multi-faceted scientific discipline. Epidemiology now includes research into the causes of chronic diseases such as cardiovascular disease and cancer. Epidemiologists often specialize in particular areas of human health, such as nutrition, occupational and environmental health, and genetics. Nevertheless, the basic epidemiologic principles have changed little since the time of Snow and his colleagues.

Epidemiology has been used to investigate the possible associations between disease and exposures in both the workplace and in the environment. Occupational epidemiologic studies established the associations between asbestos and lung cancer, vinyl chloride and angiosarcoma of the liver, benzene and leukemia, repetitive trauma, and carpal tunnel syndrome, as well as many other occupational exposure–human health effects. Environmental epidemiologic studies have investigated the associations between methyl mercury exposure and severe neurologic disease near Minamata Bay (Japan), the effects of radiation in atomic bomb survivors, and the possible carcinogenic effects of electromagnetic fields. The advantages and limitations of research in these two overlapping areas of epidemiology are discussed below.

21.2 EPIDEMIOLOGIC CAUSATION

In science, proof that a given exposure causes human health effects is established by a hierarchy of evidence. This evidence could be the existence of a medical literature with multiple individual case reports, which associates human disease with a particular exposure. There could be toxicologic evidence in experimental animals in which the particular exposure causes diseases in animals similar to those seen in humans. Regardless, epidemiologic studies are considered to be the highest level of scientific evidence for proving an association between a particular toxic exposure and human health effects.

In epidemiology, proof of causality (or the association of a particular exposure with a particular disease) is based on a variety of criteria. These criteria were first expounded by Hill in 1965, with subsequent refinement and embellishment. These criteria include a temporal relation, plausibility, consistency, strength, a dose–response relationship, and reversibility and/or preventability. In addition, consideration must be given to the appropriateness of the design and to limitations, such as sample size, in each individual epidemiologic study. Ultimately, evidence of causality is the body of epidemiologic studies meeting all of these criteria.

When considering the possibility of an association between an exposure and a disease, the exposure must precede the onset of disease. Evidence of a dose–response relationship is necessary; with an increased dose of a chemical, the risk of disease is increased. The association between the exposure and the disease must make scientific sense (e.g., have biological plausibility).

Statistical significance does not in itself signify a true association; the association must be biologically plausible as well as statistically significant. If possible, the association should be reproducible in toxicologic studies with laboratory animals and other systems such as *in vitro* systems.

The association must be shown repeatedly in different studies of different populations. The association should be preferably strong, as determined by a measure of risk. Finally, ideally, if the exposure is removed, the amount of disease (i.e., the incidence) should decrease and/or new disease should be prevented.

As stated above, a disease–exposure association is considered established if there are repeated similar findings in both toxicologic and in multiple epidemiologic studies. Further proof would be toxicologic and epidemiologic studies which show that when the exposure is removed, the amount of the particular disease decreases or disappears.

Obviously disease–exposure connections are much easier to prove in the case of acute, as opposed to chronic, health effects in both humans and laboratory animals. An illustration is that although the acute effects of carbon monoxide, such as death by asphyxiation, have been easy to establish, the long-term effects of carbon monoxide exposure associated with heart toxicity have been much more difficult to prove. The reason for this is that animals or people must be followed for longer periods of time and may be affected by many other concurrent exposures during that time. In addition, since humans have longer lifespans than many other animals, as well as subtle differences in enzymatic systems and often different routes of exposure, the extrapolation between diseases found in laboratory animals to human disease in the general human population associated with a pollutant exposure is problematic, especially for chronic diseases such as cancer.

Ultimately, if the findings disagree between epidemiologic studies with regard to a possible association between a particular exposure and a human health effect, the interpretation of these epidemiologic studies must depend on the "weight of evidence." In other words, issues such as the validity of the individual studies, the biological plausibility of the association, and the existence or absence of supporting toxicologic and other scientific evidence must all be taken into account.

21.3 TYPES OF EPIDEMIOLOGIC STUDIES: ADVANTAGES AND DISADVANTAGES

Different types of epidemiologic studies have been conducted (Table 21.2). Although predominantly an observational discipline, epidemiologic principles are used in experimental situations such as clinical trials. In observational epidemiologic studies, the study population is not manipulated. In experimental epidemiology, like toxicologic studies, population members are intentionally distributed to different groups to evaluate the effect of a particular intervention.

TABLE 21.2 Types of Epidemiologic Studies

- Observational
  - Descriptive
    - Case series
    - Surveillance
    - Ecologic
  - Analytical
    - Prevalence/cross-sectional studies
    - Case control
    - Cohort
      - Retrospective
      - Prospective
    - Nested/synthetic case control
- Experimental/intervention
  - Clinical trials/randomized controlled trials
  - Field and community trials

Source: Adapted from Beaglehole et al. (1993).

Observational studies can be divided into those that are predominantly descriptive (e.g., a description of a possible disease–exposure association) and those that are analytical (e.g., an analysis of not only the possible disease–exposure association, but of other relationships). Descriptive studies range from a report of a series of unusual cases in the medical literature to surveillance of diseases in a particular industrial population to reports of the health statistics of an entire country. With the exception of ecologic studies (in which using grouped exposure data groups in one area are compared with groups in another area, such as Japanese Hawaiians compared to native Japanese), descriptive studies only include implicit comparisons. Descriptive studies are important for formulating possible hypotheses concerning disease–exposure associations; they are not in themselves epidemiologic proof of such connections.

Analytical observational studies involve the collection of often detailed data in order to make comparisons of exposure and disease in populations. The simplest type is a cross-sectional study. A cross-sectional or prevalence study is performed at a single point in time to compare the risk for disease in unexposed and exposed populations or the risk for exposure in diseased and well populations. Again, cross-sectional studies can only suggest an association since these studies are done at a single point in time, the temporal relationship is not known, specifically, whether the exposure came before and thus caused the disease, or came after and is not truly associated. As such, cross-sectional studies are considered to be hypothesis-generating rather than able to prove an etiologic hypothesis.

A more complicated study, which can suggest an association more strongly, is the case control study in which persons with the disease (the cases) and without the disease (controls) are investigated for their different risk of reported exposures. If the exposure is associated with the disease, then the cases will be more likely to report this exposure in the past than the controls. Cohort studies are considered the *sine qua* non of analytical observational studies. In cohort studies, a group of people (the cohort) is followed in time, distinguished by their exposure (usually exposed and nonexposed), and examined for the diseases which develop over time. The population can be followed into the future in a prospective cohort study or evaluated in the past based on existing records in a retrospective cohort study. If a particular exposure is associated with a particular disease, then the exposed group would be expected to have many more cases of the particular disease than the nonexposed group.

The experimental equivalent of a toxicologic study (i.e., controlled laboratory studies in animals) in epidemiology is the experimental or intervention study. In a clinical trial, a group of people who are randomly assigned to receive either a particular exposure (usually medication) or not (the placebo), are followed to observe the effect of the exposure or the lack of the exposure. This type of epidemiologic study is rarely performed in the study of toxic exposures in humans for obvious ethical reasons. However, intervention studies which remove or mitigate a toxic exposure, then follow the population for disease reduction, can be performed.

21.4 EXPOSURE ISSUES

In order to evaluate an exposure–disease association in epidemiology, the exposure of interest must be defined. In the workplace, this has traditionally taken place by indirect means such as using a job title as a surrogate for exposure (e.g., a "pipefitter" means asbestos exposure) and more recently, by using actual workplace air measurements (both room and personal monitoring). However, these exposure measures are inadequate estimations of the dose received by a particular individual. More recently, biomonitoring (such as blood lead levels) is being used to give more accurate measures of individual exposure. These biomarkers of exposure are not without problems since the correlation of disease with many biomarkers has not been established.

The routes of exposure can determine both the presence and/or the type of human health effect. Toxic exposures in the occupational setting are generally through inhalation and/or through the skin, whereas community exposures are predominantly through contaminated water (such as at Woburn, Massachusetts) and/or the food chain (such as fish consumption at Minamata Bay, Japan). In community environmental exposures, the possibility of both inhalation of airborne pollutants and ingestion from the foodchain and water as routes of exposure must be considered.

Particular to the food chain are issues of bioaccumulation and bioconcentration (as seen with mercury as well as with many lipophilic chemicals). This means that as a chemical goes up the food chain, the concentrations of the chemical increases due to increased storage; therefore, humans, often at the top of the foodchain, will receive the highest doses. Furthermore, within human beings, the ability of certain chemicals to concentrate in fat (i.e., lipophilic) means that increased doses of a chemical can be delivered to the fetus and to the nursing child from the mother, as was shown with DDT and other organochlorine chemicals. In addition, repeated low-dose lifetime exposure can be provided with slow but continual release from fat-stored lipophilic chemicals even when external exposure has ceased.

Traditionally, many disease–exposure connections in humans were established by the evaluation of workers and their occupational diseases. This is because, with rare exceptions (such as methyl mercury exposure in Minamata, Japan), workers tend to have much higher exposures to chemicals in the workplace than the general public. However, for issues such as community exposure to hazardous waste incineration, occupational exposure information may not be entirely appropriate for extrapolation to community exposure and chronic health effects. The reasons are that community exposures are usually much lower and may occur over the entire lifetime of a person, not just during a 40-h workweek. To determine low-level exposures in communities in epidemiologic studies, large populations of people must be followed for long periods of time to see any disease effects; these studies are exceedingly difficult to perform and interpret.

Another exposure issue relevant to community exposures and human health effects is the effect of brief and/or intermittent exposures. For example, much of the existing hazardous waste literature concerns brief "accidental" exposures such as seen in Seveso (Italy); however, the relevance of the health outcomes seen with these single exposures to the more likely scenario of chronic, low-level exposures is unknown. Recently, interest in the scientific community has focused on the issue of indoor as well as outdoor exposures, since most persons in the "developed" nations spend the majority of their time (over 90 percent) indoors; in some studies, indoor exposures to various chemicals exceeded outdoor exposures by 10–100-fold.

Mixed low-level exposure with multiple different chemicals, as would be expected from community exposures to industrial processes, are another difficult exposure issue. Mixed exposure incidents are difficult to classify, to quantify, and even to measure. Furthermore, the particular mixture of exposure incidents in one community is probably not the same mixture found in another community. In addition, it is possible that mixed exposure incidents may cause more or less health effects than exposures to single chemicals since there is the possibility of synergism and/or antagonism of chemicals within the organism.

21.5 DISEASE AND HUMAN HEALTH EFFECTS ISSUES

Besides the issues of exposure, disease or human health effects must be defined and measured. In the past, completely developed end-stage diseases were considered as the most important human health effects to be studied. But because of issues of prevention, as well as the ability to detect subtler physiologic changes which are possibly reversible before the development of full-blown disease, human health effects can no longer be studied in this fashion. For example, in the past, lead toxicity was defined as encephalopathy and even death; lead levels being set are now based on the prevention of subtle neurologic effects of lead on the cognition of fetuses and young children. As discussed with biomonitoring, these subtle measures of health effect, or biomarkers of effect, have their own inherent problems since their prognostic value have not been determined.

In the past, epidemiology focused on acute infectious disease, which is relatively easy to both notice and define. However, evaluation of subchronic and chronic diseases, often months to years after initial exposure, can be very difficult. For example, in asbestos-related cancer, the exposure can be 20–40 years prior to the diagnosis of the cancer. This span of time from initial exposure to disease onset is called latency. In general, the longer the latency, the more difficult it is to establish a disease–exposure connection. In addition, chronic diseases such are heart disease may be due to different exposures in different individuals, yet appear to be the same disease. This can be particularly difficult to evaluate unless large studies of large populations are used. For example, lung cancer can be associated with exposure to smoking alone, asbestos alone, and to a combination of the two; for many years, lung cancer due to asbestos exposure without smoking was not accepted until sufficiently large studies were performed.

21.6 POPULATION ISSUES

As was discussed above, many disease–exposure connections in humans were established by the evaluation of workers and their occupational diseases. However, working populations are usually young and healthy while communities are composed of young and old people, healthy and sick.

Another important issue is that the health effects of chemicals on fetuses and growing children can be devastating at levels which are relatively tolerated by adults (as with the example of the neurologic effects of lead and mercury). Other sensitive populations identified include persons with immunosuppression, asthma, and even multiple-chemical sensitivity.

All the populations described above raise the issue of the generalizability of the results of an epidemogic study in a particular population to another population. Can epidemiologic studies that suggest a disease–exposure connection, or more to the point, a "safe" level of exposure in one human population, be extrapolated to another population?

21.7 MEASUREMENT OF DISEASE OR EXPOSURE FREQUENCY

As noted above, a basic function of epidemiology is to measure the rates or risk of disease in exposed populations (or risk of exposure in diseased populations) for comparison purposes. Rates consist of a numerator and a denominator in which the numerator is the number of people with the disease or exposure and the denominator is the total number of people at risk for this disease or exposure over a set period of time. Without evaluating the number of deaths or cases of a particular disease by the total number of people at risk in that particular population, it would be impossible to compare the risk of death or disease in that population with the risk of another population. Comparing only the number of deaths or cases between two populations can be misleading, if one population is much larger or substantially different than the other.

To increase the comparability of rates between populations, these measures of risk have been further refined by adjusting them for various population characteristics. For example, typical rates will be adjusted for age, sex, socioeconomic class, and/or race and/or ethnic group. The reason for this is that various subpopulations can experience differing risks. For example, the risk of breast cancer is at least 10 times higher in women than in men, or the risk of cancer is higher in older than younger populations.

The basic measure of disease or exposure frequency in populations is the *prevalence* of a particular disease in the population. The prevalence is defined as the number of cases of a particular disease in a specific population at a single point in time. Prevalence is expressed as percent because it is a proportion not a rate. For example, if five asbestos workers have lung cancer in a worker population of 1000 people, the prevalence of lung cancer in this worker population is 5:1000 or 0.5 percent. Prevalence includes all persons who have the particular disease at a given point in time, regardless of when they developed the disease. Therefore, the prevalence of prolonged chronic diseases such as arthritis are usually much higher than the prevalence of acutely fatal diseases such as meningitis. Prevalence is the measure of disease and exposure frequency for cross-sectional studies.

Most measurements of disease or exposure frequency in epidemiologic studies are expressed as rates: numbers of persons who developed a disease in a given population at risk during a period of time. The *incidence rate* is the classic measure of disease rates in an epidemiologic study. For example, if four out of the five lung cancer cases developed in the asbestos-exposed group of workers discussed above over the course of one year, then the incidence rate for lung cancer in that worker population would be 4/1000 per year or 0.4. The incidence rate is different from the prevalence since only persons who develop a disease during a set period of time are included. Therefore, incidence rates and prevalence cannot be directly compared. Incidence rates are the measure of disease frequency in cohort studies.

21.8 MEASUREMENT OF ASSOCIATION OR RISK

In epidemiologic studies, populations with different frequencies of disease or exposure are often compared to suggest or establish an association between particular diseases and exposure. This is also described as the risk of disease from a particular exposure. For example, if the incidence rate of lung cancer in a worker population without asbestos exposure is 1/1000 workers per year, this can be compared to the rate of lung cancer in an asbestos exposed population of 4/1000 per year. The comparison shows that the rate of lung cancer is higher in the asbestos-exposed population than the unexposed population. This suggests an association between asbestos exposure and lung cancer, or that there is a risk of lung cancer with asbestos exposure.

As mentioned above, prevalence is the measurement of disease or exposure frequency in cross-sectional studies. The risk of disease can be examined using the prevalences of disease in the exposed and unexposed populations by formulating the rate ratio and the rate difference (Table 21.3). The *rate ratio* is the ratio of the prevalence of a particular disease in the exposed population to the prevalence of the unexposed population. The *rate difference* is the difference between the prevalence of a particular disease between the exposed population and the unexposed population. A rate ratio greater than or less than one indicates increased or decreased risk of disease; a rate difference greater than 0 indicates increased risk of disease. These measures of risk can be tested for statistical significance using the chi-square (χ^2) test and a confidence interval.

The *incidence rate* is the measurement of disease or exposure frequency in *cohort studies*. The risk of disease in the exposed and unexposed population is examined with incidence rates of disease in the exposed and unexposed populations using the rate ratio and the rate difference (Table 21.4). The *rate ratio* is the ratio of the incidence of a disease in the exposed population to the incidence in the unexposed population. The *rate difference* is the difference between the incidence of a particular disease between the exposed population and the unexposed population. These measures of risk can also be tested for statistical significance using the χ^2 test and a confidence interval.

Rate ratios are often standardized to create a *standardized rate ratio* (SRR) so that rates from different populations are compared after eliminating confounding (see text below). For example, if one of the populations is older, then age as a confounder can lead to a false conclusion when comparing cancer disease rates in the two populations; the older population will inevitably have a higher cancer

TABLE 21.3 Prevalence Data: Rate Ratio and Rate Difference

		Disease	
		+	–
Exposure	+	*A*	*B*
	–	*C*	*D*

Prevalence of disease in exposed population = $A/(A + B)$
Prevalence of disease in the unexposed population = $C/(C + D)$
Rate ratio = $A/(A + B)$ divided by $C/(C + D)$
Rate difference = $A(A + B) - C/(C + D)$

TABLE 21.4 Incidence Data: Rate Ratio and Rate Difference

		Disease	
		+	–
Exposure	+	*A*	*B*
	–	*C*	*D*

Incidence of disease in exposed population = *A*/PY (person-year)
Incidence of disease in unexposed population = *C*/PY
Rate ratio = *A*/PY divided by *C*/PY
Rate difference = (*A*/PY – *C*/Py)

rate that is not due to the exposure. Therefore, standardization for age is necessary prior to making comparisons of the cancer rates between the two different populations.

A variation on the SRR used extensively in Occupational Epidemiology is the *standardized mortality ratio* (SMR) with regard to mortality or standardized morbidity ratio with regard to morbidity. First, the person years of exposure in the study group are multiplied times the rate of death from a particular disease in the reference or standard population (i.e., number of cases/number of persons in the standard population/year) to generate an expected number of cases of death from the disease. The SMR is the number of cases of death in the study population divided by the number of cases expected from the standard population (SMR = observed/expected × 100). In the example described above, the incidence rate of lung cancer in worker populations without asbestos exposure is 1/1000 workers/year while the rate of lung cancer in an asbestos-exposed population is 4/1000 workers/year with 1000 person-years. The expected number of lung cancer cases in the exposed worker group, using the unexposed worker population as the reference or standard population, is 1 (i.e., 1000 person-years × 1/1000 per year). Therefore, the SMR for lung cancer with asbestos exposure would be 4/1 × 100 or 400. A SMR greater than or less than 100 indicates increased or decreased risk of disease. SMRs and SRRs can be tested for statistical significance using the χ^2 test and a confidence interval.

In occupational epidemiologic studies, the *proportional mortality ratio* (PMR) is frequently used to measure the risk or association between exposure and disease. The PMR is used in settings when only the population and the cause of death (or disease) are known, not the number of person-years of exposure. In a PMR study, the expected number of deaths due to a particular disease is computed using the proportion of these deaths in a standard population (such as the general population), rather than the death rate as in the SMR study. The PMR study results approximate those of a SMR study when the particular cause of death is relatively uncommon. A PMR greater than or less than 100 indicates an increased or decreased risk of disease or death.

In case–control studies, the total number and proportions of persons with and without a disease are defined by the study design, rather than the true proportions in the total population. As such, usually true rates of disease cannot be derived in case–control studies. Therefore, the rate ratio is not an appropriate measure of risk; the *odds ratio* is used as an approximation of the rate ratio for the risk of disease associated with exposure (Table 21.5). An odds ratio greater than or less than 1 indicates an

TABLE 21.5 Case–Control Data: Odds Ratio

		Disease	
		+	–
Exposure	+	*A*	*B*
	–	*C*	*D*

Rate ratio = odds ratio = *AD*/*BC*

increased or decreased risk of disease. The odds ratio can be tested for statistical significance using the χ^2 square test and a confidence interval.

21.9 BIAS

Bias is the systematic difference in the study from reality. This can happen at any time during an epidemiologic study. Bias is any trend in the collection, analysis, interpretation, or review of data that can lead to conclusions that are systematically different from the truth. In general, bias affects the comparability of the data.

Epidemiologic studies are inherently biased because they are studying human populations, often based on data and records collected in the past. Much of bias can be prevented by careful study planning or avoided by the use of various statistical analysis techniques. There are several types of bias which are well described and should be evaluated in reviewing epidemiologic studies.

Selection bias is introduced during the beginning of a study when the study population is divided into different groups (either exposed or diseased). If the diseased group has different criteria for exposure than the control group, selection bias is present and the two groups are not comparable. *Observation bias* is often seen in studies where the interviewer is aware of the disease or exposure state of the study subject (i.e., not blinded); this can lead to differing amounts of information being obtained from the two groups, and thus, the two groups are not comparable. *Recall bias*, seen particularly in case–control studies, is due to the increased recall of past events (including exposures) by persons who have a disease as compared to the recall of those who do not have the disease (i.e., the controls). For example, a woman who has had a child with a birth defect is more likely to remember every medication taken during the pregnancy than a woman whose child was born without a birth defect.

Another bias commonly seen in occupational epidemiology studies has been dubbed "*the healthy worker effect*." This occurs when rates of disease in working populations are compared with those in the general population. Often, working populations in this context will show little or no increased risk of disease. However, the general population is not an appropriate comparison population for working populations. As was stated above, working populations are in general much healthier and younger than the general population, which consists of people of all ages and stages of health. Therefore, the appropriate comparison population for an exposed working population would be another unexposed working population.

Confounding is a particular form of bias which is very common in epidemiologic studies. Confounding results when the association between an exposure and a disease is really due to another exposure. For example, smoking is associated with lung cancer and with heavy alcohol intake. If exposure to smoking is not controlled for in either the study design or the analysis (see discussion below), then it could appear that lung cancer is due to exposure to heavy alcohol intake. Smoking is a common confounding variable, as well as age, sex, and socioeconomic class.

Although many factors are assumed to be confounders (age, sex, socioeconomic class, smoking, etc.), an important component of the statistical analysis is to assess for the possibility of confounding. If a particular variable is associated (i.e., has a rate ratio or its equivalent measure greater than or less than one) separately with both the disease and the exposure of interest, then that variable is considered to be a confounder.

There are a variety of ways to control for confounding in epidemiologic studies. One method, in case control studies, is to match the cases and controls on well-known confounders, such as age and sex; then, these cannot be confounding variables since the proportion of the population with the confounding variable is now the same in both the cases and controls. Another method used in clinical trials is the randomization of subjects to the treatment and nontreatment groups so that possible confounding variables are randomly distributed. Both randomization and matching are methods to control for confounding, which must be implemented in the design phase of the epidemiologic study. Usually, confounding is controlled for during the statistical analysis at the end of the epidemiologic study. One method is to stratify the data by the confounding variable, and then report the risk measure by each subpopulation. For example, different rate ratios would be reported for different age groups if age is considered a confounding variable. The most common method of controlling for confounding is the standardization of the rates of disease. An external or internal reference or standard population is used to derive standard proportions for the confounding variable; these proportions are applied to the different subpopulations so that these subpopulations are equivalent with respect to the confounding variable. Finally, if statistical modeling such as multivariate analysis is to be done, then confounding variables can be controlled by including them in the model.

21.10 OTHER ISSUES

Disease clusters are apparent increases of disease or death in time and/or space. *Epidemics* are disease clusters. In order to be epidemiologically meaningful, the individuals in the disease cluster must have a historic shared or similar exposure (also known as an *aggregation*). Many accepted occupational diseases were discovered originally as disease clusters. In environmental clusters, sometimes these disease increases are due to the same and/or shared toxic exposure in a given community, but more often the increases are due to chance. In other words, as a result of randomness, a group of individuals who live in the same geographic area has the same disease for different reasons. Therefore, drawing conclusions from a single epidemiologic study that indicates a positive disease–exposure connection is difficult.

With increased public and agency awareness, disease clusters are being reported more and more frequently. The natural assumption of individuals in communities is that a disease cluster is due to some local toxic environmental exposure. Not only is this rarely the situation, but it can be very difficult to investigate the environmental disease cluster due to issues of the small numbers of cases, poor exposure measurement, and the lack of comparability of study populations. These limitations and the possibility that the cluster is due purely to chance are difficult concepts to explain to worried community residents.

A further problem plaguing epidemiologic research of human health effects in communities with environmental exposures is the relatively small populations studied. The small numbers render valid statistical analysis and generalizability of the study conclusions questionable. Not only are the numbers of people small, but again these populations are often studied in crisis (such as with Love Canal, New York) with considerable anxiety as well as legal involvement; this atmosphere makes objective scientific investigation very difficult.

Nevertheless, as exposures decrease in occupational settings and as environmental pollution is identified and its possible human health effects questioned, more epidemiologic studies will be performed in communities in the future, despite the limitations.

21.11 SUMMARY

Epidemiology is the study of the distribution and determinants of disease or death in human populations. Since epidemiology is predominantly an observational rather than experimental science, it relies heavily on data collected in records and questionnaires. In the case of the environment or the workplace, epidemiology attempts to determine associations between a chemical exposure and particular human health effects. Using measures of risk, comparing unexposed and exposed populations for disease risk or comparing diseased and well populations for exposure risk, epidemiologic studies can associate a disease risk with a particular exposure.

REFERENCES AND SUGGESTED READING

Baker, E. L., and T. P. Matte, "Surveillance for Occupational Hazards and Disease," in *Textbook of Clinical Occupational and Environmental Medicine*, L. Rosenstock and M. R. Cullen, eds., Saunders, Philadelphia, 1994, pp. 61–67.

Beaglehole, R., R. Bonita, and T. Kjellstrom, *Basic Epidemiology*, World Health Organization, Geneva, 1993.

Centers for Disease Control (CDC), "Guidelines for investigating clusters of health events," *Morbid. Mortal. Weekly Rep.* **39**(RR-11): 1–23 (1990).

Checkoway, H., N. E. Pearce, and D. J. Crawford-Brown, *Research Methods in Occupational Epidemiology*, Oxford Univ. Press, New York, 1989.

Checkoway, H., N. Pearce, and J. M. Dement, "Design and conduct of occupational epidemiology studies: I–IV," *Am. J. Ind. Med.* **15**: 363–416 (1989).

Checkoway, H., "Epidemiology," in *Textbook of Clinical Occupational and Environmental Medicine*, L. Rosenstock, and M. R. Cullen, eds., Saunders, Philadelphia, 1994, pp. 150–168.

Eisen, E. A., and D. H. Wegman, "Epidemiology," in *Occupational Health: Recognizing and Preventing Work-Related Disease*, 3rd ed., B. S. Levy and D. H. Wegman, eds., Little, Brown, Boston, 1994, pp. 103–125.

Fleming, L. E., A. M. Ducatman, and S. L. Shalat, "Disease clusters: A central and on-going role in occupational health," *J. Occup. Med.* **33**: 818–825 (1991).

Fox, A. J., and P. F. Collier, "Low mortality rates in industrial cohort studies due to selection of work and survival in the industry," *Br. J. Prev. Soc. Med.* **30**: 225–230 (1976).

Goldsmith, J. R., *Environmental Epidemiology: Epidemiological Investigation of Community Environmental Health Problems,* CRC Press, Boca Raton, FL, 1986.

Hennekens, C. H., J. E. Buring, and S. L. Mayrent, *Epidemiology in Medicine,* Little, Brown, Boston, 1987.

Hernberg, S., *Introduction to Occupational Epidemiology,* Lewis Publishers, Philadelphia, 1992.

Hill, A. B., "The environment and disease: Association or causation?" *Proc. Roy. Soc. Med.* **58**: 295–300 (1965).

Kelsey, J. L., W. D. Thompson, and A. S. Evans, *Methods in Observational Epidemiology*, Oxford Univ. Press, New York, 1986.

Last, J. M., *A Dictionary of Epidemiology*, Oxford Univ. Press, New York, 1988.

Leaverton, P. E., *Environmental Epidemiology*, Praeger Publishers, New York, 1982.

Lilienfeld, D. E., and P. D. Stolley, *Foundations of Epidemiology*, Oxford Univ. Press, New York, 1994.

MacMahon, B., and T. F. Pugh, *Epidemiology: Principles and Methods*, Little, Brown, Boston, 1980.

Marsh, G. M., "Epidemiology of occupational diseases," in *Environmental and Occupational Medicine*, 2nd ed., W. N. Rom, ed., Little, Brown, Boston, 1992, pp. 35–50.

Monson, R. R., *Occupational Epidemiology*, 2nd ed., CRC Press, Boca Raton, FL, 1990.

Rothman, K. J., *Modem Epidemiology*, Little, Brown, Boston, 1986.

Schlesselman, J. J., *Case Control Studies*, Oxford Univ. Press, New York, 1982.

Snow, J., *On the Mode of Communication of Cholera*, Churchill, London, 1855. (reprinted in *Snow on Cholera: A Reprint of Two Papers*, Hafner, New York, 1965.)

22 Controlling Occupational and Environmental Health Hazards

PAUL J. MIDDENDORF and DAVID E. JACOBS

The purpose of this chapter is to provide an overview of the practices of individual hygiene as an application of the principles of toxicology, to protect workers and persons in the community from the adverse effects of chemicals. This chapter will

- Discuss chemical exposure limits, which are an application of toxicology to exposed populations
- Describe the roles of industrial hygiene program management in protecting occupational health
- Detail work practices, administrative, and engineering exposure control methods
- Provide examples in a variety of environments of successful control strategies

22.1 BACKGROUND AND HISTORICAL PERSPECTIVE

Efforts to control occupational health hazards intensified greatly in the twentieth century. Prior to this time, occupational health focused on identification of causative relationships between agents and harmful effects, while virtually ignoring the prevention of occupational diseases. Aristotle, Paracelsus, Agricola, Ramazzini, and Percival Potts were all involved in identification of occupational diseases, but recommended relatively few remedial actions. There were some attempts to use animal bladders as respirators for protection against toxic dusts in mines, and better ventilation, but little progress was made in prevention of occupational disease. Dr. Alice Hamilton's work in the early twentieth century documenting and describing the extent of occupational diseases in the United States is considered to be the beginning of modern industrial hygiene. She became a strong advocate for the prevention of occupational disease. In her autobiography, *Exploring the Dangerous Trades*, she described her experience with the National Lead Company and the difficulties in convincing managers of the hazards associated with lead. Eventually she was able to convince them and began a program for prevention of lead poisoning.

Two organizations were formed to advance the cause of worker health and safety. The American Conference of Governmental Industrial Hygienists (ACGIH), a private organization composed largely of industrial hygienists employed in academia and government, was formed in 1938. The American Industrial Hygiene Association was formed in 1939 and has come to include many members in both private industry and the public sector. Both groups brought practitioners together, provided means to disseminate information, exchange experiences, and conduct symposia. In addition, the American Public Health Association has had an Occupational Health and Safety Section since 1914. However, it was not until the growing public awareness of the presence and potentially harmful effects of chemicals during the 1960s that further regulation of the workplace occurred. Rachel Carson gripped the nation with her account of the effects of pesticides in the environment in her book *Silent Spring*

Principles of Toxicology: Environmental and Industrial Applications, Second Edition, Edited by Phillip L. Williams, Robert C. James, and Stephen M. Roberts.
ISBN 0-471-29321-0 © 2000 John Wiley & Sons, Inc.

(1962), and labor unions began to lobby for better protective measures in workplaces. These events resulted in a number of legislative initiatives at the end of the decade. Along with other environmental laws, the Occupational Safety and Health Act (OSHA) was passed in 1970. The OSHAct assured that every employee "has the right to a workplace free of recognized hazards." Since that time, the field of industrial hygiene has grown dramatically, both in size and sophistication.

The basic philosophies of industrial hygiene are well-recognized: protection of worker health is a worthwhile goal; the burden of protecting the worker is primarily the employer's, but the worker also has the responsibility to exercise self-protection. The hierarchy of control measures is, in order of effectiveness, elimination of toxic chemical use, substitution of less toxic material for more toxic substances, engineering controls, administrative controls, and personal protective equipment. Education and training of workers is increasingly recognized as a key aspect of industrial hygiene, as evidenced by the proliferation of "right to know" laws that have swept the country. Essentially, the idea is that an informed worker is a safer worker; if employees understand why a safety rule has been established, they are more likely to follow it.

The control of occupational health hazards is dependent on the anticipation of potential health hazards in new facilities and their recognition in existing facilities. Once the potential health hazards have been identified, they are evaluated by a process of observation of the work place and work practices and quantification of the risk associated with the hazard. When these risks reach unacceptable levels, methods of controlling the exposures must be identified and implemented to minimize the potential for harm to the worker. The identification and evaluation phases are sometimes referred to as *risk assessment*, while the control phase is referred to as risk reduction or risk management.

In practice, the industrial hygienist plays the major role in the risk assessment process. On the other hand, risk reduction involves many different players with a variety of backgrounds: the supervisors and workers who know the process and must work with any new equipment or procedures that are installed; engineers who are familiar with control technologies and understand the production process; toxicologists who may identify adverse effects from chemical exposure; physicians capable of designing medical surveillance programs to help identify emerging adverse health effects in individuals; the industrial hygienist who understands how contaminants behave in both the environment and the human body and understands when to use the various control strategies; and managers who must allocate sufficient resources to pay for controls. Each player brings special abilities and knowledge to the problems associated with minimizing health risk.

As an applied science, industrial hygiene relies on the basic sciences of biology, chemistry, physics, and statistics, as well as the social sciences—sociology, and political science—to achieve its goals.

22.2 EXPOSURE LIMITS

The bridge between toxicology and industrial hygiene has traditionally been the development of exposure limits. Exposure limits are reference values that should not be exceeded. They have been developed to protect the large majority of an exposed population from the development of environmentally induced diseases. The level of protection varies with the type and severity of the expected outcome as well as the sensitivity of the exposed population.

Air-Contaminant Exposure Limits

The threshold limit values (TLVs®) for chemical substances have been published by the American Conference of Industrial Hygienists (ACGIH) since 1946 and have increased in number since that time. Generally, the TLVs® for many substances have been lowered as the identification of more subtle adverse health effects has become possible. Mastromatteo (1988) documented the history and progression of TLVs®. ACGIH bases the TLVs® on information gathered by the ACGIH TLV® Committee from human experience in the workplace and experimental human and animal studies.

The TLVs® are designed to provide guidance to industrial hygienists in determining the potential for harm associated with air contaminant exposures. They must never be viewed as a sharp dividing line between "safe" and "unsafe" conditions, and they certainly do not preclude the appearance of adverse health effects in individuals with greater sensitivity, either through genetic or environmentally-induced mechanisms.

For chemicals, the TLVs® address exposure primarily from the inhalation route and take several forms. The TLV®-TWA is an eight hour time-weighted average (TWA) exposure to which the ACGIH TLV® Committee believes most workers may be exposed for 8-h workdays, 40 hours per week for a working lifetime without becoming ill. Some chemicals also have a short-term exposure limit (TLV®-STEL), which is a 15-min TWA exposure limit to prevent acute effects such as irritation, chronic or irreversible tissue damage, or narcosis, which could impair judgement or performance. A ceiling limit (TLV®-C) is a concentration above which employees should not be exceeded at any time.

Some compounds are considered to be carcinogens based on epidemiology studies, toxicology studies, or case histories. The qualitative differences in research results lead the ACGIH to place a carcinogen into one of five categories: (1) Confirmed Human Carcinogen, (2) Suspected Human Carcinogen, (3) Confirmed Animal Carcinogen, (4) Not Classifiable as a Human Carcinogen, and (5) Not Suspected as a Human Carcinogen. In some cases the carcinogen has an exposure limit assigned, while in others no exposure limit is assigned. For carcinogens with an exposure limit, this does not imply that the compound has a threshold, but exposures at or below the exposure limit should not result in a measurable incidence of cancer or mortality. Exposures to suspected carcinogens should be controlled to a level as low as reasonably achievable below the listed TLV®. For confirmed human carcinogens without an exposure limit, engineering controls should be installed, and workers should wear personal protective equipment that will eliminate exposure to the greatest extent possible.

The workplace has gone through many changes in the 40 years since the TLVs® were first adopted. For example, the number of women in the workplace has increased drastically. Although men and women frequently exhibit little difference in the quantitative and qualitative responses to chemical exposures, their responses to some chemicals differ. There are also more older workers in the workforce today whose sensitivities to chemicals may be increased because of age-related physiological changes. Also significant, the presence of women in the workplace has brought another part of our population, the fetus, into the workplace. The sensitivities and responses of the fetus may be much different than the sensitivities of the mother or other adult workers. As more information on these effects becomes available, the TLVs® for substances with teratogenic potential are likely to incorporate these effects and may be substantially reduced. In the interim, programs that are intended to minimize reproductive effects are likely to be implemented. The message from the Supreme Court in a legal case, the "Johnson Controls" case, in which the company policy was to exclude fertile women of child-bearing age from certain jobs where workers were exposed to lead was that these programs will be difficult to administer in ways that do not discriminate.

The Occupational Safety and Health Administration adopted the 1968 TLVs® as legally enforceable permissible exposure limits (PELs). When these were adopted, many people not familiar with the TLV® concept mistakenly interpreted these levels as "safe" for all workers since the intent of the OSHAct is "to assure as far as possible every working man and woman in the nation safe and healthful working conditions." That was never the intent of ACGIH, and the most recent version of the ACGIH TLV® booklet (1998) states

> These limits are not fine lines between safe and dangerous concentration nor are they a relative index of toxicity. They should not be used by anyone untrained in the discipline of industrial hygiene.

Before 1988, OSHA was able to promulgate only a handful of new or changed standards for air contaminants. In part, this was caused by the detailed and costly (but democratic) process prescribed for adoption of new standards, which is designed to ensure that all groups have a say in how the standard is formulated. One important concept included in many of the standards promulgated by OSHA was the "action level." "Action levels" are typically one-half the PEL, and certain actions are required to detect potential adverse health effects, such as periodic air monitoring and medical surveillance programs. Because exposures can vary greatly from one day to another with minor changes in processes, work practices, and environmental conditions, the "action level" concept is intended to identify, with 95 percent confidence, processes that are likely to exceed the occupational exposure limit on 5 percent or more of the work days. An underlying assumption of the "action level" is that exposures below the occupational exposure limit cause minimal or acceptable risk to the worker. As exposures increase above the action level, the number of days the exposure limit will likely be exceeded increases, and the probability of adverse health outcomes increases. An additional advantage of the "action level" is that workers more susceptible to the effects of a chemical are more likely to be identified through the required medical surveillance programs.

Between 1968 and 1988, ACGIH lowered many of the TLVs® and many practicing industrial hygienists referenced the TLV®s rather than the PELs (1968 TLVs®) to provide better protection to workers. The National Institute of Occupational Safety and Health (NIOSH), an agency created by the OSHAct to recommend standards to OSHA, among other responsibilities, developed many *recommended exposure levels* (RELs), which were also increasingly referenced by practicing industrial hygienists. But these were largely ignored by OSHA in the rulemaking process, in part because they were based strictly on health criteria, while OSHA was required to include feasibility considerations in their standards. NIOSH has recently stated their intent to develop RELs that do include feasibility.

In 1989, OSHA again adopted many of the ACGIH TLVs®, this time from 1987/88, as well as several NIOSH RELs. Many industrial hygienists agreed with updating the PELs because of the additional protection it afforded workers. Unfortunately, this represented a reaffirmation that the exposure limits are "safe" for all workers rather than merely guidelines. As a result of legal challenges from both organized labor and industry groups, the updated PELs were struck down by the 11th Circuit Court of Appeals in 1992, and after several continuances, OSHA accepted the decision in 1993, and the PELs reverted to those adopted in 1971. The dilemma of setting new exposure limits in a timely, but scientifically rigorous, fashion within the established regulatory framework is a difficult one. One option to speed up the process is to periodically establish a prioritized group of chemicals for which new health effects data has become available and propose new exposure limits for each compound in the group. Other alternatives may be possible if OSHA reform efforts are successful.

Another potential problem associated with adoption of TLVs® is their use by other governmental agencies as a basis for environmental exposure limits. The ACGIH considers this use generally inappropriate because of differences in the exposed populations (adults in the occupational setting versus all age groups in the general environment), exposure patterns (8 h/day, 40 h/week, 40-year working lifetime vs. 24 h/day for 70+ years), and exposure routes (inhalation versus ingestion and inhalation). Nevertheless, occupational limits are often used by environmental agencies because no other standards are available. The environmental exposure limits, usually referred to as *acceptable ambient-air concentration guidelines or standards*, usually reduce the OEL by an uncertainty factor. The uncertainty factor is based on the averaging time, usually 1, 8, or 24 h or 1 year; the duration of exposure; and the type of adverse health outcomes. Although several formulas are used, a typical one is:

$$\text{Uncertainty factor} = (\text{safety factor/TLV}^{®}) \times (\text{hours per day}/8) \times (\text{days per week}/5)$$

The safety factor is usually 10, 100, or 1000, depending on the health effects. However, Williams et al. compared environmental exposure limits developed by this method with some developed from "scratch," and found little differences. While the scratch method may be the best, as a means of providing interim protection, the use of TLVs® as a basis for environmental exposure limits may be reasonable.

Another group has proposed using the USEPA *Integrated Risk Information System* (IRIS) (database) to establish air contaminant standards, referred to as *workplace allowable air concentrations* (WACs), which are based exclusively on systemic and carcinogenic effects of chronic and subchronic exposure. The resulting exposure limits are generally lower (frequently by several orders of magnitude) than the TLVs® and PELs, which are often based on prevention of acute irritant effects. WACs are modified "reference doses" to reflect shorter daily exposure durations and the working lifetime instead of the entire lifetime. However, the current set of WACs has been developed primarily from oral exposure data rather than inhalation data, and they do not take into account the feasibility of meeting the limits. In the future, this approach with appropriate reference to inhalation data may be of tremendous value in helping to determine exposure limits, particularly integrated exposure limits, which are discussed below. For the present, however, they have been largely ignored by the industrial hygiene community.

One deficiency of the TLVs® and PELs is that they address exposure only from the inhalation route. Many chemicals may be absorbed through the skin. For chemicals that have low vapor pressure and are lipid soluble, such as many pesticides, skin absorption may be the primary route of exposure. Chemicals which are expected to have substantial exposure by the skin route are noted as a part of the TLV®, but the notation does not provide quantitative information to help with interpretation of exposure data. Ingestion is another route of exposure that the TLVs® do not address, but that, for some compounds, can lead to substantial doses in working populations. Toxicologically relevant exposures to lead and other metals often occur through poor personal hygiene and frequent hand-to-mouth contact, resulting in ingestion.

While the exposure limits are usually based on 8 h/day and 40 h/week work schedules; in many cases, work schedules are not limited to the traditional schedule. Some work schedules call for four 10-h days each week, while other novel work schedules may include a longer rotation where workers are exposed 7 days out of a week, such as the schedule where workers work 56 eight-hour workdays and then have 21 days off. Many workers have overtime and may work more than 8 h in a given day and more than 40 h in a given week. Brief and Scala argue that because these types of work schedules increase the exposure time and the amount of the chemical that reaches the target organ and because the time for clearance of the chemical from the body after exposure ends is reduced, some modification of exposure limits may be necessary. Repeated exposure may also increase stress on detoxification mechanisms and alternate metabolic pathways may be initiated, possibly leading to greater toxicity.

The Brief and Scala (1975) model for adjustment of the TLVs® is easy to use and is based strictly on the increased exposure time and decreased clearance time. For both acute and cumulative toxicants, except those whose only adverse effects are sensory irritation, the TLV® is adjusted on an 8-h basis.

$$\text{TLV}^{\circledR}_{\text{adj}} = \text{TLV}^{\circledR} \times (8\ \text{h/hours worked per day}) \times (24\ \text{h} - \text{hours worked per day})/16\ \text{hrs})$$

When the work schedule includes a 7-day workweek, such as the 56/21 schedule described above, the adjustment to the TLV® is driven by the 40-h workweek, rather than the 8-h workday:

$$\text{TLV}^{\circledR}_{\text{adj}} = \text{TLV}^{\circledR} \times (40\ \text{h/hours worked per week}) \times (168\ \text{h} - \text{hours worked per day})/128\ \text{h}$$

An advantage of this simple model is that it does not require knowledge of the biologic half-life or the mechanism of action. However, it is considered a first approximation only and does not guarantee an adequate level of protection. Additional medical surveillance is necessary to assure that workers are protected.

Although OSHA does not currently adjust their PELs except for lead, OSHA has used a model for adjustment of the PELs, outlined in Table 22.1. Substances are divided into various categories based

TABLE 22.1 Prolonged Work Schedule Categories

Category	Classification	Adjustment Criteria
1A	Ceiling standard	None
1B	Irritants	None
1C	Technologic limitations	None
2	Acute toxicants	Daily
3	Cumulative toxicants	Weekly
4	Acute and cumulative toxicants	Daily or weekly (the more protective)

on the adverse effects in which the PEL is intended to prevent. Substances in Categories 1A, 1B, and 1C are not adjusted based on a number of arguments. OSHA reasons that compounds with ceiling standards (category 1A) should not be adjusted because the adverse effect is prevented when the ceiling is not exceeded and is independent of the length of the work shift. A substance whose exposure limit is based on preventing irritation (category 1B) should not be adjusted because no cumulative effects are observed at exposures near the PEL. Some substances are regulated by OSHA at levels, which are based on the technologic feasibility of controlling exposures to specific levels rather than based on preventing adverse health effects. Exposure limits for these substances (category 1C) are not adjusted in the OSHA scheme.

For substances or their toxic metabolites that are acutely toxic, adjustments to the PEL are made to ensure that the buildup of the chemical or its toxic metabolites do not accumulate to levels that could cause adverse health effects during a single day of exposure. The adjustment to the PEL is based on the time of exposure in a single day:

$$PEL_{adj} = PEL \times (8 \text{ h/hours worked per day})$$

Substances or their toxic metabolites that may accumulate in the body over a longer period of exposure are adjusted based on the weekly exposure, rather than their daily time of exposure:

$$PEL_{adj} = PEL \times (40 \text{ h/hours worked per week})$$

For substances which have both cumulative and acute effects, the adjustment that provides the greater protection should be used.

Both the Brief–Scala and the OSHA adjustment models are generally more conservative than pharmacokinetic data indicate are necessary. When the pharmacokinetics of a substance are thoroughly known, the information can be used along with the exposure pattern to determine exposure limits for unusual exposure schedules. The available pharmacokinetic models are based on the goal of limiting the maximum body burden to the level expected following an 8-h/day, 40-h/week exposure regimen. The pharmacokinetic models also account for the form of the material, rate of metabolism, and excretion and distribution of the material in the body. Another advantage of pharmacokinetic models is that they can identify work schedules where no adjustment is necessary. The major disadvantages are that most of the models assume that the body acts as a single compartment, while the distribution of many substances is known to be better described by a two- or three-compartment model and that pharmacokinetic information is not available for many substances. A mathematical treatment of these models is beyond the scope of this chapter but is thoroughly described by Paustenbach (1985).

The process by which TLVs® are developed and the use of TLVs® at all have been criticized. Although the criticisms have some validity and exposure limits have many unresolved problems, their use should not be discontinued. Instead, the exposure limits should be regarded as our current best estimates, and, when necessary, they should be revised. They also provide a benchmark for specifying feasible control technology, and for evaluating how well those controls operate in practice.

Compliance with exposure limits is not an acceptable excuse for failing to reduce exposures to the fullest extent possible. In all cases, exposures should be reduced as much as possible and TLVs® should be considered upper limits of allowable exposure. As epidemiological, toxicological, and analytical techniques have improved, many exposure limits have been reduced and are likely to be reduced in the future. If the conceptual differences between exposure limits as guidelines and exposure limits as "guarantees of safety" are fully grasped, then the frequently changing exposure limits can be accepted as the norm, rather than the exception. The complex nature of exposures and the TLV® concepts require a trained, knowledgeable individual to interpret exposure data and workplace conditions in terms of the potential for disease.

Biological Exposure Indices

One reason why skin absorption and ingestion are not routinely addressed is because they are very difficult or impossible to quantitatively evaluate. They are frequently evaluated by a subjective observation process, followed by recommendations and control measures. Even if the recommended control measures are taken, it is not clear that the workers are adequately protected. To overcome this deficiency, as well as for other reasons, biological exposure indices (BEIs) have been proposed, and, for some chemicals, have been adopted by ACGIH. In addition, OSHA requires biological monitoring in the form of measurement of blood lead levels as part of its lead standard; cadmium and 2-microglobulin in urine, as well as cadmium in the blood as part of the cadmium standard, and, in emergency situations, urinary levels of phenol after exposure to benzene.

BEIs are reference values for biological samples from exhaled air, urine, or blood. Theoretically, hair, nails, teeth, fat deposits, and other tissues could be used, but they are either more difficult to obtain or the chance of uncontrollable sample contamination may be too great. Each BEI has a specific determinant and biological specimen to be monitored, as well as a time to monitor relative to the exposure period. The determinant can be the chemical itself, a metabolite, or a characteristic reversible biochemical change induced by the chemical. The time of monitoring is determined by the elimination half-life of the determinant. Timing becomes less critical as the elimination half-life increases.

BEIs are similar to TLVs® in that they are intended to indicate exposure level, and do not provide sharp distinctions between "safe" and "unsafe" exposures. The major advantage of BEIs is that, at least theoretically, they are capable of integrating all exposure routes to yield a better estimate of dose. However, BEIs cannot distinguish between nonoccupational and occupational exposure sources. Some BEIs are non-specific and may be applicable to several chemicals so that in the event of multiple exposures, interpretation of biological determinants becomes more difficult.

Biological exposure indices are subject to a variety of sources of variability. The same airborne, ingested, and skin exposures do not necessarily produce the same levels of the determinant in different individuals. Absorption, distribution, and metabolic processes (pharmacokinetics) vary from one individual to another. The choice of a determinant is based on the highest correlation with the level of exposure, but the determinant may not best represent the dose to the target organ, which is the most useful indicator for evaluating the health risk to the individual worker.

The use of BEIs is likely to increase with time, particularly as the techniques become more routine. Biological monitoring is not likely to completely replace the more traditional exposure assessment techniques. Rather, the two approaches will be used in tandem to complement each other and to minimize the respective weaknesses associated with each.

Integrated Exposures

Because exposures occur over a lifetime from many different sources, and sensitivities to a chemical may vary at different stages in life, the eventual public health goal should be to integrate occupational exposures into an overall strategy for limiting chemical exposures throughout the lifetime of an individual. Such a scheme would likely involve limits at varying ages and from varying sources: diet, air, and water as well as from occupational exposures. Computer models, which incorporate many of these considerations, are already well underway for specific chemicals such as lead. However, progress is likely to be slow and tedious because of the great amount of biological variation, large uncertainties in dose–response data (particularly at low doses) for different substances, and disparate political/economic forces.

Another factor to be considered is chemical interaction. Although synergism, potentiation, and antagonism are known to occur for specific chemical combinations, in current practice, exposures to chemicals which have similar effects are considered to be additive relative to their respective exposure limits, because little information is available to the contrary. This is especially troubling because workers and even the general population are routinely exposed to many chemicals simultaneously, and most commercial products are mixtures of chemicals, not pure substances. Without more specific information, additivity is strictly an estimate of the combined effects and should be viewed with considerable suspicion. In these cases increased observation and monitoring of the physical well-being of the workers and surveillance for the presence of symptoms of exposure should be done in addition to routine exposure monitoring.

22.3 PROGRAM MANAGEMENT

An industrial hygiene program can be divided into four parts: (1) management commitment and planning, (2) hazard identification, (3) hazard correction and control, and (4) training. Each of these is discussed below. Since training is a necessary component of each, it is not treated separately.

Management Commitment

All levels of management in a company must be committed to the establishment and maintenance of a safe and healthful workplace for an industrial hygiene program to function and fulfill its goals. A program must have organizational and financial backing to succeed. Top and middle level managers must establish the program policy and ensure that the company operates in the best interest of the workers' health. They must commit the resources, manpower, time, and money to the full implementation of the program. They must also provide the authority and demand accountability for the health protection of the employees.

Commitment throughout an organization to the program originates with the manager of the industrial hygiene program. The manager must maintain a commitment to the goal of protection of worker health above all other organizational goals. Managers who are perceived as "straddling the fence" are largely ineffective because other managers and workers will not take them seriously. To be most effective, the manager must develop and maintain lines of communication with all levels within the workplace.

The seemingly disparate goals of production and efficiency versus worker protection are not mutually exclusive. The challenge is to find controls that protect workers and also increase the overall productivity and, therefore, the profitability of the company. In fact, controls that interfere with productivity are frequently ineffective at protecting the workers because workers find them inconvenient and circumvent them. The true costs of operating without effective controls includes factors that are frequently overlooked by managers:

- The costs of workers' compensation
- Additional health insurance costs
- Lost workdays and the resulting loss of productivity through inefficiency

For larger operations, a well-trained and highly committed staff of industrial hygienists is also important to the success of the industrial hygiene operation. This group of professionals must fully implement the spirit as well as the letter of the program. These are the people who will carry the message of the program to distant sites. At these distant sites the attitude toward the program will be determined by how the industrial hygienist is perceived. The industrial hygienist should be involved from the beginning with any planned changes in location of processes or installation of new processes. Being involved at early stages of planning can prevent possible overexposures to chemicals and also prevent costly retrofitting of engineering controls.

The success of a program relies heavily on the line supervisor. The line supervisor is in the workplace, sees the operating conditions, and deals with the workers on a routine basis. If the line supervisor does not implement the full measures and intent of the program, it will not work.

The worker is the focus of the entire program and is also involved in the process. Employees are obligated to themselves and their families, as well as to the company, to work in a manner that will not cause illness. However, workers need four things before they can be expected to act in ways that will protect themselves:

- Complete information on the chemicals they use including the effects, how they enter the body, and the types of protective equipment they should wear

- Hazard-specific training to ensure they know how to work with the chemicals in ways which will minimize their exposures
- Proper labeling of chemical containers, including the contents and the potential health hazards, which enables workers to make the appropriate decisions
- Effective engineering controls and proper personal protective equipment

Hazard communication, or "right to know," is considered the most far-reaching standard OSHA has enacted, and, if fully implemented, will substantially increase the knowledge of toxic substances and working conditions. Already, the right-to-know concept has been extended into the community setting, and also, to some extent, into the public sector. However, although this standard has been in effect for over 15 years, failure to comply with this rule is one of the most commonly cited OSHA violations.

Before the rule was in place, labels and MSDSs (Material Safety Data Sheets) were notoriously incomplete, and substantial deficiencies remain today. For example, one recent MSDS limited the toxic effects of a lead compound to "eye and skin irritation." Indeed, since the labels and MSDSs are often prepared by those marketing the toxic substance, there may be a short-term incentive to minimize the degree of hazard as advertised on the MSDS. The opposite also occurs. The preparer includes so much information and always specifies the protective measures for worst-case situations in which the readers have difficulty determining the real hazards and appropriate levels of protection. In spite of these problems, many labels and MSDSs have gradually improved to the point where they are excellent quick reference guides to the current state of knowledge on a particular substance, including the latest animal testing and epidemiological work. Ideally, the labels should be an abstract of what appears on the MSDS, listing the important acute and chronic toxicity information (including target organs), exposure routes, necessary personal protective equipment, and the manufacturer.

The right-to-know concept has been extended to include the more fundamental, and controversial, concept of allowing employees to change the way they do their jobs to reduce a hazard. A Canadian regulation, dubbed the "right to act," extends the right-to-know concept accordingly.

Hazard Assessment

Thus far in this chapter industrial hygiene has been described broadly as the applied science devoted to understanding the interaction between exposure to chemicals and the potential hazardous effects. In the environment the manifestation of adverse health effects is minimized by eliminating or reducing exposure as much as possible. The multistep process of reducing exposures involves several distinct phases, including anticipation, recognition, evaluation, and control of exposures.

Anticipation Anticipation of adverse health effects can be difficult. Successful anticipation usually involves an examination of the production process while it remains in the design phase. However, it can also be applied to modifications of existing process, changes in ventilation characteristics, or increases in production levels. Control measures are often most cost-effective at the stage of new process design, since disruptive retrofitting measures are avoided. Downtime is eliminated, and machinery modifications can be included in space and other resource allocations. Opportunities for substituting a less toxic substance for a more toxic one are also usually more realistic at this stage, because changes in established, successful processes are resisted and capital costs may be high. In fact, substitution of less toxic compounds for more toxic substances is likely an important means for future improvements in both public health and environmental quality.

But how does one anticipate a potential overexposure? The answer is to ask questions of the other disciplines involved in the design work. Engineers, architects, economists, and other planners are typically focused on designing the most cost-effective means of production possible. Since the adverse effects of chemical exposures are often not immediately obvious and stretch out over a number of years, they are usually overlooked in preliminary designs. Changes in process are usually dictated by needed changes in the end product, and the industrial hygienist is called upon only after problems and complaints become apparent.

For example, an electronics component manufacturing company, which was in the midst of building a new plant, intended to use a chlorinated solvent to clean circuit boards as the final step in the production process. An industrial hygienist successfully anticipated the problems in the use of the particular solvent and asked the design engineers to investigate other cleaning options. Another chlorinated solvent of lower toxicity could be used, but the designers concluded that this would not clean the circuit boards effectively. The use of a water-based detergent was an alternative, but was also considered ineffective. Finally, the use of ultrasonic agitation in distilled water was chosen. This involved fitting a rather large piece of equipment into the production line, which would have been difficult had the process been built as originally conceived. In addition, costs were dramatically reduced, since water was much cheaper than the chlorinated solvent. Although the initial capital expenditure for the ultrasonic agitator was large, the cost was recovered in a short time by the reduced cleaner cost. Since no exhaust ventilation was required, heating and cooling costs for the plant were also reduced. In this instance, the industrial hygienist simply asked questions. The industrial hygienist had no knowledge of ultrasonic water cleaning for this particular production process. However, by explaining the need to examine other possibilities, the potential health hazard was successfully prevented.

By asking questions of those with more specialized knowledge, a potential health hazard can be completely eliminated before it appears. All designs and design changes should be reviewed by an industrial hygienist before implementation to avoid having to redress a problem later after exposures and possible injuries have occurred.

Recognition Unfortunately, the opportunity to participate in design formulation remains the exception rather than the rule for health professionals. In practice, health hazards are often recognized in established work settings.

In 1984, the National Research Council (NRC) examined testing needs for toxic substances and estimated that out of 5 million identified chemicals there were

- 48,523 chemicals in commerce
- 3350 pesticides
- 1815 drugs
- 8627 food additives
- 3410 cosmetics

Eliminating duplicates, the total was 53,500. By taking a randomized subset, the NRC stated that in order to form a complete health hazard assessment, further toxicological testing was needed for

- 82 percent of all drugs
- 90 percent of all pesticides
- 95 percent of all food additives
- 98 percent of all cosmetics
- Nearly 100 percent of all commercial chemicals not included above

The prospect of recognizing the hazardous effects of all chemicals appears daunting as the number of identified chemicals continues to grow. The American Chemical Society announced the identification of the ten millionth chemical in 1990. OSHA has set PELs for only about 400 of these chemicals; health standards for specific substances often take 10 years to promulgate, and the agency has been able to promulgate only 13 in its first 25 year history. Thus, neither legal limits nor existing knowledge can always be relied on to protect health. That is not to say existing knowledge should be ignored.

Considerable information on the high-volume chemicals now produced is available. Since OSHA passed its Hazard Communication Rule two useful sources of information are the container label and the Material Safety Data Sheet (MSDS).

In studying a given work setting, a review of all MSDSs and labels on hand should be completed as part of the "recognition" phase; the review should be supplemented by other sources to the extent necessary.

The MSDS and label provide basic information about the potentially hazardous agents in a given workplace, but the adequacy of the information must be assessed. Both managers and workers routinely rely on these sources for a "working knowledge" of what they should and should not do in the performance of their jobs. If the MSDSs and labels are seriously deficient, the employee's and manager's knowledge about the necessity for controlling exposures will also be wanting. Similarly, the link between symptoms and exposure to a particular substance may not be grasped. For example, welders working on galvanized steel may not make the connection between the fever they experience in the evening after their workshift and exposure to zinc oxide, the well-described "metal fume fever."

The recognition phase of the occupational hygiene survey should focus on extensive interviews with employees and managers to uncover any work-related health problems. To help assure candid and truthful responses from both, it is often necessary to conduct interviews with managers and workers separately, with guarantees that the results will remain confidential.

In many respects, the interview is similar to a medical history, in that it must be taken with care, sensitivity, and completeness. One physician has developed a medical history format for use by nonmedical personnel. The need for extensive discussions with employees, engineers, and managers is especially important given the fact that physicians seldom have access to, or fail to fully consider, information on occupational exposures when arriving at a diagnosis.

In the course of these interviews and discussions, both managers and workers frequently present information on work practices and process changes that deviate substantially from the standard operating procedures devised by engineers. In other words, the way in which the work is supposed to be done differs markedly from the way in which it is actually conducted. The changes in procedures may lead to unnecessary exposures. Understanding the employee's reasons for changing the procedures can lead to recommendations for process changes which will increase productivity and reduce exposures.

Final preparation for the visit to the plant includes gaining a general understanding of the production process. Flowcharts, blueprints, and process designs can indicate potential exposure points to specific agents.

Armed with all of the information generated above, a preliminary inspection of the work area, often known as the "walk-through survey," can be undertaken. This is often the first chance to obtain first-hand knowledge of how the work process is actually practiced. Careful observation of each task performed is the key to identifying areas and operations where further study is needed. Short interviews with front-line supervisors and workers during the walk-through survey often help determine the normal operating procedures, as well as the conditions for worst-case exposures. Copious field notes are mandatory for both the walkthrough and the evaluator's survey.

Evaluation After the walk-through survey has been completed, the field notes are reviewed, followed by a period of reflection and analysis. This should result in identification of those areas where exposure monitoring and control system evaluation are needed most.

There are several reasons for quantifying exposures for specific jobs. Documentation of compliance with exposure limits has, for better or worse, emerged as the driving force for much of the exposure monitoring conducted. However, monitoring is also critical in determining whether control measures already in place are effective in keeping exposures as low as possible. Determination of exposure levels is also critical to support epidemiological research. Finally, documentation of exposures often permits an informed scientific response to litigation.

The purpose of conducting exposure monitoring needs to be determined in each case because the strategy and techniques may differ if one is attempting to demonstrate compliance rather than control effectiveness. For example, personal exposure information, determined from air samples taken in the breathing zone of workers, is often needed to demonstrate compliance. Stationary area samples are required to determine compliance with some standards, such as cotton dust, but usually they are located near potential exposure sources to determine sources of exposure with the intent to design the most efficient control systems.

Monitoring is also performed for hazardous physical agents as well as toxic chemicals. Noise, ionizing and nonionizing radiation, and heat and cold can all be monitored and compared with exposure limits.

Today, the most widely-practiced form of exposure monitoring for toxic chemicals is air sampling. This method assesses exposure by the inhalation route only. Personal air samples are collected by attaching a sampling apparatus to the worker so that the concentration of the contaminant inside the individual's breathing zone is quantified. Typically, a small battery-powered air sampling pump is attached to the worker's belt. Tubing is routed from the pump to sampling media, which is attached to the shirt lapel so that it is within a one-foot radius of the head (i.e., the "breathing zone"). Figure 22.1 shows such an arrangement.

The choice of sampling media, flow rate, and duration of sampling depend on the substance to be collected and how it is used. The specific choice of sampling media needs to be made under the supervision of an experienced industrial hygienist and in consultation with the analytical laboratory. Certain conditions on the day of the survey, such as temperature and relative humidity, may affect the choice of sampling media.

Usually, the sampling media is analyzed to determine the total mass of the contaminant present (in milligrams). Since the pumps are set at a calibrated flowrate, the total volume of air sampled is known (in cubic meters). The exposure results are then reported in either milligrams of analyte per cubic meter (mg/m^3) of air or parts per million (ppm) by volume, both of which are time-weighted average (TWA) concentrations for the period of time the sampling was conducted.

One purpose of the walk-through survey should be to determine how many personal and/or area air samples need to be collected to adequately characterize exposure patterns. The exposed employees should be assigned to homogeneous exposure groups, and then the exposure patterns of the groups should be assessed. Statistics provides guidance on how many air samples need to be collected in order to achieve stated degrees of confidence. For example, when exposure levels are near the exposure limit and the variability in the exposure during a day is large, many samples may be needed to be assured that the exposure is less than the exposure limit with 95 percent confidence. If smaller numbers of samples are collected, the 95 percent confidence interval is larger, and may include the exposure limit within the range, so there is less confidence that the actual exposures or concentrations are below the exposure limit.

Unfortunately, statistical analysis of air sampling data is an often-neglected exercise in industrial hygiene. One reason for this has to do with cost and available resources; it is often simply too time-consuming and expensive to collect the large numbers of air samples necessary to minimize the standard errors of the measurements, particularly when variability is high. Another reason is that variations between days is typically greater than variations within a day. Therefore, multiple days of sampling may be necessary to fully characterize exposure patterns. One potential way around this quandry is to take extraordinary steps to ensure that the samples that are collected are representative. A conservative (i.e., health-protective) definition of a representative sample is one, which is collected under the "worst-case scenario," but realistic conditions. Thus, "representative" should not be confused with "average." If exposures have been measured under a worst-case scenario, some degree of confidence can be achieved that exposures during routine operations are not higher. Determining what constitutes a representative sample is often a most difficult task. The results of the interviews are especially helpful in arriving at a working definition of representative conditions. Of course, an employee should never be intentionally overexposed for the purposes of collecting an air sample.

Figure 22.1

Consider the example of an employee working 3 feet away from an open container of a solvent. On questioning, the employee states that he uses 5 pints/day on Mondays and Fridays, but only 2 pints/day on the other days of the week. Clearly, if sampling were conducted on Tuesday, it cannot be considered a "worst-case scenario." Similarly, if the sample is collected directly over the top of the container, it cannot be considered "realistic," since the employee does not inhale the air directly over the container. Thus, personal air samples should be collected on a Monday or Friday to document compliance with government standards. However, suppose the container is equipped with a slot hood exhaust system to prevent vapors from escaping into the workplace. Then we may wish to collect an area air sample directly over the top of the container to determine the effectiveness of the control system. In the latter case, we are measuring the degree of control present, not the employee's exposure.

To this point we have described the choice of sampling strategy in terms of area versus personal sampling, statistical considerations, and what a "representative" sample means. Sampling techniques, to be distinguished from sampling strategy, fall into two broad categories: direct-reading instrumentation and laboratory-based analytical methods.

Direct-reading instruments (also known as *real-time analysis*) provide an immediate measurement of levels of the given air contaminant. They are widely used in confined space entry programs, emergency response situations and hazardous-waste sites, and as supporting evidence for the laboratory-based analytical methods. Examples of direct-reading instruments include electrochemical cells, metal oxide cells, infrared gas analyzers, portable gas chromatographs, and detector tubes (also called *length-of-stain colorimetric tubes*). While some of these instruments can be worn by employees to obtain an 8-h TWA, most are used to conduct area sampling. They are excellent for determining leaks and other sources of air contaminants. They usually need to be calibrated with a test gas of known concentration immediately before and after use. The chief disadvantage of this family of techniques is that the minimum level of detection is much higher than the laboratory-based techniques. This is due to the availability of more sensitive equipment in the laboratory and the ability to "preconcentrate" the contaminant onto some type of sampling media.

Sampling techniques that require laboratory analysis are most often employed where low limits of detection are required and when integrated exposure information is acceptable. For example, activated charcoal is used to capture organic solvents, which are then analyzed by desorbing the charcoal in carbon disulfide or another appropriate solvent, taking an aliquot of the solution, and injecting it into a gas chromatograph for quantification. Other forms of laboratory analysis include ion chromatography, atomic absorption (for metal dusts), X-ray diffraction (for silica), inductively complex plasma emission spectrometry, mass spectrometry (particularly useful for identifying unknown air contaminants), scanning electron transmission microscopy (for asbestos), and a host of others.

One final form of air sampling relies on passive diffusion, instead of active pumping of air through detectors or media. These devices are often marketed as "badges" or "tubes," and can rely on colorimetric techniques or postsampling laboratory analysis. They are relatively easy to use and inexpensive. However, they are susceptible to error in areas of excessive air turbulence. Additionally, they may fail to record short bursts or peak exposures due to the longer time interval required by diffusion. Few have been validated independently.

The final stage of evaluation involves the analysis of interviews, observation of work practices, and air or physical agent monitoring results. In its simplest form, evaluation means comparing the measured exposure levels with OSHA PELs or the current TLVs®. At best, this only demonstrates possible compliance with OSHA PELs. At worst, it can provide a false sense of security because exposures below published exposure limits are not guarantees that an individual will not suffer adverse health effects.

Thorough evaluation includes an analysis of operating procedures, work practices, and "the numbers" to arrive at recommendations for reducing exposures to the lowest feasible level. For example, air monitoring of a parts cleaning operation might indicate that vapor concentrations inside workers' breathing zones are below the TLV® for the solvent. To some, this may suggest that no further measures are needed. To a competent industrial hygienist, however, the observations that the parts are much smaller than the tank and that workers always leave open the lid on the solvent tank suggests simple changes in design and work practices that could lower exposure levels. By splitting the lid so that the size of the lid is only as large as needed for the largest parts, keeping the other side closed at all times, and by instructing employees to close the lid when not in use, exposure levels can be reduced substantially at very little cost. In short, even though the level was below published exposure limits, control measures should still be specified based on direct observation of job performance.

Control Several types of control measures have already been described earlier in the discussions of anticipation, recognition, and evaluation. In fact, the choice of various control measures is usually a logical extension of the earlier efforts.

Important compromises are frequently made when workable solutions are finally identified. Truly effective control measures are those which do not excessively interfere with the performance of the job to the point where workers refuse to use them, and that are not so costly that they threaten the fundamental viability of the work process. In some cases, however, an outright ban of the final product may be warranted. In other words, some processes may seem so dangerous that they cannot be used safely even with extensive control measures in place. The United States' experiences with asbestos (now banned from nearly all commercial use), nuclear power plants (no new power plants have been ordered for a number of years), and chlorofluorocarbons (CFCs, which destroy the protective layer of ozone in the upper atmosphere) are all cases where current control technologies were perceived to be inadequate to permit further use.

In practice, the family of control measures has been relatively well described and proved. They have come to be known as the "hierarchy of controls," because some are more preferable than others. The controls are listed below in order of desirability, although in practice, a combination of control techniques is often used.

1. Substitution
2. Process modification
3. Source isolation
4. Worker isolation
5. Local exhaust ventilation
6. Dilution ventilation
7. Work practice modification
8. Administrative controls
9. Personal protective equipment (e.g., respirators)

Substitution and Process Modification Substituting a less toxic substance for one that is more toxic, or perhaps altogether doing away with the need to use a toxic substance, is the best of all possible solutions. The degree of hazard is reduced or eliminated, and further controls may not be needed. Examples of this type of control include:

- The use of water-based fountain solutions instead of isopropyl alcohol for printing operations
- The substitution of nonleaded gasoline for leaded gasoline
- The development of more efficient polymerization reactions to reduce (and practically eliminate) the offgassing of vinyl chloride monomer in the production of polyvinyl chloride plastic
- The substitution of cellulose and bimetallic compounds for former asbestos insulation and brake applications
- The use of toluene or xylene instead of benzene in certain solvent applications

One note of caution: Substitution may result in simply exchanging a known hazard for an unknown one. For example, some firms have substituted glutaraldehyde and some quaternary ammonia compounds for formaldehyde disinfection applications. However, the substitutes have not been well researched. As of this writing, no chronic toxicity testing has been done for glutaraldehyde, and only subacute testing has been completed for the quaternary ammonia compounds.

Process modification may range from the simple to the complex. One manufacturer used a chlorinated solvent to wipe grease from newly extruded plastic buckets. By adopting a more rigorous preventive maintenance program to correct grease and oil leaks when they first appeared, the cleaning step became largely unnecessary. A refrigerator manufacturer flame-sealed the copper tubing of the closed-loop refrigerant system after the CFC was added. The flame generated CFC decomposition products such as phosgene and hydrogen chloride. The process was modified so that the refrigerant was added after the system was flame-sealed, and exposure was eliminated. On the complex side, the

U.S. Air Force has been conducting research on the use of robotic lasers to strip paint from its aircraft to replace the current method of using methylene chloride–based chemical strippers.

In the late 1990s, some legislation was passed at the state level requiring industry to reduce the use of toxic substances on a certain schedule. Whether these targets are realistic remains to be seen. However, these initiatives could radically alter the prevailing conception of what is and is not feasible in the development of new processes and new substitutes.

Source Isolation Source isolation can effectively prevent exposure, and thus reduce the hazard. A potential hazard still remains, however, since a leak can develop and result in exposure. The Union Carbide incident at Bhopal, India in 1993 may be the best-known example. Backup alarm systems warning of a breech in the isolating mechanism are usually specified for this type of control measure, and emergency response procedures must be established. Specific requirements for process safety management for processes which use highly hazardous chemicals have been established by OSHA.

Enclosures with their own dedicated exhaust ventilation systems are another type of source enclosure. In poultry hatcheries, formaldehyde is used to disinfect the eggs, which are isolated inside an incubation chamber. Planning must include specification of those work processes that will require entry into the enclosure, and the types of protective measures, such as the type of respirator to be used, and how much ventilation of the enclosure will be needed after production has ceased before entry can be safely completed. Maintenance activities and emergency-response operations often require entry into such enclosure systems. In the case of the hatchery, at least 60 min of purge time at 20 air changes per hour or greater must be allowed before a worker enters the booth. To provide adequate mixing within the booth, velocities through air inlet doors should be at least 500 feet per minute (fpm), and velocities through large access doors should be at least 100 fpm.

Worker Isolation If the toxic substance cannot be isolated, then perhaps the worker can be. For example, it is not feasible to enclose a large railcar coal dumping station at a large power plant. However, a small operator's booth, with its own filtered source of fresh, tempered air, is certainly feasible. Again, plans should include provisions for emergency exit from the enclosure into the hazardous atmosphere.

Many worker isolation enclosures fail because they do not provide the necessary comfort. If heated or cooled fresh air is not supplied to the extent necessary, the employee is likely to open the door, resulting in potential overexposure. These provisions are sometimes regarded as an unnecessary "creature comfort" expense. In reality, they need to be viewed as an integral part of the control mechanism, for without it, the control fails. Even if employees are trained in the hazards of compromising the enclosure's integrity, on hot summer days the worker may view the immediate problem of baking inside the enclosure as worse than being exposed to an undetectable toxicant. This is a flaw in the design, not "human nature."

Local Exhaust and Dilution Ventilation If it is not feasible to substitute for the chemical, modify the process, or prevent the release of the air contaminant into the workers' environment, then local exhaust or dilution ventilation will be needed. Of the two, local exhaust ventilation is more desirable, because more complete capture of the contaminant is possible in most cases and smaller amounts of air will need to be moved, resulting in energy cost savings.

Local exhaust ventilation systems consist of a hood, ductwork, a fan, a pollution-control device, and an exhaust stack. Hoods can be of the external or internal variety. The external hood is designed to capture air contaminants released some distance in front of the hood, while the internal hood controls the contaminant inside some type of partial enclosure. Figure 22.2 shows a welding fume extractor and a laboratory hood, illustrating the differences between the two types of hood. Generally, a greater degree of control can be exerted by an internal hood, since the chances of cross-drafts and other sources of turbulence are less. However, even the best designed laboratory hood will exhibit some degree of leakage.

Training of employees in the proper use of ventilating equipment is crucial. Laboratory workers need to understand the consequences of placing too many bottles inside their hoods or not placing the

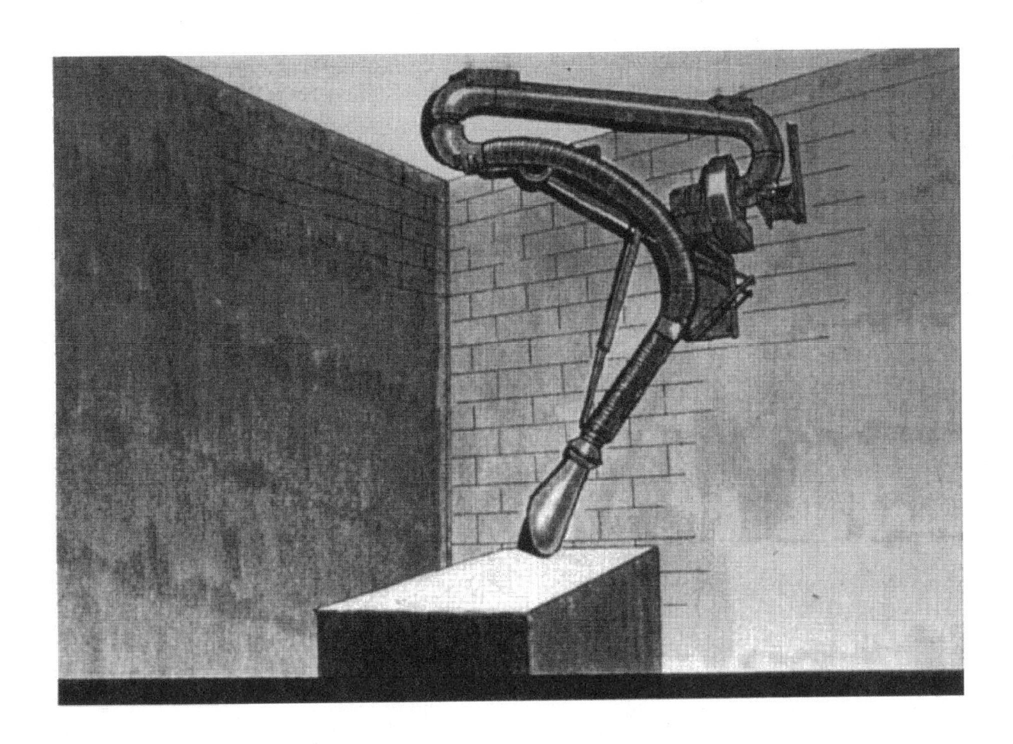

Figure 22.2.

reaction vessels at least six inches behind the hood face. Similarly, welders need to know that if they place the fume extractor too close, their shielding gas will be exhausted, resulting in an unacceptable weld; if they place it too far away, it will not prevent welding fumes from entering their breathing zones. In both cases, on-the-job training is essential. Preventive maintenance is also critical to the continued effectiveness of ventilation systems. Fans, bearings, and belts wear out with use; collectors become full and flow rates are reduced. Scheduled routine checks of the overall system and system component function is necessary to ensure the continued effectiveness of the system. Some components have lifespans that are known fairly closely. These components should be replaced at regular intervals before they fail.

If local exhaust ventilation is not feasible, then dilution ventilation (also called "general" ventilation) can be employed. Here, the contaminants are not actively removed from an employee's breathing zone, but diluted to acceptable levels. In order to be effective, dilution ventilation can only be used under the following restrictions:

- The substance is of relatively low toxicity.
- The amount used is small and is used uniformly throughout the day.
- The worker is some distance away from the source.

In many cases, the rate of use of particular substances is not uniform and is difficult to quantify. Formulas to determine the amount of dilution ventilation use an arbitrary safety factor, and the selection of the appropriate safety factor is sometimes quite difficult. Nevertheless, dilution ventilation has numerous applications and is probably the most widespread method of controlling air contaminants.

Most office buildings employ general dilution ventilation. Many cases of poor indoor air quality have been found to be caused by low rates of dilution ventilation (i.e., insufficient fresh air). Dilution ventilation is also useful in control of hot environments, confined spaces, and nuisance odors.

Specification of airflow rates, hood shape, and other factors have been developed for a number of processes.

Administrative Controls Administrative controls describe a family of measures that reduce exposures through planning and allocation of appropriate resources. One example of administrative controls involves planning for hazardous work to be completed during off-shift work hours in order to minimize the number of people potentially exposed. Another involves rotating workers in and out of hazardous areas so that 8-h TWA exposures remain low. In this second example, more people are exposed, but to lower levels. This may or may not be an improvement. For acutely toxic compounds (i.e., those with ceiling exposure limits), there is no advantage in rotating workers. If more sensitive individuals experience adverse health effects due to exposure that they would not have had otherwise, then the job rotation plan would have to be judged a failure. Job rotation can also be a tremendous burden to manage effectively. Ensuring that workers spend only the allowable time in an exposed job can be difficult. Usually a large number of workers must be available for the rotation plan. If workers are not at work because of vacation or sick leave, the number of workers available may not be enough for the rotation plan. For these and other reasons, administrative controls rank among the least preferred.

Respirators and Other Personal Protective Equipment While widely used, this family of controls is generally the least effective. Although respirators are acceptable for temporary work sites and for routine short-term exposures, OSHA does not regard the use of respirators to be a permanent means of complying with air contaminant exposure limits. The basic reason is that once the toxic agent gets past the personal protective equipment, it is available for absorption, rendering further controls ineffective. In short, these controls represent the last line of defense.

The use of respirators requires a company to implement a complete respiratory protection program. A respiratory protection program includes

- Selection and use of appropriate respirators

- Physical assessment and fit testing of affected employees
- Employee training in proper use and limitations of respirators
- Respirator storage
- Respirator inspection, cleaning and disinfection, and replacement as needed
- Regular inspection and monitoring/surveillance of work area conditions and employee exposure and stress
- Periodic program evaluation

To effectively administer a comprehensive program requires a substantial amount of time, technical competence, and money. Some program elements are a major concern, such as respirator fit and medical evaluation. Respirator manufacturers began producing multiple-size facepieces available and soft rubber facepieces only in the late 1990s to accommodate the variety of sizes and facial shapes. Respirators must be fitted to each individual either in quantitative or qualitative tests, which have used established protocols involving the use of challenge agents such as corn oil or irritant smoke. Finally, some individuals are unable to wear respirators at all, because of facial shape or a preexisting medical problem, such as dentures or heart disease. A particularly sticky concern associated with wearing respirators is that there are no definitive medical tests, which can lead a physician to determine whether a worker can wear a respirator without suffering adverse health effects.

In spite of all these limitations, respirators have saved many lives. They must be used when other feasible methods of control have been exhausted. And of course they are basic to emergency response procedures where there is no time to use other controls, or when one is facing an unknown situation that requires immediate action.

The same basic principles apply to other forms of personal protection: they all increase the stress burden on workers, but with proper training, they can form an effective barrier between the toxic substance and the body. For example, protective clothing can cause heat stress unless employees are provided with additional breaks and water intake. Face shields can protect the eyes and face from corrosive substances, but also reduce visibility. Gloves protect the hands, but at the cost of reduced dexterity. Hardhats and steel-toed shoes provide protection from safety hazards, but also add weight that the worker must carry around during the workshift.

All forms of personal protection require expert selection. The correct respirator cartridge must be used for the particular chemical, concentration, and physical state. For example, a dust filter will provide no protection against gases or vapors. Protective clothing that is not impermeable to the chemical can actually worsen skin exposure by providing a continuous reservoir of unevaporated liquid directly in contact with the skin. Guidance in the selection of appropriate protective equipment may be available on the MSDSs. Other sources include the NIOSH *Pocket Guide to Chemical Hazards* and *Guidelines for the Selection of Chemical Protective Clothing*.

Summary of Control Technologies In any given situation, it is unlikely that any single specific control technique will prevent exposures to toxic substances. The challenge in occupational hygiene is to arrive at a logical blend of all the alternatives outlined above to provide a workplace free of recognized health hazards and at least a good head start on those occupational health hazards we have yet to discover.

22.4 CASE STUDIES

Herbicide Application in the Forest

The first case study involves the exposure of herbicide applicators in the management of forests. This example demonstrates the bridge between industrial hygiene and toxicology. It also demonstrates that exposures occur from both inhalation and dermal routes, and that industrial hygiene is a valuable tool in the risk assessment process.

Herbicides are routinely applied by foresters as a management practice to maximize yields of crop species. Weed species that compete against the crop species for nutrients and space within the developing forest are eliminated by the application of herbicides.

Toxicology studies in the laboratory were performed to describe the oral and dermal pharmacokinetics of the herbicide as well as to determine a skin absorption factor for the herbicide. Volunteers ingested known amounts of the herbicide in juice, and the portion of the total dose that was excreted in the urine was determined. The time it took to excrete the material was also noted, and half-lives were calculated. A correction factor to account for the uncollected herbicide was also calculated. Known quantities were applied to the skin of volunteers for 8 h, and then washed off. Their urine was collected for 3 days after exposure and analyzed to determine the total amount that was excreted in the urine. The correction factor was applied to the amounts found in the urine to determine the total amount absorbed.

A variety of toxicological studies looking at various organs and biochemical pathways were also completed, and the most sensitive effect was competition with an organic acid excretion pathway. The no-observed-effect level (NOEL) was determined as 2.5 mg/kg per day.

To assess the occupational risks associated with use of the herbicide, a field study was initiated to determine exposure and dose to the herbicide. The study was conducted at four different sites, and five or six workers were monitored for both inhalation and dermal exposure at each site. The workers also collected their urine for 5 days: the day before the survey, the day of the survey, and 3 days after the survey. The collected urine was analyzed for the herbicide, and the total dose was estimated.

For application, workers mix the herbicide into backpacks which can be slightly pressurized by a hand pump. The herbicide mix is sprayed onto vegetation through a wand several feet long, or through a gunjet which looks like a small gun, releasing herbicide under pressure by squeezing the trigger. The spray pattern and size of the droplets is controlled by the tip inserted into the opening at the end of the applicator. When the herbicide is released, some of it remains airborne, causing some exposure by inhalation and deposition onto clothing and skin surfaces.

While applying the herbicide, applicators must walk near and sometimes brush against the vegetation that has already been sprayed, exposing the skin. Since many of the herbicides are formulated to penetrate through the waxy surface of the leaves, they are lipid-soluble, and, therefore, can more easily penetrate the skin.

Evaluation of the potential inhalation exposure was fairly routine for the herbicides. An air-sampling pump was worn by each worker for the duration of the mixing and application period, and the pump pulled ambient air from the worker's breathing zone through two filters in series: a mechanical filter to collect the mist and a solid sorbent to collect vapors.

Dermal exposure evaluation was not as straightforward, however. The workers wore patches on their clothing that were intended to absorb the herbicide they contacted. The patches were placed to represent various portions of the body. Some additional patches were worn under the clothing to determine the penetration through the clothing. In addition to the patches, hands were washed in a soap solution and rinsed with water. The soapy wash water was collected to determine dermal exposure to the hands.

The amount of herbicide that actually penetrated the skin (the dermal dose) was estimated by

- Adding the surface deposition on exposed body areas (DE), such as the face, hands, and neck, to the product of the deposition on clothed body surfaces (DC), such as the chest, arms, and legs, and the average clothing penetration (CP)
- Multiplying this sum by the skin absorption factor (SAF), the fraction of material that impinged on the skin that was expected to penetrate the skin.

$$\text{Estimated dermal dose} = [\text{DE} + (\text{DC} \times \text{CP})] \times \text{SAF}$$

The study was designed to minimize exposure by the ingestion route; workers were required to wash their hands before eating, smoking, or putting a plug of chewing tobacco in their mouths. However, dose from the ingestion route was estimated by subtracting the estimate from a dermal absorption pharmacokinetic model from the corrected amount found in the urine and subtracting away the estimated inhalation dose.

In addition to the herbicide, creatinine was measured in the urine samples to assess whether volunteers collected all their urine. Creatinine is a byproduct of metabolism and is excreted fairly uniformly from one day to the next by the same individual. Excretion levels will vary between individuals. The herbicide was quantified in each of the samples and then fitted to a pharmacokinetic model developed from the laboratory study. The model estimate of dose and the sum of the amounts found in the urine samples were compared, and the higher number was used for further calculations.

A number of differences had been noted at each site. The first application site had low growth, and the undergrowth was sparse. The second site had moderate vegetation. The third site had much higher vegetation and was fairly dense. The fourth site was similar to the second site.

Another difference between sites involved the method used to mix the herbicide. The workers at the first and fourth sites mixed the herbicide and water into a 50-gallon container (nurse tank). Backpacks were filled by gravity feed through a hose. This allowed only one worker to handle the concentrate, and he had to handle it only once. Workers at the second site mixed by adding the concentrate into each of the 3-gallon backpacks at each fill-up and handled the concentrate each time the backpacks were filled (about 6 times per day). The third group mixed into a 5-gallon container and then poured from the container into the backpacks; they had to handle the concentrate at each fill-up, and then had to pour the mix into the backpacks.

The equipment used at the three sites may also have affected exposure. Workers that used piston pumps, which generate a higher pressure inside the backpack, had a higher rate of leaks at seals, and clothing was soaked with herbicide. Their exposure levels were greater than those of workers who used diaphragm pumps, which generate a lower pressure. None of the diaphragm pumps leaked. The workers at the first site used wands to apply the material, while gunjets were used at the other two sites. The potential for exposure appeared to be much greater using the gunjet since it was held much closer to the body when sprayed. During walking, it could rub against the thigh where droplets from the tip could soak into the clothing. The gunjets also malfunctioned more often so workers had to handle them while contaminated with herbicide. A comparison of these and other factors is presented in Table 22.2. Of particular interest is that lack of training appeared to substantially increase exposure levels.

TABLE 22.2 Comparison of Herbicide Doses Across Sites

Source	Number	Geometric Mean of Dose (g)
Backpack		
Diaphragm	19	0.09
Piston	3	2.34
Handle concentrate		
Yes	11	1.26
No	11	0.93
Use tobacco		
Yes	11	0.93
No	11	1.35
Regularly apply		
Yes	20	0.91
No	2	8.71
Applicator		
Wand	13	0.71
Gunjet	9	1.51

The urine results for each individual were evaluated for complete collection, and a predicted course of elimination determined. The results from a typical individual are given in Figure 22.3. The margin of safety (MOS) was calculated from the biomonitoring data, as follows:

$$\text{MOS} = 2.5 \text{ mg/kg·day} \times \text{body weight (kg)} \div \text{dose from biomonitoring data (mg/day)}$$

The geometric means of the calculated margins of safety for each site are presented in Table 22.3.

From these results, a number of recommendations were made. Work practices were instituted requiring the use of a "nurse tank" to minimize the time and number of workers exposed to the concentrated herbicide and the mix. Workers handling the concentrate were required to wear a glove that had low permeability to the herbicide. Applicators were required to wear the same type of gloves while applying in the field and loading their backpacks. Workers were required to have a change of clothing available to change into if their clothes became contaminated with the mix or concentrate. It was not clear that the wands provided greater protection than the gunjets since there was only a small difference between sites 1 and 2. The clearest difference between exposures were the sites themselves. Directed foliar application with backpacks should be done only when the height of the vegetation is generally less than 6 feet tall.

Health Care Facilities

Ironically, the places we go to get well, hospitals, have many potential health hazards associated with them. These include

- Biological hazards such as AIDS, tuberculosis, and hepatitis

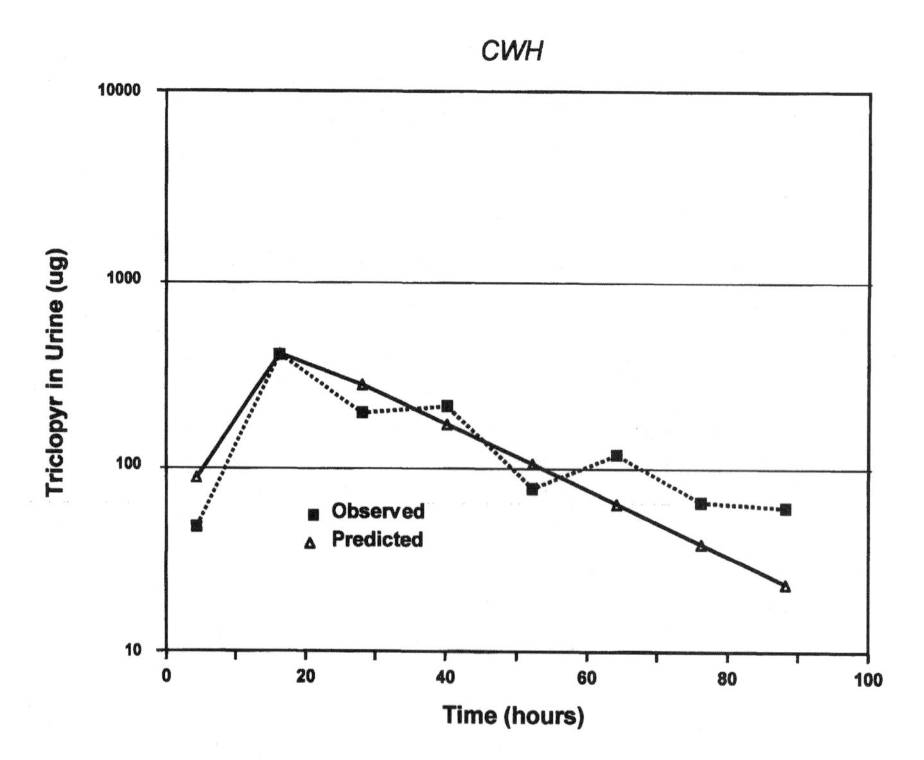

Figure 22.3.

TABLE 22.3 Margins of Safety at Forest Sites

Location	Margin of Safety
Site 1	296
Site 2	209
Site 3	61
Site 4	183

- Ergonomic problems
- Antineoplastic drugs
- Formaldehyde
- Waste anesthetic gases (nitrous oxide and fluorinated hydrocarbons)
- Ethylene oxide (a gas used to sterilize certain instruments)

In a survey of one hospital operating suite, nitrous oxide was monitored with a miniature infrared analyzer (MIRAN). The TLV® for nitrous oxide is 50 ppm 8-h TWA and the NIOSH REL is 25 ppm (operating procedure TWA). In the operating rooms themselves, the concentrations were kept well below these levels. Only once, and for a very short period, did the level rise above 50 ppm. The hose connected from the nitrous oxide wall mount receptacle to the gas-mixing unit was accidentally kicked free. Concentrations in the room quickly went to 100 ppm. As soon as the levels went up, the hose was noticed, replaced, and levels quickly returned to previous levels. The low levels in the operating rooms were expected, since they have 15–17 air changes per hour, and 100 percent of the air is fresh, sterile air. However, in the recovery rooms, the situation was not as well controlled.

In the recovery room, patients who were anesthetized exhale nitrous oxide–laden breath. Since nitrous oxide is not very soluble in blood, it quickly comes off from the blood in the lungs. Nurses must frequently bend over the patient's head to talk with them and assess the conscious level of the patient, which places their breathing zone in the breath exhalation area of the patient. Concentrations in the exhaled breath of the patients were measured up to several thousand parts per million.

General recommendations were made to minimize exposure by bringing as much fresh air as possible into the recovery room. Because patients already were complaining of feeling cold (in part due to the anesthesia), it would be expensive to condition all the air that enters the room during the winter. Another recommendation was to locate a local exhaust duct near the head of the patient to remove nitrous oxide from around the head. A difficulty is that many patients just coming out from under sedation would not understand and recognize a duct near their heads, which could cause additional stress.

In the hospital operating room example an additional condition that allowed for low-level exposure was that nitrous oxide was delivered by placing a tight-fitting mask over the patient's mouth and nose, or anesthetic could be delivered via intra-tracheal intubation, thus minimizing leakage at the point of delivery.

Dental operations do not have this luxury. A study of over 30 dental offices revealed that levels of nitrous oxide exceeded 50 ppm by wide margins. Several reasons were observed. The pipes in the nitrous oxide delivery system frequently leaked, the mixing/delivery units were overpressurized and leaked, and the scavenging system was overwhelmed by the delivered volume of gas and exhalation of the patient.

The recommendations to reduce exposure include minimizing the use of nitrous oxide to patients that truly need the sedation it affords; its routine use should be eliminated. The delivery flow rates should be reduced to the minimal effective flow rate (which varies from one patient to another). A dam should be placed in the back of the mouth to minimize the nitrous oxide, which is exhaled through the mouth. The scavenger flow rate should be set at a level, which would effectively remove nitrous oxide as it is exhaled. The scavenger flowrate should also maintain a slight negative pressure inside the deliver/scavenging nosepiece. The nosepiece should seal against the face of the wearer to minimize loss from around the face to facepiece seal. A prototype local exhaust ventilation system was developed for use in the dental operatory. With proper positioning it was found to reduce exposure levels to nitrous oxide to below 25 ppm (procedure TWA). Further studies indicated that positioning of the ventilation system is critical. The total volumetric flow rate must be kept to less than about 600 cfm (cubic feet per meter) to prevent backflow down chimneys in home-type offices, while the capture velocity must be kept above 80 fpm at the mouth to efficiently capture nitrous oxide emitted from the mouth.

Fiberglass Layup Operation

Many of the boats, golf carts, and even some of U.S. cars are made from fiberglass, which reinforces a polystyrene outer coating. The process involves spraying a polyester resin in a styrene monomer solvent, called the *gel coat*, onto a mold. The spray gun mixes the resin/solvent at the tip with MEK peroxide, a catalyst used to hasten the polymerizing reaction. The gel coat is allowed to dry and then fiberglass is applied along with a mixture of polyester resin and styrene monomer, MEK peroxide, and acetone (to liquify the surface and allow the fiberglass mixture to adhere to the gel coat), through a "chopper gun," which is much like a spray-paint gun, except that it has an additional blade, which rotates around the opening to cut the long fiberglass strands. After the fiberglass mixture dries, the product is separated from the mold, and surfaces are smoothed by sanding. Some pieces require actual cutting and grinding to provide the proper fit.

In one such operation, the gel coat sprayer was located in a separate room with an exhaust fan located in the wall. The sprayer wore a full-body covering, synthetic rubber gloves, and an air-supplied hood while spraying gel coat. His 8-h TWA exposure to styrene vapor outside the hood was 82 ppm, which was below the current OSHA PEL and the then-current ACGIH TLV® of 100 ppm. However, at the time the ACGIH was in the process of lowering the TLV® from 100 to 50 ppm, with a STEL of 100 ppm. An evaluation of the wall fan indicated that it had an average exhaust rate of 1000 cubic feet per minute (cfm) and produced 17 air changes per hour, which should have been more than adequate to reduce levels in the room to well below 50 ppm. However, when the air flow patterns in the room were observed with smoke tubes, the majority of flow into the room was observed to come from a doorway at the end of the room, which did not dilute the air in the sprayer's breathing zone; the flow of fresh air was short-circuited.

Recommendations for this area included moving the wall fan to an area farther from the side door and closer to the spraying area, which would place the sprayer between the front door and the fan; closing off the lower part of the side door to increase resistance of air movement in that direction; and instructing the sprayer to stay between the fan and the front door, spraying toward the fan so that the overspray would not be pulled back into his breathing zone. The company decided not to accept the recommendations, since the operator was wearing a supplied air respirator.

Airborne exposure of the three chopper gun operators ranged from 65 to 103 ppm. Exposure variation was caused by the individual work practices and the location relative to the vane axial fan located on one side of the spray area, which workers used to cool themselves. An exhaust ventilation system with a single slot about 3 feet from the floor was in place at the back of the layup area. Observations with smoke tubes indicated that at more than one foot from the slot there was virtually no air movement attributable to the ventilation system. Spraying was done eight feet from the ventilation hood, so there was little chance that air contaminants were being moved out of the employee's breathing zone. Visualization of the air movements with the smoke tubes indicated the airflow was across the work area from the floor fan toward an open window. The workers nearest the window were exposed to the higher levels. In addition to the air exposures, one chopper gun operator was observed to have reddened hands, apparently caused by the defatting action of acetone and styrene on the hands. Further investigation revealed that acetone was used to clean her hands of overspray material.

To reduce exposures in the layup area, the ventilation system was remodeled. A larger fan was installed to increase the airflow, the hood was angled to have greater width at the bottom than at the top, and numerous slots with smaller widths at the top than at the bottom were placed in the hood to make the airflow more uniform throughout the work area. The floor fan was removed to eliminate cross-flow, which would increase turbulence and reduce the capture efficiency of the hood. Workers were instructed and required to wear synthetic rubber gloves to minimize hand contact with the styrene and reduce the need to use acetone to clean the hands. Exposures in the area had been reduced to less than 40 ppm as a result of these measures.

Exposure to Carbon Dioxide in a Meat-Processing Industry

The rapidly growing demand for meat products in the fast-food industry has resulted in an increased use of dry ice (solid carbon dioxide) in many meat-processing plants. Contrary to conventional wisdom, carbon dioxide is not a harmless substance. It can cause a variety of health problems at relatively high exposure levels. If exposures are high enough, the results can be fatal.

The toxicity of carbon dioxide is fairly well established. It has been classified as both a stimulant and depressant of the central nervous system, an asphyxiant, and a potent respiratory stimulant. Rapid breathing, increased heart rate, headache, sweating, visual disturbances, convulsions, and death are among the symptoms related to carbon dioxide overexposure. The gas can be weakly narcotic at 30,000 ppm, and intoxication can be produced by a 30-min exposure to 50,000 ppm. Because of the extreme sensitivity of various chemoreceptors to CO_2, its high solubility in tissue fluids (20 times that of oxygen), and the permeability of the blood–brain barrier to CO_2, the effects on the respiratory and central nervous systems are rapid.

Carbon dioxide poisonings have been reported in aircraft transporting frozen food, meat-processing plants, farm silos, fermentation tanks, shipping, mining, and firefighting. Both the OSHA PEL and 1993-4 ACGIH TLV® are 5000 as an 8-h TWA. The "immediately dangerous to life and health" level set by NIOSH is 50,000 ppm.

This study describes an occupational hygiene study in three different meat-processing plants, which used dry ice to refrigerate packages, and documents how a change in production techniques (i.e., the increased use of dry ice) resulted in a significant health hazard and how the hazard can be controlled. Preliminary interviews with workers and managers revealed that several workers had been hospitalized for dizziness, hyperventilation, vomiting, and headaches. The interviews also revealed that the amount of dry ice used from one day to another varied greatly. Scheduling the full-day survey so that representative worst-case exposure levels would be obtained proved to be difficult.

Levels of exposure to carbon dioxide were initially determined with short-term detector tubes. Care was taken not to include exhaled air while sampling inside workers' breathing zones, since exhaled breath can contain as much as 59,000 ppm carbon dioxide. Normal outdoor air contains about 350–400 ppm of carbon dioxide.

The short-term detector tubes were used to determine where full-shift samples were needed. The 8-h TWA was determined using a bag sampling procedure and gas chromatography (NIOSH Analytical Method 5249). A previous attempt to measure TWA exposures using long-term detector tubes showed that the color change was not distinguishable from the background color of the medium. Therefore, long-term detector tubes were rejected as an analytical method for this study. The manufacturer of the long-term detector tubes was notified of the findings.

Samples were collected in Tedlar bags at a nominal flow rate of 20 cm^3/min. The concentration in the bag was determined in three ways: (1) short-term detector tubes were used to measure the concentration of CO_2 inside the bag, (2) an aliquot from the bag was transferred to a vacuum sampler and then shipped to the laboratory (this was considered necessary because of the possibility of bag breakage during shipment to the lab), and (3) finally, the bags themselves were shipped and analyzed. On return, the bags were checked for leaks, which were found to be common.

Generally, the direct analysis of the bags in the laboratory gave the lowest results, perhaps because of leakage during shipment. Laboratory analysis of the Vacu-Sampler cans gave the highest results. On-site analysis of bag air using short-term detector tubes gave results that were only slightly less than those of the cans. In short, acceptable TWA sampling results can be obtained at low cost by using bags followed by on-site analysis using short-term detector tubes.

The results of the sampling in all three plants showed that the highest concentrations were found in the holding coolers, and that they can exceed the IDLH level of 50,000 ppm. In one plant, workers spent nearly the entire shift working on a palletizing operation inside a relatively large holding cooler. In another plant, entry into a smaller holding cooler was confined to forklift drivers, whose time inside was relatively short. Ventilation in both areas was relatively poor, since incursion of fresh outdoor air was minimized to maintain proper refrigeration.

The rate of generation of carbon dioxide gas from dry ice in these settings is dependent upon a number of variables, including the quantity of dry ice present, the temperature, the degree of outdoor air infiltration, the size of the room, and the length of time the meat packages are held before being loaded onto trucks. Table 22.4 shows that exposures can be highly variable when measuring concentrations with short-term detector tubes. Table 22.5 shows the results of TWA exposure measurements. All workers were exposed to levels above the TLV® of 5000 ppm, and excursions well above the allowable levels were commonplace.

Several control alternatives were considered. Substitution of other methods of quick freezing offer the best method of controlling the hazard, since no carbon dioxide is present. Freeze tunnels or blast tunnels which use low-temperature air or nitrogen can be used. The meat is frozen and then packaged; this method requires more rigorous control of temperatures inside holding coolers and trucks, since no refrigerant is present inside the package itself. Nitrogen poses the potential hazard of displacement of oxygen, although it is preferable to carbon dioxide, since it poses no toxicity other than acting as an asphyxiant. The chief drawback to blast tunnels is that they occupy a great deal of floor space, and are thus difficult to fit into existing facilities. Spiral tunnels, which occupy less floor space are now available.

Another control method involves the use of local exhaust ventilation to exhaust fugitive carbon dioxide emissions from the machines that deliver the dry ice (which actually is applied in a pressurized liquid form) to the poultry package. Determination of the adequacy of the exhaust ventilation system often can be determined visually, since the cold CO_2 gas is visible. The local exhaust systems found in the three plants studied all had inadequate hood designs and airflow rates that failed to capture the CO_2 generated during package charging. Recommendations involving more complete enclosure (i.e., a better hood design) and increased exhaust air flow rates were made and found to be feasible.

TABLE 22.4 Initial Short-Term Detector Tube Sampling at Four Poultry Processing Plants

			Concentration (ppm)	
Plant	Area	Number of Employees	Range	Average
1	Breathing zone inside freezer	9	8,000–29,000	18,000
	Loading dock	2	5,000–6,500	5,750
	Dry Ice delivery to poultry packages	3	8,000–11,000	8,000
2	Holding cooler and palletizing area	3	12,000	12,000
	Loading dock	2	12,000–13,000	12,500
	Dry ice delivery to poultry packages	6	5,000–8,000	6,400
3	Holding cooler	5	23,000–60,000	33,000
	Dry Ice delivery to poultry packages (local exhaust present)	6	2,700–5,000	3,700
4	Holding cooler	3	5,000–26,000	18,000
	Palletizing area	2	11,000–30,000	21,000
	Dry ice delivery to poultry packages	4	8,000–22,000	12,000

TABLE 22.5 Comparison of Time-Weighted Average Breathing Zone Samples Using Bag Sampling, Vacuum-Sampling Cans, and Short-Term Detector Tubes[a]

Plant	Location	TWA Bag Concentration Measured with Detector Tube on Site (ppm)	TWA Bag Concentration Measured Using Vacu-Sampler and Gas Chromatography in Laboratory (ppm)	TWA Bag Concentration Measured Directly Using Gas Chromatography in Laboratory (ppm)
1	Holding cooler worker	4,900	5,800	3,700
	Palletizing line (outside holding cooler)	4,500	5,200	3,300
	Dry ice packaging, worker 1	4,500	6,300	3,500
	Dry ice packaging, worker 2	12,700	13,000	800[b]
3	Holding cooler worker	5,600	6,400	1,900[b]
	Dry ice packaging, worker 1	5,900	6,500	500[b]
	Dry ice packaging, worker 2	9,700	11,500	8,000
	Dry ice packaging worker 3	6,800	7,800	4,600
4	Dry ice packaging, worker 1	9,700	10,800	10,300
	Dry ice packaging, worker 2	14,000	15,100	12,800
	Dry ice packaging, worker 3	20,000	21,800	25,000
	Holding cooler, worker 1 (palletizing operation)	9,000	10,400	9,600
	Holding cooler, worker 2 (palletizing operation)	14,000	15,300	14,300

[a]Sampling times were approximately 300 min.

[b]Bags clearly leaked during shipment to laboratory.

Local exhaust ventilation systems obviously are not appropriate for controlling exposure levels inside the holding coolers where concentrations are greatest. Dilution ventilation rates were calculated, along with the cost of cooling the incoming fresh air. One way of reducing the energy costs involved the specification of an air-to-air heat exchanger to recover the energy in the cool, contaminated air about to be exhausted. Alarm systems were also specified for holding coolers to warn of dangerous atmospheres.

Administrative controls were also found to be effective. Workers involved in the palletizing operation inside the holding cooler were simply relocated to the outside plant area, where there was greater dilution. In another plant, workers were rotated in and out of the holding coolers on an hourly basis. This reduced TWA exposure but increased the chances of a sensitive worker being adversely affected. Some entry into holding coolers by forklift drivers was still necessary, but the overall exposures were dramatically reduced, and no further hospitalizations were reported.

Training efforts were also conducted to dispel the myth that dry ice is a harmless substance. Workers became skilled in recognizing the various signs of carbon dioxide intoxication in both themselves and in others. Material Safety Data Sheets were made available, and areas which could contain high levels of carbon dioxide were placarded with warning signs. Workers were trained in how to respond to the alarm systems which were installed.

Residential Lead-Based Paint Risk Assessment

Industrial hygiene includes evaluation and control of hazards in the community as well as the workplace. This case study describes how hazards associated with childhood lead poisoning were evaluated in a public housing authority. The results of the study were also used to control exposures to lead experienced by maintenance and renovation workers.

The risk assessment was conducted to permit the housing authority to acquire a lead-poisoning insurance policy. To manage the risks and potential claims from poisoned resident children, the insurance company required an evaluation of both immediate and long-term risks borne by the children residing in housing units owned by the authority and, more importantly, a practical program to control exposures.

A review of the childhood lead poisoning literature indicated that exposures occur through a number of pathways. These pathways include ingestion of housedust through hand-to-mouth contact and normal mouthing behavior in young children, ingestion of lead paint chips originating from deteriorated paint films, and ingestion of soil. Inhalation and dermal transfer are not considered to be important routes of exposure for young children. Drinking water was being evaluated under a citywide EPA-mandated program and was not repeated here. There were also no identifiable nearby sources of lead air emissions. Therefore, air sampling and water sampling were not performed as part of this risk assessment, which was aimed at evaluating the most likely sources of lead for the population under consideration.

As a measure of the near-term risk, paint chips from deteriorated painted surfaces, soil, and wipe dust samples were collected and analyzed for lead. Visual assessment of the condition of all paint films was completed. The quality of house cleaning was also noted. Recommendations were provided to control lead dust and soil hazards and to stabilize deteriorated leaded paint films temporarily until full-scale abatement could be accomplished.

As a measure of the long-term risk, management and maintenance practices were studied and modified so that intact lead-based paint was properly maintained. For example, the work order system was changed so that trained maintenance workers would be informed when their jobs required them to disturb surfaces coated with known or suspected lead-based paint. Interviews with various management and maintenance personnel, with regard to occupational health and safety issues, were also conducted. Methods for relocating resident children from units where dusty repair work was being completed were examined.

The risk assessment did *not* include measurement of all painted surfaces for lead, which was undertaken in a separate effort as a way of determining *potential* hazards. Ideally, both immediate and potential exposure to resident children and to maintenance workers should be evaluated through a lead-based paint risk assessment and a lead-based paint inspection, respectively.

Wipe sampling was performed in accordance with the procedure in "Guidelines for the Evaluation and Control of Lead-Based Paint Hazards in Housing." Diaper towel wipes were used as the wipe media. Ordinarily used for qualitative wipe sampling, Whatman filters were found to be insufficiently durable on housing surfaces, requiring the use of the more durable diaper wipes. Recent research has shown that settled lead dust levels measured in this fashion can be correlated to blood lead levels in resident children. Disposable gloves were used for each sample. The results of the wipe samples were compared to settled lead dust standards adopted by HUD.

Composite soil samples were collected from bare areas along the building foundations and in play areas. A single sample typically included about 50 mL of soil, which was collected as follows. The top centimeter of soil from at least five distinct spots was drawn into a 50-mL centrifuge tube. The spots were in a straight line immediately next to the building foundation, or parallel to the building face. In those units where bare soil in play areas was found, samples were collected from a random grid in the play area. Some soil samples were also collected in a line parallel to sidewalks or parking lot curbs to determine whether previous gasoline emissions were a local problem.

Of the 75 dwelling units, 15 were sampled; one of these units was unoccupied. Since it was not feasible to sample or conduct walk-through examinations of all 75 units, several criteria were developed to identify those units that were most likely to be in worst-case condition to conservatively estimate the risks in all 75 units. The units actually sampled met as many of the following criteria as possible:

- Presence of a resident child with an elevated blood lead level
- Housing or building code violations
- Chronic maintenance problems

- Dwellings with the most children
- Units reported by the housing authority to be in poor condition or where housecleaning practices were known to be deficient
- Dwellings in which at least one of the sampled units would be a vacant unit recently repaired and repainted

Dust wipe samples were collected in the following locations:

- Principal play area (living room)
- Kitchen
- Two children's bedrooms

Within each room, two wipe samples were collected, one from the midpoint or entryway floor and the other from either a window sill or a window well.

Paint chip samples were collected by using a heat gun to soften the paint, followed by use of a razor-sharp scraper to remove all layers of paint (lead-based paint is often the oldest paint layer). Composite soil samples were collected using a 50-mL polypropylene centrifuge tube.

The results indicated that levels of lead dust on the floors were well below the HUD (Department of Housing and Urban Development) clearance standard of 200 g/ft^2. However, levels of lead dust in exterior window wells and interior window sills were substantially higher, with the maximum over 9000 g/ft^2 and nearly all above the HUD standard of 500 g/ft^2 for window sills and 800 g/ft^2 for window wells. This suggested that windows contained the highest levels of lead. Although children may not contact window wells with as much frequency as floors, recent studies have indicated that lead dust levels in window wells are correlated with blood lead levels.

Soil lead levels were all below 130 ppm, which is unlikely to pose a substantial risk, given the USEPA Superfund cleanup guidance of 500—1000 ppm.

Deteriorated paint films were observed on exterior railings, exterior doors, and exterior window frames. No interior paint was in a deteriorated condition. Results of paint chip analysis from the three exterior surfaces indicated that all contained lead above 5000 ppm (the current HUD standard).

Since all surfaces had not yet been tested for the presence of lead-based paint, a management system geared to keeping lead-based paint in a nonhazardous condition was needed. The current work order system contained no warning for maintenance workers. Workers were not trained to work with lead-based paint, had not had their exposure levels evaluated for routine maintenance tasks, were not provided with respirators or protective clothing, and were not included in a medical surveillance plan.

The window wells and interior window sills were subsequently cleaned on a routine basis by residents and at least annually by trained work crews. Trained work crews cleaned all windows before residents were instructed to include them in their routine cleaning practices. The work crew cleaning consisted of an initial pass with a *high-efficiency particulate air* (HEPA) vacuum cleaner followed by a wet phosphate detergent (or other equivalent detergent) wash followed by a final pass with a HEPA vacuum. Repeated wipe samples were collected annually to make certain that lead dust levels did not reaccumulate to hazardous levels.

The deteriorated paint films on the exterior railings, doors, and windows were temporarily stabilized by wet scraping and repainting. The condition of these surfaces were visually monitored annually. Residents were encouraged to report cases of deteriorated paint to the housing authority and to wet clean all surfaces on a routine basis. Residents were informed that they should *not* attempt to remove any deteriorated or intact paint. If the paint is known to contain lead (or is suspected to contain lead), the paint film should be stabilized on an emergency repair basis by trained work crews following temporary relocation of residents during the work.

Employees involved in doing cleanup work of this sort on a routine basis wore half-mask air-purifying respirators equipped with HEPA cartridges and were included in respirator and medical surveillance programs. (Ideally, protective clothing should also be worn, although laundered uniforms can also be used as long as the uniforms are not taken home and are not worn in worker's automobiles.)

Worker's shoes were left at the maintenance headquarters and were not worn home or into cars, since they can track lead dust into locations that may be accessible to their own children. Workers were required to thoroughly wash their hands and faces before all breaks and at the end of the shift; this rule was rigorously enforced. Full showers were preferable. These precautions were necessary until worker exposures could be characterized. If exposure levels were sufficiently low, it might have been possible to reduce or eliminate some of these protective measures.

Even though levels of lead in soil were found to be relatively low, in this case all bare soil areas should be covered with sod or planted with grass seed, especially play areas in the yards of housing units. This will minimize the potential for ingestion of lead-contaminated soil.

The Housing Authority developed a formal written occupational safety and health program for its employees and a lead-based paint interim control plan. This included a brief written statement from the executive director indicating the importance of observing safe work practices within the organization.

Specific elements of the program included

- A written hazard communication program (see 29 CFR 1910.1200).
- A respirator program (see 29 CFR 1910.134). No worker should be required to wear a respirator unless the program elements are in place. Training on the limitations of the particular respirator is especially important. For example, employees should know that a respirator with a HEPA cartridge will provide absolutely no protection against solvent vapors or oxygen deficiency.
- Disciplinary procedures if safety rules are disregarded.
- Special training on the hazards of lead, asbestos, and other hazardous substances that maintenance workers might encounter.
- Emergency response procedures.
- Hazardous-waste regulations.
- Establishment of a health-and-safety committee.
- Name of the management staffer who has the authority to coordinate all lead-hazard control work.

These programs were implemented and remained in effect until all lead-based paint in the housing development was either removed through building component replacement or enclosed with durable physically fastened construction materials. These measures were implemented during a gut rehabilitation of the housing project.

No cases of children with elevated blood lead levels were reported. Thus, risks associated with the presence of lead-based paint, leaded dust, and leaded soil were managed on a practical interim basis until they could be eliminated in the course of demolition and rehabilitation activity.

22.5 SUMMARY

Industrial hygiene is devoted to the prevention of environmentally-induced disease. As such it uses a process of anticipation, recognition, and evaluation to assess the need for and types of controls necessary to protect people.

Much of the work of industrial hygiene is based on reference to exposure limits derived in part from toxicological testing as well as epidemiologic evidence, work experience, and by reference to other chemicals. The most common exposure limits are

- ACGIH TLVs®
- OSHA PELs
- Environmental exposure limits, many of which are derived from the TLVs®

- Anticipation and recognition, which involve identification of the presence of chemicals known or suspected of causing adverse health effects and understanding of potential exposure routes while handling the chemicals

Evaluation involves assessment of exposures using a variety of means, including

- Direct-reading instrumentation
- Collection devices for later analysis
- Observation of actual practices to determine exposure routes

Control of exposure may be accomplished in a number of ways, including

- Substitution or process modification
- Isolation of the source or the worker
- Ventilation
- Administrative controls
- Work practice modification
- Personal protective equipment (e.g., respirators)

A number of case studies were presented to illustrate the science and art of industrial hygiene.

REFERENCES AND SUGGESTED READING

American Conference of Governmental Industrial Hygienists, *TLVs® and BEIs*, ACGIH, Cincinnati, 1999.

American Conference of Governmental Industrial Hygienists, *Industrial Ventilation: A Manual of Recommended Practice*, ACGIH, Cincinnati, 1988.

AIHA, *A Strategy for Occupational Exposure Assessment*, N. Hawkins, S. Norwood, and J. Rock, eds., American Industrial Hygiene Association, Washington, DC, 1991.

ASHRAE, *Ventilation Handbook Series*, American Society of Heating, Refrigeration, and Air Conditioning Engineers. Atlanta, 1985.

Brief, R., and R. Scala, "Occupational exposure limits for novel work schedules," *Am. Ind. Hygiene Assoc. J.* **36**: 467–469 (1975).

Carmichael, N., R. Nolan, J. Perkins, R Davies, and S. Warrington, "Oral and dermal pharmacoki netics of triclopyr in human volunteers," *Human Toxicol.* **8**: 431–437 (1989).

Carson, R., *Silent Spring,* Houghton-Mifflin, Boston, 1962.

Castleman, B., and G. Ziem, "Corporate influence on Threshold Limit Values," *Am. J. Ind. Med.* **13**: 531–559 (1988).

Commoner, B., "Let's get serious about pollution prevention," *EPA J.* **15**: 15–19 (1989).

Cunningham, K., *A Comparison of PELs and TLVs to Health-Based Exposure Limits Derived from the IRIS Database,* New Jersey Department of Health, 1988.

DiNardi, S. R., Ed., "The Occupational Environment—Its Evaluation and Control," *Am. Ind. Hygiene Assoc.* Fairfax, VA, 1997.

Droz, P., "Biological monitoring I: Sources of variability in human response to chemical ex posures," *Appl. Ind. Hygiene* **4**: F20–F24 (1989).

Fiserova-Bergerova (Thomas), V., "Development of biological exposure indices (BEIs) and their implementation," *Appl. Ind. Hygiene* **2**: 87–92 (1987).

Hamilton, A., *Exploring the Dangerous Trades*, Little, Brown, Boston, 1943.

Hickey, J., and P. Reist, "Application of occupational exposure limits to unusual work sch edules," *Am. Ind. Hygiene Assoc. J.* **38**: 613–62 (1977).

Jacobs, D., and P. Middendorf, "Control of nitrous oxide exposures in dental operatories using local exhaust ventilation: A pilot study," *Anesthesia Prog.* **21**: 235–242 (1986).

Jacobs, D., *The OSHA Cancer Policy: Generic vs. Substance—Specific Regulation in an Area of Scientific Uncertainty*, thesis, Georgia Institute of Technology, 1988.

Jacobs, D., and M. Smith, "Exposures to carbon dioxide in the poultry processing industry," *Am. Ind. Hygiene Assoc. J.* **49**: 624–629 (1988).

Leidel, N., K. Busch, and J. Lynch, *Occupational Exposure Sampling Strategy Manual*, USDHEW/NIOSH Publication 77–173, 1977.

Lowman, S., "Hazard communication—worker perspective," in *Conference on Occupational Health Aspects of Advanced Composite Technology in the Aerospace Industry, Applied Industrial Hygiene*, special issue, 1989.

Mastromatteo, E., "Threshold Limit Values Committee for chemical substances in the work environment," *Appl. Ind. Hygiene* **3**: F12–F15 (1988).

Mickelson, R. L., D. E. Jacobs, P. A. Jensen, P. J. Middendorf, D. M. O'Brien, T. J. Fishbach, and A. A. Beasley, "Design and evaluation of a nitrous oxide local exhaust ventilation system for the dental suite," *Appl. Occup. Environ. Hygiene* **8**: 564–570 (1993).

Middendorf, P., D. Jacobs, K. Smith, and D. Mastro, "Occupational exposure to nitrous oxide in dental operatories," *Anesthesia Prog.* **33**: 91–97 (1986).

Middendorf, P., C. Timchalk, D. Rick, and B. Kropscott, "Forest worker exposure to Garlon™ 4 herbicide," *Appl. Occup. Environ. Hygiene* **9**: 589–594 (1994).

NRC (National Research Council), *Toxicity Testing: Strategies to Determine Needs and Priorities,* National Academy Press, Washington, DC, 1984.

NJDOPH (New Jersey Department of Public Health), *A Guide to Workplace Evaluation for Occupational Disease,* unpublished, 1989.

OHSAct (Occupational Health and Safety Act), *Statutes of Ontario*, Chapter 83, Section 23, 1978, as amended 1990.

OHRC (Occupational Health Resource Center), *Warning Properties of Industrial Chemicals*, Oregon Lung Association, Portland, OR, 1983.

Paustenbach, D., "Occupational exposure limits, pharmacokinetics, and unusual work schedules," in *Patty's Industrial Hygiene and Toxicology*, Vol. IIIA, 2nd ed., L. J. Cralley and L. V. Cralley, eds., Wiley, New York, 1985.

Radian Corp., *National Air Toxics Information Clearinghouse*, NATICH database report on state, local, and EPA air toxics activities, EPA-450/5-87-006, 1987.

Ross, D., "Industrial hygiene program management: Policy, scope, and responsibilities," *Appl. Ind. Hygiene* **3**: F30–F34 (1988).

Schwope, A., P. Costas, J. Jackson, J. Stull, and D. Weitzman, *Guidelines for the Selection of Chemical Protective Clothing*, 3rd ed., American Conference of Governmental Industrial Hygienists, Cincinnati, 1987.

Smith, K., P. Williams, P. Middendorf, and N. Zakraysek, "Kidney dialysis: Ambient formaldehyde levels," *Am. Ind. Hygiene J.* **45**: 48–50 (1984).

Tarlau, E., "Industrial hygiene with no limits," *AIHA J.* **51**: A9–A10 (1990).

USDHHS/NIOSH, *NIOSH Pocket Guide to Chemical Hazards,* Publication 90–117, 1990.

USDHUD, Guidelines for the Evaluation and Control of Lead-Based Paint Hazards in Housing, HUD Publication HUD-1539-LBP, Washington, DC, 1995.

USDOL/OSHA, *Compliance Officers Field Manual*, Occupational Safety and Health Administration, Washington, DC, 1979.

USDOL/OSHA, 29 CFR 1910.1200, *Hazard Communication*, 1999a.

USDOL/OSHA, 29 CFR 1910.134, *Respiratory Protection*, 1999b.

USDOL/OSHA, 29 CFR 1910.1025, *Lead*, 1999c.

USDOL/OSHA, 29 CFR 1910.1027, *Cadmium*, 1999d.

USDOL/OSHA, 29 CFR 1910.1028, *Benzene*, 1999e.

USDOL/OSHA, 29 CFR 1910.119, *Process Safety Management*, 1999f.

Williams, C., H. Jones, R. Freeman, M. Wernke, P. Williams, S. Roberts, and R. James, "The EPC approach to estimating safety for exposure to environmental chemicals," *Regul. Toxicol. Pharmacol.* **20**: 259–280 (1994).

Williams, P., M. Luster, and P. Middendorf, "Problems with selecting sample sites," *Occup. Health Safety Mag.* **52**: 21–24 (1983).

Glossary

absorption The movement of a chemical from the site of initial contact with the biologic system across a biologic barrier and into either the bloodstream or the lymphatic system.

accumulative effect of a chemical The effect of a chemical on a biologic system when the chemical has been administered at a rate that exceeds its elimination from the system. Sufficient accumulation of the chemical in the system can lead to toxicity.

acetylation The introduction of an acetyl group, CH_3CO-, onto the molecule of an organic compound having either $-OH$ or $-NH_2$ groups.

acetylator An individual with a phenotype of rapid metabolic acetylation; common in American Indians and those of Asian descent.

acetylcholine An acetic acid ester of choline normally present in many parts of the body and having important physiologic functions, such as playing a role in the transmission of an impulse from one nerve fiber to another across a synaptic junction.

acetylcholinesterase An enzyme present in nervous tissue and muscle that catalyzes the hydrolysis of acetylcholine to choline and acetic acid.

acidosis A pathologic condition resulting from accumulation of acid in, or loss of base from, the body.

action potential A momentary change in electrical potential on the surface or a nerve or muscle cell that takes place when it is stimulated, especially by the transmission of a nerve impulse.

acute toxic Adverse effects caused by a toxic agent and occurring within a short period of time following exposure.

adduct A chemical addition product (i.e., a chemical bound to an important cellular macromolecule like DNA or protein).

adenocarcinoma A malignant tumor originating in glandular tissue.

adenoma A benign epithelial tumor having a glandular origin and structure.

administrative control A method of controlling employee exposures to contaminants by job rotation or work assignment within a single workshift.

aflatoxins Toxic metabolites produced by some strains of the fungus *Aspergillus flavus*. They are widely distributed in foodstuffs, especially peanut meals.

albuminuria Presence of serum albumin in the urine; proteinuria.

alcohol An organic compound in which a hydrogen atom attached to a carbon atom in a hydrocarbon is replaced by a hydroxyl group (OH). Depending on the environment of the –C–OH grouping, they may be classified as primary, secondary, or tertiary alcohols.

aldehyde A broad class of organic compounds having the generic formula RCHO.

alicyclic Organic compounds characterized by arrangement of the carbon atoms in closed ring structures.

Principles of Toxicology: Environmental and Industrial Applications, Second Edition, Edited by Phillip L. Williams, Robert C. James, and Stephen M. Roberts.
ISBN 0-471-29321-0 © 2000 John Wiley & Sons, Inc.

aliphatic Organic compounds characterized by a straight- or branched-chain arrangement of the constituent carbon atoms.

alkane See **paraffin**.

alkyl A chemical group obtained by removing a hydrogen atom from an alkane or other aliphatic hydrocarbon.

alkylation The introduction of one or more alkyl radicals (e.g., methyl, CH_3–; ethyl, C_2H_5–; propyl, $CH_3CH_2CH_2$–; etc.) by addition or substitution into an organic compound.

allele Either of the pair of alternative characters or genes found at a designated locus on a chromosome. Chromosome pairing results in expression of a single allele at each locus.

allergy General or local hypersensitive reactions of body tissues of certain persons to certain substances (allergens) that, in similar amounts and circumstances, are innocuous to other persons. Allergens can affect the skin (producing urticaria), the respiratory tract (asthma), the gastrointestinal tract (vomiting and nausea), or may result from injections into the bloodstream (anaphylactic reaction). See also **anaphylactic-type reaction**.

alveolar macrophages Actively mobile, phagocytic cells that process particles ingested into the lung. They originate outside the lungs from precursor cells (promonocytes) in the bone marrow and from peripheral blood monocytes. They enter the alveolar interstices from the bloodstream and are able to migrate to terminal bronchioles and lymphatic vessels.

alveolus (pl. alveoli) In the lungs, small outpouchings along the walls of the alveolar sacs, alveolar ducts, and terminal bronchioles, through the walls of which gas exchange takes place between alveolar air and pulmonary capillary blood.

amelia The congenital absence of a limb or limbs. See also **phocomelia**.

amidase An enzyme that catalyzes the breakdown of an amide compound to a carboxylic acid and ammonia.

Ames assay A screening test capable of revealing mutagenic activity through reverse mutation in *Salmonella typhimurium*. Mammalian metabolism can be simulated by addition of S9 liver enzyme to the bacterial growth medium.

amide A nitrogenous compound with the general formula $RNH_2C{=}O$, related to or derived from ammonia. Reaction of an alkali metal with ammonia yields inorganic amides (e.g., sodium amide, $NaNH_2$)–. Organic amides are closely related to organic acids and are often characterized by the substitution of one or more acyl groups (RCO) for an H atom of the ammonia molecule (NH_3).

amine An organic compound formed from ammonia (NH_3) by replacement of one or more of the H atoms by hydrocarbon radicals.

amyotrophic lateral sclerosis (ALS) A disease marked by progressive degeneration of the neurons that give rise to the corticospinal tract and of the motor cells of the brainstem and spinal cord, resulting in a deficit of upper and lower motor neurons; the disease is usually fatal within 2–3 years.

anaphylatic-type reaction One of four types of allergic reaction. A violent allergic reaction to a second dose of a foreign protein or other antigen to which the body has previously been hypersensitized. Symptoms include severe vasodilation, urticaria or edema, choking, shock, and loss of consciousness. Can be fatal.

angiosarcoma Malignant tumor of vascular system arising from endothelial cells.

anoxia A complete reduction in the oxygen concentration supplied to cells or tissues.

anthropogenic Produced or caused by the actions of humans.

antibody An immunoglobulin molecule that has a specific amino acid sequence that causes it to interact only with the antigen that induced its synthesis, or with antigens closely related to it.

antigen A substance that, when introduced into the body, is capable of inducing the formation of antibodies and, subsequently, of reacting in a recognizable fashion with the specific induced antibodies.

antipyretic An agent that relieves or reduces fever.

aplasia Lack of development of an organ or tissue, or of the cellular products of an organ or tissue.

aplastic anemia A form of anemia generally unresponsive to specific antianemia therapy, in which the bone marrow may not necessarily be acellular or hypoplastic but fails to produce adequate numbers of peripheral blood elements; term is all-inclusive and probably encompasses several chemical syndromes.

apnea Cessation of breathing; asphyxia.

aromatic A major group of unsaturated cyclic hydrocarbons containing one or more rings. These are typified by benzene, which has a six-carbon ring containing three double bonds. These are also known as *arene compounds*.

arrhythmia Any variation from the normal rhythm of heart beat, including sinus arrhythmia, premature beat, heart block, atrial fibrillation, atrial flutter, pulsus altemans, and paroxysmal tachycardia.

arteriosclerosis A disease of the arteries characterized by thickening, loss of elasticity, and calcification of arterial walls, resulting in a decreased blood supply particularly to the cerebrum and lower extremities; it often develops with aging, and in hypertension and diabetes.

arthralgia Neuralgic pain in a joint or joints.

arthroosteolysis Dissolution of bone; the term is applied especially to the removal or loss of calcium from the bone; the condition is attributable to the action of phagocytic kinds of cells.

asbestosis A bilateral, diffuse, interstitial pulmonary fibrosis caused by fibrous dust of the mineral asbestos; also referred to as asbestos pneumoconiosis.

asphyxiant A substance capable of producing a lack of oxygen in respired air, resulting in pending or actual cessation of apparent life.

asthmatic response Condition marked by recurrent attacks of paroxysmal dyspnea, with wheezing caused by spasmodic contractions of the bronchi; the response is a reaction in sensitized persons.

ataxia Failure of muscular coordination; irregularity of muscular action.

atherosclerosis A form of arteriosclerosis characterized by the deposition of atheromatous plaques containing cholesterol and lipids on the innermost layer of the walls of large and medium-sized arteries.

atresia A congenital absence of closure of a normal body orifice or tubular organ; the absence or closure of a normal body orifice or tubular passage such as the anus, intestine, or external ear canal.

atrophy A decrease in the size and activity of cells, resulting from such factors as hypoxia, decreased work, and decreased hormonal stimulation.

atropine An alkaloid forming white crystals, $C_{17}H_{23}NO_{31}$, soluble in alcohol and glycerine. Used as an anticholinergic for relaxation of smooth muscles in various organs, to increase the heart rate by blocking the vagus nerve.

autonomic nervous system The part of the nervous system that regulates the activity of cardiac muscle, smooth muscle, and glands.

auxotroph Any organism (e.g., a bacterium) that, as a result of mutation, can no longer synthesize a substance that is necessary for its own nutrition (usually an amino acid), and thus requires an external supply of that substance.

autosome Any chromosome that is not a sex chromosome.

B cell An immunocyte produced in the bone marrow. B cells are responsible for the production of immunoglobulins but do not play a role in cell-mediated immunity. They are short-lived.

bactericidal Destructive to bacteria.

basophil A granular leukocyte with an irregularly shaped, relatively palestaining nucleus that is partially constricted into two lobes; cytoplasm contains coarse, bluish black granules of variable size.

benign tumor A new tissue growth (tumor) composed of cells that, although proliferating in an abnormal manner, are not invasive—that is, do not spread to surrounding, normal tissue; benign tumors are contained within fibrous enclosures.

bilirubin A bile pigment; it is a breakdown product of heme formed from the degradation of erythrocyte hemoglobin in reticuloendothelial cells, but is also formed by the breakdown of other heme pigments. Normally bilirubin circulates in plasma as a complex with albumin, and is taken up by the liver cells and conjugated to form bilirubin deglucuronide, which is the water-soluble pigment excreted in bile.

biologic half-life The time required to eliminate one-half of the quantity of a particular chemical that is in the system at the time the measurement is begun.

biomarker A specific physical or biochemical trait used to measure or indicate the effects or progress of a disease or condition.

biotransformation The series of chemical alterations of a foreign compound that occur within the body, as by enzymatic action. Some biotransformations result in less toxic products while others result in products more toxic than the parent compound.

bradycardia A slowness of the heart beat, as evidenced by a slowing of the pulse rate to less than 60 beats per minute.

bronchitis Inflammation of one or more bronchi, the larger air passages of the lungs.

byssinosis Respiratory symptoms resulting from exposure to the dust of cotton, flax, and soft hemp. Symptoms range from acute dyspnea with cough and reversible breathlessness and chest tightness on one or more days of a workweek to permanent respiratory disability owing to irreversible obstruction of air passages.

cancer A process in which cells undergo some change that renders them abnormal. They begin a phase of uncontrolled growth and spread. See also Malignant tumor.

carbamate A compound based on carbamic acid, NH_2COOH, which is used only in the form of its numerous derivatives and salts; as pesticides, carbamates are reversible inhibitors of cholinesterase. Inhibition of the enzyme is reversed largely by hydrolysis of the carbamylated enzyme and to a lesser extent by synthesis of a new enzyme.

carcinogen Any cancer-producing substance.

carcinoma A malignant tumor that arises from embryonic ectodermal or endodermal tissue.

cathartic An agent that stimulates bowel movement; a strong laxative.

cardiomyopathy General diagnostic term designating primary myocardial disease, often of obscure or unknown etiology.

catecholamine Any of a group of amines derived from catechol that have important physiological effects as neurotransmitters and hormones and include epinephrine, norepinephrine, and dopamine.

cell-mediated immunity Specific acquired immunity in which the role of small lymphocytes of thymic origin is predominant; the kind of immunity that is responsible for resistance to infectious diseases caused by certain bacteria and viruses, certain aspects of resistance to cancer, delayed hypersensitivity reactions, certain autoimmune disease, and allograft rejections, and that plays a part in certain allergies. See also **T cell**.

cellular immunity Immunity mediated by T lymphocytes. It can be transferred to a naïve individual with cells but not by serum or plasma.

cephalosporidine A broad-spectrum antibiotic of the cephalosporin group, which are penicillinase-resistant antibiotics.

chelate A chemical compound in which a metallic ion is sequestered and firmly bound with the chelating molecule; used in chemotherapeutic treatments for metal poisoning.

chemotaxis The characteristic movement or orientation of an organism or cell along a chemical concentration gradient either toward or away from the chemical stimulus.

chloracne An acne-like skin eruption caused by exposure to halogenated compounds, especially the polyhalogenated naphthalenes, biphenyls, dibenzofurans, and dioxins.

cholestasis Stoppage or suppression of the flow of bile.

cholestatic Pertaining to or characterized by cholestasis.

cholinesterase An enzyme found chiefly at nerve terminals that inactivates the neurotransmitter acetylcholine by hydrolyzing it to form acetic acid and choline.

chromhidrosis The secretion of colored sweat.

chromophore A chemical group capable of selective light absorption resulting in the coloration of certain organic compounds.

chromosome aberrations Structural mutations (breaks and rearrangements of chromosomes) or changes in number of chromosomes (additions and deletions).

chronic toxicity Adverse effects occurring after a long period of exposure to a toxic agent (with animal testing this is considered to be the majority of the animals life). These effects may be permanent or irreversible.

clastogenic Giving rise to or inducing a breakage or disruption of chromosomes.

cocarcinogen Any chemical capable of increasing the observed incidence of cancer if applied with a carcinogen, but not itself carcinogenic.

collagen The fibrous protein constituent of bone, cartilage, tendon, and other connective tissue. It is converted into gelatin by boiling.

comedo A plug in an excretory duct of the skin, containing microorganisms and desquamated keratin; a blackhead.

competitive inhibition Inhibition of enzyme activity in which the inhibitor (substrate analog) competes with the substrate for binding sites on the enzymes; such inhibition is reversible since it can be overcome by increasing the substrate concentration.

complement A complex system of proteins found in normal blood serum that combines with antibodies to destroy pathogenic bacteria and other foreign cells.

conformational change A change in the particular shape of a molecule.

conjugation The addition by drug metabolizing enzymes of hydrophilic moieties to xenobiotics to hasten their excretion from the body.

conjunctivitis Inflammation of the conjunctiva, the delicate mucous membrane that lines the eyelids and covers the exposed surface of the eye.

contact dermatitis See **dermatitis**.

contraindication Any condition, especially one of disease, that renders some particular line of treatment improper or undesirable.

corpus luteum A yellow glandular mass in the ovary formed by an ovarian follicle that has matured and discharged its ovum.

cryptorchism A development defect marked by the failure of the testes to descend into the scrotum.

cutaneous sensitization Immune reaction characterized by local skin rashes, urticaria (hives), erythema, edema, and itching. Cutaneous sensitization is thought to be initiated by the release of histamine.

cyanosis A bluish discoloration, especially of skin and mucous membranes, owing to excessive concentration of reduced hemoglobin in the blood.

cystitis Inflammation of the urinary bladder.

cytochrome oxidase An oxidizing enzyme containing iron and a porphyrin, found in mitochondria and important in cellular respiration as an agent of electron transfer from certain cytochrome molecules to oxygen molecules.

cytochrome enzymes See **mixed-function oxidase system (MFO)**.

cytokinesis The division of the cytoplasm of a cell following the division of the nucleus.

cytoplasm The protoplasm of a cell exclusive of the nucleus, consisting of a continuous aqueous solution (cytosol) and the organelles and inclusions suspended in it (phaneroplasm); the site of most of the chemical activities of the cell.

cytosol The liquid medium of the cytoplasm (i.e., cytoplasm minus organelles and nonmembranous insoluble components).

dalton A unit of mass, one-twelfth the mass of the carbon-12 atom. Carbon-12 has a mass of 12.011, and thus the dalton is equivalent to 1.0009 mass units, or 1.66×10^{-24} g. Also called the atomic mass unit (amu).

denaturation The destruction of the usual nature of a substance, usually the change in the physical properties of proteins caused by heat or certain chemicals.

depolarize Loss of the ionic gradient across a nerve cell membrane, resulting in an action potential and propagation of a nerve impulse.

dermatitis Inflammation of the skin. Contact dermatitis is a delayed allergic skin reaction resulting from contact with an allergen. Irritant dermatitis describes irritation of the skin accompanying exposure to a toxic substance.

detoxification The metabolic process by which the toxic qualities of a poison or toxin are reduced by the body.

diethylstilbestrol (DES) A synthetic estrogenic compound, $C_{18}H_{20}O_2$, prepared as a white odorless crystalline powder.

dimethyl sulfoxide (DMSO) An alkyl sulfoxide, C_2H_6OS, practically colorless in its purified form. As a highly polar organic liquid, it is a powerful solvent, dissolving most aromatic and unsaturated hydrocarbons, organic compounds, and many other substances.

diplopia A condition in which a single object is perceived as two objects; double vision.

direct carcinogen See **primary carcinogen**.

dissociation constant The equilibrium constant for the reaction by which a weak acid compound is dissociated into hydrogen ions and a conjugate base, in solution. See also ***pK***.

distal alveolar region The part of the lung composed of the alveoli, or tiny air sacs, through which gas exchange between alveolar air and blood takes place.

DMSO See **dimethyl sulfoxide**.

dose The amount of a drug needed at a given time to produce a particular biologic effect. In toxicity studies it is the quantity of a chemical administered to experimental animals at specific time intervals. The quantity can be further defined in terms of quantity per unit weight or per body surface area of the test animal. Sometimes the interval of time over which the dose is administered is part of the dose terminology. Examples are: grams (or milligrams) per kilogram of body weight (or per square meter of body surface area).

dose–response relationship One of the most basic principles of both pharmacology and toxicology. It states that the intensity of responses elicited by a chemical is a function of the administered dose (i.e., a larger dose produces a greater effect than a smaller dose, up to the limit of the capacity of the biologic system to respond).

drug-induced toxicity Toxicities that are "side effects" to the intended beneficial effect of a drug. They represent pharmacologic effects that are undesirable but that are known to accompany therapeutic doses of the drug.

dyscrasia A morbid general state resulting from the presence of abnormal material in the blood.

dysplasia Abnormal development or growth of tissues, organs, or cells.

dyspnea Difficult or labored breathing.

dysrhythmia Disturbances of rhythm, such as speech, brain waves, and heartbeat.

eczema A superficial inflammatory process involving primarily the epidermis; characterized early by redness, itching, minute papules and vesicles, weeping of the skin, oozing, and crusting, and later by scaling, lichenification, and often pigmentation.

ED_{50} The dose of a particular substance that elicits an observable response in 50 percent of the test subjects.

edema The presence of abnormally large amounts of fluid in intercellular spaces within a tissue.

electrophile A chemical compound or group that is attracted to electrons and tends to accept electrons.

elimination The removal of a chemical substance from the body. The rate of elimination depends on the nature of the chemical and the mechanisms that are used to remove the chemical from the organism. Examples of mechanisms include expiration from the lungs, excretion by the kidneys by way of the urinary system, excretion in the sweat or saliva, and chemical alteration by the organism and subsequent excretion by any of these mechanisms. See **excretion**.

emphysema Literally, an inflation or puffing up; a condition of the lung characterized by an increase, beyond the normal, in the size of air spaces distal to the terminal bronchiolus.

encephalopathy Any degenerative disease of the brain.

endoplasmic reticulum An ultramicroscopic organelle of nearly all cells of higher plants and animals, consisting of a more or less continuous system of membrane-bound cavities that ramify throughout the cell cytoplasm.

endothelial Pertaining to the layer of flat cells lining blood and lymphatic vessels.

endotoxin A toxin produced by certain bacteria and released on destruction of the bacterial cell.

engineering control A method of controlling exposures to contaminants by modifying the source or reducing the quantity of contaminants released into the environment.

enterohepatic circulation The recurrent cycle in which the bile salts and other substances excreted by the liver pass through the intestinal mucosa and become reabsorbed by the hepatic cells, and then are reexcreted.

environmental toxicology That branch of toxicology that deals with exposure of biologic tissue (more specifically, human life) to chemicals that are basically contaminants of the biologic environment, or of food, or of water. It is the study of the causes, conditions, effects, and limits of safe exposure to such chemicals.

eosinophil A structural cell or histologic element readily stained by eosin; especially, a granular leukocyte containing a nucleus usually with two lobes connected by a slender thread of chromatin, and having cytoplasm containing coarse, round granules that are uniform in size.

epidemic Spreading rapidly and extensively by infection and affecting many individuals in an area or a population at the same time: an epidemic outbreak of influenza.

epidermal tumor A tumors arising from the skin (dermal) epithelial layer.

epidermis The outermost and nonvascular layer of skin. It derives from embryonic ectoderm.

epistaxis Nosebleed.

epithelioma Any tumor developing in the epithelium, which is the kind of tissue that covers internal and external surfaces of the body.

epoxide An organic compound containing a reactive group comprising a ring formed by an oxygen atom joined to two carbon atoms, having the structure at right.

O
△

erethism Excessive irritability or sensitivity to stimulation, particularly with reference to the sexual organs, but including any body parts. Also a psychic disturbance marked by irritability, emotional instability, depression, shyness, and fatigue, which are observed in chronic mercury poisoning.

erythema The redness of the skin produced by congestion of the capillaries.

erythropoiesis The production of erythrocytes (red blood cells).

erythropoietic stimulating factor (ESF) A factor or substance that stimulates the production of erythrocytes; may be the same as erythropoietin.

erythropoietin A protein that enhances erythropoiesis.

ester A compound formed from an alcohol and an acid by removal of water.

esterase Any of various enzymes that catalyze the hydrolysis of an ester.

ether A colorless, transparent, mobile, very volatile liquid, highly inflammable, and with a characteristic odor; many ethers are used by inhalation as general anesthetics; the usual anesthetic forms are diethyl ether or ethyl ether.

excretion The process whereby materials are removed from the body to the external environment. If a chemical is in solution as a gas at body temperature, it will appear in the air expired from the animal; if it is a nonvolatile substance, it may be eliminated by the kidney via the urinary system, or it may be chemically altered by the animal and then excreted by means of any of the mechanisms available to the animal, such as excretion in the urine, in the sweat, or in the saliva. See **elimination**.

fibrosis The formation of excessive fibrous tissue, as in a reparative or reactive process.

follicle-stimulating hormone One of the gonadotropic hormones of the anterior pituitary, which stimulates the growth and maturation of graafian follicles in the ovary, and stimulates spermatogenesis in the male.

forensic toxicology The medical aspects of the diagnosis and treatment of poisoning and the legal aspects of the relationships between exposure to and harmful effects of a chemical substance. It is concerned with both intentional and accidental exposures to chemicals.

gastric lavage The process of washing out the stomach with saline solution using a lavage tube, to remove poisons taken orally.

gastritis Inflammation of the stomach.

gene The basic unit of inheritance, recognized through its variant alleles; a segment of DNA coding a designated function (or related functions).

genotoxicity A measure of the potency of adverse effect of a substance directly of DNA.

genotype The entire allelic composition of an individual (or genome), or of a certain gene or set of genes.

germ cell An ovum or a sperm cell of one of its developmental precursors.

gingivitis Inflammation of the gums of the mouth.

glutathione A naturally occurring tripeptide, serving as a biological redox agent or substrate for certain conjugation reactions in chemical metabolism.

glycoside Any of a group of organic compounds, occurring abundantly in plants, that yield a sugar and one or more nonsugar substances on hydrolysis.

gonadotropin A hormone that stimulates the growth and activity of the gonads, especially any of several pituitary hormones that stimulate the function of the ovaries and testes.

granulocyte Any cell containing granules, especially a leukocyte containing neutrophil, basophil, or eosinophil granules in its cytoplasm.

halogenation The incorporation of one of the halogen elements, usually chlorine or bromine, into a chemical compound.

heinz body Microscopic bodies noted in red blood cells with enzyme deficiencies, identified as either cholesterinolein-based or as dead cytoplasm resulting from oxidative injury to and precipitation of hemoglobin.

hematopoietic Pertaining to or affecting the formation of blood cells; an agent that promotes the formation of blood cells.

hematuria Blood in the urine.

hemolytic Pertaining to, characterized by, or producing hemolysis. The liberation of hemoglobin; the separation of hemoglobin from the red cells and its appearance in the plasma.

hemolytic anemia Anemia owing to shortened *in vivo* survival of mature red blood cells, and inability of the bone marrow to compensate for their decreased life span.

hemoptysis The coughing or spitting up of blood from the respiratory tract.

hemorrhagic cystitis Urinary bladder inflammation compounded with bleeding.

hemosiderosis A general increase in iron stores in tissues without tissue damage.

hepatomegaly Enlargement of the liver.

hepatotoxin A toxin destructive of liver cells.

histamine A physiologically active amine, $C_5H_9N_3$, found in plant and animal tissue. It is released from cells of the immune system in human beings as part of an allergic reaction.

hives A skin condition characterized by intensely itching welts and caused by an allergic reaction to internal or external agents, an infection, or a nervous condition. Also called *urticaria.*

homolog One of a series of compounds, each of which is formed from the one before it by the addition of a constant element or a constant group of elements, as in the homologous series C_NH_{2N+2}, compounds of which would be CH_4, C_2H_6, C_3H_8, or similar.

humoral immunity The component of the immune response involving the transformation of B lymphocytes into plasma cells that produce and secrete antibodies to a specific antigen.

hydrocarbon An organic compound consisting exclusively of the elements carbon and hydrogen. Derived principally from vegetable sources, petroleum, and coal tar.

hydrolysis Decomposition of a chemical compound by reaction with water, such as the catalytic conversion of starch to glucose.

hydrophilic Readily absorbing water; hygroscopic.

hydroxylation An oxidative reaction that introduces one or more hydroxyl groups into an organic compound.

hyperalgesia A heightened or excessive sensitivity to pain.

hyperemia An excess of blood in some part of the body.

hyperkeratosis Overgrowth of the corneous layer of the skin, or any disease characterized by that conditions.

hyperpigmentation Abnormally increased pigmentation.

hyperplasia Abnormal multiplication or increase in the number of normal cells in normal arrangement in a tissue.

hypersensitivity A state of extreme sensitivity to an action of a chemical; for example, the individuals of a test population who fit into the "low end" of an ED_{50} or LD_{50} curve (i.e., those individuals who react to a very low dose as opposed to the median effective dose).

hypokinesis Abnormally decreased mobility; abnormally decreased motor function or activity.

hyposensitivity The state of decreased sensitivity; for example, the individuals of a test population who fit into the "high end" of an ED_{50} or LD_{50} curve (i.e., those individuals who respond only to a very high dose as compared to the median effective dose).

hypoxia A partial reduction in the oxygen concentration supplied to cells or tissues.

immune response See **sensitization reaction**.

incidence An expression of the rate at which a certain event occurs, as the number of new cases of a specific disease occurring during a certain period.

inclusion body An abnormal structure in a cell nucleus or cytoplasm having characteristics staining properties and associated especially with certain viral infections, such as rabies and smallpox.

infarct An area of necrosis in a tissue caused by local lack of blood resulting from obstruction of circulation to the area.

inhalation route The movement of a chemical from the breathing zone, through the air passageways of the lung, into the alveolar area, across the epithelial cell layer of the alveoli and the endothelial cell layer of the capillary wall, and into the blood system.

inotropic Affecting the force of muscular contraction, especially in the heart muscle.

interleukins A generic term for a group of protein factors that affect primary cells and are derived from macrophages and T cells that have been stimulated by antigens or mitogens.

interleukin-1 Any of a group of protein substances, released by macrophages and other cells, that induce the production of interleukin-2 by helper T cells and stimulate the inflammatory response.

interleukin-2 A lymphokine that is released by helper T cells in response to an antigen and interleukin-1 and stimulates the proliferation of helper T cells. It has been used experimentally to treat cancer.

intraperitoneal Within the peritoneal cavity; an intraperitoneal injection is one in which a chemical is injected into the abdominal fluid of an animal.

ionization The dissociation of a substance in solution into ions.

irritant dermatitis See **dermatitis**.

ischemia Deficiency of blood owing to a functional constriction or actual obstruction of a blood vessel.

isotonic Describing a solution with the same solute concentration as another solution (e.g., tissue culture media and cellular cytoplasm of cultured cells).

isozyme A member of a family of proteins with related structure and function.

kepone Insecticide and fungicide having the formula $C_{10}Cl_{10}O$; causes excitability, tremor, skin rash, opsoclonus, weight loss, and in some cases (in animals) testicular atrophy.

keratoacanthoma A rapidly growing papular lesion, with a crater filled with a keratin plug, which reaches maximum size and then resolves spontaneously within 4–6 months from onset.

keratosis Any horny growth, such as a wart or callosity.

ketone Any compound containing the carbonyl group C=O and having hydrocarbon groups attached to its carbonyl carbon.

LD_{50} That dose of a particular substance that, administered to all animals in a test, is lethal to 50 percent of the animals. It is that dose of a compound which will produce death in 50 percent of the animals-hence, the median lethal dose. The values of LD_{50} should be reported in terms of the duration over which the animals were observed. If a time is not given, it is assumed they were observed for 24 h.

lacrimation The secretion and discharge of tears.

laryngitis Inflammation of the larynx, a condition attended with dryness and soreness of the throat, hoarseness, cough, and dysphagia (difficulty in swallowing).

leukocyte A white blood cell or corpuscle; classified as either granular or nongranular.

leukocytosis A transient increase in the number of leukocytes in the blood, resulting from various causes, such as hemorrhage, fever, infection, or inflammation.

leukopenia Lower-than-normal number of leukocytes in the blood; the normal concentration is 4000–11,000 leukocytes in 1 ml of blood.

Leydig's cells The interstitial cells of the testes (between the seminiferous tubules), believed to furnish the male sex hormone.

lichen planus An inflammatory skin disease characterized by the appearance of wide, flat, violaceous, itchy, polygonal papules, occurring in circumscribed patches, and often very persistent. The hair follicles and nails may become involved, and the buccal mucosa may be affected.

lipid peroxidation Interaction of free radicals with the lipid constituents of a membrane, resulting in alterations of structure and function of the membrane.

lipophilicity Having an affinity for fats.

lipoprotein Any of a group of conjugated proteins in which at least one component is a lipid. Lipoproteins, classified according to their densities and chemical qualities, are the principal means by which lipids are transported in the blood.

locus of action (site of action) The part of the body (organ, tissue, or cell) where a chemical acts to initiate the chain of events leading to a particular effect.

luteinizing hormone A gonadotropic hormone of the anterior pituitary, which acts with the follicle-stimulating hormone to cause ovulation of mature follicles and secretion of estrogen by thecal and granulosa cells.

lymphocyte A mononuclear leukocyte with a deep-staining nucleus containing dense chromatin and a pale-blue-staining cytoplasm. Chiefly a product of lymphoid tissue. Participates in humoral and cell-mediated immunity. See also **B cell**; **T cell**.

lymphokine Any of various substances released by T cells that have been activated by antigens. They function in the immune response through a variety of actions, including stimulating the production of nonsensitized lymphocytes and activating macrophages.

macrophage Any of the large phagocytic cells of the reticuloendothelial system.

makeup air In workplace ventilation, air introduced into an area to replace the air that has been removed.

malignant tumor Relatively autonomous growth of cells or tissue. Each type of malignant tumor has a different etiology and arises from a different origin. The condition tends to become progressively worse and to result ultimately in death. There are many common properties of malignant tumors, but the invasion of surrounding tissue and the ability to metastasize are considered the most characteristic.

margin of safety The magnitude of the range of doses involved in progressing from a noneffective dose to a lethal dose. Consequently, the slope of the dose–response curve is an index of the margin of safety of a compound.

megakaryocyte A giant cell found in bone marrow, containing a greatly lobulated nucleus from which mature blood platelets originate.

mesenchymal cells (tissue) The meshwork of embryonic connective cells or tissue in the mesoderm from which are formed the connective tissues of the body, the blood vessels, and the lymphatic vessels.

mesothelioma A tumor developed from the mesothelial tissue—the simple squamous-celled layer of the epithelium, which covers the surface of all true serous membranes (lining the abdominal cavity, covering the heart, and enveloping the lungs).

metabolism The biochemical reactions that take place within an organism. It involves two processes: anabolism (assimilation or constructive processes) and catabolism (disintegration or destructive processes). All metabolic processes involve energy transfer.

metallothionein An inducible metal-binding protein involved in trafficking and detoxification mechanisms for various heavy metals.

metaplasia The transformation of cells from a normal to an abnormal state.

metastasis The establishment of a secondary growth site, distant from the primary site. One of the primary characteristics of a malignant tumor.

methemoglobin A compound formed from hemoglobin by oxidation of iron in the ferrous state to the ferric state. Methemoglobin does not combine with oxygen.

methemoglobinemia Presence of methemoglobin in the blood, resulting in cyanosis.

microsomes The fragments of the smooth reticular endothelium. This is the source of the microsomal enzymes that are capable of catalyzing a variety of biotransformation reactions, including hydroxylation, dealkylation, deamination, alkyl side-chain oxidation, hydrolysis, and reduction.

miosis Contraction of the pupil of the eye.

mitochondria Small spherical or rod-shaped components (organelles) found in the cytoplasm of cells, enclosed in a double membrane. They are the principal sites of energy generation (ATP) and they contain the enzymes of the Krebs and fatty acid cycles and the respiratory pathways. Mitochondria contain an extranuclear source of DNA and have genetic continuity.

mitosis The process in cell division by which the nucleus divides, typically consisting of four stages—prophase, metaphase, anaphase, and telophase—followed by cytokinesis and normally resulting in two new cells, each of which contains a complete copy of the parental chromosomes.

mixed-function oxidase system (MFO) A nonspecific, multienzyme complex on the smooth endoplasmic reticulum of cells in the liver and various other tissues. These enzymes constitute the important enzyme system involved in phase I reactions (i.e., oxidation/reduction reactions). Also called *cytochrome P450 enzymes*.

monclonal antibody A homogeneous antibody that is produced by a clone of antibody-forming cells and that binds with a single antigenic determinant.

monocyte A mononuclear phagocytic leukocyte with an ovoid or kidney-shaped nucleus, containing lacy, linear chromatin, and abundant gray-blue cytoplasm fitted with fine, reddish and azure granules.

morbidity The rate of sickness or ratio of sick persons to well persons in community.

multiple myeloma A malignant proliferation of plasma cells in bone marrow causing numerous tumors and characterized by the presence of abnormal proteins in the blood.

muscarine A highly toxic alkaloid $C_9H_{20}NO_2$, related to the cholines, derived from the red form of the mushroom *Amanita muscaria*.

mutagen Any substance causing genetic mutation.

mutagenesis The induction of those alterations in the information content (DNA) of an organism or cell that are not due to the normal process of recombination. Mutagenesis is irreversible and is cumulative, in the event of increased mutation rates or decreased selection pressures.

mutagenic tests Test of an agent to determine effects on the faithful replication of genetic material. The genetic damage can occur in both somatic and germinal cell lines.

mutation A permanent offspring-transmissible change in genetic material or structure. Such changes may manifest themselves as altered morphology or altered ability to direct the synthesis of proteins.

myalgia Muscular pain or tenderness, especially when diffuse and nonspecific.

myasthenia gravis A disease characterized by progressive fatigue and generalized weakness of the skeletal muscles, especially those of the face, neck, arms, and legs, caused by impaired transmission of nerve impulses following an autoimmune attack on acetylcholine receptors.

myelin sheath The insulating envelope of myelin that surrounds the core of a nerve fiber or axon and facilitates the transmission of nerve impulses. In the peripheral nervous system, the sheath is formed from the cell membrane of the Schwann cell and, in the central nervous system, from oligodendrocytes. Also called *medullary sheath.*

myeloid leukemia Leukemia arising from myeloid tissue (bone marrow) characterized by unrestrained growth of the granular, polymorphonuclear leukocytes and their precursors.

myelotoxin A cytotoxin that causes destruction of bone marrow cells.

myoclonus A sudden twitching of muscles or parts of muscles, without any rhythm or pattern, occurring in various brain disorders.

myotonia Tonic spasm or temporary rigidity of one or more muscles, often characteristic of various muscular disorders.

narcosis A condition of deep stupor or unconsciousness produced by a drug or other chemical substance.

nasopharyngeal region The part of the pharynx lying above the level of the soft palate (also known as the *postnasal space*).

necrosis Death of one or more cells, or of part of a tissue or organ, generally owing to irreversible damage.

nematocyst A capsule within specialized cells of certain coelenterates, such as jellyfish, containing a barbed, threadlike tube that delivers a paralyzing sting when propelled into attackers and prey.

neoplasm Literally, new growth, usually characterized by a random abnormal "immature," meiosis-type cell division and proliferation.

nephritis Inflammation of the kidney; a focal or diffuse proliferative or destructive process, which may involve the glomerulus, tubule, or interstitial renal tissue.

neurodermatitis A nonspecific pruritic skin disorder presumed to result from prolonged vigorous scratching, rubbing, or pinching, sometimes forming polymorphic lesions.

neuroendocrine Or, relating to, or involving the interaction between the nervous system and the hormones of the endocrine glands.

neurofibril One of the delicate threads running in every direction through the cytoplasm of the body of a nerve cell and extending into the axon and dendrites of the cell.

neuromuscular endplate A flattened discoid expansion at the neuromuscular junction, where a myelinated motor nerve fiber joins a skeletal muscle fiber.

neuropathy General term denoting functional disturbances and/or pathologic changes in the nervous system.

neutropenia A decrease in the number of neutrophilic leukocytes in the blood.

neutrophil A granular leukocyte having a nucleus with three to five lobes connected by slender threads of chromatin and cytoplasm, containing fine, inconspicuous granules.

nicotinic effect Poisoning by nicotine or a compound related in structure or action, characterized by stimulation (low doses) and depression (high doses) of the central and autonomic nervous systems. In extreme cases, death results from respiratory paralysis. Also referred to as *nicotinism*.

nitrosamine Any of a group of *n*-nitroso derivatives of secondary amines. Some show carcinogenic activity.

NOEL See **no-observable-effect level**.

noncompetitive inhibition Inhibition of enzyme activity by inhibitors that combine with the enzyme on a site other than that utilized by the substrate; such inhibition may be irreversible or reversible.

nonspecific chemical action The action of a chemical, such as a strong acid or base or concentrated solution of organic solvent, which occurs in all cells in direct proportion to the concentration in contact with the tissue. This is a nonselective effect and its intensity is directly related to the concentration of the chemical.

nonspecific receptor Secondary receptor within the body, which combine with or react with a chemical; however, the function of the cell is not influenced by the product that is formed. Such receptors are usually combining sites on proteins.

no-observable-effect level (NOEL) A measure of the toxicity of a substance, established by the U.S. Environmental Protection Agency (USEPA); the level of a substance that, when administered to a group of experimental animals, does not produce those effects observed at higher levels, and at which no significant differences between the exposed animals and the unexposed or control animals are observed.

olefin A class of unsaturated aliphatic hydrocarbons having one or more double bonds. Also called *alkene*.

oncogenic Giving rise to tumors or causing tumor formation.

opsoclonus A condition characterized by rapid, irregular, nonrhythmic horizontal and vertical oscillations of the eyes, observed in various disorders of the brainstem or cerebellum.

optic neuritis Inflammation of the optic nerve; it may affect the part of the nerve within the eyeball, or the portion behind the eyeball.

oral route The entry of a chemical into the body by way of the gastrointestinal tract. Although absorption to some extent takes place throughout the tract, the majority of the absorption takes place in the area of the villi of the small intestine.

organic acid Any acid, the radical of which is a carbon derivative; a compound in which a hydrocarbon radical is joined to COOH (carboxylic acid) or to SO_3H (sulfonic acid).

organochlorine pesticides These compounds are extremely stable and persistent in the environment. They are efficiently absorbed by ingestion, and act on the central nervous system to stimulate or depress it. Signs and symptoms of toxicity vary with the specific chemical. In general, mild poisoning cases cause symptoms such as dizziness, nausea, abdominal pain, and vomiting. In chronic poisoning, weight loss and loss of appetite, temporary deafness, and disorientation can occur.

organophosphate pesticides These are irreversible inhibitors of cholinesterase, thus allowing accumulating of acetylcholine at nerve endings. They are rapidly absorbed into the body by ingestion, through intact skin, including the eye, and by inhalation. Poisoning symptoms range from headache, fatigue, dizziness, vomiting, and cramps in mild cases, to the rapid onset of unconsciousness, local or generalized seizure, and other manifestations of a cholinergic crisis in severe cases.

osteomalacia A condition of softening of the bones characterized by pain, tenderness, loss of weight, and muscular weakness.

osteoporosis Abnormal rarefaction of bone, seen most commonly in the elderly.

osteosclerosis Hardening or abnormal density of bone.

ototoxic Having a toxic effect on the structures of the ear, especially on its nerve supply.

pancytopenia A form of anemia in which the capacity of the bone marrow to generate red blood cells is defective. This anemia may be caused by bone marrow disease or exposure to toxic agents, such as radiation, chemicals, or drugs.

paraffin A class of aliphatic hydrocarbons characterized by a straight or branched carbon chain afid having the generic formula C_nH_{2n+2}; also called *alkane*.

paranoid schizophrenia A psychotic state characterized by delusions of grandeur or persecution, often accompanied by hallucinations.

parasympathetic Craniosacral division of the autonomic nervous system. These cholinergic nerves are associated with normal body functions (e.g., smooth muscle in blood vessels, salivary glands, and GI tract).

paresthesia (also **paraesthesia**) A skin sensation, such as burning, prickling, itching, or tingling, with no apparent physical cause.

Parkinsonism A group of neurologic disorders characterized by hypokinesia, tremor, and muscular rigidity.

PEL See **permissible exposure limit**.

percutaneous absorption The transfer of a chemical from the outer surface of the skin through the horny layer (dead cells), through the epidermis, and into the systemic circulation. A variety of factors, such as pH, extent of ionization, molecular size, and water and lipid solubility govern transfer of chemicals through the skin.

perinatal toxicology The study of toxic responses to occupationally or environmentally encountered substances during a woman's exposure to those substances from the time of conception through the neonatal period.

peripheral neuritis Inflammation of the nerve ending or of terminal nerves.

permissible exposure limit (PEL) A measure of the toxicity of a substance, established by the U.S. Department of Labor, Occupational Safety and Health Administration (OSHA); an 8-h, time-weighted average (TWA) limit of exposure is most commonly used. The limit is commonly expressed as the concentration of a substance per unit of air volume (mg/m^3, ppm, $fibers/cm^3$, etc.)

pernicious anemia The progressive, megaloblastic anemia resulting from lack of vitamin B_{12}, sometimes accompanied by degeneration of the postier an lateral columns of the spinal cord.

peroxidase Any of a group of enzymes that occur especially in plant cells and catalyze the oxidation of a substance by a peroxide.

personal protective equipment Any devices worn by individuals as protection against hazards in the environment or the workplace, including respirators, gloves, goggles, and earmuffs.

pesticide Any substance used to destroy or inhibit the action of plant or animal pests. See **carbamate**; **organochlorine pesticides**; **organophosphate pesticides**.

petechiae Tiny, nonraised, perfectly round, purplish red spots caused by intradermal or submucosal hemorrhaging.

pH A value taken to represent the acidity or alkalinity of an aqueous solution. It is defined as the logarithm of the reciprocal of the hydrogen-ion concentration of a solution:

$$pH = \ln \frac{1}{[H^+]}$$

pharmacokinetics The field of study concerned with the techniques used to quantify the absorption, distribution, metabolism, and excretion of drugs or chemicals in animals, as a function of time.

pharmacology The unified study of the properties of chemical agents (drugs) and living organisms and all aspects of their interactions. An expansive science encompassing areas of interest germane to many other disciplines.

phenothiazine A green, tasteless compound with the formula $C_{12}H_9NS$, prepared by fusing diphenylamine with sulfur; also, a group of tranquilizers resembling phenothiazine in molecular structure.

phocomelia A developmental anomaly characterized by the absence of the proximal portion of a limb or limbs, such that hands or feet are attached to the trunk of the body by a single small, irregularly shaped bone.

phospholipase Any enzyme that catalyzes the hydrolysis of a phospholipid.

photophobia Abnormal sensitivity, usually of the eyes, to light.

photosensitivity reactions Undesirable reactions in the skin of persons exposed to certain chemicals when the skin is also exposed to sunlight (in some cases, to artificial light). Dermatologic lesions form, which vary from sunburn-like responses to edematous, vesiculated lesions or bullae.

phototoxicity Capacity of a chemical to (nonimmunlogically) sensitize the skin to a light-induced reaction.

pilosebaceous units Relating to the hair follicles and sebaceous glands.

pK The acidic dissociation constant of a compound; the pH of an aqueous solution of an acid or base at which equal concentrations of each are present, at the point at which dissociation is half-complete. The negative logarithm of the ionization constant K_a.

pneumoconiosis Accumulation of dusts in the lungs and the tissue reaction to the presence of such dust.

pneumonitis Inflammation of lung tissue.

point mutation An alteration in a single nucleotide pair in the DNA molecule, usually leading to a change in only one biochemical function.

poison The term used to describe those materials or chemicals that are distinctly harmful to the body.

polymorphism The occurrence of different genetic forms or types of a protein that produce phenotypically distinct populations.

polymorphonuclear Having a nucleus deeply lobed or so divided as to appear to be multiple.

polyneuritis Inflammation of several nerves simultaneously, as in lead palsy.

polyneuropathy A disease involving several nerves.

porphyrin Any of a group of iron-free or magnesium-free cyclic tetrapyrrole derivatives occurring universally in protoplasm. They form the basis of the respiratory pigments of animals and plants.

potency A comparative expression of chemical or drug activity measured in terms of the dose required to produce a particular effect of given intensity relative to a given or implied standard of reference. If two chemicals are not both capable of producing an effect of equal magnitude, they cannot be compared with respect to potency.

potentiation A condition whereby one substance is made more potent in the presence of another chemical that alone produces no response.

pressure, static The potential pressure exerted in all directions by a fluid at rest.

pressure total The algebraic sum of static and velocity pressures, representing the total energy in the system.

pressure, velocity The kinetic pressure exerted in the direction of flow necessary to cause a fluid at rest to flow at a given velocity.

primary carcinogens Chemicals that act directly and without biotransformation. Also called *direct carcinogens.*

primary irritants Chemicals that induce local, minor to severe inflammatory response, or even extreme necrosis, of cells of a tissue, in direct relation to the concentration available to the tissue. This is termed a *nonspecific chemical action*, the toxicity of which may be manifested at the site of exposure (e.g., skin or in the respiratory tract). Examples of these types of chemicals are strong acids or bases, ammonia, and acrolein.

Probenecid A white, odorless crystalline powder, with the formula $C_{13}H_{19}NO_4S$, soluble in dilute alkali, alcohol, and acetone; used to increase serum concentrations of certain antibiotics, as well as being an agent to promote uric acid secretion in the urine.

procarcinogen Chemicals that require metabolism to another, more reactive or toxic chemical form before their carcinogenic action can be expressed.

proerythropoietin A precursor of erythropoietin.

psoriasis A chronic, hereditary, recurrent, papulosquarnous dermatitis, the distinctive lesion of which is a vivid red macula, papule, or plaque covered almost to its edge by silvery lamellated scales. It usually involves the scalp and extensor surfaces of the limbs, especially the elbows, knees, and shins.

pyrethroid Any of several synthetic insecticidal compounds similar to the nature pyrethrums extracted from crushed chrysanthemums.

Raynaud's phenomenon Intermittent attacks of severe pallor of the fingers or toes and sometimes of the ears and nose, brought on characteristically by cold and sometimes by emotion.

receptors See **specific receptor**; **nonspecific receptor**.

renal osteodystrophy A condition resulting from chronic kidney disease. The onset early in childhood is characterized by impaired renal function, elevated serum phosphorus and low or normal serum calcium levels, and stimulation of parathyroid function. The resultant bone disease includes a variety of symptoms, including osteitis fibrosa cystica, osteomalacia, osteoporosis, and osteosclerosis. Renal dwarfism may result from childhood onset.

reproduction tests Tests that determine (or estimate) the effects of an agent on fertility, gestation, and offspring; usually conducted on more than one generation of test animals. Toxicity in either parent may affect fertility as the direct result of altered gonadal function, estrus cycle, mating behavior, and conception rates. Effects on gestation concern the development of the fetus. Effects on offspring concern growth, development, and sexual maturation; and effects on the mother concern lactation and acceptance of the offspring.

resorption The loss of substance in the mucous lining of the uterus.

reticuloendothelial system Phagocytic macrophages present in linings of sinuses and in reticulum of various organs and tissues. A functionally important body defense mechanism; the phagocytic cells have both endothelial and reticular attributes and the ability to take up particles of colloidal dyes.

risk assessment A methodologic approach in which the toxicities of a chemical are identified, characterized, and analyzed for dose–response relationships, and a mathematical model is applied to the data to generate a numerical estimate that can serve as a guide to allowable exposures.

risk estimation Mathematical modeling of the animal and/or human toxicity data, combined with evaluation of human exposures, so as to estimate the probability or incidence of effects on human health.

risk management The process of applying a risk assessment to the conditions that exist in society, so as to balance exposures to toxic agents against needs for products and processes that may be inherently hazardous.

safety factor A factor that presumably reflects the uncertainties inherent in the process of extrapolating data about toxic exposures (i.e., intraspecies and interspecies variations). With this approach, an allowable human exposure to a compound can be determined by dividing the no-observable-effect level (NOEL) established in chronic animal toxicity studies by some safety factor. Also called *uncertainty factor.*

sarcoma A cancer that arises from mesodermal tissue (supporting or connective tissue).

sclerodermatous skin change A chronic hardening and shrinking of the connective tissues of any part of the body, including the skin, heart, esophagus, kidney, and lung. The skin may become thickened and hard, and the condition may be generalized and rigid, and pigmented patches may occur, limited to the distal parts of the extremities and face, or to the digits, or localized to oval or linear areas a few centimeters in diameter.

sclerosis A thickening or hardening of a body part, as of an artery, especially from excessive formation of fibrous interstitial tissue.

sensitization reaction An immunologic response to a chemical. The mechanism of immunization involves the following events: initial exposure of an animal to a chemical substance, an induction period in the animal; and the production of a new protein termed an *antibody*. The initial exposure does not result in cellular damage but causes the animal to be "sensitized" to subsequent exposure to the chemical. Exposure of the animal to the same chemical on a subsequent occasion will lead to the formation of sensitized antigen, which will react with the preformed antibodies and lead to a response in the tissues in the form of cellular damage.

SGOT Serum glutamic–oxaloacetic transaminase. An enzyme found in the liver and muscle tissue and used to detect early membrane permeability as part of a test of the activity of enzymes present in liver cells.

SGPT Serum glutamic–pyruvic transaminase. An enzyme used in the identification and measurement of the activity of enzymes present in liver cells. SGPT is found in the liver and heart tissues. It is an indicator of early membrane permeability, as is SGOT.

silicosis A type of pneumocomosis due to the inhalation of the dust of stone, sand, or flint containing silicon dioxide. It results in the formation of generalized nodular fibrotic changes in both lungs.

site of action See **locus of action**.

SNARL Suggested no-adverse-response level. A measure of toxicity established by the National Research Council.

somatic cell Any cell of a plant or an animal other than a germ cell. Also called *body cell.*

specific receptor Macromolecular constituent of tissue capable of combining reversibly with a compound by means of chemical bonds; the tissue element with which a compound interacts to provide its characteristic biologic effect.

spermatozoa The mature fertilizing gamete of a male organism, usually consisting of a round or cylindrical nucleated cell, a short neck, and a thin, motile tail. Also called *sperm cell.*

spirometer An instrument for measuring the air taken into and exhaled from the lungs.

spirometry The measurement of the breathing capacity of the lungs.

squamous cell carcinoma Carcinoa developing from squamous epithelium (composed of flattened, platelike cells) and characterized by cuboid cells.

stereoisomers Two substances of the same composition differing only in the relative spatial positions of their constituent atoms and/or groups.

steric hindrance The nonoccurrence of an expected chemical reaction owing to inhibition by a particular atomic grouping.

sulfhemoglobin An abnormal greenish form of hemoglobin containing sulfur that is bound to heme.

sulfotransferase An enzyme that catalyzes the transfer of sulfate from a donor molecule to an acceptor.

sympathetic Thorocolumbar division of the autonomic nervous system mainly involved with homeostasis (e.g., vasoconstriction, glucose mobilization, adrenaline release).

sympathomimetic Mimicking the effects of impulses conveyed by adrenergic postganglionic fibers in the sympathetic nervous system.

synapse The anatomical relation of one nerve cell to another; the region of junction between processes of two adjacent neurons, forming the place where a nervous impulse is transmitted from one neuron to another.

synergism The situation in which the combined effects on a biologic system of two chemicals acting simultaneously is greater than the algebraic sum of the individual effects of these chemicals.

tachycardia A rapid heart rate, especially one above 100 beats per minute in an adult.

tachypnea Rapid breathing.

T cell Thymus-dependent lymphocytes; these pass through the thymus or are influenced by it on their way to the tissues; they can be suppliers or assist the stimulation of antibody production in B cells in the presence of antigen, and can kill such cells as tumor and transplant-tissue cells. T cells are responsible for all cell-mediated immunity and immunologic memory.

TD_{50} That dose of a substance that, administered to all animals in a test, produces a toxic response in 50 percent of them. The toxic response may be any adverse effect other than death.

teratogen Any substance capable of causing malformation during development of the fetus.

thalidomide A sedative and hypnotic drug commonly used in Europe in the early 1960s. It was discovered to be the cause of serious congenital anomalies in the fetus, notably amelia and phocomelia.

threshold dose (ThD) The minimal dose effective in prompting an all-or-none response.

threshold limit value (TLV) A term for exposure limits established by the American Conference of Governmental Industrial Hygienists (ACGIH). That concentration of any airborne substance to which it is believed, through animal toxicity testing and human exposure data, that workers can be exposed to 8 h per day, 40 h per week for a working lifetime, without suffering adverse health effects or significant discomfort. TLV measurements are usually based on 8-h time-weighted average (TWA) exposures but may be expressed as ceiling values.

time-weighted average (TWA) A method of combining multiple air-sample results collected on one individual during a workshift, so as to derive the overall average exposure for the entire shift (or exposure period). Measurements of chemical exposure can be made in each phase, and the exposure estimate is calculated according to the formula

$$E = \frac{C_1T_1 + C_2T_2 + \ldots C_NT_N}{(T_1 + T_2 \ldots T_N)}$$

where E exposure, C concentration measured in phase N, and T_N = duration of phase N.

TLV See **threshold limit value**.

tolerance The ability of an organism to show less response to a specific dose of a chemical than it showed on a prior occasion when subjected to the same dose.

totipotency The ability of a cell, such as an egg, to give rise to unlike cells and thus to develop into or generate a new organism or part.

toxicity A relative term generally used in comparing the harmful effect of one chemical on some biologic mechanism with the effect of another chemical.

toxicology The scientific study of poisons and their actions and detection, and treatment of the condition produced by them. Also the study of the effects of chemicals on biologic systems, with emphasis on the mechanisms of harmful effects of chemicals and the conditions under which harmful effects occur. Thus, toxicology is a multidisciplinary science.

tracheitis Inflammation of the trachea.

tumor An abnormal mass of tissue, the growth of which exceeds and is uncoordinated with that of normal tissue. The basic types are benign and malignant. See also **benign tumor**; **malignant tumor**.

tumorigenicity The process or quality of producing or giving rise to a tumor.

Type I reaction (immediate) A hypersensitive reaction that is manifested in allergic asthma, hay fever, and excema, developing within minutes after exposure to an antigen.

type II reaction (antibody-mediated) A hypersensitive reaction that is caused by an antibody reaction to cell surface antigens; occurs when red blood cells are destroyed in transfusion reactions.

ubiquinone A quinone compound that serves as a electron carrier between flavoproteins and in cellular respiration.

uncertainty factor See **safety factor**.

unction An ointment; the application of an ointment or salve.

uropathy Any pathologic change in the urinary tract.

Urticarria See **hives**.

ventricular fibrillation Arrhythmia characterized by fibrillary contractions of the ventricular muscle owing to rapid repetitive excitation of myocardial fibers without coordinated contraction of the ventricle.

xenobiotic A chemical foreign to the biologic system (i.e., chemicals that are not normal endogenous compounds for the biologic system).

INDEX